NOBEL LECTURES IN PHYSIOLOGY OR MEDICINE 1981–1990

NOBEL LECTURES

INCLUDING PRESENTATION SPEECHES
AND LAUREATES' BIOGRAPHIES

PHYSICS

CHEMISTRY

PHYSIOLOGY OR MEDICINE

LITERATURE

PEACE

ECONOMIC SCIENCES

NOBEL LECTURES

INCLUDING PRESENTATION SPEECHES
AND LAUREATES' BIOGRAPHIES

PHYSIOLOGY
OR
MEDICINE

1981–1990

EDITOR-IN-CHARGE

TORE FRÄNGSMYR

Uppsala University
Uppsala, Sweden

EDITOR

JAN LINDSTEN

The Karolinska Medico-Chirurgical Institute
Stockholm, Sweden

World Scientific
Singapore • New Jersey • London • Hong Kong

Published for the Nobel Foundation in 1993 by

World Scientific Publishing Co. Pte. Ltd.
P O Box 128, Farrer Road, Singapore 9128
USA office: Suite 1B, 1060 Main Street, River Edge, NJ 07661
UK office: 73 Lynton Mead, Totteridge, London N20 8DH

NOBEL LECTURES IN PHYSIOLOGY OR MEDICINE (1981–1990)

ISBN 981-02-0792-1
ISBN 981-02-0793-X (pbk)

Printed in Singapore by Continental Press Pte Ltd

Foreword

Since 1901 the Nobel Foundation has published annually "Les Prix Nobel" with reports from the Nobel Award Ceremonies in Stockholm and Oslo as well as the biographies and Nobel lectures of the laureates. In order to make the lectures available to people with special interests in the different prize fields the Foundation gave Elsevier Publishing Company the right to publish in English the lectures for 1901–1970, which were published in 1964–1972 through the following volumes:

Physics 1901–1970	4 vols.
Chemistry 1901–1970	4 vols.
Physiology or Medicine 1901–1970	4 vols.
Literature 1901–1967	1 vol.
Peace 1901–1970	3 vols.

Elsevier decided later not to continue the Nobel project. It is therefore with great satisfaction that the Nobel Foundation has given World Scientific Publishing Company the right to bring the series up to date beginning with the Prize lectures in Economics in 2 volumes 1969–1990. Thereafter the lectures in all the other prize fields will follow.

The Nobel Foundation is very pleased that the intellectual and spiritual message to the world laid down in the laureates' lectures, thanks to the efforts of World Scientific, will reach new readers all over the world.

Lars Gyllensten
Chairman of the Board

Stig Ramel
Executive Director

Stockholm, June 1991

Preface

The Nobel Week is an annual, international event the highlight of which is the Nobel Prize Award Ceremony in the Concert Hall of Stockholm on December 10th, that is on the death-day of Alfred Nobel. It is a fascinating occurrence from both the scientific and social points of view, mainly because it puts the significance of science for society into bright focus. Only one item on the entire agenda is compulsory for the Laureates during this week, the Nobel Lecture. Thus, each Laureate has to deliver a lecture on the topic for which the prize has been awarded. The lectures are generally given on December 8th, that is before the Prize Award Ceremony, so that the Laureates can enjoy the festivities in a more relaxed way.

The Laureates in Physiology or Medicine present their lectures at the Karolinska Institute, that is on the ground of the Prize Awarding Institution. This gives the scientists and students at the Institute a unique possibility to enjoy presentations of some of the most significant contributions to biomedical science and also to meet the Laureates personally. That is why these lectures have such a special atmosphere (video tape recordings of the lectures are kept in the Nobel Archives at the Karolinska Institute).

The Nobel Lectures are published in an annual book series, "Les Prix Nobel". Each volume in this series contains, in addition to the lectures given by all Laureates for a specific year, short biographical notes and portraits of the Laureates as well as reports on the prize presentation ceremonies in Stockholm and Oslo.

In the present two volumes the Nobel Lectures given by the Laureates in Physiology or Medicine for the years 1971–1990 have been reprinted. Since these lectures are time bound documents, only minor changes, for instance of printing errors, have been made.

As a member of the Nobel Assembly and secretary-general of the Nobel Assembly and the Nobel Committee at the Karolinska Institute during the years 1979–1990, I have had the pleasure of meeting most of the Laureates from the time period covered in these two volumes. It has indeed been a pleasure to reread their lectures with the perspective that time now has given them.

Jan Lindsten
M.D., Professor
September 1992

Contents

1981
Physiology or Medicine

ROGER W. SPERRY

"for his discoveries concerning the functional specialization of the cerebral hemispheres"

and

DAVID H. HUBEL &
TORSTEN N. WIESEL

"for their discoveries concerning information processing in the visual system"

THE NOBEL PRIZE FOR PHYSIOLOGY OR MEDICINE

Speech by Professor DAVID OTTOSON of the Karolinska Institute.
Translation from the Swedish text.

Your Majesties, Your Royal Highnesses, Ladies and Gentlemen,

One day in October 1649, René Descartes, the French philosopher and mathematician acknowledged as the greatest brain researcher of the period, arrived in Stockholm at the pressing invitation of Queen Christina. It was with much hesitation that Descartes went to Sweden as he wrote "the land of bears between rocks and ice". In the letters to his friends, he complained bitterly that he was obliged to present himself at the Royal Palace at five o'clock each morning to instruct the young queen in philosophy, so avid was she for knowledge. Modern brain research scientists and followers in the Cartesian footsteps are not faced with the same demands as winners of the Nobel Prize, but they are met with other tribulations — and expectations.

Descartes, with the help of philosophy, sought to find the answer to his questions of the functions of the mind. Later research has had other means at its disposal and has tried to feel its way forward by other methods. Sperry has succeeded with sophisticated methods to extract from the brain some of its best guarded secrets and has allowed us to look into a world which until now has been nearly completely closed to us. Hubel and Wiesel have succeeded in breaking the code of the message which the eyes send to the brain and have thereby given us insight into the neuronal processes underlying our visual experiences.

The brain consists of two halves, hemispheres, which are structurally identical. Does this mean that we have two brains or that the two hemispheres have different tasks? The answer to this question can appear impossible to find because the brain halves are united by millions of nerve threads and, therefore, work in a complete functional harmony. However, it has been known for more than a hundred years that despite their similarity and close linkage the two hemispheres have in part different tasks to fulfill. The left hemisphere is specialized for speech and has, therefore, been considered absolutely superior to the right hemisphere. For the right hemisphere it has been difficult to find a role and it has generally been regarded as a "sleeping partner" of its left companion. In a way the roles of the two hemispheres were somewhat like those of man and wife of an old-time marriage.

In the beginning of the 1960s Sperry had the occasion to study some patients in whom the connections between the two hemispheres had been severed. The surgical intervention had been undertaken as a last resort to alleviate the epileptic seizures from which the patients suffered. In most of them an improvement occurred and there was a decrease in the frequency of their epileptic fits. Otherwise, the operation did not appear to be accompanied by any changes in the personality of the patients. However, Sperry was able, using brilliantly designed test methods to demonstrate that the two hemispheres in these patients had each its own stream of conscious awareness, perceptions, thoughts, ideas and memories, all of which were cut off from the corresponding experiences in the opposite hemisphere.

The left brain half is, as Sperry was able to show, superior to the right in abstract thinking, interpretation of symbolic relationships and in carrying out detailed analysis. It can speak, write, carry out mathematical calculations and in its general function is rather reminiscent of a computer. Furthermore, it is the leading hemisphere in the control of the motor system, the executive and in some respects the aggressive brain half. It is with this brain half that we communicate. The right cerebral hemisphere on the other hand is mute and in essence lacks the possibility to reach the outside world. It cannot write and can only read and understand the meaning of simple words in noun form and does not grasp the meaning of adjective or verb. It almost entirely lacks the ability to count and can only carry out simple additions up to 20. It completely lacks the ability to subtract, multiply and divide. Because of its muteness, the right brain half gives the impression of being inferior to the left. However, Sperry in his investigations was able to reveal that the right hemisphere in many ways is clearly superior to the left. Foremost, this concerns the capacity for concrete thinking, the apprehension and processing of spatial patterns, relations and transformations. It is superior to the left hemisphere in the perception of complex sounds and in the appreciation of music; it recognizes melodies more readily and also can accurately distinguish voices and tones. It is, too, absolutely superior to the left hemisphere in perception of nondescript patterns. It is with the right hemisphere we recognize the face of an acquaintance, the topography of a town or landscape earlier seen.

It is soon 50 years since Pavlov, the great Russian physiologist, put forward the suggestions that mankind can be divided into thinkers and artists. Pavlov was perhaps not entirely wrong in making this proposal. Today we know from Sperry's work that the left hemisphere is cool and logical in its thinking, while the right hemisphere is the imaginative, artistically creative half of the brain. Perhaps it is so that in thinkers the left hemisphere is dominant whereas in artists it is the right.

Hubel and Wiesel came in the mid-50s to the laboratory of the neurophysiologist S. W. Kuffler in Baltimore. Kuffler had at this time completed a series of investigations marked by an extraordinary experimental elegance in which he demonstrated how the picture that falls into the eyes is processed by the cells of the retina. Kuffler, who passed away a year ago, had by his work indicated the lines on which to continue analysis of the information processing of the visual system. This is, therefore, a fitting occasion on which to pay tribute to the memory of Kuffler for his important contribution.

The signal message that the eye sends to the brain can be regarded as a secret code to which only the brain possesses the key and can interpret the message. Hubel and Wiesel have succeeded in breaking the code. This they have achieved by tapping the signals from the nerve cells in the various cell layers of the brain cortex. Thus, they have been able to show how the various components of the retinal image are read out and interpreted by the cortical cells in respect to contrast, linear patterns and movement of the picture over the retina. The cells are arranged in columns, and the analysis takes place in a strictly ordered sequence from one nerve cell to another and every nerve cell is responsible for one particular detail in the picture pattern.

Hubel and Wiesel in their investigations were also able to show that the ability of the cortical cells to interpret the code of the impulse message from the retina develops early during a period directly after birth. A prerequisite for this development to take place is that the eye is subjected to visual experiences. If during this period one eye is sutured even for a few days, this can result in permanently impaired vision because the capacity of the brain to interpret the picture has not developed normally. For this to take place it is not only essential that the eye is reached by light but also that a sharp image is formed on the retina and that retinal image has a pattern of contours and contrasts. This discovery reveals that the brain has a high degree of plasticity at an early stage immediately after birth.

Hubel and Wiesel have disclosed one of the most well guarded secrets of the brain: the way by which its cells decode the message which the brain receives from the eyes. Thanks to Hubel and Wiesel we now begin to understand the inner language of the brain. Their discovery of the plasticity of the brain cortex during an early period of our life has implications reaching far beyond the field of visual physiology and proves the importance of a richly varied sensory input for the development of the higher functions of the brain.

Dr. Sperry, Dr. Hubel and Dr. Wiesel,

You have with your discoveries written one of the most fascinating chapters in the history of brain research. You, Dr. Sperry, have given us more profound insights into the higher functions of the brain than all the knowledge acquired in the twentieth century. You, Dr. Hubel and Dr. Wiesel, have translated the symbolic calligraphy of the brain cortex. The deciphering of the hieroglyphic characters of the ancient Egyptians has been denoted as one of the greatest advances in the history of philology. By breaking the code of the enigmatic signals of the visual system you have made an achievement which for all time will stand out as one of the most important in the history of brain research.

It is a privilege and pleasure for me to convey to you the warmest congratulations of the Nobel Assembly of the Karolinska Institute and to invite you to receive your Nobel Prize from the hands of His Majesty the King.

Roger Sperry

ROGER SPERRY

Birthplace and Family: Born August 20, 1913, in Hartford, Connecticut to Francis Bushnell and Florence Kraemer Sperry of Elmwood, a small suburb. Father was in banking; mother trained in business school and after dad's death, when I was 11 years old, she became assistant to the principal in the local high school. One brother, Russell Loomis, a year younger, went into chemistry. I was married to Norma Gay Deupree, December 28, 1949. We have one son, Glenn Michael (Tad), born October 13, 1953 and one daughter, Janeth Hope, born August 18, 1963.

Education: My early schooling was in Elmwood, Connecticut and William Hall High School in West Hartford, Connecticut. I attended Oberlin College on a 4 year Amos C. Miller Scholarship. After receiving the AB in English in 1935, I stayed on 2 years more in Oberlin for an MA in Psychology, 1937, under Professor R.H. Stetson. I then took an additional third year at-large at Oberlin to prepare for a switch to Zoology for Ph.D. work under Professor Paul A. Weiss at the University of Chicago. After receiving the Ph.D. at Chicago in 1941, I did a year of postdoctoral research as a National Research Council Fellow at Harvard University under Professor Karl S. Lashley.

Professional positions: Biology research fellow, Harvard University, at Yerkes Laboratories of Primate Biology (1942–46); Assistant professor, Department of Anatomy, University of Chicago (1946–52); Associate professor of psychology, University of Chicago (1952–53); Section Chief, Neurological Diseases and Blindness, National Institutes of Health (1952–53); Hixon professor of psychobiology, California Institute of Technology (1954–present).

Awards and Honors: Amos C. Miller Scholarship, Oberlin College (1931–35); National Research Council Fellowship (1941–42); Distinguished Alumni Citation; Oberlin College (1954); Elected National Academy of Sciences (1960); Elected American Academy of Arts and Sciences (1963); Howard Crosby Warren Medal, Society of Experimental Psychologists (1969); Distinguished Scientific Contribution Award, American Psychological Association (1971); California Scientist of the Year Award (1972); Co-recipient William Thomson Wakeman Research Award, National Paraplegia Foundation (1972); Honorary Doctor of Science degree, Cambridge University (1972); Passano Award in Medical Science (1973); Elected American Philosophical Society (1974); Elected Honorary Member American Neurological Association (1974); Co-recipient Claude Bernard Science Journalism Award (1975); Karl Lashley Award of American Philosophical Society (1976); Elected Foreign Member of Royal Society (1976); Honorary Doctor of Science Degree, University of Chicago (1976); Elected member of Pontifical Academy of Sciences (1978); Honorary

Doctor of Science Degree, Kenyon College (1979); Wolf Prize in Medicine (1979); Ralph Gerard Award of the Society of Neurosciences (1979); International Visual Literacy Association Special Award (1979); Albert Lasker Medical Research Award (1979); Honorary Doctor of Science Degree, The Rockefeller University (1980); American Academy of Achievement Golden Plate Award (1980).

A vocational and anti-brain-strain: Collected and raised large American moths in grade school. Ran trap line and collected live wild pets during junior high school years. Three-letter man in varsity athletics in high school and college. Through middle life continued evening and weekend diversionary activities including sculpture, ceramics, figure drawing, sports, American folk dance, boating, fishing, snorkeling, water colors, and collecting unusual fossils—among which we have a contender for the world's 3rd largest ammonite.

Selected Bibliography

1. The problem of central nervous reorganization after nerve regeneration and muscle transposition. R.W. Sperry. Quart. Rev. Biol. 20: 311–369 (1945).
2. Regulative factors in the orderly growth of neural circuits. R.W. Sperry. Growth Symp. 10: 63–67 (1951).
3. Cerebral organization and behavior. R.W. Sperry. Science 133: 1749–1757 (1961).
4. Chemoaffinity in the orderly growth of nerve fiber patterns and connections. R.W. Sperry. Proc. Nat. Acad. Sci. USA 50: 703–710 (1963).
5. Interhemispheric relationships: the neocortical commissures; syndromes of hemisphere disconnection. R.W. Sperry, M.S. Gazzaniga, and J.E. Bogen. In *Handbook Clin. Neurol.* P.J. Vinken and G.W. Bruyn (Eds.), Amsterdam: North-Holland Publishing Co. 4: 273–290 (1969).
6. Lateral specialization in the surgically separated hemispheres. R.W. Sperry. In *Neurosciences Third Study Program.* F. Schmitt and F. Worden (Eds.), Cambridge: MIT Press 3: 5–19. (1974).
7. Mind-brain interaction: mentalism, yes; dualism, no. R.W. Sperry. Neuroscience 5: 195–206 (1980). Reprinted in *Commentaries in the Neurosciences.* A.D. Smith, R. Llanas and P.G. Kostyuk (Eds.), Oxford: Pergamon Press, pp. 651–662 (1980).
8. *Science and moral priority: merging mind, brain and human values.* R.W. Sperry. Vol. 4 of Convergence, (Ser. ed. Ruth Anshen) New York: Columbia University Press (1982).

SOME EFFECTS OF DISCONNECTING THE CEREBRAL HEMISPHERES

Nobel lecture, 8 December 1981

by

ROGER SPERRY

Division of Biology, California Institute of Technology, Pasadena, California U.S.A.

INTRODUCTION: CLASSIC VIEW OF CEREBRAL DOMINANCE

To start by looking back a little, recall that even a small brain lesion, if critically located in the left or language hemisphere, may selectively destroy a person's ability to read, while at the same time sparing speech and the ability to converse. The printed page continues to be seen, but the words have lost their meaning. This condition typically follows from focal damage to the angular gyrus in the left hemisphere. It also results from lesions interrupting the neural input to this left angular gyrus from the visual or calcarine cortical areas (1, 2). It is natural to conclude in such cases that the left hemisphere is responsible for reading while the undamaged right hemisphere, in contrast, must be 'word-blind' or incapable of seeing meaning in the printed word.

The same applies with respect to the capacity to comprehend *spoken* words. Focal lesions within Wernicke's area near the base of the left temporal lobe or, again, lesions that disconnect this area from its input arriving from the auditory receiving centers of the cortex have been shown to regularly abolish the capacity to understand spoken language (2). Speech continues to be heard but the meaning is lost. Again, such cases seem to tell us that word comprehension is confined to the left hemisphere and that the spared right hemisphere must be word-deaf, as well as word-blind.

The accumulation of many observations of this kind where left, but not right, focal damage destroys the comprehension, as well as the expression, of language helped to give rise over the years to the so-called classic view in neurology of a dominant or major, left, language hemisphere and a subordinate, or minor, nonlanguage hemisphere. The minor hemisphere in addition to being unable to talk, and unable to write, and word-deaf and word-blind, was inferred by extrapolation to be typically lacking also in the higher cognitive faculties associated with language and symbolic processing.

This classic view of cerebral dominance was further reinforced by parallel findings on apraxia in which disorders of learned volitional movement were reported to follow predominantly lesions on the left side. The left hemisphere accordingly came to be regarded as being also the leading motor executive for the direction and control of higher volitional movements and the major repository for the cerebral engrams of motor learning (3, 4). Evidence for left dominance extended further to calculation and arithmetic reasoning (5). Thus, with few exceptions, the bulk of the collected lesion evidence up through the 1950s into the early '60s converged to support the picture of a leading, more highly evolved and intellectual left hemisphere and a relatively retarded right hemisphere that by contrast, in the typical righthander brain, is not only mute and agraphic but also dyslexic, word-deaf and apraxic, and lacking generally in higher cognitive function.

CONTRASTING EVIDENCE FROM COMMISSUROTOMY

It thus came as a considerable surprise in the early 1960s when tests on commissurotomy or 'split-brain' patients seemed to indicate the presence in the right, so-called 'minor' hemisphere of a considerable capacity for cognitive understanding and the comprehension of language, both written and spoken. These were patients of the neurosurgeons Joseph Bogen and his chief, Phillip Vogel of the White Memorial Medical Center in Los Angeles. The patients had undergone a midline surgical section of the corpus callosum and other forebrain commissures in a last resort effort to control severe, intractable epilepsy. The operation severed all neural cross connections for direct communication between the two hemispheres. From experience with this operation in human patients (6) and from nearly 10 years of split-brain animal studies (7), it could be predicted that the effect would not be seriously incapacitating as far as ordinary daily activities were concerned, and this proved to be the case. Given six months to a year for recovery, and in the absence of other major brain pathology, a person with complete section of the forebrain commissures would go undetected as a rule in a casual first meeting or conversation or even through an entire routine medical exam.

Our early studies with Michael Gazzaniga (8, 9, 10) on these patients seemed to show from the start that the disconnected right hemisphere was by no means word-deaf as anticipated, nor either word-blind. Lateralized testing for linguistic abilities showed the right hemisphere to be largely mute and agraphic, but nevertheless able to comprehend, at a moderately high level, words spoken aloud by the examiner. The disconnected right hemisphere also was able to read printed words flashed to the left visual field—as demonstrated manually in each case by selective retrieval or by pointing to corresponding objects or pictures in a choice array. The commissurotomy patients were also able with the right hemisphere to choose correct written or spoken words to match presented objects or pictures and to go correctly from spoken to printed words and vice versa. Correct tactual retrieval by the right hemisphere was achieved for objects not directly named but only described with complex spoken phrases like "a measuring instrument", "container for liquids", etc. With the disconnected right hemisphere, these patients could also spell three and four letter words with cutout letters and could read such words presented tactually. These semantic capabilities of the right hemisphere have more recently been affirmed and extended in a comprehensive series of experiments by Zaidel (11) using his improved scleral lens technique that allows prolonged viewing. So strong was contemporary neurological doctrine to the contrary in the early sixties that Dr. Bogen felt obliged in good conscience to withdraw his name from our initial papers on language.

Our own conviction that the answers on these language tests had to be coming from the right and not from the left half of the brain was based on lateralized testing procedures in which the speaking left hemisphere could be shown, by follow-up verbal questions, to have remained incognisant or quite unaware of the answers and performances being ascribed to the right hemisphere. Each disconnected hemisphere behaved as if it were not conscious of cognitive events in the partner hemisphere—just as had been the case in our split-brain animal studies of the 1950s, started by Ronald Myers (12) at the University of Chicago. Each brain half, in other words, appeared to have its own, largely separate, cognitive domain with its own private perceptual, learning and memory experiences, all of which were seemingly oblivious of corresponding events in the other hemisphere. Although the basic hemisphere deconnection syndrome in man (10) proved to be essentially similar to that worked out earlier in cats and monkeys, its manifestation was much more dramatic in the human subjects. The speaking hemisphere in these patients could tell us directly in its own words that it knew nothing of the inner experience involved in test performances correctly carried out by the mute partner hemisphere. Lateralization of brain functions could be inferred, not only from the deficiency or absence of function on one side but also from its concurrent presence on the other.

RIGHT HEMISPHERE LANGUAGE CONTROVERSY

The unexpected language capacities found in the right hemisphere after commissurotomy posed some controversial issues the answers to which are still not entirely resolved. Very simply, the problem raised is the following: Why is it that the right hemisphere is able to do things following commissurotomy, such as reading, that it fails to do in the presence of focal damage in the left hemisphere? It has been suggested in answer (13,14,15) that the commissurotomy evidence may be misleading because of an atypical bilateral spread of language into the right hemisphere correlated with the long-term epilepsy and associated pathology. A further criticism has invoked individual variation in view of the small patient group involved.

We have favored another interpretation which suggests conversely that it is the unilateral lesion evidence that has been misleading. The reasoning here says that left lesions in the presence of the commissures act to prevent the expression of latent function, actually present but suppressed, within the undamaged right hemisphere (10). This interpretation assumes that the two halves of the brain, when connected, work closely together as a functional unit with the leading control being in one or the other. When this unitary function is rendered defective by a one-sided lesion, the resultant impaired function prevails with respect to both hemispheres. That is, the two continue to operate as an integral, though defective, functional unit. Only after the intact right hemisphere is released from its integration with the disruptive and suppressive influence of the damaged hemisphere, as effected by commissurotomy, can its own residual function become effective.

This interpretation found support also in the limited hemispherectomy data available (16). The same reasoning has seemed to apply as well to phenomena of unilateral neglect and apraxia neither of which proved to be nearly so severe in lateralized tests after commissurotomy as one might have expected from the lateral lesion findings. Although the final word on these various issues is not yet in, the foregoing interpretation has received considerable support in subsequent commissurotomy studies which reveal the presence in the disconnected right hemisphere of additional superior cognitive capacities that can hardly be ascribed either to an atypical bilateralization of language or, any longer, to individual variation. There is reason to think that these other faculties also had gone unrecognized because of complexities that inevitably prevail in the presence of the commissures.

RIGHT HEMISPHERE SPECIALIZATION

Earlier indications of right hemisphere specialization in the lateral lesion data, such as in facial recognition, dressing, making block designs, drawing three-dimensional cubes, etc., had been ascribed to asymmetry in the sensory and motor-executive realms primarily rather than in higher central cognitive levels. These right hemisphere functions were referred to as 'visuospatial', 'constructional', or 'praxic'. In keeping with conventional conceptions of cerebral dominance, any higher cognitive processing that might be involved in such activities could be assumed to be contributed from the left hemisphere via the commissures. Our own initial interpretations of these activities did not depart substantially from the classic view (17).

By 1967, however, the collected observations on the commissurotomy subjects were being taken to uphold the conclusion (18) that each of the disconnected hemispheres, not only the left, has its own higher gnostic functions. Each hemisphere in the lateralized testing procedures appeared to be using its own percepts, mental images, associations and ideas. As in the split-brain animal studies, each could be shown to have its own learning processes and its own separate chain of memories, all of course, essentially inaccessible to conscious experience of the other hemisphere.

Added evidence for involvement of the right hemisphere in higher intellectual processing came from study of a case of congenital absence of the corpus callosum with an above-average verbal IQ and in whom speech was found to be present in the right as well as the left hemisphere (19, 20). The scholastic records of this college student with callosal agenesis were fair to good for courses that involved language and verbal facility, but contrastingly poor for subjects such as geometry and geography that involved spatial and related nonverbal faculties which we now commonly associate with the right hemisphere. The extra language in the right hemisphere had apparently been attained at the expense of the usual nonverbal cognitive faculties that otherwise normally develop there.

More direct, controlled evidence for right hemisphere superiority in tasks requiring higher cognitive ability came from studies by Jerre Levy (21, 22) aimed specifically at cognitive specialties of the right hemisphere. She found that the mental capacity to make intermodal spatial transformations from three-dimensional to unfolded, two-dimensional forms was much better developed in the right hemisphere. Also where items in the test series showed higher scores by the left hemisphere there was a corresponding drop in right hemisphere performance suggesting a left-right polarity in cognitive abilities.

From these data, taken in conjunction with available clues from the literature, Levy proposed that left and right hemispheres are characterized by inbuilt, qualitatively different and mutually antagonistic modes of cognitive processing, the left being basically analytic and sequential, the right spatial and synthetic. A rationale was added for the evolution of cerebral asymmetry (23) based on the functional advantages of having the two cognitive modes develop in separate hemispheres in order to minimize mutual interference.

In succeeding years thinking evolved rapidly along these lines and became strengthened and refined through a series of studies (24–31) in which it proved possible to demonstrate further that the so-called subordinate or minor hemisphere, which we had formerly supposed to be illiterate and mentally retarded and thought by some authorities to not even be conscious, was found to be in fact the superior cerebral member when it came to performing certain kinds of mental tasks. The right hemisphere specialities were all, of course, nonverbal, nonmathematical and nonsequential in nature. They were largely spatial and imagistic, of the kind where a single picture or mental image is worth a thousand words. Examples include reading faces, fitting designs into larger matrices, judging whole circle size from a small arc, discrimination and recall of nondescript shapes, making mental spatial transformations, discriminating musical chords, sorting block sizes and shapes into categories, perceiving wholes from a collection of parts, and the intuitive perception and apprehension of geometrical principles. The emphasis meantime became shifted somewhat from that of an intrinsic antagonism and mutual incompatibility of left and right processing to that of a mutual and supportive complementarity.

In many cases the observed left-right cognitive differences were rather subtle and qualitative in nature, such that they would easily be obscured in lateral lesion studies by individual variation and background pathology. Under the conditions of commissurotomy where background factors are equalized and where close left-right comparisons become possible within the same subject working the same problem, even slight lateral differences become significant. The same individual can be observed to employ consistently one or the other of two distinct forms of mental approach and strategy, much like two different people, depending on whether the left or right hemisphere is in use.

FURTHER EXTENSIONS

Further developments from other sources have advanced in many directions through study of various normal, brain-damaged and other select populations (32, 33), exploring correlations with handedness, gender, occupational preferences and ability, special innate talents, genetic variations like Turner's syndrome, congenital dyslexia, endocrine pathology, autism, dreaming, hypnosis, inverted writing—and others. In some cases the conclusions along with the growing wave of semipopular extrapolations and speculations concerning "left-brain" vs. "right-brain" functions call for a word of caution. The left-right dichotomy in cognitive mode is an idea with which it is very easy to run wild. Qualitative shifts in mental control may involve up-down, front-back, or various other organizational changes as well as left-right differences. Furthermore, in the normal state the two hemispheres appear to work closely together as a unit, rather than one being turned on while the other idles. Much yet remains to be settled in all these matters. Even the main idea of differential left and right cognitive modes is still under challenge in some quarters in favor of the view that the right hemisphere specialities are primarily praxic or 'manipulospatial' in character and that higher cognition and self awareness are associated mainly with language in the left hemisphere (34, 35).

Regardless of remaining uncertainties concerning laterality, one beneficial outcome that appears to hold up is an enhanced awareness, in education and elsewhere, of the important role of nonverbal components and forms of intellect. Another broadly relevant outcome, that derives from evidence involving familial, mutational, sexual and other innate variations, is a growing recognition of, and respect for the inherent individuality in the structure of human intellect. The more we learn, the more complex becomes the picture for predictions regarding any one individual and the more it seems to reinforce the conclusion that the kind of unique individuality in our brain networks makes that of fingerprints or facial features appear gross and simple by comparison. The need for educational tests and policy measures to selectively indentify, accommodate, and maximize the differentially specialized forms of individual intellectual potential becomes increasingly evident.

SELF CONSCIOUSNESS AND SOCIAL AWARENESS

Earlier contentions that the right hemisphere is not even conscious largely gave way by the mid seventies to an intermediate position conceding that the mute hemisphere may be conscious at some lower elemental levels, but claiming that it lacks the higher, reflective, self-conscious kind of inner awareness that is special to the human mind and is needed, so it is said, to qualify the right conscious system as a "self" or "person" (36,37). Self awareness in particular is reported, on the basis of mirror tests mainly, to be a predominantly human attribute and is rated by developmental as well as by evolutionary standards to be a highly advanced phase of conscious awareness.

Accordingly we undertook to test the right hemisphere more specifically for the presence of self recognition and related forms of self and social awareness. With perception of pictorial stimuli confined to one hemisphere by the scleral contact lens occluder developed by Eran Zaidel (38), the subject merely had to point to select items in a multiple choice array in answer to various kinds of leading questions regarding his or her knowledge and feelings concerning the content of the pictures. Subject's responses included also differential emotional expressions, thumbs-up, thumbs-down evaluations, exclamations, replies to 20-question type prompting and spontaneous remarks relevant to the emotional aspects of affect-laden stimuli.

The results (39) revealed that the disconnected right hemisphere readily recognizes and identifies him or herself among a choice array of portrait photos, and in doing so, generates appropriate emotional reactions and displays a good sense of humor requiring subtle social evalulations. Similar findings were obtained for pictures of the immediate family, relatives, acquaintances, pets, personal belongings, familiar scenes and also political, historical and religious figures, as well as television and screen personalities. The relatively inaccessible inner world of the nonspeaking hemisphere was thus found to be surprisingly well developed. The general level of performance on these tests was in good accord with that obtained from the left hemisphere of the same subject or in free vision. Results to date suggest the presence of a normal and well developed sense of self and personal relations along with a surprising knowledgeability in general.

Similar projective procedures were used to explore for a sense of time in the right hemisphere and the presence of concern for the future with thus far no evidence of abnormal deficit. The nonvocal hemisphere appears to be quite cognisant of the person's daily and weekly schedules, the calendar, seasons, and important dates of the year. The right hemisphere also makes appropriate discriminations that show concern with regard to the thought of possible future accidents and personal or family losses. The need for life, fire, and theft insurance, for example, seems to be properly appreciated by the extensively tested mute hemisphere of these patients.

Unlike other aspects of cognitive function, emotions have never been readily confinable to one hemisphere. Though generated by lateralized input, the emotional effects tend to spread rapidly to involve both hemispheres, apparently through crossed fiber systems in the undivided brain stem. In the above tests for self consciousness and social awareness it was found that even subtle shades of emotion or semantic connotations generated in the right hemisphere could be quite helpful to the left hemisphere in its efforts to guess the nature of a stimulus known only to the right hemisphere. The results suggested that this affective, connotational or semantic component could play an extremely important role in cognitive processing generally.

The more structured and specific informational components of cognitive processing were shown to be separable from the emotional and connotational components. The former remained confined within the hemisphere in which it was generated, whereas the emotional overtones leaked across to influence neural processing in the other hemisphere. The evidence of this separability is in itself significant in regard to questions of the organization of the neural mechanisms of cognition. Also, since the affective component appears to be an eminently conscious property, the fact that it crosses at lower brainstem levels is of interest in reference to the structural basis of consciousness. A major thrust in our current work is aimed at determining more precisely what shades of emotional, connotational or semantic content are able to cross through the brainstem and how they affect cognitive processing on the other side. In these studies we are using a new technique just developed for lateralizing vision (40,41). It allows prolonged viewing without attachments to the eye.

PROGRESS ON MIND–BRAIN PROBLEM

In closing it remains to mention briefly that one of the more important things to come out of the split-brain work, as an indirect spin-off, is a revised concept of the nature of consciousness and its fundamental relation to brain processing (42,43,44). The key development here is a switch from prior non-causal, parallelist views to a new causal, or "interactionist" interpretation that ascribes to inner experience an integral causal control role in brain function and behavior. In effect, and without resorting to dualist views, the mental forces and properties of the conscious mind are restored to the brain of objective science from which they had long been excluded on materialist-behaviorist principles.

Acceptance of the revised "causal view" and the reasoning involved, now becoming widespread, carries important implications for science and for scientific views of man and nature. Cognitive introspective psychology and related cognitive science can no longer be ignored experimentally, or written off as "a science of epiphenomena", nor either as something that must, in principle, reduce eventually to neurophysiology. The events of inner experience, as emergent properties of brain processes, become themselves explanatory causal constructs in their own right, interacting at their own level with their own laws and dynamics. The whole world of inner experience (the world of the humanities) long rejected by 20th century scientific materialism, thus becomes recognized and included within the domain of science.

Basic revisions in concepts of causality are involved in which the whole, besides being "different from and greater than the sum of the parts", also causally determines the fate of the parts, without interfering with the physical or chemical laws for the subentities at their own level. It follows that physical science no longer perceives the world to be reducible to quantum mechanics or to any other unifying ultra element or field force. The qualitative, holistic properties at all different levels become causally real in their own form and have to be included in the causal account. Quantum theory on these terms no longer replaces or subsumes classical mechanics but rather just supplements or complements.

The results add up to a fundamental change in what science has long stood for throughout the materialist-behaviorist era (45). The former scope of science, its limitations, world perspectives, views of human nature, and its societal role as an intellectual, cultural and moral force all undergo profound change. Where there used to be a chasm and irreconcilable conflict between the scientific and the traditional humanistic views of man and the world (46,47), we now perceive a continuum. A unifying new interpretative framework emerges (48) with far reaching impact not only for science but for those ultimate value-belief guidelines by which mankind has tried to live and find meaning.

ACKNOWLEDGMENTS

Our split-brain studies could hardly have succeeded without the contributions of a long line of very able graduate students and postdoctoral associates. I am particularly grateful to Ronald Myers who started the animal work in his doctoral research at the University of Chicago; Michael Gazzaniga, first to work with the human subjects and Jerre Levy, first to demonstrate superior cognitive processing in the right hemisphere. All contributed immensely to these respective innovations as did others to more specific aspects of the program. We are deeply indebted to Drs. Joseph Bogen and Phillip Vogel for generously making their patients available for study, and to the patients themselves without whose long and willing cooperation the human work would not have been possible.

Our work has been dependent for funding since the late 1950s on successive federal grants conferred mainly by the National Institutes of Mental Health. My chair at Caltech was made possible and has been sustained throughout by the F. P. Hixon Fund of the California Institute of Technology donated to bring to the Institute research bearing on "the 'why' of human behavior."

For research assistance in the human studies we owe much to the dedicated efforts of Dahlia Zaidel over a 9 year period beginning in 1967, and to those also of Lois MacBird in both the animal, and more recently, the human work extending over a 25 year period to the present. Our research progress has been dependent in no small measure on the consistent support received on all sides at Caltech. My own efforts could not have prospered without the constant help and understanding of Norma, my wife, whose competence and willingness in handling matters of our home and family has freed me over the years to give added time to problems of the laboratory.

REFERENCES

1. Greenblatt, S. H. Neurosurgery and the anatomy of reading: A practical review. *Neurosurgery*, 1977, 1, 6–15.
2. Brown, J. W. *Aphasia; Apraxia and Agnosia; Clinical and theoretical aspects*. Springfield, III. Charles C. Thomas, Publisher, 1972.
3. Liepmann, H. Der weitere Krankheitsverlauf bei dem einseitig Apraktischen und der Gehirnbegund auf Grund von Serienschnitten. *Monatschrift für Psychiatrie und Neurologie*, 1906, 19, 217–243.

4. Geschwind, N. The apraxias: neural mechanisms of disorders of learned movement. *Am. Scient.*, 1975, 63, 188–195.
5. Hecaen, H. Clinical symptomatology in right and left hemispheric lesions. In: V. B. Mountcastle (Ed.), *Interhemispheric Relations and Cerebral Dominance.* Baltimore: The Johns Hopkins Press, 1962, 215–243.
6. Akelaitis, A. J. A study of gnosis, praxis, and language following section of the corpus callosum. *J. Neurosurg.*, 1944, 1, 94–102.
7. Sperry, R. W. Cerebral organization and behavior. *Science,* 1961, 133, 1749–1757.
8. Gazzaniga, M. S. & Sperry, R. W. Language after section of the cerebral commissures. *Brain,* 1967, 90, (I), 131–148.
9. Sperry, R. W. & Gazzaniga, M. S. Language following disconnection of the hemispheres. In: C. H. Millikan & F. L. Darley (Eds.), *Brain Mechanisms Underlying Speech and Language.* New York: Grune & Stratton, Inc., 1967, 177–184.
10. Sperry, R. W., Gazzaniga, M. S. & Bogen, J. E. Interhemispheric relationships: the neocortical commissures; syndromes of hemisphere disconnection. In: P. J. Vinken & G. W. Bruyn (Eds.), *Handbook of Clinical Neurology.* Amsterdam: North-Holland Publishing Company, 1969, 4, 177–184.
11. Zaidel. E. Lexical structure in the right hemisphere. In: P. Buser and A. Rougeul-Buser (Eds.) *Cerebral Correlates of Conscious Experience.* Elsevier North-Holland Biomedical Press, Amsterdam, 1978, 177–197.
12. Myers, R. E. Interocular transfer of pattern discrimination in cats following section of crossed optic fibers. *J. comp. physiol. Psychol.*, 1955, 48.
13. Geschwind, N., "Discussion". p. 222 in *Les Syndromes de disconnexion calleuse chez l'homme.* In: F. Michel and B. Schott (Eds.) Lyon Hospital Neurologique, 1974.
14. Selnes, O. A. A note on "On the representation of language in the right hemisphere of right-handed people." *Brain and Language,* 1976, 583–590.
15. Whitaker, H. A. & Ojemann, G. A. Lateralization of higher cortical functions: a critique. In: S. Dimond & D. A. Blizard (Eds.), *Evolution and Lateralization of the Brain,* Annals New York Academy of Sciences, 1977, 299, 459–473.
16. Smith, A. Speech and other functions after left (dominant) hemispherectomy. *J. Neurol. Neurosurg. & Psychiat.*, 1966, 29, 467–471.
17. Bogen, J. E. & Gazzaniga, M. S. Cerebral commissurotomy in man. Minor hemisphere dominance for certain visuospatial functions. *J. of Neurosurg.*, 1965, 394–399.
18. Sperry, R. W., Vogel, P. J. & Bogen, J. E. Syndrome of hemisphere deconnection. In: P. Bailey and R. E. Foil (Eds.), *Proceedings 2nd Pan-Am Congress of Neurology,* Puerto Rico, 1970, 195–200.
19. Saul R. & Sperry, R. W. Absence of commissurotomy symptoms with agenesis of the corpus callosum. *Neurology,* 1968, 18, 307.
20. Sperry, R. W. Plasticity of neural maturation. *Developmental Biology Supplement,* 1968, 2, 306–327.
21. Levy, J. Information processing and higher psychological functions in the disconnected hemispheres of commissurotomy patients. Unpublished doctoral dissertation, California Institute of Technology, 1970. (Ann Arbor, Mich.: University Microfilms No. 70–14, 844).
22. Levy-Agresti, J. & Sperry, R. W. Differential perceptual capacities in major and minor hemispheres. *Proc. Nat. Acad. Sci.*, 1968, 61, 1151.
23. Levy, J. Possible basis for the evolution of lateral specialization of the human brain. *Nature,* 1969, 224, 614–615.
24. Zaidel, D. & Sperry, R. W. Performance on the Raven's Colored Progressive Matrices test by subjects with cerebral commissurotomy. *Cortex,* 1973, 9, 34–39.
25. Nebes, R. D. Superiority of the minor hemisphere in commissurotomized man for the perception of part-whole relations. *Cortex,* 1971, 7, 333–349.
26. Nebes, R. D. Dominance of the minor hemisphere in commissurotomized man on a test of figural unification. *Brain,* 1972, 95, Part III, 633–638.
27. Gordon, H. W. Hemispheric asymmetries in the perception of musical cords. *Cortex,* 1971, 6, 387–398.

28. Kumar, S. Lateralization of concept-formation in human cerebral hemispheres. *Caltech Biology Annual Report,* California Institute of Technology, 1971, No. 136, 118–119.
29. Milner, B. & Taylor, L. Right-hemisphere superiority in tactile pattern-recognition after cerebral commissurotomy: evidence for nonverbal memory. *Neuropsychologia,* 1972, 10, 1–15.
30. Levy, J., Trevarthen, C. & Sperry, R. W. Perception of bilateral chimeric figures following hemispheric deconnexion. *Brain,* 1972, 95, 61–78.
31. Franco, L. & Sperry, R. W. Hemisphere lateralization for cognitive processing of geometry. *Neuropsychologia,* 1977, 15, 107–114.
32. *Hemisphere function in the human brain.* S. Dimond & J. G. Beaumont (Eds.) London: Paul Eleck, Ltd., 1974.
33. *Evolution and Laterialization of the Brain.* S. T. Dimond and D. A. Blizard (Eds.) Annals of the New York Academy of Sciences, 299. The New York Academy of Sciences, New York, 1977.
34. LeDoux, J. E., Wilson, D. H. & Gazzaniga, M. S. Manipulo-spatial aspects of cerebral lateralization: clues to the origin of lateralization. *Neuropsychologia,* 1977, 15, 743–749.
35. Eccles, J. C. The Human Psyche. (The Gifford Lectures). Springer-Verlag: Berlin, 1980.
36. DeWitt, L. Consciousness, mind and self: The implications of the split-brain studies. *Brit. J. Philos. Sci.,* 1975, 26, 41–47.
37. Popper, K. & Eccles, J. *The self and its brain:* An argument for interactionism. Berlin: Springer International, 1977.
38. Zaidel, E. A technique for presenting lateralized visual input with prolonged exposure. *Vision Res.,* 1975, 15, 283–289.
39. Sperry, R. W., Zaidel, E. & Zaidel, D. Self recognition and social awareness in the deconnected minor hemisphere. *Neuropsychologia,* 1979, 17, 153–166.
40. Sperry, R. W. & Myers, J. J. A simplified technique for lateralizing visual input. *Caltech Biology Annual Report,* No. 231, 1981.
41. Myers, J. J. & Sperry, R. W. A simple technique for lateralizing visual input that allows prolonged viewing. *Behav. Res. Meth. Instrum.* 1982 in press.
42. Sperry, R. W. Mind, brain and humanist values. In: J. R. Platt (Ed.), *New Views of the Nature of Man.* 71–92. Chicago: University of Chicago Press. 1965.
43. Sperry, R. W. A modified concept of consciousness. *Psychological Review,* 1969, 76, 532–536.
44. Sperry, R. W. Mind-brain interaction: Mentalism, Yes; Dualism, No. *Neuroscience,* 1980, 5, 195–206.
45. Sperry, R. W. Changing Priorities. *Ann. Rev. Neurosci.,* 1981, 4, 1–15.
46. Snow, C. P. *The two cultures and the scientific revolution.* New York: Cambridge University Press, 1959.
47. Jones, W. T. *The Sciences and the Humanities: Conflict and Reconciliation.* University of California Press, Berkeley and Los Angeles, 1965.
48. Sperry, R. W. *Science and Moral Priority: Merging Mind, Brain and Human Values.* Vol. stet 4 of *Convergence* (Ser. ed. R. N. Anshen) New York: Columbia University Press, 1982.

competition that is found in graduate students today; it was possible to take more long-shots without becoming panic stricken if things didn't work out brilliantly in the first few months. We were not free from financial worries, as graduate students in biology by and large are now; until I entered the army my income was close to zero, and I owe a huge debt to my wife Ruth for supporting us through those lean and exploited years of residency and fellowship training.

Scientifically, I could hardly have chosen a better place than Walter Reed. In the neuropsychiatry division David Rioch had assembled a broad and lively group of young neuroscientists, notably M.G.F. Fuortes and Robert Galambos in neurophysiology, Walle Nauta in neuroanatomy, Joseph Brady and Murray Sidman in experimental psychology and John Mason in chemistry. As in Montreal, the focus was on the entire nervous system, not on a subdivision of biological subject matter based on methods. I worked under the supervision of Fuortes. We began by collaborating for six months on a spinal cord project, and it was then that I had my only apprenticeship in experimental neurophysiology. Fuortes had a genuine feel for biology that was rare among neurophysiologists in those days. I also learned and benefited much from a most able and helpful research assistant, Calvin Henson. My main project while at Walter Reed was a comparison of the spontaneous firing of single cortical cells in sleeping and waking cats. I began by recording from the visual cortex: it seemed most sensible to look at a primary sensory area, and the visual was easiest, there being less muscle betweeen that part of the brain and the ouside world. It was first necessary to devise a method for recording from freely moving cats and to develop a tungsten microelectrode tough enough to penetrate the dura. That took over a year, but in the end it was exciting to be able to record from a single cell in the cortex of a cat that was looking around and purring.

In 1958 I moved to the Wilmer Institute, Johns Hopkins Hospital, to the laboratory of Stephen Kuffler, and there I began collaboration with Torsten Wiesel. A year later Kuffler's entire laboratory (nine families) moved to Harvard Medical School in Boston, at first as part of the Department of Pharmacology under Otto Krayer, who was largely responsible for bringing Kuffler to Harvard. Five years later, in a move unprecedented for Harvard, we became the new Department of Neurobiology. My daily contacts with Stephen Kuffler (until his death a year ago) and with Edwin Furshpan, Edward Kravitz, David Potter and Simon LeVay have been both fun and enriching. During the past twenty two years, besides working with Torsten, I have collaborated briefly with Ursula Dräger, Helga Ginzler, and Ann Graybiel. At present I am working with Margaret Livingstone.

Since the age of five I have spent a disproportionate amount of time on music, for many years the piano, then recorders, and now the flute. I do woodworking and photography, own a small telescope for astronomy, and I ski and play tennis and squash. I enjoy learning languages, and have spent untold hours looking up words in French, Japanese and German dictionaries. In the laboratory I enjoy almost everything, including machining, photography, computers, sugery—even neurophysiology.

This is perhaps a suitable place to express my deep gratitude to the Eye Institute of the National Institutes of Health, to the U.S. Air Force, the Klingenstein Foundation, and to the Rowland Foundation for their generous support of our research. Also the Faculty of Harvard University deserves my thanks for tolerating such a truculent colleague.

EVOLUTION OF IDEAS ON THE PRIMARY VISUAL CORTEX, 1955–1978: A BIASED HISTORICAL ACCOUNT

Nobel lecture, 8 December 1981

by
DAVID H. HUBEL

Harvard Medical School, Department of Neurobiology,
Boston, Massachusetts, U.S.A.

INTRODUCTION

In the early spring of 1958 I drove over to Baltimore from Washington, D.C., and in a cafeteria at Johns Hopkins Hospital met Stephen Kuffler and Torsten Wiesel, for a discussion that was more momentous for Torsten's and my future than either of us could have possibly imagined.

I had been at Walter Reed Army Institute of Research for three years, in the Neuropsychiatry Section headed by David Rioch, working under the supervision of M.G.F. Fuortes. I began at Walter Reed by developing a tungsten microelectrode and a technique for using it to record from chronically implanted cats, and I had been comparing the firing of cells in the visual pathways of sleeping and waking animals.

It was time for a change in my research tactics. In sleeping cats only diffuse light could reach the retina through the closed eyelids. Whether the cat was asleep or awake with eyes open, diffuse light failed to stimulate the cells in the striate cortex. In waking animals I had succeeded in activating many cells with moving spots on a screen, and had found that some cells were very selective in that they responded when a spot moved in one direction across the screen (e.g. from left to right) but not when it moved in the opposite direction (1) (Fig. 1). There were many cells that I could not influence at all. Obviously there was a gold mine in the visual cortex, but methods were needed that would permit the recording of single cells for many hours, and with the eyes immobilized, if the mine were ever to begin producing.

I had planned to do a postdoctoral fellowship at Johns Hopkins Medical School with Vernon Mountcastle, but the timing was awkward for him because he was remodeling his laboratories. One day Kuffler called and asked if I would like to work in his laboratory at the Wilmer Institute of Ophthalmology at the Johns Hopkins Hospital with Torsten Wiesel, until the remodeling was completed. That was expected to take about a year. I didn't have to be persuaded; some rigorous training in vision was just what I needed, and though Kuffler himself was no longer working in vision the tradition had been maintained in his laboratory. Torsten and I had visited each other's laboratories and it was clear that we had common interests and similar outlooks. Kuffler

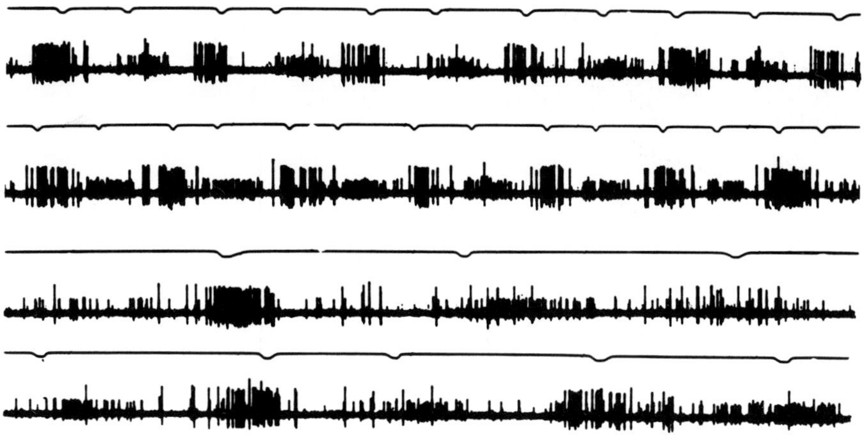

Figure 1. Continuous recording from striate cortex of an unrestrained cat. In each dual trace the lower member shows the microelectrode oscilloscope recording from two cells, one with large impulses, the other smaller ones. The stimulus was small to-and-fro hand movements in front of the cat. Each movement interrupted a light beam falling on a photoelectric cell, producing the notches in the upper beam. The upper two pairs of records represent fast movements, the lower ones slower movements. Each line represents 4 seconds. (1)

suggested that I come over to discuss plans, and that was what led to the meeting in the cafeteria.

It was not hard to decide what to do. Kuffler had described two types of retinal ganglion cells, which he called "on-center" and "off-center". The receptive field of each type was made up of two mutually antagonistic regions, a center and a surround, one excitatory and the other inhibitory. In 1957 Barlow, FitzHugh and Kuffler had gone on to show that as a consequence retinal ganglion cells are less sensitive to diffuse light than to a spot just filling the receptive-field center (2). It took me some time to realize what this meant: that the way a cell responds to any visual scene will change very little when, for example, the sun goes behind a cloud and the light reflected from black and white objects decreases by a large factor. The cell virtually ignores this change, and our subjective assessment of the objects as black or white is likewise practically unaffected. Kuffler's center-surround receptive fields thus began to explain why the appearance of objects depends so little on the intensity of the light source. Some years later Edwin Land showed that the appearance of a scene is similarly relatively independent of the exact color composition of the light source. The physiological basis of this color independence has yet to be worked out.

The strategy (to return to our cafeteria) seemed obvious. Torsten and I would simply extend Stephen Kuffler's work to the brain; we would record from geniculate cells and cortical cells, map receptive fields with small spots, and look for any further processing of the visual information.

My reception in Kuffler's office the first day was memorable. I was nervous and out of breath. Steve at his desk, rotated around on his chair and said "Hi, David! Take off your coat. Hang up your hat. Do up your fly." His laboratory was informal! But it took me a month, given my Canadian upbringing, to force myself to call him Steve. For the first three months no paycheck arrived and finally I screwed up the courage to go in and tell him. He laughed and laughed, and then said "I forgot!"

Torsten and I didn't waste much time. Within a week of my coming to Hopkins (to a dark and dingy inner windowless room of the Wilmer Institute basement, deemed ideal for visual studies) we did our first experiment. For the time being we finessed the geniculate (at Walter Reed I had convinced myself that geniculate cells were center-surround) and began right away with cortex. The going was rough. We had only the equipment for retinal stimulation and recording that had been designed a few years before by Talbot and Kuffler (3). A piece of apparatus resembling a small cyclotron held the anesthetized and paralyzed cat with its head facing almost directly upwards. A modified ophthalmoscope projected a background light and a spot stimulus onto the retina. The experimenter could look in, see the retina with its optic disc, area centralis and blood vessels, and observe the background light and the stimulus spots. Small spots of light were produced by sliding 2 cm × 5 cm metal rectangles containing various sizes of holes into a slot in the apparatus, just as one puts a slide into a slide projector. To obtain a black spot on a light background one used a piece of glass like a microscope slide, onto which a black dot had been glued. All this was ideal for stimulating the retina and recording directly from retinal ganglion cells, since one could see the electrode tip and know where to stimulate, but for cortical recording it was horrible. Finding a receptive field on the retina was difficult, and we could never remember what part of the retina we had stimulated. After a month or so we decided to have the cat face a projection screen, as I had at Walter Reed and as Talbot and Marshall had in 1941 (4). Having no other head holder, we continued for a while to use the ophthalmoscope's head holder, which posed a problem since the cat was facing directly up. To solve this we brought in some bed sheets which we slung between the pipes and cobwebs that graced the ceiling of the Wilmer basement, giving the setup the aura of a circus tent. On the sheets we projected our spots and slits. One day Vernon Mountcastle walked in on this scene, and was horror struck at the spectacle. The method was certainly inconvenient since we had to stare at the ceiling for the entire experiment. Then I remembered having seen in Mountcastle's laboratory a Horsley-Clarke head holder that was not only no longer being used but also had the name of the Wilmer Institute engraved on it. It was no other than the instrument that Talbot had designed for visual work when he and Marshall mapped out visual areas I and II in the cat, in 1941 (4). For years Vernon had used it in his somatosensory work, but he had recently obtained a fancier one. Torsten and I decided to reclaim the Wilmer instrument, not without some trepidation. To give ourselves confidence we both put on lab coats, for the first and last times in our lives, and looking very professional walked over to Physiology. Though Mountcastle was his usual friendly and generous self, I suspect he was loath to part with this treasure, but the inscription on the stainless steel was not to be denied and we walked off with it triumphantly. It is still in use (now at Harvard: we literally stole it from the

Wilmer), and has probably the longest history of uninterrupted service of any Horsley-Clarke in the world.

A short while before this adventure we had gone to a lecture by Vernon (this was a few years after his discovery of cortical columns) (5) in which he had amazed us by reporting on the results of recording from some 900 somatosensory cortical cells, for those days an astronomic number. We knew we could never catch up, so we catapulted ourselves to respectability by calling our first cell No. 3000 and numbering subsequent ones from there. When Vernon visited our circus tent we were in the middle of a 3-unit recording, cell Nos. 3007, 3008, and 3009. We made sure that we mentioned their identification numbers. All three cells had the same receptive-field orientation but neither Vernon nor we realized, then, what that implied.

At times we were peculiarly inept. Our first perfusion of a cat was typical. One morning at about 2:00 a.m. we had arranged two huge bottles on an overhead shelf, for saline and formalin, and were switching over from saline to formalin when the rubber tubing came off the outlet of the formalin bottle and gave us an acrid early morning cold shower. We did not relish being preserved at so young an age! The reference to 2:00 a.m. perhaps deserves some comment, because neurophysiologists, at least those who study animals, have the reputation of doing experiments that last for days without respite. We soon found that such schedules were not for us. I knew we were losing traction in an experiment when Torsten began to talk to me in Swedish; usually this was around 3:00 a.m. The longest experiment we ever did was one in which I arrived home just as my family was sitting down for breakfast. I had almost driven off the road on the way back. At the risk of becoming what Mountcastle termed "part-time scientists" we decided to be more lenient with ourselves, giving the deteriorating condition of the animal as the official reason for stopping early.

Our first real discovery came about as a surprise. We had been doing experiments for about a month. We were still using the Talbot-Kuffler ophthalmoscope and were not getting very far; the cells simply would not respond to our spots and annuli. One day we made an especially stable recording. (We had adapted my chronic recording system, which made use of Davies' idea of a closed chamber (6), to the acute experimental animals, and no vibrations short of an earthquake were likely to dislodge things.) The cell in question lasted 9 hours, and by the end we had a very different feeling about what the cortex might be doing. For 3 or 4 hours we got absolutely nowhere. Then gradually we began to elicit some vague and inconsistent responses by stimulating somewhere in the midperiphery of the retina. We were inserting the glass slide with its black spot into the slot of the ophthalmoscope when suddenly over the audiomonitor the cell went off like a machine gun. After some fussing and fiddling we found out what was happening. The response had nothing to do with the black dot. As the glass slide was inserted its edge was casting onto the retina a faint but sharp shadow, a straight dark line on a light background. That was what the cell wanted, and it wanted it, moreover, in just one narrow range of orientations.

This was unheard of. It is hard, now, to think back and realize just how free we were from any idea of what cortical cells might be doing in an animal's daily life. That the retinas mapped onto the visual cortex in a systematic way was of course well known, but it was far from clear what this apparently unimaginative remapping was good for. It seemed inconceivable that the information would enter the cortex and leave it unmodified, especially when Kuffler's work in the retina had made it so clear that interesting transformations took place there between input and output. One heard the word "analysis" used to describe what the cortex might be doing, but what one was to understand by that vague term was never spelled out. In the somatosensory cortex, the only other cortical area being closely scrutinized, Mountcastle had found that the cells had properties not dramatically different from those of neurons at earlier stages.

Many of the ideas about cortical function then in circulation seem in retrospect almost outrageous. One has only to remember the talk of "suppressor strips", reverberating circuits, or electrical field effects. This last notion was taken so seriously that no less a figure than our laureate-colleague Roger Sperry had had to put it to rest, in 1955, by dicing up the cortex with mica plates to insulate the subdivisions, and by skewering it with tantalum wire to short out the fields, neither of which procedures seriously impaired cortical function (7, 8). Nevertheless the idea of ephaptic interactions was slow to die out. There were even doubts as to the existence of topographic representation, which was viewed by some as a kind of artifact. One study, in which a spot of light projected anywhere in the retina evoked potentials all over the visual cortex, was interpreted as a refutation of topographic representation, but the result almost certainly came from working with a dark-adapted cat and a spot so bright that it scattered light all over the retina. It is surprising, in retrospect, that ideas of non-localization could survive in the face of the masterly mapping of visual fields onto the cortex in rabbit, cat and monkey done by Talbot and Marshall far back in 1941 (4).

It took us months to convince ourselves that we weren't at the mercy of some optical artifact, such as anyone can produce by squinting one's eyes and making vertical rays emanate from street lights. We didn't want to make fools of ourselves quite so early in our careers. But recording in sequence in the same penetration several cells with several different optimal orientations would, I think, have convinced anyone. By January we were ready to take the cells we thought we could understand (we later called them "simple cells") and write them up. Then as always what guided and sustained us was the attitude of Stephen Kuffler, who never lectured or preached but simply reacted with buoyant enthusiasm whenever he thought we had found something interesting, and acted vague and noncommittal when he found something dull. Neither of us will ever forget writing our first abstract, for the International Congress of Physiology in 1959 (9). We labored over it, and finally gave a draft to Kuffler. The following day when I came in Torsten was looking more glum than usual, and said "I don't think Steve much liked our abstract". It was clear enough that Kuffler wasn't quite satisfied: his comments and suggestions contained

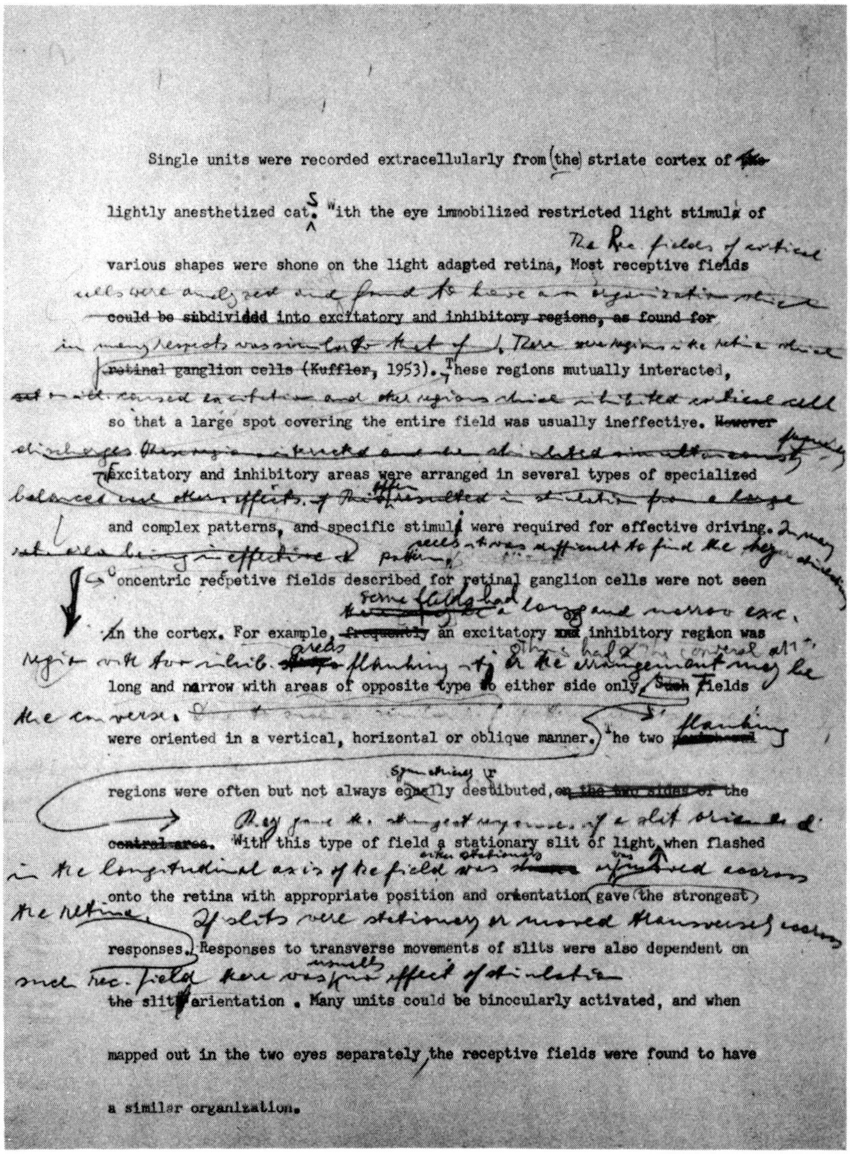

Single units were recorded extracellularly from the striate cortex of the lightly anesthetized cat. With the eye immobilized restricted light stimuli of various shapes were shone on the light adapted retina. Most receptive fields could be subdivided into excitatory and inhibitory regions, as found for retinal ganglion cells (Kuffler, 1953). These regions mutually interacted, so that a large spot covering the entire field was usually ineffective. However excitatory and inhibitory areas were arranged in several types of specialized and complex patterns, and specific stimuli were required for effective driving. Concentric recpetive fields described for retinal ganglion cells were not seen in the cortex. For example, frequently an excitatory and inhibitory region was long and narrow with areas of opposite type to either side only. Such fields were oriented in a vertical, horizontal or oblique manner. The two regions were often but not always equally destibuted, on the two sides of the central area. With this type of field a stationary slit of light when flashed onto the retina with appropriate position and orientation gave the strongest responses. Responses to transverse movements of slits were also dependent on the slit orientation. Many units could be binocularly activated, and when mapped out in the two eyes separately the receptive fields were found to have a similar organization.

Figure 2. First draft our first abstract (8), showing comments by Kuffler.

more words than our original (Fig. 2)! Writing, it may be added, did not come easy to either of us at the beginning, and our first paper, in 1959 (10) went through eleven complete reworkings.

HIERARCHY OF VISUAL CELLS

During the years 1959–62, first at the Wilmer Institute and then at Harvard Medical School, we were mainly concerned with comparing responses of cells in the lateral geniculate body and primary visual cortex of the cat. In the lateral geniculate we quickly confirmed my Walter Reed finding that the receptive fields are like those of retinal ganglion cells in having an antagonistic concentric center-surround organization. But now we could compare directly the responses of a geniculate cell with those of a fiber from an afferent retinal ganglion cell, and we found that in geniculate cells the power of the receptive-field surround to cancel the input from the center was increased. This finding was subsequently confirmed and extended in a beautiful set of experiments by Cleland, Dubin and Levick (11), and for many years remained the only known function of the lateral geniculate body.

In the cat striate cortex it soon became evident that cells were more complex than geniculate cells, and came in several degrees of complexity (12). One set of cells could be described by techniques similar to those used in the retina by Kuffler; we called these "simple". Their receptive fields, like the fields of retinal ganglion cells and of lateral geniculate cells, were subdivided into antagonistic regions illumination of any one of which tended to increase or decrease the rate of firing. But simple cells differed from retinal ganglion cells and lateral geniculate cells in the striking departure of their receptive fields from circular symmetry; instead of a single circular boundary between center and surround the antagonistic subdivisions were separated by parallel straight lines whose orientation (vertical, horizontal or oblique) soon emerged as a fundamental property (Fig. 3a). The optimal stimulus, either a slit, dark bar or edge, was easily predictable from the geometry of the receptive field, so that a stationary line stimulus worked optimally when its boundaries coincided with the boundaries of the subdivisions (Fig. 3c), and displacing the line to a new position parallel to the old one generally resulted in a sharp decline in the response. Perhaps most remarkable of all was the precise nature of the spatial distribution of excitatory and inhibitory effects: not only did diffuse light produce no response (as though the excitatory and inhibitory effects were mutually cancelling with the precision of an acid-base titration), but any line oriented at 90° to the optimal was also without effect, regardless of its position along the field, suggesting that the subpopulations of receptors so stimulated also had precisely mutually cancelling effects.

In the cat, simple cells are mostly found in layer IV, which is the site of termination of the bulk of the afferents from the lateral geniculate body. The exact connections that lead to orientation specificity are still not known, but it is easy to think of plausible circuits. For example, the behavior of one of the commonest kinds of simple cells may be explained by supposing that the cell receives convergent excitatory input from a set of geniculate cells whose on-centers are distributed in overlapping fashion over a straight line (Fig. 3b). In the monkey, the cells of Layer IVc (where most geniculate fibers terminate) seem all to be concentric center-surround, and the simple cells are probably mainly in the layers immediately superficial to IVc. No one knows why this extra stage of center-surround cells is intercalated in the monkey's visual pathway.

The next set of cells we called "complex" because their properties cannot be derived in a single logical step from those of lateral geniculate cells (or, in the monkey, from the concentric cells of layer IVc). For the complex cell, com-

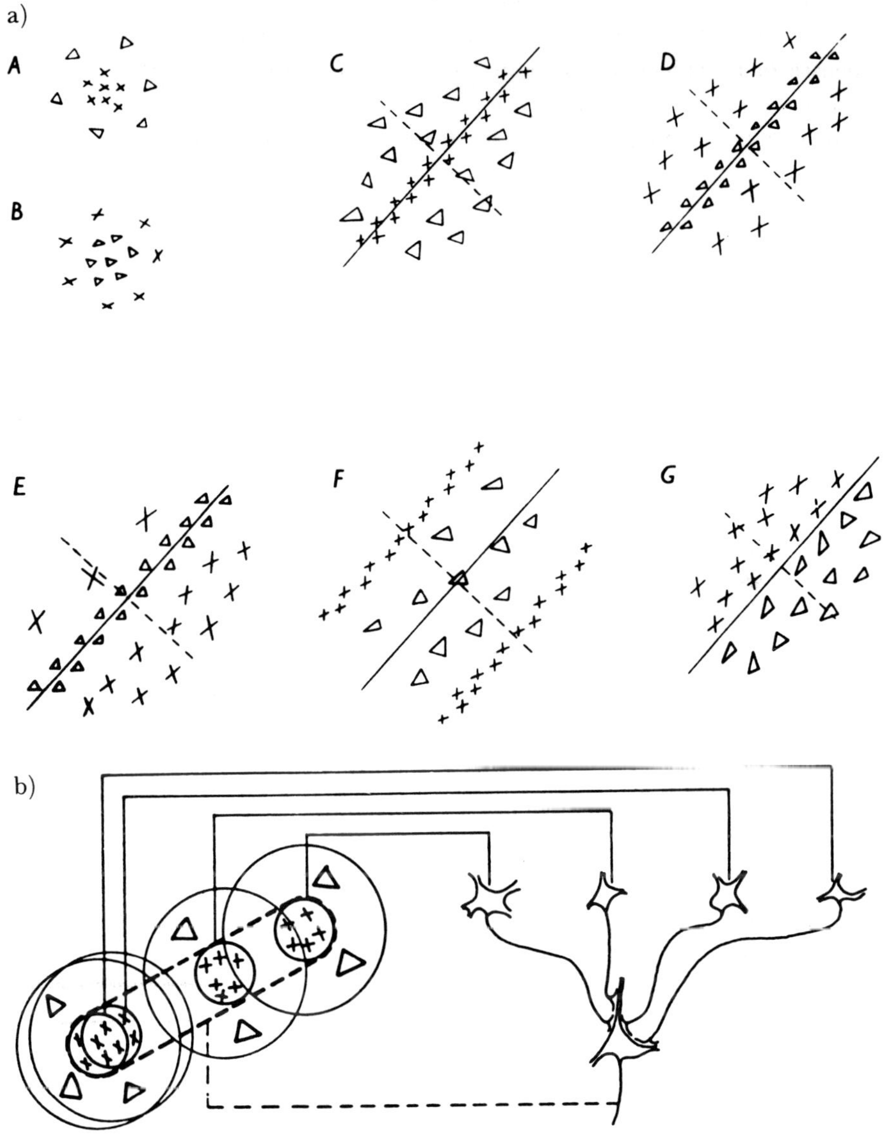

Figure 3. a) Common arrangement of lateral geniculate (A and B) and simple cortical (C−G) receptive fields. X, areas giving excitatory responses ('on' responses); △, areas giving inhibitory responses ('off' responses). Receptive-field orientations are shown by continuous lines through field centers; in the figure these are all oblique, but each arrangement occurs in all orientations (Fig. 2 (12)).

b) Possible scheme for explaining the organization of simple receptive fields. A large number of lateral geniculate cells, of which four are illustrated in the right in the figure, have receptive fields with 'on' centers arranged along a straight line on the retina. All of these project upon a single cortical cell, and the synapses are supposed to be excitatory. The receptive field of the cortical cell will then have an elongated 'on' center indicated by the interrupted lines in the receptive-field diagram to the left of the figure (Fig. 19 (12)).

c) Responses to shining a rectangular slit of light 1 × 8°, so that center of slit is superimposed on center of receptive field, in various orientations, as shown. Receptive field is of type C (see part (a) of this Figure), with axis vertically oriented (Fig. 3 (10)).

c)

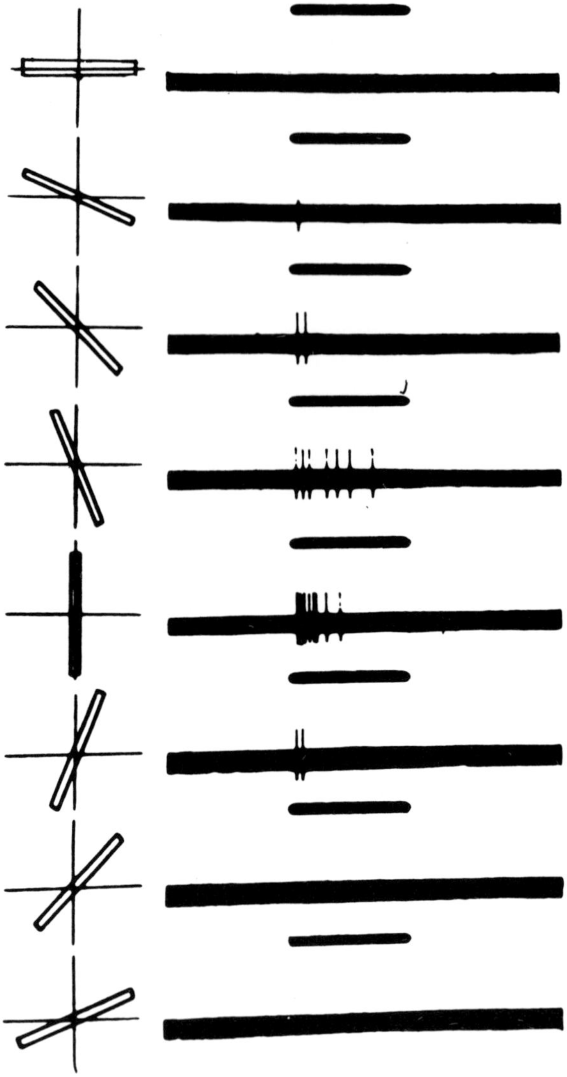

pared to the simple cell, the position of an optimally oriented line need not be so carefully specified: the line works anywhere in the receptive field, evoking about the same response wherever it is placed (Fig. 4a). This can most easily be explained by supposing that the complex cell receives inputs from many simple cells, all of whose receptive fields have the same orientation but differ slightly in position (Fig. 4b). Sharpness of tuning for orientation varies from cell to cell, but the optimal orientation of a typical complex cell in layer II or III in the monkey can be easily determined to the nearest 5–10°, with no more equipment than a slide projector.

For a complex cell, a properly oriented line produces especially powerful responses when it is swept across the receptive field (Fig. 4c). The discharge is generally well sustained as long as the line keeps moving, but falls off quickly if the stimulus is stationary. About half of the complex cells fire much better to one direction of movement than to the opposite direction, a quality called

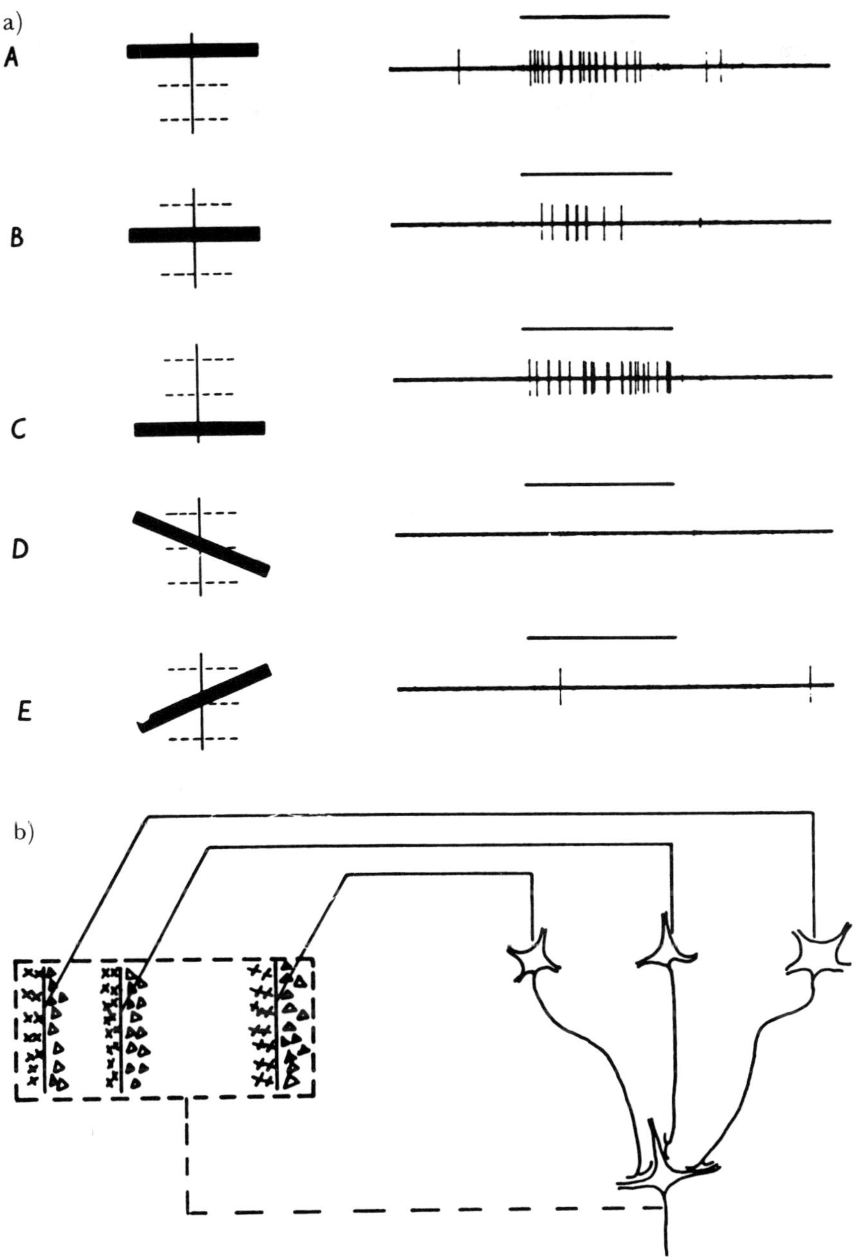

Figure 4. a) Complex cell responding best to a black horizontally oriented rectangle placed anywhere in the receptive field (A–C). Tilting the stimulus rendered it ineffective (D, E).

b) Same cell, showing response to a moving horizontal bar, downward movement better than upward (A), and no response to a moving vertical bar (B). Time 1 sec. (Figs. 7 and 8 (12)).

c) Possible scheme for explaining the organization of complex receptive fields. A number of cells with simple fields, of which three are shown schematically, are imagined to project to a single cortical cell of higher order. Each projecting neuron has a receptive field arranged as shown to the left: an excitatory region to the left and an inhibitory region to the right of a vertical straight-line boundary. The boundaries of the fields are staggered within an area outlined by the interrupted lines. Any vertical-edge stimulus falling across this rectangle, regardless of its position, will excite some simple-field cells, leading to excitation of the higher-order cell. (Fig. 20 (12)).

c)

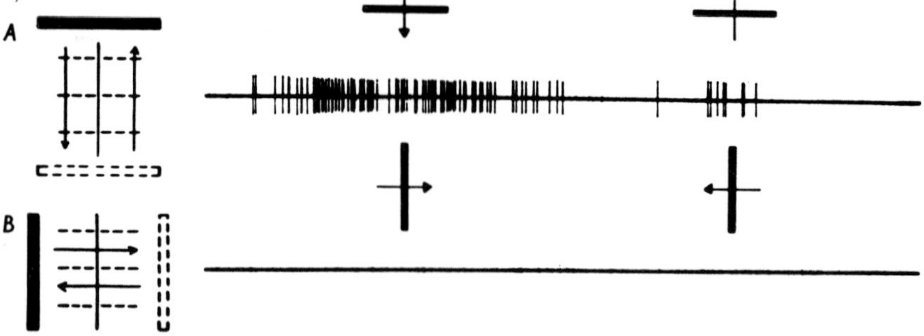

"directional selectivity", which probably cannot be explained by any simple projection of simple cells onto complex cells, but seems to require inhibitory connections with time delays of the sort proposed by Barlow and Levick for rabbit retinal ganglion cells (13).

HYPERCOMPLEX CELLS

Many cells, perhaps 10–20% in area 17 of cat or monkey, respond best to a line (a slit, edge or dark bar) of limited length; when the line is prolonged in one direction or both, the response falls off. This is called "end stopping". In some cells the response to a very long line fails completely (Fig. 5) (14). We originally called these cells "hypercomplex" because we looked upon them as next in an ordered hierarchical series, after simple and complex. We saw hypercomplex cells first in areas 18 and 19 of the cat, and only later in area 17. Dreher subsequently found cells, in all other ways resembling simple cells, that showed a similar fall-off in response as the length of the stimulus exceeded some optimum (15). It seems awkward to call these cells hypercomplex; they are probably better termed "simple end-stopped" in contrast to "complex end-stopped".

Complex cells come in a wide variety of subtypes. Typical cells of layer II and III have relatively small receptive fields, low spontaneous activity, and in the monkey may not only be highly orientation selective but also fussy about wave length, perhaps responding to red lines but not white. They may or may not be end-stopped. Cells in layers V and VI have larger fields. Those in V have high spontaneous activity and many respond just as well to a very short moving line as to long one. Many cells in layer VI respond best to very long lines (16). These differences are doubtless related to the important fact, first shown with physiologic techniques by Toyama, Matsunami and Ohno (17) and confirmed and extended by anatomical techniques, that different layers project to different destinations—the upper layers mainly to other cortical regions, the fifth layer to the superior colliculus, pons and pulvinar, and VI back to the lateral geniculate body and to the claustrum.

In the last 10 or 15 years the subject of cortical receptive-field types has become rather a jungle, partly because the terms 'simple' and 'complex' are used differently by different people, and undoubtedly partly because the categories themselves are not cleanly separated. Our idea originally was to empha-

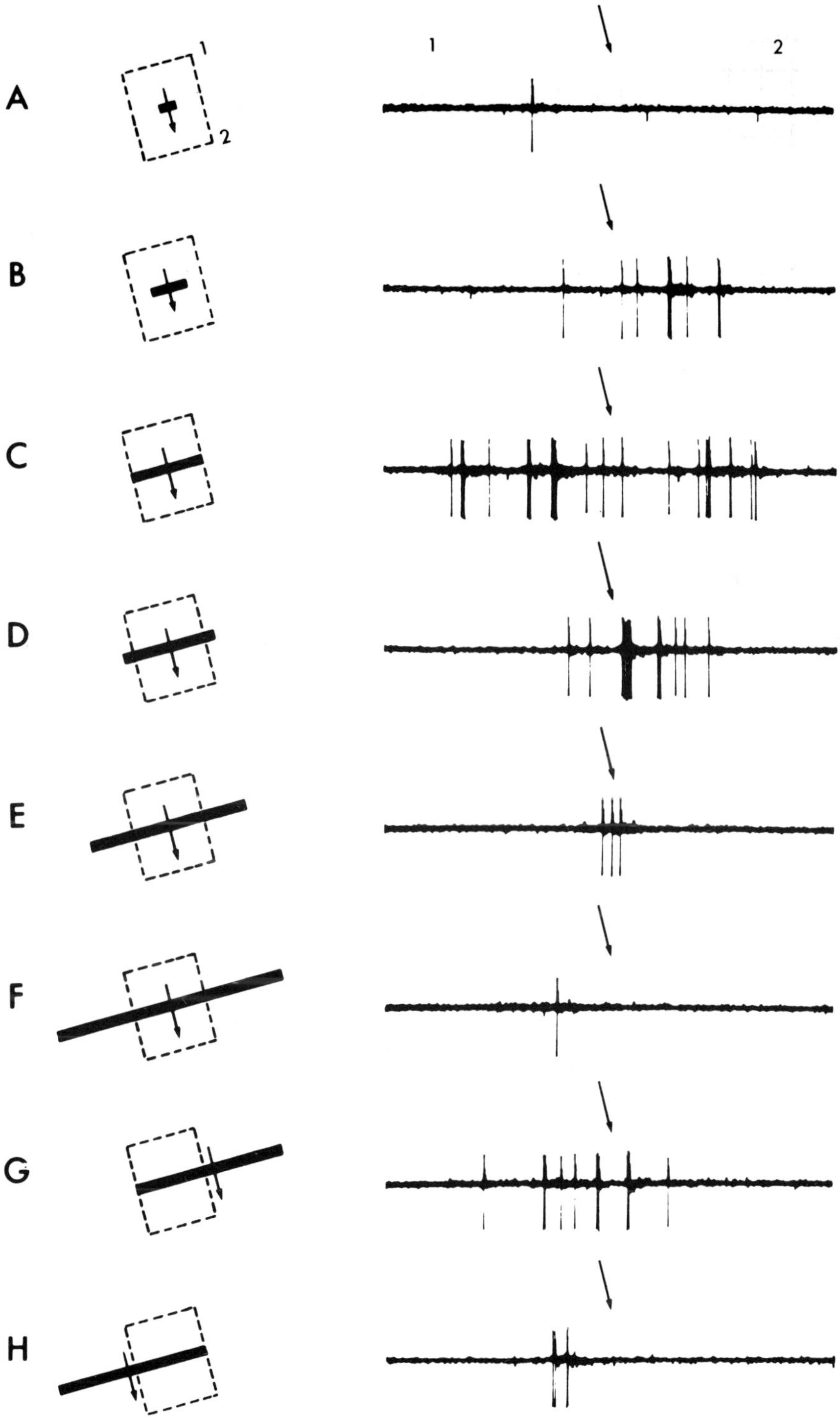

Figure 5. Hypercomplex cell, responding to a black bar oriented 2:30—8:30, moving downward. Optimum response occurred when stimulus swept over area outlined (C); stimulating more than this region (D–H) or less (A, B) resulted in a weaker response. Sweep duration 2.5 sec. (Fig. 19 (14)).

size the tendency toward increased complexity as one moves centrally along the visual path, and the possibility of accounting for a cell's behavior in terms of its inputs. The circuit diagrams we proposed were just a few examples from a number of plausible possibilities. Even today the actual circuit by which orientation specificity is derived from center-surround cells is not known, and indeed the techniques necessary for solving this may still not be available. One can nevertheless say that cells of different complexities, whose receptive fields are in the same part of the visual field and which have the same optimal orientation, are likely to be interconnected, whereas cells with different optimal orientations are far less likely to be interconnected. In the monkey a major difficulty with the hierarchical scheme as outlined here is the relative scarcity of simple cells, compared with the huge numbers of cells with concentric fields in IVc, or compared with the large number of complex cells above and below layer IV. The fact that the simple cells have been found mainly in layer IVb also agrees badly with Jennifer Lund's finding that layer IVcβ projects not to layer IVb but to layer III. One has to consider the possibility that in the monkey the simple-cell step may be skipped, perhaps by summing the inputs from cells in layer IV on dendrites of complex cells. In such a scheme each main dendritic branch of a complex cell would perform the function of a simple cell. All such speculation serves only to emphasize our ignorance of the exact way in which the properties of complex cells are built up.

Knowing how cortical cells respond to some visual stimuli and ignore others allows us to predict how a cell will react to any given visual scene. Most cortical cells respond poorly to diffuse light, so that when I gaze at a white object on a dark background, say an egg, I know that those cells in my area 17 whose receptive fields fall entirely within the boundaries of the object will be unaffected. Only the fields that are cut by the borders of the egg will be influenced, and then only if the local orientation of a border is about the same as the orientation of the receptive field. A slight change in position of the egg without changing its orientation will produce a dramatic change in the population of activated simple cells, but a much smaller change in the activated complex cells.

Orientation-specific simple or complex cells "detect" or are specific for the direction of a short line segment. The cells are thus best not thought of as "line detectors": they are no more line detectors than they are curve detectors. If our perception of a certain line or curve depends on simple or complex cells it presumably depends on a whole set of them, and how the information from such sets of cells is assembled at subsequent stages in the path, to build up what we call "percepts" of lines or curves (if indeed anything like that happens at all), is still a complete mystery.

ARCHITECTURE

When I began my training in neurophysiology at Walter Reed I was lucky enough to be influenced by new and vigorous traditions of experimental neuroanatomy, represented by Walle Nauta, and by a new blend of neuroanatomy and neurophysiology represented at Walter Reed, Johns Hopkins, and the National Institutes of Health by (among others) Jerzy Rose, Vernon Mountcastle, and Robert Galambos. One day very near the beginning of my term at Walter Reed, Jerzy Rose, on the steps of the Research Institute, very sternly told me that I had better make it my business to know exactly where my recording electrode was. I subsequently began to use the Hopkins – Walter Reed technique of making one electrode track or several parallel tracks through cortex, recording as many cells as possible in each track and then reconstructing the tracks from the histology. This made it possible to work out the response properties of single cells and also to learn how they were grouped. It was put to use most dramatically by Vernon Mountcastle, whose discovery of columns in the somatosensory cortex was surely the single most important contribution to the understanding of cerebral cortex since Cajal. Our addition to the reconstruction technique was the strategy of making multiple small (roughly 100 μm diameter) electrolytic lesions along each track by passing small currents through the tungsten electrodes. I worked out this method at Walter Reed by watching the coagulation produced at the electrode tip on passing currents through egg white. The lesions made it possible to be sure of the positions of at least several points along a track; other positions were determined by interpolating depth readings of the microelectrode advancer.

By the early 1960s our research had extended into four different but overlapping areas. Closest to conventional neurophysiology was the working out of response properties (i.e. receptive fields) of single cells. We became increasingly involved with architecture, the grouping of cells according to function into layers and columns, studied by track reconstructions. This led in turn to experiments in which single-cell recording was combined with experimental anatomy. It began when one day James Sprague called to tell us that his chief histological technician, Jane Chen, was moving to Boston and needed a job: could we take her? Luckily we did, and so, despite our not possessing anatomical union cards, we acquired an expert in the Nauta method of making lesions in nervous tissue and selectively staining the degenerating axons . It seemed a terrible waste not to use this method and we soon got the idea of working out detailed pathways by making microelectrode lesions that were far smaller than conventional lesions and could be precisely placed by recording with the same electrodes. It became possible to make lesions in single layers of the lateral geniculate body, with results to be discussed shortly. Finally, still another phase of our work involved studies of newborn animals' postnatal development, and the effects of distorting normal sensory experience in young animals. This began in 1962 and grew steadily. Torsten Wiesel will discuss these experiments.

Having mentioned Jane Chen, this is perhaps as good a place as any to acknowledge our tremendous debt to many research assistants who have helped us over the past 22 years, especially to Jane and to Janet Wiitanen and Bea Storai, and also to Jaye Robinson, Martha Egan, Joan Weisenbeck, Karen Larson, Sharon Mates, Debra Hamburger, Yu-Wen Wu, Sue Fenstemaker, Stella Chow, Sarah Kennedy, Maureen Packard and Mary Nastuk. For photographic assistance I am grateful to Sandra Spinks, Carolyn Yoshikami and Marc Peloquin. In electronics and computers David Freeman has continued to amaze us with his wizardry for 12 years. And for secretarial help and preservation of morale and sanity I want to thank Sheila Barton, Pat Schubert and Olivia Brum.

ORIENTATION COLUMNS

What our three simultaneously recorded cells, Nos. 3009, 3010 and 3011, mapped out on the overhead sheet in September 1958, with their parallel orientation axes and separate but overlapping field positions, were telling us was that neighboring cells have similar orientations but slightly different

6a)

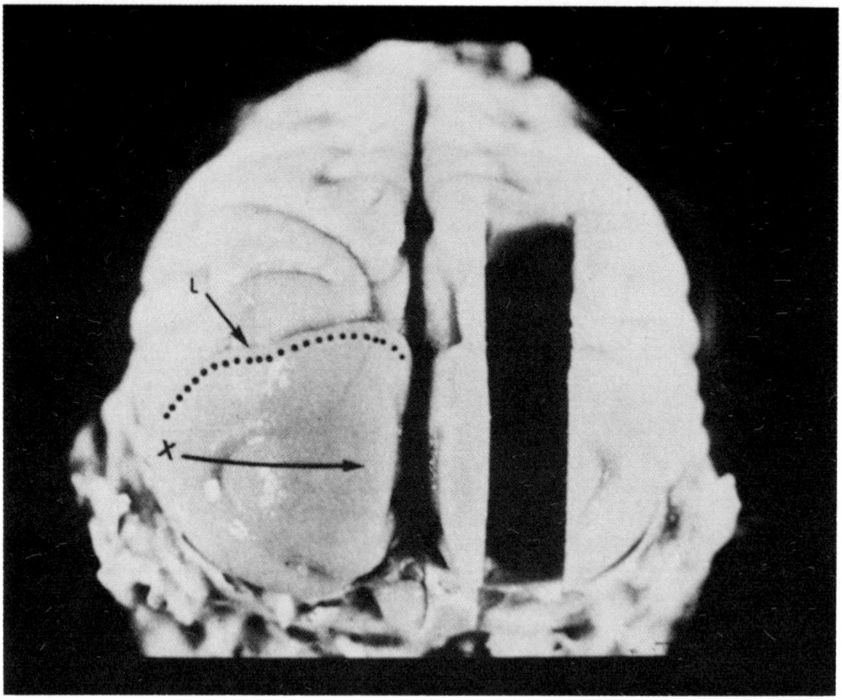

Figure 6. a) Brain of a macaque monkey, perfused with formalin, viewed from above and behind. The occipital lobe is demarcated in front by the lunate sulcus (L) and consists mainly of the striate cortex, area 17, which occupies most of the smooth surface, extending forward to the dotted line (the 17–18 border). If followed medially area 17 curves around the medial surface of the brain and continues in a complex buried fold, a part of which lies underneath the convexity and parallel to it. X marks the projection of the fovea; movement in the direction of the arrow corresponds to movement along the horizon; movement along the dotted line, to movement down along the vertical midline of the visual field. The brain on removal from the skull does not, of course, look exactly like this: the groove in the right hemisphere was made by removing a parasagittal block of tissue to produce the cross section of Fig. 6b. (Fig. 6a (29)).

b) Low power Nissl-stained section from a parasagittal block such as that of Fig. 6a. It is what would be seen if one could stand in the groove of 6a and look to the left. A marks the outer convexity; B the buried fold, and arrows indicate the 17–18 borders, the upper right one of which is indicated by the dotted line in Fig. 6a. (Fig. 6b (29)).

c) Cross section through monkey striate cortex showing conventional layering designations. W, white matter. Deeper layers (VI, V) of the buried fold of cortex are shown in the lower part of the figure (compare Fig. 6b). Cresyl violet. (Fig. 10 (29)).

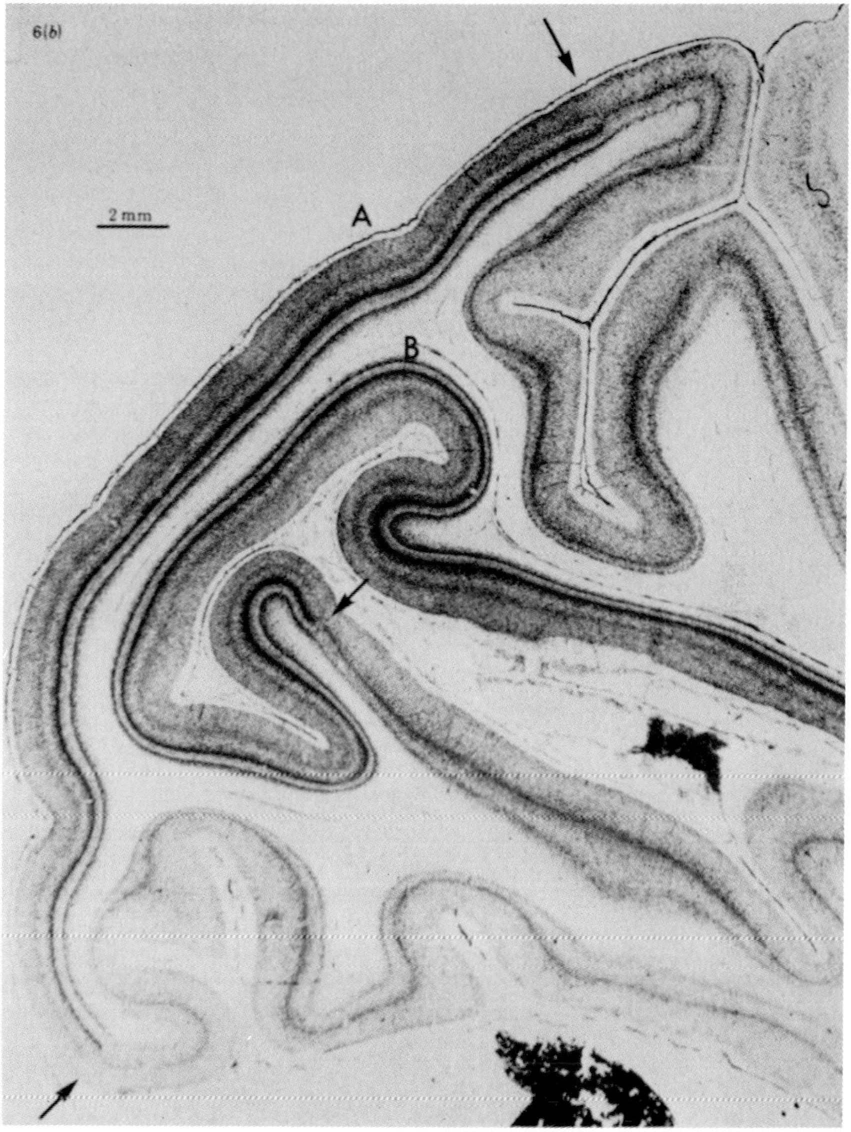

receptive-field positions. We of course knew about our visitor Mountcastle's somatosensory colums, and we began to suspect that cells might be grouped in striate cortex according to orientation; but to prove it was not easy.

Our first indication of the beauty of the arrangements of cell groupings came in 1961 in one of our first recordings from striate cortex of monkey, a spider monkey named George. In one penetration, which went into the cortex at an angle of about 45° and was 2.5 mm long, we were struck right away by something we had only seen hints of before: as the electrode advanced the orientations of successively recorded cells progressed in small steps, of about 10° for every advance of 50 μm. We began the penetration around 8:00 p.m.; five hours later we had recorded 53 successive orientations without a single

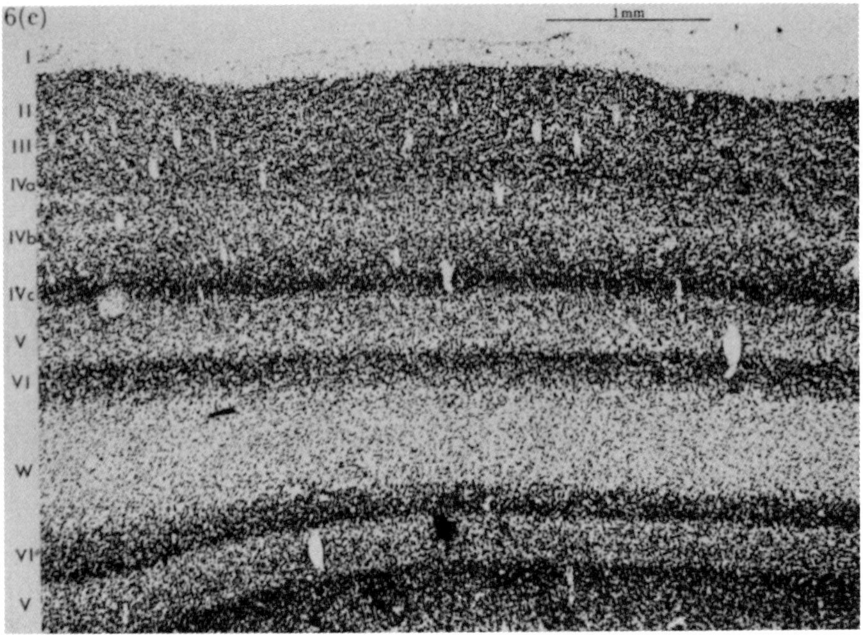

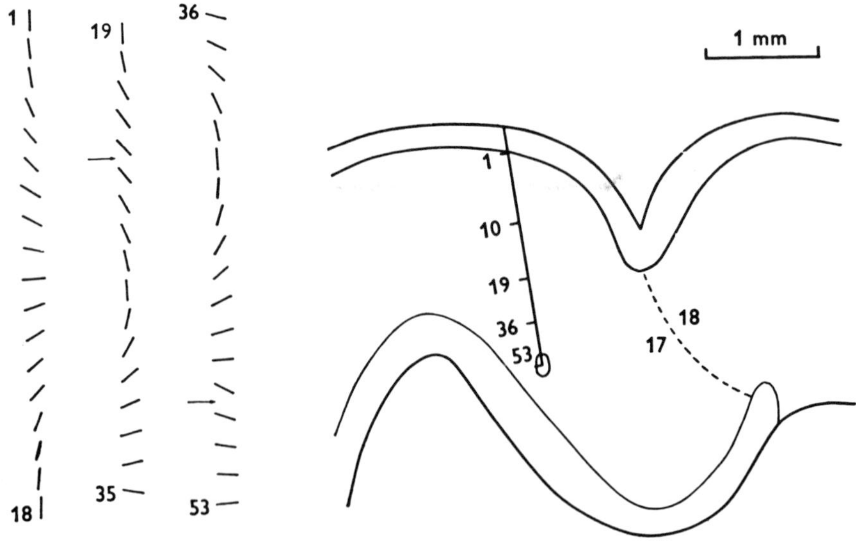

Figure 7. a) Reconstruction of a penetration through striate cortex about 1 mm from 17–18 border, near occipital pole of a spider monkey called George. To the left of the figure the lines indicate receptive-field orientations of cells in the columns traversed; each line represents one or several units recorded against a rich unresolved background activity. Arrows indicate reversal of directions of shifts in orientation (32).

b) Graph of stimulus orientation in degrees vs. distance along electrode track in mm, in the experiment shown in (a). Vertical is taken as 0°, clockwise is positive, anticlockwise negative.

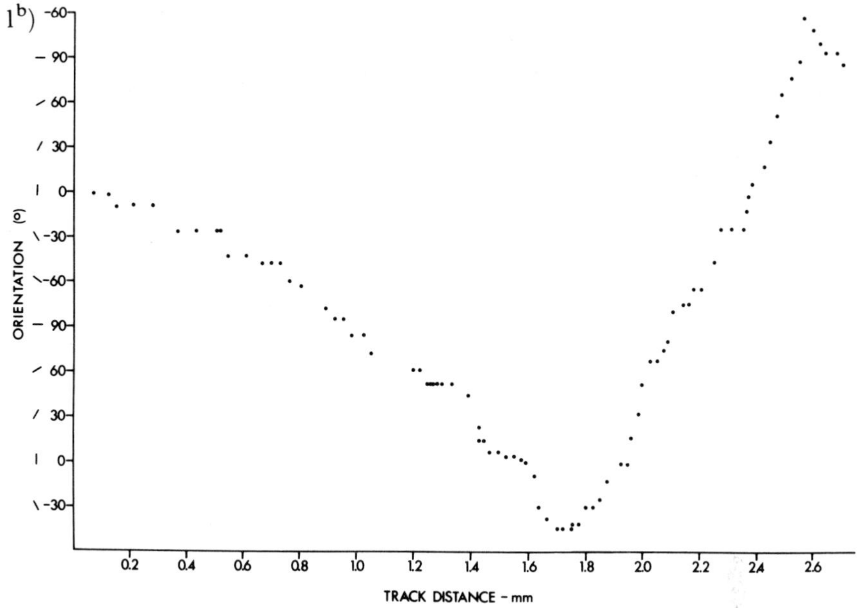

large jump in orientation (Fig. 7). During the entire time, in which I wielded the slide projector and Torsten mapped the fields, neither of us moved from our seats. Fortunately our fluid intake that day had not been excessive! I have shown this illustration many times. So far only one person, Francis Crick, has asked why there was no interruption in layer IVc, where according to dogma the cells are not orientation-specific. The answer is that I don't know.

In the cat we had had occasional suggestions of similar orderliness, and so we decided to address directly the problem of the shape and arrangement of the groupings (18). By making several closely-spaced oblique parallel penetrations we convinced ourselves that the groupings were really columns in that they extended from surface to white matter and had walls that were perpendicular to the layers (Fig. 8). We next made multiple close-spaced penetrations, advancing the electrode just far enough in each penetration to record one cell or a group of cells. To map a few square mm of cortex this way required 50–100 penetrations, each of which took about 10–15 minutes. We decided it might be better to change careers, perhaps to chicken farming. But although the experiments were by our standards exhausting they did succeed in showing that orientation columns in the cat are not generally pillars but parallel slabs that intersect the surface either as straight parallel stripes or swirls (Fig. 9).

Reversals in direction of orientation shift, like those shown in Fig. 7, are found in most penetrations. They occur irregularly, on the average about once every millimeter, and not at any particular orientation such as vertical or horizontal. We still do not know how to interpret them. Between reversals the plots of orientation vs. electrode position are remarkably linear (19). I once, to exercise a new programmable calculator, actually determined the *coefficient of linear correlation* of such a graph. It was 0.998, which I took to mean that the line must be very straight indeed.

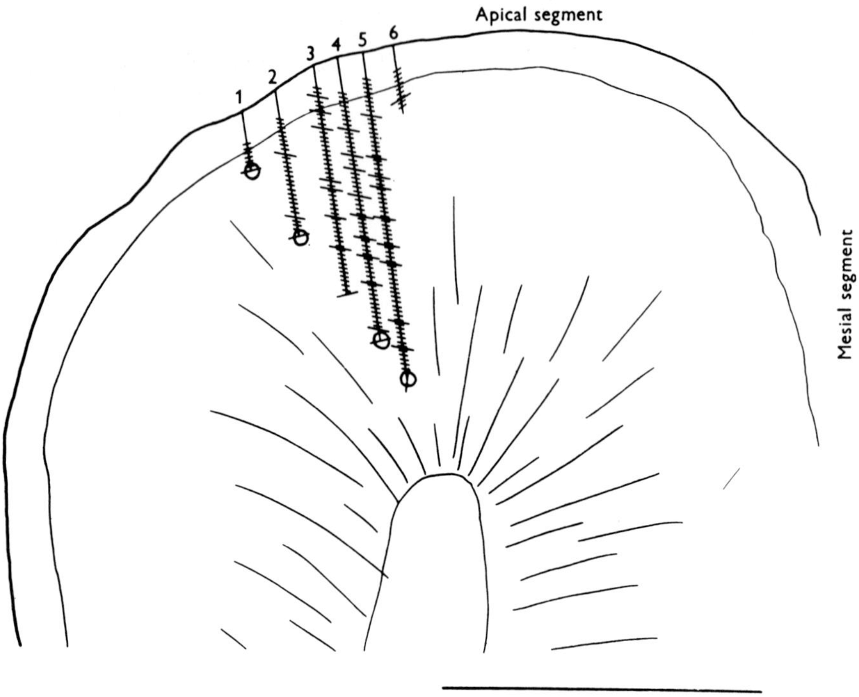

Figure 8. Coronal section through cat visual cortex showing reconstructions of 6 parallel microelectrode penetrations. (Nos. 1, 2, 4 and 5 end with lesions shown as circles.) Short lines perpendicular to tracks indicate receptive-field orientation; lines perpendicular to tracks represent horizontal orientation. (The longer of these lines represent single cells, the shorter ones, multiple unit recordings). Most of the territory traversed by penetrations 2–5 is in one orientation column, whose left hand border lies at the ends of tracks 2–4. Scale 1 mm. (Fig. 2 (18)).

For some years we had the impression that regular sequences like the one shown in Fig. 7 are rare—that most sequences are either more or less random or else the orientation hovers around one angle for some distance and then goes to a new angle and hovers there. Chaos and hovering do occur but they are exceptional, as are major jumps of 45–90°. It took us a long time to realize that regularity is the rule, not the exception, probably because only around the mid-1970s did we begin making very oblique or tangential penetrations. Also for these experiments to be successful requires electrodes coarse enough to record activity throughout a penetration, and not simply every 100 μm or so. Such electrodes look less aesthetically pleasing, a fact that I think has happily tended to keep down competition.

Our attempts to learn more about the geometry of orientation columns in the monkey by using the 2-deoxyglucose technique (20) suggest that iso-orientation lines form a periodic pattern but are far from straight, being full of swirls and interruptions. Experiments done since then (21) suggest that the deoxyglucose is possibly also labelling the cytochrome blobs (see below). Similar work in the tree shrew by Humphrey (22) has shown a much more regular pattern and Stryker, Wiesel and I have seen more regularity in the cat (unpublished). Both tree shrew and cat lack the cytochrome blobs.

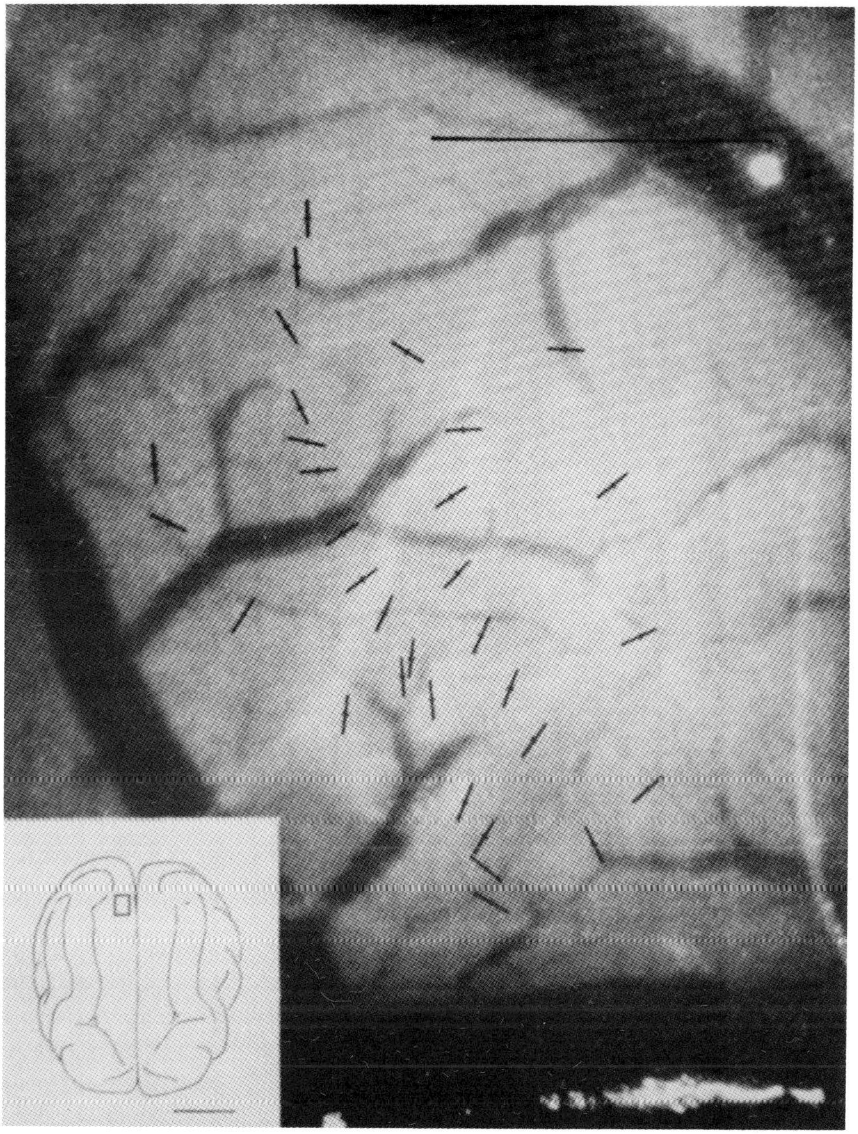

Figure 9. Surface map of a small region of cat visual cortex. Receptive field orientations are shown for 32 superficial penetrations. Regions of relatively constant orientation run more or less right-to-left in the figure, or medio-lateral on the brain (see inset). Going from above down, in the figure, or from posterior to anterior of the brain, is associated with anticlockwise rotation. Scale 1 mm; inset scale 1 cm. (Plate 2 (18)).

OCULAR DOMINANCE COLUMNS

A major finding in our 1959 and 1962 papers (10, 12), besides the orientation selectivity, was the presence in the striate cortex of a high proportion of binocular cells. Since recordings from the lateral geniculate body had made it clear that cells at that stage are for all practical purposes monocular, this answered the question of where, in the retinogeniculocortical pathway, cells first received convergent input from the two eyes. More interesting to us than the mere binocularity was the similarity of a given cell's receptive fields in the two eyes, in size, complexity, orientation and position. Presumably this forms the basis of the fusion of the images in the two eyes. It still seems remarkable that a cell should not only be wired with the precision necessary to produce complex or hypercomplex properties, but should have a duplicate set of such connections, one from each eye. (That this is hard wired at birth will form some of the material for Torsten Wiesel's lecture.) Though the optimum stimulus is the same for the two eyes, the responses evoked are not necessarily equal; for a given cell one eye is often better than the other. It is as if the two sets of connections were qualitatively similar but, for many cells, different in density. We termed this relative effectiveness of the two eyes "eye preference" or "relative ocular dominance".

In the macaque it was evident from the earliest experiments that neighboring cells have similar eye preferences. In vertical penetrations the preference remains the same all the way through the cortex. Layer IVc, in which cells are monocular, is an exception; here the eye that above and below layer IV merely dominates the cells actually monopolizes them. In penetrations that run parallel to the layers there is an alternation of eye preference, with shifts roughly

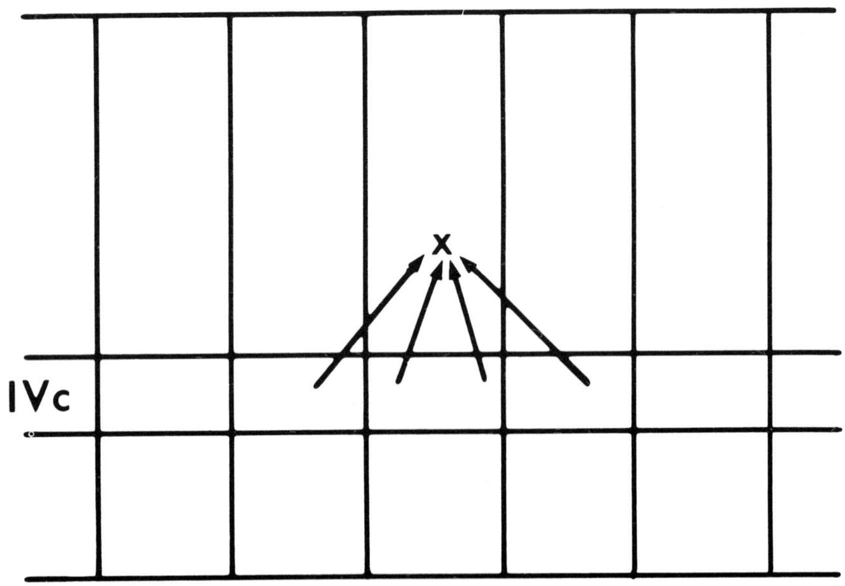

Figure 10. Scheme to illustrate the wiring of a binocular cell in a layer above (or below) layer IVc. The bulk of the afferents to the striate cortex from the lateral geniculate body, themselves monocular, are in the macaque monkey strictly segregated by eye affiliation in layer IVc, and thus the cells in this layer are strictly monocular. A cell outside of IVc, labelled X in the diagram, receives its input connections, directly or indirectly by one or more synapses, from cells in IVc (to some extent also, perhaps, from layers IVa and VI). Cells in IVc will be more likely to feed into X the closer they are to it; consequently X is likely to be binocular, dominated by the eye corresponding to the nearest patch in IVc. The degree of dominance by that eye is greater, the closer X is to being centered in its ocular dominance column, and cells near a boundary may be roughly equally influenced by the two eyes. (Fig. 12 (29)).

every 0.5 mm. The conclusion is that the terminals from cells of the lateral geniculate distribute themselves in layer IVc according to eye of origin, in alternating patches about 0.5 mm wide. In the layers above and below layer IV horizontal and diagonal connections lead to a mixing that is incomplete, so that a cell above a given patch is dominated by the eye supplying that patch but receives subsidiary input from neighboring patches (Fig. 10).

The geometry of these layer-IV patches interested us greatly, and was finally determined by several independent anatomical methods, the first of which involved the Nauta method and modifications of the Nauta method for staining terminals worked out first by Fink and Heimer and then by a most able and energetic research assistant, Janet Wiitanen (23). By making small lesions in single geniculate layers we were able to see the patchy distribution of degenerating terminals in layer IV, which in a face-on view takes the form not of circumscribed patches but of parallel stripes. We also showed that the ventral (magnocellular) pair of layers projects to the upper half of IVc, (subsequently called IVc α by Jennifer Lund), whereas the dorsal 4 layers project to the lower half (IVc β), and that the line of Gennari (IVb), once thought to receive the strongest projection, is actually almost bereft of geniculate terminals.

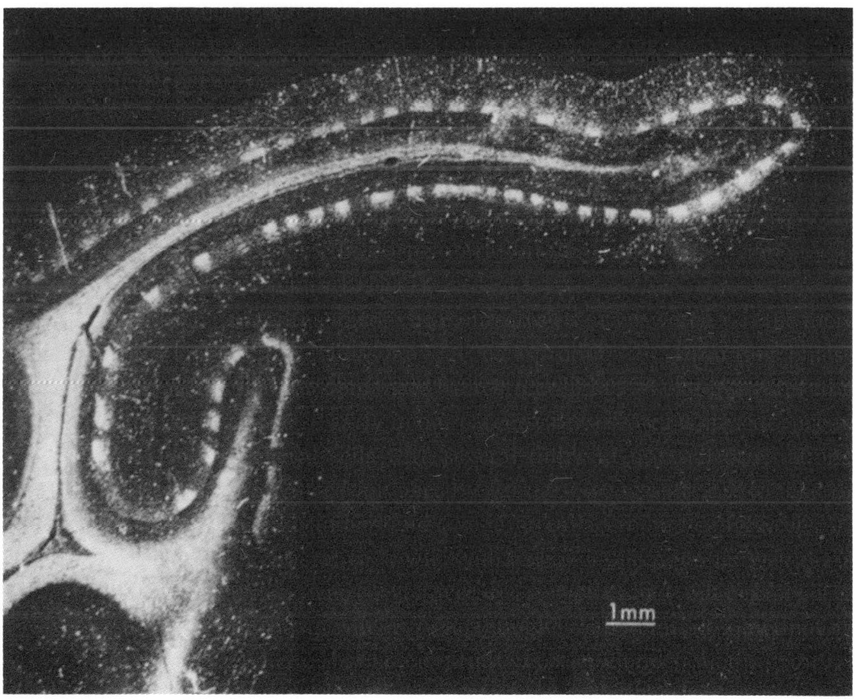

Figure 11. Dark-field autoradiograph of striate cortex in an adult macaque in which the ipsilateral eye had been injected with tritiated proline-fucose 2 weeks before. Labelled areas show as white. Section passes in a plane roughly perpendicular to the exposed surface of the occipital lobe, and to the buried part immediately beneath (roughly, through the arrow of Fig. 6a). In all, about 56 labelled patches can be seen. (Fig. 22 (29)).

While the Nauta studies were still in progress we read a paper in which Bernice Grafstein reported that after injecting a radioactive aminoacid into the eye of a rat, radioactive label could be detected in the contralateral visual cortex, as though transneuronal transport had taken place in the geniculate (24). (The rat retinogeniculocortical pathway is mainly crossed.) It occurred to us that if we injected the eye of a monkey we might be able to see label autoradiographically in area 17. We tried it, but could see nothing. Soon after, while visiting Ray Guillery in Wisconsin, I saw some aminoacid transport autoradiographs which showed nothing in light field but in which label was perfectly obvious in dark field. I rushed back, we got out our slides, borrowed a dark-field condenser, and found beautiful alternating patches throughout all the binocular part of area 17 (25) (Fig. 11). This method allowed us to reconstruct ocular dominance columns over much wider expanses than could be mapped with the Nauta method (Fig. 12). It led to a study of the pre- and postnatal visual development of ocular dominance columns, and the effects of visual deprivation on the columns, which Torsten will describe.

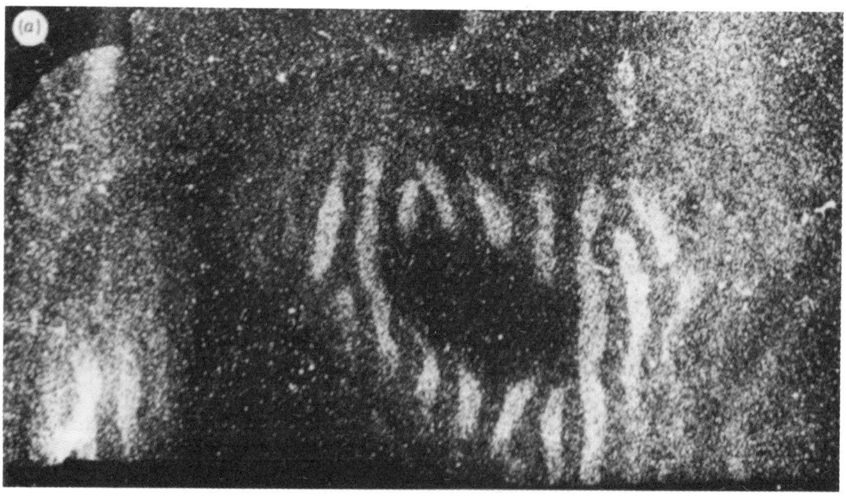

Figure 12. Autoradioagraphs from the same (normal) animal as Fig. 11, but hemisphere contralateral to the injected eye (dark field).

a) A section tangential to the exposed dome-like surface of the occipital lobe, just grazing layer V, which appears as an oval, surrounded by layer IVc, which appears as a ring containing the labelled parallel bands; these appear as light against the dark background.

b) A composite made by cutting out layer IVc from a number of parallel sections such as the one shown in (a), and pasting them together to show the bands over an area some millimeters in extent.

c) Reconstruction of layer IVc ocular dominance columns over the entire exposed part of area 17 in the right occipital lobe, made from a series of reduced-silver sections (33). The region represented is the same as the part of the right occipital lobe shown in Fig. 6a. Dotted line on the left represents the midsagittal plane where the cortex bends around. Dashed c-shaped curve is the 17–18 border, whose apex, to the extreme right, represents the fovea. Every other column has been blackened in, so as to exhibit the twofold nature of the set of subdivisions. Note the relative constancy of column widths.

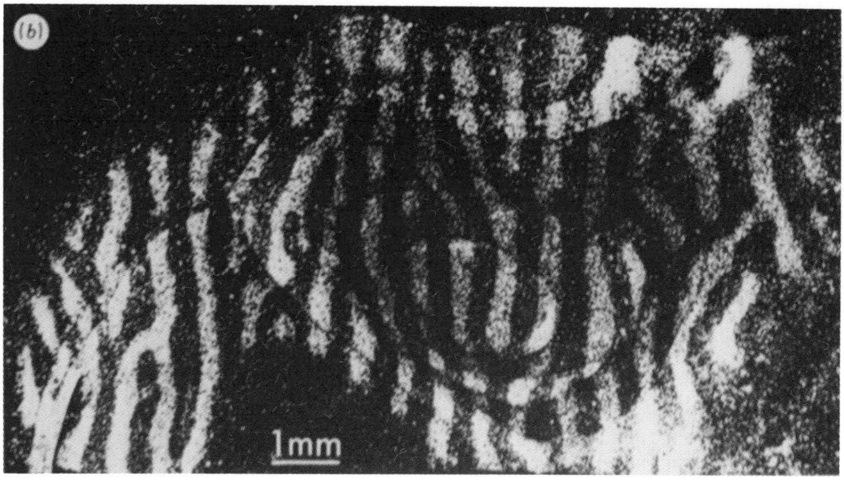

(c)

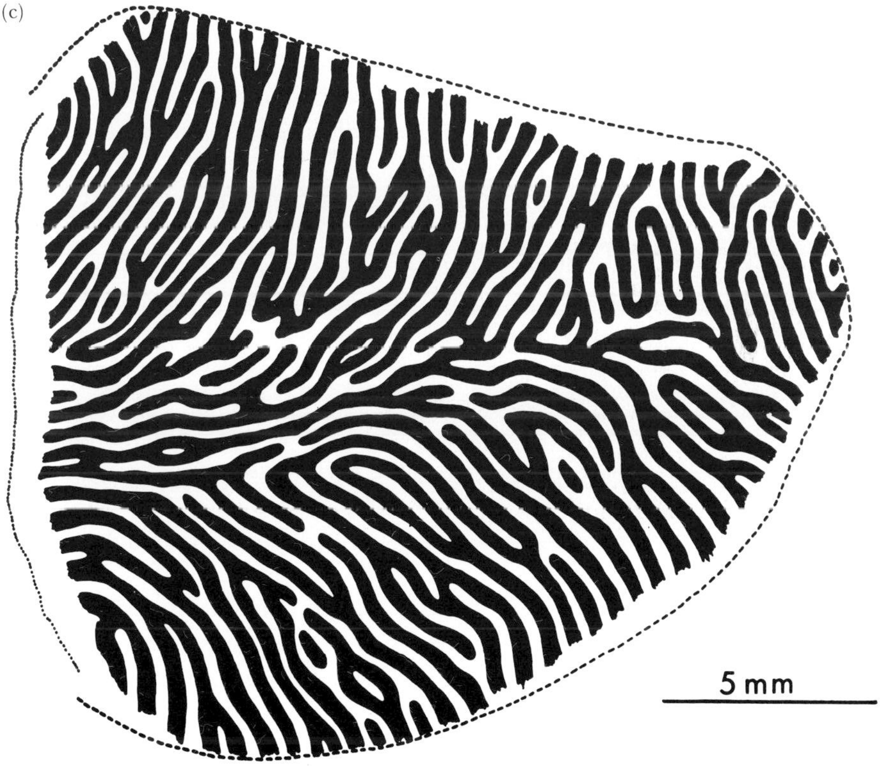

RELATIONSHIP BETWEEN COLUMNS, MAGNIFICATION AND FIELD SIZE

To me the main pleasures of doing science are in getting ideas for experiments, doing surgery, designing and making equipment, and above all the rare moments in which some apparently isolated facts click into place like a Chinese puzzle. When a collaboration works, as ours has, the ideas and the clicking into place often occur simultaneously or collaboratively; usually neither of us has known (or cared) exactly where or from whom ideas came from, and sometimes one idea has occurred to one of us, only to be forgotten and later resurrected by the other. One of the most exciting moments was the realization that our orientation columns, extending through the full thickness of the cat cortex, contain just those simple and complex cells (later we could add the hypercomplex) that our hierarchical schemes had proposed were interconnected (12). This gave the column a meaning: a little machine that takes care of contours in a certain orientation in a certain part of the visual field. If the cells of one set are to be interconnected, and to some extent isolated from neighboring sets, it makes obvious sense to gather them together. As Lorente de Nó showed (26), most of the connections in the cortex run in an up-and-down direction; lateral or oblique connections tend to be short (mostly limited to 1 to 2 mm) and less rich. These ideas were not entirely new, since Mountcastle had clearly enunciated the principle of the column as an independent unit of function. What was new in the visual cortex was a clear function, the transformation of information from circularly symmetric form to orientation-specific form, and the stepwise increase in complexity.

A similar argument applies to the ocular dominance columns, a pair of which constitutes a machine for combining inputs from the two eyes—combining, but not completely combining, in a peculiar grudging way and for reasons still not at all clear, but probably related in some way to stereopsis. (Whatever the explanation of the systematically incomplete blending, it will have to take into account the virtual but not complete absence of dominance columns in squirrel monkeys.) If the eyes are to be kept functionally to some extent separate, it is economical of connections to pack together cells of a given eye preference.

To my mind our most aesthetically attractive and exciting formulation has been the hypercolumn (not, I admit, a very attractive term!) and its relation to magnification. The idea grew up gradually, but took an initial spurt as a result of a question asked by Werner Reichardt during a seminar that I gave in Tübingen. I had been describing the ordered orientation sequences found in monkeys like George, when Werner asked how one avoided the difficulty arising from the fact that as you move across the cortex visual field position is changing, in addition to orientation. Could this mean that if you looked closely you would find, in one small part of the visual field, only a small select group of orientations represented? The answer seemed obvious: I explained that in any one part of the visual field all orientations are represented, in fact probably several times over. Afterwards the question nagged me. There must be more to it than that. We began to put some seemingly isolated facts together. The visual fields map systematically onto the cortex but the map is distorted: the fovea is disproportionately represented, with 1 mm about equivalent to 1/6° of visual field. As one goes out in the visual field the representation falls off, logarithmically, as Daniel and Whitteridge had shown (27), so that in the far periphery the relationship is more like 1 mm = 6°. Meanwhile the average size of receptive fields grows from center of gaze to periphery. This is not unexpected when one considers that in the fovea our acuity is very much higher than in the periphery. To do the job in more detail takes more cells, each looking after a smaller region; to accommodate the cells takes more cortical surface area. I had always been surprised that the part of the cortex representing the fovea is not obviously thicker than that representing the periphery: the surprise, I suppose, comes from the fact that in the retina near the fovea the ganglion cell layer is many times thicker than in the periphery. The cortex must be going out of its

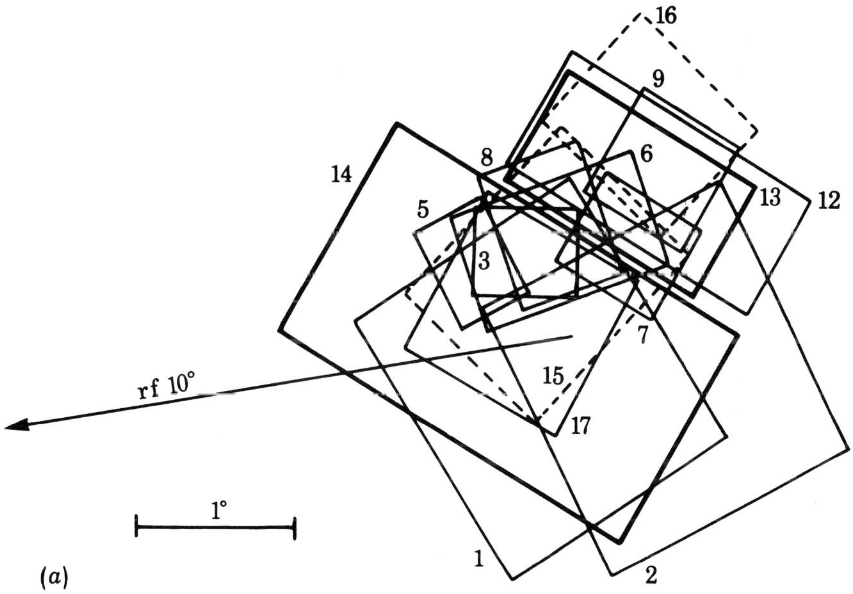

Figure 13. a) Receptive-field scatter: Receptive-field boundaries of 17 cells recorded in a penetration through monkey striate cortex in a direction perpendicular to the surface. Note the variation in size, and the more or less random scatter in the precise positions of the fields. The penetration was made in a part of the cortex corresponding to a visual field location 10° from the center of gaze, just above the horizontal meridian. Fields are shown for one eye only. Numbers indicate the order in which the cells were recorded. (Fig. 1 (28)).

b) Receptive-field drift: Receptive fields mapped during one oblique, almost tangential penetration through striate cortex, in roughly the same region as in (a). A few fields were mapped along each of four 100 μm segments, spaced at 1 mm intervals. These four groups of fields are labelled 0, 1, 2 and 3. Each new set of fields was slightly above the other, in the visual field, as predicted from the direction of movement of the electrode and from the topographic map of visual fields onto cortex. Roughly a 2 mm movement through cortex was required to displace the fields from one region to an entirely new region. (Fig. 2 (28)).

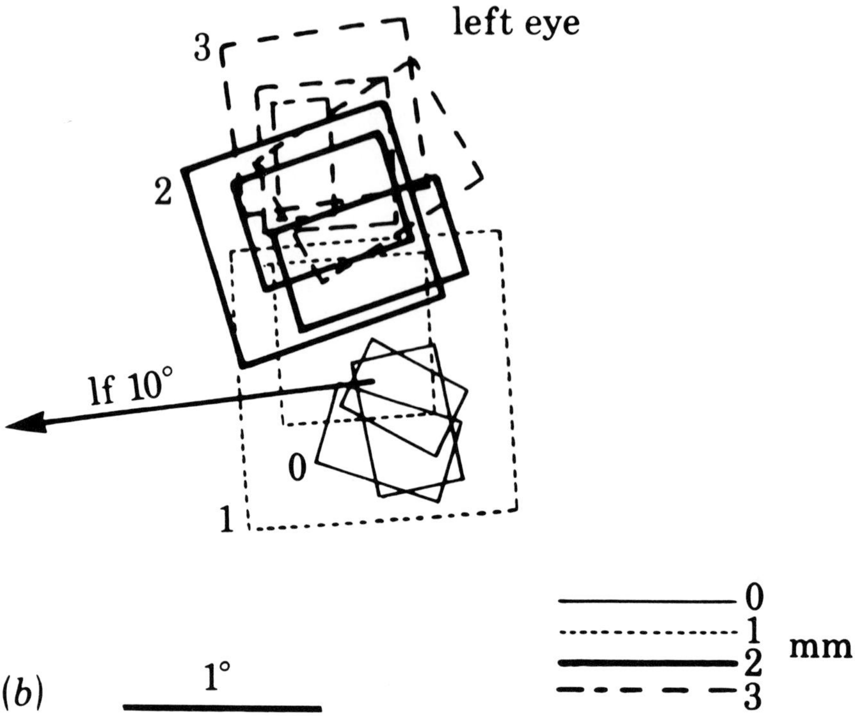

(b)

way to keep its uniformity by devoting to the detailed tasks more area rather than more thickness.

We decided to look more carefully at the relationship between receptive-field size and area of cortex per unit area of visual field (28). When an electrode is pushed vertically through the cortex and encounters a hundred or so cells in traversing the full thickness, the receptive fields vary to some extent in size, and in a rather random way in position, so that the hundred maps when superimposed cover an area several times that of an average receptive field (Fig. 13a). We call this the "aggregate receptive field" for a particular point on the cortex. On making a penetration parallel to the surface there is a gradual drift in field position, superimposed on the random staggering, in a direction dictated by the topographic map (Fig. 13b). We began to wonder whether there was any law connecting the rate of this drift in aggregate position and the size of the fields. It was easy to get a direct answer. It turned out that a movement of about 2 mm across the cortex is just sufficient to produce a displacement, in the visual field, out of the region where one started and into an entirely new region. This held everywhere across the striate cortex (and consequently in the visual field). In the fovea the displacement was tiny and so were the fields. As one went out, both increased in size, in parallel fashion (Fig. 14). Now things indeed seemed to mesh. George and other monkeys had taught us that a 1−2 mm movement across cortex is accompanied by an angular shift in receptive-field orientation of 180−360°, more than one full complement of orientations. We have termed such a set of orientation columns (180°) a "hypercolumn". Meanwhile, the

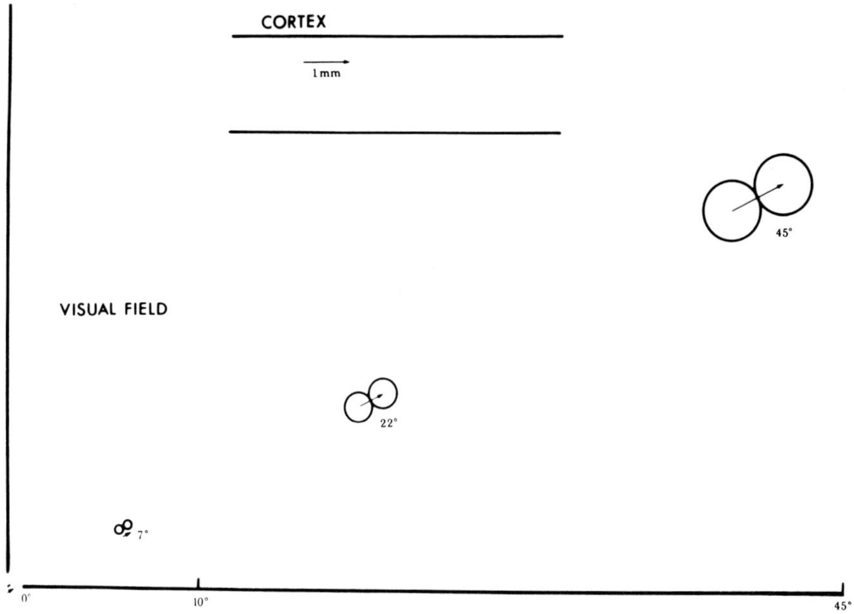

Figure 14. Variation of receptive-field drift with eccentricity: The diagram represents one quadrant of the field of vision, and the circles represent aggregate receptive fields, the territory collectively occupied by respective fields of cells encountered in a microelectrode penetration perpendicular to the cortical surface. Each pair of circles illustrates the movement in aggregate receptive field accompanying a movement along the cortex of 1–2 mm. Both the displacement and the aggregate field size vary with distance from the fovea (eccentricity), but they do so in parallel fashion. Close to the fovea the fields are tiny, but so is the displacement accompanying a 1–2 mm movement along the cortex. The greater the distance from the fovea, the greater the two become, but they continue to remain roughly equal (28).

ocular dominance shifts back and forth so as to take care of both eyes every millimeter – a hypercolumn for ocular dominance. Thus, in one or two square millimeters there seems to exist all the machinery necessary to look after everything the visual cortex is responsible for, in a certain small part of the visual world. The machines are the same everywhere; in some parts the information on which they do their job is less detailed, but covers more visual field (Fig. 15).

Uniformity is surely a huge advantage in development, for genetic specifications need only be laid down for a 1–2 mm block of neural tissue, together with the instruction to make a thousand or so.

We could, incidentally, have called the entire machine a hypercolumn, but we did not. The term as we define it refers to a complete set of columns of one type. I mention this because uniformity has obvious advantages, not just for the cortex but also for terminology. Perhaps one could use "module" to refer to the complete machine.

There are two qualifications to all of this. I do not mean to imply that there need really be 2,000 separate definable entities. It need not matter whether one begins a set of orientation columns at vertical, horizontal or any one of the

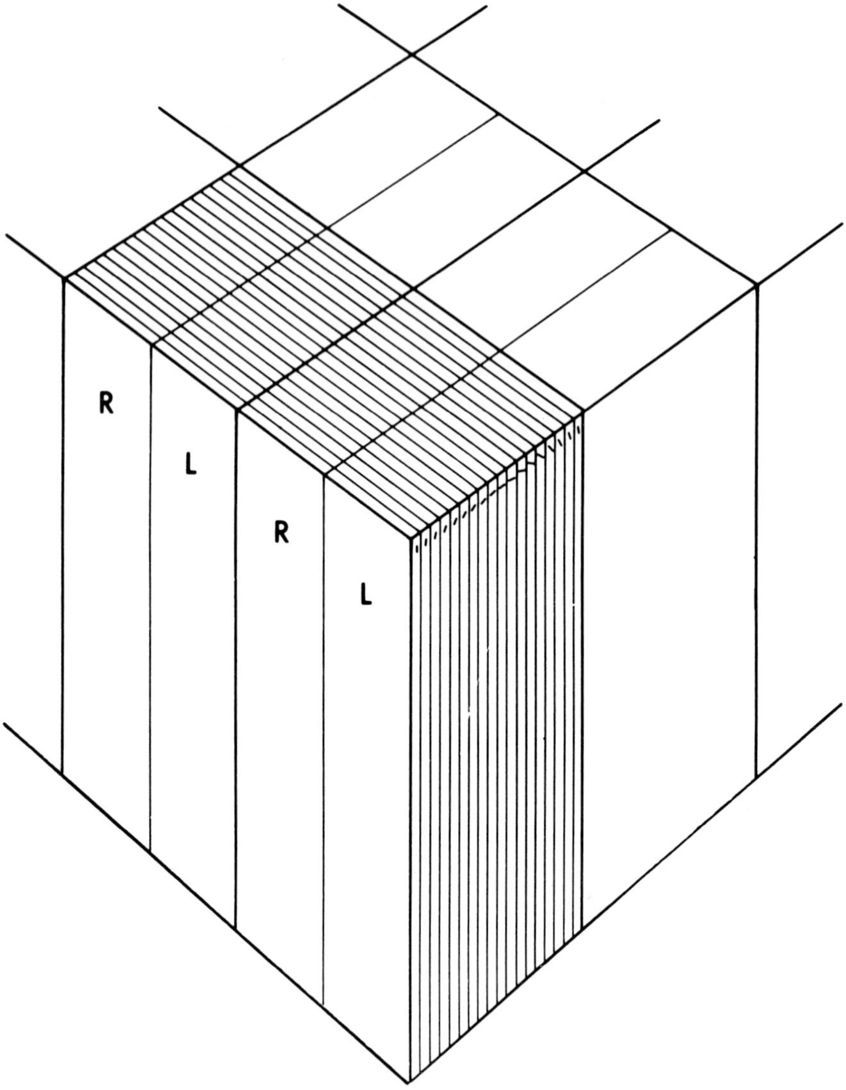

Figure 15. Model of the striate cortex, to show roughly the dimensions of the ocular dominance slabs (L,R) in relation to the orientation slabs and the cortical thickness. Thinner lines separate individual columns; thicker lines demarcate the two types of hypercolumn: two pairs of ocular dominance columns and two sets of orientation columns. The placing of these hypercolumn boundaries is of course arbitrary; one could as well begin the orientation hypercolumn at horizontal or any of the obliques. The decision to show the two sets of columns as intersecting at right angles is also arbitrary, since there is at present no evidence as to the relationship between the two sets. Finally, for convenience the slabs are shown as plane surfaces, but whereas the dominance columns are indeed more or less flat, the orientation columns are not known to be so, and may when viewed from above have the form of swirls. (Fig. 27 (29)).

obliques; the decision is arbitrary. One requires two dominance columns, a left and a right, and it makes no difference which one begins with. (In fact, as I will soon point out, it now looks as though the blocks of tissue may really be discrete, to a degree that we could not have imagined two years ago.) Second, there may well be some differences in cortical machinery between the center and periphery of the visual field. Color vision and stereopsis, for example, probably decline in importance far out in the visual fields. I say this not to be obsessively complete but because in the next few years someone will probably find some difference and pronounce the concept wrong. It may of course be wrong, but I hope it will be for interesting reasons.

I should perhaps point out that the retina *must* be nonuniform if it is to do a more detailed job in the center. To have more area devoted to the center than to the periphery is not an option open to it, because it is a globe. Were it anything else the optics would be awkward and the eye could not rotate in its socket.

A few years ago, in a Ferrier Lecture (29), Torsten and I ended by saying that the striate cortex is probably now (was, then) in broad outline, understood. This was done deliberately: one did not want the well to dry up. When one wants rain the best strategy is to leave raincoat and umbrella at home. So the best way to guarantee future employment was to declare the job finished. It certainly worked. Two years ago Anita Hendrickson and her coworkers and our laboratory independently discovered that monkey striate cortex, when sectioned parallel to the surface and through layers II and III and stained for the enzyme cytochrome oxidase, shows a polka-dot pattern of dark blobs quasi-regularly spaced 1/2−1 mm apart (Fig. 16) (30,21). It is as if the animal's brain had the measles. The pattern has been seen with several other enzymatic stains, suggesting that either the activity or the machinery is different in the blob regions. The pattern has been found in all primates examined, including man, but not in any nonprimates. In macaque the blobs are clearly lined up along ocular dominance columns (19). Over the past year Margaret Livingstone and I have shown that the cells in the blobs lack orientation selectivity, resembling, at least superficially, cells of layer IVc (31). They are selectively labeled after large injections of radioactive proline into the lateral geniculate body, so it is clear that their inputs are not identical to the inputs to the rest of layers II and III. Thus, an entire system has opened up whose existence we were previously quite unaware of and whose anatomy and functions we do not yet understand. We are especially anxious to learn what, if any, the relationship is between the cytochrome blobs and the orientation columns.

Things are at an exciting stage. There is no point leaving the umbrella home; it is raining, and raining hard.

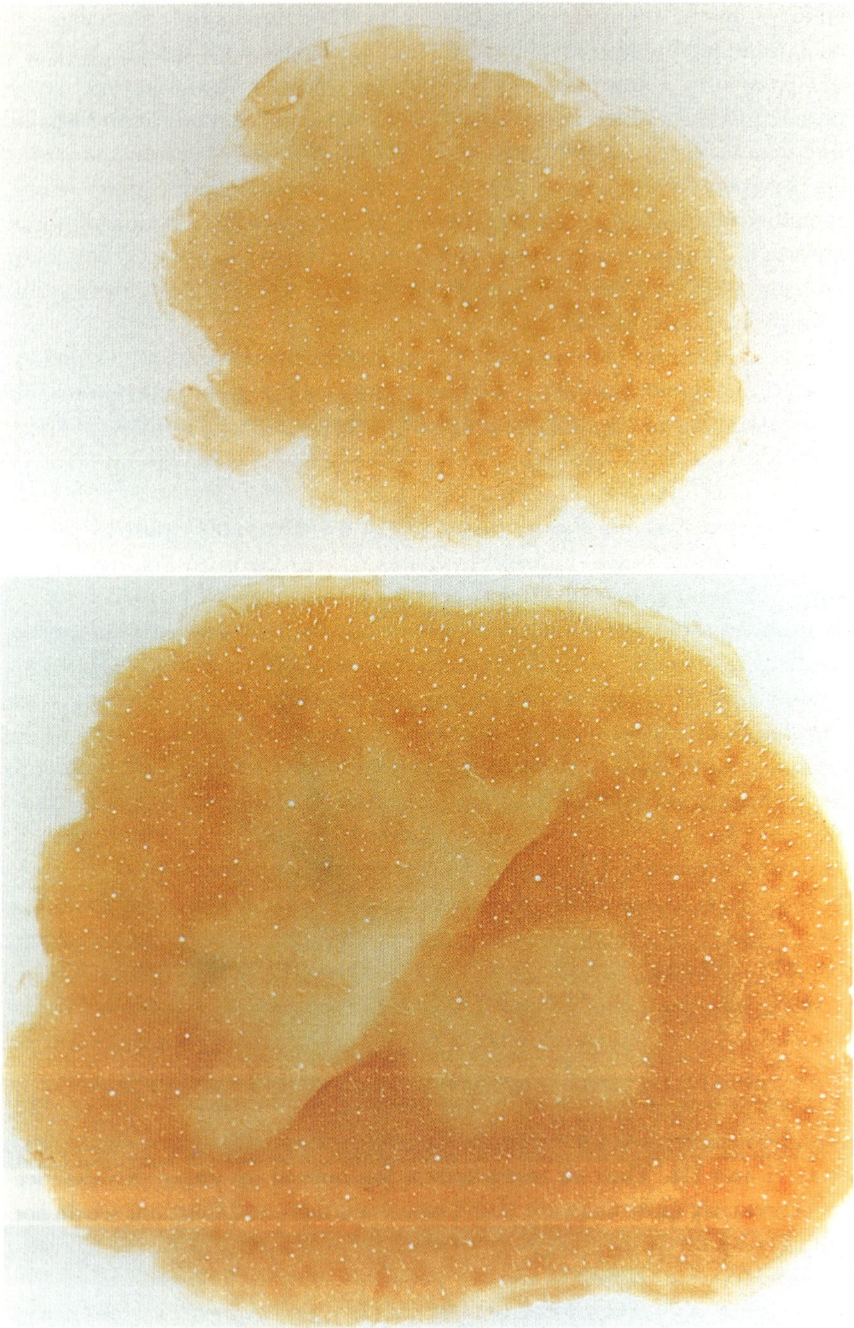

Figure 16. Tangential sections through the visual cortex of the squirrel monkey; cytochrome oxidase stain. The sections pass through the 17−18 border, which runs obliquely in the figure with area 17 below and to the right and 18 above and left. (Hubel and Livingstone, unpublished) The left-hand section passes through layer III, and the blobs can be seen easily in area 17. The right-hand section is tangential to layer V where blobs can be again seen, though faintly; these lie in register with the upper-layer blobs. The coarse pattern in area 18 is now under study and promises to be interesting.

REFERENCES

1. Hubel, D. H., (1958) Cortical unit responses to visual stimuli in nonanesthetized cats. Amer. J. Ophthal. 46: 110–122.
2. Barlow, H. B., FitzHugh, R. and Kuffler, S. W., (1957) Dark adaptation, absolute threshold and Purkinje shift in single units of the cat's retina. J. Physiol. 137: 327–337.
3. Talbot, S. A. and Kuffler, S. W., (1952) A multibeam ophthalmoscope for the study of retinal physiology. J. Opt. Soc. Am. 42: 931–936.
4. Talbot, S. A. and Marshall, W. H., (1941) Physiological studies on neural mechanisms of visual localization and discrimination. Am. J. Ophthal. 24: 1255–1263.
5. Mountcastle, V. B., (1957) Modality and topographic properties of single neurons of cat's somatic sensory cortex. J. Neurophysiol. 20: 408–434.
6. Davies, P. W., (1956) Chamber for microelectrode studies in the cerebral cortex. Science 124: 179–180.
7. Sperry, R. W., Miner, N., and Meyers, R. E., (1955) Visual pattern perception following subpial slicing and tantalum wire implantations in the visual cortex. J. Comp. Physiol. Psych. 48: 50–58.
8. Sperry, R. W. and Miner, N., (1955) Pattern perception following insertion of mica plates into visual cortex. J. Comp. Physiol. Psych. 48: 463–469.
9. Hubel, D. H. and Wiesel, T. N., (1959) Respective field organization of single units in the striate cortex of cat. XXI Int. Congr. Physiol. Sci., Buenos Aires, p. 131.
10. Hubel, D. H. and Wiesel, T. N., (1959) Receptive fields of single neurones in the cat's striate cortex. J. Physiol. 148: 574–591.
11. Cleland, B. G., Dubin, M. W., and Levick, W. R., (1971) Simultaneous recording of input and output of lateral geniculate neurones. Nature New Biol. 231: 191–192.
12. Hubel, D. H. and Wiesel, T. N., (1962) Receptive fields, binocular interaction and functional architecture in the cat's visual cortex. J. Physiol. 160: 106–154.
13. Barlow, H. B. and Levick, W. R., (1965) The mechanism of directionally selective units in rabbit's retina. J. Physiol. 178: 477–504.
14. Hubel, D. H. and Wiesel, T. N., (1965) Receptive fields and functional architecture in two non-striate visual areas (18 and 19) of the cat. J. Neurophysiol. 28: 229–289.
15. Dreher, B. (1972) Hypercomplex cells in the cat's striate cortex. Invest. Ophth. 11: 355–356.
16. Gilbert, C. D. (1977) Laminar differences in receptive field properties of cells in cat visual cortex. J. Physiol. 268: 391–421.
17. Toyama, K., Matsunami, K., and Ohno, T., (1969) Antidromic identification of association, commissural and corticofugal efferent cells in cat visual cortex. Brain Res. 14: 513–517.
18. Hubel, D. H. and Wiesel, T. N., (1963) Shape and arrangement of columns in cat's striate cortex. J. Physiol. 165: 559–568.
19. Hubel, D. H. and Wiesel, T. N., (1974) Sequence regularity and geometry of orientation columns in the monkey striate cortex. J. Comp. Neur. 158: 267–294.
20. Sokoloff, L., Reivich, M., Kennedy, C., DesRosiers, M. H., Patlak, C. S., Pettigrew, K. D., Sakurada, O. and Shinohara, M., (1977) The [^{14}C] deoxyglucose method for the measurement of local cerebral glucose utilization: theory, procedure, and normal values in the conscious and anesthetized albino rat. J. Neurochem. 28: 897–916.
21. Horton, J. C. and Hubel, D H., (1981) Regular patchy distribution of cytochrome oxidase staining in primary visual cortex of macaque monkey. Nature 292: 762–764.
22. Humphrey, A. L., Skeen, L. C., and Norton, T. T., (1980) Topographic organization of the orientation column system in the striate cortex of the tree shrew (*Tupaia glis*). II. Deoxyglucose mapping. J. Comp. Neur. 192: 549–566.
23. Hubel, D. H. and Wiesel, T. N., (1972) Laminar and columnar distribution of geniculo-cortical fibers in the macaque monkey. J. Comp. Neur. 146: 421–450.
24. Grafstein, B. (1971) Transneuronal transfer of radioactivity in the central nervous system. Science 172: 177–179.

25. Wiesel, T. N., Hubel, D. H., and Lam, D. M. K., (1974) Autoradiographic demonstration of ocular-dominance columns in the monkey striate cortex by means of transneuronal transport. Brain Res. 79: 273–279.
26. Lorente de Nó, R. (1949) Cerebral cortex: architecture, intracortical connections, motor projections. Chapt. 15 in Fulton, J. F.: *Physiology of the Nervous System*, 3rd edition, Oxford University Press, New York and London.
27. Daniel, P. M. and Whitteridge, D., (1961) The representation of the visual field on the cerebral cortex in monkeys. J. Physiol., Lond. 159: 203–221.
28. Hubel, D. H. and Wiesel, T. N., (1974) Uniformity of monkey striate cortex: a parallel relationship between field size, scatter, and magnification factor. J. Comp. Neur. 158: 295–306.
29. Hubel, D. H. and Wiesel, T. N., (1977) Ferrier Lecture. Functional architecture of macaque monkey visual cortex. Proc. R. Soc. Lond. B. 198: 1–59.
30. Hendrickson, A. E., Hunt, S. P., and Wu, J.-Y., (1981) Immunocytochemical localization of glutamic acid decarboxylase in monkey striate cortex. Nature 292: 605–607.
31. Hubel, D. H. and Livingstone, M. S., (1981) Regions of poor orientation tuning coincide with patches of cytochrome oxidase staining in monkey striate cortex. Neurosci. Abst. 11th Ann. Meeting, Los Angeles, 118.12.
32. Hubel, D. H. and Wiesel, T. N., (1968) Receptive fields and functional architecture of monkey striate cortex. J. Physiol. 195: 215–243.
33. LeVay, S., Hubel, D. H., and Wiesel, T. N., (1975) The pattern of ocular dominance columns in macaque visual cortex revealed by a reduced silver stain. J. Comp. Neur. 159: 559–576.

Torsten Wiesel

TORSTEN N. WIESEL

I was born in Uppsala Sweden in 1924, the youngest of five children. My father, Fritz S. Wiesel, was chief psychiatrist and head of Beckomberga Hospital, a mental institution located on the outskirts of Stockholm. We were brought up by my mother, Anna-Lisa (b. Bentzer), at the hospital and were sent by bus to Whitlockska Samskolan, a coeducational private school in the city. I was a rather lazy, mischievous student, interested mainly in sports. My election as president of the high school's athletic association was my only memorable achievement during that period. Suddenly, at the age of 17, I became a serious student and I did reasonably well as a medical student. My curiosity about the workings of the nervous system was stimulated by the lectures of Carl Gustaf Bernhard and Rudolf Skoglund, my professors in neurophysiology. Because of my background I was also interested in psychiatry, and I spent one year while I was a medical student working with patients in different mental hospitals.

When my studies were completed I returned to Professor Bernhards's laboratory at the Karolinska Institute in 1954 to do basic neurophysiological research. The following year I had the good fortune to be invited to the United States as a postdoctoral fellow in Dr. Stephen Kuffler's laboratory at the Wilmer Institute, Johns Hopkins Medical School. Dr. Kuffler had just published his now classical study of the receptive field arrangements of cat retinal ganglion cells. This was an important extension of the pioneering work of Drs. Hartline and Granit, for which they received the 1967 Nobel Prize. David Hubel joined the laboratory in 1968, and the two of us decided to explore the receptive field properties of cells in the central visual pathways. This marked the beginning of our twenty year collaboration.

In 1959 Dr. Kuffler was invited to become a professor of pharmacology at Harvard Medical School, and he brought a group of young and enthusiastic investigators with him from Johns Hopkins Medical School. The effectiveness of this group of neuroscientists in research and teaching, and the foresight of Dr. Ebert, then the Dean of the Medical School, led to the formation of the Department of Neurobiology with Stephen Kuffler as the chairman. In addition to David Hubel and myself, the original group of emigres from Johns Hopkins included Edwin Furshpan and David Potter; together with Edward Kravitz we became the original faculty of the new department. David and I now had the opportunity to continue our work in a stimulating environment. Our collaboration continued until the late seventies. In the past several years I worked with Charles Gilbert, a young investigator in the Department. In 1973 I was asked to be head of the Department of Neurobiology. Dr. Kuffler, who meant so much to all of us, continued his work as a University Professor until he died suddenly in 1980. My only regret is that he could not join David and me in the celebration of the Nobel Prize.

I was married to Teeri Stenhammar 1956–1970 and Ann Yee 1973–1981. My daughter Sara Elisabeth was born in 1975. Aside from my work my interests lie in the arts and in world affairs.

Honors & Awards:

1967 A.M. (Hon.), Harvard University

1971 The Dr. Jules C. Stein Award, presented by the Trustees for Research to Prevent Blindness

1972 The Lewis S. Rosenstiel Prize, presented by Brandeis University

1972 Ferrier Lecture (Royal Society of London)

1975 The Freidenwald Award, presented by the Trustees of the Association for Research in Vision and Ophthalmology, Inc.

1976 The Grass Lecture (Society for Neuroscience)

1977 The Karl Spencer Lashley Prize, presented by the American Philosophical Society

1978 The Louisa Gross Horwitz Prize, presented by Columbia University

1979 The Dickson Prize, presented by the University of Pittsburgh

1980 The Ledlie Prize, Harvard University

1980 Society for Scholars (Johns Hopkins University)

1981 The Nobel Prize in Physiology or Medicine

THE POSTNATAL DEVELOPMENT OF THE VISUAL CORTEX AND THE INFLUENCE OF ENVIRONMENT

Nobel lecture, 8 December 1981

by

TORSTEN N. WIESEL

Harvard Medical School, Department of Neurobiology,
Boston, Massachusetts, U.S.A.

INTRODUCTION

In the early sixties, having begun to describe the physiology of cells in the adult cat visual cortex,[1] David Hubel and I decided to investigate how the highly specific response properties of cortical cells emerged during postnatal development. We were also interested in examining the role of visual experience in normal development, a question raised and discussed by philosophers since the time of Descartes. The design of these experiments was undoubtedly influenced by the observation that children with congenital cataract still have substantial and often permanent visual deficits after removal of the cataract and proper refraction.[2] Also, behavioral studies had shown that animals raised in the dark or in an environment devoid of contours have a similar impairment of their visual functions.[3,4]

Because of the difficulties associated with raising kittens in total darkness, we decided to fuse the lids by suture. This procedure prevented any form vision without completely depriving the animal of light. We expected this to be an effective procedure because cortical cells respond to contours and are insensitive to changes in levels of diffuse light.[5] Initially we raised kittens with only one eye closed, using the other eye as a control. This design turned out to be fortunate, because—as shown below—the effects of single eye closure on the visual cortex are more dramatic than the results obtained from animals raised with both eyes occluded or kept in the dark.

Our initial findings were that kittens with one eye occluded by lid suture during the first three months of life were blind in the deprived eye, and that in the striate cortex the majority of the cells responded only to stimulation of the normal eye.[6] This defect appeared to be localized to the visual cortex, perhaps at the site of interaction between geniculate afferents and cortical cells.[7] From another series of experiments, we found that the properties of orientation specificity and binocularity developed through innate mechanisms.[8] This result, taken together with the monocular deprivation experiment, indicated that neural connections present early in life can be modified by visual experience.

Such neural plasticity was not observed in the adult cat, but existed only during the first three postnatal months.[9]

The early experiments were done in the cat, but we soon turned our attention to the rhesus macaque monkey. After having demonstrated that cells in the monkey visual cortex also respond selectively to lines of different orientations and often are binocular[10] we showed that the monkey was also susceptible to visual deprivation,[11] a finding subsequently confirmed and extended.[12, 13, 14, 15] Further advances in our understanding of the nature of and mechanism underlying the deprivation phenomena depended on working out some of the functional architecture of the visual cortex. This was done through further physiological experiments in the normal animal and by using newly developed anatomical methods.[16, 17, 18, 19, 20] Over the years we have pursued the normal and developmental studies in parallel, and this has accelerated our progress in both areas. For example, while the deprivation experiments depended on the understanding of the functional architecture of the normal adult animal, we were alerted to the existence of ocular dominance columns in the cat by experiments we had done in strabismic animals.[21]

In this lecture I will present our current understanding of the development of the monkey visual cortex and the role of visual experience in influencing neural connections. Rather than attempting to discuss in any detail the now very extensive literature in the field, my emphasis will be on the work carried out in our laboratory (for reviews see references 22, 23 and 24). David Hubel and I did much of this work in collaboration with Simon LeVay.

MONOCULAR DEPRIVATION

The procedure of suturing a monkey's eyelid shut creates a condition similar to a cataract, since though the light reaching the retina through the closed lid is only slightly attenuated (by factor of 3), the forms of objects are no longer visible. As mentioned above, when the deprived eye is opened after months of deprivation, the animal is unable to see with it; there are no obvious changes in the ocular media, the retina or the LGN that can explain this deficit; instead marked changes have occurred at the level of the primary visual cortex (striate cortex). Even if the ocular media are clear, the occluded eye develops with time a marked axial length myopia (5−12 D over a 1 year period).[25]

One way of seeing the change is to record from cells in the striate cortex and determine their ocular preference.[7] In the monkey there is normally a fairly even balance between cells driven preferentially by one eye and cells driven preferentially by the other.[10] In layer IV most cells are strictly monocular, and outside of layer IV they are usually binocular, though they still tend to respond more strongly to stimulation of one eye than to the other. There are about as many cells preferring stimulation of the left eye as cells preferring stimulation of the right (Fig. 1, left). Under conditions of monocular deprivation, however, the great majority of cortical cells are driven exclusively by the non-deprived eye (Fig. 1, right).[11, 12, 13] One could ask whether this can be accounted for in terms of changes occurring at the level of the lateral geniculate nucleus.

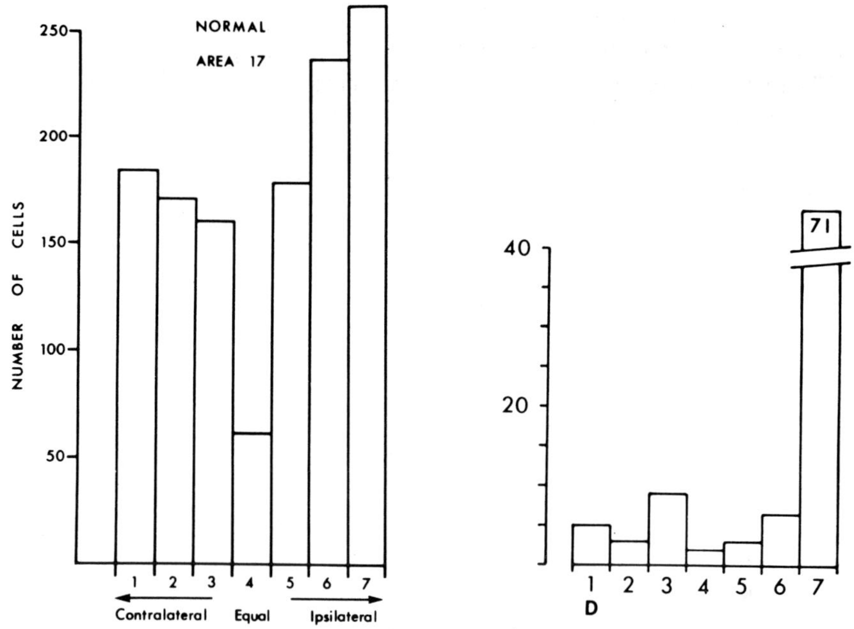

Fig. 1. Ocular dominance histograms in normal and monocularly deprived rhesus macaque monkeys.

Left histogram: 1256 cells recorded from area 17 in normal adult or juvenile rhesus monkeys.[13]

Right histogram: obtained from a monkey in which the right eye was closed at 2 weeks for 18 months.[15] It shows the relative eye preference of 100 cells recorded from the left hemisphere. The letter D indicates the side of the histogram corresponding to dominance by the deprived eye.

Cells in layer IVC are excluded in this figure and in histograms of all other figures. Cells in group 1 are driven exclusively from the contralateral eye, those in group 7 from the ipsilateral eye, group 4 cells are equally influenced, and the remaining group are intermediate.

Although the cells lying in the geniculate layers that receive input from the deprived eye are smaller than those in the non-deprived layers (Fig. 2), they are present in normal numbers, respond briskly to stimulation of the deprived eye and have normal receptive fields. Since geniculate cells are functionally normal and the cortical cells are altered in their properties there must be some change in the effectiveness of the geniculocortical connections. We were interested in investigating whether there were any structural changes associated with this abnormality.

The first aspect of cortical organization to be examined is the pattern of input of the geniculate afferents to the cortex. This can be done using the autoradiographic technique for tracing neuronal connections, transynaptically from the eye[18, 20] or by a fiber stain[19] (see also lecture of David Hubel). When in the normal monkey the input from the lateral geniculate nucleus reaches layer IV of the cortex, the information from the two eyes is still segregated. The input from each eye is distributed into a series of branching and anastomosing bands, which are about 0.5 millimeter wide, and alternate with similar bands serving the other eye (Fig. 3A). This pattern of innervation forms the anatomical basis for ocular dominance columns. Cells in the superficial and deep layers, while

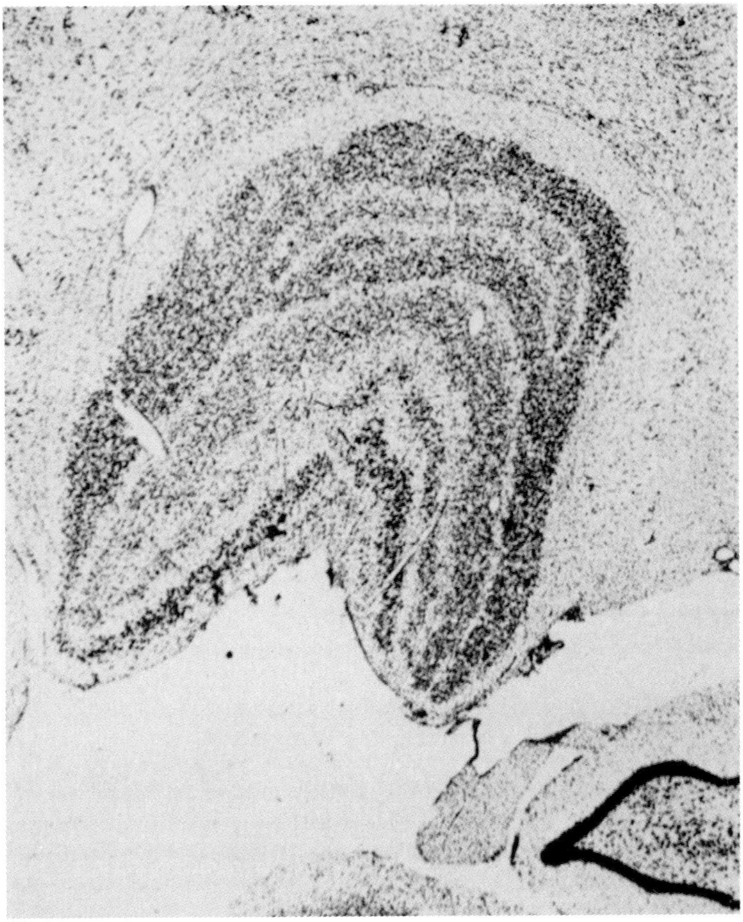

Fig. 2. Coronal section through the right lateral geniculate nucleus of the monkey with right eye closed at 2 weeks for 18 months (Fig. 1, right). Note the atrophy in the layers receiving input from the deprived eye (indicated by arrows). Stained with cresyl violet; frozen section.[13]

tending to be more binocular than cells in layer IV, still are more strongly influenced by the eye that provides input to the column in which they reside. The relative influence of the two eyes is shown by making tangential electrode penetrations through different cortical layers. Such penetrations in the normal monkey show regular changes in eye preference as expected from the columnar arrangements (Fig. 3A; Fig. 4, top).

In an animal that has undergone monocular deprivation, the geniculate terminals with input from the non-deprived eye take over much of the space that would normally have been occupied by terminals from the deprived eye (Fig. 3B). [13, 15] The deprived eye input has shrunken down to occupy the small strips lying between the terminals of the non-deprived eye input. Tangential electrode penetrations through cortical layers reveal long expanses of cells driven by the non-deprived eye interrupted by small patches of cells that are either unresponsive or driven by the deprived eye (Fig. 4, middle). As will be

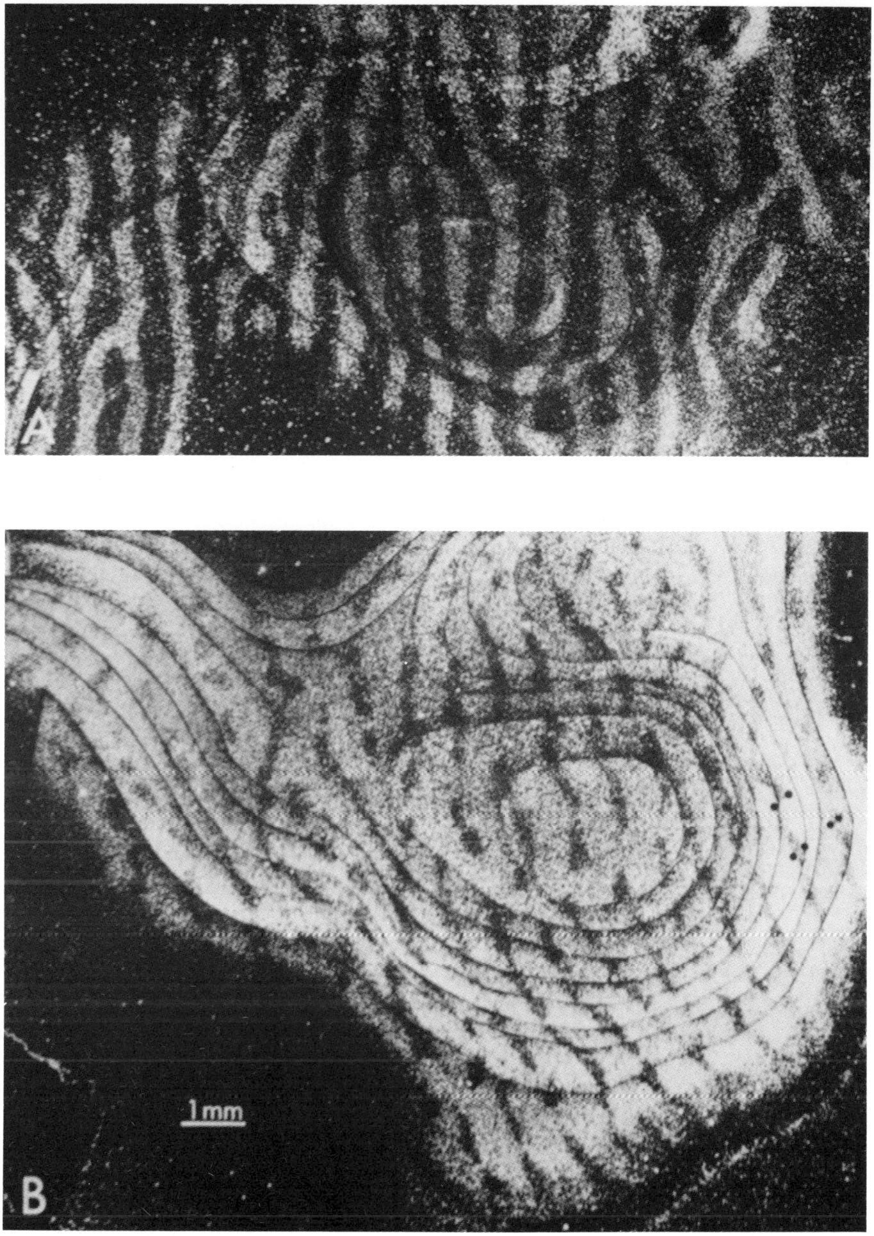

Fig. 3. Dark field autoradiographs of monkey striate cortex following injection of ^{3}H-proline in the vitreous of one eye 2 weeks before.

A: Normal monkey, a montage of a series of tangential sections through layer IVC. The light stripes, representing the labelled eye columns, are separated by gaps of the same width representing the other eye. *B*: Monocularly deprived monkey, again a montage from a series of tangential sections through layer IVC. Same monkey as in Fig. 1, right, and Fig. 2, which had the right eye closed at 2 weeks for 18 months. The input from the normal eye is in form of expanded bands which in places coalesce, obliterating the narrow gaps which represent the columns connected to the closed eye.

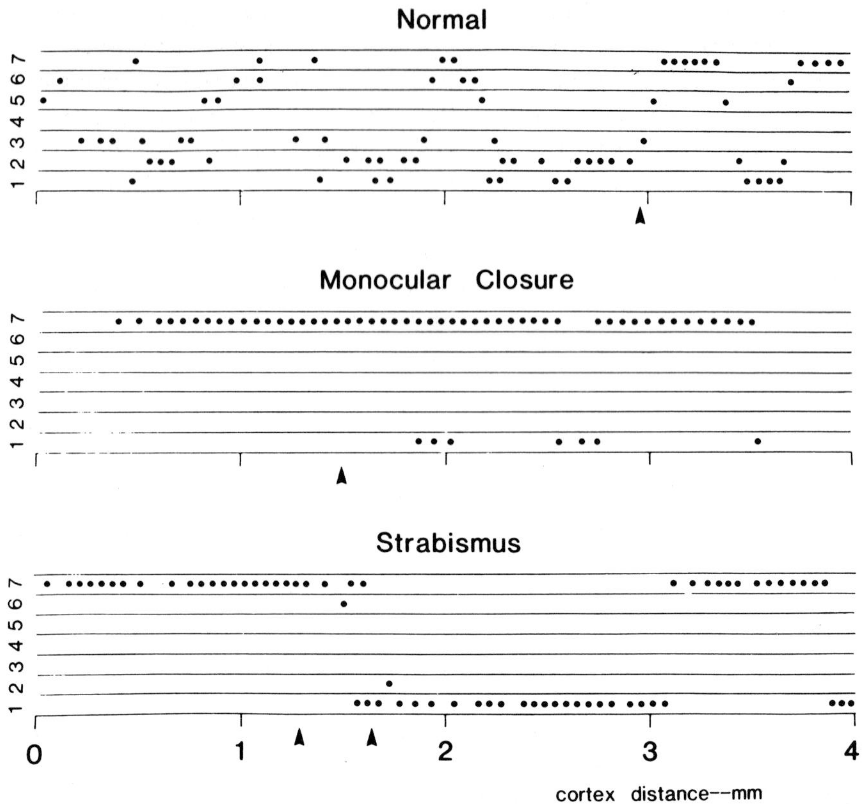

Fig. 4. Eye preference of cells recorded in oblique penetrations through the cortex in a normal monkey, a monocularly deprived monkey, and a monkey raised with strabismus. Ocular dominance categories (1–7) are shown relative to the distance the electrode penetrated through the cortex.

Top: Penetration in a normal monkey which shows the sinusoidal shift in eye preference with distance. The arrowhead indicates that the electrode entered layer IVC, in which cells are monocular, and there are abrupt shifts of dominance from one eye to the other.

Middle: Oblique penetration in a monkey raised with monocular closure (same monkey as in Figs. 1–3). Note that outside layer IV all cells are driven only by normal eye (7), and in layer IVC (see arrow) there are only short stretches of cells with input from the deprived eye. The observed overlap of input from left and right eye is not present in the normal monkey.

Bottom: A monkey with a 10° convergent squint produced by sectioning of the lateral rectus at three weeks. (Same animal as in Fig. 11, left histogram). The illustration shows the result of oblique penetration through the striate cortex made when the animal was $3^1/_2$ years old. Even outside of layer IV cells were monocular, with equal stretches of cortex dominated by either eye.

shown later in this paper, this expansion of the input from the non-deprived eye occurs at the level of single geniculate afferents. Cells in the deprived layers of the geniculate are smaller than normal. One reason for this is that their shrunken cortical arbors may require a smaller soma to maintain them, as originally proposed by Guillery and Stelzner (1970).[26]

Morphological examination of the lateral geniculate nucleus in these animals showed that there is a good relationship (r = 0.91) between the relative size of normal and deprived cells and the relative size of normal and deprived ocular dominance columns in layer IVC.[15] Thus, measuring geniculate cell sizes is yet another means of evaluating the effects of monocular closure.

From the histogram shown in Fig. 1 one cannot tell whether many cells have changed allegiance from the deprived to the non-deprived eye or have simply become unresponsive. The autoradiographic labelling of the afferents in layer IV (Fig. 3B) shows that a greater proportion of the cells in layer IV receive direct input from the non-deprived eye. The consequence of this change is that cells at later stages have shifted their allegience from the deprived to the non-deprived eye, rather than becoming unresponsive. This conclusion is supported by the physiological findings that the large majority of cells in superficial and deep layers respond only to the stimulation of the normal eye (Fig. 4, middle).

THE CRITICAL PERIOD

Having observed these dramatic effects of monocular suture early in an animal's life, we wanted to determine if there was a period over which the cortex retained its plasticity.

Our experiments in adult cats and monkeys[6, 15] showed that long periods of monocular lid suture did not result in the sort of changes in the visual cortex described above. Instead, we found that there is a definite period of time, early in life, during which the visual system shows this lability. We termed this the "critical period." The permanent visual deficits observed in children with congenital cataracts are therefore most likely a result of changes in the visual cortex that occurred during the critical period. Adult humans suffering from cataracts for many years will have normal vision when the cataracts have been removed presumably because they are well past their critical period at the onset of the disease.

The critical period in the monkey was estimated by closing one eye at different ages and keeping it closed for several months or longer.[15] The deprivation effect was gauged by the relative influence of the two eyes on single cortical cells (ocular dominance distribution), by the distribution of the input from the two eyes in layer IV (using the autoradiographic technique shown in Fig. 3), and by comparing the cell sizes in deprived and non-deprived layers of the LGN. The physiological results in monkeys with one eye closed at 2 weeks, 10 weeks, 1 year of age, and in the adult are illustrated in terms of ocular dominance histograms in Fig. 5. The earliest closure produced the most severe shift of preference toward the normal eye. The same degree of shift could be seen up to an age of 6 weeks. At that age the animal's susceptibility to monocular deprivation began to decline, but as is shown in the figure, it was still pronounced at 10 weeks and was detectable at one year. As indicated above, there were no cortical changes when the closure was done in the adult.

The changes occurring in the geniculo-cortical innervation were in general agreement with the physiology, though the time course was somewhat different (Fig. 6). Animals with a closure at 2 and $5\,^1/_2$ weeks showed the expected expansion of the non-deprived geniculate terminals; closure at 10 weeks showed a more moderate expansion, and at one year the pattern was indistin-

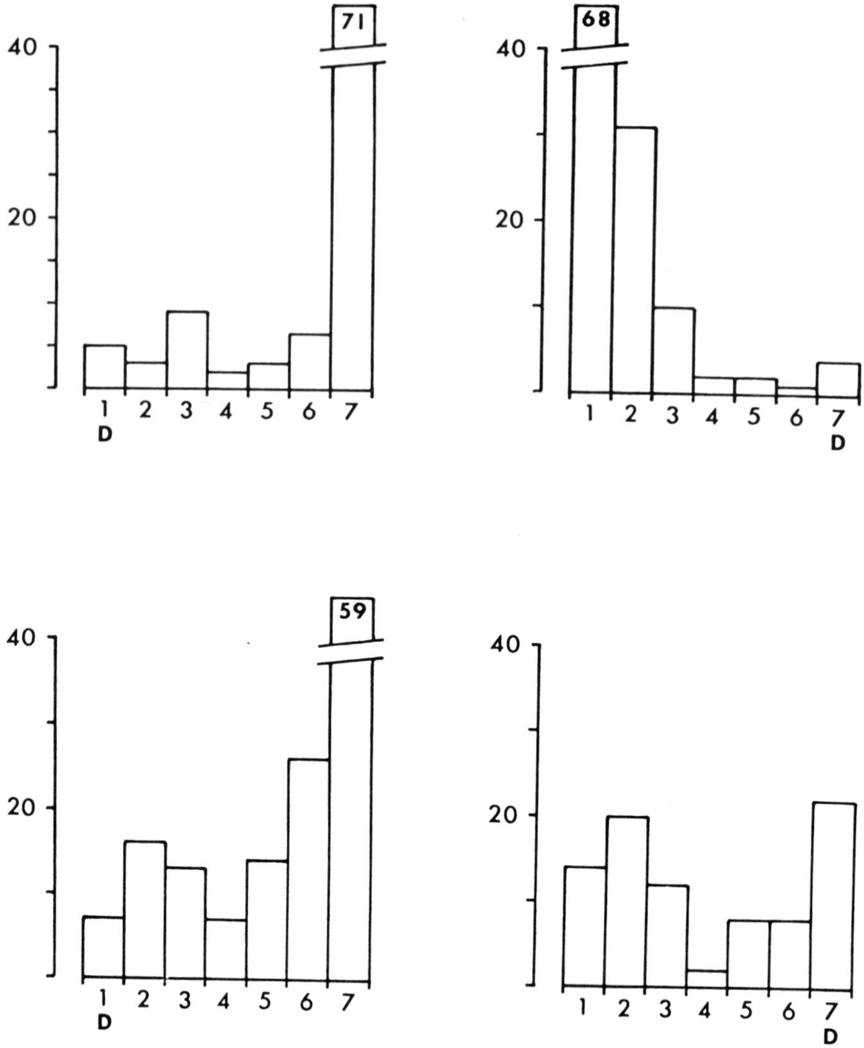

Fig. 5. Ocular dominance histograms of monkeys with one eye occluded by lid suture at different ages and examined after relatively long periods of closure.[15]

Upper left: Same monkey as in Figs. 1–4. Right eye closed at 2 weeks for 18 months (cf. Fig. 1, right).

Upper right: A monkey with right eye closed at 10 weeks for 4 months. Strong dominance of normal eye but not as pronounced as at earlier closure. Duration of deprivation relatively short, but our experience is that at this age the main changes in eye preference occur within the first few months of closure.

Lower left: A monkey with right eye closed at 1 year for a period of 1 year. A moderate shift in preference toward the non-deprived eye. This was particularly true for cells in layers II and III.

Lower right: Adult monkey (6 years old) with one eye occluded for $1^1/_2$ years. There was no obvious difference in eye preference from that observed in the normal monkey.

guishable from that in the adult. Geniculate cell sizes in the deprived layers changed in a parallel fashion, showing marked shrinkage at early closures, moderate reduction in closure at 10 weeks and no change when closed at one year. Since in the closure at one year we observed physiological changes in the

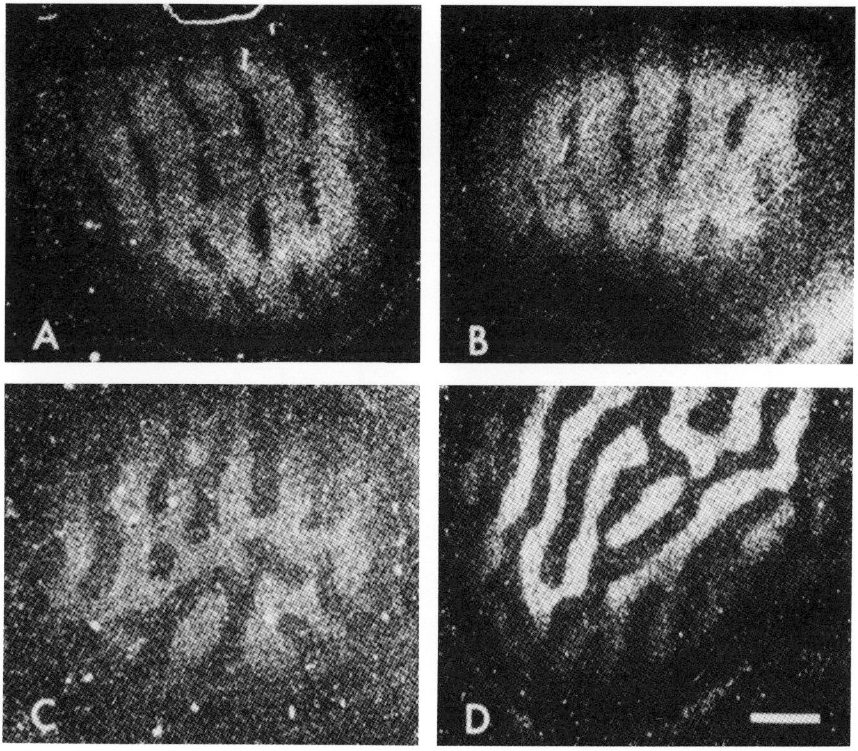

Fig. 6. Autoradiographic labelling patterns from the striate cortex of four monocularly deprived monkeys illustrating the distribution of geniculate terminals in layer IVC after closures at different ages. In all cases the normal (left) eye was injected with ^{3}H-proline thereby labelling the non-deprived geniculate terminals.[15]

A: Right eye closed at 2 weeks for 18 months. Same animal as in Fig. 1 (right), Fig. 4 (middle) and Fig. 5 (upper left).

B: Right eye closed at $5^1/_2$ weeks for 16 months.

C: Right eye closed at 10 weeks for 4 months. Same animal as in Fig. 5 (upper right).

D: Right eye closed at 14 months for 14 months. The unlabelled bar is 1 mm.

absence of a change in the pattern of geniculate innervation, there must presumably be changes occurring at subsequent levels in the cortical circuit.[13, 14] At any rate, in the adult even this "higher level" of plasticity disappears.

The high degree of susceptibility to deprivation at early ages is also apparent from experiments in which one eye in monkeys was closed for short periods. Before 6 weeks of age, it was sufficient to close an eye for a few days to obtain substantial change in eye preference. The ocular dominance histogram from a monkey with one eye closed for 12 days is shown in Fig. 7 (left). During the subsequent several months a marked change required several weeks of closure, and during the second year any change required months of closure.

From these and similar experiments by us[13, 15] and others[14, 27] we conclude that the macaque monkey is highly susceptible to monocular deprivation during the first six weeks of life, at which age the sensitivity declines progressively, so that at $1\,^1/_2$ to 2 years the monkey loses this type of neural plasticity.

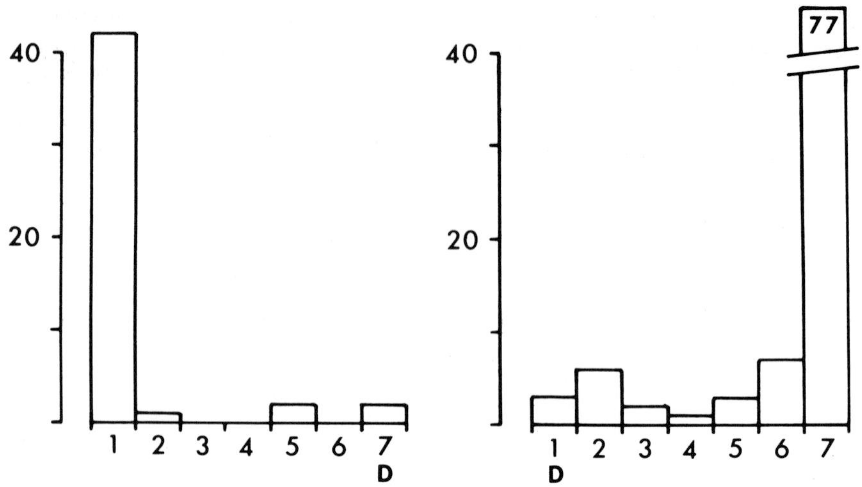

Fig. 7. *Left*: Ocular dominance histogram for 47 neurons recorded in a 20-day-old monkey whose right eye had been closed since 8 days of age. The physiological picture is similar to that seen after months of deprivation.[15]
Right: Ocular dominance histogram of 99 cells recorded in a monkey whose right eye was closed from 21 to 30 days of age. In spite of a subsequent 4 years of binocular vision, most cortical neurons were still unresponsive to stimulation of the right eye.[15]

The length of the critical period varies between species. In cats it is 3 to 4 months,[9, 28] and from clinical observations in humans it may extend up to 5–10 years, though the susceptibility to deprivation appears to be most pronounced during the first year and declines with age.[29, 30, 31]

RECOVERY FROM DEPRIVATION

Monocular closure during the entire critical period in cats and monkeys leads to permanent blindness.[32, 33, 34] Presumably there is no recovery of vision after the eye is opened because the pattern of geniculate innervation and the eye preference of cortical cells can no longer be modified. During the period of high susceptibility partial recovery of vision in the deprived eye is possible after brief periods of monocular closure.[15, 34, 35, 36] This was shown in a monkey with one eye closed between days 21–30, after which the monkey lived with both eyes open for a period of 4 years. Initially the animal appeared blind in the deprived eye but, with time, it slowly regained the use of the eye and the final acuity was 20/80–100 as compared to 20/40 in the non-deprived eye. The recordings from the striate cortex showed a marked dominance of the non–deprived eye (Fig. 4, right). If there was an increase in the number of cells driven by the once deprived eye, it was not very obvious. There was a marked narrowing of the deprived columns and a corresponding widening of the non-deprived ones. Thus, even if there had been some behavioral recovery, these results demonstrate that a few days of monocular closure had caused clear physiological and anatomical changes in the striate cortex.

These results are relevant to observations in children who have been monocularly deprived for short periods of time. When tested later, some children were found to have reduced acuity in the once patched eye and the degree of deficit depended on how young the child was at the time of patching.[38, 39] The experience in children with cataract removal indicates that surgery must be performed very early in the critical period in order to prevent the appearance of any deficit.[40, 41]

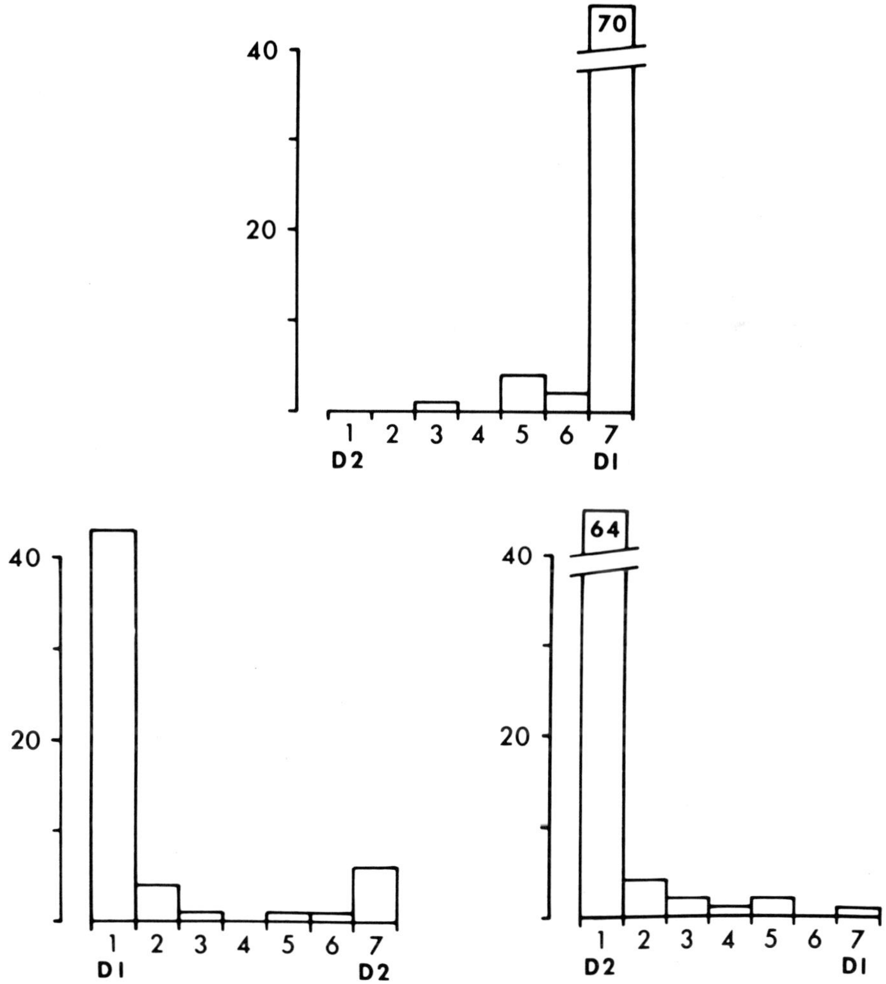

Fig. 8. Ocular dominance histograms of cortical cells recorded in three monkeys in which reversed suture was done at various ages.[15]
Upper histogram: 77 cells recorded from the right striate cortex of a monkey with the right eye (D_1) closed at 2 days for 3 weeks and the left eye (D_2) closed at 3 weeks for about 8 months. Nearly all neurons responded only to the initially deprived right eye (D_1).
Lower left: 56 neurons recorded from the left striate cortex. Right eye (D_1) closed at 3 days for 6 weeks; left eye (D_2) closed at 6 weeks for $4^1/_2$ months. Again nearly all neurons were driven exclusively by the initially deprived right eye (D_1).
Lower right: 74 neurons recorded from the left striate cortex. Right eye (D_1) was closed at 7 days for 1 year; left eye (D_2) was closed at 1 year for $2^1/_2$ years. In this case there was no effect of the reversal; nearly all the calls responded only to the initially open eye.

A procedure commonly used in children with strabismic amblyopia is to place a patch over the good eye to improve vision in a weak eye. In monocularly deprived animals it was possible to open the sutured eye and close the normal eye, here termed "reverse suture". Both in the cat and monkey, reverse suture led to a complete switch in eye preference if it was done within the early part of the critical period.[15, 27, 28, 37] The geniculate innervation of layer IVC also reversed so that the shrunken regions controlled by the initially closed eye expanded at the expense of the other eye, and consequently the cortical cells switched eye preference in favor of the eye closed first.[15, 37, 42] An example is shown in Figure 8 (top) in which the eye reversal was done at 3 weeks, and the recordings done 8 months later. The ocular dominance histo-

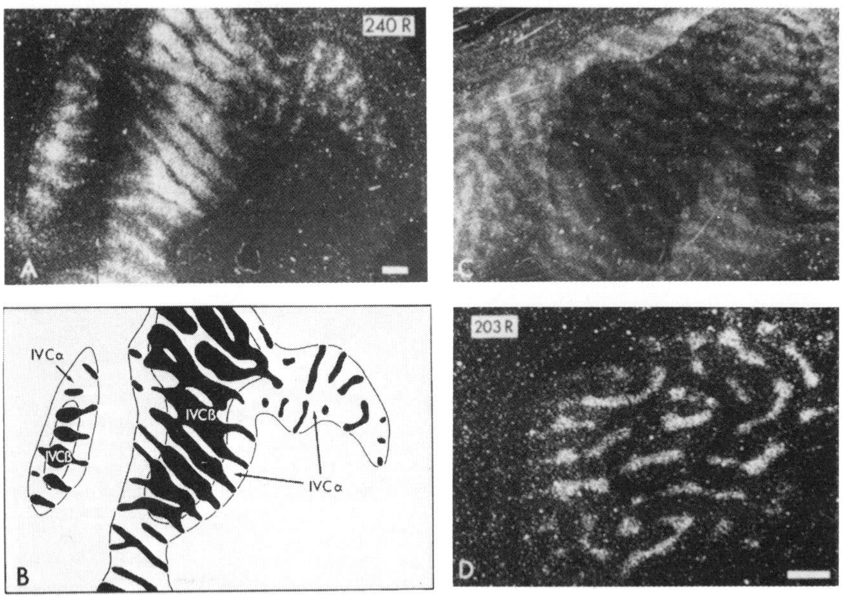

Fig. 9. Effect of reversed suture at various ages on the labelling pattern of geniculate terminals in layer IVC.

A: Same monkey as in the upper part of Fig. 8, in which reversed suture was done at 3 weeks of age. The initially deprived (right) eye was injected with ^{3}H-proline. A single tangential section. In the central region the labelled bands are expanded. In the surrounding belt the bands are contracted. These two regions correspond to the β and α sublaminae of layer IVC. Unlabelled bar is 1 mm.

B: Key to A, showing the distribution of label (in black) and the boundaries of the sublaminae IVCα and IVCβ, traced from an adjacent section stained with cresyl violet. Note that the thin labelled bands in IVCα run into the centers of the enlarged bands in IVCβ, meaning that the two sets of bands, though of very different width, are still in register with each other just as they are in normal animals.

C: The autoradiographic montage of the labelling pattern in monkey with reversed suture at 6 weeks (cf. lower left, Fig. 8). The initially deprived right eye was injected. The labelled bands are of about normal width indicating a recovery from the effects of the early deprivation, but not a complete reversal. Most of the montage shows layer IVCβ. In layer IVCα the labelled columns remained shrunken (not illustrated).

D: Autoradiographic montage from the monkey with reverse suture at 1 year of age (Fig. 8, lower right). The labelled columns (for the initially deprived right eye) remain shrunken, indicating that the late reversal did not permit any anatomical recovery. Scalemarker = 1 mm.

gram shows that the initially closed eye, which at the time of eye reversal would have influenced very few cortical cells, now became strongly dominant. The autoradiography (Fig. 9A) shows a marked expansion in layer IVCβ of the initially deprived geniculate terminals. When reversal was done at 6 weeks the physiology indicated a complete reversal, with a strong dominance of the initially deprived eye (Fig. 8, lower left). Such a marked shift was not reflected in the innervation of layer IV, in which the initially deprived eye had succeeded only in regaining its normal territory (Fig. 9C). This indicates that a significant part of the changes were occurring at the level of intrinsic cortical connections. Finally, reversing at one eye failed to produce any restoration of the function of the initially deprived eye (Fig. 8, lower right; Fig. 9D). Though it is possible to cause changes by monocular deprivation at one eye (Fig. 5, lower left), it appears to be more difficult to repair connections that have already been changed once.

Looking more closely at the autoradiography of the geniculate input to layer IVC in the monkey with the reversal at 3 weeks (Fig. 9 A and B), one sees a surprising result. The initially deprived eye took over much of the area of innervation of the lower part of layer IVC (IVCβ), but failed to reverse the dominance of the other eye in the upper part of layer IVC (IVCα). Apparently, the eye preference of the majority of cortical cells is determined primarily by the cells in IVCβ (Figs. 8 top, 9A and B). Layer IVCβ is innervated by cells in the dorsal part of the lateral geniculate nucleus (parvocellular layers) and IVCα by cells in the ventral part of the same nucleus (magnocellular layers). The result of the eye reversal experiment indicates that the critical period is different for the two cell types. Whereas the critical period is over for the magnocellular input at 3 weeks, the parvocellular input apparently begins to lose its ability to expand at 6 weeks (Fig. 9C), a time when intracortical connections still show considerable plasticity. This suggests that each functional unit has a unique program of development throughout the brain.

MECHANISM: DISUSE VERSUS COMPETITION

These experiments demonstrate that when a binocular cortical cell is not stimulated by a given eye, then the input from that eye drops out. Other forms of visual deprivation have shed some light on the mechanism of the effect of monocular deprivation. For example, if disuse were an important factor, one might expect that with both eyes closed, cortical cells would not be driven by either eye. Experiments in cats and monkeys raised under conditions of binocular deprivation showed, however, that cells were readily driven by the two eyes.[28, 43, 44] The cortex in the monkey was nonetheless altered in a very substantial way, in that very few cells were binocularly responsive.[44] This is illustrated in Fig. 10, in which a monkey had both eyes sutured from birth to 4 weeks of age. Except for the obvious lack of binocular cells, the cortex seemed quite normal. The cells were briskly responsive, showed a high degree of orientation selectivity, and had regular sequences of shifts in orientation preference. From tangential electrode penetrations we were also able to see a clear

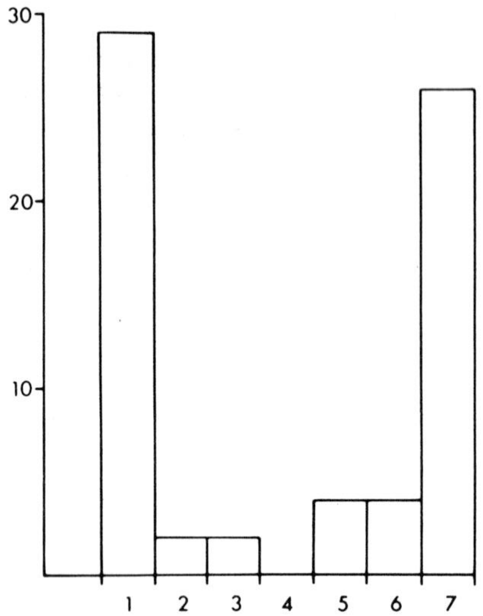

Fig. 10. Ocular dominance histogram of a monkey with binocular lid suture from birth to 30 days of age. Note the low number of binocular cells.[44]

segregation of the cells into unusually distinct ocular dominance columns, even outside of layer IV. When monkeys are kept in the binocularly deprived condition for many months a considerable fraction of the cells are unresponsive or respond only sluggishly, and often show lack of orientation preference. Binocular closures in kittens had a similar effect except that neither short nor long term deprivation led to an obvious loss of binocular cells.[28, 43]

Evidence for competitive mechanisms has also been found by measurements of geniculate cell sizes in an ingenious set of experiments.[26, 45, 46] First it was demonstrated in monocularly occluded kittens that deprived cells in the monocular segment of the nucleus were of normal size, whereas those in the binocular segment showed marked shrinkage.[26] Next Guillery produced a monocular region in the zone of binocular overlap by making a local retinal lesion in the normal eye of monocularly occluded kittens.[45] Again the deprived geniculate cells with no competitive input from the other eye were of normal size, and those outside the topographical area corresponding to the retinal lesion showed the usual shrinkage. Finally, it could be demonstrated that binocular closure in kittens apparently did not lead to a reduction in geniculate cell size[46] as originally reported by us.[43] These experiments lend strong support to the hypothesis that competitive mechanisms rather than disuse are prime factors in producing the changes observed under conditions of monocular deprivation.

Because many cortical cells are binocular from birth, the loss in the monkey of binocular cells at early times after closure suggests that in order for cortical cells to sustain a binocular input the two eyes must work together. Another situation that interrupts coordinated activity from the two eyes is strabismus.[47] One way of producing experimental strabismus is to section an extraocular muscle. Sectioning the lateral rectus causes the eye to deviate inward (convergent strabismus), whereas sectioning the medial rectus produces an outward deviation of the eye (divergent strabismus). After surgery the sectioned muscle usually reattaches behind the original site, so that except for the misalignment, normal eye movements are restored. In four monkeys with convergent strabismus the operation was performed between 3–5 weeks.[48] When the animals were examined after a year or more, three of them had normal acuity in both eyes but lacked the ability to fuse the images in the two eyes. The striate cortex of these animals had normal single unit activity, but there was a striking absence of binocular cells (Fig. 11, left). Tangential penetrations showed that the monocular cells were grouped in the usual regular columnar pattern (Fig. 4, bottom), suggesting that binocular cells had lost the input from the non-dominant eye. The fourth monkey had low acuity in one eye and fewer cortical cells were driven from that eye than from the normal eye (Fig. 11, right). When in five additional monkeys a strabismus was produced at later times during the critical period, there was an increase in the proportion of binocularly driven cells. From these results and experiments in cats and monkeys reported ear lier[49, 50, 51] it seems that the period during which the cortex can be influenced by the artificially induced strabismus is comparable in duration and sensitivity to that observed with monocular deprivation.

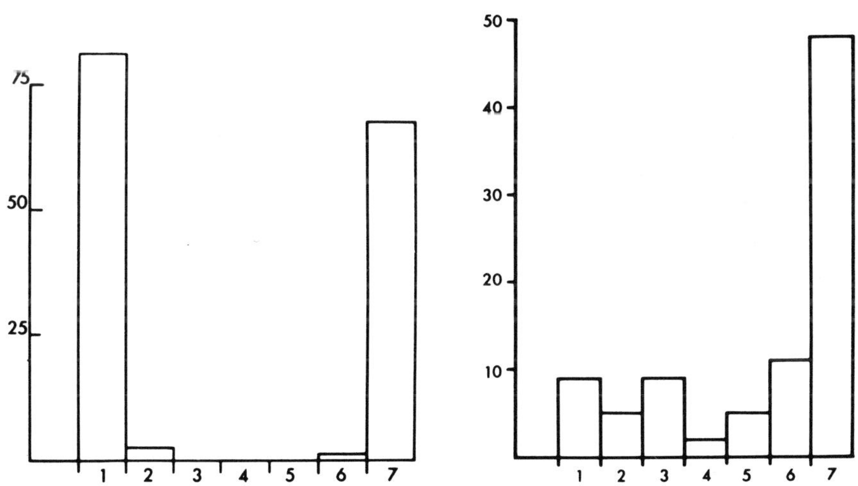

Fig. 11. Ocular dominance histograms of cells recorded in the striate cortex of two strabismic rhesus monkeys.[48]

Left: Histogram shows the eye preference of cells recorded in a 3 year old monkey in which the lateral rectus of the right eye was sectioned at 3 weeks of age. There is a nearly complete absence of binocular cells; the cells are driven exclusively either by the right or the left eye. As shown in Fig. 4 (lower) cells are clustered in a columnar fashion. The monkey had a 10° convergent strabismus, normal acuity in both eyes, but could not fuse images presented separately to the two eyes.

Right: A monkey with lateral rectus muscle of the right eye sectioned at 3 weeks. The animal had a convergent strabismus. Behavioral testing showed normal acuity in the left eye (20/30) and lower acuity in right eye (20/60 to 20/120). There was no difference in refraction between the two eyes. The histogram shows that the amblyopic eye influenced fewer neurons in the superficial and deep layers of the striate cortex. The ocular dominance columns in layer IVC had normal appearance when examined in tangential sections stained with a reduced silver method (Liesegang).[19]

The binocular deprivation and strabismus experiments support the notion that competition, rather than disuse, is the main cause of the observed changes.[43] The right circumstances must exist, however, for the competition to occur, since cells in the normal monkey tend to be dominated by one eye or the other,[10] and the dominant eye does not take over the cell completely. The difference between normal and deprived animals is that under normal conditions a cell receives input synchronously from the two eyes, whereas in monocularly deprived, strabismic, or binocularly deprived animals the two eyes do not act together. The maintenance of a given input may depend on the rate of firing of the postsynaptic cell while that input is active[3] so that in normal animals the non-dominant input is maintained by the activity of the dominant input. Carefully designed experiments by Singer et al[52] and Wilson et al[53] have provided support for the notion that it is crucial to activate the postsynaptic cell in order to change ocular dominance (for a more general discussion of synapse formation and stabilization see references 54 and 55).

In addition to providing insight into the mechanisms of development and plasticity in the visual cortex, the strabismus experiments may be of direct clinical relevance. A common situation in children with strabismus is that they have good vision in both eyes, but cannot fuse the images in the two eyes. These children often use the two eyes alternately, fixating and attending first with one eye and then with the other. The lack of binocular cells in strabismic animals is perhaps the physiological basis of this condition.[47, 12] Another common consequence of strabismus in children is a loss of acuity in one eye (strabismic amblyopia). The physiological mechanism of this condition is less well understood, even if our experiments (Fig. 11, right) and those of others[12, 56] indicate that one eye has been weakened in its ability to drive cortical cells, as is seen in monocular deprivation.

As mentioned above, late monocular deprivation in the monkey (see Fig. 5, lower left) and reversal experiments (Fig. 8, lower left, and Fig. 9C) caused alterations in the cortical circuit at stages subsequent to the input from the lateral geniculate nucleus. Another series of experiments illustrated this point quite dramatically. The approach is a variation of the original experiments by Hirsch and Spinelli[57] and by Blakemore and Cooper[58] in which kittens were raised viewing only stripes of one orientation. In our experiments we allowed a monkey to see vertical stripes through one eye only.[59] The other eye was deprived by lid suture. This effectively produced a different condition of deprivation for different populations of cortical cells: those with vertical orientation preference were monocularly deprived, and those with horizontal orientation preference were binocularly deprived. We recorded from the striate cortex after 57 hours of exposure (between days 12–54 after birth) and found normal levels of activity and cells of all orientations with their usual regular columnar arrangement.[17] There was an overall dominance of the open eye, but when we produced separate ocular dominance histograms for vertically and horizontally oriented cells (Fig. 12), it became clear that horizontally oriented cells tended to be driven monocularly by either eye (a picture typical of binocular deprivation), and vertically oriented cells tended to be driven monocularly by the exposed eye only (a picture typical of monocular deprivation). Thus, these findings again demonstrate that in addition to influencing the thalamocortical input, deprivation can alter the connections in the cortex without causing changes in that input. This question has also been addressed in the kitten by somewhat different approaches but with essentially the same results.[60, 61]

In looking at the effects of various forms of deprivation one gets certain insights into the processes that govern the balance between different inputs, enabling the visual cortex to integrate information in the appropriate manner. We have learned that competition and synchronization of inputs are important factors in forming and maintaining this balance. If these processes are disturbed early in life, the system can be permanently altered.

NORMAL DEVELOPMENT

We cannot properly evaluate the experiments on visual deprivation without having detailed knowledge about the normal development of the visual system. To assess the relative importance of the genetic program and the visual environment, it is necessary to evaluate the capabilities of the visual cortex at birth. The monkey is visually alert at or very soon after birth and cells in the

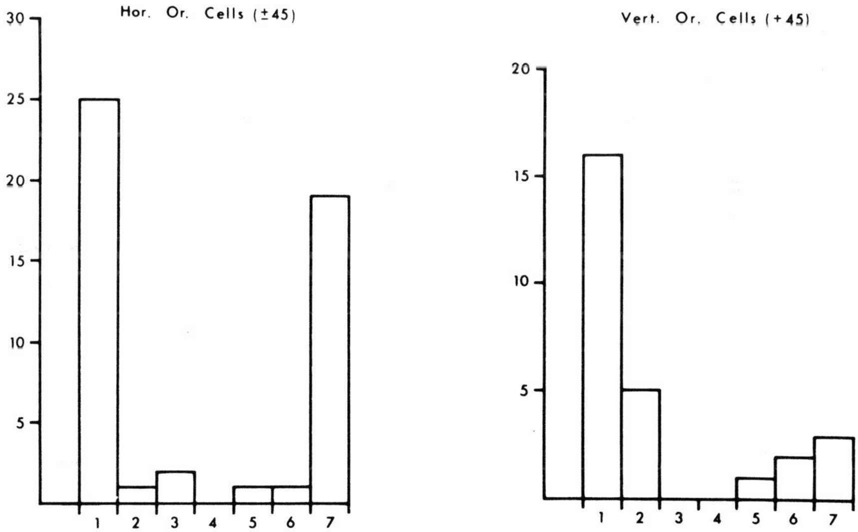

Fig. 12. Ocular dominance histograms of cells recorded from a monkey with the right eye closed at 12 days of age, then dept in the dark except for 57 hours of self-exposure to vertical stripes during the subsequent 42 days.
Left histogram: 48 cells recorded in the right striate cortex with preferred orientation within ± 45° of the horizontal axis. There were few binocular cells and a good number of monocular cells responding to stimulation of either the left or the right eye. Similar distribution of eye preference to that seen in binocularly deprived animals (cf. Fig. 10).
Right histogram: 27 cells with preferred orientation within ± 45° of the vertical axis. Majority of the recorded cells responded to the open eye producing a histogram similar to that seen after monocular deprivation.[59]

visual cortex show orientation preference and binocularity, as in the adult monkey. This was shown by single cell recordings in neonatal monkeys with no experience of contours or forms.[44]

Compared to the monkey, the kitten is far less well developed at birth; the eyes do not open until the second week and kittens tend to spend their first three weeks mainly eating and sleeping. In the visual cortex cells tend to give weak or erratic responses during the early postnatal period and come to respond like adult cells at about 3–4 weeks of age.[9, 28, 62] During the same time period cortical cells differentiate and active synapse formation takes place.[63, 64] Whether cells in the cat visual cortex develop normally without visual experience, as was originally reported for binocularly sutured kittens,[8] was questioned at first[65] but subsequently confirmed in several studies.[62,66,67,68] Whether in the kitten all cortical cells can develop fully through innate mechanisms is not entirely clear, since animals raised in the dark or with binocular lid closures seem to have a certain fraction of unresponsive or unoriented cortical cells.[62, 66, 68]

The newborn animal does differ from the adult in one significant respect, relating to the segregation of the afferents from the two eyes in layer IVC. In the newborn monkey we were able to show by eye injection of ^{3}H-proline that the inputs from the two eyes are strongly overlapping with only a mild fluctuation in eye dominance in a bandlike pattern.[13] Sokoloff and his colleagues[69] confirmed this observation using the 2-deoxyglucose method. In the monkey foetus Rakic showed that initially the left and right-eye afferents overlap completely, and not until a few days before birth do they begin to sort out into ocular dominance columns.[70] We followed this process of segregation postnatally; it was completed by 4–6 weeks of age (Fig. 13).[13] Recordings also indicated an initial overlap, followed by separation of the inputs from the two eyes in layer IVC and the time course of the events were similar. The process of segregation did not require visual experience, since it also occurred in an animal raised in the dark.[15] In kittens ocular dominance columns are formed much as they are in monkeys, showing a sequence of initial overlap and segregation during the first few months of life,[71] even though in this species visual experience appears necessary for their normal development.[72]

In the kitten it has been possible to examine the segregation of ocular dominance columns at a single cell level. In the early postnatal period, a single geniculate afferent gives off numerous branches innervating without interruption an area covering several future ocular dominance columns without interruption (Fig. 14, top).[73] As the axon matures, there appears to be a selective loss of branches, so that ultimately it innervates ocular dominance columns serving one eye and leaves gaps for the columns serving the opposite eye (Fig. 14, middle).[74, 76] In a cat monocularly deprived during its critical period, a geniculocortical afferent with input from the normal eye is shown in the bottom of Fig. 14.[77] It appears to have innervated an area that normally would have been occupied by the other eye.

Both the autoradiography (Fig. 3) and the single cell reconstructions (Fig. 14) suggest a mechanism for the expansion and contraction of ocular domin-

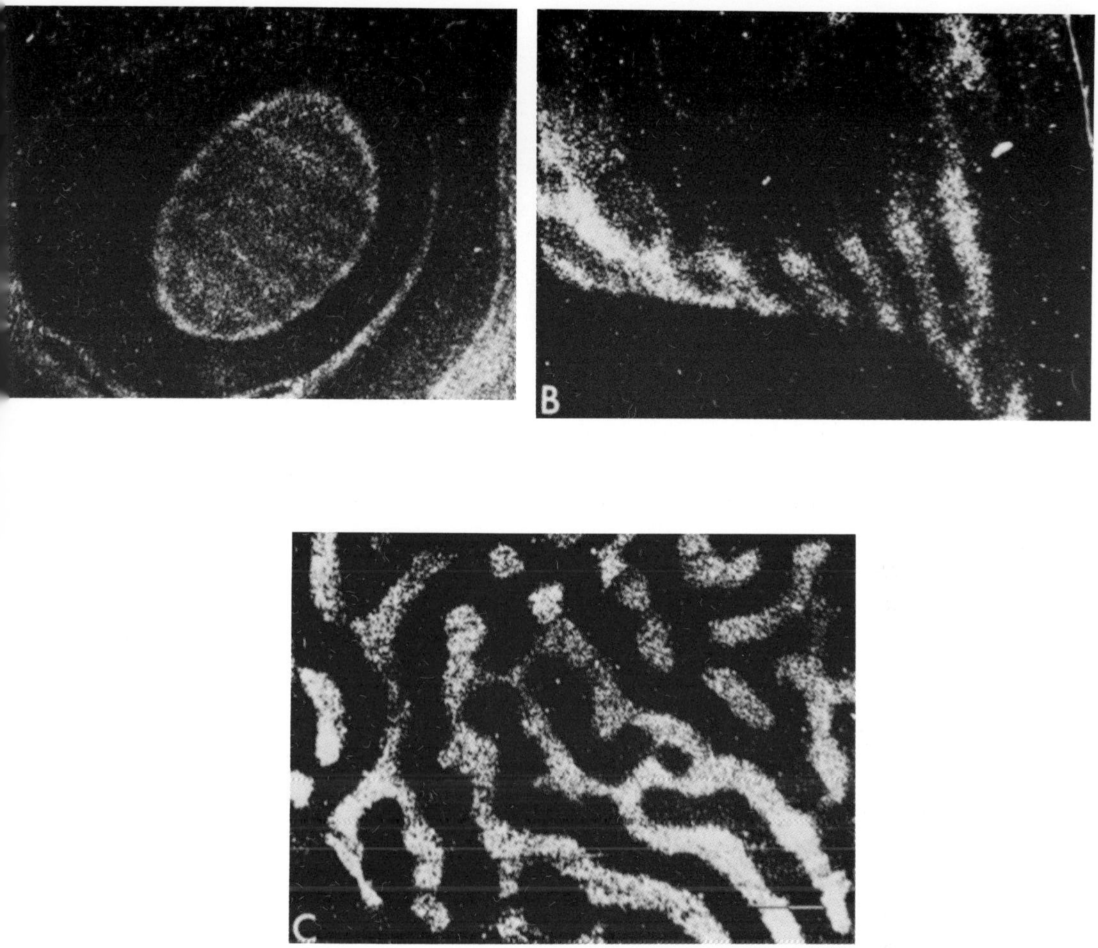

Fig. 13. Darkfield autoradiographs of geniculate afferent terminals in the striate cortex of normal neonatal monkeys in which the right eye had been injected with a radioactive tracer 1–2 weeks earlier.[15]

A: A normal 6-day-old monkey; single section from the left hemisphere that graces IVC tangentially in the central oval region. Silver grains are distributed continuously over layer IVC, but there are bands of alternating higher and lower grain density, indicating that afferents for the two eyes are in the process of columnar segregation.

B: A normal 3-week-old monkey; a single section through layer IVC of the left hemisphere which shows a clear columnar pattern but with a slight blurring of the margin of labelled bands, suggesting a modest intermixing of left and rigth eye afferents at the borders of ocular dominance columns.

C: A 6-week-old normal monkey; autoradiographic montage of the geniculate labelling pattern in layer IVC of the right striate cortex. The ocular dominance columns appear as sharply defined as in the adult monkey. The unlabelled bar is 1 mm. Adapted from Reference 15.

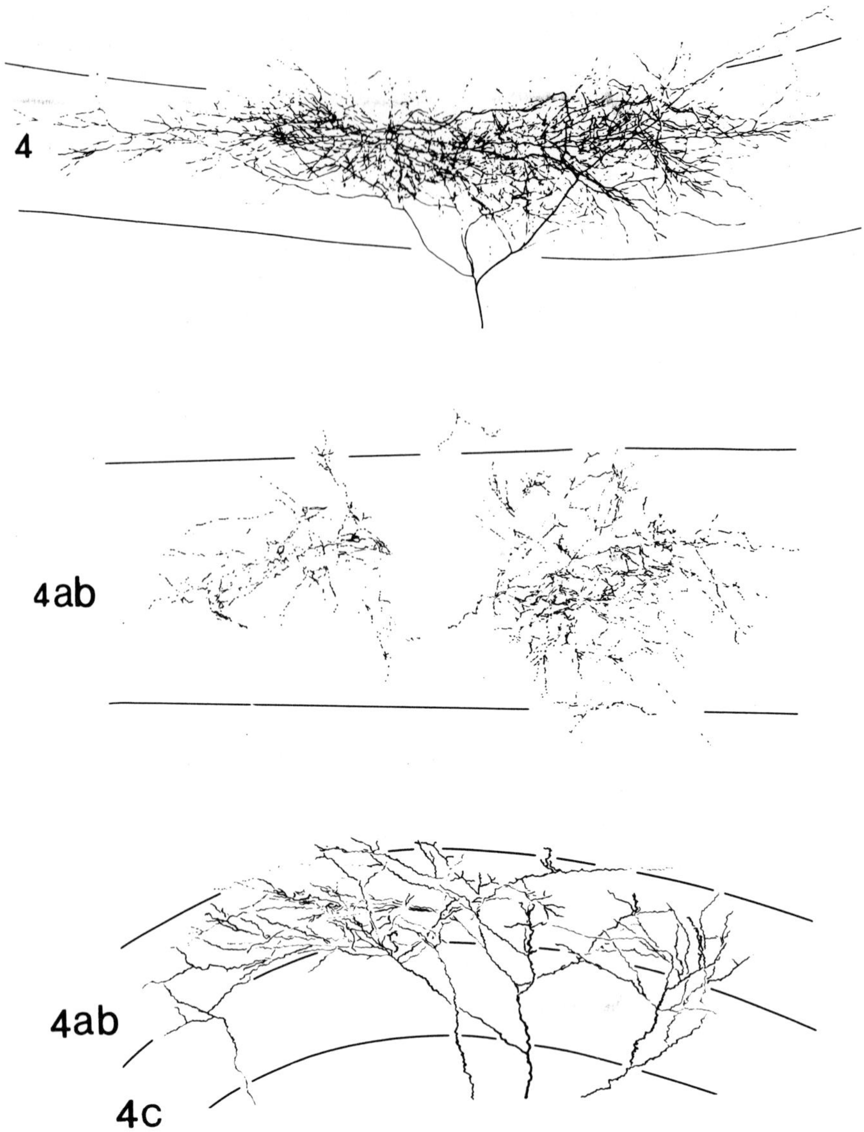

Fig. 14. Patterns of arborization of single geniculate axons in layer 4 of cat visual cortex.
Upper: A 17-day-old kitten; the arborization of a single afferent is shown at an age prior to the columnar segregation.[71] The axon, impregnated in its entirety with the Rapid Golgi method, arborizes profusely and uniformly over a disc-shaped area that is more than 2 mm in diameter. The use of this illustration is gratefully acknowledged (LeVay, S. and Stryker, M.P.).[73]
Middle: Normal adult cat; off-center geniculate afferent (Y-type)[75] injected intra-axonally with horse radish peroxidase (HRP) in the striate cortex. The arborization is entirely within layer 4 ab and forms two patches separated by a terminal-free gap. Presumably this pattern corresponds to the segregation of the input from the two eyes in a columnar fashion.[76]
Lower: Monocularly deprived cat (2 weeks – more than 1 year); a non-deprived Y-type afferent with an on-center receptive field injected intra-axonally with HRP. The arborization is primarily within layer 4 ab but does not have the normal patchy distribution of terminals. The absence of a terminal free region indicates that non-deprived geniculate branches are present in the territory that normally belongs only to the other eye.[77]

ance columns in monocular closure. The terminals with input from the normal eye continue to occupy the territory which normally they would have relinquished, while the deprived terminals are trimmed to an abnormal extent. Another mechanism appears to operate at slightly later stages: when the ocular dominance columns are fully segregated (at six weeks of age in the monkey), monocular closure still causes an expansion and a contraction of the columns similar to that seen after earlier closure (Fig. 6B). This argues for a mechanism of sprouting by one axon into the territory originally occupied exclusively by the deprived eye. Perhaps sprouting occurs from the axon branches that traverse the ocular dominance columns for the other eye (Fig. 14, middle). Reverse suture experiments also indicate that both trimming and sprouting are involved in plastic changes of the geniculocortical pathway (Figs. 9A, B, C). We know little of the biochemical mechanisms underlying these changes, except for the intriguing observation of the possible role of norepinephrine in neural plasticity.[78] Since the critical period seems to vary in onset and duration between different brain regions and even between layers of an individual cortical area (cf. IVCα and IVCβ in monkey striate cortex,[15] Fig. 9A and B) the control of plasticity appears to be specific and localized, not a phenomenon controlled by diffuse processes.

CONCLUSIONS

Innate mechanisms endow the visual system with highly specific connections, but visual experience early in life is necessary for their maintenance and full development. Deprivation experiments demonstrate that neural connections can be modulated by environmental influences during a critical period of postnatal development. We have studied this process in detail in one set of functional properties of the nervous system, but it may well be that other aspects of brain function, such as language, complex perceptual tasks, learning, memory and personality, have different programs of development. Such sensitivity of the nervous system to the effects of experience may represent the fundamental mechanism by which the organism adapts to its environment during the period of growth and development.

ACKNOWLEDGMENTS

I wish to express my gratitude and affection to the staff, faculty and students in the Department of Neurobiology. For over two decades this has been a unique place because of its blend of scientific excellence and compassion. The late Stephen Kuffler played a very special role in the creation of this environment and his spirit and attitude will always serve as an inspiration. My scientific career began in a serious way at Johns Hopkins Medical School and developed further during the past twenty years at Harvard Medical School. I am indebted to both of these institutions for providing me with invaluable opportunities for scientific training and research. I want also to recognize Robert Winthrop who generously provided the resources for my endowed professorship. It is obvious that our work could not have been carried out all these years without federal support; it is a pleasure to acknowledge the steady and generous support from the National Eye Institute. Finally, I wish to thank Drs. Anne Houdek and Charles Gilbert for their help in preparing the manuscript.

REFERENCES

1. Hubel, D. H. and Wiesel, T. N., J. Physiol. 160: 106–154 (1982).
2. von Senden, M., Space and Sight. The Free Press, Glencoe, 1960.
3. Hebb, D. O., The Organization of Behavior. Wiley, New York, 1949.
4. McCleary, R. A., Genetic and Experiential Factors in Perception. Scott Foresman and Company, Glenview, 1970.
5. Hubel, D. H. and Wiesel, T. N., J. Physiol. 148: 574–591 (1959).
6. Wiesel, T. N. and Hubel, D. H., T. N., J. Neurophysiol. 26: 1003–1017 (1963).
7. Wiesel, T. N. and Hubel, D. H., J. Neurophysiol. 26: 978–993 (1963).
8. Hubel, D. H. and Wiesel, T. N., J. Neurophysiol. 26: 994–1002 (1963).
9. Hubel, D. H. and Wiesel, T. N., J. Physiol. 206: 419–436 (1970).
10. Hubel, D. H. and Wiesel, T. N., J. Physiol. 195: 215–243 (1968).
11. Wiesel, T. N. and Hubel, D. H., Int. Union of Physiol. Sciences (I.U.P.S.) ABSTRACTS P 118–119 (1971).
12. Baker, F. H., Grigg, P. and von Noorden, G. K., Brain Res. 66: 185–208 (1974).
13. Hubel, D. H., Wiesel, T. N. and LeVay, S., Phil. Trans. Soc. London B. 278: 377–409 (1977).
14. Blakemore, C., Garey, L. J. and Vital-Durand, F., J. Physiol. 283: 223–262 (1978).
15. LeVay, S., Wiesel, T. N. and Hubel, D. H., J. Comp. Neurol. 191: 1–51 (1980).
16. Hubel, D. H. and Wiesel, T. N., J. Comp. Neurol. 146: 421–450 (1972).
17. Hubel, D. H. and Wiesel, T. N., J. Comp. Neurol. 158: 267–294 (1974).
18. Wiesel, T. N., Hubel, D. H. and Lam, D., Brain Res. 79: 273–279 (1974).
19. LeVay, S., Hubel, D. H. and Wiesel, T. N., J. Comp. Neurol. 159: 559–576 (1975).
20. Hubel, D. H. and Wiesel, T. N., Proc. R. Soc. Lond. B. 198: 1–59 (1977).
21. Hubel, D. H. and Wiesel, T. N., J. Neurophysiol. 28: 229–289 (1965).
22. Barlow, H. B., Nature 258: 199–204 (1975).
23. Pettigrew, J. D. In: Neuronal Plasticity, ed. C. Cotman. pp. 311–330, New York, Raven (1978). Symposium presentation, Berlin.
24. Movshon, J. A. and von Sluyter, R. C., Ann. Rev. Psychol. 32: 477–552 (1981).
25. Wiesel, T. N. and Raviola, E., Nature 266: 66–68 (1977).
26. Guillery, R. W. and Stelzner, D. J., J. Comp. Neurol. 139: 413–422 (1970).
27. Crawford, M. L. J., Blake, R., Cool, S. J. and von Noorden, G. K., Brain Res. 84: 150–154 (1975).
28. Blakemore, C. and von Sluyter, R. C., J. Physiol. (London) 237: 195–216 (1974).
29. Arden, G. B. and Barnard, W. M., Trans. Ophthalmol. Soc. U. K. 99: 419–431 (1979).
30. Vaegan, and Taylor, D., Trans. Ophthalmol. Soc. U. K. 99: 432–439 (1979).
31. von Noorden, G. K., Am. J. Ophthalmol. 92: 416–421 (1981).
32. Dews, P. B. and Wiesel, T. N., J. Physiol. 206: 437–455 (1970).
33. Ganz, L., Hirsch, H. V. B. and Tieman, S. B., Brain Res. 44: 547–568 (1972).
34. Wiesel, T. N., Unpublished results.
35. Mitchell, D. E., Cynader, M. and Movshon, J. A., J. Comp. Neurol. 176: 53–64 (1977).
36. Olson, C. R. and Freeman, R. D., J. Neurophysiol. 41: 65–74 (1978).
37. Blakemore, C., Vital-Durand, F. and Garey, L. J., Proc. R. Soc. Lond. B 213: 399–423 (1981).
38. Awaya, S., Sugawara, M. and Miyake, S., Trans. Ophthalmol. Soc. U. K. 99: 447–454 (1979).
39. Odom, J. V., Hoyt, C. S. and Marg, E., Arch. Ophthalmol. 99: 1412–1416 (1981).
40. Beller, R., Hoyt, C. S., Marg, E. and Odom, J. V., Amer. J. Ophthal. 91(5): 559–565 (1981).
41. Jacobson, S. G., Mohindra, J. and Held, R., Brit. J. Ophthalmol. 65: 727–735 (1981).
42. Swindale, N. V., Vital-Durand, F. and Blakemore, C., Proc. R. Soc. Lond. B 213: 435–450 (1981).
43. Wiesel, T. N. and Hubel, D. H., J. Neurophysiol. 28: 1029–1040 (1965).

44. Wiesel, T. N. and Hubel, D. H., J. Comp. Neurol. 158: 307–318 (1974).
45. Guillery, R. W., J. Comp. Neurol. 144: 117–130 (1972).
46. Guillery, R. W., J. Comp. Neurol. 148: 417–422 (1973).
47. Hubel, D. H. and Wiesel, T. N., J. Neurophysiol. 28: 1041–1059 (1965).
48. Wiesel, T. N. and Hubel, D. H., Unpublished results.
49. Yinon, U., Exp. Brain Res. 26: 151–157 (1976).
50. Crawford, M. L. J. and von Noorden, G. K., Invest. Ophthalmol. 18: 496–505 (1979).
51. Jacobson, S. G. and Ikeda, H., Exp. Brain Res. 34: 11–26 (1979).
52. Singer, W., Rauschecker, J. and Werth, R., Brain Res. 134: 568–572 (1977).
53. Wilson, J. R., Webb, S. V., Sherman, S. M., Brain Res. 136: 277–287 (1977).
54. Stent, G. S., Proc. Natl. Acad. Sci. U.S.A. 70: 997–1001 (1973).
55. Changeux, J.-P. and Danchin, A., Nature 264: 705–712 (1976).
56. Ikeda, H. and Wright, M. J., Exp. Brain Res. 25: 63–77 (1976).
57. Hirsch, H. V. B. and Spinelli, D. N., Science 168: 869–871 (1970).
58. Blakemore, C. and Cooper, G. F., Nature 228: 477–478 (1970).
59. Wiesel, T. N., Carlson, M. and Hubel, D. H., In preparation.
60. Cynader, M. and Mitchell, D. E., Nature 270: 177–178 (1977).
61. Rauschecker, J. P. and Singer, W., Nature 280: 58–80 (1979).
62. Buiserret, P. and Imbert, M., J. Physiol. 255: 511–525 (1976).
63. Marty, R., Archiv. D'Anatomie Microscopique et de Morphologie Experimentale 52: 129–264 (1962).
64. Cragg, B. G., J. Comp. Neurol. 160: 147–166 (1975).
65. Pettigrew, J. D., J. Physiol. 237: 49–74 (1974).
66. Blakemore, C. and von Sluyter, R. C., J. Physiol. 248: 663–716 (1975).
67. Sherk, H. and Stryker, M. P., J. Neurophysiol. 39: 63–70 (1976).
68. Singer, W. and Tretter, F., J. Neurophysiol. 39: 613–630 (1976).
69. Des Rosiers, M. H., Sakurada, O., Jehle, J., Shinohara, M., Kennedy, C. and Sokoloff, L., Science 200: 447–449 (1978).
70. Rakic, P., Philos. Trans. R. Soc. London, Ser. B. 278: 245–260 (1977).
71. LeVay, S., Stryker, M. P. and Shatz, C. J., J. Comp. Neurol. 179: 223–244 (1978).
72. Swindale, N. V., Nature 290: 332–333 (1981).
73. LeVay, S. and Stryker, M. P., Soc. Neurosci. Symp. 4: 83–98 (1979).
74. Ferster, D. and LeVay, S., J. Comp. Neurol. 182: 923–944 (1978).
75. Enroth-Cugell, C. and Robson, J. G., J. Physiol. 182: 923–944 (1966).
76. Gilbert, C. and Wiesel, T. N., Nature 280: 120–125 (1979).
77. Gilbert, C. and Wiesel, T. N., Unpublished observations.
78. Kasamatsu, T. and Pettigrew, J. D., J. Comp. Neurol. 185: 139–162 (1979).

1982
Physiology or Medicine

SUNE K. BERGSTRÖM
BENGT I. SAMUELSSON and
SIR JOHN R. VANE

"for their discoveries concerning the prostaglandins and related biologically active substances"

THE NOBEL PRIZE FOR PHYSIOLOGY OR MEDICINE

Speech by Professor BENGT PERNOW of the Karolinska Institute.
Translation from the Swedish text

Your Majesties, Your Royal Highnesses, Ladies and Gentlemen,

Hippocrates, the Father of Medicine, taught us that good health requires the four humors – blood from the heart, phlegm from the brain, and the yellow and black bile from the liver and spleen – to be in harmony with one another, or to put it simply, that we have sound humors. The teachings of Hippocrates also speak of the body's efforts to stay healthy and its constant battle against disease.

This more than 2.000-year-old concept may be said to prevail even today. Within the complex system that makes up the human body, it is essential that there be a balance between not only the organs but the individual cells. Their ability to interact determines our actions and our attitudes at any given time. A disturbance of this balance leads to undermined health. To maintain the balance and to prevent it from being upset by external or internal stress factors, nature has provided us with a number of regulatory systems. Prostaglandins and related biologically active substances constitute one of these systems.

It was Professor *Ulf von Euler* who opened up this area when he showed, almost 50 years ago, that seminal fluid in man and animals contains a substance that influences blood vessels and muscle fibers. He called the newly discovered substance Prostaglandin. Von Euler was awarded the Nobel Prize in 1970 for his fundamental discoveries within the fascinating world of hormones and signal substances.

The great breakthrough in prostaglandin research came in the late 1950s when *Sune Bergström* purified the first prostaglandins and determined their structure. This marked the beginning of the discovery of a hitherto entirely unknown biological system that regulates several vital life processes and that intervenes when the body's normal balance is disturbed. The widespread involvement of this system is indicated by the fact that almost all cells in our body have the ability to form one or several of its components. *John Vane* has called them defense hormones. Time permits me to give only two examples of how the system works.

Blood is constantly flowing in our blood vessels. But the blood cells have a tendency to bunch up to form blood clots that stop blood flow. To prevent this from happening in healthy and intact vessels, substances that belong to the prostaglandin system are constantly being formed within the vessel wall and the blood cells. These substances see to it that the blood can circulate without interruption. If this system is upset, formation of blood clots is inevitable.

Our white blood cells form a defense barrier against infection. Therefore, it is important that these blood cells converge on an inflamed tissue to attack and, if possible, destroy the injurious intruders. The latest addition to this biological system – the leukotrienes, recently discovered by *Bengt Samuelsson* – play an important role because of their ability to lure white blood cells to the site of injury and bind them to the vessel wall.

Thus, this entire array of substances indeed warrants the name defense hormones.

However, here as elsewhere, moderation is essential. In certain situations, such as in allergic reactions, prostaglandins and leukotrienes are overproduced and give rise to the entire set of symptoms that characterize allergies. When, for example, an asthma sufferer is exposed to the specific substances to which he or she is over-sensitive, leukotrienes are produced in large quantities in the lungs and trigger the severe asthma attack.

Thus, the knowledge acquired of how this system is formed and how it functions has to a significant degree increased our understanding of how our bodies can keep our humors sound. But it has also taught us to understand the mechanisms behind a number of widespread diseases – allergies, inflammations and vascular diseases, to name a few. This new knowledge has, in turn, provided the prerequisites for enabling us to more specifically, and thus more effectively, threaten these disorders and prevent their occurrence.

It is no wonder that prostaglandin research is now being carried on all over the world. Nevertheless, the researchers who once were the pioneers and still continue to lead development in the field are Bergström, Samuelsson and Vane.

As mentioned, *Sune Bergström* laid the groundwork for the current development by isolating the first prostaglandins and showing that it was not a matter of one substance but of a whole system. Bergström showed, too, that unsaturated fatty acids form the parent substance for this system, and thereby focused research attention on these acids.

Bergström's student, *Bengt Samuelsson,* is responsible for the chemical development since the 1960s. He has isolated and determined the structure of several of the most significant components within the system. More thoroughly than anyone else, he has explained for us how this complicated system is built up, and has helped us to understand the relationship between its various components.

Among *John Vane's* contributions may be mentioned the discovery of an important component of the system – prostacyclin. Vane has also revealed that the secret of the pain-relieving and fever-reducing properties of acetylsalicylic acid is its inhibition of the formation of prostaglandins. We are more familiar with acetylsalicylic acid under the name aspirin. All at once Vane was able to explain how this most widely used medicine in the world, for whose existence we have all had reason to be grateful at one time or another, actually functions in our bodies. By this discovery, Vane also provided an important weapon for those continuing the research directed towards establishing the functional role of the prostaglandin system.

The scientific contributions we are rewarding today have inspired researchers all over the world, and prostaglandin research may now be said to have entered its most dynamic phase. At the same time it may be said to be a beautiful example of how support to basic research has turned out to be a very good investment for society as a whole.

Drs. Bergström, Samuelsson and Vane,

Your discoveries have revealed for us a previously unknown biological system that not only plays a decisive role in normal life processes but also contributes to the imbalance that characterizes several diseases. Proceeding from varying points of departure you have, together, clarified the structure, biological properties and fundamental functions of this system. Your discoveries have stimulated an entire world in their intensive research aimed at finding the causes and cures for a number of diseases of importance to mankind.

On behalf of the Karolinska Institute and its Nobel Assembly I have the privilege to convey to you our warmest congratulations, and I now ask you to receive your Nobel Prize from the hands of His Majesty the King.

SUNE K. D. BERGSTRÖM

I was born January 10th, 1916 in Stockholm, Sweden.

Degrees:

1944	D. Med. Sci., Biochemistry, Karolinska Institutet, Stockholm
1944	M. D., Karolinska Institutet, Stockholm
1944	Docent of Physiological Chemistry, Karolinska Institutet, Stockholm

Appointments:

1938	Research Fellowship, London University, London
1940–41	Research Fellowship, Columbia University, New York
1941–42	Squibb Institute for Medical Research, New Brunswick, N.J.
1944–47	Assistant at the Biochemical Department, The Medical Nobel Institute, Karolinska Institutet, Stockholm
1946–47	Research Fellowship, Basel University, Basel
1947–58	Professor of Physiological Chemistry, The University of Lund, Lund
1958–80	Professor of Chemistry, Karolinska Institutet, Stockholm
1963–66	Dean of the Medical Faculty, Karolinska Institutet, Stockholm
1969–77	Rector of Karolinska Institutet, Stockholm
1975–	Chairman of the Board of Directors, The Nobel Foundation, Stockholm
1977–82	Chairman of the WHO Global Advisory Committee on Medical Research, Geneva
1983	President of the Royal Swedish Academy of Sciences

Memberships and Honorary Memberships:

1952–58,	Swedish Medical Research Council, Stockholm
1964–70	Swedish Medical Research Council, Stockholm
1955–62	Swedish Natural Science Research Council, Stockholm
1965	Royal Swedish Academy of Sciences, Stockholm
1965	Swedish Academy of Engineering Sciences, Stockholm
1965	American Academy of Arts and Sciences, Boston
1973	National Academy of Sciences, Washington, D.C.
1973	American Society of Biological Chemists
1976	Academy of Sciences USSR, Moscow
1977	Academia Leopoldina, Halle, DDR

1978	Institute of Medicine, NAS, Washington, D.C.
1980	Royal Society of Edinburgh, Edinburgh
1982	Medical Academy USSR, Moscow
1982	Finska Vetenskaps-Societeten, Helsingfors
1982	Swedish Society of Medical Sciences, Stockholm

Honors and Awards:

Doctor h. c., University of Basel, Basel, 1960
Doctor h. c., University of Chicago, Chicago, 1960
"La Madonnina", Lectureship, Milan, 1972
Dunham Lecturer, Harvard University, Boston, 1972
Anders Jahre Medical Prize, Oslo, 1972
The Gairdner Award, University of Toronto, Toronto, 1972
Dohme Lecturer, Johns Hopkins University, Baltimore, 1972–73
Merrimon Lecturer, University of North Carolina, Chapel Hill, 1973
The V. D. Mattia Lectureship of the Roche Institute, USA, 1974
Harvey Lecture, The Harvey Society, New York, 1974
The Louisa Gross Horwitz Prize, Columbia University, New York, 1975
The Francis Amory Prize, American Academy of Arts and Sciences, 1975
Doctor h. c., Harvard University, Boston, 1976
Doctor h. c., Mount Sinai Medical School, New York, 1976
Doctor h. c., Medical Academy of Wrocław, Poland, 1976
The Albert Laser Basic Medical Research Award, New York, 1977
General Amir Chand Oration, All India Institute, New Delhi, 1978
Cairlton Lecture, University of Texas Health Science Centre, Dallas, 1979
The Robert A. Welch Award in Chemistry, Houston, 1980
Nobel Laureate in Physiology or Medicine, Stockholm 1982

THE PROSTAGLANDINS: FROM THE LABORATORY TO THE CLINIC

Nobel lecture, 8 December 1982

by

SUNE BERGSTRÖM

Karolinska Institutet
S-10401 Stockholm, Sweden

The chemistry and biochemistry of the prostaglandins are now described in textbooks, and I have reviewed the early work several times (1-4). I will instead discuss the background of the early prostaglandin work here in Sweden and also some recent developments in the clinical fields that I have been associated with.

Now let me digress for a moment. My scientific work started with Dr Erik Jorpes in 1934 when I participated in his early heparin work. He is more responsible than anybody else for purifying heparin and introducing it as a drug into the clinic. At that time Professor Einar Hammarsten was Director of the Chemistry Department at the Karolinska, then one of the leading laboratories in the world also in the field of nucleic acids and of peptide hormones, i.e. secretin and cholecystokinine.

Dr Jorpes always maintained that it was too bad that nobody worked on lipids or steroids in Sweden. He financed a trip for me to England in 1938 where I spent a few months working on bile acids with Dr G.A.D. Haslewood at Hammersmith Postgraduate Medical School. The following year I got a fellowship from the British Council to work for a year at Dr Marrian's laboratory in Edinburgh. It was cancelled when the war broke out, but I was then lucky enough to get a Swedish American Fellowship and worked for a year and a half at Columbia University and at the Squibb Institute with Dr Oskar Wintersteiner on cholesterol autoxidation 1940-1942. After returning home I started working on the autoxidation of linoleic acid and identified the structure of the main reaction products. It was found that a conjugation of the double bonds took place and that oxygen was introduced as an hydroperoxide group at carbon atoms 9 or 13. I also found that the lipoxygenase enzyme of soy beans, just described by Dr Sumner, yielded the same products as the heavy metal catalyzed autoxidation. At that time I was working in Dr Hugo Theorell's laboratory, and we started to cooperate on the purification of the soy bean lipoxygenase enzyme.

My involvement with the prostaglandins started at the Meeting of the Physiological Society of Karolinska Institutet, October 19th, 1945 where I reported on my work on the oxidation of linoleic acid. Dr Hugo Theorell was chairman, Dr Yngve Zotterman secretary and Dr Ulf von Euler signed the minutes of the meeting.

After this meeting von Euler asked me if I might be interested to study the small amount of his lipid extracts of sheep vesicular glands that he had stored since before the war. I agreed, and we began a most stimulating and pleasant cooperation.

The first observation that there was some biologically active compounds in human semen had been done in the Department of Obstetrics and Gynecology at Columbia University in New York. In 1930 Kurzrok and Lieb with the technical cooperation of Dr Sara Ratner (5) reported that when they were doing artificial inseminations in women, they sometimes got violent contractions, sometimes relaxation, of the uterus.

An interesting coincidence is that I met Dr Kurzrok and worked at the same laboratory at Columbia University as Dr Sara Ratner for a year but never heard anything about prostaglandin.

In 1933 the British pharmacologist Goldblat (6, 7) had reported that human semen contained a factor that reduced blood pressure and stimulated smooth muscle.

At about the same time Dr von Euler was making a thorough study of the occurrence of compound P in various organs, the peptide he had discovered a few years earlier. In semen and in extracts of "prostate" or vesicular glands of monkeys, sheep and goats there was a strong blood pressure decreasing factor that also stimulated smooth muscles. He showed that the factor was different from P, that it was lipid soluble and he gave it the name prostaglandin. Dr Theorell had shown that the activity behaved as an acid in his electrophoresis appartus (8—10).

In the early work the rabbit intestinal strip test was used to follow the activity. We used mainly countercurrent extractions to purify von Euler's crude extract.

I had just brought home one of Dr Lyman Craig's stainless steel counter current extraction machines that proved ideal for working with these small amounts. We managed to purify the crude extract about 500 times. The most active fractions consisted of unsaturated hydroxy acids that were free of nitrogen (11).

The work was interrupted for a few years due to my appointment to the chair of physiological chemistry at the University of Lund in 1948.

A practically empty institution had to be rebuilt, reequipped and staffed, but it was a very fortunate time (1948) for the biomedical sciences in Sweden. The large buildup of resources for basic biomedical research at the Swedish Universities, initiated by the former Prime Minister Tage Erlander, had just started. The Swedish Medical Research Council had also just been started. An additional advantage during the following decade was that the National Institutes of Health (NIH) in the U.S.A. started their unique international programme of support of biomedical research.

We were fortunate to receive quite sizeable grants for our work in the field of steroid and bile acid metabolism from NIH for a number of years.

A group of able graduate students could then be trained in these fields— Bengt Borgström, Jan Sjövall, Sven Lindstedt, Henry Danielsson, Bengt Samulesson and Rolf Blomstrand. They have all more or less directly contributed to the early prostaglandin development that was started up again in the fifties.

Collection of sheep glands was organized in Sweden and Norway. Using counter current fractionations and partition chromatography two crystalling compounds prostaglandin E_1 and $F_{1\alpha}$ (12–14) were isolated in small amounts in 1957.

The correct empirical formulas could be determined in cooperation with Wolfgang Kristen who had developed an ultramicro carbon-hydrogen determination in Dr Einar Stenhagen's laboratory in Uppsala. Together with the molecular weight, determined by Dr Ragnar Ryhage in his mass spectrometer, the formulas were found to be $C_{20}H_{34}O_5$ and $C_{20}H_{36}O_5$, respectively.

Collection of sheep glands was then expanded in Sweden, Norway, Iceland, Greenland and also in the U.S.A. through the Horniel Institute. These activities were generously supported by the Upjohn Co., U.S.A.

In 1958–1959 the whole research group moved from Lund to the Department of Chemistry at the Karolinska Institute in Stockholm. What really played a decisive role in the prostaglandin work was the mass spectrometer development that Dr Ragnar Ryhage was doing there. He had built the first

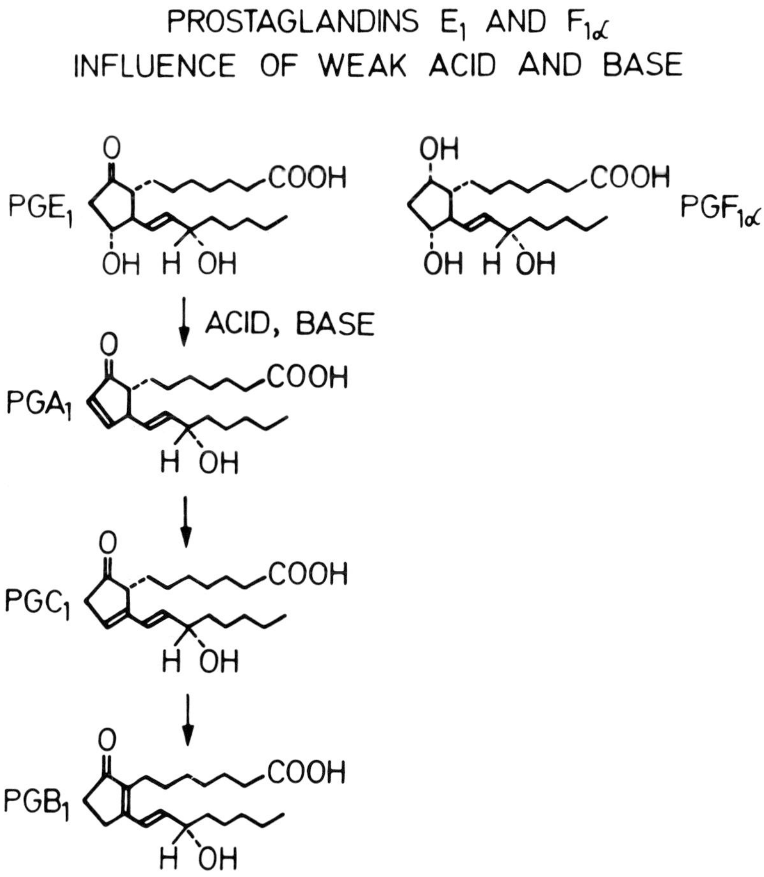

Figure 1. The influence of weak acid and base on PGE_1 and $PGF_{1\alpha}$.

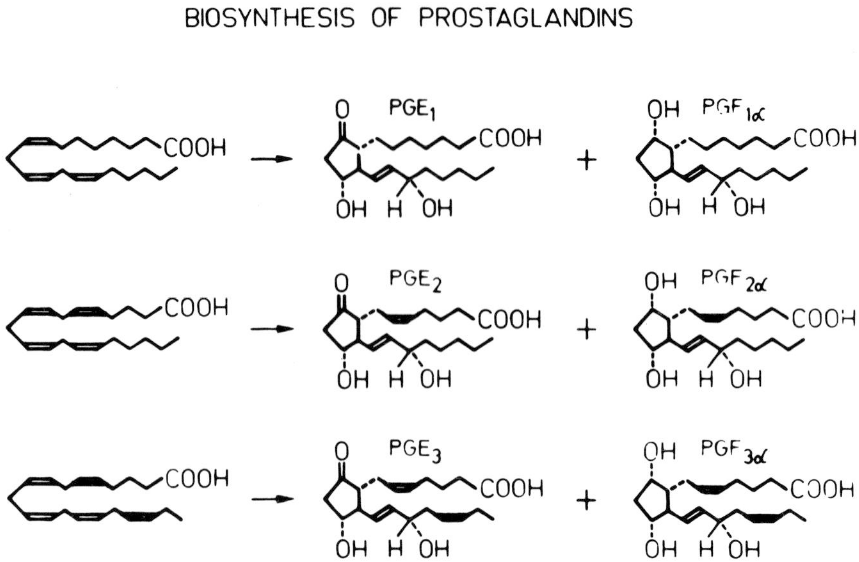

Figure 2. Biosynthesis of prostaglandins with homogenates of sheep vesicular glands.

functioning combination instrument of a gas chromatograph and a mass spectrometer. The structure of PGE_1 and $F_{1\alpha}$ was then deduced largely from mass spectrometric identification of the products formed by oxidative degradation of these prostaglandins and the compounds formed from PGE_1 by treatment with a weak acid or base indicated in Fig. 1. The structure was then confirmed and the stereochemistry determined by Dr Sixten Abrahamsson in Gothenburg on a derivative of $PGF_{1\alpha}$. It took him about one year to do this using the technology of those days. Today it would take less than three weeks. The absolute configuration was finally settled (25, 26).

By 1962 the six prostaglandins as listed in Fig. 2 had been isolated and their structures determined (15–24). It had also been found that these prostaglandins occur in many other tissues outside the male reproductive organs.

These 20-carbon prostaglandins have cis-double bonds located as in certain essential fatty acids, when counting from the carboxyl. This made us suspect that these naturally occurring acids might be precursors.

I then telephoned Dr David van Dorp at Unilevers Research Laboratories in Holland where these acids had been synthesized and isotopically labelled earlier. When I asked him if we could get some isotopically labelled C_{20} acids, it turned out that he was also planning to start investigating if they were precursors, and he had just prepared isotopically labelled dihomogamma-linoleic acid for this purpose. He was most generous and sent us samples of the labelled acid. We started simultaneously incubating the labelled acids with homogenates of sheep glands and could inform each other after two days that indeed labelled prostaglandin E_1 and $F_{1\alpha}$ were produced in good yield. We published our findings simultaneously in 1964. This work was then followed up with the labelled tetraenoic arachidonic acid and the pentaenoic acid shown in Fig. 2 which yielded prostaglandins of the 2 and 3 series, respectively (27–30). This discovery was also made independently at the Upjohn Co. using glands from bulls.

As the yields could be up to 70% these enzymatic methods were of great practical importance for getting material for further biochemical, physiological and pharmacological work.

We prepared several grammes of various prostaglandins whereas Unilever made considerably more, and the Upjohn Co. must have prepared hundreds of grammes.

In spite of the biosynthetic method the supply situation became critical around 1970 when prostaglandins were needed for clinical work before large scale total syntheses had been developed. An unexpected discovery was then made by Weinheimer and Spraggins (31) who had been studying the biochemistry of marine organisms in the Mexican Gulf. They found that up to one and one half percent of the dry weight of a Gorgonia coral consisted of 15 epiprostaglandin A_2 and related compounds. For several years the Upjohn Co. isolated these prostaglandins from corals which they transformed into the prostaglandins supplied for clinical trials. But from about 1973 the supplies have been prepared by total synthesis.

When the structure of the prostaglandins was published, many projects had been started on their total synthesis.

Dr E.J. Corey, who had collaborated with us in studies of the biosynthesis of bile acids with his stereospecifically labelled cholesterol, was the first to succeed. In a series of classical papers in 1968–69 (32–37) he described several elegant methods that still form the basis for most synthetic work in the prostaglandin field.

By that time many pharmaceutical industries had become interested in the field and large synthetic programmes to synthesize prostaglandin analogues were started. These synthetic efforts were in part guided by the knowledge of the metabolism of prostaglandins accumulated in Stockholm. More than 5000 prostaglandin analogues have now been prepared and tested in one way or another.

When the biosynthesis from the precursor acids was done with homogenates from lungs, it had been found that the prostaglandins formed were extensively further metabolized to inactive compounds (38–41) as indicated in Fig. 3, i.e. by saturation of the double bond between carbon 13 and 14 and dehydrogenation at C_{15}. These reactions are so effective *in vivo* that these prostaglandins are usually almost completely inactivated during one passage through the lungs. In later work *in vivo* it was found that these inactive compounds are further beta- and gammaoxidized before excretion. A possible way to block the first reactions is to replace the C_{15} hydrogen with a methyl group. The 15-methyl prostaglandins that were prepared by Bundy et al. at the Upjohn Co. (42) turned out to be very useful for clinical work and 15-methyl $F_{2\alpha}$ is now registered as a drug.

Some of the different analogues that have been studied in the clinic are listed

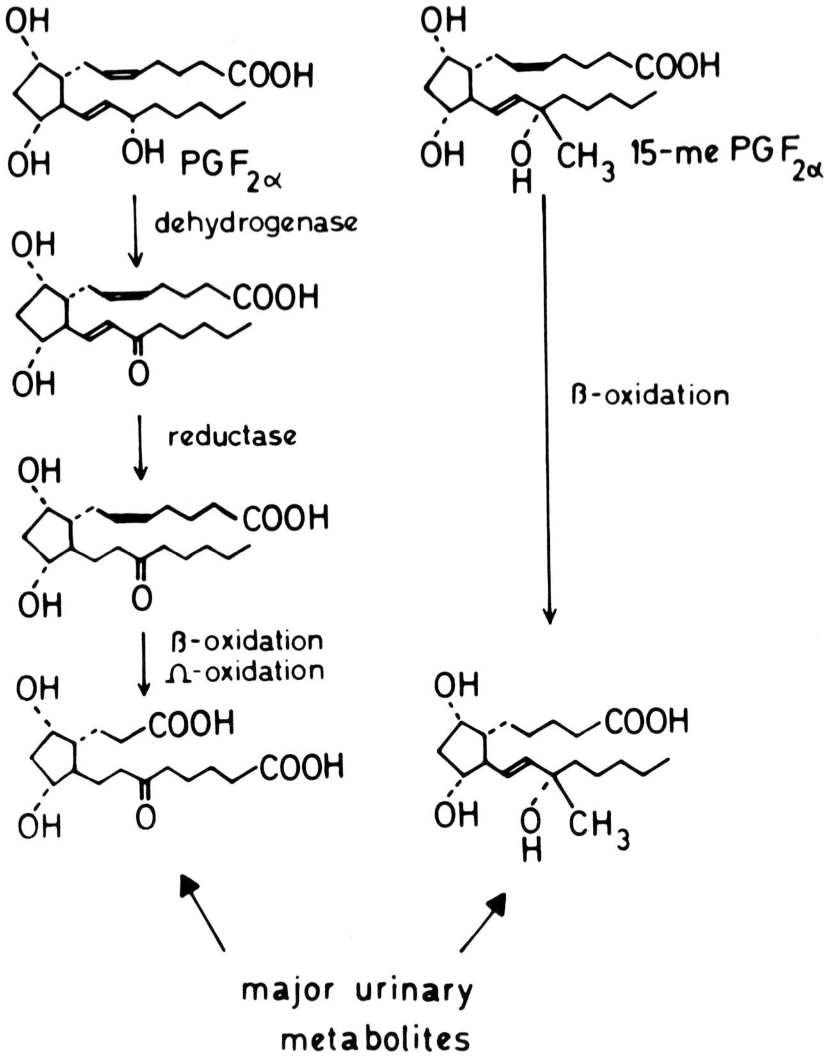

Figure 3. Main metabolic pathway of $PGF_{2\alpha}$ and its inhibition by the methyl group at C-15.

in Fig. 4. All are prepared by the Upjohn Co. unless indicated. These efforts to get compounds with longer lasting or more specific actions are continuing.

In the early sixties we had some difficulty trying to convince our colleagues to study the properties of the prostaglandins in their various pharmacological and physiological systems. This changed expecially after early studies done in cooperation with Drs Dan Steinberg, Martha Vaughan and Jack Orloff at the National Institutes of Health (43–47). They were working with rat adipose tissue that releases glycerol and fatty acids due to increased lipolysis when stimulated with epinephrine, norepinephrine, glucagon or ACTH. It was found that PGE_1 and PGE_2 strongly inhibited the effects of all these compounds. PGE_1 and PGE_2 were also found to counteract both the lipolysis and the blood

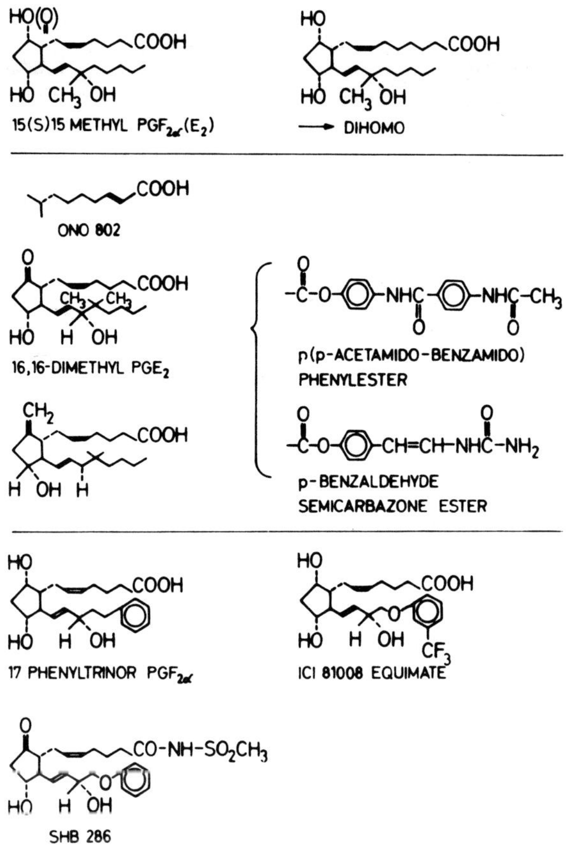

Figure 4. Some prostaglandin analogues that have been studied in clinical trials.

pressure effect caused by epinephrine *in vivo.* It was later found that these effects were mediated by a decrease in the level of cyclic AMP in the fat cells. In this case the level was decreased. In most other cell types the E prostaglandins raise the level of cAMP. In some cases the F prostaglandins have been shown to influence the level of cyclic GMP. I am not going further into this field, but these findings explain why various prostaglandins can influence the metabolism of so many different cell types. A comprehensive summary of the early studies of the biological effects of the prostaglandin was published in 1968 (48).

It turned out that there were not only large differences between the effects of different prostaglandins and their analogues but also between different animal species. Obviously one had to turn to clinical studies in order to explore the physiological role and therapeutic potential of these compounds.

The first study on the cardiovascular effects of the pure prostaglandins in humans was done at the Karolinska (49, 50). In humans PGE_1 caused a fall in blood pressure and an increase in heart rate. My colleagues Drs Lars A. Carlson, Bengt Pernow and Lennart Kaijser and their collaborators have done much of the early, fundamental work on the effects of various analogues on the human cardiovascular system (51–54).

Dr Carlson and his collaborators demonstrated that 1/100 of a milligram of

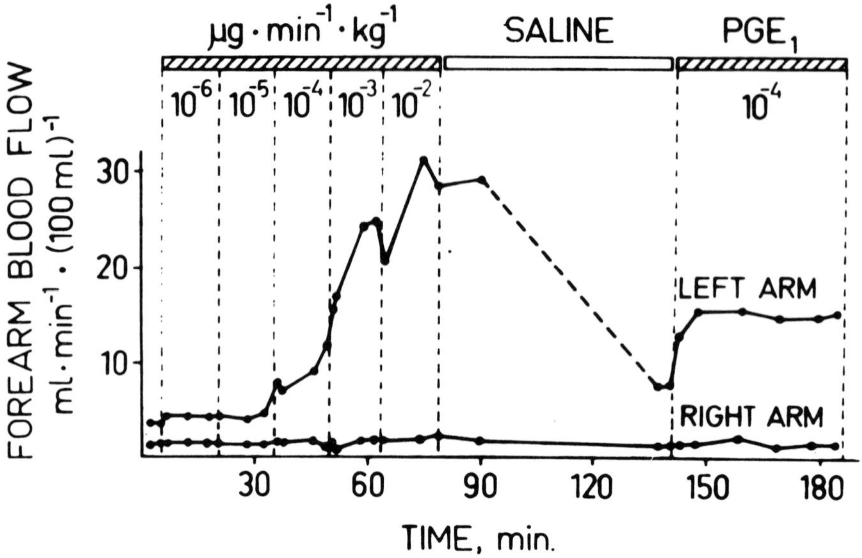

Figure 5. The effect on forearm blood flow of continuous intraarterial infusion of PGE_1 into left brachial artery.

PGE_1 can increase the blood flow through an arm tenfold if it is given intraarterially (Fig 5). No change was seen in the other arm as the compound is inactivated during the passage through the lungs (55). Together with Dr Anders Olsson he has developed this observation into a therapy for peripheral vascular diseases; in some cases this therapy has a dramatic effect.

Another therapeutic utility relates to the patency of *Ductus arteriosus*. It normally closes soon after birth. In certain cases of congenital malformations the E prostaglandins have found clinical use to keep the ductus open for a few days so that corrective surgery can be performed a few days later with less danger to the newborn baby.

On the other hand, the closure of a ductus that remains open can be supported by administration of a cyclooxygenase inhibitor like aspirin or indomethacin etc. that inhibits the local prostaglandin synthesis.

I will now turn to the effects of prostaglandins on the gastrointestinal tract, a field in which Dr Robert of the Upjohn Co. has been very active. In animal experiments he found that the gastric secretion could be inhibited by oral administration or injections of prostaglandins of the E-type. The methyl analogues 15(S) 15-methyl PGE_2 and 16, 16-dimethyl PGE_2 were many times more active than PGE_2 itself (58). Early clinical studies were done here in Stockholm by the late Drs Sven Andersson and B. Nylander. They found that an oral dose of only 80 microgrammes of 16, 16-dimethyl PGE practically blocked the pentagastrin stimulated gastric secretion in humans for several hours (Fig. 6). This work has been continued and expanded upon by Dr Catja Johansson and her collaborators. Extensive clinical trials in many countries have been conducted that have demonstrated a healing effect on ulcers with several analogues. For a recent summary see (58).

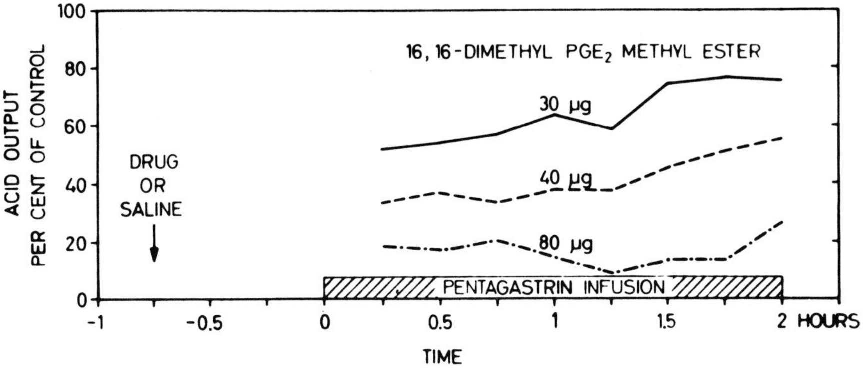

Figure 6. The effect of 16, 16-dimethyl prostaglandin E_2 on gastric secretion.

Another important observation that Dr Robert made was a phenomenon that he called "cytoprotection". If large doses of indomethacin or aspirin are administered to rats or if you give boiling water, strong acid or base orally to anesthesized rats, extensive erosions and/or bleeding of the gastric mucosa occurs and the animals usually die in a few days. But if less than a microgramme of 16,16-dimethyl PGE_2 is administered orally 30 minutes beforehand, the mucosa is protected. An interesting observation by Robert et al. (59) is also that mild irritants like dilute ethanol can effect a certain protection presumably by stimulating the mucosal cells to an increased biosynthesis of prostaglandins. Cytoprotection can be observed at doses and with prostaglandins that do not inhibit acid secretion. Some prostaglandins stimulate secretion of mucus and bicarbonate. The mechanism of cytoprotection is not clear but it is certainly a reality (58). The effect is not limited to the gastrointestinal mucosa. Liver and kidney cells can also be protected to some extent from necrosis caused by carbontetrachloride, etc. by administration of 16, 16-dimethyl PGE_2 (60).

It was logical at this point to look into the effect of cytoprotective prostaglandins on some clinical side effects of indomethacin and aspirin. Normally there is an invisible bleeding of a fraction of one ml of blood from the gut per day. The usual therapeutic dose of indomethacin increases this amount to 3 to 5 ml and in a few cases causes serious bleedings. It was demonstrated here at the Karolinska for the first time that only a third of a milligram of PGE_2 or 40 microgrammes × 3 of 15(R) 15-methyl PGE_2 (61, 62) completely protected the patient from this bleeding caused by indomethacin (Fig. 7). It is still too early to evaluate the therapeutic utility of the administration of prostaglandins to reduce the side effects of these drugs (NSAID).

The last aspects that I will discuss are related to fertility and the use of prostaglandins in obstetrics and gynecology. That is where the prostaglandins were discovered, and that is where they have found their greatest utility so far.

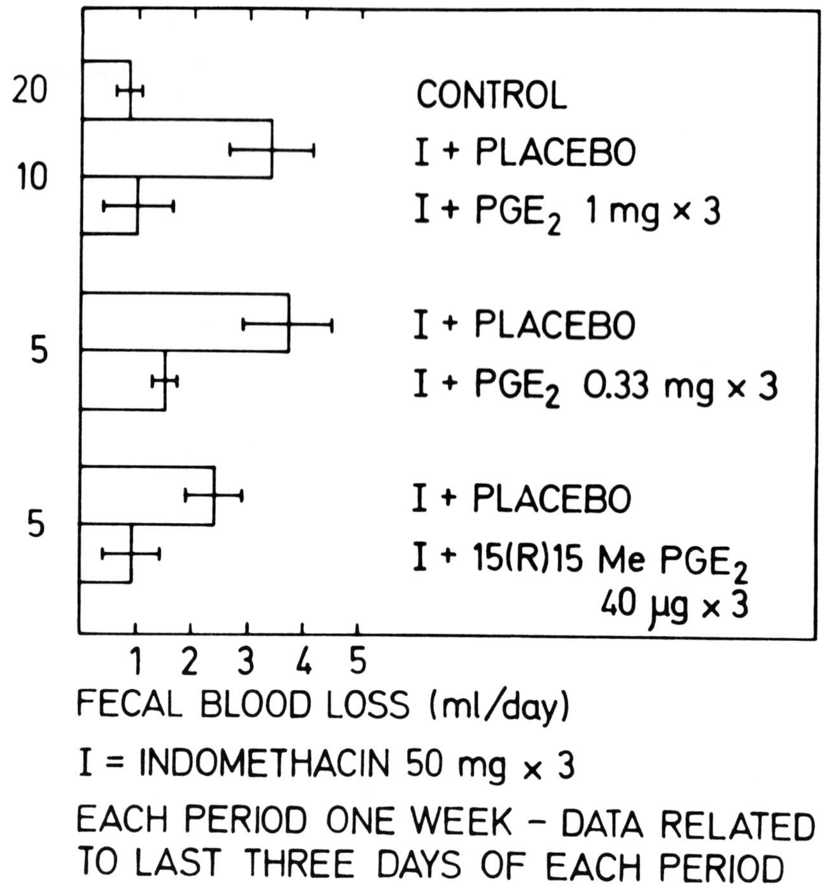

Figure 7. The effect of prostaglandins on the fecal blood loss during indomethacin administration.

Dr Marc Bygdeman had just started his thesis work with Dr von Euler when the pure prostaglandins became available. He studied the effects of the different prostaglandins first *in vitro* and then *in vivo* in Dr Ulf Borell's Department at the Karolinska together with Dr Nils Wiquist. They first demonstrated clinically that PGE_1 causes uterine contractions when injected in small amounts. This of course led to the expectation that administration of prostaglandins might initiate labor or cause pregnancy interruption (63–65).

Dr Sultan Karim was the first to report on the clinical use of prostaglandins to initiate labor.

The first therapeutic abortion with $PGF_{2\alpha}$ was done in May 1969 at the Karolinska Institute. During the remainder of the year the scientists at the Karolinska and also Drs Karim and Filshie in Uganda conducted further studies with $PGF_{2\alpha}$ and PGE_2. The results of both groups were published in the same number of Lancet in January 1970 (66, 67).

An intense activity then started all around the world exploring these findings with normal prostaglandins and later with some of the analogues as shown in Fig. 4.

APPROXIMATE DOSES FOR PREGNANCY INTERRUPTION

	$PGF_{2\alpha}$	PGE_2	15(S) 15 Me $F_{2\alpha}$	16,16 diMe E_2
INTRA VENOUS	~ 100 mg (50–100 μg/min)	~5 mg 5 μg/min	~5 mg 5 μg/min	~ 0.5 mg 0.5 μg/min
INTRA AMNIOTIC	40 – 50 mg (1×)	5–10 mg (1×)	2.5 mg (1×)	
EXTRA AMNIOTIC	~ 5 mg (~ 9×0.75 mg)	1–2 mg (6×)	1 mg (1×)	
I. VAGINAL SUPPOSITORIES	250 mg (5 × 50 mg)	60 – 80 mg (3-4 × 20 mg)	~3 mg (Me-ester) (1 × 3 mg)	~4 mg (4 × 1 mg)
I. MUSCULAR	—	—	2 mg (6 × 0.3)	—
ORAL	—	—	—	0.2 –0.6 mg (60% Success)

Figure 8. Summary of early clinical trials indicating approximate doses for interruption of pregnancy.

The early clinical data that were obtained in Stockholm are summarized in Fig. 8 (66–73).

The two natural prostaglandins PGE_2 and $F_{2\alpha}$ were first studied extensively. When they were administered i. v. they caused very pronounced side effect. When administered intraamniotically or extraamniotically the side effects were tolerable.

The 15(S)15-methyl $PGF_{2\alpha}$ that was supplied by the Upjohn Co., was then studied and found to be much more active and giving much less side effects. It could also be injected i. m., and we developed vaginal suppositories containing the methyl ester. It is now a registered drug in many countries.

16,16-dimethyl PGE_2 was even more active—sometimes also very effective after oral administration of less than a milligramme. It caused, however, more side effects and furthermore had stability problems. Recently an analogue of 16,16-dimethyl PGE_2, in which the carbonyl group at C9 has been replaced by a methylene group is being extensively studied as it combines high activity with stability and very low side effects (74).

At that time an important development was initiated by SIDA (the Swedish International Development Authority). In the sixties they had become heavily involved in supporting family planning in many developing countries. However, they found that the methods available left much to be desired and therefore were considering how best to stimulate and support research and development in the field.

The Director, Mr Ernst Michanek and Mr Carl Wahren had developed advanced plans to start an international research foundation located here at the Karolinska for this purpose. However, for various reasons, it was decided to make a feasability study of an alternative arrangement together with the Ford Foundation and WHO, in which I had the honor to participate. This resulted in the creation of WHO's "Special programme" for research on human reproduction in 1971–72. The work should be focused on the needs of developing countries. The voluntary contributions to the programme soon exceeded ten million US dollars annually of which more than half were provided from Swedish sources during the seventies.

One of the "Task Forces" of the programme was devoted to exploring the potential of the prostaglandins to interrupt pregnancy. During the first five years I had the stimulating assignment as chairman of this group of outstanding experts. The exploratory work done at the Karolinska and by Dr Karim's group formed the basis for large international coordinated clinical trials. Figure 9 indicates the locations of the cooperating clinics and Fig. 10 the number of cases completed. The Task Force has continued its activities under the chairmanship of Dr Bygdeman.

The most important new development has been the interruption of pregnancy during the "postconceptional" period, i.e. the first three weeks after a missed menstrual period.

I have a very vivid memory of a late evening in Bombay when Dr Borell, Dr Bygdeman and myself were compiling results of abortions during different weeks of pregnancy and found a success rate of practically 100 per cent complete abortions during this period (Fig. 11). This observation has been studied extensively, and Dr Bygdeman et al. has just completed a successful trial with a hundred cases in Stockholm in which the patients even administered the drug themselves in their home. This method obviously has a very great potential especially in developing countries.

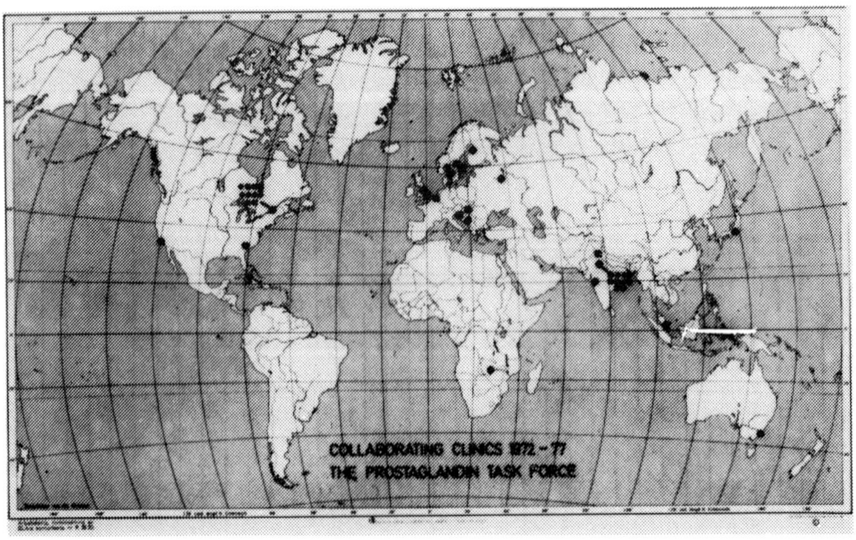

Figure 9. Participating clinics in coordinated trials of the WHO Prostaglandin Task Force.

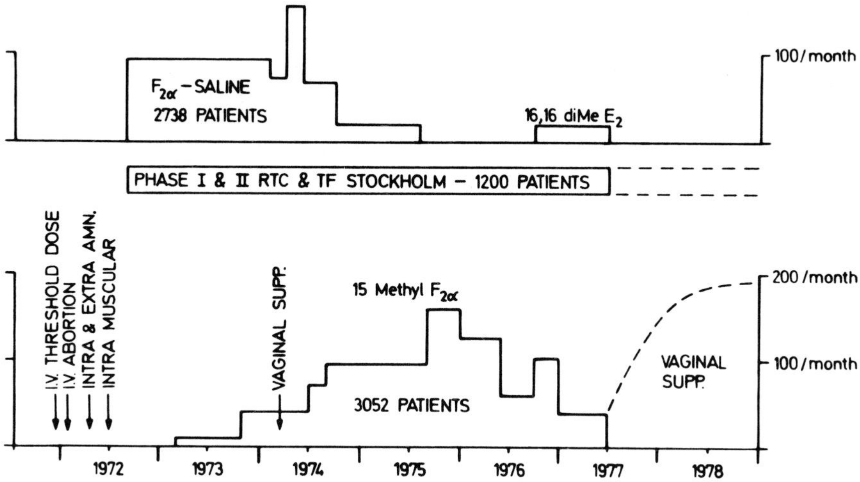

Figure 10. Number of patients participating in coordinated trials of the WHO Prostaglandin Task Force.

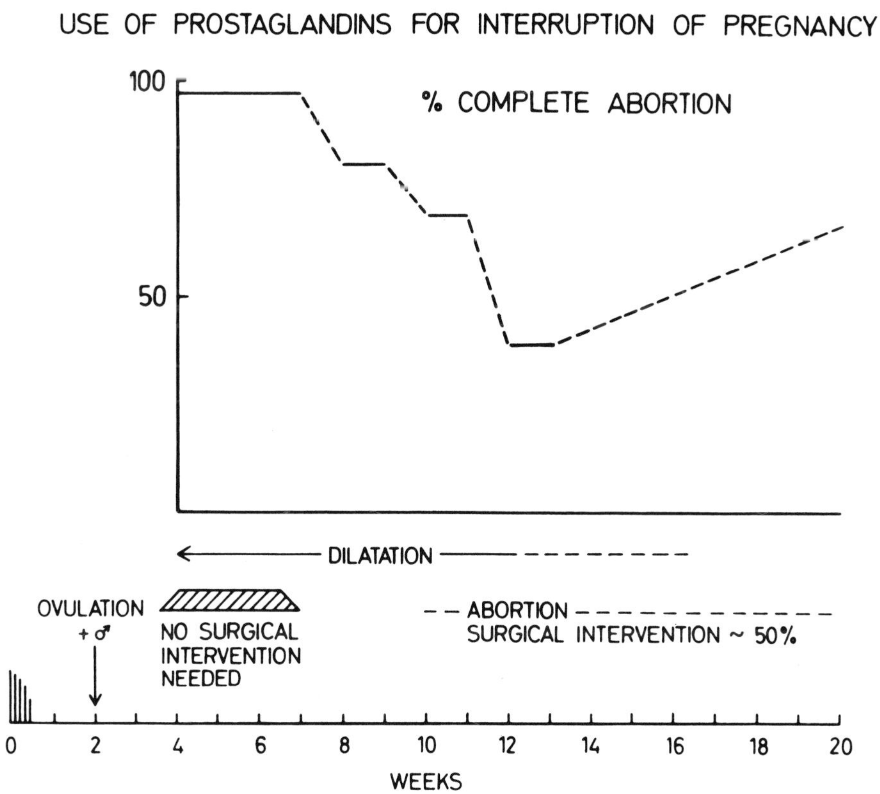

Figure 11. Per cent of "complete" abortions after interruption with prostaglandins during different weeks of pregnancy.

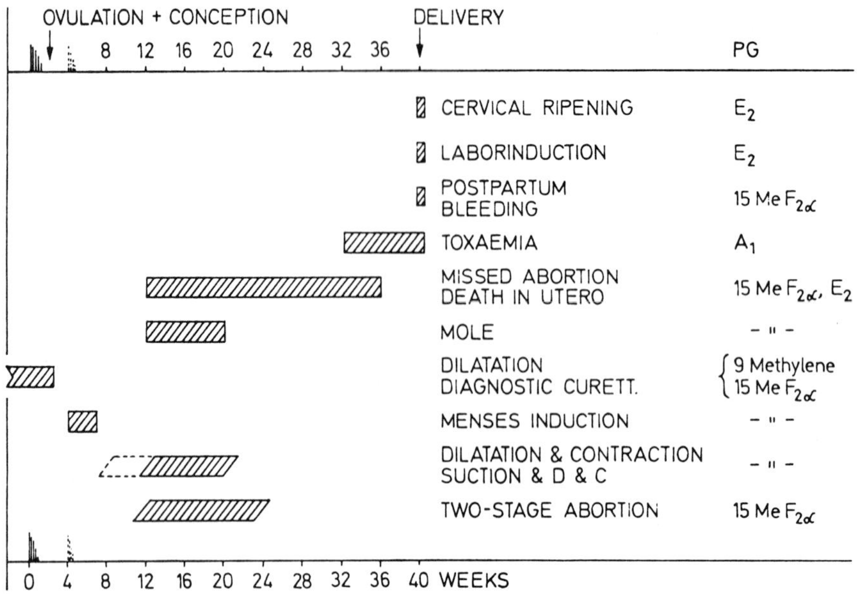

Figure 12. Summary of therapeutic utility of prostaglandins in human fertility.

The therapeutic use of prostaglandins is, however, not limited to interruption of pregnancy. It is being used for labor induction, treatment of serious bleeding after delivery and dilatation for diagnostic curettage. Promising results have been reported of treating eclampsia with PGA_1 by Dr M. Toppozada in Alexandria.

A summary of the use of prostaglandins in the fertility area is given in Fig. 12.

Prostaglandins have also found extensive use in animal husbandry especially for "synchronization" of herds of cattle, i.e. after injection of cows with 25 mg of $PGF_{2\alpha}$ at an appropriate time, their cycles are interrupted and the whole herd can be inseminated three days later.

My presentation has of necessity been very short and sketchy and will be complemented by my colleagues.

The prostaglandin precursors and the enzymic systems appear to be present in practically all nucleated animal cells. They can biosynthesize characteristic mixtures depending on cell type and conditions on appropriate stimulation. As the highest concentration is found in the lowest animal species one can speculate that the ease of autoxidation of the percursor acids—a reaction that has even been reported to produce prostaglandins in minute amounts (75) — has led to their utilization in metabolism early on in development and for many different functions.

They are apparently mainly playing a role as local regulators even if $PGF_{2\alpha}$ functions as a classical hormone in sheep, where it is produced in the uterus at the end of the cycle and transported in the blood to the ovary where it causes luteolysis.

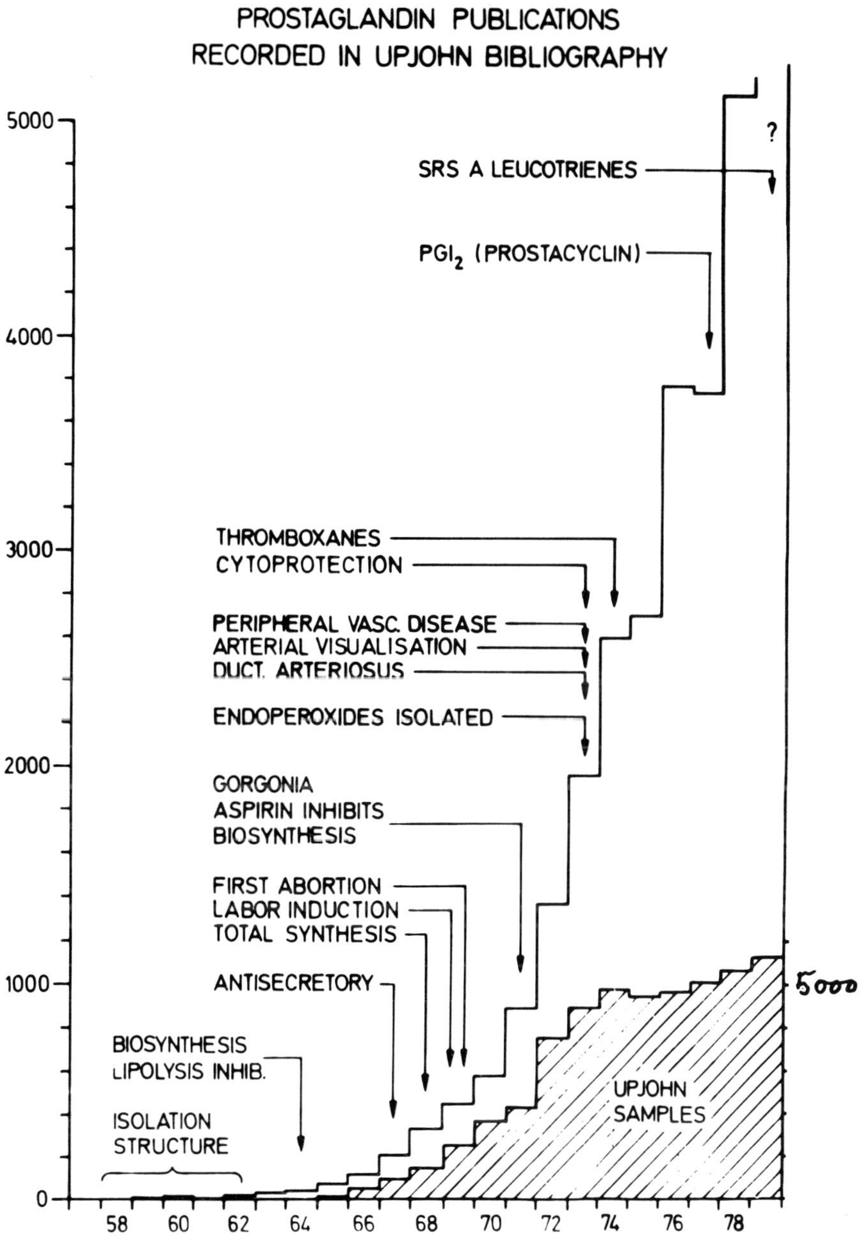

Figure 13. Annual publications related to prostaglandins in relation to the scientific development of the field. Free samples of prostaglandins distributed by the Upjohn Co., Kalamazoo.

The complex pattern of the arachidonic acid metabolism makes studies of the roles of prostaglandins in the complex regulatory mechanisms in organs like the kidney, lung etc. a difficult and challenging task.

Figure 13 illustrates how the field has developed from the early publications on the isolation, structure and biosynthesis to the present level of more than five thousand papers annually.

'How did all these scientists get their prostaglandins?' It is illustrated in Fig. 13. The Upjohn Company has sent out something like 75 000 free samples during this period, and you can see that there is a correlation between the number of publications and samples up to about 1970. At that point the prostaglandin containing coral was found in the Mexican Gulf. A number of pharmaceutical industries collected corals, started synthetic programmes and then also supplied samples to scientists.

At this moment it is appropriate to point out that the whole prostaglandin field had been very much slower in developing without the oustanding research and development work and generous supply policy that was initiated and organized at the Upjohn Co. by Dr David Weisblat.

The prostaglandin story again illustrates the importance of interaction between the pharmaceutical industry and the academic biomedical scientists. In this case the special programme for human reproduction of the World Health Organization has also greatly contributed to the speed of development and to the buildup of research and development capabilities in this field in developing countries.

REFERENCES

1. Bergström S. The prostaglandins. Recent Progr. Hormone Res. 22: 153–175, 1966.
2. Bergström S. Prostaglandins: Members of a new hormonal system. Science (N.Y.) 157: 382–391, 1967.
3. Bergström S. Isolation, structure and action of the prostaglandins. In: Prostaglandins, Proc. 2nd Nobel Symp., Stockholm, June 1966, ed. by S. Bergström and B. Samuelsson, pp. 21–30, Almqvist and Wiksell, Stockholm; Interscience, New York, 1967.
4. Bergström S., and Samuelsson B. Prostaglandins. Annu. Rev. Biochem. 34: 101–108, 1965.
5. Kurzrok R., and Lieb C.C. Biochemical studies of human semen. II. The action of semen on the human uterus. Proc. Soc. Exp. Biol. Med. 28: 268–272, 1930.
6. Goldblatt M.W. A depressor substance in seminal fluid. J. Soc. Chem. Ind. (London) 52: 1056–1057, 1933.
7. Goldblatt M.W. Properties of human seminal plasma. J. Physiol. (London) 81: 208–218, 1935.
8. Euler U.S. von. Zur Kenntnis der pharmakologischen Wirkungen von Nativsekreten und Extrackten männlicher accessorischer Geschlechtsdrusen. Arch. Exp. Pathol. Pharmakol. 175: 78–81, 1934.
9. Euler U.S. von. A depressor substance in the vesicular gland. J. Physiol. (London) 84: 21P, 1935.
10. Euler U.S. von. On the specific vasodilating and plain muscle stimulating substances from accessory genital glands in man and certain animals (prostaglandin and vesiglandin). J. Physiol. (London) 88: 213–234, 1937.
11. Bergström S. Prostaglandinets kemi. Nord. Med. 42: 1465–1466, 1949.
12. Bergström S., and Sjövall J. The isolation of prostaglandin. Acta Chem. Scand. 11: 1086, 1957.
13. Bergström S., and Sjövall J. The isolation of prostaglandin F from sheep prostate glands. Acta Chem. Scand. 14: 1693–1700, 1960.
14. Bergström S., and Sjövall J. The isolation of prostaglandin E from sheep prostate glands. Acta Chem. Scand. 14: 1701–1705, 1960.
15. Bergström S., Dressler F., Krabisch L., Ryhage R., and Sjövall J. The isolation and structure of a smooth muscle stimulating factor in normal sheep and pig lungs. Arkiv Kem. 20: 63–66, 1962.
16. Bergström S., Dressler F., Ryhage R., Samuelsson B., and Sjövall J. The isolation of two further prostaglandins from sheep prostate glands. Arkiv Kem. 19: 563–567, 1962.
17. Bergström S., Krabisch L., Samuelsson B., and Sjövall J. Preparation of prostaglandin F from prostaglandin E. Acta Chem. Scand. 16: 969–974, 1962.
18. Bergström S., Krabisch L., and Sjövall J. Smooth muscle stimulating factors in ram semen. Acta Chem. Scand. 14: 1706–1710, 1960.
19. Bergström S., Ryhage R., Samuelsson B., and Sjövall J. The structure of prostaglandin E, F_1 and F_2. Acta Chem. Scand. 16: 501–502, 1962.
20. Bergström S., Ryhage R., Samuelsson B., and Sjövall J. Degradation studies on prostaglandins. Acta Chem. Scand. 17: 2271–2280, 1963.
21. Bergström S., Ryhage R., Samuelsson B., and Sjövall J. The structures of prostaglandin E_1, $F_{1\alpha}$ and $F_{1\beta}$. J. Biol. Chem. 238: 3555–3564, 1963.
22. Bergström S., and Samuelsson B. Isolation of prostaglandin E_1 from human seminal plasma. J. Biol. Chem. 237: PC3005–PC3006, 1962.
23. Samuelsson B. The structure of prostaglandin E_3. J. Amer. Chem. Soc. 85: 34P, 1963.
24. Samuelsson B. Isolation and identification of prostaglandins from human seminal plasma. J. Biol. Chem. 238: 3229–3234, 1963.
25. Abrahamsson S., Bergström S., and Samuelsson B. The absolute configuration of prostaglandin F_{2-1}. Proc. Chem. Soc. 332, 1962.
26. Nugteren D.H., and Dorp D.A. van, Bergström S., Hamberg M., and Samuelsson B. Absolute configuration of the prostaglandins. Nature (London) 212: 38–39, 1966.
27. Bergström S., Danielsson H., Klenberg D., and Samuelsson B. The enzymatic conversion of essential fatty acids into prostaglandins. J. Biol. Chem. 239: PC4006–PC4008, 1964.

28. Bergström S., Danielsson H., and Samuelsson B. The enzymatic formation of prostaglandin E_1 from arachidonic acid. Biochim. Biophys. Acta 90: 207–210, 1961.
29. Dorp D.A. van, Beerthuis R.K., Nugteren D.H. and Vonkeman H. The biosynthesis of prostaglandins. Biochim. Biophys. Acta 90: 204–207, 1964.
30. Dorp D.A. van, Beerthuis R.K., Nugteren D.H., and Vonkeman H. Enzymatic conversion of all-cis-polyunsaturated fatty acids into prostaglandins. Nature (London) 203: 839–941, 1964.
31. Weinheimer A.J., and Spraggins R.L. The occurrence of two new prostaglandin derivatives in the gorgonian Plexora Homomalla. Chemistry of Coelentrates. XV. Tetrahedron Letters 5185–5188, 1969.
32. Corey E.J., Vlattas I., Andersen N.H., and Harding K. A new total synthesis of prostaglandins of the E_1 and F_1 series including 11-epiprostaglandins. J. Amer. Chem. Soc. 90: 3247, 1968.
33. Corey E.J., Andersen N.H., Carlson R.M., Paust J., Vedejs E., Vlattas I., and Winter R.E.K. Total synthesis of prostaglandins. Synthesis of the pure dl-E_1, -$F_{1\alpha}$, -$F_{1\beta}$, -A_1 and -B_1 hormones. J. Amer. Chem. Soc. 90: 3245–3247, 1968.
34. Corey E.J., Weinshenker N.M., Shaaf, T.K., and Huber W. Stereo-controlled synthesis of prostaglandins $F_{1\alpha}$ and E_2 (dl). J. Amer. Chem. Soc. 91: 5675–5677, 1969.
35. Corey E.J., Vlattas I., and Harding K. Total synthesis of natural (levo) and enantiometric (dextro) forms of prostaglandin E_1. J. Amer. Chem. Soc. 91: 535–536, 1969.
36. Corey E.J., Arnold Z., and Hutton J. Total synthesis of prostaglandins E_2 and $F_{2\alpha}$ (dl) via a tricarbocyclic intermediate. Tetrahedron Letters 307–310, 1970.
37. Corey E.J., Noyori R., and Schaaf T.K. Total synthesis of prostaglandins $F_{1\alpha}$, E_1, $F_{2\alpha}$, and E_2 (natural forms) from a common synthetic intermediate. J. Amer. Chem. Soc. 92: 2586–2587, 1970.
38. Änggård E., Green K., Samuelsson B. Synthesis of tritium-labeled prostaglandin E_1 and studies on its metabolism in guinea pig lung. J. Biol. Chem. 240: 1932–1940, 1965.
39. Änggård E., and Samuelsson B. Biosynthesis of prostaglandins from arachidonic acid in guinea pig lung. J. Biol. Chem. 240: 3518–3521, 1965.
40. Änggård E., and Samuelsson B. The metabolism of prostaglandins in lung tissue. In: Prostaglandins, Proc. 2nd Nobel Symp., Stockholm, June 1966, ed. by S. Bergström and B. Samuelsson, pp. 97–106, Almqvist and Wiksell, Stockholm; Interscience, New York, 1967.
41. Änggård E., and Samuelsson B. Metabolism of prostaglandin E_1 in guinea pig lung: The structure of two metabolites. J. Biol. Chem. 239: 4097–4102, 1964.
42. Bundy G.L., Lincoln F.H., Nelson N.A., Pike J.E., and Schneider W.P. Novel prostaglandin syntheses. Annal. N.Y. Acad. Sci., Prostaglandins 180: 76–79, 1971.
43. Steinberg D., Vaughan M., Nestel P., and Bergström S. Effects of prostaglandin E opposing those of catecholamines on blood pressue and on triglyceride breakdown in adipose tissue. Biochem. Pharmacol. 12: 764–766, 1963.
44. Steinberg D., Vaughan M., Nestel P.J., Strand O., and Bergström S. Effects of the prostaglandins on hormone-induced mobilization of free fatty acids. J. Clin. Invest. 43: 1533–1540, 1964.
45. Orloff J., Handler J.S., and Bergström S. Effect of prostaglandin (PGE_1) on the permeability response of the toad bladder to vasopressin, theophyrline and adenosine 3', 5' monophosphate. Nature (London) 205: 397–398, 1965.
46. Orloff J., and Grantham J. The effect of prostaglandin (PGE_1) on the permeability response of rabbit collecting tubules to vasopressin. In: Prostaglandins, Proc. 2nd Nobel Symp., Stockholm, June 1966, ed. by S. Bergström and B. Samuelsson, pp. 143–146, Almqvist and Wiksell, Stockholm; Interscience, New York, 1967.
47. Orloff J., and Handler J. The role of adenosine 3,5-phosphate in the action of antidiuretic hormone. Amer. J. Med. 47: 757–768, 1967.
48. Bergström S., Carlson L.A., and Weeks J.R. The prostaglandins: A family of biologically active lipids. Pharmacol. Rev. 20: 1–48, 1968.
49. Bergström S., Dunér H., von Euler U.S., Pernow B., and Sjövall J. Observations on the effects of infusion of prostaglandin E in man. Acta Physiol. Scand. 45: 145–151, 1959.
50. Bergström S., Eliasson R., von Euler U.S., and Sjövall J. Some biological effects of two crystalline prostaglandin factors. Acta Physiol. Scand. 45: 133–144, 1959.

51. Carlson L.A. Metabolic and cardio-vascular effects in vivo of prostaglandins. In: Prostaglandins, Proc. 2nd Nobel Symp., Stockholm, June 1966, ed. by S. Bergström and B. Samuelsson, pp. 123–132, Almqvist and Wiksell, Stockholm; Interscience, New York, 1967.
52. Carlson L.A., Ekelund L.-G. and Orö L. Clinical and metabolic effects of different doses of prostaglandin E_1 in man. Acta Med. Scand. 183: 423–430, 1968.
53. Carlson L.A., and Hallberg D. Basal lipolysis and effects of noradrenaline and prostaglandin E_1 on lipolysis in human subcutaneous and omental adipose tissue. J. Lab. Clin. Med. 71: 368–377, 1968.
54. Carlson L.A., Irion E., and Orö L. Effect of infusion of prostaglandin E_1 on the aggregation of blood platelets in man. Life Sci 7: 85–90, 1968.
55. Bevegård S., and Orö L. Effect of prostaglandin E_1 on forearm blood flow. Scand. J. clin. Lab. Invest. 23: 347–353, 1969.
56. Robert A. The inhibitory effects of prostaglandins on gastric secretion. Progress in Gastroventerol. Vol. III. G.B.J. Glass (ed.). Grune and Stratton, New York 1977, 777–801.
57. Nylander B., Andersson S. Gastric secretory inhibition induced by three methyl analogs of prostaglandin E_2 administered intragastrically to man. Scand. J. Gastroent. 9: 751–758, 1974.
58. Johansson C., and Bergström S. Prostaglandins and protection of the gastroduodenal mucosa. Scand. J. Gastroenterol. Suppl. nr. 77: 21–46, 1982.
59. Robert A., Lancaster C., Hanchar A.J., and Nezamis J.E. Mild irritants prevent gastric necrosis through prostaglandin formation: histological study. Gastroenterology 74: 1086, 1978.
60. Ruwart M.J., Rush B.D., Friedle N.M., Piper R.C., Kolaja G.J. Protective effects of 16,16 dimethyl PGE_2 on the liver and kidney. Prostaglandins 21: 97–102, 1981.
61. Johansson C., Kollberg B., Nordemar R., Samuelsson K., and Bergström S. Protective treatment with prostaglandin E_2 in the gastrointestinal tract during indomethacin treatment of rheumatic patients. Gastroenterology 76: 479–483, 1980.
62. Cohen M.M., Cheung G., and Lyster D.M. Prevention of aspirin-induced faecal blood loss by prostaglandin E_2. Gut 21: 602–606, 1980.
63. Bygdeman M., and Eliasson R. A comparative study on the effect of different prostaglandin compounds on the motility of the isolated human myometrium. Medicina experimentalis 9: 409–415, 1963.
64. Bygdeman M. The effect of different prostaglandins on human myometrium in vitro. Acta Physiol. Scand., 63, Suppl. 242, 1–78, 1964.
65. Bygdeman M., Kwon S., and Wiqvist N. The effect of prostaglandin E_1 on human pregnant myometricom in vivo. In: Prostaglandins, Proc. 2nd Nobel Symp., Stockholm, June 1966, ed. by S. Bergström and B. Samuelsson, pp. 93–96, Almqvist and Wiksell, Stockholm; Interscience, New York, 1967.
66. Roth-Brandel U., Bygdeman M., Wiqvist N., and Bergström S. Prostaglandins for induction of therapeutic abortion. Lancet 1: 190–191, 1970.
67. Karim S.M.M., and Filshie G.M. Therapeutic abortion using prostaglandin $F_{2\alpha}$. Lancet 1: 157, 1970.
68. Wiqvist N., and Bygdeman M. Therapeutic abortion by local administration of prostaglandins. Lancet 2: 716–717, 1970.
69. Bygdeman M., Beguin N., Toppozada M., Wiqvist N., and Bergström S. Intra-uterine administration of 15(S)-15 Methyl-$PGF_{2\alpha}$ for induction of abortion. Lancet I: 13–36, 1972.
70. Bygdeman M., Green K., Toppozada M., Wiqvist N., and Bergström S. The influence of prostaglandin metabolites on the uterine response to $PGF_{2\alpha}$. A clinical and pharmacokinetic study. Life Sci. 14: 521–531, 1974.
71. Bygdeman M., Martin J.N., Wiqvist N., Gréen K., and Bergström S. Reassessment of systematic administration of prostaglandins for induction of midtrimester abortion. Prostaglandins 8: 157–169, 1974.
72. Bygdeman M., Gréen K., Lundström V., Ramadan M., Fotiou S., and Bergström S. Induction of abortion by vaginal administration of 15(S)15-methyl prostaglandin $F_{2\alpha}$ methyl ester. A comparison of two delivery systems. Prostaglandins 12: 27–51, 1976.

73. Bygdeman M., Ganguli A., Kinoshita K., and Lundström V. Development of a vaginal suppository suitable for single administration of interruption of second trimester pregnancy. Contraception 15: 129–141, 1977.
74. Gréen K., Vesterqvist O., Bygdeman M., Christenssen N.J., and Bergström S. Plasma levels of 9-deoxo-16,16-dimethyl-9-methylene-PGE_2 in connection with its development as an abortifacient. Prostaglandins 24: 451–466, 1982.
75. Nugteren D.H., Vonkeman H., and van Dorp D.A. Non-enzymatic conversion of all-cis 8,11,14-eicosatrienoie acid into prostaglandin E_1. Receil Travaux Chim. Pays-Bas 86: 1237–1245, 1967.

Bengt Samuelsson

BENGT SAMUELSSON

I was born in Halmstad, Sweden, on may 21, 1934 to Anders and Kristina Samuelsson. After attending public schools I studied medicine at the University of Lund where I met my wife Karin (Bergstein). We have one son (Bo) and two daughters (Elisabet and Astrid).

After a few years in Lund I moved to Karolinska Institutet in Stockholm in order to do graduate work in biochemistry in parallel with medical studies. In 1960 I finished my dissertation and became docent in medical chemistry. A year later I also obtained my MD degree from Karolinska Institutet. After a year as research fellow in the Department of Chemistry at Harvard University, Cambridge, Mass., U.S.A., I returned to Karolinska Institutet. In 1967 I was appointed professor of medical chemistry at the Royal Veterinary College in Stockholm, and after a few years I moved back to Karolinska Institutet to become professor and chairman of the department of physiological chemistry. Concurrently with my research positions I have also held administrative posts. I was dean of the medical faculty from 1978 to 1983, and is presently rector of Karolinska Institutet.

My research interests were originally in cholesterol metabolism with emphasis on reaction mechanisms. After the structural work on prostaglandins with Sune Bergström in 1959–1962 I have mainly been interested in transformation products of arachidonic acid. This has led to the discovery of endoperoxides, thromboxanes and the leukotrienes, and my group has mainly been involved in studying the chemistry, biochemistry and biology of these compounds and their role in biological control system. The research has implications in several clinical areas, particularly in thrombosis, inflammation and allergy.

Appointments

Assistant professor of medical chemistry, Karolinska Institutet 1961–1966

Research fellow, Department of Chemistry, Harvard University, Cambridge, Mass., U.S.A., 1961–1962.

Professor of medical chemistry, Royal Veterinary College, Stockholm, Sweden, 1967–1972.

Professor of medical and physiological chemistry, Karolinska Institutet, Stockholm, Sweden, 1973–

Chairman of the Department of Chemistry, Karolinska Institutet, Stockholm, Sweden, 1973–

Visiting professor in chemistry, Harvard University, Cambridge, Mass., U.S.A., spring term 1976.

Dean of the Medical Faculty, Karolinska Institutet, Stockholm, Sweden, July 1, 1978–June 30, 1983

Rector of Karolinska Institutet, July 1, 1983 –

Memberships, Awards and Honors

Swedish Medical Association's Jubilee Award, Stockholm, Sweden (1968)
Anders Jahres Award, Oslo University, Oslo, Norway (1970)
Louisa Gross Horwitz Award, Columbia University, New York, U.S.A. (1975)
Honorary Member American Society of Biological Chemists (1976)
Intrascience Medalist, Santa Monica, California, U.S.A. (1976)
Albert Lasker Basic Medical Research Award, New York, U.S.A. (1977)
Honorary Degree of Doctor of Science, University of Chicago, Chicago, Illinois, U.S.A. (1978)
Ciba Geigy Drew Award in Biomedical Research, Madison, New Jersey, U.S.A. (1980)
Member of the Royal Swedish Academy of Sciences (1981)
Lewis S. Rosenstiel Award in Basic Medical Research, Brandeis University, Boston, Mass., U.S.A. (1981)
Swedish Medical Association's Jubilee Award, Stockholm, Sweden (1981)
The Gairdner Foundation Award, Toronto, Canada (1981)
Heinrich Wieland Prize, Munich, West Germany (1981)
The Bror Holmberg Medal of the Swedish Chemical Society (1982)
Honorary Member Association of American Physicians (1982)
Member of the Mediterranean Academy, Catania, Italy (1982)
Foreign Honorary Member of the American Academy of Arts and Sciences (1982)
American Chemical Society Division of Medical Chemistry Award (1982)
Waterford Bio-Medical Science Award, La Jolla, California, U.S.A. (1982)
International Association of Allergology & Clinical Immunology Award, London, Great Britian (1982)
Honorary Member, Swedish Medical Association, Stockholm, Sweden (1982)
Honorary Degree of Doctor of Science, University of Illinois, U.S.A. (1983)

FROM STUDIES OF BIOCHEMICAL MECHANISMS TO NOVEL BIOLOGICAL MEDIATORS: PROSTAGLANDIN ENDOPEROXIDES, THROMBOXANES AND LEUKOTRIENES

Nobel Lecture, December 8, 1982

by

BENGT SAMUELSSON

Deparment of Physiological Chemistry, Karolinska Institutet,
S-104 01 Stockholm, Sweden

INTRODUCTION

Following the completion of the structural work on the prostaglandins with Sune Bergström and co-workers (for reviews, see refs. 1—3) I was very fortunate in being able to spend a year in the Chemistry Department at Harvard University, Cambridge, Mass. During this stay I had the opportunity to study both theoretical and synthetic organic chemistry. The year in Cambridge had a profound effect on my future research. At this time Konrad Bloch, E.J. Corey, Frank Westheimer, Robert B. Woodward and several other prominent scientists were among the faculty members of the department, and it was indeed a stimulating place for a young M.D. interested in chemistry. I was working in E.J. Corey's laboratory, and we have continued to collaborate in several areas since then.

In 1964 it was established that there was a biogenetic relationship between the polyunsaturated fatty acids and prostaglandins (4,5). This finding was of considerable biological interest and since the mechanisms of the reactions involved were unknown, I decided to study this problem in the laboratory I had established at that time.

MECHANISM OF BIOSYNTHESIS OF PROSTAGLANDINS

The conversion of 8, 11, 14-eicosatrienoic acid into prostaglandin E_1 (PGE_1) involves the introduction of two hydroxyl groups and one keto group. Incubation of 8, 11, 14-eicosatrienoic acid in an atmosphere of $^{18}O_2$ showed that the oxygen atoms of the hydroxyl groups were derived from oxygen gas whereas the keto oxygen did not contain any ^{18}O (6,7). That this was due to exchange between the keto oxygen of the water was shown in later experiments (8). In these studies the keto group was reduced immediately with borohydride and the resulting trihydroxy acid derivative was shown to contain three atoms of ^{18}O. These experiments were extended by carrying out the reaction in a mixture of $^{18}O_2$ (8). The reduced product was converted into the trimethoxy derivative, and the side chain carrying a hydroxyl group was cleaved off by oxidation with permanganate

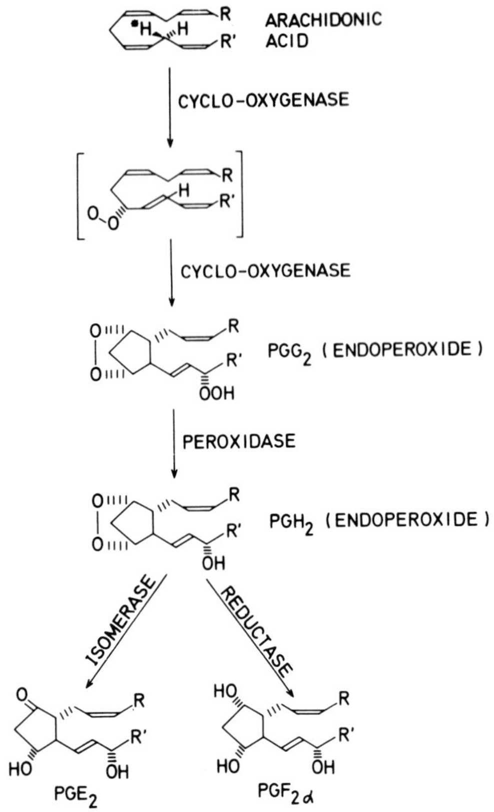

Fig. 1. Mechanism of prostaglandin biosynthesis.

periodate. The resulting dicarboxylic acid ester contained the two oxygens that were introduced into the five-membered ring during the biosynthesis. Analysis of this molecule by mass spectrometry showed that it contained either two atoms of ^{18}O or two atoms of ^{16}O in the ring and that molecules with one atom of ^{18}O and one atom of ^{16}O in the ring were virtually absent. The experiment demonstrated that the oxygen atom of the hydroxyl group at C-11 and of the keto group at C-9 originated in the same molecule of oxygen (Fig. 1).

It was also shown that the hydrogens at C-8, C-11, and C-12 are retained in their original positions during the conversion of 8,11,14-eicosatrienoic acid into PGE_1 which is in agreement with the mechanism proposed (see below). Experiments with hypothetical intermediates, namely, 15-hydroperoxy-8,11,13-eicosatrienoic acid and 15-hydroxy-8,11,13-eicosatrienoic acid indicated that the initial reaction consisted of the introduction of the oxygens in the ring (9,10). However, two different mechanisms for the incorporation of the oxygen molecule seemed possible (9,10,11). In one of the pathways leading from 8,11,14-eicosatrienoic acid to the proposed intermediate, the oxygen is added across carbon atoms C-9 and C-11 with concomitant formation of the new carbon-carbon bond between C-8 and C-12. The other pathway involves a lipoxygenase-like reaction with formation of 11-peroxy-8,12,14-eicosatrienoic acid as the initial step. In both of the pathways leading to the cyclic peroxide intermediate one hydrogen at C-13 is removed. In the latter pathway the hydrogen at C-13 is most likely removed as the initial step whereas the removal of this hydrogen occurs later in the first pathway. It was thus conceivable that the removal of the hydrogen was the rate-determining step and that substitution of tritium for hydrogen in the precursor would give a kinetic isotope effect. The resulting enrichment should appear in the precursor acid in the second pathway whereas the first pathway should produce enrichment of tritium in an oxygenated intermediate. Precursor acids, which were stereospecifically labeled with tritium at C-13 and labeled with ^{14}C at C-3, were therefore synthesized (10).

The conversion of the doubly labeled acids into PGE_1 was catalyzed by a vesicular gland preparation and the $^{3}H/^{14}C$ ratio of the precursor, product and the precursor remaining after the reaction was determined. It was found that 13D-^{3}H-3-^{14}C 8,11,14-eicosatrienoic acid retained the tritium label during the conversion to PGE_1. The 13L-^{3}H-3-^{14}C-8,11,14-eicosatrienoic acid, however, was transformed into PGE^1 with essentially complete loss of tritium. The precursor isolated after 75 % conversion was significantly enriched (284 % retention) with respect to tritium. Thus, the initial step in the transformation of 8,11,14-eicosatrienoic acid into prostaglandin is the stereospecific elimination of the 13L-hydrogen. This reaction is followed by introduction of oxygen at C-11 in a lipoxygenase-like reaction to give 11-peroxy-8,12,13-eicosatrienoic acid (10,12). It is of interest in this context that soybean lipoxygenase removes the same hydrogen both specifically (ω8) and stereospecifically (L) (9). However, the plant lipoxygenase introduces the oxygen molecule in ω6 position whereas the lipoxygenase, which is a component of the prostaglandin synthetase, introduces the oxygen in ω10 position (9). The 11-peroxy-8,12,13-eicosatrienoic acid visualized to be formed in the initial oxygenation is subsequently transformed into an endoperoxide (8) by a concerted reaction involving addition of oxygen at C-15, isomerization of the Δ^{13} double bond, formation of the new carbon-carbon bond between C-8 and C-12 and attack by the oxygen radical at C-9. This is shown in Fig. 1. Indirect evidence indicates a free radical mechanism (13,14). The endoperoxide is transformed into PGE_1 by removal of hydrogen at C-9 or into $PGF_{1\alpha}$ by a reductive cleavage of the peroxide.

When washed microsomes were used as enzyme source, eicosatrienoic acid gave rise to other products (15,16,17). These cannot act as precursors in the biosynthesis of prostaglandins; however, their structures and the fate of ^{3}H in labeled precursors during their formation provided additional evidence for the proposed scheme of the transformation. The monohydroxy acid fraction from an incubation with eicosatrienoic acid was shown to consist of 11-hydroxy-8,12,14-eicosatrienoic acid, 15-hydroxy-8,11,13-eicosatrienoic acid and 12-hydroxy-8(*trans*),10(*trans*)-heptadecadienoic acid. The mechanism of formation of the C_{17} acid was studied by using 3-^{14}C-eicosatrienoic acid, which also contained tritium label at C-9, C-10 or C-11 (16). These experiments showed that the conversion resulted in loss of the tritium label in these three positions. Furthermore, malonaldehyde was identified by condensation with L-arginine to give δ-N-2-(pyrimidinyl)-L-ornithine or with urea to form 2-hydroxypyrimidine. The derivative of malonaldehyde contained the ^3H label from the 9-^3H and 11-^3H labeled precursor whereas ^3H originally in position C-10 was lost by enolization of the malonaldehyde.

It was further found that a compound with chromatographic properties similar to those of PGE_1 and which was transformed into $PGF_{1\alpha}$ bý borohydride reduction was formed from eicosatrienoic acid (17). The structure of the new product was found to be 9α, 15-dihydroxy-11-ketoprost-13-enoic acid (11-dehydro-$PGF_{1\alpha}$ (18). In the conversion of 9-^3H,3-^{14}C- and 11-^3H, 2-^{14}C-8,11,14-eicosatrienoic acid to 11-dehydro-$PGF_{1\alpha}$, the latter precursor lost the ^3H label, whereas the 9-^3H label was retained. These experiments on the structures of the various products from eicosoatrienoic acid and the fate of the ^3H labels in their formation provided strong evidence for the existence of the endoperoxide intermediate.

ISOLATION AND CHARACTERIZATION OF PROSTAGLANDIN ENDOPEROXIDES

Subsequently it was possible to detect and isolate an endoperoxide from short-time incubations of arachidonic acid with the microsomal fraction of homogenates of sheep vesicular glands (19). The incubation mixtures were treated with stannous chloride in ethanol in order to reduce endoperoxide into $PGF_{2\alpha}$. This was followed by sodium borodeuteride reduction and determination of the resulting PGF_2 species by multiple-ion analysis. This method made it possible to assay PGE_2 as well as 11-dehydro-$PGF_{2\alpha}$ and $PGF_{2\alpha}$. It was of particular interest that a peak of $PGF_{2\alpha}$ appeared in the initial phase of the incubation period. No metabolic transformation of PGF compounds had been observed in preparations of sheep vesicular gland, and thus it was unlikely that the $PGF_{2\alpha}$ peak could be ascribed to enzymatic formation of $PGF_{2\alpha}$ followed by rapid metabolic degradation. Furthermore, when the $SnCl_2$ and sodium borodeuteride reduction was omitted, the peak of $PGF_{2\alpha}$ disappeared, indicating that $PGF_{2\alpha}$ was formed by chemical reduction of an oxygenated derivative present in the initial phase of the incubation. That an oxygenated intermediate was formed and temporally accumulated was also suggested by the finding that the rate of PGE_2 formation was slower than the rate of oxygenation of the precursor acid.

Additional support for the existence of an oxygenated intermediate that was convertible into $PGF_{2\alpha}$ by $SnCl_2$ reduction came from experiments in which reduced glutathione or p-mercuribenzoate was added to the microsomal suspension. The former agent increased the rate of PGE_2 formation and suppressed the peak of $PGF_{2\alpha}$, whereas the latter agent decreased the rate of PGE_2 formation with a simultaneous increase in the height and duration of the $PGF_{2\alpha}$ peak. The oxygenated intermediate detected by these experiments was also isolated. On thin-layer radiochromatographic analysis of the product (methyl esters) isolated after a 30-second incubation of labelled arachidonic acid with microsomes in the presence of p-mercuribenzoate a new radioactive derivative appeared. This derivative was identified as the methyl ester of the earlier postulated endoperoxide. In an extension of these studies the endoperoxide described above was obtained as the free acid; in addition, an endoperoxide carrying a hydroperoxy group at C-15 was isolated (20,21). We suggested the trivial names PGG_2 for the less polar endoperoxide (15-hydroperoxy-9α, 11α-peroxidoprosta-5,13-dienoic acid) and PGH_2 for the more polar endoperoxide (15-hydroxy-9α,11α-peroxidoprosta-5,13-dienoic acid). The structure of PGG_2 was established by three separate experiments. Treatment of PGG_2 with mild reducing agents such as $SnCl_2$ and triphenylphospine gave $PGF_{2\alpha}$ as the major product. This showed the presence of a peroxide bridge between C-9 and C-11 but did not discriminate between a hydroxy and a hydroperoxy group at C-15 since the agents used would reduce the latter group into the former. In a second experiment, PGG_2 was treated with lead tetraacetate in benzene followed by triphenylphosphine. In this case 15-keto-$PGF_{2\alpha}$ was the major product. Lead tetraacetate causes dehydration in hydroperoxides into ketones, and therefore, formation of a 15-keto-prostaglandin from PGG_2 by this treatment strongly indicated the presence of a hydroperoxy group at C-15. The isomerization of PGG_2 into 15-hydroperoxy-PGE_2 in aqueous medium gave independent evidence for a peroxide group at C-15 (Fig. 1).

Two reactions are involved in the conversion of PGG_2 into PGE_2, i.e., reduction of the hydroperoxy group at C-15 into a hydroxy group (peroxidase) and isomerization of the endoperoxide structure into a β-hydroxyketone (endoperoxide isomerase) (Fig. 1). The endoperoxide isomerase was found to be almost entirely associated with the microsomal fraction. The enzymic activity was stimulated by reduced glutathione.

The endoperoxides were quite unstable ($t_{1/2} = 5$ min). However, if they were stored under anhydrous conditions in acetone they could be kept for several weeks. When their biological activity was tested on *in vitro* preparations it was found that the effects of the endoperoxides on gastrointestinal smooth muscle were comparable to those of PGE_2 and $PGF_{2\alpha}$. On the other hand, the effects on vascular (rabbit aorta) and airway (guinea pig trachea) smooth muscle were considerably greater than those of PGE_2 and $PGF_{2\alpha}$, respectively (22). Both endoperoxides were potent contractors of the isolated human umbilical artery (23). Administration of PGG_2 and PGH_2 intravenously to guinea pigs (22) produced an increase in insufflation pressure, which was more marked than that caused by corresponding doses of $PGF_{2\alpha}$. The cardiovascular effects of the endoperoxides showed a complex pattern. The blood pressure response was triphasic, i.e., a transient fall consistently followed by a shortlasting rise and then by a sustained reduction. These studies on vascular and airway smooth muscle demonstrated that the endoperoxides had effects that could not be attributed to conversion into the stable prostaglandins.

Additional studies in our laboratory showed that the two endoperoxides also had unique effects on platelets. Thus, PGG_2 and PGH_2 induced rapid and irreversible aggregation of human platelets (19,20,24).

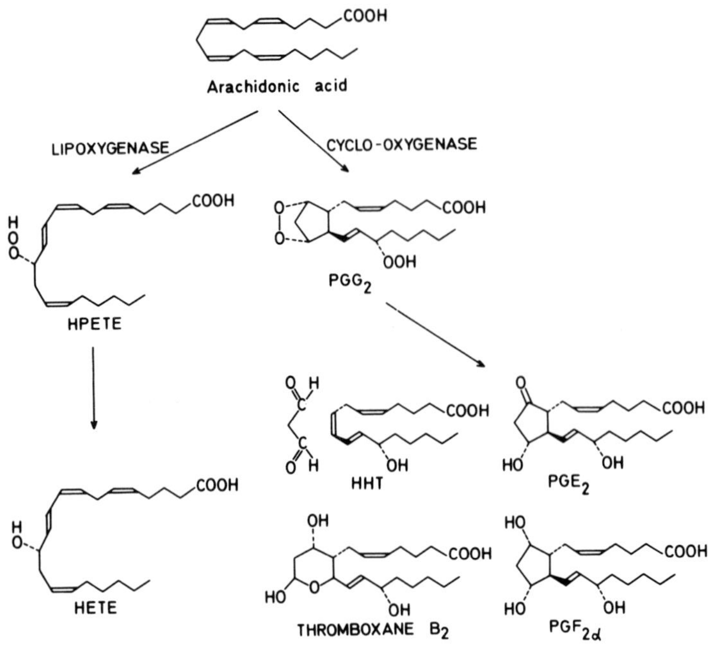

Fig. 2. Transformation of arachidonic acid in human platelets.

DISCOVERY OF AN UNSTABLE AGGREGATING FACTOR AND THE THROMBOXANES

The biological effects of the pure endoperoxides were of particular interest in relation to other studies which demonstrated that arachidonic acid caused aggregation when added to human platelets (25,26) and that labile aggregating material (LASS) was formed from this acid when it was incubated with preparations of sheep vesicular glands (27,28,29). Furthermore, the potency of the endoperoxides in causing contractions of the isolated rabbit aorta was of particular interest in relation to the so-called rabbit aorta contracting substance (RCS) (30). RCS was reported to be formed in guinea pig lung during anaphylaxis, and was later suggested to be due to the endoperoxide intermediate in prostaglandin biosynthesis (31). We found that material with similar biological properties was formed after addition of arachidonic acid to human platelets. However, the rabbit aorta contracting substance from guinea pig lung and platelets was found to consist of one major component with a $t_{1/2}$ of about 30 seconds and a minor component of PGG_2 and/or PGF_2 with a $t_{1/2}$ of 4—5 minutes (32). The short-lived major component of RCS could be generated by addition of arachidonic acid to platelets.

We therefore incubated 1-^{14}C-arachidonic acid with suspensions of washed human platelets in order to obtain structural information about RCS. Three major metabolites were isolated (33). One of them was found to be 12L-hydroxy-5,8,10,14-eicosatetraenoic acid (12-HETE) (Fig. 2) The corresponding hydro-

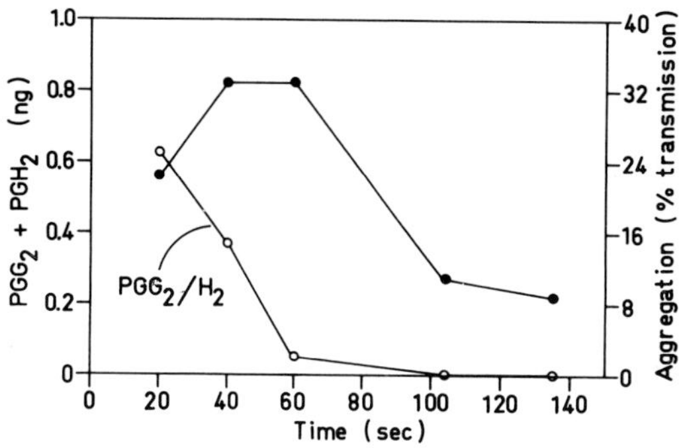

Fig. 3. Maximum aggregation induced by 0.1 ml of suspensions of washed platelets incubated for different times with 120 ng of arachidonic acid (●—●). The content of PGG_2 /H_2 in these samples is also given (○—○). The platelet suspension in the aggregometer tube was preincubated for 2 min with 1.4×10^5 M indomethacin.

peroxide (HPETE) could be isolated after incubation of arachidonic acid with sonicated platelets. Formation of 12-HETE from arachidonic acid was also reported to occur in bovine platelets (34). A more polar metabolite was identified as 12L-hydroxy-5,8,10-heptadecatrienoic acid (HHT) whereas a third component was found to be the hemiacetal derivative of 8(1-hydroxy-3-oxopropyl)-9-12L-dihydroxy-5,10-heptadecadienoic acid (thromboxane B_2, PHD). [1-^{14}C]-PGG_2 added to suspensions of human platelets was rapidly converted into HHT and thromboxane B_2.

All of the identified metabolites of arachidonic acid were stable compounds and could therefore not be identical to the very unstable ($t_{1/2}$ = 30 seconds) RCS. Additional biological work with the platelets involving characterization of the material formed from arachidonic acid was therefore carried out. When arachidonic acid was incubated with washed platelets and an aliquot of the incubate was transferred to a suspension of platelets preincubated with indomethacin, aggregation took place. This was not due to PGG_2 or PGH_2 since the amounts found were only about one per cent of those required to explain the response. A more detailed analysis of the appearance of the aggregating factor and the endoperoxides showed that the amount of endoperoxides was highest in the very early phase of the incubation period, whereas the aggregating factor had a maximum later (35) (Fig. 3). Experiments using filtrates of incubates prepared as described above showed that the aggregating factor was very unstable. When the log dose (arbitrary units) was plotted against time of incubation at 37°C, a linear relationship was obtained. The half-life of the aggregating factor was 33—46 seconds. A factor with similar properties was also generated from the endoperoxide, PGG_2. In addition to inducing irreversible aggregation, the unstable factor also caused release of serotonin from platelets.

Further work involving $^{18}O_2$ experiments suggested that thromboxane B_2 was formed from PGG_2 by rearrangement and subsequent incorporation of one molecule of H_2O (33). It was therefore conceivable that if the rearranged inter-

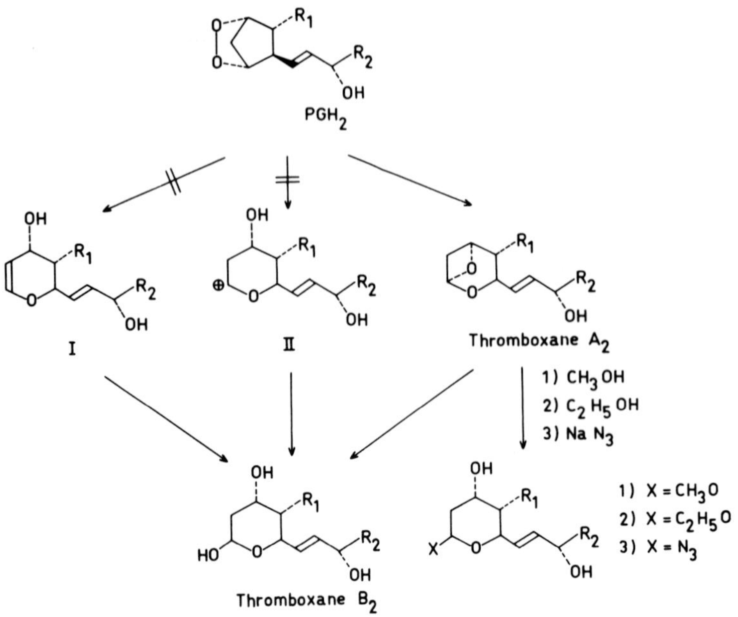

Fig. 4. Scheme of transformations of PGH_2 into thromboxane derivatives.

mediate had an appreciable lifetime it should be trapped in the presence of nucleophilic reagents (36). This was found to be the case. Additon of 25 volumes of methanol to washed platelets incubated with arachidonic acid for 30 seconds gave two derivatives that were less polar than thromboxane B_2. The mass spectral data indicated that the two compounds obtained by addition of methanol were epimers of thromboxane B_2 methylated at the hemiacetal hydroxyl group (Fig. 4). The two epimers also appeared when methanol was added to platelets incubated with PGG_2 for 30 seconds. Addition of ethanol to platelets incubated with arachidonic acid for 30 seconds similarly gave rise to epimers of thromboxane B_2 ethylated at the hemiacetal hydroxyl group. Finally, addition of 5 volumes of 5M sodium azide to platelets incubated with arachidonic acid gave an azido alcohol, i.e. a derivative of thromboxane B_2 in which the hemiacetal hydroxyl group was replaced by an azido group. The trapping experiments showed the existence of a very unstable intermediate in the conversion of PGG_2 into thromboxane B_2.

In order to determine the half-life of the intermediate, the platelet suspension was incubated with [1-^{14}C]-arachidonic acid for 45 seconds and the reaction was stopped by filtration. The clear, essentially platelet-free filtrate was kept at 37°C, and aliquots were removed after different times and immediately added to 25 ml of methanol containing tritium-labelled mono-O-methylthromboxane B_2. A linear relationship between the logarithms of the $^{14}C/^{3}H$ ratios of the purified methyl ester of mono-O-methyl-thromboxane B_2 and the times of incubation was obtained. The half-life thus obtained was 32+2 (SD) seconds.

Fig. 4 shows the proposed structure of the unstable intermediate. The acetal carbon atom binding two oxygens should be susceptible to attack by nucleophiles, e.g. H_2O (giving thromboxane B_2) as well as CH_3OH, C_2H_5OH and N_3^- (giving derivatives of thromboxane B_2 described above). Addition of CH_3O^2H to platelets incubated with arachidonic acid led to formation of mono-O-methylthromboxane B_2 lacking carbon bound ^2H. This finding excluded an alternative structure of the unstable intermediate, i.e. an unsaturated oxane ((I) in Fig. 4). Furthermore, the $t_{1/2}$ of thromboxane A_2 seemed to exclude a carbonium ion structure ((II) in Fig. 4), which in aqueous medium should be considerably less stable.

The available evidence indicated that the aggregating factor and RCS were due to the same compound. Thus, they were both derived from arachidonic acid or PGG_2, their formation from arachidonic acid was blocked by indomethacin, and their half-lives were similar. It was proposed that this material is identical with the unstable intermediate detected chemically in platelets (Fig. 4) because of similar properties.

The new oxane derivatives were named thromboxanes because of their structure and origin. Thromboxane A_2 is the highly unstable bicyclic compound, and thromboxane B_2 is the stable derivative provisionally named PHD. The subscript indicates the number of double bonds, as in the prostaglandin nomenclature. The structure of thromboxane B_2 has been confirmed by synthesis (37,38,39,40).

Thromboxane A_2 has been shown to possess a variety of strong biological effects. The best known of these are induction of platelet aggregation and the release reaction (36,40) as well as strong constricting effects on vascular smooth muscle. The first vessel to be studied in this respect was the rabbit aorta (30,32); later similar contractile responses were observed in other vessels as well, such as coronary arteries (42,43,44,45,46), the mesenteric and celiac arteries (47,48), the umbilical artery (23,49), and others.

These dual effects of TXA_2, induction of vasoconstriction and platelet aggregation both come into operation after a vessel has been injured. They indicate that TXA_2 probably plays a role in normal hemostasis *in vivo* as well as in pathological conditions with an increased tendency to vasospasm and/or thrombosis. Several reviews have been written on the biological effects and possible roles of thromboxanes *in vivo* (e.g. Refs. 41,50). Thromboxane A_2 has also potent contractile effects on airways demonstrable both *in vitro* and *in vivo* (51).

Following the isolation of the endoperoxides and the discovery of thromboxane A_2 it was found that arterial tissue converts the endoperoxide into a product with opposite effects (52). Structural work demonstrated that it was an enol ether derivative (53). This vasodilator and antiaggregating compound was named PGI_2, or prostacyclin. Thromboxane A_2 and prostacyclin probably form a hemostatic mechanism for control of the tonus of blood vessels and the aggregation of platelets *in vivo*. These platelet and vessel wall reactions, which are of considerable interest in antithrombotic therapy, are summarized in Fig. 5. The main focus is now on the development of specific thromboxane synthetase inhibitors for this purpose. For a recent review see ref. 54.

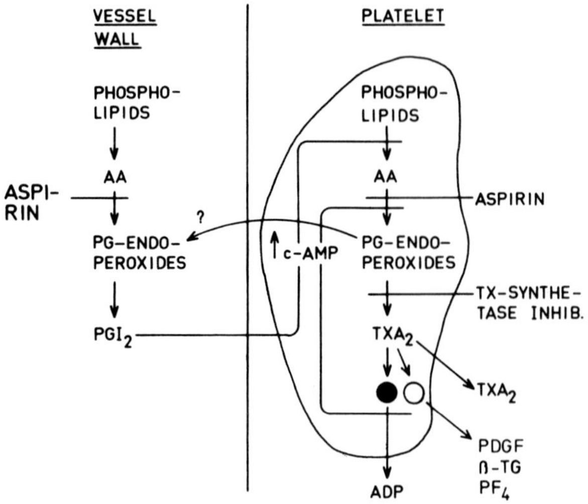

Fig. 5. Interaction between platelets and the vessel wall.

DISCOVERY OF THE LEUKOTRIENES

The role of prostaglandins in inflammation was brought into focus with the discovery by Vane and collaborators that non-steroidal anti-inflammatory drugs like aspirin inhibit the enzyme (cyclo-oxygenase) responsible for conversion of arachidonic acid into prostaglandins and thromboxanes (55). Anti-inflammatory steroids also inhibit the formation of prostaglandins; however, the mechanism of action is different. The steroids inhibit the formation of prostaglandins by blocking the release of arachidonic acid from the phospholipids. Since aspirin-like drugs and steroids have different anti-inflammatory effects it seemed conceivable to us that some of these differences might be due to formation of additional pro-inflammatory derivatives of arachidonic acid. Studies of the transformation of arachidonic acid in leukocytes, which were carried out to test this hypothesis, have recently resulted in the recognition of a novel group of compounds, the leukotrienes. These compounds seem to be of importance in both immediate hypersensitivity reactions and inflammation.

When arachidonic acid was incubated with polymorphonuclear leukocytes it was found that the major metabolite was a new lipoxygenase product, viz. 5(S)-hydroxy-6,8,11,14-eicosatetraenoic acid (5-HETE) (56). Additional studies also demonstrated the formation of 5(R), 12(R)-dihydroxy-6,8,10,14-eicosatetraenoic acid (major product) (leukotriene B_4, c.f. below), two additional 5(S),12-dihydroxy-6,8,10-*trans*,14-*cis*-eicosatetraenoic acids, epimeric at C-12, and two isomeric 5,6-dihydroxy-7,9,11,14-eicosatetraenoic acids (Fig. 6) (57,58).

Stereochemical studies, demonstrating formation of two acids with all trans conjugated trienes and epimeric at C-12 and one major isomer (12R) with different configuration of the triene raised the question of the mechanism of

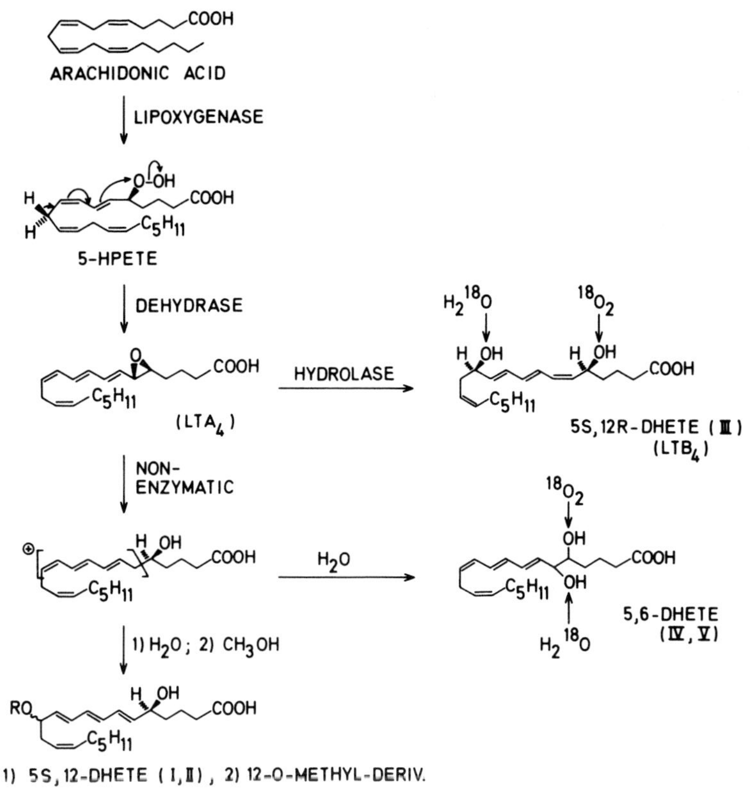

Fig. 6. Formation of dihydroxy derivatives via unstable intermediate. Origin of oxygen and trapping experiments.

formation (58). With isotopic oxygen it was demonstrated that the oxygen of the alcohol group at C-5 originated in molecular oxygen, whereas the oxygen of the alcohol group at C-12 was derived from water (Fig. 6) (59). These observations led us to develop the hypothesis that leukocytes generated an unstable intermediate which would undergo nucleophilic attack by water, alcohols, and other nucleophiles. Leukocytes were therefore incubated for 30 seconds with arachidonic acid followed by addition of 10 volumes of methanol, 10 volumes of ethanol or 0.2 volumes of N HCl. Trapping with methanol (or ethanol) yielded two new less polar compounds, which were present in equal amounts and which had ultraviolet spectra identical to those of compounds I and II (Fig. 6). Infrared spectrometry indicated that the conjugated trienes had all-trans geometry. Mass spectrometric analyses of the two compounds showed that they were isomeric and carried hydroxyl groups at C-5 and methoxy groups at C-12. The alcohol groups at C-5 had (S) configuration and it was obvioius that the compounds were the C-12 epimers of 5(S)-hydroxy,12-methoxy-6,8,10,14(E.E.E.Z)-eicosatetraenoic acid (Fig.6).

These experiments demonstrated that leukocytes generated a metabolite of arachidonic acid, which can undergo a facile nucleophilic reaction with alcohols. Analysis of samples obtained from trapping experiments performed under different conditions always showed inverse relationships between the amount of compounds I and II and their 12-0-alkyl derivatives. This result suggested that compounds I and II and the 12-0-alkyl derivatives were formed nonenzymatically from the same intermediate.

The stability of the intermediate was determined by incubating leukocytes with arachidonic acid for 45 seconds followed by addition of 1 volume of acetone in order to inactivate the enzyme. After different time intervals aliquots of the mixture were transferred to flasks containing 15 volumes of methanol. Analysis by chromatography showed that tht $t_{1/2}$ of the intermediate, measured as the 12-0-methyl derivative was 3—4 min. Concomitantly with the decrease in the concentration of the intermediate the concentrations of compounds I, II, IV and V increased whereas the concentrations of compounds III and 5-hydroxy-6,8,11,14-eicosatetraenoic acid remained constant. These data suggested that the epimeric 5,6- and 1,12-dihydroxy acids (compounds I, II, IV and V) are formed by non-enzymatic hydrolysis of a common unstable intermediate, whereas compound III is generated by enzymatic hydrolysis of the same intermediate (Fig. 6). Similar experiments performed at acid and alkaline pH showed that the intermediate was acid labile and considerably more stable under alkaline conditions. It was also found that the two 5,6-dihydroxy-derivatives (IV and V) were formed non-enzymatically) from the same intermediate as the enzymatic product, 5S,12R-dihydroxy-eicosatetraenoic acid, and that ^{18}O from molecular oxygen was exclusively retained at C-5 of these derivatives whereas ^{18}O from water was introduced at C-6 or C-12. On basis of the experimental data described above, the structure 5,6- oxido-7,9,11,14-eicosatetraenoic acid (leukotriene A_4, LTA_4) (Fig. 6) was proposed for the intermediate (59).

The formation of compounds I—V from the epoxide intermediate is shown in Fig. 6. With the exeption of compound III these are formed by chemical hydrolysis of the epoxide through a mechanism involving a carbonium ion. This derivative added hydroxyl anion preferentially at C-6 and C-12 to yield four isomeric products which contain the stable conjugated triene structure. The formation of compound III is enzymatic since it is not racemic at C-12 and since it is only formed by non-denatured cell preparations.

The structure, 5,6-oxido-7,9,11,14-eicosatetraenoic acid (leukotriene A_4, c.f. below) (59) proposed for the intermediate has subsequently been confirmed by chemical synthesis and the stereochemistry has been elucidated (60). The 5S,12R-dihydroxy acid formed enzymatically was earlier shown to contain one *cis* and two *trans* double bonds in the conjugated triene. The location of the *cis* double bond (Δ^6-position) was recently established by synthetic methods (61). The allylic epoxide intermediate can exist in free form since it has been isolated from human polymorphonuclear leukocytes (62). The suggested mechanism for the biosynthesis of the epoxide from arachidonic acid (Fig. 6) involves initital formation of 5-hydroperoxy-6,8,11,14-eicosatetraenoic acid (5-HPETE). The epoxide is formed from 5-HPETE and subsequent abstraction of the pro-R hydrogen at C-10, and elimination of hydroxyl anion from the hydroperoxy group (63). The dehydration reaction has been found to be catalyzed by a soluble enzyme, which was recently isolated from leukocytes (64).

SLOW REACTING SUBSTANCE OF ANAPHYLAXIS (SRS-A)

The occurrence of a smooth muscle stimulating factor (SRS) appearing in the perfusate of guinea pig lung treated with cobra venom was described in 1938 (65). The factor was shown to be released also by immunological challenge (66). Biological studies of SRS suggested that it might be an important mediator of anaphylactic and other immediate hypersensitivity reactions (67,68). Characterization of SRS indicated that it was a polar lipid with UV-absorption and that it might contain sulfur (69,70,71). Studies with labelled arachidonic acid suggested that this acid might be incorporated into SRS (72,73).

Studies in our laboratory showed that treatment of human neutrophils with the ionophore A23187 resulted in stimulation of the synthesis of the 5,12-dihydroxy acid (LTB_4) described above (58). On the basis of the stimulatory effect of the ionophore on both SRS production (74) and LTB_4 formation, the UV-absorbance data and other considerations we developed the hypothesis that there was a biogenetic relationship between the unstable allylic epoxide intermediate in neutrophils and SRS generated in a variety of systems.

For production of relatively large amounts of SRS we found that murine mastocytoma cells (CXBGABMCT-1) stimulated with the calcium ionophore A23187 were more suitable than previously described systems (75). The SRS was purified by high pressure liquid chromatography. The purified material showed an absorbance maximum at 280 nm and gave a typical contraction of guinea pig ileum, which was reversed by FPL-55712 (75). The ultraviolet characteristics resembled those of the dihydroxy acids. However, the maximum was shifted to a 10 nm higher wavelength. This was in agreement with a sulfur substituent α to a conjugated triene. Labelled arachidonic acid and cysteine were incorporated into the product.

Degradation of SRS by Raney nickel desulfurization gave 5-hydroxy-arachidic acid, indicating that the arachidonic acid derivative and cysteine were linked by a thioether bond (Fig. 7). This finding also supported the hypothesis that there was a biogenetic relationship between the 5-lipoxygenase pathway in leukocytes and SRS. The positions of the double bonds in SRS were determined by reductive ozonolysis. The isolation of 1-hexanol among the products indicated that the Δ^{14}-double bond of arachidonic acid had been retained. The approach used for locating the conjugated triene was based on previous studies in our laboratory that had shown that arachidonic acid and related fatty acids containing two methylene interrupted *cis* double bonds at the $\omega 6$ and $\omega 9$ positions are oxygenated to give derivatives with isomerization of the $\omega 6$ double bond to $\omega 7$. Incubation of the isolated SRS with lipoxygenase resulted in isomerization of the Δ^{14}-double bond into conjugation with the conjugated triene (forming a tetraene) since there was a bathochromic shift of 30 nm. This finding indicated that SRS contained a Δ^{11}-*cis* double bond and additional double bonds at $\Delta 7$ and $\Delta 9$. The structural work at this stage showed that SRS was a derivative of 5-hydroxy-7,9,11,14-eicosatetraenoic acid with a cysteine containing substituent in the thioether linkage at C-6. Derivatization of cysteine was suggested by the failure to isolate alanine after desulfurization.

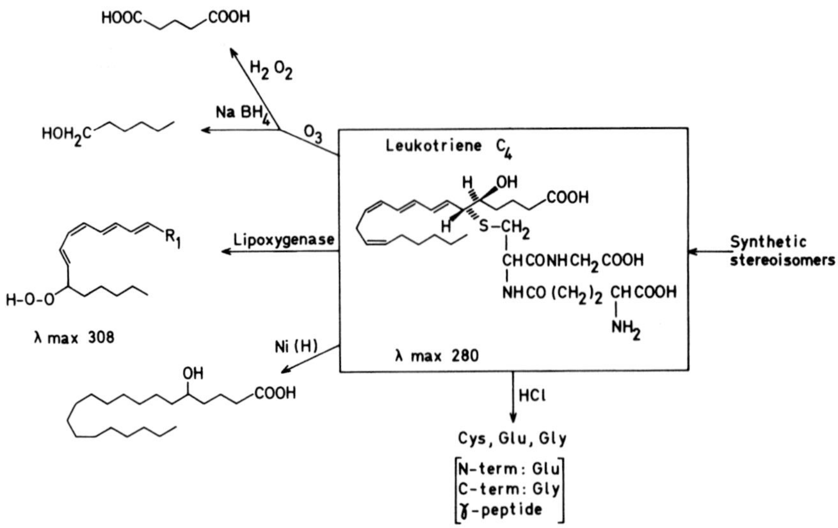

Fig. 7. Some transformations in structural studies on SRS.

The cysteine containing substituent was therefore referred to as RSH in the reports of this work (75,76,77). Further studies involving amino acid analyses of acid hydrolyzed SRS demonstrated that in addition to cysteine, one mol of glycine and one mol of glutamic acid were present per mol of SRS. End group (dansyl method and hydrozinolysis) and sequence analyses (dansyl-Edman procedure) of the peptide showed that it was γ-glutamylcysteinylglycine (glutathione). The structure of the SRS from murine mastocytoma cells was therefore 5-hydroxy-6-S-glutathionyl-7,9,11,14-eicosatetraenoic acid, leukotriene (LT)C_4 (c.f. below) (Fig. 7) (78). The structure was confirmed by comparison with synthetic material. This represented the first structure determination of an SRS-A (78). The preparation and some properties of corresponding cysteinylglycine derivative (LTD_4) and cysteinyl derivative (LTE_4) were also reported at the same time (78). These compounds have later been isolated from natural sources (see below). The proposed stereochemistry for LTC_4 was confirmed and unambiguously assigned by total synthesis including preparation of stereoisomers of LTC_4 (79). The synthetic work was carried out by E.J. Corey and co-workers. LTC_4 is thus 5(S)-hydroxy,6(R)-*S*-glutathionyl-7,9-*trans*-11,14-*cis*-eicosatetraenoic acid.

Later studies using a different cell type, the RBL-1 cells, demonstrated that the major slow reacting substance was LTD_4 (5(S)-hydroxy,6(R)-*S*-cysteinylglycine-7,9-*trans*-11,14-*cis*-eicosatetraenoic acid) (80).

Following the structure determination of SRS from mastocytoma cells (75,78) and synthetic preparation of LTC_4, LTD_4 and LTE_4 (78) all of these cysteine containing leukotrienes (Fig. 8) have been found in a variety of biological systems using comparison with synthetic material or partial characterization by chemical or physical methods for identification (Table 1). SRS-A is thus a

Table 1. Identification of leukotrienes from different sources

Source	LTA_4	LTB_4	LTC_4	LTD_4	LTE_4	References
Rabbit peritoneal leukocytes	+	+				57, 59
Human peripheral leukocytes	+	+	+			86, 119, 120
Mouse mastocytoma cells	+		+			75, 77, 121
Rat basophilic leukemia cells				+	+	80, 122, 123
Rat peritoneal cells			+	+	+	124, 125, 126
Rat leukocytes		+				127, 128
Rat macrophages		+				129
Mouse macrophages			+			130
Human lung			+	+		131
Guinea pig lung				+		132
Cat paws				+	+	133

mixture of leukotrienes containing cysteine, i.e. the parent compound LTC_4 and the metabolites LTD_4 and LTE_4.

Transformation of LTA_4 into LTC_4 by enzymatic addition of glutathione has been demonstrated in both mastocytoma cells and human leukocytes pretreated with the inhibitor of arachidonic acid metabolism,BW755 (81). These studies confirm the originally proposed pathway for biosynthesis of SRS, i.e. formation of LTA_4 from arachidonic acid via 5-HPETE followed by addition of glutathione to LTA_4 with opening of the epoxide at the allylic position C-6 to give LTC_4 (75).

The biological significance of the biosynthetic pathways described and the cumbersome systematic names of the compounds involved suggested the introduction of a trivial name for these entities (76). The term "leukotriene" was chosen because the compounds were discovered in leukocytes and the common structural feature is a conjugated triene. Various members of the group have been designated alphabetically: leukotrienes A are 5,6-oxido-7,9-*trans*-11-*cis*; leukotriene B, 5(S),12(R)-dihydroxy-6-*cis*-8,10-*trans*; leukotrienes C, 5(S)-hydroxy-6(R)-γ-glutamyl-cysteinyl-glycyl-7,9-*trans*-11-*cis*; and leukotrienes E, 5(S)-hydroxy-6(R)-*S*-cysteinyl-7,9-*trans*-11-*cis*-eicosapolyenoic acids. Since precursor acids containing the Δ^5 double bond system (i.e. 5,8,11-eicosatrienoic acid, arachidonic acid and 5,8,11,14,17-eicosapentaenoic acid) can be converted to leukotrienes containing 3—5 double bonds, a subscript denoting this number is used (82). Leukotriene A_4 is thus the epoxy derivative of arachidonic acid which can be further trnsformed to leukotrienes B_4, C_4 and E_4.

Leukotriene C_4 is metabolized to leukotriene D_4 by enzymatic elimination of glutamic acid by γ-glutamyl transpeptidase (80). The remaining peptide bond in leukotriene D_4 is hydrolyzed by a renal dipeptidase to give leukotriene E_4 (83). It has recently been found that LTE_4 can also function as acceptor of γ-glutamic acid forming a γ-glutamyl, cysteinyl derivative, named LTF_4 (84,85) (Fig. 8).

In addition to the 5-lipoxygenase, leukocyte preparations contain enzymes catalyzing introduction of oxygen at C-12 and C-15 (86). Recently, evidence has been obtained for leukotriene formation after initial oxygenation at either of these positions (87,88,89,90,91).

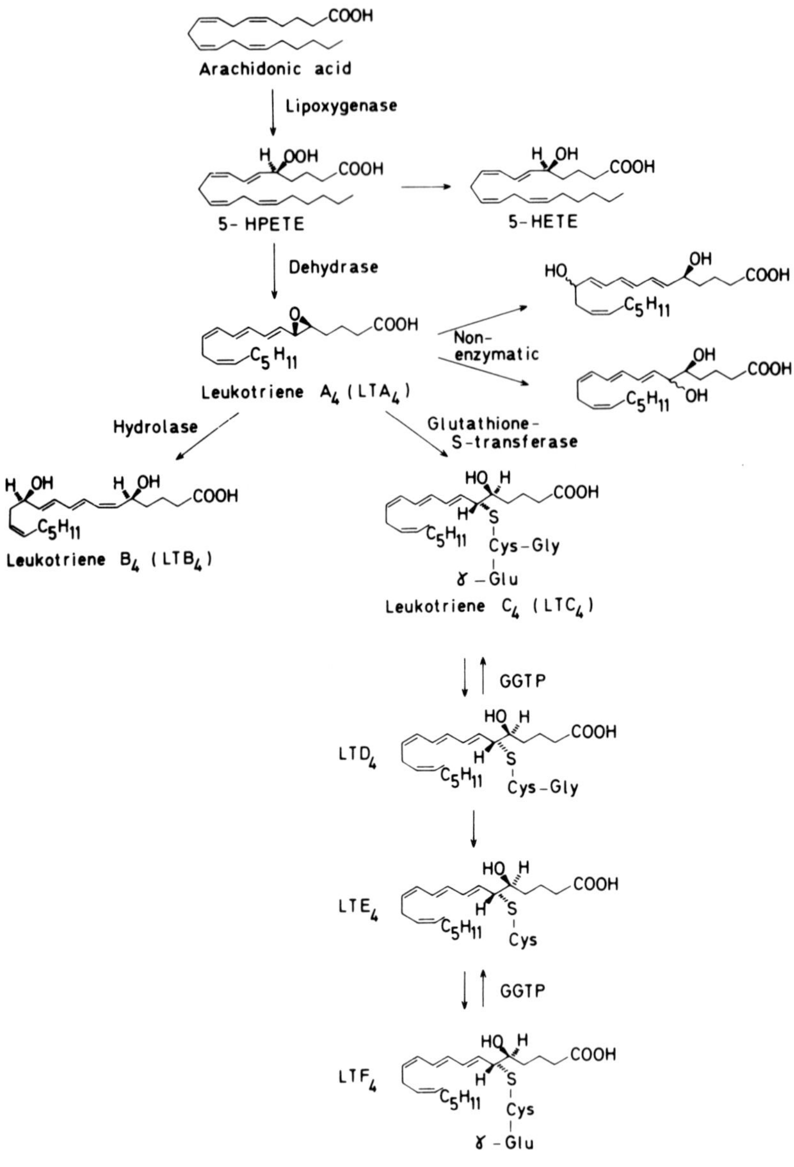

Fig. 8. Formation of leukotrienes via the 5-lipoxygenase pathway.

BIOLOGICAL EFFECTS OF THE LEUKOTRIENES

Studies with pure leukotrienes have provided detailed information about the effects of this group of compounds in different biological systems. The leukotrienes containing cystein (LTC_4, D_4 and E_4) are potent bronchoconstrictors in several species including humans, and they seem to have specific effects on the peripheral airways (92,93,94,95,96,97,98,99). They also show potent vasoconstrictor activity and have negative ionotropic effects on the cardiac contractions (100).

In recent studies using bronchi from atopic patients sensitive to birch pollen (101) the relative importance of the leukotrienes as mediators of anaphylaxis has been demonstrated. Treatment of the preparation with a histamine antagonist, mepyramine, and cyclo-oxygenase inhibitor, indomethacin, did not reduce the response to the specific allergen. However, benoxaprofen and a prostacyclin derivative (U-60257), both of which block leukotriene formation (102,103), inhibited the anaphylactic contraction in bronchi from asthmatics induced by birch pollen. Incubation of the atopic lung tissue with antigen resulted in a release of LTC_4, LTD_4 and LTE_4 which could be inhibited by the prostacyclin derivative U-60257 (101). These studies indicate that the leukotrienes containing cystein (LTC_4, LTD_4 and LTE_4) are major mediators of airway anaphylaxis; the finding that inhibition of leukotriene formation blocks ascaris induced asthma in monkeys also indicates that leukotriene antagonists or inhibitors of their formation could be of therapeutic value in the treatment of bronchial asthma (102).

When injected intradermally into guinea pigs LTC_4 and LTD_4 cause extravasation of Evan's blue (92,96,98). More recent studies involving intravital microscopy of the cheek pouch of the hamster (Mesocricetus auratus) have demonstrated specific effects of these leukotrienes on the permeability of the post-capillary venules (104). According to dose-response curves, LTC_4 and LTD_4 both induced a significant increase of vascular permeability at much lower concentrations than histamine. Leukotriene C_4 was approximately 5000 times more potent than histamine in this respect. The cysteinyl containing leukotrienes seem to increase the vascular permeability by a direct action on the vessel wall, since it occurs rapidly and does not require release of histamine or prostaglandins or the participation of polymorphonuclear leukocytes. Leukotriene B_4 also causes extravasation of plasma, although at higher concentrations. The reaction occurs with some latency and requires adhering leukocytes. Administration of a vasodilator together with leukotrienes potentiates the increase in plasma leakage caused by a submaximal dose of leukotrienes, as has been reported in the guinea pig for PGE_2 and LTD_4 (105) and in the guinea pig, rabbit and rat for PGE_2 and LTB_4 (106,107).

When LTB_4 was administered to the hamster cheek pouch in the same dose-range as LTC_4 it caused a dramatic increase in the adhesion of leukocytes to the endothelium in small venules (104). Increased adherence of human leukocytes caused by LTB_4 has also been demonstrated *in vitro* using a column of nylon fibers (108).

During superfusion with LTB_4 (6—10 min) the number of interstitial white cells increased. This finding is consistent with the chemotactic stimulant property of LTB_4. This effect of LTB_4 has been demonstrated *in vitro* using either the Boyden chamber technique or migration under agarose (109,110,111,112). *In vivo* this effect has been monitored by determining white cell accumulation in the peritoneal cavity of guinea pigs following intraperitoneal injection of LTB_4 (113). The studies described above indicate that LTB_4 might be a mediator in the migration of leukocytes from the blood to areas of inflammation. Recent work has also demonstrated that LTB_4 activates neutrophils. Addition of nano-

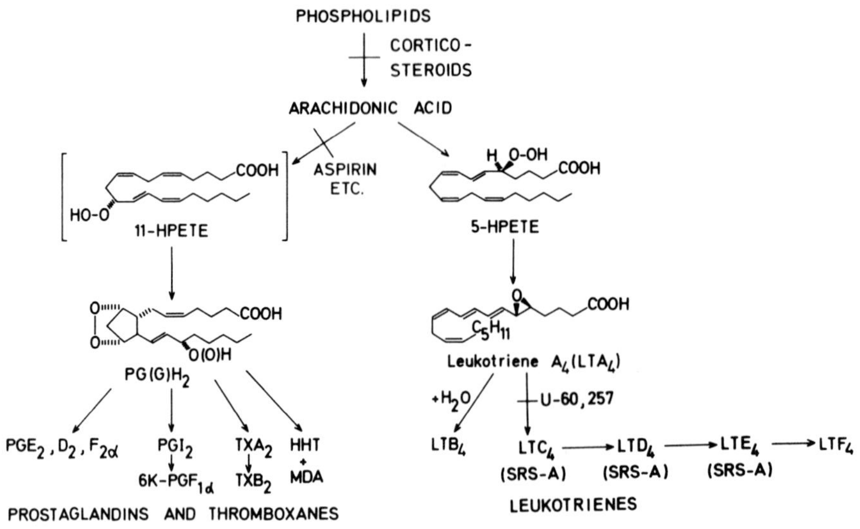

Fig. 9. Formation of prostaglandins, thromboxanes and leukotrienes.

molar concentrations of LTB_4 to the cells results in rapid aggregation, degranulation, superoxide generation and mobilization of membrane associated calcium (114,115,116).

Studies on the mechanism of action of anti-inflammatory steroids indicate that they inhibit the release of the precursor acid, arachidonic acid, whereas cyclo-oxygenase inhibitors as aspirin block the transformation of this acid into prostaglandins and thromboxanes (Fig. 9). The steroid induced inhibition of arachidonic acid release, proposed to be due to formation of peptide inhibitors of phospholipase A_2, prevents formation of not only prostaglandins and thromboxanes but also leukotrienes and other oxygenated derivatives (117,118). Some of the therapeutic effects of steroids which are not shared by aspirintype drugs might therefore be due to inhibition of leukotriene formation. The recent increased knowledge about the transformation of arachidonic acid and the biological effects of the metabolites seem to provide new possibilities to develop novel and more specific therapeutic agents.

ACKNOWLEDGEMENTS

It is a pleasure to acknowledge the important contributions of my associates who participated in the research described in this article.

The work from the author's laboratory was supported by the Swedish Medical Research Council (project 03X-217) and the Knut and Alice Wallenberg Foundation.

REFERENCES

1. Bergström, S. and Samuelsson, B.: Ann. Rev. Biochem. *34*, 101 (1965).
2. Samuelsson, B.: Angew. Chemie. Int. *4*, 410 (1965).
3. Bergström, S.: Les Prix Nobel, 1982.
4. Bergström, S., Danielsson H. and Samuelsson, B.: Biochim. Biophys. Acta, *90*, 207 (1964).
5. Van Dorp, D.A., Beerthius, R.K., Nugteren, D.H. and Vonkeman, H.: Biochim. Biophys. Acta, *90*, 104 (1964).
6. Nugteren, D.H. and Van Dorp, D.A. Biochim. Biophys. Acta, *98*, 645 (1965).
7. Ryhage, R. and Samuelsson, B.: Biochem. Biophys. Res. Commun. *19*, 279 (1965).
8. Samuelsson, B.: J. Am. Chem. Soc. *87*, 3011, (1965).
9. Hamberg, M. and Samuelsson, B.: J. Biol. Chem. *242*, 5329 (1967).
10. Hamberg, M. and Samuelsson, B.: J. Biol. Chem. *242*, 5344 (1967).
11. Samuelsson, B.: Progr. Biochem. Pharmacol. 3, 59–70 (1967).
12. Hamberg, M.: Eur. J. Biochem. *6*, 135, (1958).
13. Nugteren, D.H., Beerthuis, R.K. and van Dorp, D.A.: *In*: Nobel Symposium 2, Prostaglandins. Eds. S. Bergström and B. Samuelsson. Almqvist & Wiksell, Stockholm, p. 45–50 (1967).
14. Samuelsson, B., Granström, E. and Hamberg, M.: *In*: Nobel Symposium 2, Prostaglandins. Eds. S. Bergström and B. Samuelsson. Almqvist & Wiksell, Stockholm, P. 51–66 (1967).
15. Hamberg, M. and Samuelsson, B.: J. Am. Chem. Soc. *88*, 2349 (1966).
16. Hamberg, M. and Samuelsson, B.: J. Biol. Chem. *24*, 5344 (1967).
17. Nugteren, D.H., Beerthuis, R.K. and van Dorp, D.A.: Res. Trav. Chim. *85*, 405 (1966).
18. Granström, E., Lands, W.E.M. and Samuelsson, B.: J. Biol. Chem. *243*, 4104 (1968).
19. Hamberg, M. and Samuelsson, B.: Proc. Natl. Acad. Sci. USA, *70*, 899 (1973).
20. Hamberg, M., Svensson, J., Wakabayashi, T. and Samuelsson, B.: Proc. Natl. Acad. Sci. USA, *71*, 345 (1974).
21. Nugteren, D.H. and Hazelhof, E.: Biochim. Biophys. Acta, *326*, 448 (1973).
22. Hamberg, M., Hedqvist, P., Strandberg, K., Svensson, J. and Samuelsson, B.: Life Sci. *16*, 451 (1975).
23. Tuvemo, T., Strandberg, K., Hamberg, M. and Samuelsson, B.: Acta Physiol. Scand. *96*, 145 (1976).
24. Samuelsson, B. and Hamberg, M.: *In*: Prostaglandin Synthetase Inhibitors. Eds. H.J. Robinson and J.R. Vane. Raven Press, New York, p. 107 (1974).
25. Vargaftig, B.B. and Zirinis, P.: Nature (Lond.) New. Biol. *244*, 114 (1973).
26. Silver, M.J., Smith, J.B., Ingerman, C. and Kocsis, J.J.: Prostaglandins, *4*, 863 (1973).
27. Willis, A.L.: Science, *183*, 325, (1974).
28. Willis, A.L.: Prostaglandins, *5*, 1 (1974).
29. Willis, A.L., Vane, F.M., Kuhn, D.C., Scott, C.G. and Petrin, M.: Prostaglandins, *8*, 453 (1974).
30. Piper, P.J. and Vane, J.R.: Nature (Lond.), *223*, 29 (1969).
31. Gryglewski, R.J. and Vane, J.R.: Br. J. Pharmacol. *45*, 37 (1972).
32. Svensson, J., Hamberg, M. and Samuelsson, B.: Acta Physiol. Scand. *94*, 222 (1975).
33. Hamberg, M. and Samuelsson, B.: Proc. Natl. Acad. Sci. USA, *71*, 3400 (1974).
34. Nugteren, D.H.: Biochim. Biophys. Acta *380*, 299 (1975).
35. Samuelsson, B.: *In*: Advances in Prostaglandin and Thromboxane Research. Eds. B. Samuelsson and R. Paoletti. Raven Press, New York. Vol. 1, p. 1 (1976).
36. Hamberg, M., Svensson, J. and Samuelsson, B.: Proc. Natl. Acad. Sci. USA, *72*, 2994 (1975).
37. Nelson, N.A. and Jackson, R.W.: Tetr. Lett. *37*, 3275 (1976).
38. Kelly, R.C., Schettler, I. and Stein, S.J.: Tetr. Lett. *37*, 3279 (1976).
39. Schneider, W.P. and Morge, R.A.: Tetr. Lett. *37*, 3283 (1976).
40. Corey, E.J., Schibasaki, M. and Knolle, J.: Tetrahedron Lett. *19*, 1625 (1977).
41. Moncada, S. and Vane, J.R.: Pharmacol. Rev. *30*, 293 (1979).
42. Ellis, E.F., Oelz, O., Roberts, L.J., II, Payne, N.A., Sweetman, B.J., Nies, A.S. and Oates, J.A.: Science, *193*, 1135 (1976).

43. Needleman, P., Kulkarni, O.S. and Raz, A.: Science, *195*, 4090 (1977).
44. Svensson, J. and Hamberg, M.: Prostaglandins, *12*, 943 (1976).
45. Terashita, Z., Fukkui, H., Nishikawa, K., Hirat, M. and Kikuchi, S.: Eur. J. Pharmacol. *53*, 49 (1978).
46. Wang, H.H., Kulkarni, P.S. and Eakins, K.E.: Eur. J. Pharmacol. *66*, 31 (1980).
47. Bunting, S., Moncada, S. and Vane, J.R.: Br. J. Pharmacol. *57*, 462 (1976).
48. Moncada, S., Ferreira, S.H. and Vane, J.R.: *In*: Advances in Prostaglandin and Thromboxane Research. Ed. J.C. Frölich. Raven Press. New York, Vol. 5, p. 211 (1978).
49. Tuvemo, T., Strandberg, K., Hamberg, M. and Samuelsson, B.: *In:* Advances in Prostaglandin and Thromboxane Research. Eds. B. Samuelsson and R. Paoletti. Raven Press, New York. Vol. 1, p. 425–428 (1976).
50. Gryglewski, R.J., Dembinska-Kiec, A. and Korbut, R.: Acta Biol. Med. Ger. *37*, 715 (1978).
51. Svensson, J., Strandberg, K., Tuvemo, T. and Hamberg, M.: Prostaglandins, *14*, 425 (1977).
52. Moncada, S., Gryglewski, R.J., Bunting, S. and Vane, J.R.: Nature (Lond.) *263*, 663 (1976).
53. Johnson, R.A., Morton, D.R., Kinner, J.H., Gorman, R.R., McGuire, J.C., Sun, F.F., Whittaker, N., Bunting, S., Salmon, J., Moncada, S. and Vane, J.R.: Prostaglandins, *12*, 915 (1976).
54. Granström, E., Diczfalusy, U., Hamberg, M., Hansson, G., Malmsten, C. and Samuelsson, B.: *In*: Advances in Prostaglandin, Thromboxane and Leukotriene Research. Ed. J.A. Oates. Raven Press, New York. Vol. 10, p. 15 (1982).
55. Vane, J.R.: *In*: Advances in Prostaglandins and Thromboxane Research. Eds. B. Samuelsson and R. Paoletti. Raven Press, New York. Vol. 2, p. 791 (1976).
56. Borgeat, P., Hamberg, M. and Samuelsson, B.: J. Biol. Chem. *251*, 7816 (1976).
57. Borgeat, P. and Samuelsson, B.: J. Biol. Chem. *254*, 2643 (1979).
58. Borgeat, P. and Samuelsson, B.: J. Biol. Chem. *254*, 7865 (1979).
59. Borgeat, P. and Samuelsson, B.: Proc. Natl. Acad. Sci. USA, *76*, 3213 (1979).
60. Rådmark, O., Malmsten, C., Samuelsson, B., Clark, D.A., Giichi, G., Marfat, A. and Corey, E.J.: Biochem. Biophys. Res. Commun. *9*, 954 (1980).
61. Corey, E.J., Marfat, A., Goto, G. and Brion, F.: J. Am. Chem. Soc. *102*, 7984 (1981).
62. Rådmark, O., Malmsten, C., Samuelsson, B., Goto, G., Marfat, A. and Corey, E.J.: J. Biol. Chem. *255*, 11828 (1980).
63. Panossian, A., Hamberg, M. and Samuelsson, B.: FEBS Lett. *150*, 511 (1982).
64. Rådmark, O. and Shimizu, T., in press 1982.
65. Feldberg, W. and Kellaway, C.H.: J. Physiol. (Lond.), *94*, 187 (1938).
66. Kellaway, C.H. and Trethewie, E.R.: Q. J. Exp. Physiol. *30*, 121 (1940).
67. Austen, K.F.: J. Immunol. *121*, 793 (1978).
68. Brocklehurst, W.E.: J. Physiol. (Lond.) *120*, 16P (1953).
69. Morris, H.R., Taylor, G.W., Piper, P.J., Sirois, O. and Tippins, J.R.: FEBS Lett. *87*, 203 (1978).
70. Orange, R.P., Murphy, R.C., Karnovsky, M.L. and Austen, K.F.: J. Immunol. *110*, 760 (1973).
71. Strandberg, K. and Uvnäs, B.: Acta Physiol. Scand. *82*, 359 (1971).
72. Bach, M.K., Brashler, J.R. and Gorman, R.R.: Prostaglandins, *14*, 21 (1977).
73. Jakschik, B.A., Falkenhein, S. and Parker, C.W.: Proc. Natl. Acad. Sci. USA, *74*, 4577 (1977).
74. Conroy, M.C., Orange, R.P. and Lichtenstein, L.M.: J. Immunol. *116*, 1677 (1976).
75. Murphy, R.C., Hammarström, S. and Samuelsson, B.: Proc. Natl. Acad. Sci. USA, *76*, 4275 (1979).
76. Samuelsson, B., Borgeat, P., Hammarström, S. and Murphy, R.C.: Prostaglandins, *17*, 785 (1979).
77. Samuelsson, B., Borgeat, P., Hammarström, S. and Murphy, R.C.: *In*: Advances in Prostaglandin and Thromboxane Research. Eds. B. Samuelsson, P. Ramwell and R. Paoletti. Raven Press, New York, Vol. 6, p. 1 (1980).
78. Hammarström, S., Murphy, R.C., Samuelsson, B., Clark, D.A., Mioskowski, C. and Corey, E.J.: Biochem. Biophys. Res. Commun. *91*, 1266 (1980).

79. Hammarström, S., Samuelsson, B., Clark, D.A., Goto, G., Marfat, A., Mioskowski, C. and Corey, E.J.: Biochem. Biophys. Res. Commun. *92*, 946 (1980).
80. Örning. L., Hammarström, S. and Samuelsson, B.: Proc. Natl. Acad. Sci. USA, *77*, 2014 (1980).
81. Rådmark, O., Malmsten, C. and Samuelsson, B.: FEBS Lett. *110*, 213 (1981).
82. Samuelsson, B. and Hammarström, S.: Prostaglandins, *19*, 645 (1980).
83. Bernström, K. and Hammarström, S.: J. Biol. Chem. *256*, 9579 (1981).
84. Anderson, M.E., Allison, R.D. and Meister, A.: Proc. Natl. Acad. Sci. USA, *79*, 1088 (1982).
85. Uehara, N., Ormstad, K., Orrenuis, S., Örning, L. and Hammarström, S.: *In*: Advances in Prostaglandin, Thromboxane and Leukotriene Research. Eds. B. Samuelsson, P. Ramwell and R. Paoletti. Raven Press, New York. Vol. 11, p. 147 (1983).
86. Borgeat, P. and Samuelsson, B.: Proc. Natl. Acad. Sci. USA, *76*, 2148 (1979).
87. Jubiz, W., Rådmark, O., Lindgren, J.Å., Malmsten, C. and Samuelsson, B.: Biochem. Biophys. Res. Commun. *99*, 976 (1981).
88. Lindgren, J.Å. and Samuelsson, B.: in press (1983).
89. Lundberg, U., Rådmark, O., Malmsten, C. and Samuelsson, B.: FEBS Lett. *126*, 127 (1981).
90. Rådmark, O., Lundberg, U., Jubiz, W., Malmsten, C. and Samuelsson, B.: *In*: Advances in Prostaglandin, Thromboxane and Leukotriene Research. Eds. B. Samuelsson and R. Paoletti. Raven Press, New York. Vol. 9, p. 61 (1982).
91. Maas, R.L., Brash, A.R. and Oates, J.A.: *In*: Advances in Prostaglandin, Thromboxane and Leukotriene Research. Eds. B. Samuelsson and R. Paoletti. Raven Press, New York. Vol. 9, p. 29 (1983).
92. Dahlén, S.-E., Hedqvist, P., Hammarström, S. and Samuelsson, B.: Nature, *288*, 484 (1980).
93. Hedqvist, P., Dahlén, S.-E., Gustafsson, L., Hammarström, S. and Samuelsson, B.: Acta Physiol. Scand. *110*, 331 (1980).
94. Lewis, R.A., Austen, F.K., Drazen, J.M., Soter, M.A., Figueiredo, J.C. and Corey, E.J.: *In*: Advances in Prostaglandin, Thromboxane and Leukotriene Research. Eds. B. Samuelsson and R. Paoletti. Raven Press, New York. Vol. 9, p. 137 (1982).
95. Lewis, R.A., Lee, C.W., Levine, L., Morgan, R.A., Weiss, J.W., Drazen, J.M., Oh, H., Hoover, D., Corey, E.J. and Austen, K.F.: *In*: Advances in Prostaglandin, Thromboxane and Leukotriene Research. Eds. B. Samuelsson, P. Ramwell and R. Paoletti. Raven Press, New York. Vol. 11, p. 15 (1983).
96. Piper, P.J. and Tippins, J.R.: *In*: Advances in Prostaglandin, Thromboxane and Leukotriene Research. Eds. B. Samuelsson, P. Ramwell and R. Paoletti. Raven Press, New York. Vol. 9, p. 183 (1982).
97. Smedegård, G., Hedqvist, P., Dahlén, S.-E., Revenäs, B., Hammarström, S. and Samuelsson, B.: Nature, *295*, 327 (1982).
98. Drazen, J.M., Austen, F.K., Lewis, R.A., Clark, D.A., Goto, G., Marfat, A. and Corey, E.J.: Proc. Natl. Acad. Sci. USA, *77*, 4354 (1980).
99. Weiss, J.W., Drazen, J.M., Coles, N., McFadden, E.R., Weller, P.W., Corey, E.J., Lewis, R.A. and Austen, K.F.: Science, *216*, 196 (1982).
100. Levi, R., Burke, J.A. and Corey, E.J.: *In*: Advances in Prostaglandin, Thromboxane and Leukotriene Research. Eds. B. Samuelsson and R. Paoletti. Raven Press, New York. Vol. 9, p. 215 (1982).
101. Hansson, G., Björck, T., Dahlén, S.-E., Hedqvist, P., Granström, E. and Dahlén, B.: *In*: Advances in Prostaglandin, Thromboxane and Leukotriene Research. Eds. B. Samuelsson, P. Ramwell and R. Paoletti. Raven Press, New York. Vol. 12, p. 153 (1983).
102. Bach, M.K., Brashler, J.R., Fritzpatrick, F.A., Griffin, R.L., Iden, S.S., Johnson, H.G., McNee, M.L., McGuire, J.C., Smith, H.W., Smith, R.J., Sun, F.F. and Wasserman, M.A.: *In*: Advances in Prostaglandin, Thromboxane and Leukotriene Research. Eds. B. Samuelsson, P. Ramwell and R. Paoletti. Vol. 11, p. 39 (1983).
103. Dawson, W., Boot, J.R., Harvey, J. and Walker, J.R.: Eur. J. Rheumatol. and Infl. *5*, 61 (1982).
104. Dahlén, S.-E., Hedqvist, P., Hammarström, S. and Samuelsson, B.: Nature, *288*, 484 (1980).
105. Peck, M.J., Piper, P.J. and Williams, T.J.: Prostaglandins, *20*, 863 (1980).

106. Bray, M.A., Cunningham, F.M., Ford-Hutchinson, A.W. and Smith, M.J.H.: Br. J. Pharmacol. *72*, 483 (1981).
107. Wedmore, C.V. and Williams, T.J.: Nature, *289*, 658 (1981).
108. Palmblad, J., Malmsten, C.L., Udén, A.-M., Rådmark, O., Engstedt, L. and Samuelsson, B.: Blood, *58*, 658 (1981).
109. Ford-Hutchinson, A.W., Bray, M.A., Doig, M.V., Shipley, M.E. and Smith, M.J.H.: Nature *286*, 264 (1980).
110. Malmsten, C.L., Palmblad, J., Udén, A.-M., Rådmark, O., Engstedt, L. and Samuelsson, B.: Acta Physiol. Scand. *110*, 449 (1980).
111. Palmer, R.M.J., Stephney, R.J., Higgs, G.A. and Eakins, K.E.: Prostaglandins, *20*, 411 (1980).
112. Goetzl, E.J. and Pickett, W.C.: J. Immunol. *125*, 1789 (1980).
113. Smith, M.J.H., Ford-Hutchinson, A.W. and Bray, M.A.: J. Pharm. Pharmacol. *32*, 517 (1980).
114. Serhan, C.N., Fridovich, J., Goetzl, E.J., Dunham, P.B. and Weissmann, G.: J. Biol. Chem. *257*, 4746 (1982).
115. Feinmark, S.J., Lindgren, J.Å., Claesson, H.-E., Malmsten, C. and Samuelsson, B.: FEBS Lett. *136*, 141 (1981).
116. Serhan, C.N., Radin, A., Smolen, J.E.: Korchak, H., Samuelsson, B. and Weissmann, G.: Biochem. Biophys. Res. Commun. *107*, 1006 (1982).
117. Blackwell, G.J., Garnuccion, R., Di Rosa, M., Flower, R.J., Parente, L. and Persico, P.: Nature, *287*, 147 (1980).
118. Hirata, F., Shiffmann, E., Venkatasubramanian, K., Salomon, D. and Axelrod, J.: Proc. Natl. Acad. Sci. USA, *77*, 2533 (1980).
119. Rådmark, O., Malmsten, C.L.: Samuelsson, B., Goto, G., Marfat, A. and Corey, E.J.: J. Biol. Chem. *255*, 11828 (1980).
120. Hansson, G. and Rådmark, O.: FEBS Lett, *122*, 87 (1980).
121. Hammarström, S. and Samuelsson, B.: FEBS Lett. *122*, 83 (1980).
122. Morris, H.R., Taylor, G.W., Piper, P.J., Samhoun, M.N. and Tippins, J.R.: Prostaglandins, *19*, 185 (1980).
123. Parker, C.W., Falkenhein, S.F. and Huber, M.M.: Prostaglandins, *20*, 863 (1980).
124. Bach, M.K., Brashler, J.R., Hammarström, S. and Samuelsson, B.: J. Immunol. *125*, 115 (1980).
125. Bach, M.K., Brashler, J.R., Hammarström, S. and Samuelsson, B.: Biochem. Biophys. Res. Commun. *93*, 1121 (1980).
126. Lewis, R.A., Drazen, J.M., Austen, K.F., Clark, D.A. and Corey, E.J.: Biochem. Biophys. Res. Commun. *96*, 271 (1980).
127. Ford-Hutchinson, A.W., Bray, M.A., Cunningham, F.M., Davidson, E.M. and Smith, M.J.H.: Prostglandins, *21*, 143 (1981).
128. Siegel, M.I., McConnell, R.T., Bonser, R.W. and Cautrecasas, P.: Prostaglandins, *21*, 123 (1981).
129. Doig, M.V. and Ford-Hutchinson, A.W.: Prostaglandins, *20*, 1007 (1980).
130. Rouzer, C.A., Scott, W.H., Cohn, Z.A., Blackburn, P. and Manning, J.M.: Proc. Natl. Acad. Sci. USA, *77*, 4928 (1980).
131. Lewis, R.A., Austen, K.F., Drazen, J.M., Clark, D.A., Marfat, A. and Corey, E.J.: Proc. Natl. Acad. Sci. USA, *77*, 3710 (1980).
132. Morris, H.R., Taylor, G.W., Piper, P.J. and Tippins, J.R.: Nature, *285*, 104 (1980).
133. Hoglum, J., Pai, J.-K., Atrache, V., Sok, K.-E. and Sih, C.J.: Proc. Natl. Acad. Sci. USA, *77*, 5688 (1980).

John Vane.

JOHN R. VANE

I was born in Tardebigg, Worcestershire, on the 29th March 1927, one of three children, with an elder sister and brother. My father, Maurice Vane, was a son of immigrants from Russia and my mother, Frances Vane, came from a Worcestershire farming family.

We lived in a suburb of Birmingham where I attended the local state school from the age of five. I then went on to King Edward VI High School in Edgbaston, Birmingham. However, the war was beginning and the whole school was evacuated into the countryside, alongside Repton School in Derbyshire. The expected bombings did not take place, and early in 1940 the school moved back to Birmingham. The air raids then started, and for the next four years, my school and home life were coloured by the trappings of war. With my family, I spent nights in the air-raid shelter at the bottom of the garden and at school we firewatched and trained as (or pretended to be) young soldiers.

At the age of 12, my parents gave me a chemistry set for Christmas and experimentation soon became a consuming passion in my life. At first, I was able to use a Bunsen burner attached to my mother's gas stove, but the use of the kitchen as a laboratory came to an abrupt end when a minor explosion involving hydrogen sulphide spattered the newly painted decor and changed the colour from blue to dirty green!

Shortly afterwards, my father, who ran a small company making portable buildings, erected a wooden shed for me in the garden, fitted with bench, gas and water. This became my first real laboratory, and my chemical experimentation rapidly expanded into new fields.

At High School I progressed through the pure sciences, and in 1944 it seemed natural to move to the University of Birmingham (which was just across the road from the school) to study Chemistry. However, the enthusiasm with which I had approached experimentation in Chemistry in the garden shed was soon dampened, for at university experimentation was nonexistent. The only unknown in the practical class was the percentage yield in the chemical synthesis involved. It was, I suppose, at this stage that I began to realise that my interest lay not in chemistry but more in experimentation. Thus, when Maurice Stacey, the Professor of Chemistry, asked me what I wanted to do when I graduated, I said "anything but chemistry". Stacey then told me that he had received a letter that morning from Professor Harold Burn in Oxford asking whether he could recommend another young chemist (he had sent one the previous year) to go to Oxford to be trained in pharmacology. Without hesitation I grasped the opportunity and immediately went to the library to find out what pharmacology was all about! That brief exchange with Stacey reshaped my whole career.

I went to Burn's department in 1946. I had no biological training of any sort and very little motivation. I found inspiration in working with him and caught his enthusiasm for pharmacology. If anyone can be said to have moulded the subject of pharmacology around the world, it is he. He did this through his particular style of research, through the lucidity of his writings, but most of all through the school which he founded. Young, impressionable scientists from various disciplines and older, less impressionable pharmacologists all came to work with him. His laboratory gradually became the most active and important centre for pharmacological research in the U.K. and the main school for training of young pharmacologists. It was his energy and inspiration that set my career into one of adventure in the fields of bioassay and pharmacology. It was Burn who reinforced for me the essence of experimentation and that is, never to ignore the unusual.

After qualifying for a B.Sc. in pharmacology, I spent a few months in Sheffield University as a research worker in the pharmacology department but then went back to Oxford to the Nuffield Institute for Medical Research in order to study for a D. Phil. with Dr. Geoffrey Dawes. In 1951 I was awarded the Stothert Research Fellowship of The Royal Society and this enabled me to complete my doctorate in 1953. Oxford was also an important milestone for it was there that my wife and I made our first home, and it was there that my daughters Nicola and Miranda were born.

In 1953, we all went to Newhaven, Connecticut where, at the invitation of Dr. Arnold Welch, who was then Chairman, I joined the Department of Pharmacology at Yale University as Assistant Professor in Pharmacology. That was a lively and bustling department, but after 2 years we returned to the U.K., where I started work with Professor W. D. M. Paton at the Institute of Basic Medical Sciences of the University of London in the Royal College of Surgeons of England. This was an unusual department, for the teaching was only for graduates, and was not time consuming, thus offering plenty of time for research. I stayed there for 18 years, progressing from Senior Lecturer to Reader to Professor of Experimental Pharmacology. From 1961 to 1973, Professor G. V. R. Born, a close friend from my Oxford days, was the Chairman of the Department and we enjoyed a strong symbiotic relationship, each maintaining an active group of graduate students and research workers. Interestingly, our fields of research endeavour (platelets and prostaglandins) only coalesced in a significant way after we had both moved on.

It was here that I developed, together with my group, the cascade superfusion bioassay technique for measurement of, dynamically and instantaneously, the release and fate of vasoactive hormones in the circulation or in the perfusion fluid of isolated organs. In the mid-1960's, our attention was focused on prostaglandins, leading in 1971 to the forging of the link between aspirin and the prostaglandins.

In 1973, I was offered the position of Group Research and Development Director for The Wellcome Foundation. In making my decision, I was conscious that Henry Wellcome, seventy years before, had recruited Henry Dale to work in (and soon to direct) the Wellcome Physiological Research Laboratories, the forerunners of the present Research and Development Directorate. When Henry Dale, then at Cambridge, first received the offer from Wellcome, he hesitated over accepting it. "Friends to whom I mentioned this approach" he said, "were almost unanimous in advising me to have nothing to do with it. I should be selling my scientific soul for a mess of commercial potage". Nevertheless, he accepted and had no regrets. I also found amongst a few of my friends a resistance to the idea of me entering into industrial science. It was as if to say that good science can only be promulgated in academia. Those friends were wrong; like Dale I accepted and had no regrets. I took with me from the Royal College of Surgeons a nucleus of colleagues, and this has expanded over the last few years into a Prostaglandin Research department under the leadership of Dr. Salvador Moncada. It was in this department that prostacyclin was discovered and its pharmacology developed.

Fellowships

1973 Honorary Member of the Polish Pharmacological Society
1973 Fellow of the Institute of Biology
1974 Fellow of the Royal Society
1977 Walter C. McKenzie Visiting Professorship, University of Alberta, Edmonton, Canada
1978 Honorary Fellowship of the American College of Physicians
1978 Member of the Royal Academy of Medicine of Belgium
1979 Foreign Member of the Royal Netherlands Academy of Arts & Sciences
1979 Visiting Professor, Harvard University, Cambridge, Mass., U.S.A.
1980 Foreign Member of the Polish Academy of Sciences
1982 Foreign Honorary Member of the American Academy of Arts and Sciences, U.S.A.
1982 Honorary Fellowship of the Swedish Society of Medical Sciences
1983 Foreign Associate of the National Academy of Sciences, U.S.A.

Honorary degrees

1977 D. Med. (Hon. Causa) Copernicus Academy of Medicine, Cracow
1978 Doctor Hon Causa René Descartes University, Paris
1980 Doctor of Science (Hon. Causa) Mount Sinai Medical School, City University of New York, U.S.A.
1983 Doctor of Science, Aberdeen University

Medals, prizes and awards

1977 Baly Medallist of the Royal College of Physicians
1977 Albert Lasker Basic Medical Research Award
1979 Joseph J. Bunim Medal of the American Rheumatism Association
1980 Peter Debye Prize, University of Maastricht, Holland
1980 Nuffield Lecture & Gold Medal, Royal Society of Medicine, England
1980 Feldberg Foundation Prize
1980 Ciba Geigy Drew Award, Drew University, U.S.A.
1981 Dale Medallist, Society for Endocrinology

ADVENTURES AND EXCURSIONS IN BIO-ASSAY: THE STEPPING STONES TO PROSTACYCLIN

Nobel lecture, 8 December, 1982

by

JOHN R. VANE

Wellcome Research Laboratories, Langley Court,
Beckenham, Kent, U.K.

Physiology has spawned many biological sciences, amongst them my own field of pharmacology. No man has made a more important contribution to the fields of physiology and pharmacology than Sir Henry Dale (1875–1968, Nobel Laureate in Physiology or Medicine in 1936). Dale had a great influence not only on British pharmacology in general but also on my own scientific endeavours. Indeed, I can put forward a strong case for considering myself as one of Dale's scientific grandchildren. My early days as a pharmacologist were influenced not only by Dale himself but also by his school of colleagues, including Burn, Gaddum and von Euler. It was Burn who taught me the principles and practice of bioassay. Some of Gaddum's first publications were on the development of specific and sensitive methods for biological assay and he maintained a deep interest in this subject for the rest of his life (1). In 1964 he said "the pharmacologist has been a 'jack of all trades' borrowing from physiology, biochemistry, pathology, microbiology and statistics – but he has developed one technique of his own, and that is the technique of bioassay" (2).

Expensive, powerful and sophisticated chemical methods, such as gas chromatography and mass spectrometry, have been developed and perfected for detection and quantification of prostaglandins (PGs) and related substances. One should not forget, however, that starting with the discovery and isolation of prostaglandins by von Euler (3; see also Bergström in this volume), biological techniques and bioassay have contributed very substantially to the development of the field. Bioassay has provided crucial information on the role of the lungs in the removal of circulating prostaglandins (4), the participation of prostaglandins in inflammatory reactions (5, 6), the contribution of prostaglandins to the autoregulation and maintenance of blood flow to the kidney (7–9), the inhibitory effect of aspirin-like drugs on the biosynthesis of prostaglandins (10–12), the mediation of pyrogen fever by prostaglandins (13), and the release of rabbit aorta-contracting substance (RCS; now identified as thromboxane A_2, TXA_2) from lungs during anaphylaxis (14, 15). Moreover, in 1976 bioassay made possible the discovery of PGX, now renamed prostacyclin (PGI_2), the latest member of the prostaglandin family (16–19). Indeed, it is doubtful whether the biological significance of any of the unstable products of arachidonic acid metabolism would have been recognised without bioassay techniques. With extraordinary simplicity and convenience, by its very nature, bioassay distinguishes between the important biologically active compounds and their closely related but biologically unimportant metabolites.

In this review I shall discuss the development of the cascade superfusion bioassay technique and some of the discoveries and concepts which arose from its application, leading up to the discovery and development of prostacyclin. The effects of prostacyclin in man and its clinical assessment (another application of bioassay) will also be discussed.

1. CASCADE SUPERFUSION BIOASSAY

a. Development

Most uses of bioassay involving smooth muscle demand high sensitivity and specificity. These aspects have been achieved first, by limiting the volume of fluid bathing the isolated tissue and second, by using an assay organ sensitive to, and relatively specific for, the test substances under study. Further specificity can be achieved by using a combination of several tissues which present a characteristic pattern of response to the test substance or substances. This takes advantage of the principle of parallel pharmacological assay, regarded by Gaddum (20) as strong evidence for the identity of a compound.

Magnus (21) introduced the idea of suspending an isolated portion of smooth muscle in a chamber containing a nutrient fluid and measuring changes in tissue tone. The organ baths used today are modified versions of that used by Dale (22). Gaddum (23) applied the experimental design developed by Finkleman (24) to the assay of minute amounts of biologically active substances. He called his technique "superfusion" in contrast to perfusion. This consisted of bathing an assay tissue with a stream of fluid which was momentarily stopped at the moment of addition of the test substance. Vane (25) introduced the idea of superfusing several tissues in cascade (generally up to six, arranged in two banks) (Fig. 1). Besides being useful for the parallel assay of individually injected samples, this arrangement also allows parallel analysis of the active components present in a fluid stream (most commonly Krebs' solution) taken from the outflow of a perfused organ.

Another innovation introduced by Vane (25) was to use blood as the superfusion medium (the blood-bathed organ technique). The anaesthetised animal is heparinised, and blood is continuously removed at a constant rate of 10–15 ml/min (dogs, cats and rabbits). Lower rates can be used from guinea pigs (26). The blood (either from a vein or from an artery) superfuses the assay tissues and is then returned by gravity to a large vein.

Plainly, when perfusate from an organ or blood from an animal is used for superfusion, substances can reach the assay tissues within a few seconds of generation or release. This element of "instantaneity" is an important aspect of cascade superfusion bioassay in that it detects the biological activity of chemically unstable compounds whose activity would otherwise be lost in an extraction process. Another important feature of the method is that it gives the maximum opportunity for serendipity. The dynamic nature of the assay also allows the measurement of inactivation of an infused substance across a particular vascular bed. Further modifications of the bioassay technique have been developed by Collier (27), Ferreira and Souza Costa (28) and Gryglewski and colleagues (29).

b. Choice of bioassay tissues

It is usually possible to find a piece of smooth muscle which is particularly sensitive to the hormone under investigation and relatively insensitive to other substances. Indeed, think of any part of the body and you can be sure that the pharmacologist has cut it out, put it into an isolated organ bath, or perfused its vessels in order to study the effects of drugs. For bioassay, segments of the gastro-intestinal tract or spirally cut strips of vascular tissue have mainly been used. Such procedures are the backbone not only of bioassay but also of classical pharmacology.

Figure 2 depicts the reactions of some superfused isolated tissues to various endogenous substances in concentrations likely to be found in circulating blood. It should be remembered that when blood is used as the superfusing medium, some smooth muscle preparations (but not others) exhibit an increased resting tone often lasting for the duration of the experiment. Such increased tone reduces sensitivity to substances which cause contraction but increases sensitivity to those which induce relaxation.

For detection of the classical prostaglandins, the most useful combination of assay tissues is the rat stomach strip, rat colon and chick rectum. For PG endoperoxides and later for prostacyclin, vascular tissues were added such as strips of coeliac or mesenteric artery (30). Strips of bovine coronary artery (31) are especially useful for they contract to PGE_2 but relax to prostacyclin (Fig. 2).

The specificity of a bioassay can be increased still further by the use of antagonists. For instance, contractions of the rat stomach strip induced by 5-hydroxytryptamine can be abolished by a specific antagonist such as methysergide, thereby leaving the preparation more specifically sensitive to the prostaglandins. The rat colon is relaxed by catecholamines but contracted by angiotensin II; when both are present in the superfusion fluid, the catecholamines reduce the contraction produced by angiotensin. This unwanted interference is prevented by blocking the actions of catecholamines with a β-receptor antagonist. When blood is used as the superfusion fluid, the antagonist can be perfused through the closed lumen (Fig. 1) of, say, the rat colon (32) thus localising the blocking agent to the assay tissue and minimising its effects on the whole animal. For a detailed discussion of the limitations of the cascade superfusion bioassay technique, the reader is referred to Vane (33) and Moncada, Ferreira and Vane (34).

c. Measurement of substances by cascade superfusion bioassay

The technique is well suited for measuring substances released into the circulation, such as catecholamines or angiotensin II, and also for determining the fate of substances released or infused into different parts of the circulation.

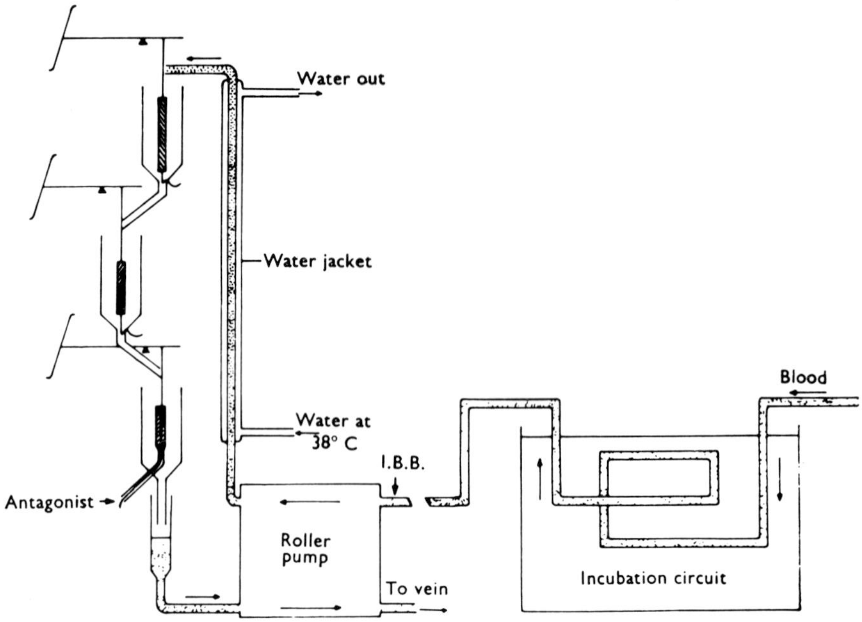

Figure 1. Diagram of the blood-bathed organ technique. Blood is continuously withdrawn from a convenient vessel by a roller pump, kept at 37° C by a water jacket and then allowed to superfuse a series of isolated organs, the longitudinal movements of which are recorded. The blood is then collected in a reservoir and returned to the animal. In some experiments the blood flows through a length of silicone tubing in a water bath (incubating circuit) before superfusing the isolated tissues. Drugs can be applied directly to the isolated tissues by infusions or injections into the bathing blood (I. B. B.) or with a time delay into the incubating circuit (from Vane 1969, reference 33, by permission of The Macmillan Press Ltd.).

i. Release of substances in response to stimuli

Release of catecholamines from the adrenal medulla can be detected and quantitated by use of a rat stomach strip and chick rectum. This technique was used to demonstrate that circulating catecholamines appeared to play little or no part in arterial baroreceptor reflexes (25) and also that catecholamines are released into the circulation during anaphylaxis (35). Of the substances released during the anaphylactic reaction histamine, bradykinin and slow reacting substance in anaphylaxis (SRS-A) will all in turn release adrenaline when injected intravenously, although there may be species differences in the mechanisms of action and in the sensitivity of the adrenal medulla (36–38).

An early use of the blood-bathed organ technique was to show the sequential release of angiotensin II and catecholamines during haemorrhage (39). Release of bradykinin into the blood stream by the intravenous injection of kallikrein or by contact of the blood with glass was easily demonstrated (40), but we consistently failed to demonstrate with this technique the endogenous release of bradykinin by physiological manipulation. However, circulating kinins were demonstrated during hypotension due to haemorrhage in the dog, and the concentrations detected in the bloodstream were sufficient to lower a normal blood pressure (41).

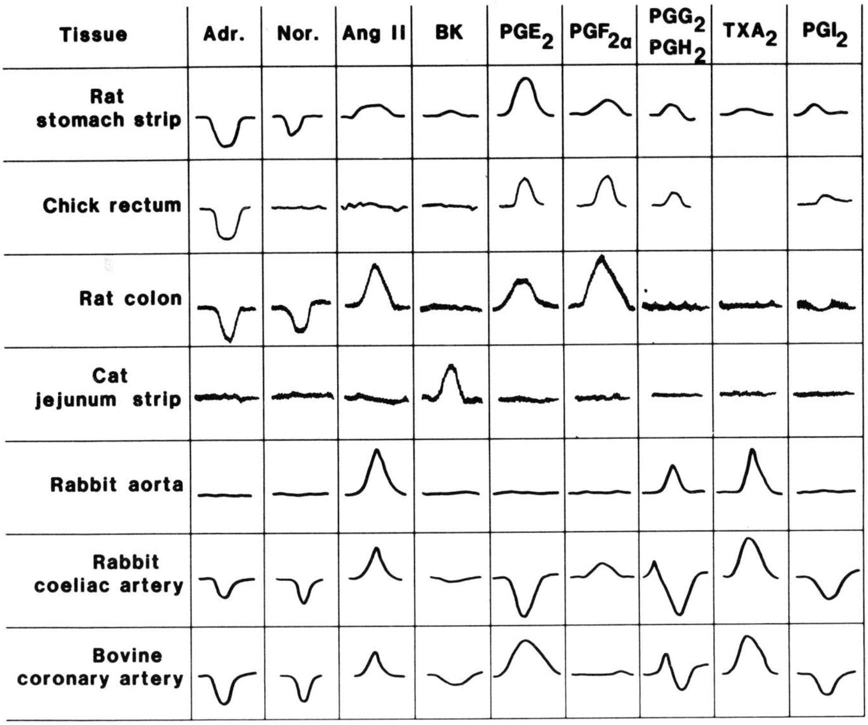

Figure 2. Diagram showing the reactions of some blood bathed organs to various endogenous substances in concentrations of 0.1–5.0 ng/ml. The actions of the catecholamines can be abolished by treatment with suitable blocking agents.

ii. Inactivation of circulating vasoactive substances

By comparing the effects on the blood-bathed assay organs of infusions of a substance over several minutes into a particular vascular bed with the effects of similar infusions given into the effluent of the vascular bed, it is possible to assay the percentage of the substance disappearing in one circulation through that vascular bed (Fig. 3). This technique allowed the demonstration of the inactivation of several vasoactive substances as they passed through vascular beds such as the hind legs, the liver and the lungs.

Angiotensin II, for example, is unaffected by passage through the pulmonary circulation, either *in vivo* or in isolated lungs of all species studied, including rats, dogs, guinea pigs, cats and man (for review see Bakhle and Vane, 42). However, the same substance is substantially inactivated (50–70 %) in one passage through peripheral vascular beds such as the liver, the kidneys and the hind legs (43).

We also studied the fate of adrenaline and noradrenaline and found in both cats and dogs that 70–95 % of an intra-arterial infusion of adrenaline or noradrenaline disappeared in one passage through the hind quarters. The lungs, however, inactivated up to 30 % of an infusion of noradrenaline without interfering with the passage of adrenaline (44, 45).

Bradykinin is fairly rapidly destroyed in blood and has a half-life in the blood stream of cat or dog of about 17 seconds. Ferreira and Vane (46) showed that whereas the liver inactivated about 50 %, the lungs inactivated about 80 % of the bradykinin infusion.

It was observations such as these that drew our attention to the metabolic and pharmacokinetic function of the pulmonary circulation. The selectivity of the pulmonary inactivation mechanism is strikingly demonstrated by the way in which the lungs inactivate bradykinin but allow other peptides such as eledoisin, substance P, physalaemin, vasopressin and oxytocin to pass through without change (42).

The metabolism of prostaglandins in the pulmonary circulation *in vivo* was first studied by Vane and his colleagues (4, 47) who showed that almost all of an infusion of PGE_1, PGE_2 or $PGF_{2\alpha}$ was inactivated in one passage through the lungs. McGiff et al. (48) confirmed that PGE_1 and PGE_2 were avidly removed by dog lung *in vivo* but further showed that PGA_1 and PGA_2 survived the passage through the lungs without change. Thus, within this very closely related group of substances, the inactivation process can distinguish between the individual members. Interestingly, after the discovery of prostacyclin, we also found (49) that prostacyclin (unlike PGE_2, see Fig. 4) survived passage through the pulmonary circulation without change. In other vascular beds, the inactivation of prostacyclin in a single passage (50–70 %) was comparable to that of PGE_2. Thus, the hind quarters, and particularly the liver, removed some of the prostacyclin which reaches those beds.

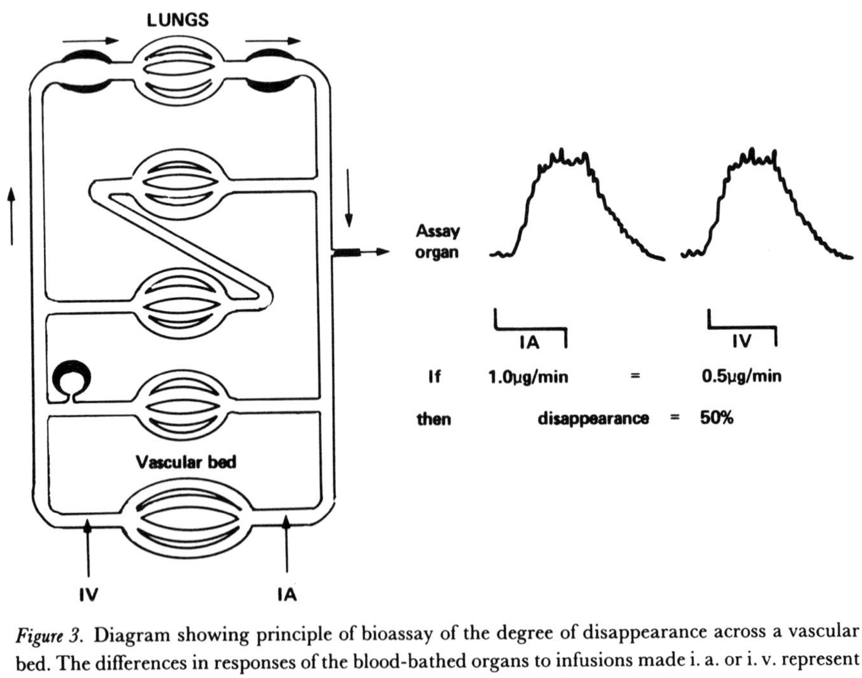

Figure 3. Diagram showing principle of bioassay of the degree of disappearance across a vascular bed. The differences in responses of the blood-bathed organs to infusions made i. a. or i. v. represent the degree of removal of the substance in one passage through the vascular bed.

Overall, the inactivation of prostacyclin in one circulation was about 50 % giving a metabolic half life of one circulation time (*c* 30 sec) as compared with the chemical half life of 2–3 min. The inactivation mechanisms for PGE_2 and prostacyclin (50) are similarly dependent on PG 15-hydroxydehydrogenase, (PGDH). However, the disappearance of prostaglandins in the pulmonary circulation depends on two mechanisms, namely uptake and enzyme attack by PGDH, so our results suggested that prostacyclin is not a substrate for the PG uptake mechanism.

From the differential removal of vasoactive hormones by the pulmonary circulation we proposed that they could be divided into at least two types— "local" and "circulating" hormones. The local hormones are those which are largely removed by the lungs and if they have a physiological function, it is probably localised at or near to the site of release. It is intriguing to think that venous blood may be full of potent, as yet unidentified, chemicals released by peripheral vascular beds but removed by the lungs before they can cause effects in the arterial circulation. Interestingly, in 1970, Gryglewski and Vane described the release of an unidentified substance into the venous blood after infusion of isoprenaline into the hind legs of dogs (51). The pattern of activity of this substance on the blood-bathed assay tissues was unlike that of any prostaglandin known at that time, but it can now be clearly identified as prostacyclin.

Circulating hormones are those which pass through the lungs, either unchanged (adrenaline, histamine, vasopressin, prostacyclin) or with an actual increase in activity. One such demonstration of an increase in activity on passage through the pulmonary circulation was associated with the renin-angiotensin system. We showed (52) that, contrary to popular belief, conversion of angiotensin I to angiotensin II did not take place in the bloodstream, but was largely accomplished in the pulmonary circulation. This was demonstrated both *in vivo* (52–54) and *in vitro* (Fig. 5).

iii. Our studies of substances released during anaphylaxis

Piper and Vane (14, 55, 56) made a series of studies in which they investigated the release of mediators from perfused lungs isolated from sensitised guinea pigs. We found, as expected, that there was a large release of histamine when the lungs were challenged with antigen. As also expected, we found a release of SRS-A. However, we were excited at that time to find the release of three other substances which had not previously been associated with anaphylaxis (Fig. 6). We detected the release of prostaglandin-like substances with our bioassay system and were later able to show by thin layer chromatography that prostaglandins E_2 and $F_{2\alpha}$ were present in the effluent. Even more exciting was the detection of the release of a previously undescribed substance which caused a strong contraction of strips of rabbit aorta and which we called, because of this effect "rabbit aorta contracting substance" or RCS. Two properties of RCS intrigued us. First, it was chemically unstable and if we introduced a delay coil of a few minutes before the lung effluent reached the assay tissues the activity had disappeared, although that of histamine, PGE_2, $PGF_{2\alpha}$ and SRS-A was still present. Second, we found that the release of RCS during anaphylaxis was

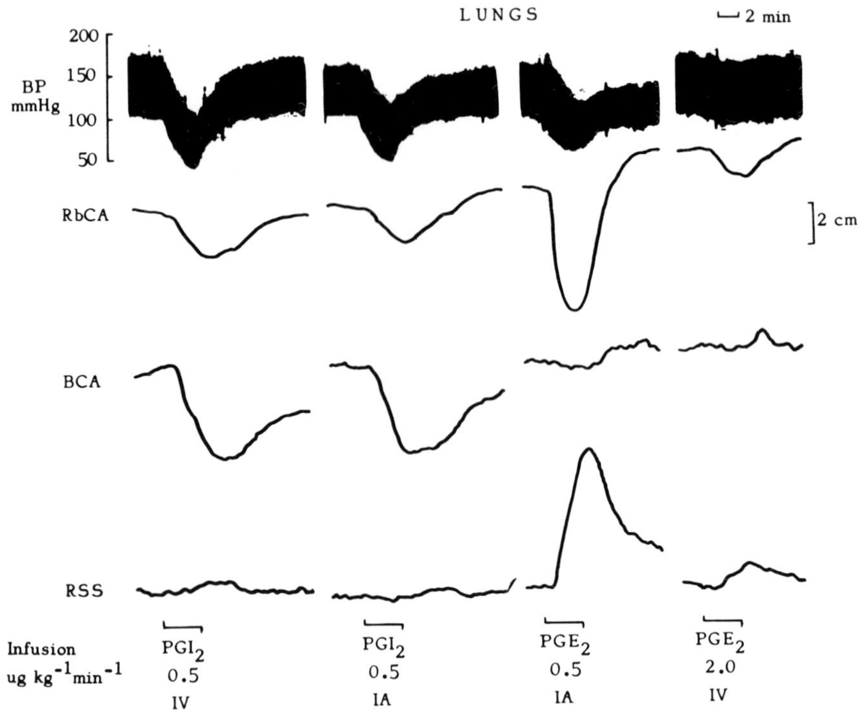

Figure 4. Passage of prostacyclin (PGI_2) and prostaglandin E_2 (PGE_2) across the lungs. Spiral strips of rabbit coeliac artery (RbCA), bovine coronary artery (BCA) and rat stomach strip (RSS) were bathed in arterial blood from a dog. PGI_2 infused intravenously (i. v.) caused similar effects on the bioassay tissues and blood pressure (BP) as infusion into the root of the aorta (i. a.) indicating that PGI_2 did not disappear across the lungs. In contrast, much more than 75 % of PGE_2 was inactivated in passage through the pulmonary circulation. (From Dusting, Moncada and Vane, 1978 (reference 49), by permission of The Macmillan Press Ltd.).

selectively prevented by aspirin and other similar compounds. Piper and Vane (14) postulated that RCS may be involved in causing those symptoms which aspirin relieves. Isolated lungs also released prostaglandins into the perfusate when particles (up to 120 microns) were infused into the pulmonary artery (57).

At this time we were also becoming interested in prostaglandin release from other tissues. Mammalian cells of all types disgorge prostaglandins at the slightest provocation, but the tissue content of prostaglandins is very low compared with the release. This is well illustrated in the dog spleen, from which less than 1 μg (or 7 ng per g wet weight) can be extracted; however, the spleen can release up to 10 μg of prostaglandin (assayed as PGE_2) per minute when it is stimulated. Horton and his colleagues (58) were the first to demonstrate that PGE_2 was released into splenic venous blood following splenic nerve stimulation in the dog. We became interested in the characteristics of this release and further showed that the output (PGE_2 and $PGF_{2\alpha}$) was associated with contraction of the spleen, for it could be induced by adrenaline and

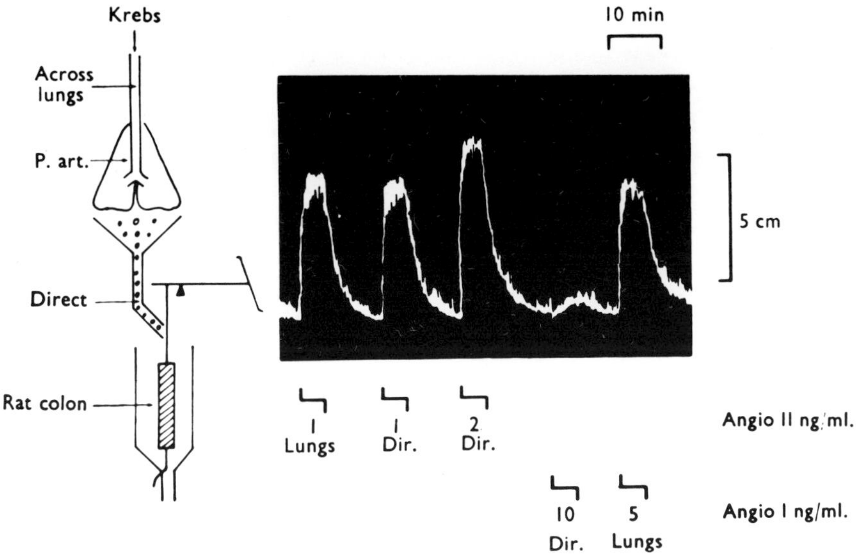

Figure 5. Increased activity of angiotensin I when passed through guinea pig isolated lungs. The diagram on the left shows the experimental procedure. Infusions of angiotensin II (Angio II) through the lungs or direct to the rat colon (Dir.) gave the same response of the rat colon, showing that there was no destruction in the lungs. An infusion of angiotensin I (Angio I) at 10 ng/ml direct to the rat colon gave a minimal response, but when half this concentration (5 ng/ml) was infused through the lungs there was a much greater contraction of the rat colon. Time, 10 min, vertical scale 5 cm. (From Vane, 1969, reference 33, by permission of The Macmillan Press Ltd.)

prevented by α-adrenoreceptor antagonists. Interestingly, as in the lungs, we found that prostaglandins were released by the spleen in response to infusions of particles (59).

2. ASPIRIN AND PROSTAGLANDIN BIOSYNTHESIS

In research there is always a "climate" of experience which acts as a background to important discoveries. I have tried to indicate the "climate" in our laboratory at The Royal College of Surgeons of England around the year 1970. We had a major interest in the release and fate of vasoactive hormones and were pursuing this with especial reference to the lungs. We had discovered RCS as an unstable substance released from lungs during anaphylaxis, and we knew that its release was inhibited by aspirin and other asprini-like drugs. We had become interested in the prostaglandins and had come to the idea that any tissue which was distorted or disturbed or traumatised would release prostaglandins. In this context, it seemed to me that each distension of the lungs might cause a prostaglandin release which could help to adjust the regional pulmonary blood flow. Indeed, the idea that a prostaglandin release might be important in controlling regional blood flow in the lungs had also been suggested by Liljestrand (60). With over-distension, any prostaglandins released might be detected in the arterial blood stream. I started a series of experiments using the blood-bathed organ technique to test this hypothesis. In anaesthetised dogs it was easy to show that when they were hyperventilated, there was an output into the arterial blood from the pulmonary circulation of an RCS-like substance and of PGE_2 and $PGF_{2\alpha}$ (Vane, unpublished). It was at this time that I was impressed by the effects of an infusion of aspirin into the hyperventilated dog, for not only was the associated hypotension reduced, but there was also a strong inhibition of the prostaglandin release. It was this experiment that led me to the idea (over a weekend) that aspirin might be interfering with prostaglandin biosynthesis. On the Monday morning I said to Sergio Ferreira and Priscilla Piper "I think I know how aspirin works" and set about doing an experiment. Änggård and Samuelsson (61) had described a preparation of guinea pig lungs in which a crude cell-free homogenate was used to convert arachidonic acid into PGE_2 and $PGF_{2\alpha}$. Although I was inexperienced in working with biochemical techniques, for I have always believed in using whole animals or organs whenever possible, I homogenised some guinea pig lungs, spun off the cell debris, divided the supernatant into test-tubes, added arachidonic acid and measured by bioassay the amounts of PGE_2 and $PGF_{2\alpha}$ formed. To some of the tubes was added aspirin, indomethacin or morphine. By the end of that day I was convinced that aspirin and indomethacin (but not morphine) strongly inhibited the formation of prostaglandins from arachidonic acid (see Fig. 7).

During the time that I was confirming and extending this first experiment, Ferreira and Moncada (12) began to study the effects of aspirin and indomethacin on prostaglandin release from the spleen (Fig. 8). Independently of these observations, Smith and Willis were using platelets to measure the effects of aspirin on prostaglandin formation. The results of these three studies were published simultaneously in 1971 (10–12). Vane (10) developed the hypothesis that this biochemical intervention in prostaglandin formation by the aspirin-like drugs is the basis of their therapeutic action.

An explanation of the therapeutic action of aspirin and its congeners had long been sought in terms of inhibition of a specific enzyme or biological function. Although these drugs inhibited a wide variety of enzymic reactions *in vitro* no convincing relationship could be established between such inhibition and their known anti-inflammatory, antipyretic and analgesic actions. This was largely because of the high concentrations needed for enzyme inhibition. At the time that we discovered that aspirin-like drugs inhibited the biosynthesis of prostaglandins in low concentrations, there was some evidence that the prostaglandins participated in the pathogenesis of inflammation and fever, and this reinforced the suggestion that inhibition of prostaglandin biosynthesis could explain the clinical action of these drugs. In the years which have elapsed since the original observations, a considerable body of evidence has accumulated which supports this hypothesis. Our knowledge about the inflammatory process has also increased, and the way in which prostaglandins participate in this process has been considerably clarified.

Inhibition of prostaglandin biosynthesis is a property peculiar to the aspirin-like drugs, since many otherwise pharmacologically active agents are inactive

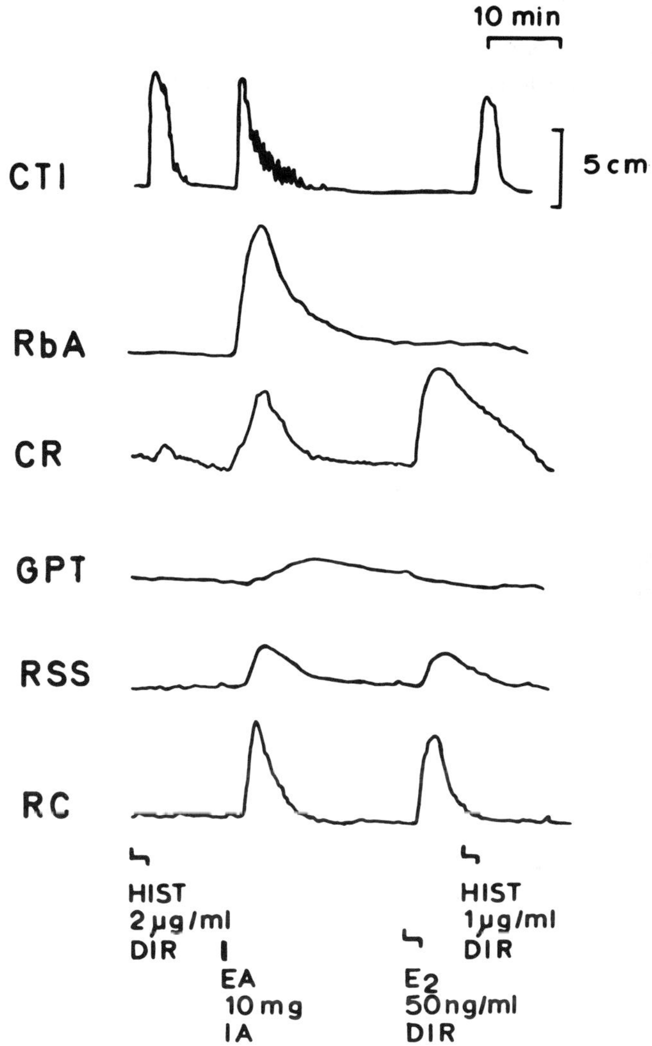

Figure 6. Release of mediators from isolated lungs of sensitised guinea pigs. The lungs were perfused through the pulmonary artery with Krebs' solution and the effluent superfused a cat terminal ileum (CTI), rabbit aorta spiral strip (RbA), chick rectum (CR), guinea pig trachea (GPT), rat stomach strip (RSS) and rat colon (RC). All tissues except CTI were blocked with antagonists to 5HT, catecholamines and histamine. Infusions of histamine (2 μg and 1 μg/ml DIR) and of prostaglandin E_2 (E_2, 50 ng/ml DIR) directly to the assay tissues demonstrated the selective sensitivity of the assay system. Anaphylaxis was induced in the lungs by injecting ovalbumen intra-arterially (EA 10 mg I.A.). Contractions of CTI demonstrated release of histamine, RbA release of RCS, GPT release of SRS-A and CR, RSS and RC release of prostaglandins. Time 10 min; vertical scale 5 cm. (Piper and Vane, published in Vane, 1971, reference 163, by permission of The Ciba Foundation).

against this enzyme system, including the opiates, antihistamines, α-and β-adrenoreceptor blocking agents and antagonists of acetylcholine and 5HT. The anti-inflammatory steroids are also inactive against this enzyme although they

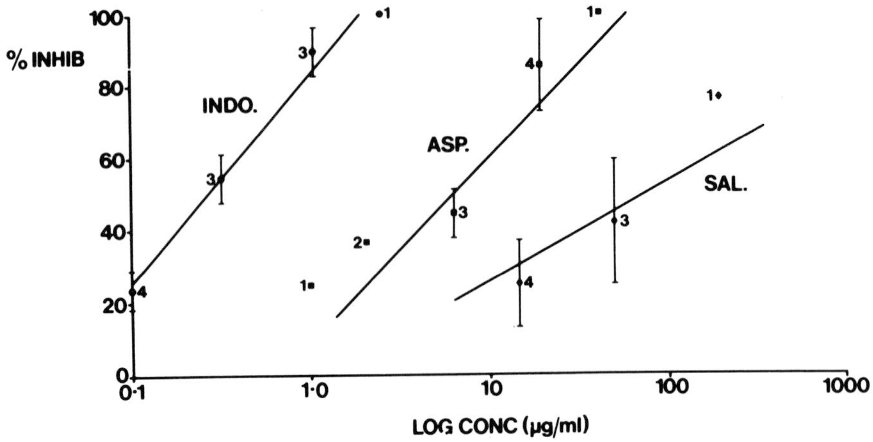

Figure 7. Concentration (μg/ml) of indomethacin (●), aspirin (■) and salicylate (◆) plotted on a log scale against the percentage inhibition of prostaglandin synthesis (assayed as $PGF_2\alpha$ on rat colons). The lines are those calculated for best fit. Numbers by the points indicate number of experiments. When three or more estimates were averaged, the standard error of the mean is shown. (From Vane, 1971, reference 10. Reprinted by permission from *Nature*. Copyright (c) 1971 Macmillan Journals Limited).

can reduce prostaglandin production by inhibition of phospholipase A_2 (for review see Flower, Blackwell, Di Rosa and Parente, 62).

Any hypothesis which purports to explain the action of a drug in terms of an anti-enzyme action must satisfy at least two basic criteria. First, the free concentrations achieved in plasma during therapy must be sufficient to inhibit the enzyme in question. Second, there must be a reasonable correlation between the level of anti-enzyme activity and the therapeutic potency. Clearly, there is abundant evidence to show that both these criteria are satisfied and there is also good evidence that therapeutic dosage reduces prostaglandin biosynthesis in man (for review, see Vane, Flower and Salmon, 63).

3. PROSTACYCLIN

a. The advent of the prostaglandin endoperoxides and the discovery of prostacyclin

The isolation by Samuelsson and others of the prostaglandin endoperoxides in the early 1970's was a major step forward in prostaglandin research (see Samuelsson in this volume). The demonstration that the endoperoxides caused platelet aggregation and that they were transformed in platelets to TXA_2 led us and others to the conclusion that most of the activity associated with RCS was due to TXA_2 (64, 65) (Fig. 9).

From Samuelsson's work we knew that TXA_2 could be released by platelets. We isolated the enzyme from the "microsomal" fraction of platelets and showed by our bioassay techniques that endoperoxides, when incubated with this fraction (even at 0°C), were rapidly transformed into TXA_2 which potently contracted rabbit aorta and induced platelet aggregation. This enzyme, which

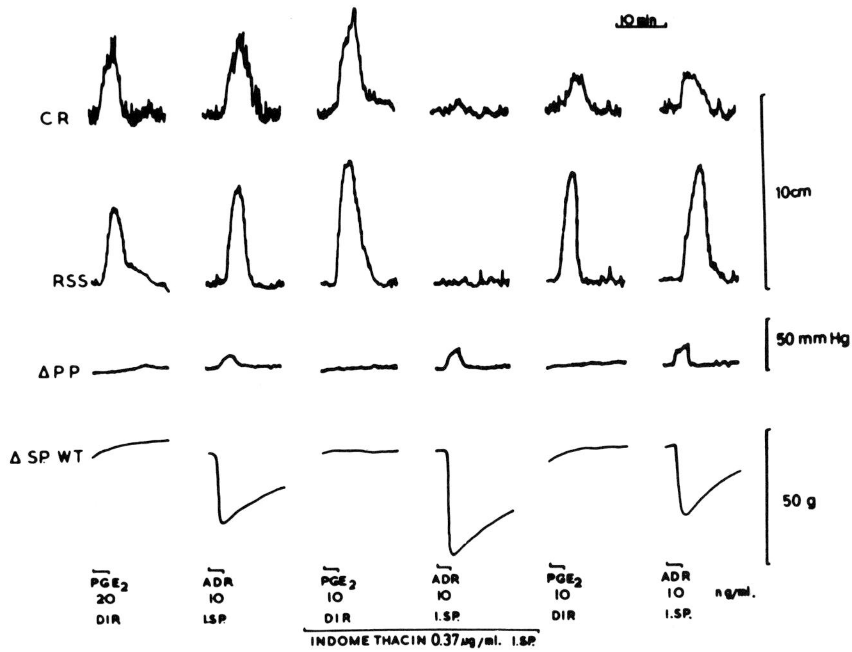

Figure 8. A spleen from a dog was perfused with Krebs'-dextran solution at a rate of 20 ml/min. A continuous sample (10 ml/min) of the splenic outflow, with antagonists to histamine, 5HT and catecholamines added, was used to superfuse the assay tissues. The figure shows the effects of prostaglandins on a chick rectum (CR; top) and a rat stomach strip (RSS). The next two tracings (bottom) show changes in perfusion pressure (PP) and spleen weight (SP.wt.). Except when infused into the spleen indomethacin was added to the splenic outflow to give a concentration of 0.37 μg/ml. The first panel shows contractions of CR and RSS induced by prostaglandin E_2 (20 ng/ml DIR). Next an adrenaline infusion into the spleen (ADR 10 ng/ml I. SP) induced a rise in perfusion pressure, a fall in spleen weight and an output of prostaglandins equivalent to PGE_2 at about 20 ng/ml. Indomethacin (0.37μg/ml) was then infused into the spleen. During the next 25 min the assay tissues relaxed (not shown) and were then more sensitive to PGE_2 (10 ng/ml DIR). Adrenaline (40 min after start of indomethacin) now caused a greater increase in perfusion pressure, a greater decrease in spleen weight, but no output of prostaglandin. After stopping the indomethacin infusion into the spleen, the reactivity of the assay tissues gradually decreased and the output of prostaglandin induced by adrenaline infusion into the spleen gradually returned. The adrenaline stimulation shown was made 70 min after stopping the indomethacin. (From Ferreira, Moncada and Vane, 1971, reference 12. Reprinted by permission from *Nature*. Copyright (c) 1971 Macmillan Journals Limited).

we called "thromboxane synthetase" (66, 67) is now an important therapeutic target for the development of compounds with anti-thrombotic potential.

Moncada, Gryglewski and Bunting then began to look at other tissues to determine whether they also could generate TXA_2. To do this, they took microsomal fractions of several different tissues and measured, again with the superfusion cascade bioassay, formation of either the classical stable prostaglandins E_2 and $F_{2\alpha}$, or of TXA_2. It was Moncada's suggestion that we should look into the biosynthetic system of vascular tissue, since vascular endothelium and platelets might share some structural features. Indeed, after several weeks of work we found that microsomal fractions of pig aorta incubated with the

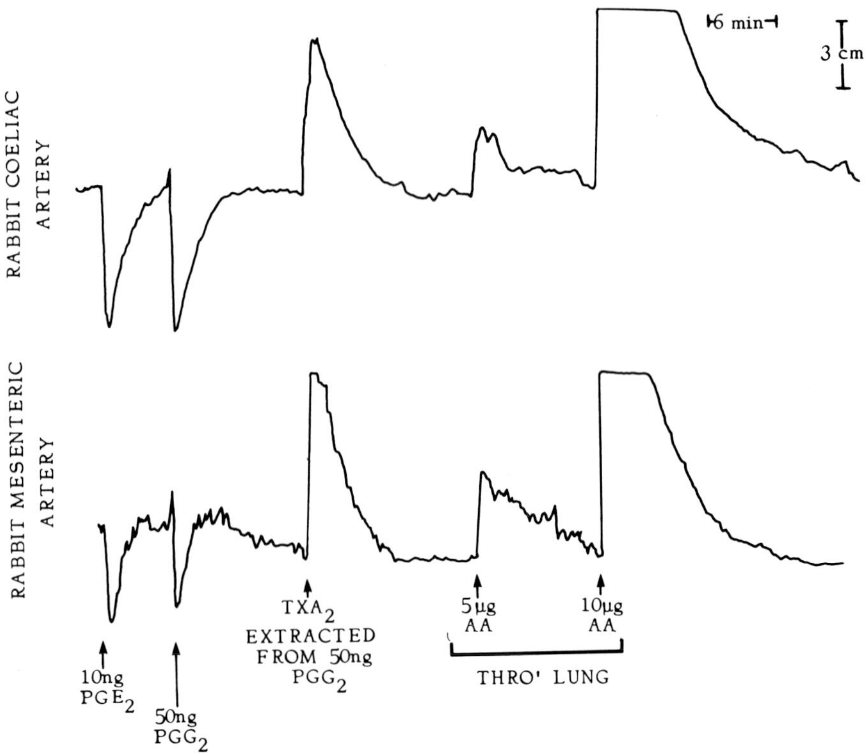

Figure 9. Rabbit coeliac and mesenteric artery strips were superfused with the outflow of a pair of guinea pig lungs perfused with Krebs' solution at 10 ml/min. Prostaglandin E_2 (10 ng) relaxes the two tissues as does PGG_2. Thromboxane A_2 generated from 50 ng PGG_2 produced a contraction. Challenge of the lungs with arachidonic acid (AA 5 and 10 µg) produces a dose-dependent release of TXA_2-like material. (From Moncada and Vane, 1977, reference 65, by permission of Academic Press Inc [New York]).

endoperoxide did not generate classical prostaglandins even though the endoperoxide activity (measured as RCS) disappeared. We eventually came to the conclusion that the endoperoxide was being transformed into an unknown prostaglandin and began to refer to this substance as PGX. By using recently-developed bioassay tissues such as the rabbit coeliac and mesenteric arteries (30) we were able to distinguish between the endoperoxides (which caused a biphasic effect) and PGX (which only relaxed them). Importantly, PGX, in contrast to TXA_2, inhibited the clumping of platelets. Like TXA_2, it was also unstable, with a half life of two minutes at 37°C. Boiling the solution for 15 seconds destroyed all measured activity.

The first paper on PGX was published by Moncada, Gryglewski, Bunting and Vane in Nature in October 1976 (16). Although the structure was then unknown, many of the characteristics of PGX were described, together with some important concepts. PGX was different from the other products of PG endoperoxides and its biological activity on isolated tissues, its instability and its potent anti-aggregatory activity, distinguished it from PGD_2, PGE_2, $PGF_{2\alpha}$,

TXA_2 and TXB_2. PGX relaxed strips of rabbit mesenteric and coeliac arteries, but contracted rat stomach strip, chick rectum, guinea pig tracheal chain and guinea pig ileum, although its contractile potency on these tissues was less than that of the classical prostaglandins. The rat colon was not contracted by PGX: indeed, spontaneous movement was decreased.

In this first paper, the transformation of the PG endoperoxides by platelets to TXA_2, which caused platelet aggregation and vascular contraction, was contrasted with their transformation by blood vessel microsomes to PGX, which had potent anti-aggregatory properties and relaxed vascular strips. Thus, the concept was suggested that a balance between the amounts of TXA_2 formed by platelets and PGX formed by blood vessel walls might be critical for thrombus formation. Indeed, in the light of the discovery of this anti-thrombotic property associated with arterial walls, we recalled the pre-Lister vitalistic view that in some way the arteries kept the blood fluid.

We also developed the concept that platelets attempting to stick to vessels may release endoperoxides which are then used by the blood vessel wall to generate PGX, thus limiting or preventing further platelet clumping. We also suggested that plaque formation on the arterial wall could hinder access of platelet endoperoxides to the PGX generating system. These important properties and concepts are now well established and developed (for review, see Moncada and Vane, 68).

The structure of PGX was established through a collaborative research programme between scientists at the Upjohn Company in Kalamazoo and The Wellcome Foundation Ltd. in Beckenham (19). This also led to the first chemical syntheses of the substance which was renamed prostacyclin (PGI_2) (19, 69).

There is now a plethora of names for prostacyclin. The chemical name is 5-{(1*S*,3*Z*,5*R*,6*R*,7*R*)-7-hydroxy-6-[(1*E*,3*S*)-3-hydroxy-1-octenyl]-2-oxabicyclo [3.3.0]oct-3-ylidene pentanoic}acid. As a freeze-dried pharmaceutical preparation, the approved name is Epoprostenol and the trade names are Flolan (Wellcome) and Cyclo-prostin (Upjohn). To maintain consistency in the scientific literature, the trivial name, prostacyclin, should be used whenever possible.

b. The formation and properties of prostacyclin

Prostacyclin is the main product of arachidonic acid in all vascular tissues so far tested including those of man (Fig. 10). The ability of the large vessel wall to synthesise prostacyclin is greatest at the intimal surface and progressively decreases toward the adventitia (70). Culture of cells from vessel walls also shows that endothelial cells are the most active producers of prostacyclin (71, 72).

Prostacyclin relaxes isolated vascular strips and is a strong hypotensive agent through vasodilation of all vascular beds studied, including the pulmonary and cerebral circulations. (For review, see Moncada and Vane, 73). Several authors have suggested that prostacyclin generation participates in or accounts for functional hyperaemia (74, 75).

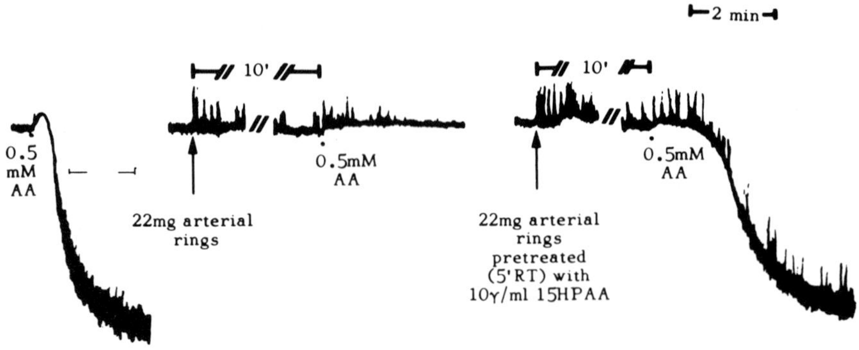

Figure 10. Inhibition of platelet aggregation by rings of human vascular tissue. Cut rings (15–30 mg), when incubated at 37° C in human platelet rich plasma (P.R.P.) for 10 min, inhibited the aggregation produced by 0.5 mmol arachidonic acid (AA). When the rings were pretreated by incubation with 15 hydroperoxyarachidonic acid (15-HPAA) (10 μg/ml) for 5 min at 22° C, and then added to the P.R.P., aggregation was once more observed after addition of 0.5 mmol AA. (From Moncada, Higgs and Vane, 1977, reference 164, by permission of The Lancet).

Prostacyclin is the most potent endogenous inhibitor of platelet aggregation yet discovered. This effect is short-lasting *in vivo,* disappearing within 30 minutes of cessation of intravenous administration. Prostacyclin disperses platelet aggregates *in vitro* (16, 76) and in the circulation of man (77). Moreover, it inhibits thrombus formation in models using the carotid artery of the rabbit (76) and the coronary artery of the dog (78), protects against sudden death (thought to be due to platelet clumping) induced by intravenous arachidonic acid in rabbits (79), and inhibits platelet aggregation in pial venules of the mouse when applied locally (80).

Prostacyclin inhibits platelet aggregation by stimulating adenylate cyclase, leading to an increase in cAMP levels in the platelets (81, 82). In this respect prostacyclin is much more potent than either PGE_1 or PGD_2, and its effect is longer-lasting. In contrast to TXA_2, prostacyclin enhances Ca^{++} sequestration in platelet membranes (83). Moreover, inhibitory effects on platelet phospholipase (84, 85) and platelet cyclo-oxygenase (86) have been described. All these effects are related to its ability to increase cAMP in platelets. Prostacyclin, by inhibiting several steps in the activation of the arachidonic acid metabolic cascade, exerts an overall control of platelet aggregability.

Prostacyclin increases cAMP levels in cells other than platelets (for review, see Moncada, 87) raising the possibility that in these cells a balance with the thromboxane system exerts a similar homeostatic control of cell behaviour to that observed in platelets. Thus, the prostacyclin/TXA_2 system may have wider biological significance in cell regulation. An example is that prostacyclin inhibits white cell adherence to the vessel wall (88, 89) to nylon fibres and to endothelial monolayers *in vitro* (90). Prostacyclin increases cAMP in the endothelial cell itself, suggesting a negative feedback control for prostacyclin production by the endothelium (91–93).

One of the functional characteristics of the intact vascular endothelium is its non-reactivity to platelets: clearly, prostacyclin generation could contribute to this thromboresistance. Moreover, prostacyclin inhibits platelet aggregation (platelet-platelet interaction) at much lower concentrations than those needed to inhibit adhesion (platelet-collagen interaction) (94). Thus, prostacyclin may permit platelets to stick to vascular tissue and to interact with it so allowing platelets to participate in the repair of a damaged vessel wall while at the same time preventing or limiting thrombus formation.

c. Prostacyclin and cytoprotection

In addition to its well-known vasodilator and anti-aggregating actions, prostacyclin shares with other prostaglandins a "cytoprotective activity", as yet not clearly defined. This activity has usually been studied on gastric ulcers (95). We have suggested (68, 96) that this third property may be important in explaining certain therapeutic effects of prostacyclin. For instance, in models of myocardial infarction, prostacyclin reduces infarct size (97–99), arrhythmias (100), oxygen demand (99) and enzyme release from infarcted areas (101). In sheep, prostacyclin protected the lungs against injury induced by endotoxin (102). There was also a beneficial effect in endotoxin shock in the dog (103) and cat (104) where prostacyclin improves splanchnic blood flow and reduces the formation and release of lysosomal hydrolases. The effects of hypoxic damage in the cat isolated perfused liver are also substantially reduced by prostacyclin (105). Canine livers can be preserved *ex vivo* for up to 48 hours and then successfully transplanted using a combination of refrigeration, Sacks' solution and prostacyclin (106).

All these effects could be related to a result obtained recently by Moncada and colleagues (107). The addition to platelets of prostacyclin during their separation from blood and subsequent washing substantially improves their immediate functionality *in vitro*. In addition, whereas platelets normally are functional for about 6 h, when prepared with the addition of prostacyclin they remain functional for more than 72 h (107). This extended viability of platelets *in vitro* is not accompanied by a prolonged increase in levels of cAMP, thus separating the effect from the classical anti-aggregating activity (108). Interestingly, there has been a study demonstrating a dissociation between anti-aggregating and cytoprotective effects of a prostacyclin analogue in a model of acute myocardial ischaemia (109).

All these results suggest that some of the therapeutic effects of prostacyclin might be related to this cytoprotective effect and point to even wider indications for prostacyclin in cell or tissue preservation *in vivo* and *in vitro*.

d. Prostacyclin and atherosclerosis

Lipid peroxides, such as 15-hydroperoxy arachidonic acid (15-HPAA), are potent and selective inhibitors of prostacyclin generation by vessel wall microsomes or by fresh vascular tissue (Fig. 10) (17, 18, 110, 111). There are high concentrations of lipid peroxides in advanced atherosclerotic lesions (112). Lipid peroxidation induced by free radical formation occurs in vitamin E deficiency, the ageing process and perhaps also in hyperlipidaemia accompanying atherosclerosis (113). Accumulation of lipid peroxides in atheromatous plaques could predispose to thrombus formation by inhibiting generation of prostacyclin by the vessel wall without reducing TXA_2 production by platelets. Moreover, platelet aggregation is induced by 15-HPAA, and this aggregation is not inhibited by adenosine or PGE_1 (114). Human atheromatous plaques do not produce prostacyclin (115, 116). In normal rabbits the production of prostacyclin by the luminal surface of the aorta is abolished by de-endothelialisation and slowly recovers with re-endothelialisation over a period of about 70 days. However, the recovery of prostacyclin formation did not occur in rabbits made moderately hypercholesterolaemic by diet (117). These results suggest that it would be worth exploring whether attempts to reduce lipid peroxide formation by inhibiting peroxidation influence the development of atherosclerosis and arterial thrombosis. Vitamin E acts as an antioxidant and perhaps its empirical use in arterial disease in the past (118–120) had, in fact, a biochemical rationale. For discussion of the implication of prostacyclin and TXA_2 in diseases other than atherosclerosis see Moncada and Vane (68).

e. Clinical applications of prostacyclin

Prostacyclin is available as a stable freeze-dried preparation (Epoprostenol) for administration to man. Intravenous infusion of prostacyclin in healthy volunteers leads to a dose-related inhibition of platelet aggregation, dispersal of circulating platelet aggregates, arteriolar vasodilatation, increases in skin temperature, facial flushing and sometimes headache (87, 121). Infusion of prostacyclin into patients susceptible to migraine or cluster headache induces, in most cases, a headache different from those usually experienced (122).

Extracorporeal circulation of blood brings it into contact with artificial surfaces which cannot generate prostacyclin. In the course of such procedures thrombocytopaenia and loss of platelet haemostatic function occur and make an important contribution to the bleeding problems following charcoal haemoperfusion and prolonged cardiopulmonary bypass in man. Formation of microemboli during cardiopulmonary bypass may also contribute to the cerebral complications which sometimes follow this procedure. Platelet damage and thrombocytopaenia were prevented by prostacyclin both in animal models of extracorporeal circulation (87, 121) and in man.

In patients with fulminant hepatic failure undergoing charcoal haemoperfusion (123) prostacyclin infusion prevented the fall in platelet count and elevation of β-thromboglobulin seen in the control patients. Gimson and his colleagues (124) have made almost 200 charcoal haemoperfusions on a daily basis using prostacyclin for platelet protection in the treatment of 76 patients with fulminant hepatic failure. Remarkable survival rates (65 %) were obtained in the 31 patients who had been referred early and in whom the serial haemoperfusions were started whilst the signs of grade III encephalopathy were still apparent (not rousable but may or may not respond to painful stimuli). The authors thought that this was probably the major factor in the improved survival rate, a reflection of the better biocompatibility of the system because

Figure 11. Filters. Electron micrographs of the downstream face of filters taken from the arterial lines of cardio-pulmonary bypass operations in man. The left hand picture (patient receiving placebo) shows formation of platelet aggregates, some clogging the pores (40 ± 5 μ) of the filter. The right hand picture (patient receiving prostacyclin) shows lack of platelet adhesion. The few cells sticking to the filter are leucocytes. (Electron micrographs kindly provided by Dr. N. Read and Mr. P. J. Astbury, Wellcome Research Laboratories).

prostacyclin was used, allowing the patients to be treated at an earlier stage. In the group treated later, with Grade IV encephalopathy already present, 20 % survived, so that the overall survival rate from the 76 patients was 38 %. These results (especially those treated early) compare favourably with a survival rate of 15 % in patients under standard intensive care measures.

Several double blind clinical trials of prostacyclin in cardiopulmonary by-pass have been published (125–131). The treatment groups showed a preservation of platelet number and function, with a reduction in the blood loss in the first 18 hours after operation. In the trial by Longmore and colleagues (130) the blood loss was halved. In that by Walker and co-workers (129), filters were used and the formation of platelet aggregates on the filters from the placebo group contrasted strikingly with the lack of platelet adhesion to those from patients treated with prostacyclin (Fig. 11). The heparin-sparing effect of prostacyclin was confirmed and the vasodilator effects were not troublesome; indeed, Nobak and colleagues (131) suggest that these effects may be utilised in controlling intra-bypass hypertension. Clearly, the use of prostacyclin or an analogue should allow improvements in the methodology of extracorporeal circulations.

Therapeutic assessment of prostacyclin is still in its infancy with many trials in progress. The results are, therefore, still preliminary, but nevertheless they point the way to conditions in which prostacyclin therapy may be useful. In open trials, prostacyclin was of benefit to patients with peripheral vascular disease both through relief of ischaemic pain and improved ulcer healing (132–137). Placebo-controlled blind trials are now in progress, and the results of the first to be analysed (138) are encouraging. In the 13 patients infused intravenously for 4 days with placebo, 3 showed reduction in rest pain at 5 days, 2 at 1 month and 1 at 6 months. After 6 months, 3 had died and 5 others had received surgical intervention. Of those 15 patients who were infused with prostacyclin for 4 days (average 7 ng/kg/min i.v.), at 5 days all had reduction in rest pain. At 1 month, 9 still showed a substantial improvement, which was also evident in 7 patients at 6 months. By this time, two other patients in the group had received surgical intervention and one had died. Zygulska-Mach and colleagues (139) infused prostacyclin into 3 patients with sudden blockage of central retinal veins. Improvement was observed in those two patients who were treated within the first 48 hours.

Prostacyclin also induces long-lasting improvements in patients with Raynaud's phenomenon. Intravenous infusion of the drug for 72 hours produced striking reductions in the frequency, duration and severity of the disease in 21 of 24 patients. The improvement lasted for a mean of 9–10 weeks, and in 3 patients, subjective improvement was still reported 6 months after the infusion. Pain relief was a striking feature presumably associated with the increased blood flow as indicated by increased temperature of the hands and fingers (140). Belch and coworkers (141) have also reported successful treatment in 4 out of 5 patients, and a double-blind clinical trial (142) has now confirmed these results in Raynaud's phenomenon. There was an overall improvement still present at 6 weeks in 6 of 7 patients receiving prostacyclin, but only in 1 of 7 receiving placebo. The prostacyclin patients had a significant fall in the number and duration of attacks over the 6 weeks period post infusion, whereas there was no change in the placebo group.

Gryglewski and his colleagues in Cracow, who first demonstrated the beneficial effects of infusion of prostacyclin in ischaemic disease of the legs, have now obtained dramatic improvements following prostacyclin infusion in 10 patients with ischaemic stroke (143). Patients with transient ischaemic attacks and haemorrhagic stroke were excluded. With prostacyclin treatment there was a reversal of symptoms strikingly sooner in all 10 patients than could have been expected and in 6 patients during the first 6 hour infusion. One patient died 2 weeks later of a second stroke, but the other 9 have maintained return of function for (so far) up to 6 months.

Prostacyclin has been successfully used in a few cases of pulmonary hypertension and is more effective than PGE_1 (144–146). Single case studies have suggested that prostacyclin may be useful in the treatment of patent ductus arteriosus (147) and pre-eclamptic toxaemia (148).

Beneficial effects of intravenous infusion of prostacyclin were obtained in 9 patients with severe congestive heart failure refractory to digitalis and diuretics (149). Mean pulmonary and systemic pressures and vascular resistances were reduced and heart rate, cardiac index and stroke index were all increased during the infusion, with facial flushing as the only side effect.

Bergman and colleagues (150) gave an intravenous infusion of prostacyclin to patients with coronary artery disease with no deleterious effects. Heart rate and cardiac index were increased and mean blood pressure, systemic and pulmonary resistance all fell. Mean atrial pacing time to angina rose from 142 to 241 seconds. They concluded that acute administration of prostacyclin was beneficial in angina, having effects similar to those of the short-acting nitrates. In 5 patients with coronary artery disease, prostacyclin was safely infused directly into diseased coronary arteries (151), and there was a beneficial effect of intravenous prostacyclin infusions in patients with unstable angina (137). However, prostacyclin had no effect on the number, severity and duration of ischaemic episodes in 8 of 9 patients with variant angina, although consistent relief was seen in the ninth patient (152).

A prostacyclin deficiency has been reported in thrombotic thrombocytopaenic purpura (TTP) (153). Infusion of prostacyclin into two patients with TTP did not produce an increase in circulating platelet count (153, 154). However, FitzGerald and colleagues (155) have reported an increase in platelet count and an improvement in the neurological status of one such patient during 18 days of prostacyclin infusion. They were sufficiently encouraged to conclude that the controlled evaluation of prostacyclin in TTP was warranted.

Infusion of prostacyclin protects transplanted kidneys from hyperimmune rejection in dogs (156) and in patients with chronic renal transplant rejection (157).

Clearly, there are many clinical conditions which may respond to prostacyclin treatment and its place (or that of chemically stable analogues) in therapeutics will be defined in the next few years. Some of these conditions are preeclamptic toxaemia (158), haemolytic uraemic syndrome (159), peptic ulceration (160), the thrombotic complications associated with transplant rejection (156), the prevention of tumour metastasis (161) and the treatment of pulmonary embolism (162).

Acknowledgments

Over the years I have had a remarkably talented and productive group of colleagues. They have come from many parts of the world, including Australia (Greg Dusting, Bob Hodge), Belgium (Arnold Herman), Brazil (Sergio Ferreira), Great Britain (Mick Bakhle, Rod Flower, Gerry and Annie Higgs, John Hughes, Robert Lowe, Priscilla Piper, John Salmon, Brendan Whittle, Ivor Williams), Holland (Franz Nijkamp), Honduras (Salvador Moncada), Italy (Domenico Regoli), Poland (Krystyna Herbaczynska-Cedro, Richard Gryglewski, Janina Staszewska-Barczak), Singapore (Kevin Ng), and the United States (Jim Aiken, Alan Block, Nobby Gilmore, Jack McGiff, Phil Needleman). Some of these have returned to my laboratory several times. Others have started as technicians and are still with me as Ph.Ds. All (and many others) have contributed to the work which I have described. In particular, the discovery of prostacyclin was made in the Prostaglandin Research Department of The Wellcome Research Laboratories under the direction of Dr. Salvador

Moncada and with the co-operation of Richard Gryglewski and Stuart Bunting. The further pre-clinical development of prostacyclin in our laboratories was also largely directed by Dr. Salvador Moncada. I would like to thank all of these friends and colleagues for their help and collaboration over the years. Their participation in the work described in this article is gratefully acknowledged.

I would also like to thank Mrs. A. Higgs and Mrs. A. Skinner for their help in the preparation of this manuscript.

REFERENCES

1. Feldberg, W. John Henry Gaddum, Biographical Memoirs of Fellows of the Royal Society, *13,* 57 (1967).
2. Gaddum, J. H. In: Drugs in our Society, pp. 17–26. The John Hopkins Press, Baltimore, Maryland. (1964).
3. von Euler, U. S. Biochim. Biophys. Acta *499,* 48 (1936).
4. Ferreira, S. H., and Vane, J. R. Nature (Lond). *216,* 868 (1967).
5. Willis, A. L., J. Pharm. Pharmacol. *21,* 126 (1969).
6. Willis, A. L. In: Prostaglandins, Peptides and Amines, pp. 31–38. Academic Press, New York. (1969).
7. Herbaczynska-Cedro, K., and Vane, J. R. Circ. Res. *33,* 428 (1973).
8. Lonigro, A. J., Itskovitz, H. D., Crowshaw, K., and McGiff, J. C. Circ. Res. *32,* 712 (1973).
9. Lonigro, A. J., Terragno, N. A., Malik, K. U., and McGiff, J. C. Prostaglandins *3,* 595 (1973).
10. Vane, J. R. Nature (New Biol). *231,* 232 (1971).
11. Smith, J. B., and Willis, A. L. Nature (New Biol). *231,* 235 (1971).
12. Ferreira, S. H., Moncada, S., and Vane, J. R. Nature (New Biol). *231,* 237 (1971).
13. Feldberg, W., Gupta, K. P., Milton, A. S., and Wendlandt, S., J. Physiol. (Lond). *234,* 279 (1973).
14. Piper, P. J., and Vane, J. R. Nature (Lond). *223,* 29 (1969).
15. Hamberg, M., Svensson, J., and Samuelsson, B. Proc. Natl. Acad. Sci. USA, *72,* 2994 (1975).
16. Moncada, S., Gryglewski, R. J., Bunting, S., and Vane, J. R. Nature (Lond). *263,* 663 (1976).
17. Gryglewski, R. J., Bunting, S., Moncada, S., Flower, R. J., and Vane, J. R. Prostaglandins *12,* 685 (1976).
18. Moncada, S., Gryglewski, R. J., Bunting, S., and Vane, J. R. Prostaglandins *12,* 715 (1976).
19. Johnson, R. A., Morton, D. R., Kinner, J. H., Gorman, R. R. McGuire, J. R., Sun, F. F., Whittaker, N., Bunting, S., Salmon, J., Moncada, S., and Vane, J. R. Prostaglandins *12,* 915 (1976).
20. Gaddum, J. H. Pharmacol. Rev. *11,* 241 (1959).
21. Magnus, R. Ergeb. Physiol. *2,* 637 (1903).
22. Dale, H. H., J. Pharmacol. Exp. Ther. *4,* 167 (1912).
23. Gaddum, J. H. Br. J. Pharmacol. Chemother. *8,* 321 (1953).
24. Finkleman, B. J. Physiol. (Lond). *70,* 145 (1930).
25. Vane, J. R. Br. J. Pharmacol. *23,* 360 (1964).
26. Palmer, M. A., Piper, P. J., and Vane, J. R. Br. J. Pharmacol. *49,* 226 (1973).
27. Collier, J. G. Br. J. Pharmacol. *44,* 383 (1972).
28. Ferreira, S. H., and Souza Costa, F. S. Eur. J. Pharmacol. *39,* 379 (1976).
29. Gryglewski, R. J., Korbut, R., and Ocetkiewicz, A. C. Nature (Lond). *273,* 765 (1978).
30. Bunting, S., Moncada, S., and Vane, J. R. Br. J. Pharmacol. *57,* 462P (1976).
31. Needleman, P., Bronson, S. D., Wyche, A., Sivakoff, M., and Nicolaou, K. C., J. Clin. Invest. *61,* 839 (1978).

32. Hodge, R. L., Lowe, R. D., and Vane, J. R., J. Physiol. (Lond). *185,* 613 (1966).
33. Vane, J. R. Br. J. Pharmacol. *35,* 209 (1969).
34. Moncada, S., Ferreira, S. H., and Vane, J. R. In: Advances in Prostaglandin and Thromboxane Research, Vol. 5. Raven Press, New York, pp. 211–236. (1978).
35. Piper. P. J., Collier, H. O., and Vane, J. R. Nature (Lond). *213,* 838 (1967).
36. Staszewska-Barczak, J., and Vane, J. R., J. Physiol. (Lond). *177,* 57P. (1965).
37. Staszewska-Barczak, J., and Vane, J. R. Br. J. Pharmac. Chemother. *25,* 728 (1965).
38. Staszewska-Barczak, J., and Vane, J. R. Br. J. Pharmac. Chemother. *30,* 655 (1967).
39. Regoli, D., and Vane, J. R., J. Physiol. *183,* 513 (1966).
40. Ferreira, S. H., and Vane, J. R. Br. J. Pharmac. Chemother. *29,* 367 (1967).
41. Berry, H. E., Collier, J. G., and Vane, J. R. Clin. Sci. *39,* 349 (1970).
42. Bakhle, Y. S., and Vane, J. R. Physiol. Rev. *54,* 1007 (1974).
43. Hodge, R. L., Ng, K. K. F., and Vane, J. R. Nature (Lond). *215,* 138 (1967).
44. Vane, J. R. Pharmac. Rev. *18,* 317 (1966).
45. Ginn, R. W., and Vane, J. R. Nature (Lond) *219,* 740 (1968).
46. Ferreira, S. H., and Vane, J. R. Br. J. Pharmac; Chemother. *30,* 417 (1967).
47. Piper, P. J., Vane, J. R., and Wyllie, J. H. Nature (Lond) *225,* 600 (1970).
48. McGiff, J. C., Terragno, N. A., Strand, J. C., Lee, J. B., Lonigro, A. J., and Ng. K. K. F. Nature *223,* 742 (1969).
49. Dusting, G. J., Moncada, S., and Vane, J. R. Br. J. Pharmacol. *64,* 315 (1978).
50. McGuire, J. C., and Sun, F. F. Arch. Biochem. and Biophysics *189,* 92 (1978).
51. Gryglewski, R., and Vane, J. R. Br. J. Pharmacol. *39,* 573 (1970).
52. Ng, K. K. F., and Vane, J. R. Nature (Lond). *216,* 762 (1967).
53. Ng, K. K. F., and Vane, J. R. Naunyn-Schmiedeberg's Arch. Pharmak. exp. Path. *259,* 2 (1968).
54. Ng, K. K. F., and Vane, J. R. Nature (Lond). *218,* 144 (1968).
55. Piper, P. J., and Vane, J. R. In: Prostaglandins, Peptides and Amines. Academic Press, London and New York, pp. 15–19. (1969).
56. Piper, P., and Vane, J. Ann. N. Y. Acad. Sciences *180,* 363 (1971).
57. Lindsey, H. E., and Wyllie, J. H. Br. J. Surg. *57,* 738 (1970).
58. Davis, B. N., Horton, E. W., and Withrington, P. G. Br. J. Pharmacol. *32,* 127 (1968).
59. Gilmore, N., Vane, J. R., and Wyllie, J. H. In: Prostaglandins, Peptides and Amines. Academic Press, London and New York. pp. 21–29. (1969).
60. Liljestrand, G. In: Nobel Symposium 2, Prostaglandins. Stockholm, pp. 107–108 (1967).
61. Änggård, E., and Samuelsson, B., J. Biol. Chem. *239,* 4097 (1964).
62. Flower, R. J., Blackwell, G., Di Rosa, M., and Parente, L. In: Mechanisms of Steroid Hormone Action. Macmillan Press, London, pp 97–114 (1981).
63. Vane, J. R., Flower, R. J., and Salmon, J. A. In: Prostaglandins and Related Lipids, vol. 2. Alan R. Liss, Inc. New York. pp. 21–45. (1982).
64. Svensson, J., Hamberg, M., and Samuelsson, B. Acta Physiol. Scand. *94,* 222 (1975).
65. Moncada, S., and Vane, J. R. In: Biochemical Aspects of Prostaglandins and Thromboxanes. Academic Press, New York. pp. 155–177 (1977).
66. Needleman, P., Moncada, S., Bunting, S., Vane, J. R., Hamberg, M., and Samuelsson, B. Nature *261,* 558 (1976).
67. Moncada, S., Needleman, P., Bunting, S., and Vane, J. R. Prostaglandins *12,* 323 (1976).
68. Moncada, S., and Vane, J. R. Harvard Medical School Bicentennial Celebration Proceedings. To be published by John Wiley. (In press 1983).
69. Whittaker, N. Tetrahedron Letters No. 32, 2805 (1977).
70. Moncada, S., Herman, A. G., Higgs, E. A., and Vane, J. R. Thromb. Res. *11,* 323 (1977).
71. Weksler, B. B., Marcus, A. J., and Jaffe, E. A. Proc. Natl. Acad. Sci. USA. *74,* 3922 (1977).
72. MacIntyre, D. E., Pearson, J. D., and Gordon, J. L. Nature *271,* 549 (1978).
73. Moncada, S., and Vane, J. R. Pharmac. Rev. *30,* 293 (1979).
74. Whittle, B. J. R. In: Gastro-intestinal Mucosal Blood Flow. Churchill Livingstone, Edinburgh, London, pp. 180–191. (1980).
75. Axelrod, L., and Levine, L. Diabetes *30,* 163 (1981).

76. Ubatuba, F. B., Moncada, S., and Vane, J. R. Thromb. Diath. Haemorrh. *41,* 425 (1979).
77. Szczeklik, A., Gryglewski, R. J., Nizankowski, R., Musial, J., Pieton, R. and Mruk, J. Pharmac. Res. Commun. *10,* 545 (1978).
78. Aiken, J. W., Gorman, R. R., and Shebuski, R. J. Prostaglandins *17,* 483 (1979).
79. Bayer, B.-L., Blass, K. E., and Forster, W. Br. J. Pharmacol. *66,* 10 (1979).
80. Rosenblum, W. I., and El Sabban, F. Stroke *10,* 399 (1979).
81. Gorman, R. R., Bunting, S., and Miller, O. V. Prostaglandins *13,* 377 (1977).
82. Tateson, J. E., Moncada, S., and Vane, J. R. Prostaglandins *13,* 389 (1977).
83. Kaser-Glanzmann, R., Jakabova, M., George, J., and Luscher, E. Biochim. Biophys. Acta *466,* 429 (1977).
84. Lapetina, E. G., Schmitges, C. J., Chandrabose, K., and Cuatrecasas, P. Biochem. Biophys. Res. Commun. *76,* 828 (1977).
85. Minkes, M., Stanford, M., Chi, M., Roth, G., Raz, A., Needleman, P., and Majerus, P. J. Clin. Invest. *59,* 449 (1977).
86. Malmsten, C., Granström, E., and Samuelsson, B. Biochem. Biophys. Res. Commun. *68,* 569 (1976).
87. Moncada, S. Br. J. Pharmacol. *76,* 3 (1982).
88. Higgs, G. A., Moncada, S., and Vane, J. R., J. Physiol (Lond). *280,* 55P (1978).
89. Higgs, G. A. In: Cardiovascular Pharmacology of the Prostaglandins. Raven Press, New York. pp. 315–325. (1982).
90. Boxer, L. A., Allen, J. M., Schmidt, M., Yoder, M., and Baehner, R. L., J. Lab. Clin. Med. *95,* 672 (1980).
91. Hopkins, N. K., and Gorman, R. R., J. Clin. Invest. *67,* 540 (1981).
92. Schafer, A. I., Gimbrone, M. A. Jr., and Handin, R. I. Biochem. Biophys. Res. Commun. *96,* 1640 (1980).
93. Brotherton, A. A. F., and Hoak, J. C. Proc. Natl. Acad. Sci. USA. *79,* 495 (1982).
94. Higgs, E. A., Moncada, S., Vane, J. R., Caen, J. P., Michel, H., and Tobelem, G. Prostaglandins *16,* 17 (1978).
95. Whittle, B. J. R. Brain Res. Bull. *5* (Suppl. 1) 7 (1980).
96. Vane, J. R. In: Advances in Prostaglandin, Thromboxane and Leukotriene Research. Vol 11. Raven Press, New York. pp. 449–456. (1983).
97. Jugdutt, B. F., Hutchins, G. M., Bulkley, B. H., and Becker, L. C. Clin. Res. *27,* 177A (1979).
98. Ogletree, M. L., Lefer, A. M., Smith, J. B., Nicolaou, K. C. Eur. J. Pharmacol. *56,* 95 (1979).
99. Ribeiro, L. G. T., Brandon, T. A., Hopkins, D. G., Reduto, L. A., Taylor, A. A., and Miller, R. R. Am. J. Cardiol, *47,* 835 (1981).
100. Starnes, V. A., Primm, R. K., Woolsey, R. L., Oates, J. A., and Hammon, J. W., J. Cardiovasc. Pharmacol. *4,* 765 (1982).
101. Ohlendorf, R., Perzborn, E., and Schrör, K. Thromb. Res. *19,* 447 (1980).
102. Demling, R. H., Smith, M., Gunther, R., Gee, M., and Flynn, J. Surgery *89,* 257 (1981).
103. Fletcher, J. R. and Ramwell, P. W. Circ. Shock *7,* 299 (1980).
104. Lefer, A. M., Tabas, J., and Smith, E. F. III, Pharmacol. *21,* 206 (1980).
105. Araki, H., and Lefer, A. M. Am. J. Physiol. *238,* H176 (1980).
106. Monden, M., and Fortner, J. G. Ann. Surg. *196,* 38 (1982).
107. Moncada, S., Radomski, M., and Vargas, J. R. Br. J. Pharmacol. *75,* 165P (1982).
108. Blackwell, G. J., Radomski, M., Vargas, J. R., and Moncada, S. Biochim. Biophys. Acta *718,* 60 (1982).
109. Schrör, K., Ohlendorf, R., and Darius, H., J. Pharmac. Exp. Ther. *219,* 243 (1981).
110. Bunting, S., Gryglewski, R., Moncada S., and Vane, J. R. Prostaglandins *12,* 897 (1976).
111. Salmon, J. A., Smith, D. R., Flower, R. J., Moncada, S., and Vane, J. R. Biochim. Biophys. Acta *523,* 250 (1978).
112. Glavind, J., Hartmann, S., Clemmesen, J., Jessen, K. E., and Dam, H. Acta Pathol. Microbiol. Scand. *30,* 1 (1952).
113. Slater, T. F. Free Radical Mechanisms in Tissue Injury. Pion. Ltd., London. (1972).
114. Mickel, H. S., and Horbar, J. Lipids *9,* 68 (1974).

115. D'Angelo, V., Villa, S., Mysliwiec, M., Donati, M. B., and De Gaetano, G. Thromb. Diath. Haemorrh. *39,* 535 (1978).
116. Sinzinger, H., Feigl, W., and Silberbauer, K. Lancet *ii,* 469 (1979).
117. Eldor, A., Falcone, D. J., Hajjar, D. P., Minick, C. R., and Weksler, B. B. Am. J. Pathol. *107,* 186 (1982).
118. Boyd, A. M., and Marks, J. Angiology *14,* 198 (1963).
119. Haeger, K. Vasc. Dis. *5,* 199 (1968).
120. Marks, J. Vitamins and Hormones *20,* 573 (1962).
121. Vane, J. R., J. Endocrin. *95,* 3P (1982).
122. Peatfield, R. C., Gawel, M. J., and Clifford Rose, F. Headache *21,* 190 (1981).
123. Gimson, A. E. S., Hughes, R. D., Mellon, P. J., Woods, H. F., Langley, P. G., Canalese, J., Williams, R., and Weston, M. J. Lancet *i,* 173 (1980).
124. Gimson, A. E. S., Braude, S., Mellon, P. J., Canalese, J., and Williams, R. Lancet *ii,* 681 (1982).
125. Bennett, J. G., Longmore, D. B., and O'Grady, J. In: Clinical Pharmacology of Prostacyclin. Raven Press, New York. pp. 201–208. (1981).
126. Bunting, S., O'Grady, J., Fabiani, J.-N., Terrier, E., Moncada, S., and Vane, J. R. In: Clinical Pharmacology of Prostacyclin. Raven Press, New York. pp. 181–193. (1981).
127. Chelly, J., Tricot, A., Garcia, A., Boucherie, J.-C., Fabiani, J.-N., Passelecq, J., and Dubost, Ch. In: Clinical Pharmacology of Prostacyclin. Raven Press, New York. pp. 209 (1981).
128. Rådegran, K., Egberg, N., and Papaconstantinou, C. Scand. J. Thoracic Cardliovasc. Surg. *15,* 263 (1981).
129. Walker, I. D., Davidson, J. F., Faichney, A., Wheatley, D., and Davidson, K. In: Clinical Pharmacology of Prostacyclin. Raven Press, New York. pp. 195–199. (1981).
130. Longmore, D. B., Bennett, J. G., Hoyle, P. M., Smith, M. A., Gregory, A., Osivand, T., and Jones, W. A. Lancet *i,* 800 (1981).
131. Noback, C. R., Tinker, J. H., Kaye, M. P., Holcomb, G. R., and Pluth, J. R. Circulation *62* (Suppl. 3), 1242 (1980).
132. Negus, D., In: Hormones and Vascular Disease. Pitman Medical. p. 181 (1981)
133. Olsson, A. G. Lancet *ii,* 1076 (1980).
134. Pardy, B. J. H., Lewis, J. D., and Eastcott, H. H. G., Surgery *88,* 826 (1980).
135. Soreide, O., Segadahl, L., Trippestad, A., and Engedal, H. Scand. J. Thoracic Cardiovasc. Surg. *16,* 71 (1982).
136. Szczeklik, A., Nizankowski, R., Skawinski, S., Szczeklik, J., Gluszko, P., and Gryglewski, R. J. Lancet *i,* 1111 (1979).
137. Szczeklik, A., and Gryglewski, R. In: Clinical Pharmacology of Prostacyclin. Raven Press, New York. pp. 159–167 (1981).
138. Belch, J. J. F., McKay, A., McArdle, B., Lieberman, P., Pollock, J. G., Lowe, G. D. O., Forbes, C. D., and Prentice, C. R. M. Lancet *i,* 315 (1983).
139. Zygulska-Mach, H., Kostka-Trabka, E., Niton, A., and Gryglewski, R. J. Lancet *ii,* 1075 (1980).
140. Dowd, P. M., Martin, M. F. R., Cooke, E. D., Bowcock, S. A., Jones, R., Dieppe, P. A., and Kirby, J. D. T. Br. J. Dermatol. *106,* 81 (1982).
141. Belch, J. J. F., Newman, P., Drury, J. K., Capell, H., Leiberman, P., James, W. B., Forbes, C. D., and Prentice, C. R. M. Thromb. Haem. *45,* 255 (1981).
142. Belch, J. J. F., Newman, P., Drury, J. K., McKenzie, F., Capell, H., Leiberman, P., Forbes, C. D., and Prentice, C. R. M. Lancet *i,* 313 (1983).
143. Gryglewski, R. J., Nowak, S., Kostka-Trabka, E., Bieron, K., Dembinska-Kiec, A., Blaszczyk, B., Kusmiderski, J., Markowska, E., and Szmatola, S. Pharm. Res. Commun. *14,* 879 (1982).
144. Watkins, W. D., Peterson, M. B., Crone, R. K., Shannon, D. C., and Levine, L. Lancet *i,* 1083, (1980).
145. Rubin, L. J., Groves, B. M., Reeves, J. T., Frosolono, M., Handel, F., and Cato, A. E. Circulation *66* (2) Part 1, 334 (1982).
146. Szczeklik, J., Szczeklik, A., and Nizankowski, R. Lancet *ii,* 1076 (1980).

147. Lock, J. E., Olley, P. M., Coceani, F., Swyer, P. R., and Rowe, R. D. Lancet *i,* 1343 (1979).
148. Fidler, J., Bennett, M. J., de Swiet, M., Ellis, C., and Lewis, P. J. Lancet *ii,* 31 (1980).
149. Yui, Y., Nakajima, H., Kawai, C., and Murakami, T. Am. J. Cardiol. *50,* 320 (1982).
150. Bergman, G., Daly, R., Atkinson, L., Rothman, M., Richardson, P. J., Jackson, G., and Jewitt, D. E. Lancet *i,* 569 (1981).
151. Hall, R. J. C., and Dewar, H. A. Lancet *i,* 949 (1981).
152. Chierchia, S., Patrono, C., Crea, F., Ciabattoni, G., de Caterina, R., Cinotti, G. A., Distante, A., and Maseri, A. Circulation *65,* 470 (1982).
153. Hensby, C. N., Lewis, P. J., Hilgard, P., Mufti, G. J., Hows, J., and Webster, J. Lancet *ii,* 748 (1979).
154. Budd, G. T., Bukowski, R. M., Lucas, F. V., Cato, A. E., and Cocchetto, D. M. Lancet *ii,* 915 (1980).
155. FitzGerald, G. A., Roberts, L. J. II., Maas, D., Brash, A. R., and Oates, J. A. In: Clinical Pharmacology of Prostacyclin. Raven Press, New York. p. 81 (1981).
156. Mundy, A. R., Bewick, M., Moncada, S., and Vane, J. R. Prostaglandins *19,* 595 (1980).
157. Leithner, C., Sinzinger, H., and Schwarz, M. Prostaglandins *22,* 783 (1981).
158. Fidler, J., Ellis, C., Bennett, M. J., de Swiet, M., and Lewis, P. J. In: Clinical Pharmacology of Prostacyclin. Raven Press, New York. pp. 141–143 (1981).
159. Webster, J., Borysiewicz, L. K., Rees, A. J., and Lewis, P. J. In: Clinical Pharmacology of Prostacyclin. Raven Press, New York. pp. 77–80 (1981).
160. Whittle, B. J. R., Kauffman, G. L., and Moncada, S. Nature *292,* 472 (1981).
161. Honn, K. V., Cicone, B., and Skoff, A. Science *212,* 1270 (1981).
162. Utsunomiya, T., Krausz, M. M., Valeri, C. R., Shepro, D., and Hechtman, H. B. Surgery *88,* 25 (1980).
163. Vane, J. R. In: Identification of Asthma. Churchill Livingstone, pp. 121–131. (1971).
164. Moncada, S., Higgs. E. A., and Vane, J. R. Lancet *i,* 18 (1977).

1983
Physiology or Medicine

BARBARA McCLINTOCK

"for her discoveries of mobile genetic elements"

THE NOBEL PRIZE FOR PHYSIOLOGY OR MEDICINE

Speech by Professor NILS RINGERTZ of the Karolinska Institute.
Translation from the Swedish text

Your Majesties, Your Royal Highnesses, Ladies and Gentlemen,

The Nobel prize in Physiology or Medicine for 1983 recognizes a great discovery about the organization of genes on chromosomes and how these genes, by changing places, can alter their function. This discovery, made while investigating blue, brown, and red spots on maize kernels, resulted in new knowledge of great medical importance — information which provides the key to problems as diverse as hospital infections, African sleeping sickness and chromosome changes in cancer cells. In order to explain this link, we must start at the beginning; namely with Barbara McClintock's investigations of coloured spots on maize kernels.

The maize cobs that we buy at the supermarket usually have yellow kernels. This is not always the case with wild forms of maize. In Central and South America where maize originated, one can still find primitive types of maize where the kernels are blue, brown or red. The colour depends on pigments in the surface layer of the kernel endosperm. The endosperm is the food store for the developing seedling. The synthesis of kernel pigments is controlled by the genes of the maize plant. In some cases one finds differently coloured kernels on the same cob. The explanation for this is that the cob is formed from a group of female flowers. Each of these female flowers may be fertilized independently by a pollen grain from a male flower. Maize cobs with differently coloured kernels arise when the pollen grains do not carry the same genes for endosperm pigments. All these phenomena can be explained on the basis of the laws of the inheritance stated by Gregor Mendel in 1866. What cannot be explained, however, and what puzzled plant breeders in the 1920's, was that maize kernels sometimes have numerous spots or dots, rather than being evenly coloured as would be expected. It was suspected that the dots on the kernels were due to the instability of genes involved in the pigment synthesis. These genes were believed to undergo mutations during the development of the kernel. Should such a mutation be inherited by several generations of daughter cells it would result in a differently coloured spot. This idea received further support when it was found that maize with variegated kernels also had broken chromosomes. The problem of variegation in maize was of slight importance from a practical point of view, but it fascinated Barbara McClintock because it evidently could not be explained on the basis of Mendelian genetics.

McClintock analyzed this phenomenon by studying chromosome changes and the results of crossing experiments in maize with different patterns of variegation. She was able to identify a series of genes on chomosome number 9 that determine pigmentation and other characteristics of the endosperm. She found that variegation occurred when a small piece of chromosome 9 moved from one place on the chromosome to another close to a gene coding for a pigment. The usual effect was to switch off the gene, and furthermore, the chromosome frequently showed a break at the site of integration. McClintock called these types of genetic material "control elements" since they clearly altered the function of neighbouring genes. In a series of very advanced experiments carried out between 1948 and 1951, McClintock mapped several families of control elements. These elements affected not only the pigmentation pattern of the maize kernels but other properties as well. She also pointed out that mobile genetic elements were probably present in insects and higher animals. In spite of this, her observations received very little attention. This was because her findings, when first presented, were overshadowed by the discovery that the DNA molecule stores the genetic information in its structure. It also became evident that mutations involving only one change in one of the building blocks in the DNA molecule could have serious effects. Under these circumstances, it is not surprising that few geneticists were prepared to accept that genes could jump in the irresponsible manner that McClintock proposed for controlling elements. The "state of the art" in molecular genetics at that time made it difficult to accept "jumping genes", and thus McClintock had to await the development of methodological tools powerful enough to verify in biochemical terms her great discovery.

In the mid-sixties, mobile genetic elements were found to play an important role in the spreading of resistance to antibiotics from resistant to sensitive strains of bacteria. This type of transferable drug resistance is a serious problem in hospitals since it causes infections that are very difficult to treat. During the 1970's, more support was found for the medical significance of mobile genetic structures. It was found, for instance, that the transposition of genes is an important step in the formation of antibodies. It has always been a mystery how the body, using a limited number of genes, can form an almost endless number of different antibodies to foreign substances. Nature has solved this problem according to the building block principle. When an individual is born, the chromosomes carry a set of mobile building blocks for antibody genes. By recombining these blocks in various ways in different cells, the body is able to generate millions of genes for antibodies.

During the last few years mobile genetic structures have attracted great interest in cancer research. In certain forms of cancer, growth regulating genes called oncogenes, are transposed from one chromosome to another. Tumour viruses in birds and mice have been found to carry oncogenes which they, in all likelihood, originally picked up from a host cell. If a virus then introduces these genes in the wrong place on the chromosomes of a normal cell, the latter is transformed into a cancer cell.

McClintock's discovery of mobile genetic elements in maize, therefore, has been found to have counterparts also in bacteria, animals and humans.

What led McClintock to devote her research to the variegation of maize kernels was that it did not fit in with Mendelian genetics. With immense perseverance and skill, McClintock, working completely on her own, carried out experiments of great sophistication that demonstrated that hereditary information is not as stable as had previously been thought. This discovery has led to new insights into how genes change during evolution and how mobile genetic structures on chomosomes can change the properties of cells. Her research has helped to elucidate a series of complicated medical problems.

Dr. McClintock,

I have tried to summarize to this audience your work on mobile genetic elements in maize and to show how basic research in plant genetics can lead to new perspectives in medicine. Your work also demonstrates to scientists, politicians and university administrators how important it is that scientists are given the freedom to pursue promising lines of research without having to worry about their immediate practical applications. To young scientists, living at a time of economic recession and university cutbacks, your work is encouraging because it shows that great discoveries can still be made with simple tools.

On behalf of the Nobel Assembly of the Karolinska Institute I wish to convey to you our warmest congratulations and I ask you to receive your Nobel prize in Physiology or Medicine from His Majesty the King.

Barbara McClintock

BARBARA McCLINTOCK

In the fall of 1921 I attended the only course in genetics open to undergraduate students at Cornell University. It was conducted by C. B. Hutchison, then a professor in the Department of Plant Breeding, College of Agriculture, who soon left Cornell to become Chancellor of the University of California at Davis, California. Relatively few students took this course and most of them were interested in pursuing agriculture as a profession. Genetics as a discipline had not yet received general acceptance. Only twenty-one years had passed since the rediscovery of Mendel's principles of heredity. Genetic experiments, guided by these principles, expanded rapidly in the years between 1900 and 1921. The results of these studies provided a solid conceptual framework into which subsequent results could be fitted. Nevertheless, there was reluctance on the part of some professional biologists to accept the revolutionary concepts that were surfacing. This reluctance was soon dispelled as the logic underlying genetic investigations became increasingly evident.

When the undergraduate genetics course was completed in January 1922, I received a telephone call from Dr. Hutchison. He must have sensed my intense interest in the content of his course because the purpose of his call was to invite me to participate in the only other genetics course given at Cornell. It was scheduled for graduate students. His invitation was accepted with pleasure and great anticipations. Obviously, this telephone call cast the die for my future. I remained with genetics thereafter.

At the time I was taking the undergraduate genetics course, I was enrolled in a cytology course given by Lester W. Sharp of the Department of Botany. His interests focused on the structure of chromosomes and their behaviors at mitosis and meiosis. Chromosomes then became a source of fascination as they were known to be the bearers of "heritable factors". By the time of graduation, I had no doubts about the direction I wished to follow for an advanced degree. It would involve chromosomes and their genetic content and expressions, in short, cytogenetics. This field had just begun to reveal its potentials. I have pursued it ever since and with as much pleasure over the years as I had experienced in my undergraduate days.

After completing requirements for the Ph.D. degree in the spring of 1927, I remained at Cornell to initiate studies aimed at associating each of the ten chromosomes comprising the maize complement with the genes each carries. With the participation of others, particularly that of Dr. Charles R. Burnham, this task was finally accomplished. In the meantime, however, a sequence of events occurred of great significance to me. It began with the appearance in the fall of 1927 of George W. Beadle (a Nobel Laureate) at the Department of Plant

Breeding to start studies for his Ph.D. degree with Professor Rollins A. Emerson. Emerson was an eminent geneticist whose conduct of the affairs of graduate students was notably successful, thus attracting many of the brightest minds. In the following fall, Marcus M. Rhoades arrived at the Department of Plant Breeding to continue his graduate studies for a Ph.D. degree, also with Professor Emerson. Rhoades had taken a Masters degree at the California Institute of Technology and was well versed in the newest findings of members of the Morgan group working with Drosophila. Both Beadle and Rhoades recognized the need and the significance of exploring the relation between chromosomes and genes as well as other aspects of cytogenetics. The initial association of the three of us, followed subsequently by inclusion of any interested graduate student, formed a close-knit group eager to discuss all phases of genetics, including those being revealed or suggested by our own efforts. The group was self-sustaining in all ways. For each of us this was an extraordinary period. Credit for its success rests with Professor Emerson who quietly ignored some of our seemingly strange behaviors.

Over the years, members of this group have retained the warm personal relationship that our early association generated. The communal experience profoundly affected each one of us.

The events recounted above were, by far, the most influential in directing my scientific life.

Born: Hartford, Connecticut, U.S.A, 16 June, 1902

Secondary Education: Erasmus Hall High School, Brooklyn, New York.

Earned Degrees:

B.S. Cornell University, Ithaca, New York, 1923
M.A. Cornell University, Ithaca, New York, 1925
Ph.D. Cornell University, Ithaca, New York, 1927

Positions held:

Instructor in botany, Cornell University, 1927–1931
Fellow, National Research Council, 1931–1933
Fellow, Guggenheim Foundation, 1933–1934
Research Associate, Cornell University, 1934–1936
Assistant Professor, University of Missouri, Columbia, Missouri, 1936–1941
Staff Member, Carnegie Institution of Washington, Cold Spring Harbor, New York, 1942–1967
Distinguished Service Member, Carnegie Institution of Washington, Cold Spring Harbor, New York, 1967 to Present
Visiting Professor, California Institute of Technology, 1954
Consultant, Agricultural Science Program, The Rockefeller Foundation, 1963–1969
Andrew D. White Professor-at-Large, Cornell University, 1965–1974

Honorary Doctor of Science:
University of Rochester, 1947
Western College for Women, 1949
Smith College, 1957
University of Missouri, 1968
Williams College, 1972
The Rockefeller University, 1979
Harvard University, 1979
Yale University, 1982
University of Cambridge, 1982
Bard College, 1983
State University of New York, 1983
New York University, 1983

Honorary Doctor of Humane Letters:
Georgetown University, 1981

Awards:
Achievement Award, Association of University Women, 1947
Merit Award, Botanical Society of America, 1957
Kimber Genetics Award, National Academy of Sciences, 1967
National Medal of Science, 1970
Lewis S. Rosenstiel Award for Distinguished Work in Basic Medical Research, 1978
The Louis and Bert Freedman Foundation Award for Research in Biochemistry, 1978
Salute from the Genetics Society of America, August 18, 1980
Thomas Hunt Morgan Medal, Genetics Society of America, June, 1981
Honorary Member, The Society for Developmental Biology, June, 1981
Wolf Prize in Medicine, 1981
Albert Lasker Basic Medical Research Award, 1981
MacArthur Prize Fellow Laureate, 1981
Honorary Member, The Genetical Society, Great Britain, April, 1982
Louisa Gross Horwitz Prize for Biology or Biochemistry, 1982
Charles Leopold Mayer Prize, Académie des Sciences, Institut de France, 1982

THE SIGNIFICANCE OF RESPONSES OF THE GENOME TO CHALLENGE

Nobel lecture, 8 December, 1983

by

BARBARA McCLINTOCK

Carnegie Institution of Washington
Cold Spring Harbor Laboratory
Cold Spring Harbor, New York, U.S.A.

I. *Introduction*

An experiment conducted in the mid-nineteen forties prepared me to expect unusual responses of a genome to challenges for which the genome is unprepared to meet in an orderly, programmed manner. In most known instances of this kind, the types of response were not predictable in advance of initial observations of them. It was necessary to subject the genome repeatedly to the same challenge in order to observe and appreciate the nature of the changes it induces. Familiar examples of this are the production of mutation by X-rays and by some mutagenic agents. In contrast to such "shocks" for which the genome is unprepared, are those a genome must face repeatedly, and for which it is prepared to respond in a programmed manner. Examples are the "heat shock" responses in eukaryotic organisms, and the "SOS" responses in bacteria. Each of these initiates a highly programmed sequence of events within the cell that serves to cushion the effects of the shock. Some sensing mechanism must be present in these instances to alert the cell to imminent danger, and to set in motion the orderly sequence of events that will mitigate this danger. The responses of genomes to unanticipated challenges are not so precisely programmed. Nevertheless, these are sensed, and the genome responds in a descernible but initially unforeseen manner.

It is the purpose of this discussion to consider some observations from my early studies that revealed programmed responses to threats that are initiated within the genome itself, as well as others similarly initiated, that lead to new and irreversible genomic modifications. These latter responses, now known to occur in many organisms, are significant for appreciating how a genome may reorganize itself when faced with a difficulty for which it is unprepared. Conditions known to provoke such responses are many. A few of these will be considered, along with several examples from nature implying that rapid reorganizations of genomes may underlie some species formations. Our present knowledge would suggest that these reorganizations originated from some "shock" that forced the genome to restructure itself in order to overcome a threat to its survival.

Because I became actively involved in the subject of genetics only twenty-one years after the rediscovery, in 1900, of Mendel's principles of heredity, and at a stage when acceptance of these principles was not general among biologists, I have had the pleasure of witnessing and experiencing the excitement created by revolutionary changes in genetic concepts that have occurred over the past sixty-odd years. I believe we are again experiencing such a revolution. It is altering our concepts of the genome: its component parts, their organizations, mobilities, and their modes of operation. Also, we are now better able to integrate activities of nuclear genomes with those of other components of a cell. Unquestionably, we will emerge from this revolutionary period with modified views of components of cells and how they operate, but only, however, to await the emergence of the next revolutionary phase that again will bring startling changes in concepts.

II. *An experiment with Zea mays conducted in the summer of 1944, and its consequences*

The experiment that alerted me to the mobility of specific components of genomes involved the entrance of a newly ruptured end of a chromosome into a telophase nucleus. This experiment commenced with the growing of approximately 450 plants in the summer of 1944, each of which had started its development with a zygote that had received from each parent a chromosome with a newly ruptured end of one of its arms. The design of the experiment required that each plant be self-pollinated. This was in order to isolate from the self-pollinated progeny new mutants that were expected to appear, and to be confined to locations within the arm of a chromosome whose end had been ruptured. Each mutant was expected to reveal the phenotype produced by a minute homozygous deficiency, and to segregate in a manner resembling that of a recessive allele in an F_2 progeny. Their modes of origin could be projected from the known behavior of broken ends of chromosomes in successive mitoses. In order to observe those mutants that might express an altered seedling character, forty kernels from each self-pollinated ear were sown in a seedling bench in the greenhouse during the winter of 1944–45.

Some seedling mutants of the type expected did segregate, but they were overshadowed by totally unexpected segregants exhibiting bizarre phenotypes. These segregants were variegated for type and degree of expression of a gene. Those variegated expressions given by genes associated with chlorophyll development were startingly conspicuous. Within any one progeny chlorophyll intensities, and their pattern of distribution in the seedling leaves, were alike. Between progenies, however, both the type and the pattern differed widely. Variegated seedlings from the different progenies were transferred to pots in order to observe the variegated phenomenon in the later developing, larger leaves. It soon became apparent that modified patterns of gene expression were being produced, and that these were confined to sharply defined sectors in a leaf. Thus, the modified expression appeared to relate to an event that had occurred in the ancestor cell that gave rise to the sector. It was this event that was responsible for altering the pattern and/or type of gene expression in descendant cells, often many cell generations removed from the event. It was soon evident that the event was related to some cell component that had been unequally segregated at a mitosis. Twin sectors appeared in which the patterns of gene expression in the two side-by-side sectors were reciprocals of each other. For example, one sector might have a reduced number of uniformly distributed fine green streaks in a white background in comparison with the number and distribution of such streaks initially appearing in the seedling and showing elsewhere on the same leaf. The twin, on the other hand, had a much increased number of such streaks. Because these twin sectors were side-by-side they were assumed to have arisen from daughter cells following a mitosis in which each daughter had been modified in a manner that would differentially regulate the pattern of gene expression in their progeny cells. After observing many such twin sectors, I concluded that regulation of pattern of gene expression in these instances was associated with an event occurring at a mitosis in which one daughter cell had gained something that the other daughter cell had lost. Believing that I was viewing a basic genetic phenomenon, all attention was given, thereafter, to determine just what it was that one cell had gained that the other cell had lost. These proved to be transposable elements that could regulate gene expressions in precise ways. Because of this I called them "controlling elements". Their origins and their actions were a focus of my research for many years thereafter. It is their origin that is important for this discussion, and it is extraordinary. I doubt if this could have been anticipated before the 1944 experiment. It had to be discovered accidently.

III. *Early observations of the effect of X-rays on chromosomes*

The 1944 experiment took place thirteen years after I had begun to examine the behavior of broken ends of chromosomes. It was knowledge gained in these years that led me to conceive of this experiment. Initial studies of broken ends of chromosomes began in the summer of 1931. At that time our knowledge of chromosomes and genes was limited. In retrospect we might call it primitive. Genes were "beads" arranged in linear order on the chromosome "string." By 1931, however, means of studying the "string" in some detail was provided by newly developed methods of examining the ten chromosomes of the maize complement in microsporocytes at the pachytene stage of meiosis. At this stage the ten bivalent chromosomes are much elongated in comparison to their metaphase lengths. Each chromosome is identifiable by its relative length, by the location of its centromere, which is readily observed at the pachytene stage, and by the individuality of the chromomeres strung along the length of each chromosome. At that time maize provide the best material for locating known genes along a chromosome arm, and also for precisely determining the break points in chromosomes that had undergone various types of rearrangement, such as translocations, inversions, etc. The usefulness of the salivary gland chromosomes of *Drosophila* for such purposes had not yet been recognized. This came several years later. In the interim, maize chromosomes were revealing, for the first time, some distinctive aspects of chromosome organization and behavior. One of these was the extraordinary effect of X-rays on chromosomes.

The publications of H. J. Muller in 1927 and 1928 (1, 2) and of Hanson in 1928 (3) reporting the use of X-rays for obtaining mutations in *Drosophila*, and similarly that of Stadler in 1928 (4) with the barley plant, produced a profound effect on geneticists. Here was a way of obtaining mutations at will. One did not need to await their spontaneous appearances. Many persons over many years continued to use X-rays for such purposes. But X-rays did not fulfill initial expectations of their usefulness. For other purposes, however, they have been most valuable, particularly for obtaining various types of structural reorganizations of the genome, from minute deficiencies to multiple rearrangements of chromosomes.

It was to observe the effects of X-rays on chromosomes of maize that brought me to the University of Missouri at Columbia in the summer of 1931. Prior to 1931 Dr. Stadler had been using X-rays to obtain mutations in maize. He had developed techniques for isolating those mutations that occur at selected gene loci. One method was to irradiate pollen grains. Pollen grains carry the haploid male gametes. The irradiated male gametes in Stadler's experiments carried wild-type alleles of known recessive mutants. Irradiated pollen was placed on the silks of ears of plants that were homozygous for one or more recessive alleles located in known linkage groups. An X-ray-induced mutation altering the expression of the wild-type allele of one of these recessives should be identifiable in an individual plant derived from such a cross. By the summer of 1931 Stadler had many plants in his field at Columbia, Missouri, from which one could choose those that exhibited one or another of these recessive phenotypes. Stadler had asked me if I would be willing to examine such plants at the meiotic stages to determine what types of events might be responsible for these recessive expressions. I was delighted to do so, as this would be a very new experience. Following my arrival at Columbia in June, 1931, plants were selected whose chromosomes were to be examined. The knowledge gained from these observations was new and impressive. Descriptions and photographs summarizing these observations appeared in a bulletin published by the University of Missouri Agricultural College and Experiment Station (5).

None of the recessive phenotypes in the examined plants arose from "gene mutation". Each reflected loss of a segment of a chromosome that carried the wild-type allele, and X-rays were responsible for inducing these deficiencies. They also were responsible for producing other types of chromosome rearrangements, some of them unexpectedly complex. A conclusion of basic significance could be drawn from these observations: broken ends of chromosomes will fuse, 2-by-2, and any broken end with any other broken end. This principle has been amply proved in a series of experiments conducted over the years. In all such instances the break must sever both strands of the DNA double helix. This is a "double-strand break" in modern terminology. That two such broken ends entering a telophase nucleus will find each other and fuse, regardless of the initial distance that separates them, soon became apparent.

After returning to Cornell University in the fall of 1931, I received a reprint from geneticists located at the University of California, Berkeley. The authors described a pattern of variegation in *Nicotiana* plants that was produced by loss of a fragment chromosome during plant development. The fragment carried the dominant allele of a known recessive present in the normal homologues. Loss of the dominant allele allowed the recessive phenotype to be expressed in the descendants of those cells that had lost this fragment. It occurred to me that the fragment could be a ring-chromosome, and that losses of the fragment were caused by an exchange between sister chromatids following replication of the ring. This would produce a double-size ring with two centromeres. In the following anaphase, passage of the centromeres to opposite poles would produce two chromatid bridges. This, I thought, could prevent the chromosome from being included in either telophase nucleus. I sent my suggestion to the geneticists at Berkeley who then sent me an amused reply. My suggestion, however, was not without logical support. During the summer of 1931 I had seen plants in the maize field that showed variegation patterns resembling the one described for *Nicotiana*. The chromosomes in these plants had not been examined. I then wrote to Dr. Stadler asking if he would be willing to grow more of the same material in the summer of 1932 that had been grown in the summer of 1931. If so, I would like to select the variegated plants to determine the presence of a ring chromosome in each. Thus, in the summer of 1932 with Stadler's generous cooperation, I had the opportunity to examine such plants. Each plant did have a ring chromosome. It was the behavior of this ring that proved to be significant. It revealed several basic phenomena. The following was noted: (*1*) In the majority of mitoses replication of the ring chromosome produced two chromatids that were completely free from each other and thus could separate without difficulty in the following anaphase. (*2*) Sister strand exchanges do occur between replicated or replicating chromatids, and the frequency of such events increases with increase in the size of the ring. These exchanges produce a double-size ring with two centromeres. (*3*) Mechanical rupture occurs in each of the two chromatid bridges formed at anaphase by passage of the two centromeres on the double-size ring to opposite poles of the mitotic spindle. (*4*) The location of a break can be at any one position along any one bridge. (*5*) The broken ends entering a telophase nucleus then fuse. (*6*) The size and content of each newly constructed ring depend on the position of the rupture that had occurred in each bridge (6, 7, 8).

The conclusion seems inescapable that cells are able to sense the presence in their nuclei of ruptured ends of chromosomes, and then to activate a mechanism that will bring together and then unite these ends, one with another. And this will occur regardless of the initial distance in a telophase nucleus that separated the ruptured ends. The ability of a cell to sense these broken ends, to direct them toward each other, and then to unite them so that the union of the two DNA strands is correctly oriented, is a particularly revealing example of the sensitivity of cells to all that is going on within them. They make wise decisions and act upon them.

Evidence from X-rays, ring chromosomes, and that obtained in later experiments (9, 10, 11, 12), gives unequivocal support for the conclusion that broken ends will find each other and fuse. The challenge is met by a programmed response. This may be necessary, as both accidental breaks and programmed breaks may be frequent. If not repaired, such breaks could lead to genomic deficiencies having serious consequences.

IV. *The entrance into a telophase nucleus of a single broken end of a chromosome*

In the mid-nineteen-thirties another event inducing chromosome rupture was discovered. It revealed why crossing-over should be suppressed between the centromere and the nucleolus organizer in organisms in which chiasmata terminalize, from the initial location of a crossover to the end of the arm of the chromosome. In maize terminalization occurs at the diplotene stage of meiosis. This is before the nucleolus breaks up, which it does at a later stage in the first meiotic prophase. It is known that the force responsible for terminalization is strong. It is enough to induce chromosome breakage should the terminalization process be blocked before the terminalizing chiasma reaches the end of the arm of a chromosome. In maize the centromere and the nucleolus organizer on the nucleolus chromosome are relatively close together. No crossovers have been noted to occur between them. However, if a plant is homozygous for a translocation that places the centromere on the nucleolus chromosome some distance from its nucleolus organizer, crossing over does occur in the interval between them (10). A chiasma so located starts its terminalization process to reach the end of the arm. It is stopped, however, at the nucleolus border. The terminalizing chromatid strands cannot pass through the nucleolus. Instead, the two strands are ruptured at this border. Fusions then occur between the ruptured ends establishing, thereby, a dicentric chromosome deficient for all of the chromatin that runs through the nucleolus and continues beyond to the end of the arm. At the meiotic anaphase, passage of the two centromeres of the dicentric chromosome to opposite poles of the spindle produces a bridge. This bridge is ruptured, and again, the rupture can occur at any one location along the bridge. Now a *single* ruptured end of a chromosome enters the telophase nucleus. How, then, does the cell deal with this novel situation?

In order to determine how a cell responds to the presence of a single ruptured end of a chromosome in its nucleus, tests were conducted with plants that were heterozygous for a relatively long inversion in the long arm of chromosome 4 of maize. It had been known for some time that a crossover within the inverted segment in plants that are heterozygous for an inversion in one arm of a chromosome would result in a dicentric chromosome, and also an acentric fragment composed of all the chromatin from the distal breakpoint of the inversion to the end of the arm. A chromatin bridge would form at the meiotic anaphase by passage of the two centromeres on the dicentric chromosome to opposite poles of the spindle. Mechanical rupture of this bridge as the spindle elongated would introduce a single broken end into the telophase nucleus, as illustrated in a to d, Fig. 1. The intent of this experiment was to observe this chromosome in the following mitotic division in order to determine the fate of its ruptured end. This could be accomplished readily by observing the first mitotic division in the microspore. Meiosis on the male side gives rise to four haploid spores, termed microspores. Each spore enlarges. Its nucleus and nucleolus also enlarge. Approximately seven days after completion of meiosis this very enlarged cell prepares for the first post-meiotic mitosis. This mitosis produces two cells, a very large cell with a large, active nucleus and nucleolus, and a small cell with compact chromatin in a small nucleus, surrounded by a

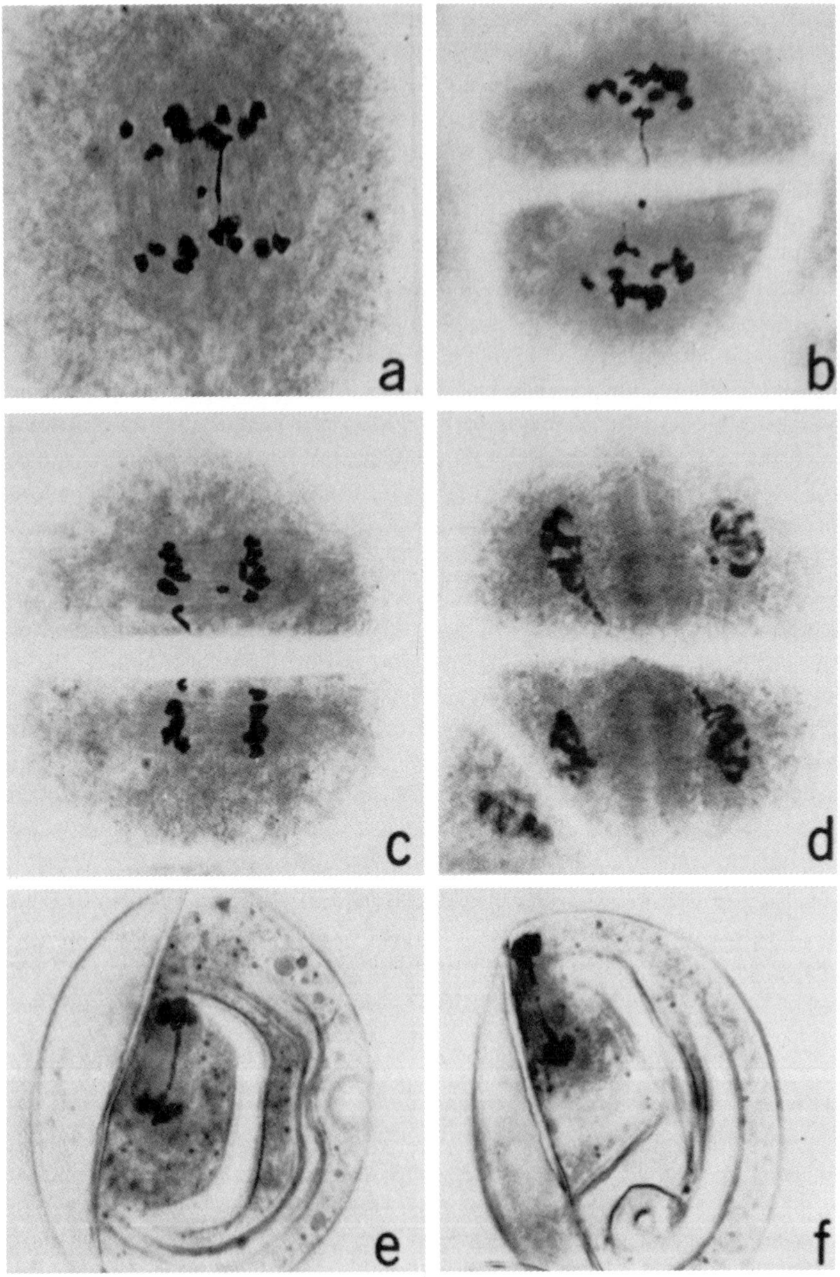

Figure 1. Photographs illustrating the behavior of a newly ruptured end of a chromosome at the meiotic mitoses in microsporocytes and in the post-meiotic mitosis in the microspore. *a.* Chromatin bridge at the first meiotic anaphase, with accompanying acentric fragment. Note the thin region in the bridge where rupture probably would have occurred at a slightly later stage. *b*. Two sister cells at a very late prophase of the second meiotic mitosis. The rupture of the chromatid bridge that occurred at the previous anaphase severed the bridge at a non-median position. The larger segment so produced appears in the upper cell and opposite to the shorter segment in the lower cell. Their locations away from the other divalent chromatids relate to late entrances into the previous telophase nuclei, caused by tension on the bridge before its rupture. Their placements show the

thin layer of cytoplasm. This is the generative cell. Sometime later it undergoes a mitosis that will produce two condensed sperm cells. With completion of this division the pollen grain is nearly ready to function. The first division in the microspore may be observed readily merely by using a squash technique. The division of the generative cell, on the other hand, is obscured by the densely packed starch grains that have accumulated during the interval between the two mitoses.

Examination of the first mitotic division in the microspore revealed a strange behavior of the single broken end that had entered a meiotic telophase nucleus. The replicated chromosome again was dicentric. The two chromatids produced by the replication process appeared to be fused at the location of the break that had occurred at the previous meiotic anaphase. In the spore, passage to opposite poles of the two centromeres of this newly created dicentric chromosome again produced a chromatid bridge that again was ruptured (Fig. 1, e, f). Thus, a newly ruptured end of the chromosome again entered each telophase nucleus. How would this newly broken end behave in subsequent mitoses? To determine this requires that the pollen grain with nuclei having such a ruptured end of a chromosome be functional. This could not be in the described instance because pollen grains whose nuclei had such a ruptured end would be deficient for a large terminal segment of the long arm of chromosome 4. Pollen grains whose nuclei have such a deficiency are unable to function.

The problem was resolved by obtaining plants having one chromosome of the maize complement with a duplication of all of its short arm in reverse orientation; its homologue had either a normal organization of its short arm, or better in the test to be performed, a short terminal deficiency of this arm that will not allow pollen grains receiving this chromosome to function. A crossover at the meiotic prophase, as shown in Fig. 2, produces a dicentric chromosome that simulates two normal chromosomes attached together at the ends of their short arms, and a fragment chromosome with telomeres at both ends. The dicentric chromosome, produced by the crossover, initiates the chromatid type

positions they occupied in these telophase nuclei. The acentric fragment is in the lower cell, close to the cell membrane that was formed at the end of the first meiotic mitosis. *c.* Anaphase of the second meiotic mitosis. The chromatid with a ruptured end in each cell is placed closer to the newly formed cell wall than are other chromatids, and for reasons given in *b.* Note dissimilar lengths of the arms with ruptured ends. The acentric fragment is near the middle of the spindle in the upper cell. *d.* Telophase of the second mitotic mitosis with extensions in two of the four nuclei pointing toward each other, one in the upper left nucleus and one in the lower right nucleus. The shapes of these nuclei reflect the off-positioning of chromatids having a newly ruptured end. Such off-positioning starts with the first meiotic telophase, continues throughout the interphase and into the prophase, metaphase, and anaphase of the second meiotic mitoses and then into the telophases, as shown here. The acentric fragment is adjacent to but not within the nucleus to the upper right. Note formation of the cell plate in each cell that anticipates the four spores that are products of meiosis. *e.* Mitotic anaphase in the microspore showing a chromatid bridge produced by "fusion" of the replicated broken end. *f.* Same as *e* but a slightly later stage showing rupture of the bridge. (Photographs adapted from Missouri Agricultural Experiment Station Research Bulletin 290, 1938.)

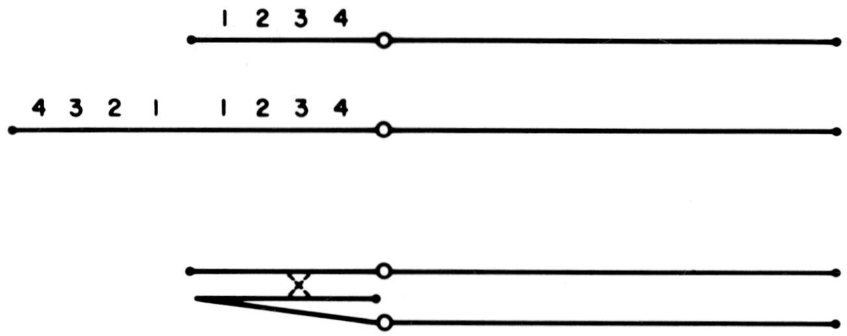

Figure 2. Stylized representation of a crossover between a chromosome 9 with a normal short arm, upper line, and one with a duplication of this arm in reverse orientation, line below. In the lower two lines an exchange between two homologously associated arms is indicated by the cross. Such an exchange would give rise to a dicentric chromosome that simulates two normal chromosomes 9 attached together at the ends of their short arms, plus a small acentric fragment composed of the short arm of chromosome 9. The open circles represent centromeres. Telomeres are depicted as small knobs at the ends of chromosomes.

of breakage-fusion-bridge cycle. This cycle, initially detected at the first mitosis in a microspore, could now be followed in subsequent mitoses. This is because the location of breaks in some of the anaphase bridges gave rise to chromosomes with at least a full complement of genes necessary for pollen functioning. Such functional pollen grains falling on the silks of ears will deliver their two sperm cells to the embryo sac inside a kernel-to-be. One sperm will contribute to the development of the embryo and the other will contribute to the development of the endosperm.

On the female side only a single cell in the kernel-to-be undergoes meiosis, and the embryo sac arises from only one of the four spores produced by the two meiotic mitoses. The other three spores degenerate. This one haploid cell, the megaspore, then undergoes three successive mitoses to form the embryo sac, or female gametophyte. Of the cells in the embryo sac, only the egg cell and the much enlarged central cell need be considered here. The very large central cell has two haploid nuclei positioned close to each other and near the egg cell. Following delivery of the two sperms to the embryo sac, one sperm nucleus fuses with the egg cell nucleus to form the diploid zygote. The other sperm nucleus and the two nuclei in the central cell fuse to form the primary endosperm nucleus, which is triploid. (The term "double fertilization" is commonly employed in referring to these events.) Thus, the embryo and endosperm are formed separately, although both share the same genes, one set from each parent for the embryo, and two sets from the female parent and one set from the male parent for the endosperm. Although developing separately, the two structures are placed side-by-side in the mature kernel, as illustrated in Fig. 3.

It was soon learned that the chromatid type of breakage-fusion-bridge cycle, initiated at a meiotic anaphase, will continue during development of the pollen grain and the embryo sac. Whenever a sperm nucleus contributes a chromo-

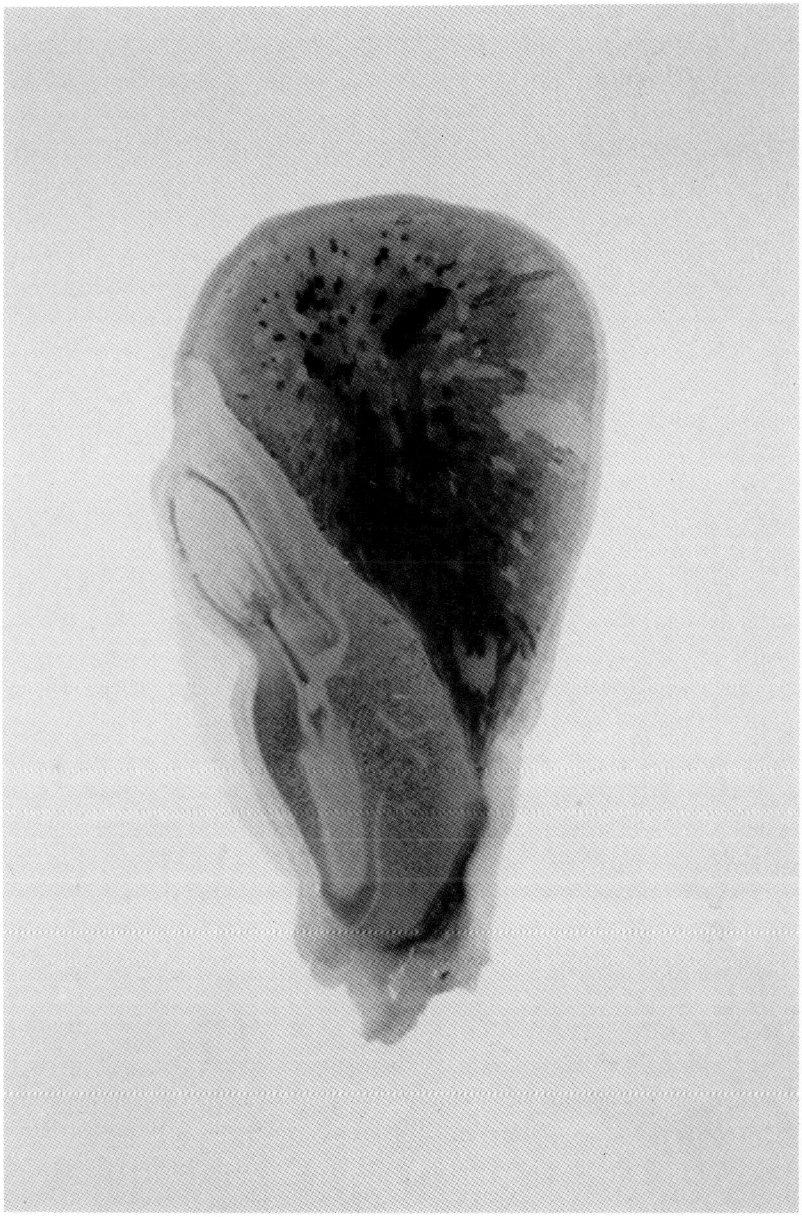

Figure 3. Longitudinal section through a mature maize kernel to show its parts. The cut surface was treated with an iodine-potassium iodide solution to stain amylose in the starch granules of individual cells. The narrow outer layer of the kernel is the pericarp, a maternal tissue. The embryo and endosperm are side-by-side but clearly delimited from each other. The endosperm is above and to the right of the embryo. In this photograph, four parts of the embryo may be noted. To the left and adjacent to the endosperm is the scutellum with its canals. The shoot, to the upper left, and the primary root below it, are connected to each other by the scutellar node. The different staining intensities in the endosperm cells reflect different amounts of amylose in them. These differences relate to the presence and action of a transposable *Ac* element at the *Wx* locus (23). The *Wx* gene is responsible for conversion of amylopectin to amylose, but only in the endosperm, not in the embryo.

some with a newly broken end to the primary endosperm nucleus, this cycle will continue throughout mitoses in the developing endosperm. Similarly, if the two nuclei in the central cell each have such a ruptured end of a chromosome, either the chromosome or chromatid type of breakage-fusion-bridge cycle will occur throughout endosperm development. When, however, a single ruptured end of a chromosome is delivered to the zygote nucleus by either the egg or the sperm nucleus, the ruptured end will "heal" subsequently; the cycle ceases in the developing embryo. Although not yet proven at the molecular level, it is altogether likely that the healing process represents the formation of a new telomere at the ruptured end. This assures that the special requirement for DNA replication at free ends of chromosomes will be satisfied. This new telomere functions normally thereafter. It is as stable in this regard as any other telomere of the maize complement, and tests of this cover many cell and plant generations.

A cell capable of repairing a ruptured end of a chromosome must sense the presence of this end in its nucleus. This sensing activates a mechanism that is required for replacing the ruptured end with a functional telomere. That such a mechanism must exist was revealed by a mutant that arose in my stocks. When homozygous, this mutant would not allow the repair mechanism to operate in the cells of the plant. Entrance of a newly ruptured end of a chromosome into the zygote is followed by the chromatid type of breakage-fusion-bridge cycle throughout mitoses in the developing plant. This suggests that the repair mechanism in the maize strains I have been using is repressed in cells producing the male and female gametophytes and also in the endosperm, but is activated in the embryo. Although all of this was known before the 1944 experiment was conducted, the extent of trauma perceived by cells whose nuclei receive a *single* newly ruptured end of a chromosome that the cell cannot repair, and the speed with which this trauma is registered, was not appreciated until the winter of 1944—45.

V. *Proof that entrance of a newly ruptured end of a chromosome into a telophase nucleus can initiate activations of previously silent genomic elements*

By 1947 it was learned that the bizarre variegated phenotypes that segregated in many of the self-pollinated progenies grown on the seedling bench in the fall and winter of 1944–45, were due to the action of transposable elements. It seemed clear that these elements must have been present in the genome, and in a silent state previous to an event that activated one or another of them. To my knowledge, no progenies derived from self-pollination of plants of the same strain, or related strains, had ever been reported to have produced so many distinctly different variegated expressions of different genes as had appeared in the progenies of these closely related plants grown in the summer of 1944. It was concluded that some traumatic event was responsible for these activations. The unique event in the history of these plants relates to their origin. Both parents of the plants grown in the summer of 1944 had contributed a chromosome with a newly ruptured end to the zygote that gave rise to each of these plants. The rupture occurred, in the first instance, at a meiotic anaphase in each parent, and the ruptured end then underwent the succession of mitotic anaphase breaks associated with the chromatid type of breakage-fusion-bridge cycle during development of the male and female gametophytes—the pollen grain and the embryo sac. I suspected that an activating event could occur at some time during this phase of the life history of the parent plants. I decided, then, to test if this might be so.

The newly activated elements, isolated from the initial experiment, were observed to regulate gene expressions following insertion of an element, or one of its derivatives, at a gene locus (13, 14, 15). In some instances the general mode of regulation resembled that produced by the *Dotted* "gene" on the standard recessive allele, *a*, of the *A* gene. This *a* allele represents the second recognized gene among a number of others whose action is required for production of anthocyanin pigment, either red or purple, in plant tissues and also in several tissues of the kernel (16). In the mid-nineteen-thirties Marcus Rhoades discovered this *Dotted* (*Dt*) element in a strain of Black Mexican sweet corn (17, 18). It behaved as a dominant gene that caused the otherwise very stable but non-functional *a* allele to mutate to new alleles that allowed anthocyanin pigment to be formed in both plant and kernel. The name *Dotted*, given to it, refers to the pattern of mutations that is expressed in plants and kernels homozygous for the *a* allele on chromosome 3 and having a *Dt* element located elsewhere in the chromosome complement. Small streaks of red or purple pigment appear in plants; the kernels have dots of this pigment distributed over the aleurone layer. (The aleurone layer is the outermost layer of the endosperm.)

Suspecting that this *Dt* had originated from activation of a previously silent element in the maize genome, and also suspecting that such silent elements must be present in all maize genomes, it was decided to test whether the breakage-fusion-bridge cycle would activate one such silent *Dt* element. My stocks that were homozygous for the *a* allele had never given any indication of *Dt* action. Therefore, these stocks were used to test if a presumed silent *Dt* element could be activated by the chromatid type of breakage-fusion-bridge cycle. Plants homozygous for *a*, and having a chromosome 9 constitution similar to that described for Fig. 2, were used as pollen parents in crosses to plants that also were homozygous for *a*. These pollen parents had the duplication of the short arm as shown in Fig. 2, but its homologue was deficient for a terminal segment of this arm that would not allow pollen grains having it to function. It was determined that 70 to 95 percent of the *functional* pollen grains produced by these plants carried sperms having a chromosome 9 with a newly ruptured end of its short arm, the initial rupture having occurred at the previous meiotic anaphase. Thus, most of the embryos and endosperms in the kernels on ears produced by the described cross started development with a newly ruptured end of the short arm of chromosome 9 in both embryo and endosperm. These kernels were searched for dots of pigment in their aleurone layer. A number of kernels had such dots. Many of these dots were confined to a restricted area of the aleurone layer, suggesting that this area represented a sector derived from a single cell in which a silent *Dt* element had been activated. One kernel had dots distributed over all of the aleurone layer, suggesting that the sperm nucleus contributing to the primary endosperm nucleus already had an activated *Dt* element in it. Tests of the plant arising from this kernel indicated that the sister sperm nucleus that had fused with the egg nucleus did not have such an activated element. Apparently, the activating event had occurred in the nucleus of only one of the two sperms. Significantly, this is only two mitoses removed from initiation of the chromatid type of breakage-fusion-bridge cycle. As mentioned earlier, this cycle continues during successive mitoses in the development of the endosperm. This continuing cycle could explain the presence in some kernels of sectors with pigmented dots, and this, in turn, would imply that activations of silent elements could occur at any time that this cycle remains in operation (19, 20, 21).

A similar test was conducted some years later by Doerschug (22), using the same constitution of the pollen parent as that just described. He obtained similar results. In his tests, however, two kernels with spots of pigment distributed over the entire aleurone layer, proved to have an activated *Dt* element in the plant grown from each of these kernels. The behavior of these two newly activated *Dt* elements was extensively studied by Doerschug. The two elements differed from each other not only in their location in the chromosome complement, but also in their mode of control of the time and place of change in *a* gene action. We now know that such differences in performance of these elements are expected.

Doerschug's two *Dt* isolates are most significant for appreciating the speed of response of a genome to entrance of a newly ruptured end of a chromosome into a telophase nucleus. Each *Dt* element must have been activated in the microspore nucleus or not later than the generative nucleus produced by division of the microspore nucleus. The unexpected event probably is sensed and acted upon from the initial entrance of a single ruptured end of a chromosome into a telophase nucleus, and in each subsequent nucleus that receives such a newly ruptured end. It is recognized that *Dt* is only one among a number of silent, but potentially transposable elements, that are present in maize genomes. Most probably some of these other silent elements were activated during the described test, but they were not able to be recognized as were activations of *Dt* elements. A similar approach could be used to detect such activations if a proper indicator stock were chosen for the test. Detection of silent elements is now made possible with the aid of DNA cloning method. Silent *Ac* (Activator) elements, as well as modified derivatives of them, have already been detected in several strains of maize (23). When other transposable elements are cloned it will be possible to compare their structural and numerical differences among various strains of maize. Present evidence suggests that wide differences may be found in this regard, as they have been found for middle and highly repetitious DNA sequences (24). In any one strain of maize the number of silent but potentially transposable elements, as well as other repetitious DNAs, may be observed to change, and most probably in response to challenges not yet recognized.

There are clear distinctions in comportment of ends of chromosomes on entering telophase nuclei. These relate to: (*1*) all chromosomes having normal ends, (*2*) two chromosomes, each with a single ruptured end, or one chromosome with both ends ruptured, and (*3*) one chromosome with a single broken end. Both ends of normal, unbroken chromosomes have a normal telomere. No difficulties are experienced. Two ruptured ends, neither with a telomere, will find each other and fuse. In these instances there is no immediate telomere problem. A single broken end has no telomere, and no other broken end with which to fuse. If the cell cannot make a new telomere, which is the case in the maize gametophytes and the endosperm, trauma must be experienced as the evidence indicates. Telomeres are especially adapted to replicate free ends of chromosomes. When no telomere is present, attempts to replicate this uncapped end may be responsible for the apparent "fusions" of the replicated chromatids at the position of the previous break as well as for perpetuating the chromatid type of breakage-fusion-bridge cycle in successive mitoses. Activation of potentially transposable elements, as well as other structural modifications of the chromosomes not considered here, are recognizable consequences of the cell's response to the continuing trauma.

VI. *Further examples of response of genomes to stress*

Cells must be prepared to respond to many sources of stress. Mishaps that affect the operation of a cell must be occurring continuously. Sensing these and instigating repair systems are essential. We are aware of some of the mishaps affecting DNA, and also of their repair mechanisms, but many others could be difficult to recognize. Homeostatic adjustments to various accidents would be required if these accidents occur frequently. Many such mishaps and their adjustments would not be detected unless some event or observation directed attention to them. Some, however, are so conspicuous that they cannot fail to be noted. For example, in *Drosophila,* some sensing device recognizes when the amount of rDNA is above or below the standard amount, and then sets in motion the system that will make the proper adjustment. Similarly, amitotic divisions of macronuclei in ciliates may result in unequal distributions of DNA to daughter nuclei. These deviations are sensed in each daughter cell. To make adjustments, one cell may respond by increasing its DNA content to reach the standard amount. The other cell may discard the excess DNA. There must be numerous homeostatic adjustments required of cells. The sensing devices and the signals that initiate these adjustments are beyond our present ability to fathom. A goal for the future would be to determine the extent of knowledge the cell has of itself, and how it utilizes this knowledge in a "thoughtful" manner when challenged.

One class of programmed responses to stress has received very little attention by biologists. The stress signal induces the cells of a plant to make a wholly new plant structure, and this to house and feed a developing insect, from egg to the emerging adult. A single Vitus plant, for example, may have on its leaves three or more distinctly different galls, each housing a different insect species. The stimulus associated with placement of the insect egg into the leaf will initiate reprogramming of the plant's genome, forcing it to make a unique structure adapted to the needs of the developing insect. The precise structural organization of a gall that gives it individuality must start with an initial stimulus, and each species provides its own specific stimulus. For each insect species the same distinctive reprogramming of the plant genome is seen to occur year-after-year. Some of the most interesting and elaborate plant galls house developing wasps. Each wasp species selects its own responding oak species, and the gall structure that is produced is special for each wasp to oak combination. All of these galls are precisely structured, externally and internally, as a rapid examination of them will show.

The galls on roots of legumes that are associated with nitrogen fixing bacteria are readily available for examination. They illustrate in their own way an example of reprogramming of the plant genome by a stimulus received from a foreign organism. Induction of such reprogrammings by insects, bacteria, fungi, and other organisms, which are not a required response of the plant genome at some stage in its life history, is quite astounding. But it is no more astounding, it would seem, than the sharing of a single genome by two brilliantly designed organisms, the caterpillar and the moth. It is becoming increasingly apparent that we know little of the potentials of a genome. Nevertheless, much evidence tells us that it must be vast.

Many known and explored responses of genomes to stress are not so precisely programmed. Activation of potentially transposable elements in maize is one of these. We do not know when any particular element will be activated. Some responses to stress are especially significant for illustrating how a genome may modify itself when confronted with unfamiliar conditions. Changes induced in genomes when cells are removed from their normal locations and placed in tissue culture surroundings are outstanding examples of this.

The establishment of a successful tissue culture from animal cells, such as those of rat or mouse, is accompanied by readily observed genomic restructuring. None of these animal tissue cultures has given rise to a new animal. Thus, the significance of these changes for the organism as a whole is not yet directly testable. The ability to determine this is a distinct advantage of plant tissue cultures. Many plant tissue cultures have developed new plants and, in some instances, many plants from a single initial cell or tissue isolate. A reason for this difference in behavior of animal and plant tissue cultures is not difficult to find. In many animals the germline is set aside early in cleavage, allowing the soma—a dead-end structure—to develop by any means, including genome restructuring and nonreversible programming. In higher plants, each fertile flower has the equivalent of a "germline" in it. The flower makes the gametes and initiates embryo formation. In this regard, consider the many flowers that may be produced by a bush or a tree. Some system must operate to reprogram the genome in those cells of the flower that will produce the gametes and establish the zygote. This implies that the specific programming sequences, earlier initiated and required for flower production, must be "erased" in order to return the genome to its very early state. If this occurs in so many places in a bush or a tree, then it is not surprising that it may occur in a plant cell or a cluster of cells not within a flower. Also in many plants such resettings are a common means of initiating new individuals from somatic cells. In these instances, however, the process of resetting is regulated, and the genome is not permanently restructured. This is not true for plants arising from many tissue cultures. The treatment, from isolation of the cell or cells of a plant, to callus formation, and then to production of new plants from the cells of these calluses, must inflict on the cells a succession of traumatic experiences. Resetting of the genome, in these instances, may not follow the same orderly sequence that occurs under natural conditions. Instead, the genome is abnormally reprogrammed, or decidedly restructured. These restructurings can give rise to a wide range of altered phenotypic expressions. Some of the altered phenotypes are readily observed in the newly produced plants themselves. Others appear in their progeny. Some initially displayed altered phenotypes do not reappear in the progeny. Their association with genomic change remains problematic. Other altered phenotypes clearly reflect genomic restructuring, and various levels of this have been observed. It may be safe to state that no two of the callus derived plants are exactly alike, and none is just like the plant that donated the cell or cells for the tissue culture. The many levels of genomic modification that already are known and expressed as changed genotypes and phenotypes could be potent sources for selection by the plant breeder, and incidentally, for theoretical ponderings by the biologist.

Modifications in gene expression may be induced in plants when they are infected with an RNA virus. Instances of this may be detected merely by viewing infected plants in the field. For example, patterns of anthocyanin pigment distribution, normally highly regulated and prominently displayed in the flowers of a plant, may appear grossly distorted in those parts of a plant that clearly reveal the virus infection. Recently, it was learned that infection of maize plants with barley stripe mosaic virus, an RNA virus, may traumatize cells to respond by activating potentially transposable elements. These, in turn, may then enter a gene locus and modify its expression (25). Such changes in expression of known genes may be exhibited in the self-pollinated progeny of infected plants. More often they are detected in later generations. Yet, no virus genome has been detected in the immediate progeny of infected plants or in those plants shown to have a transposable element newly inserted at a known gene locus.

Species crosses are another potent source of genomic modification. Plants have provided many excellent examples of this. The advantage of plants is the ease of making crosses to obtain hybrids, the simplicity of growing them, the ready availability of their chromosomes, and the ability to obtain progeny in quantities, if necessary. The alterations produced when the genomes of two species are combined reflect their basic incompatibilities. Evidence of this is the appearance of the same types of genome change whenever the same two species are crossed. Expressions of incompatibilities do differ, but their nature is always in accordance with the particular two species whose genomes are combined. The genus *Nicotiana* has a large number of species that differ from each other in chromosome number, chromosome organization, and phenotypic expressions. Genome incompatibilities have been observed in a large number of

2-by-2 combinations of species. An illustration is the behavior of chromosomes in the hybrid plant produced by the cross of *N. tabacum* by *N. plumbaginifolia*. The chromosomes of *plumbaginifolia* are lost during development of this hybrid plant. Although whole chromosome losses appear to be common, other irregularities in chromosome behavior also occur. These are chromosome fragments, chromosome bridges in somatic anaphases, and the appearance in an occasional metaphase plate of a single, very much elongated chromosome, termed a "megachromosome." The presence of one or two such hugely elongated chromosomes in some somatic metaphase plates characterizes the hybrid derived from the cross of *N. tabacum* × *N. otophora*. In this instance it is known that a heterochromatic segment in each of two chromosomes of the *otophora* set contributes to these linear amplifications (26, 27). Hybrids produced by crosses of distantly related *Nicotiana* species are known to give rise to tumors, some of which resemble teratomas. In one instance it was shown that tumor production relates to a single heritable modification which was initiated in the hybrid.

Major restructuring of chromosome components may arise in a hybrid plant and continue to arise in its progeny, sometimes over successive plant generations. The restructuring may range from apparently simple to obviously complex. These are associated with translocations, inversions, deficiencies, duplications, etc., that are simple in some instances or variously intercalated in others. New stable or relatively stable "species" or "genera" have been derived from such initial hybrids. The commercially useful plant, *Triticale*, is an example. Wheat (*Triticum*) and rye (*Secale*) were crossed and the combined set of chromosomes doubled to provide reproductive stability. Nevertheless, this genome was not altogether stable. Selections continued in later generations for better performances with considerable success, even though instabilities were not eliminated altogether. Some species of *Triticum* undoubtedly arose by a comparable mechanism as that outlined for *Triticale*, and different related genera made their contribution to some of these *Triticum* species. Evidence for this is exceptionally clear (30).

Undoubtedly, new species can arise quite suddenly as the aftermath of accidental hybridizations between two species belonging to different genera. All evidence suggests that genomic modifications of some type would accompany formation of such new species. Some modifications may be slight and involve little more than reassortments of repetitious DNAs, about which we know so little. (The adjective "slight" refers to the apparent simplicity of the restructuring mechanism rather than the significance of its consequences). Major genome restructuring most certainly accompanied formation of some species. Studies of genomes of many different species and genera indicate this. Appreciation of the various degrees of reassortment of components of a genome, that appear during and following various types of genome shock, allows degrees of freedom in considering such origins. It is difficult to resist concluding that some specific "shock" was responsible for the origins of new species in the two instances to be described below.

The organization of chromosomes in many closely related species may resemble one another at the light microscope level. Only genetic and molecular analyses would detect those differences in their genomes that could distinguish them as species. In some instances of this type, distinctions relate to the assortment of repetitious DNAs over the genome, as if a response to shock had initiated mobilities of these elements. In other instances, distinctions between related species are readily observed at the light microscope level, such as polyploidizations that are common in plants, or amplifications of DNA that do not alter chromosome number or basic metaphase morphologies. Others relate to chromosome fusions or fragmentations, or readily observed differences in the placement of specific DNA segments. The literature is full of descriptions of differences in chromosome organization among the species of a genus. Two instances of these latter differences warrant special consideration here, because the observed differences in chromosome organization suggest origins from a response to a single event. One response gave rise to extensive fusions of chromosomes. The other placed heterochromatic segments at new locations within the chromosomes of the set.

That such multiple chromosome changes may relate to some initial event occurring in a cell of the germline is proposed and defended in a review article by King (31). An example that would fit his proposal is the organization of chromosomes of the Indian muntjac deer *(Muntiacus muntjak)* (32) when compared with its closely related species, *M. reevesi,* the Chinese muntjac. The latter species has 46 chromosomes as the diploid number, whereas the Indian muntjac has 6 chromosomes in the female and seven chromosomes in the male, and these chromosomes are huge in comparison with those of the Chinese muntjac. Observations of the chromosomes in the hybrid between these two species strongly supports chromosome fusion as the mechanism of origin of the reduced number and huge size of the Indian muntjac chromosomes (33). In general, evidence of fusion of chromosomes is plentiful. When two or three chromosomes of a set appear to have arisen by fusion, the question of simultaneous or sequential events responsible for these fusions cannot be determined with certainty. In the case of the Indian muntjac it is difficult to avoid the conclusion that the fusions of so many chromosomes resulted from some initial shocking event that activated a fusion mechanism already known to exist from the fusions of individual chromosomes in many other organisms. Whatever the cause, the changed chromosome organization is stunning.

Another stunning example of differences in chromosome organization between species is reported by S. Beermann in an extraordinarily thorough and fascinating account (34). This report describes the chromosome organization in three species of the copepod genus *Cyclops.* The main differences among them to be considered here relate to distributions of conspicuous heterochromatic blocks in the chromosomes of each species. In one species, these blocks are confined to the ends of chromosomes. In another species, blocks of heterochromatin are at the ends of chromosomes, but also positioned to each side of the centromere. In the third species, blocks of heterochromatin are distributed all along the chromosomes. An additional feature of this heterochromatin is its unchanged presence in cells of the germline, in contrast to its elimination at cleavages specific for each species and in cells destined to produce the soma.

The elimination process is associated with formation of rings of DNA cut out from the heterochromatin (35). Again it is difficult to avoid concluding that these distinctive distributions of heterochromatin relate to unusual and disturbing events, and that these events activate mechanisms that can redistribute heterochromatin to specific sites.

VII. *Concluding statement*

The purpose of this discussion has been to outline several simple experiments conducted in my laboratory that revealed how a genome may react to conditions for which it is unprepared, but to which it responds in a totally unexpected manner. Among these is the extraordinary response of the maize genome to entrance of a single ruptured end of a chromosome into a telophase nucleus. It was this event that, basically, was responsible for activations of potentially transposable elements that are carried in a silent state in the maize genome. The mobility of these activated elements allows them to enter different gene loci and to take over control of action of the gene wherever one may enter. Because the broken end of a chromosome entering a telophase nucleus can initiate activations of a number of different potentially transposable elements, the modifications these elements induce in the genome may be explored readily. In addition to modifying gene action, these elements can restructure the genome at various levels, from small changes involving a few nucleotides, to gross modifications involving large segments of chromosomes, such as duplications, deficiencies, inversions, and other more complex reorganizations.

It was these various effects of an initial traumatic event that alerted me to anticipate unusual responses of a genome to various shocks it might receive, either produced by accidents occurring within the cell itself, or imposed from without, such as virus infections, species crosses, poisons of various sorts, or even altered surroundings such as those imposed by tissue culture. Time does not allow even a modest listing of known responses of genomes to stress that could or should be included in a discussion aimed at the significance of responses of genomes to challenge. The examples chosen illustrate the importance of stress in instigating genome modification by mobilizing available cell mechanisms that can restructure genomes, and in quite different ways. A few illustrations from nature are included because they support the conclusion that stress, and the genome's reactions to it, may underlie many species formations.

In the future attention undoubtedly will be centered on the genome, and with greater appreciation of its significance as a highly sensitive organ of the cell, monitoring genomic activities and correcting common errors, sensing the unusual and unexpected events, and responding to them, often by restructuring the genome. We know about the components of genomes that could be made available for such restructuring. We know nothing, however, about how the cell senses danger and instigates responses to it that often are truly remarkable.

REFERENCES

1. Muller, H.J., Science *66*, 84–87 (1927).
2. Muller, H.J., Proc. Nat. Acad. Sci. (U.S.A.) *14*, 714–726 (1928).
3. Hanson, F. B., Science *67*, 562–563 (1928).
4. Stadler, L.J., Science *68*, 186–187 (1928).
5. McClintock, B., Missouri Agr. Exp. Sta. Res. Bull. *163*, 1–30 (1931).
6. McClintock, B., Proc. Nat. Acad. Sci (U.S.A.) *18*, 677–681 (1932).
7. McClintock, B., Genetics *23*, 315–376 (1938).
8. McClintock, B., Genetics *26*, 542–571 (1941).
9. McClintock, B., Genetics *26*, 234–282 (1941).
10. McClintock, B., Cold Spring Harbor Symp. Quant. Biol. *9*, 72–80 (1941).
11. McClintock, B., Proc. Nat. Acad. Sci (U.S.A.) *28*, 458–463 (1942).
12. McClintock, B., Stadler Genetics Symp. *10*, 25–48 (1978).
13. McClintock, B., Carnegie Inst. Wash. Year Book No. *47*, 155–169 (1948).
14. McClintock, B., Carnegie Inst. Wash. Year Book No. *48*, 142–154 (1949).
15. McClintock, B., Proc. Nat. Acad. Sci (U.S.A.) *36*, 344–355 (1950).
16. Emerson, R. A., Cornell University Agr. Exp. Sta. Memoir *16*, 231–289 (1918).
17. Rhoades, M. M., Jour. Genetics *33*, 347–354 (1936).
18. Rhoades, M. M., Genetics *23*, 377–397 (1938).
19. McClintock, B., Carnegie Inst. Wash. Year Book No. *49*, 157–167 (1950).
20. McClintock, B., Carnegie Inst. Wash. Year Book No. *50*, 174–181 (1951).
21. McClintock, B., Brookhaven Symp. in Biol. *18*, 162–184 (1965).
22. Doerschug, E. B., Theor. and Appl. Genetics *43*, 182–189 (1973).
23. Fedoroff, N., Wessler S. and Shure, M. Cell *35*, 235–242 (1983).
24. Hake, S. and V. Walbot, Chromosoma (Berl.) *79*, 251–270 (1980).
25. Mottinger, J. P., Dellaporta, S. L. and Keller, P. In press.
26. Gerstel, D. U. and Burns, J. A. Chromosomes Today *1*, 41–56 (1966).
27. Gerstel, D. U. and Burns, J. A. Genetica *46*, 139–153 (1976).
28. Smith, H. H., Brookhaven Symp. in Biol. *6*, 55–78 (1954).
29. Smith, H. H., Brookhaven Lecture Series *52*, 1–8 (1965).
30. Feldman, M., In "Evolution of Crop Plants" (N. W. Simmonds, ed.) pp. 120–128 Longman, London and New York, (1976).
31. King, M., Genetica *59*, 53–60 (1982).
32. Wurster, D. H. and Benirschke, K. Science *168*, 1364–1366 (1970).
33. Shi Liming, Yingying Ye and Xingsheng, Duan. Cytogen. Cell Genet. *26*, 22–27 (1980).
34. Beermann, S., Chromosoma (Berl.) *60*, 297–344 (1977).
35. Beermann, S., and Meyer G. F. Chromosoma (Berl.) *77*, 277–283 (1980).

1984
Physiology or Medicine

NIELS K. JERNE
GEORGES J. F. KÖHLER and
CÉSAR MILSTEIN

"for theories concerning the specificity in development and control of the immune system and the discovery of the principle for production of monoclonal antibodies"

THE NOBEL PRIZE FOR PHYSIOLOGY OR MEDICINE

Speech by Professor HANS WIGZELL of the Karolinska Institute.
Translation from the Swedish text

Your Majesties, Your Royal Highnesses, Ladies and Gentlemen,

It is typical for the human mind that little thought goes to the functions of our body when we are healthy, yet acute interest frequently develops in times of disease. The immune system is a somewhat anonymous, talented and well-trained cellular society within ourselves which must function properly to maintain our health. The immune defence has the inherent capacity to rapidly recognize foreign material and can subsequently remember this contact for decades, thus creating the basis for vaccination. Through a clever usage of genetic material and large numbers of cells, the immune system within a single human being is able to produce defence molecules, antibodies, in billions of different shapes. The Nobel prize winners in physiology or medicine this year have all worked with the capacity of the immune system to produce specific antibodies.

Niels Jerne is the great theoretician in modern immunology. He entered the immunological arena comparatively late in his life and was 44 years old when in 1955 he published his first important theory of the construction of the immune system. Jerne proposed that the well-known capacity of the immune defence to recognize myriades of foreign molecules was something predetermined, already existing in the body when the very first contact with a foreign structure was made. What then happened was merely a selection amongst the naturally occurring antibody population resulting in an increase in production of exactly those antibodies which happened to have a good fit for the structure. Jerne's theory stood in great contrast to prevailing theories at that time, but it was rapidly confirmed and extended. We now know that Darwin's laws about natural selection indeed apply to the cells of the immune system: Those cells which happen to have received the property to produce a wanted antibody type will upon vaccination be rewarded with regard to proliferative capacity and survival.

Jerne took another known feature of the immune defence as the starting point for his next important theory, in 1971. The immune system always expresses a very strong defence of the private and unique features of the tissues within one individual. This behaviour creates great problems whenever attempts are made to transplant tissue from one individual to another. Jerne assumed that the molecules in the tissues that cause these reactions, the so-called transplantation antigens, must have their normal functions within the body of the individual. He proposed that one function of these molecules could be to serve as a specific driving force for the cells of the immune system, thus creating a large number of cells from which cells especially suitable to defend the tissues of the host would become selected. Special organs, such as the thymus, were assumed to serve as a combination of greenhouse and university for these cells. In this theory Jerne did predict to a great extent how the specificity of the cellmediated immunity is generated.

In the third great theory, in 1974, Niels Jerne introduced us to the mirror halls of immunology. The immune system is visualized somewhat like a gigantic computer where constant communication and regulation takes place in between the different components, the cells. The number of cells in such a network system in an adult human being exceeds 10^{12} (one million millions); and the system has through its capacity to produce billions of different forms of antibodies an enormous inbuilt richness with regard to structural variations. Jerne proposed that this should allow the creation of multiple complementary situations where certain antibodies would have select capacity to combine with their mirror images. Some antibodies would according to the theory then even mimic foreign molecules against which other antibodies would normally be produced during immunization. Jerne postulated that pairs of antibodies and their mirror images would spontaneously be produced during the development of the immune system, thus creating the possibilities for communication networks and regulatory equlibria. During immunization the foreign molecules would enter the mirror halls of immunology where the different pairs of antibodies and cells would perform their interdependent piruettes and separate the partners chasing away the mirror images. This change in equilibrium would then serve as a driving force, resulting in immunity. It is now well documented that network forces of the fascinating type that Jerne predicted do indeed exist inbuilt in our own immune systems. The theory also predicted the almost mind-boggling possibility that antibodies of the mirror image-type could replace the foreign material completely when inducing immunity. This is now a proven reality. Thus, it is for instance possible to induce a long lasting immunity against hepatitis virus by immunization with mirror image antibodies to the antibodies against hepatitis virus without ever using the virus in the vaccination.

In conclusion, Niels Jerne has via his visionary theories enabled modern immunology to make major leaps of progress. Several concepts in immunology now considered as self-evident have their roots in some of his pioneering thoughts.

In order to fully understand the importance of Georges Köhler's and César Milstein's discoveries we should first take some steps back. Sera from intentionally immunized animals or humans constitute very important tools in the hospitals as well as in research laboratories. They are used to diagnose infectious diseases, as well as to determine the concentration of a particular hormone in a sample. But every one of these immune sera contains a unique mixture of antibodies produced by a large number of different cells and their progeny and the various antibodies react in a similar yet distinctly different manner. Thus, each immune serum has to be tested to determine the special features of that particular serum with regard to its ability to distinguish between two related hormones, different bacteria, etc. Regardless of whether the immune serum is close to being perfect or not, it will always be used up, and it is then necessary to start again trying to produce a similar kind of serum. International standardizations of tests using immune sera have thus been greatly hampered.

The discovery and development of principles for production of the so-called monoclonal antibodies by the hybridoma technique by Georges Köhler and César Milstein have largely solved all the above major problems. And the story of the discovery of the technique also contains the moral of a saga, where the evil is put into the service of the good. How did this discovery take place? César Milstein is a highly prominent biochemist working for a long time in Cambridge in England. A major interest in his research has been to explore various facets of antibody production. Milstein used tumor cells which had arisen in cells of a type that normally produce antibodies. Such tumors also produce proteins which in all respects look like antibodies, although it is difficult to find suitable foreign structures to which they can bind. Milstein wanted amongst other things to study what would happen if two different tumor lines were allowed to fuse, e.g. what would happen to the production of the antibody-like proteins if for instance the tumor cells came from different species? Milstein constructed tumor cell lines allowing only hybrid cells between the two tumor cells to grow in certain defined tissue culture solutions. The systems worked and the hybrid cells produced large quantities of the antibody-like proteins, some of which at the molecular level could be shown to be hybrid molecules as well.

At the same time the young researcher Georges Köhler struggled in Basel in Switzerland to study normal antibody-producing cells in tissue culture. His research was in part frustrating as he could only get very few cells to survive for short periods of time. Köhler knew of the important studies of Milstein, and it seemed logical to see if normal antibody-forming cells could be fused with tumor cells to produce long-lived hybrid cell lines. If this was indeed possible the experiments of Milstein would indicate that they should then continue to produce their antibodies. At the same time the normally evil feature of tumor cells, the capacity to proliferate for ever, would now be turned into a very beneficial feature. Köhler went to Milstein's laboratory and together they wrestled with the problems and managed to solve them in a hectic two year period, 1975–1976. By that time they had succeeded to develop a technique allowing them at will to fish up exactly those rare antibody-producing cells that they wanted from a sea of cells. These cells were fused with tumor cells creating hybrid cells with eternal life and capacity to produce the very same antibody in high quantity. Köhler and Milstein called these hybrid cells hybridomas, and as all cells in a given hybridoma come from one single hybrid cell, the antibodies made are monoclonal.

Köhler's and Milstein's development of the hybridoma technique for production of monoclonal antibodies have in less than a decade revolutionized the use of antibodies in health care and research. Rare antibodies with a tailor-made-like fit for a given structure can now be made in large quantities. The hybridoma cells can be stored in tissue banks and the very same monoclonal antibody can be used all over the world with a guarantee for eternal supply. The precision in diagnosis is greatly improved, and entirely new possibilities for therapy have been opened up via the hybridoma technique. Rare molecules present in trace amounts in complex solution can now be purified in an efficient manner using monoclonal antibodies. In all, it is therefore correct to describe the hybridoma technique discovered by Georges Köhler and César Milstein as one of the major methodological advances in medicine during this century.

Dr. Jerne, Dr. Köhler and Dr. Milstein,

On behalf of the Nobel Assembly of the Karolinska Institute I would like to congratulate you on your outstanding accomplishments and ask you to receive the Nobel Prize in Physiology or Medicine from the hands of His Majesty the King.

Niels Jerne

NIELS K. JERNE

born 23rd December 1911, London

My parents, Hans Jessen Jerne and Else Marie Lindberg, and their ancestors (back to the seventeenth century and earlier) all lived on the island Fanø and in a small adjacent area of western Jutland in Denmark. My family moved to London in 1910, and then to Holland during the first world war. I received my Baccalaureate in Rotterdam in 1928.

After two years of studying physics at the University of Leiden, I switched to medicine at the University of Copenhagen where I presented my thesis on the avidity of antibodies in 1951.

My wife Alexandra and I married in 1964, and now live in our house near Avignon. Further details of my curriculum vitae:

Research worker at the Danish State Serum Institute (1943–1956)
Research fellow at the California Institute of Technology, Pasadena (1954–1955)
Head of the Sections of Biological Standards and of Immunology at the World Health Organization, Geneva (1956–1962)
Professor of Biophysics at the University of Geneva (1960–1962)
Professor of Microbiology and Chairman of the Department, University of Pittsburgh (1962–1966)
Professor of Experimental Therapy at the Johann-Wolfgang-Goethe-Universität, Frankfurt, and Director of the Paul-Ehrlich-Institut, Frankfurt (1966–1969)
Director of the Basel Institute for Immunology, Basel (1969–1980)
Special Immunology Adviser to the Director of the Institut Pasteur, Paris (1981–1982)
Member emeritus and Honorary Chairman of the Advisory Board of the Basel Institute for Immunology (from 1981)

Member of the WHO Advisory Committee on Medical Research (1949–1968)
Member of the Advisory Committee on Medical Research of the Panamerican Health Organization (1963–1966)
Member of the Expert Advisory Panel of Immunology of the WHO since 1962
Honorary Member of the Robert-Koch-Institut, Berlin (1966)
Foreign Honorary Member of the American Academy of Arts and Sciences (1967)
Member of the Royal Danish Academy of Sciences (1969)
Chairman, Council of the European Molecular Biology Organization (1971–1975)
Gairdner Foundation International Award, Toronto (1970)
Doctor of Science, h.c., University of Chicago (1972)

Honorary Member of the American Association of Immunologists (1973)
Foreign Associate of the National Academy of Sciences (USA) (1975)
Waterford Bio-Medical Science Award, La Jolla (1978)
Doctor of Science, h.c., Columbia University, New York (1978)
Foreign Member of the American Philosophical Society (1979)
Doctor of Science, h.c., University of Copenhagen (1979)
Marcel Benoist Prize, Bern (1979)
Fellow of the Royal Society (1980)
Doctor of Science, h.c., University of Basel (1981)
Member of the Académie des Sciences de l'Institut de France (1981)
Paul Ehrlich Prize, Frankfurt (1982)
Honorary Member of the British Society for Immunology (1983)
Doctor of Medicine, h.c., Erasmus University, Rotterdam (1983)

The work referred to in the citation for the award of the Nobel Prize is mainly included in the following papers:

"The natural selection theory of antibody formation"
Proc.Nat.Acad.Sci.USA *41*, 849–857, 1955
"Immunological speculations"
Ann.Rev.Microbiol. *14*, 341–358, 1960
"Plaque formation in agar by single antibody-producing cells"
(with Albert A. Nordin), Science *140*, 405, 1963
"The natural selection theory of antibody formation: ten years later"
in "Phage and the origins of molecular biology"
Cold Spring Harbor Lab. of Quant. Biology 301–312, 1966
"Antibodies and learning"
in "The Nerurosciences"
The Rockefeller University Press 200–205, 1967
"Waiting for the End"
Cold Spring Harbor Symp. on Quant. Biology *32*, 591–603, 1967
"The somatic generation of immune recognition"
Eur.J.Immunol. *1*, 1–9, 1971
"What precedes clonal selection ?"
in "The ontogeny of Acquired Immunity"
Ciba Foundation Symposium, Elsevier, Amsterdam 1–15, 1972
"Towards a network theory of the immune system"
Ann. Immunol. (Inst. Pasteur) 125C, 373–389, 1974
"The immune system: a web of v-domains"
Academic Press, New York, Harvey Lectures *70*, 93–110, 1976
"Idiotypic networks and other preconceived ideas"
Immunological Reviews *79*, 5–24, 1984

THE GENERATIVE GRAMMAR OF THE IMMUNE SYSTEM

Nobel lecture, 8 December 1984

by

NIELS K. JERNE

Château de Bellevue, F-30210 Castillon-du-Gard, France

Grammar is a science that is more than 2000 years old, whereas immunology has become a respectable part of biology only during the past hundred years. Though both sciences still face exasperating problems, this lecture attempts to establish an analogy between linguistics and immunology, between the descriptions of language and of the immune system. Let me first recall some of the essential elements of the immune system, with which I shall be concerned. In 1890, von Behring and Kitasato (12) were the first to discover antibody molecules in the blood serum of immunized animals, and to demonstrate that these antibodies could neutralize diphtheria toxin and tetanus toxin. They also demonstrated the *specificity* of antibodies: tetanus antitoxin cannot neutralize diphtheria toxin, and vice versa. During the first 30 years, or more, after these discoveries, most immunologists believed that all cells of our body are capable of producing antibodies, and it took until the 1950's before it became clear, and until 1960 before it was demonstrated (13), that only the white blood cells named lymphocytes can produce antibodies. The total number of lymphocytes represent a little more than 1% of the body weight of an animal. Thus, it would not be wrong to say that our immune system is an organ consisting of about 10^{12} lymphocytes

$$\text{(human) immune system} = 10^{12} \text{ lymphocytes}$$

or in a mouse that is 3000 times smaller than we

$$\text{(mouse) immune system} = 3\text{x}10^{8} \text{ lymphocytes}$$

This brief description of the immune system disregards the fact that lymphocytes interact with most other cells in the body, which in my definition do not belong to the immune system sensu strictu.

Let me draw attention to the fact that this number of lymphocytes in the immune system is at least one order of magnitude larger than the number of neurons in the nervous system. Also, we should note that lymphocytes travel among most other cells of our body, that they circulate in blood and lymph, and occur in large concentrations in the spleen, lymph nodes, appendix, thymus and bone marrow. Strangely enough, however, they seem to be excluded from the brain. The 1960's was a very fruitful decade of immunological discoveries, of which I shall name a few: In the beginning of the decade, the primary structure of antibody molecules was clarified (14, 15); then followed the demonstration that the dictum of Burnet (2, 16) was correct, namely that all antibody molecules synthesized by one given lymphocyte are identical; and finally, towards the end of that decade, lymphocytes were shown to fall into two classes, called T cells and B cells, existing in the body in almost equal numbers (17, 18, 5). Only B lymphocytes, or B cells, however, can produce and secrete antibody molecules.

Schematically, I could picture this as in Fig. 1.

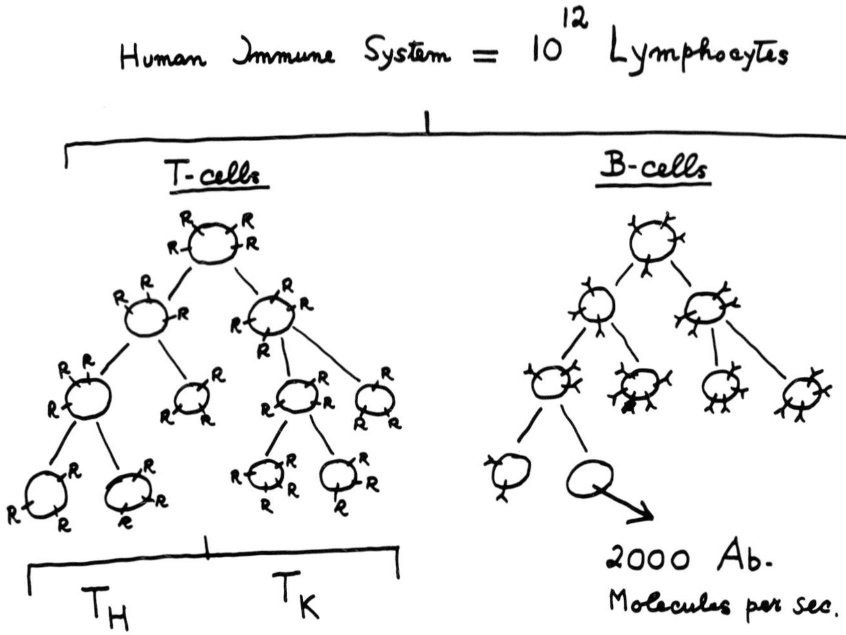

Fig. 1.

What I should like you to retain from this picture is both what we know as well as what we do not know at present. Thus, B lymphocytes are known to carry so-called receptor molecules on their surface (about 10^5 identical receptors per B cell), and when such a "resting" B cell is properly stimulated to divide and to mature, its descendants will end up excreting about 2000 antibody molecules per second, all of which are identical, and similar or identical to the receptors that the resting B cell originally displayed. This clonal nature of antibody formation was clearly demonstrated in the early 1970's (19, 20). The normal antibody response of an animal to a foreign antigen involves a large number of different clones, however, and is characterized by the polyclonal production, usually of several hundreds of different antibody molecules (21). T lymphocytes are also known to carry receptor molecules on their surface, but firstly these molecules are yet not well known because they have been discovered only during the past two years and secondly, the T cells do not excrete such molecules: these T cell receptors are antigen-recognizing molecules, but they do not contribute to the population of freely circulating antibody molecules which are purely B cell products. Furthermore, there are at least two different types of T cells, one of which is called T helper cells (T_H) (22) because they help B cells to become stimulated (or, in their absence, they deny B cells to receive a proper stimulus); the other type is called T killer cells (T_K) (23) because they are capable of killing other cells which they consider undesirable (such as virus-infected cells, or cells transplanted from another individual); moreover, as suppressor cells, they may prevent B cells from being stimulated (6).

Thus, in this picture, B cells have the single-minded desire to express their antibody language, but are subordinate to the T cells, that can either enhance or suppress this capacity.

Before looking at grammar, I must briefly describe the structure of an antibody molecule. No matter what you try to investigate in biology turns out to become increasingly complex — thus also the structure of antibody molecules. The basic element of all antibody molecules has been shown to be a Y-shaped protein structure (26) of about 150,000 daltons molecular weight. This is a three-dimensional structure, like all molecules and cells that biology has to deal with. Three-dimensionality tends to perplex our mind which is most at ease with one-dimensional, linear sequences, but I shall try to make a rough two-dimensional sketch of an antibody in Fig. 2.

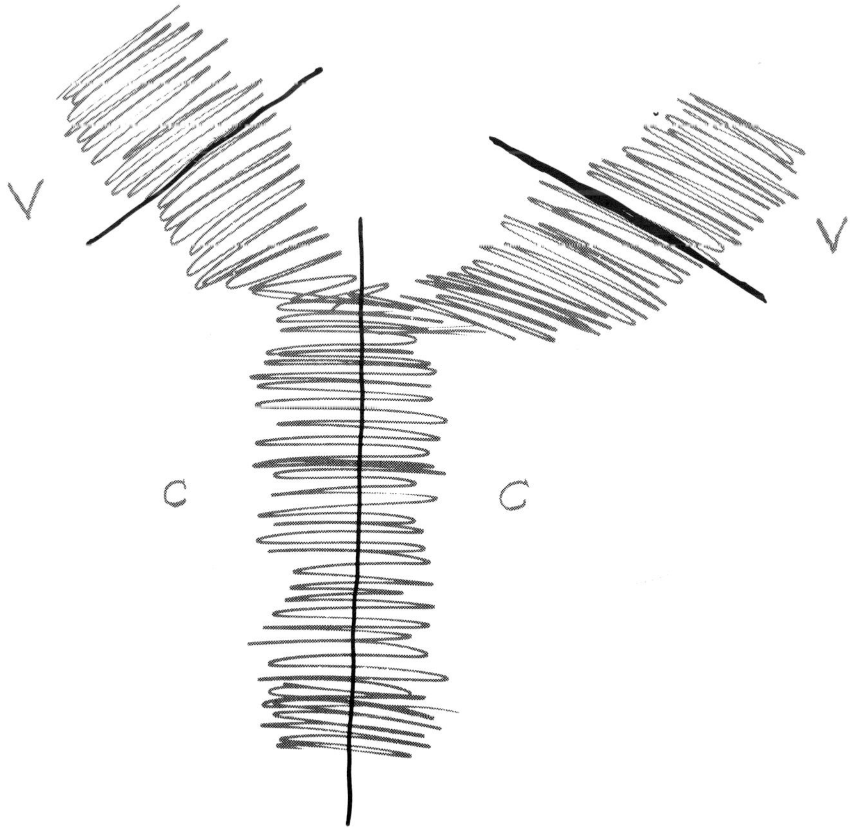

Fig. 2.

We can make some important cuts through this molecule. The vertical cut shows symmetry. It divides the antibody molecule in two identical halves. The two other cuts separate a so-called "constant" part (c) from a "variable" part (v). By constant part, we mean that molecules of different antibody specificity have this part in common (such as, for example, diphtheria antitoxin and tetanus antitoxin). By variable part, we mean that this is the region of the antibody molecule that determines its "specificity". The two variable parts are identical; that is to say, that the molecule, with regard to specificity, is divalent. The difficulty we face is not to transform this sketchy two-dimensional picture into a one-dimensional primary structure, but into a three-dimensional tertiary structure. The primary structure (14, 15) has been clarified: the half molecule is made up of a light polypeptide chain of about 214 amino acid residues, and a heavy polypeptide chain of a little more than 400 amino acid residues, as in Fig. 3.

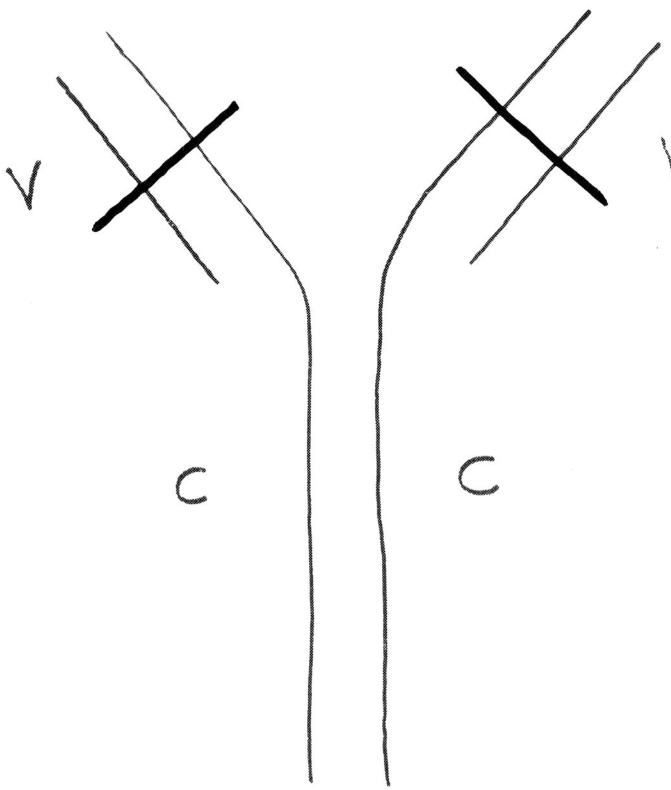

Fig. 3.

It turned out that antibody molecules of different specificity have identical amino acid sequences in their carboxyl-ended regions, but that they vary with respect to the amino acid sequences in the amino-ended regions of both heavy and light chains (24). It became obvious immediately that the great diversity of antibody molecules, the great number of different molecules which antibodies can recognize, or in other words, the great repertoire of antibody specificities, must result from an enormous number of varieties in the variable regions with respect to amino acid sequences. This insight does not solve our problems, however. It is like saying that the great variety of words or sentences in a language results from the enormous number of varieties with respect to the sequences of letters or of phonemes.

Interpretation in immunology remained practically as it had traditionally been, namely that the variable region of an antibody molecule forms a three-dimensional "combining site", and that "specificity" simply means that this combining site is complementary in shape to part of the three-dimensional profile of an antigen molecule. Antigen is the word that was given, and is still in use, for molecules that can induce the immune system to produce specific antibodies which can recognize these antigens. Traditionally, the antibody combining site was conceived of as a cleft which recognizes a protuberance on the outer shape of an antigenic molecule, and all antibodies were named after the antigens that they recognized, such as diphtheria antitoxin, antisheep red blood cell antibodies, anti-TNP, etc. (25). I shall now try to give you an impression of the size of this system of antigens and specific antibody molecules. Let us first consider macromolecules of molecular weights exceeding 10,000 daltons; they may be polysaccharides, proteins, lipoproteins, nucleic acids, viruses, bacteria – in fact any such molecule or particle existing in the world is an antigen to which the immune system can make specific antibodies. Moreover, molecules such as nitrophenol, or arsonate, or any organic or inorganic molecules you care to mention are antigenic when attached to a so-called carrier molecule, for example to a protein: the immune system will then produce antibodies that specifically recognize these molecules, even if they have been synthesized in a chemical laboratory without ever before having existed in the world (1). How is this possible? For example, the immune system of a mouse possesses no more than about 10^8 B lymphocytes, which would be the maximal available repertoire of variable regions on its antibody molecules. We realize that "recognition" need not be perfect, and that the same "combining site" might recognize, with more or less precision, a number of similar antigens.

I shall now turn to some remarkable discoveries, made during the past 25 years, showing that the variable regions of antibody molecules are themselves antigenic and invoke the production of anti-antibodies. Kunkel (27) showed that monoclonal myeloma antibodies, when injected into another animal, induce specific antibodies which recognize the particular myeloma antibodies used, but which do not recognize any other myeloma antibodies isolated from other myeloma patients. This work was extended by others, but mainly by Jacques Oudin and his colleagues in Paris, who showed that ordinary antibody molecules that arise in an immunized animal are antigenic and invoke the formation of specific anti-antibodies (28, 29, 30, 31). In other words, the variable region of an antibody molecule constitutes not only its "combining site", but also presents an antigenic profile (named its idiotype) against which anti-idiotypic antibodies can be induced in other animals. Moreover, it turned out that this antigenic, idiotypic profile of the variable region of a given antibody molecule is not a single site, but consists of several distinct sites against which a variety of different anti-idiotypic antibody molecules can be made. These individual sites are now named idiotopes, implying that the idiotype of one antibody molecule can be described as a set of different immunogenic idiotopes. And finally, it has been shown that the immune system of a single animal, after producing specific antibodies to an antigen, continues to produce antibodies to the idiotopes of the antibodies which it has itself made. The latter anti-idiotopic antibodies likewise display new idiotypic profiles, and the immune system turns out to represent a network of idiotypic interactions (7, 8, 9, 10, 11).

I shall now show, in Fig. 4, a preliminary picture of how we might vaguely try to imagine the shape of the variable region of an antibody molecule.

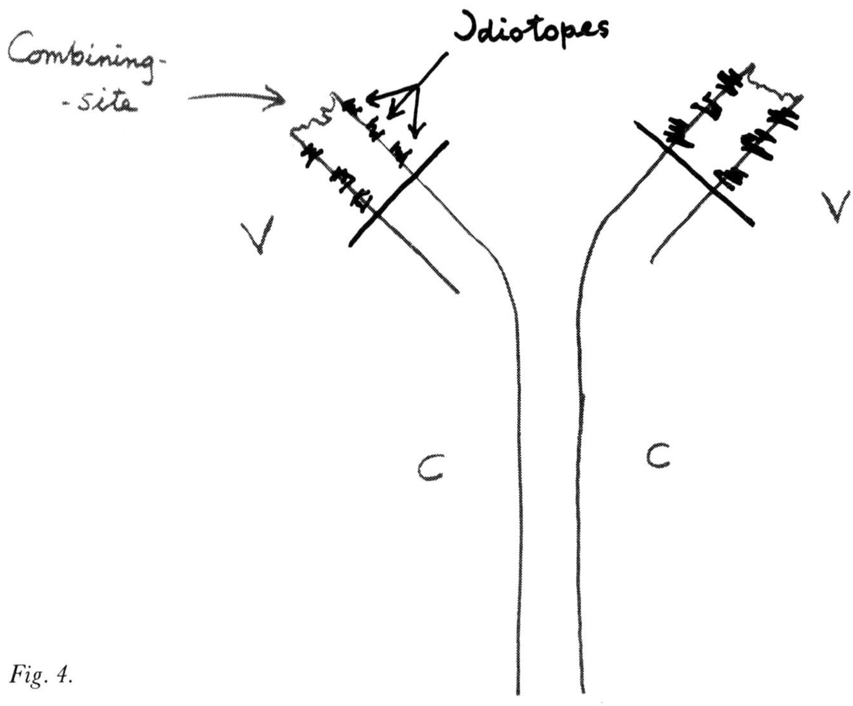

Fig. 4.

This picture is an historical compromise: we uphold our antigen-centered tradition (25) by retaining the notion of a "combining site" which enables the antibody molecule to recognize the antigenic molecule that induced its production, and we simply add a number of idiotopes on the same variable region, which are capable of inducing the production of other antibody molecules that have "combining sites" recognizing these idiotopes.

We are now getting into trouble, however, when trying to interpret our experimental results. As I said earlier, a resting B lymphocyte displays on its surface about 100,000 identical receptor molecules which are representations of the type of antibody molecules this B cell and its progeny will produce if stimulated by an antigen.

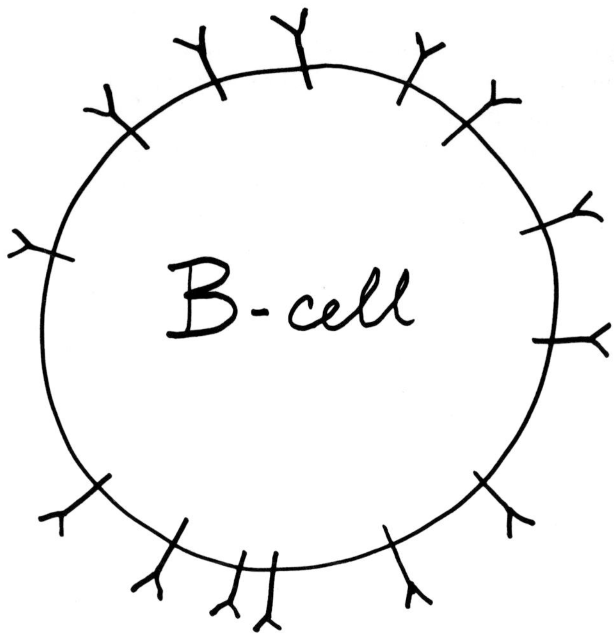

Fig. 5.

Let us return to this picture for a moment, as in Fig. 5, and let us make an enlargement, in Fig. 6, of a small part of the surface of this B cell, focusing on one receptor molecule only, making the usual cuts between constant and variable regions:

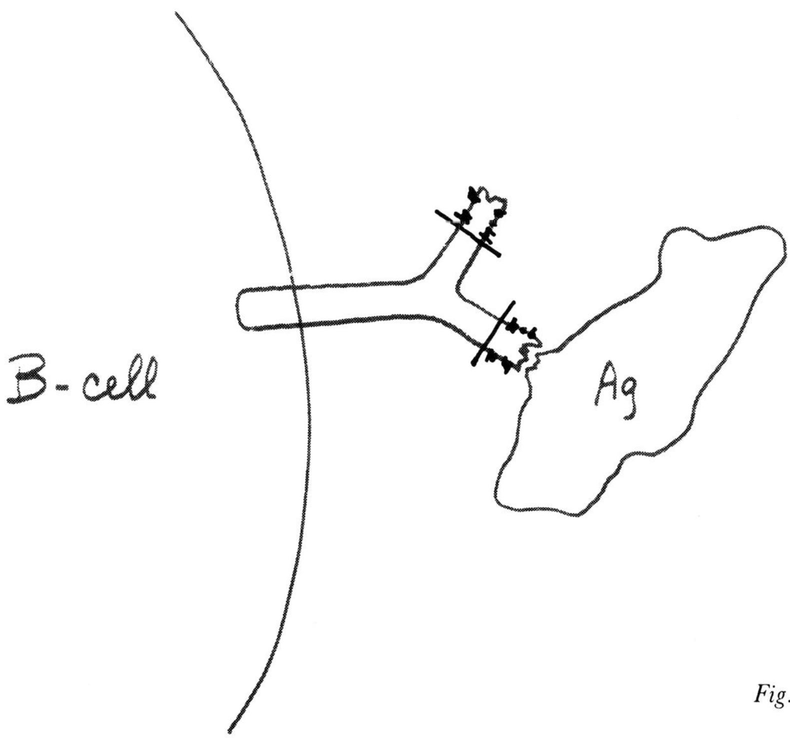

Fig. 6.

As you see, I have retained the traditional distinctions between the antigen-recognizing combining site, and a number of antigenic idiotopes. I have also added an imaginary profile of an antigenic molecule, part of which is recognized by the combining site of the B cell receptor. This is the basic picture of the selective theories of antibody formation, as most clearly formulated by Burnet (16, 2). The antigen "selects" the lymphocytes by which it is recognized, and stimulates these cells to proliferate, to mature, and to secrete antibodies with fitting combining sites. Clearly, T cell control is also involved, as well as growth factors, maturation factor, etc., but this picture remains the basic idea of antibody induction. Fig. 7 shows one of these antibody molecules, which recognizes part of the surface profile ("epitope") of the antigen.

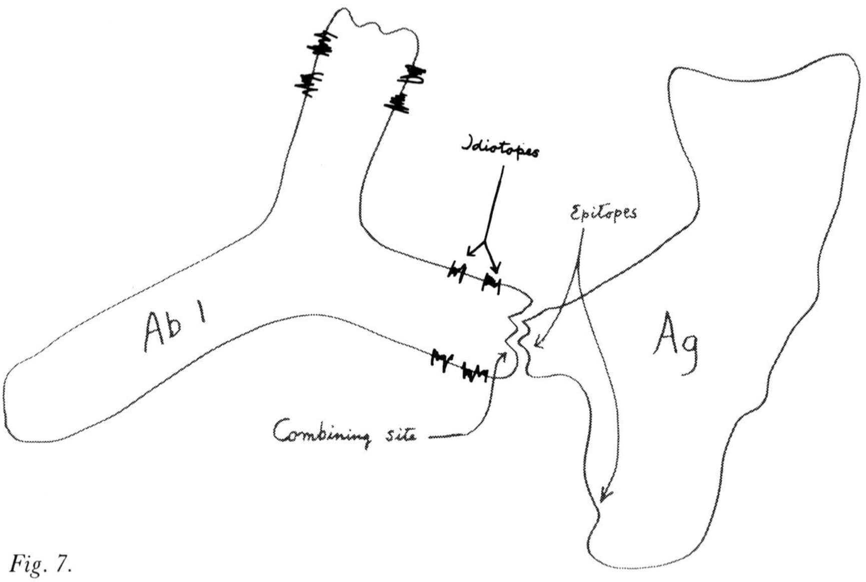

Fig. 7.

Now imagine, however, that this antibody molecule (Abl), displaying both its combining site as well as its antigenic idiotopes, is itself used as antigen. It is then possible to envisage two situations, α and β.

Fig. 8 shows, schematically, a freely circulating Abl molecule which recognizes, and sticks to, a receptor on a B cell. The figure envisages the stimulation of two different B cells by Abl molecules now acting as antigens. In the α case, the combining site of the receptor on a B cell recognizes an idiotope of Abl, and the cell may be stimulated to produce the corresponding anti-idiotopic antibodies (Ab2). In the β case, however, it is the combining site of the Ab1 molecule which recognizes an idiotope of the receptor on a B cell which may thus be stimulated to produce antibodies which possess idiotopes that have a shape that is similar to the epitope displayed by the original antigen. Experiments have shown both these situations actually to occur. For example, if the original antigen is insulin, and Ab1 is an anti-insulin antibody, then some of the anti-idiotypic antibodies (Ab2) of the β type show a similarity to insulin and have

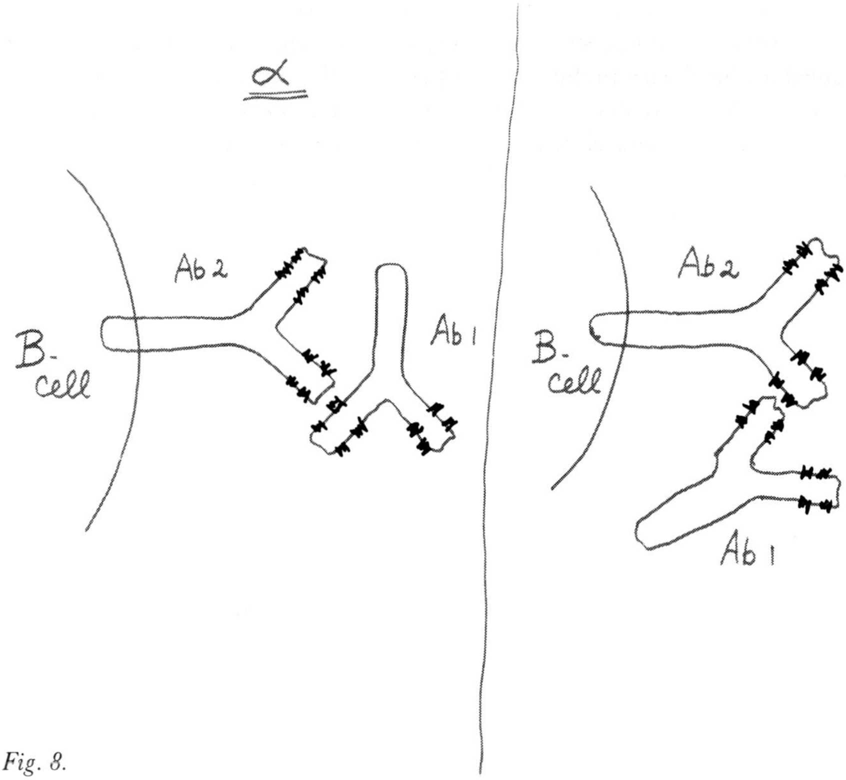

Fig. 8.

been shown, by Sege and Peterson, to function like insulin (32). Similar results in other systems have been obtained by Cazenave and Roland, by Strosberg, by Urbain, and their colleagues, and by others (33, 10, 39, 35).

The point I wish to make, however, is to consider whether the two situations α and β are fundamentally different, or not. Is there a difference between saying that Ab1 recognizes Ab2 or that Ab2 recognizes Ab1? Can we, at this three-dimensional, molecular level, distinguish between "recognizing" and "being recognized"? If not, it becomes meaningless to distinguish between idiotopes and combining sites, and we could merely say that the variable region of an antibody molecule displays several equivalent combining sites, or a set of idiotopes, and that every antibody molecule is multi-specific. I do not have to belabour this point which has been made repeatedly (36, 37, 38, 25). Instead, I should now like to introduce some numerology into this discussion. How large is the number of different antibodies that the immune system of one single animal (be it a human or a mouse) can make? This number, during the past few decades, has been estimated, on more or less slender evidence, to exceed ten millions, and this enormous diversity has been designated as the "repertoire" of the B lymphocytes. Such a "repertoire" has been characterized by Coutinho as "complete" (39). "Completeness" means that the immune system can respond, by the formation of specific antibodies, to any molecule existing in the world, including, as I said earlier, to molecules that the system has never before encountered. Immunologists sometimes use words they have borrowed from linguistics, such as "immune response". Looking at languages, we find that all of them make do with a vocabulary of roughly a hundred thousand words, or less. These vocabulary sizes are a hundred-fold smaller than the estimates of the size of the antibody repertoire available to our immune system. But if we consider that the variable region that characterizes an antibody molecule is made up of two polypeptides, each about 100 amino acid residues long, and that its three-dimensional structure displays a set of several combining sites, we may find a more reasonable analogy between language and the immune system, namely by regarding the variable region of a given antibody molecule not as a *word* but as a *sentence* or a phrase. The immense repertoire of the immune system then becomes not a vocabulary of words, but a lexicon of sentences which is capable of responding to any sentence expressed by the multitude of antigens which the immune system may encounter.

At this point, I shall make a quotation from Noam Chomsky (3) concerning linguistics: "The central fact to which any significant linguistic theory must address itself is this: a mature speaker can produce a new sentence of his language on the appropriate occasion, and other speakers can understand it immediately, though it is equally new to them ... Grammar is a device that specifies the infinite set of well-formed sentences and assigns to each of these one or more structural descriptions. Perhaps we should call such a device a *generative grammar* ... which should, ideally, contain a central syntactic component ..., a phonological component and a semantic component." That is the end of my quotation. For the size of the set of possible sentences in a language, Chomsky uses the word "open-endedness", and I now think that "open-ended" is the best description also of the "completeness" of the antibody repertoire. As for the components of a generative grammar that Chomsky mentions, we could with some imagination equate these with various features of protein structures. Every amino acid sequence is a polypeptide chain, but not every sequence will produce a well-folded stable protein molecule with acceptable shapes, hydrophobicity, electrostatics, etc. Some grammatical rules would seem to be required. It is harder, however, to find an analogy to semantics: does the immune system distinguish between meaningful and meaningless antigens? Perhaps the distinction between "self" and "non-self" is a valid example. It would seem, at first sight, that the immune response to a sentence presented by an invading protein molecule is merely to select, from amongst its enormous preformed antibody repertoire, a suitable mirror image of part of this antigenic sentence. As you will know, Leonardo da Vinci wrote his private journal in the mirror image of ordinary handwriting. It is difficult, without considerable practice, to write and read mirror handscript. Let me show you an example in Fig. 9.

Fig. 9.

On the following two figures, I shall use the device of showing ordinary letters in black, and of using greyly marked zones to indicate the mirror images of these letters.

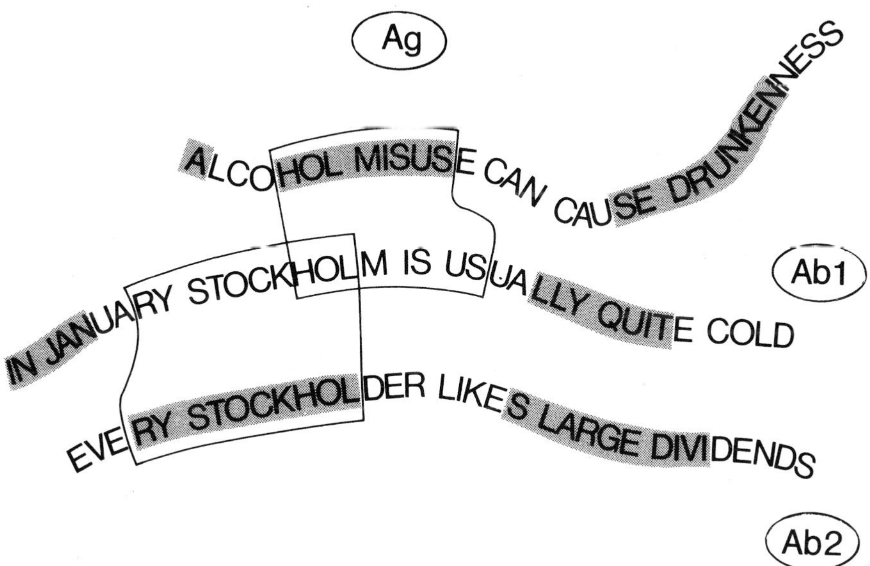

Fig. 10.

Fig. 10 shows an antigenic "sentence", part of which is mirrored by Ab1. The anti-idiotopic Ab2 mirrors part of Ab1, but bears no relation to the original antigen. Fig. 11 is a little more complex. Here, the original antigen is insulin, and the letter-sequence "OF INSULIN DE" represents its active site, which is mirrored by Ab1. Of the two anti-idiotopic antibodies shown, Ab2α and Ab2β, the latter mirrors this mirror image, and thus displays the active site of insulin (32).

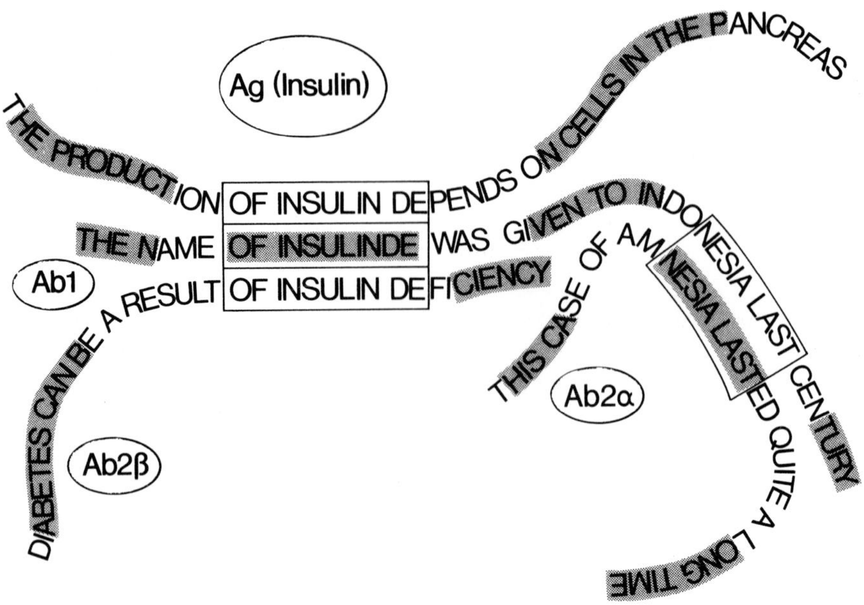

Fig. 11.

I should perhaps again emphasize that the sentences representing antibodies possess partial mirror images of an antigenic sentence. These antibodies are not echoes of the invading antigen, but were already available to the animal in its repertoire of B cells before the antigen arrived. This is the important insight that followed the introduction into immunology of the selective theories in the 1950's. Also, I must emphasize another important quantitative aspect of the situation facing the immune system. It has been estimated that one human individual produces about 10,000 different proteins, such as enzymes, hormones, cell surface proteins, etc. At the same time, we estimate that the immune system maintains a repertoire exceeding ten million different proteins, namely antibody molecules. This is a thousand-fold more than all other body proteins taken together. Man and mice, normally, have about ten milligrams of antibodies in a milliliter of their blood. Thus, a normal human possesses between about 50 to 100 grams of freely circulating antibodies, called immunoglobulins. If we divide this figure by 10^7 different specificities, we are still left with an average of 5 to 10 micrograms of every specificity in the available repertoire, representing an average of about $3x10^{13}$ monoclonal antibodies of every specificity. For mice that are 3 000 times smaller, we would have to divide these figures by 3 000, which would still leave a mouse with an average of 2 to 3 nanograms of antibodies of every of 10^7 specificities. That even such nanogram quantities of monoclonal anti-idiotypic antibodies, when introduced into mice, produce remarkable effects has been shown by Rajewsky and his colleagues (40, 41, 42).

I should therefore like to conclude that in its dynamic state our immune system is mainly self-centered, generating anti-idiotypic antibodies to its own antibodies, which constitute the overwhelming majority of antigens present in the body. The system also somehow maintains a precarious equilibrium with the other normal selfconstituents of our body, while reacting vigorously to invasions into our body of foreign particles, proteins, viruses, or bacteria, which incidentally disturb the dynamic harmony of the system.

The inheritable "deep" structure of the immune system is now known: certain chromosomes of all vertebrate animals contain DNA segments which encode the variable regions of antibody polypeptides. Furthermore, experiments in recent years have demonstrated the generative capacities of this innate system. In proliferating B lymphocytes, these DNA segments are the targets for somatic mutations, which result in the formation of antibody variable regions which differ, in amino acid sequences, from those encoded by the stem cell from which these B cells have arisen (43, 44, 45, 46, 47). The experiments showed that it was still possible, however, to identify the original stem cell genes that must have undergone these mutations. Expressed in linguistic terms, such investigations belong to the etymology of the immune system.

As immunologists, we should like to know the semantics of the inheritable gene structures. What is the meaning of the basic lexicon, or what are the specificities of the antibodies, B cell receptors and T cell receptors as encoded in the genes of our germ cells? It is known that B cells recognize the language of the T cell receptors. I have said so little about the latter because T cell receptorology is still in too early a stage of development. An immune system of enormous complexity is present in all vertebrate animals. When we place a population of lymphocytes from such an animal in appropriate tissue culture fluid, and when we add an antigen, the lymphocytes will produce specific antibody molecules, in the absence of any nerve cells (48). I find it astonishing that the immune system embodies a degree of complexity which suggests some more or less superficial though striking analogies with human language, and that this cognitive system has evolved and functions without assistance of the brain.

It seems a miracle that young children easily learn the language of any environment into which they were born. The generative approach to grammar, pioneered by Chomsky (4), argues that this is only explicable if certain deep, universal features of this competence are innate characteristics of the human brain. Biologically speaking, this hypothesis of an inheritable capability to learn any language means that it must somehow be encoded in the DNA of our chromosomes. Should this hypothesis one day be verified, then linguistics would become a branch of biology.

REFERENCES

Books

1. Landsteiner, K., The Specificity of Serological Reactions. Howard University Press, Washington (1947)
2. Burnet, F.M., The Clonal Selection Theory of Acquired Immunity. Cambridge University Press (1959)
3. Chomsky, N., Current Issues in Linguistic Theory. Janua Linguarum, Series minor, Mouton, The Hague (1964)
4. Chomsky, N., Language and Mind. Harcourt Brace Jovanovich, New York (1972)
5. Greaves, M.F., Owen, J.J.T., and Raff, M.C., T and B Lymphocytes. American Elsevier, New York (1974)
6. Golub, E.S., The Cellular Basis of the Immune Response. Sinauer Associates, Massachusetts (1977)
7. Immunoglobulin Idiotypes. Eds. Janeway, C., Sercarz, E.E., and Wigzell, H., Academic Press, New York (1981)
8. Idiotypes: Antigens on the Inside. Ed. Westen-Schnurr, I., Editiones Roche, Basel (1982)
9. Immune Networks. Eds. Bona, C.A., and Köhler, H., Annals N.Y. Acad. Sci. *418* (1983)
10. Idiotypy in Biology and Medicine. Eds. Köhler, H., Urbain, J., and Cazenave, P.-A., Academic Press, New York (1984)
11. The Biology of Idiotypes. Eds. Greene, M.J., and Nisonoff, A., Plenum Press, New York (1984)

Articles

12. v. Behring, E., and Kitasato, S., Deutsche Med. Wochenschr. *16*, 1113 (1890)
13. Gowans, J.L., and McGregor, D.D., Progr. Allergy *9*, 1 (1965)
14. Porter, R.R., Biochem. J. *73*, 119 (1959)
15. Edelman, G.M., J. Amer. Chem. Soc. *81*, 3155 (1959)
16. Burnet, F.M., Aust. J. Sci. *20*, 67 (1957)
17. Miller, J.F.A.P., and Mitchell, G.F., Proc. Natl. Acad. Sci. USA *59*, 296 (1968)
18. Raff, M.C., Immunology *19*, 637 (1970)
19. Bosma, M., and Weiler, E., J. Immunol. *104*, 203 (1970)
20. Lefkovits, I., Eur. J. Immunol. *2*, 360 (1972)
21. Fazekas de St. Groth, S., Underwood, P.A., and Scheidegger D., in: Protides of the Biological Fluids. Ed. Peeters, H., p. 559, Pergamon Press, Oxford (1980)
22. Claman, H.N., Chaperon, E.A., and Triplett, R.F., Proc. Soc. Exp. Biol. Med. *122*, 1167 (1966)
23. Govaerts, A., J. Immunol. *85*, 516 (1960)
24. Hilschmann, N., and Craig, L.C., Proc. Natl. Acad. Sci. USA *53*, 1403 (1965)
25. Coutinho, A., Forni, L., Holmberg, D., Ivars, F., and Vaz, N., Immunol. Rev., ed. Möller, G., *79*, 151 (1984)
26. Valentine, R.C., and Green, N.M., J. Mol. Biol. *27*, 615 (1967)
27. Slater, R.J., Ward, S.M., and Kunkel, H.G., J. Exp. Med. *101*, 85 (1955)
28. Oudin, J., and Michel, M., C.R. Acad. Sci. Paris *257*, 805 (1963)
29. Oudin, J., and Michel, M., J. Exp. Med. *130*, 595, 619 (1969)
30. Kunkel, H.G., Mannik, M., and Williams, R.C., Science *140*, 1218 (1963)
31. Gell, P.G.H., and Kelus, A.S., Nature *201*, 687, (1964)
32. Sege K. and Peterson P.A., Proc. Natl. Acad. Sci. USA *75*, 2443 (1978)
33. Jerne, N.K., Roland, J., and Cazenave, P.-A., EMBO J. *1*, 243 (1982)
34. Strosberg, A.D., Couraud, P.-O., and Schreiber, A., Immunol. Today *2*, 75 (1981)
35. Guillet, J.G., Kaveri, S.V., Durieu, O., Delavier, C., Hoebeke, J., and Strosberg, A.D., Proc. Natl. Acad. Sci. USA, *82*, 1781 (1985)
36. Richards, F.F., and Konigsberg, W.H., Immunochemistry *10*, 545 (1973)
37. Varga, J.M., Konigsberg, W.H., and Richards, F.F., Proc. Natl. Acad. Sci. USA *70*, 3269 (1973)

38. Jerne, N.K., Immunol. Rev., ed. Möller, G., *79*, 5 (1984)
39. Coutinho, A., Ann. Immunol. (Inst. Pasteur) *131D*, 235 (1980)
40. Kelsoe, G., Reth, M., and Rajewsky, K., Eur. J. Immunol. *11*, 418 (1981)
41. Rajewsky, K., and Takemori, T., Ann. Rev. Immunol. *1*, 569 (1983)
42. Müller, C.E., and Rajewsky, K., J. Exp. Med. *159*, 758 (1984)
43. McKean, D.M., Hüppi, K., Bell, M., Staudt, L., Gerhard, W., and Weigert, M., Proc. Natl. Acad. Sci. USA *81*, 3180 (1984)
44. Rüdikoff, S., Pawlita, M., Pumphrey, J., and Heller, M., Proc. Natl. Acad. Sci. USA *81*, 2162 (1984)
45. Bothwell, A.L.M., Paskind, M., Reth, M., Imanishi-Kari, T., Rajewsky, K., and Baltimore, D., Cell *24*, 625 (1981)
46. Sablitzky, F., Wildner, G., and Rajewsky, K., EMBO J. *4*, 345 (1985)
47. Sims, J., Rabbitts, T.H., Estess, P., Slaughter, C., Tucker, P.W., and Capra, J.D., Science *216*, 309 (1982)
48. Mishell, R.I., and Dutton, R.W., J. Exp. Med. *126*, 423 (1967)

Georges J.F. Köhler

GEORGES J. F. KÖHLER

Name: *Georges*, Jean, Franz Köhler
Born: 17.4.1946 in München. German Nationality

Education and research experience

April 1965	Abitur in Kehl, beginning of studies in Biology at the University of Freiburg.
Jan. 1971	Diploma in Biology, work on repair-deficient strains of Escherichia coli and computer assisted instruction.
April 1974	Ph.D., University of Freiburg. Thesis work on immunological studies of the enzyme β-galactosidase, carried out at the Institute for Immunology, Basel, Switzerland, under the supervision of Professor Fritz Melchers.
April 1974 to March 1976	Postdoctoral work in cell biology (lymphocyte fusion) in Dr. C. Milstein's laboratorium at the Medical Research Council, Laboratory of Molecular Biology. Work supported by an EMBO long-term fellowship. Publication: G. Köhler and C. Milstein (1975) "Continuous cultures of fused cells secreting antibody of predefined specificity". Nature 256: p.495–497.
April 1976 to present	Member of the Basel Institute for Immunology; Molecular and cellular work on lymphocyte hybrids.

Member of the European Organization of Molecular Biology (EMBO), Honory Lecturer at the University of Basel, Switzerland, Doctor honoris causa of the University of Centre Limburg, Belgium, numerous awards, becoming director of the Max Planck Institut für Immunbiologie in Freiburg, Germany.

DERIVATION AND DIVERSIFICATION OF MONOCLONAL ANTIBODIES

Nobel lecture, 8 December, 1984

by

GEORGES KÖHLER

Basel Institute for Immunology, Grenzacherstr. 487,
Postfach, Basel, Switzerland

A mouse can make ten million different antibodies, each synthesized by its own B-lymphocytes. About 1000 different antibodies are able to recognize one single antigenic determinant. When a conventional mouse response to such a determinant is analysed, only 5–10 different antibody species are seen, representing probably a random sample of the total repertoire. So even when appropriate absorptions or allogeneic immunizations have confined the antibody heterogeneity to one single determinant the sera obtained have four disadvantages: (1) the titers are low, (2) the antibodies, while specific for a single determinant, are nevertheless heterogeneous, (3) the supply is limited, and (4) the same combination of specific antibodies is impossible to reproduce in a new animal.

The method of lymphocyte fusion developed together with C. Milstein, MRC, Cambridge, England, provides a tool to overcome these limitations (Fig. 1). Mouse myeloma tumour cells are fused to spleen cells derived from a mouse which previously had been immunized with antigen. About 50 % of the hybrid cells combine the hoped for parental traits: vigorous growth in tissue culture derived from the myeloma tumour cell and antibody production coming from the splenic B cell. A relatively high proportion of the hybridoma cells secrete antibody specific to the immunizing antigen (1, 2).

The advantages of this technique are:

1) Single specificity. Each hybrid produces only one antibody.
2) Unlimited supply of antibody. The hybrids are immortal like tumour cells, can be frozen, secrete 10–50 μg/ml antibody into the culture fluid, and produce titers as high as 1–10 mg antibody per ml body fluid upon injection into mice.
3) Impure antigens lead to pure antibody reagents. The monoclonal antibody by definition characterizes only one antigen of the many injected into the mouse.
4) All specificities can be rescued. The empirical observation seems to be that if an immune response can be elicited in the mouse also specific hybridomas can be derived.
5) Enrichment of specific hybridomas. Specific B cells are rare even in the spleen cell population of an immunized mouse. They are found to be enriched 10–100 fold in the corresponding hybridoma population.

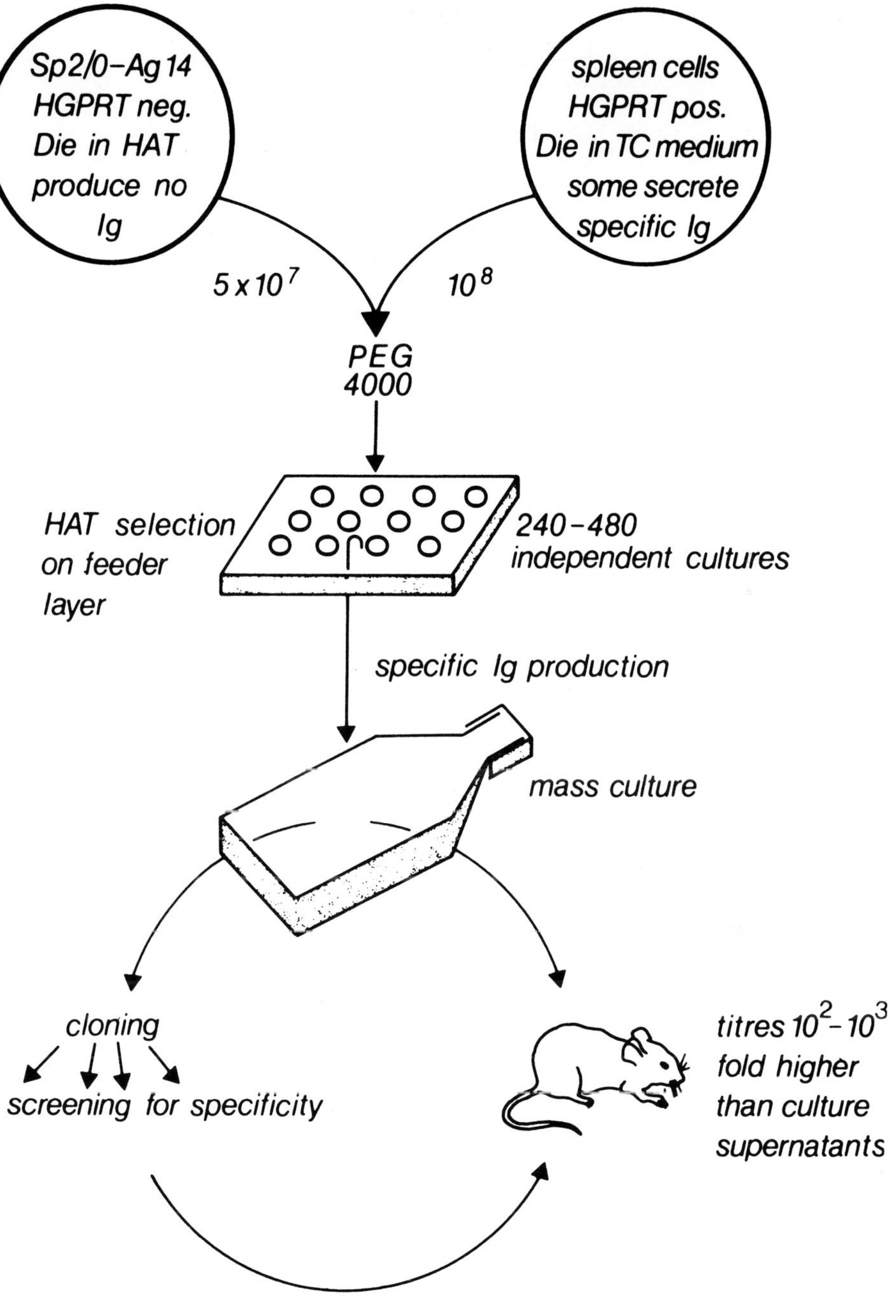

Fig. 1 – The hybridoma technique.
The fusion line Sp2/0-Ag14 comes itself from a hybridoma line and is devoid of endogeneous immunoglobulin production (61). HGPRT =hypoxanthinephosphoribosyltransferase, an enzyme necessary in DNA synthesis when the normal pathway is blocked by aminopterin. HAT = hypoxanthine, aminopterin, thymidine (1) (62,63). PEG = polyethylenglycol.

Table 1. Uses of monoclonal antibodies © 1982 John Wiley & Sons Ltd, From Genetic Engineering to Biotechnology – The Critical Transition. Edited by W. J. Whelan and Sandra Black.

Use	Application	Success, example	References
to *define* antigen	– on bacteria,	Classification	3
	viruses	becomes easier	4
	parasites		5
	– on cells	11 human T cell antigens established	6
		Histocompatibility antigen typing	7
		tumor antigens	8
purify	– factors, hormones	5000 fold enrichment of interferon	9
	– cell membrane	200 fold enrichment of Ia antigen of rat	10
detect and *quantify*	– in crude mixtures	human chorionic-gonadotropin in pregnancy tests	11
map	– epitope characterization	7 determinants on mouse μ chain (constant region)	12
modify	– infectivity	sporozoite of plasmodium berghei	13
	– toxicity	digoxin overdose	14
	– function	αLy2suppresses T cell killing	15
	– immunogenicity	αRhesus factor	16
select	– α idiotype	enhancement and suppression of anti-NP immune response	17
	– mutations	Histocompatibility antigen	18
		Influenza A virus	19
localize	– in organs	nervous system of leech	20
	– in body	tumor imaging in humans	21

6) Dominance of antibody secretion. B-hybridomas secrete high level of antibodies, irrespective whether or not the normal B cell was a high producer.
7) Antibodies become manipulable. The hybridoma cell lines can be mutated to produce antibodies not found in nature.
8) It is a general method. The only generalization so far, was to apply cell fusion for the rescue of T cell functions, where T hybridoma lines secreting different lymphokines, exerting killer function or helper and suppressor activities have been generated.

Some selected applications are summarized in Table 1.

In the past, conventional sera have been used to define, purify, detect, quantify, map, modify, select and localize antigen. Thus, with few exceptions (18, 19), the monoclonal antibodies have not led to applications unthought of previously. However, as pure chemicals of higher precision and unlimited supply they are replacing the conventional sera and thereby contributing to a worldwide standardisation of antibody mediated reactions.

One of the great advantages of the hybridoma lines lies not only in the production of monoclonal antibodies but also in the availability of the machinery which produces them. One can diversify a single antibody molecule by selecting mutant molecules, by introducing new immunoglobulin heavy and light chains into hybridoma lines, by cloning and mutating its genes by "reverse genetics" techniques and reintroducing them into cell lines or into the germline of a mouse. The variants give some insights into the structure-function relationship of immunoglobulin protein, their RNA and genes. The analysis of variants have also led to a more speculative hypothesis about the interaction of H and L chains in multigene families.

Diversification of monoclonal antibodies

Many investigators have generated variant forms of monoclonal antibodies (Table 2), with perhaps more interesting properties than the ones I will describe here. But this is a somewhat personal account of this development, and not meant to be a review. It is focused, where possible, on experiments using one hybridoma cell line, Sp6/HLGK. The line is derived from a Balb/c mouse immunized with trinitrophenyl-lipopolysaccharide (TNP-LPS). It secretes an immunoglobulin mu heavy chain associated with kappa light chain of anti-TNP specificity (designated HL in Sp6 and equivalent hybridomas) together with a gamma and kappa chain derived from the fusion myeloma line X63-Ag8 (designated GK in Sp6 and equivalent hybridomas) (2).

Table 2. Mutations and alterations in cell lines producing monoclonal antibodies. Reprinted with permission from Pergamon Press Ltd, © 1984 Pergamon Press.

Selection	Mutation/alteration	Frequency	Reference
loss of isotype	loss of H- or L-chain	$1/10^2$	22
	domain deletions	$1/10^3$	23, 24
	loss of λ-chain secretion due to point mutation	$>1/10^4$	25, 26
change in IEF	domain deletions frameshift, point mutation	$1/10^3$	27
cell sorter, positive selection for different isotype	change of isotype often associated with deletions	$\sim 1/10^5-10^7$	28, 29, 30
loss of idiotype, loss and gain of antigen binding	possible gene conversion point mutations	$10^{-2}-10^{-3}$	31, 32, 33
loss of lytic activity	deletions, frameshifts insertions	$1/10^3-10^4$	34, 35, 36
reverse genetics	chimaeric antibodies		
	$V_H-C\kappa$		37, 38
	^{V}mouse$-^C$human		39, 40
	antibody-enzyme		41
secondary hybrids	monovalent antibodies		41 a
	two specificities		41 b
	complementation of specificity		49

Table 3. H-CHAIN TOXICITY

$3H + 3L \rightarrow$	$\frac{2H + 3L}{3H + 2L} \sim \frac{2}{1}$
$2H + 2L \rightarrow$	$\frac{1H + 2L}{2H + 1L} \sim \frac{8}{1}$
$1H + 1L \rightarrow$	$\frac{0H + 1L}{1H + 0L} \gtrsim \frac{100}{1}$
2H + 3L	
1H + 2L	L-loss as frequent as H-loss
0H + 1L	

Immunoglobulin chain loss variants lead to the heavy chain toxicity hypothesis

The first hybridomas all produced two immunoglobulins, the specific lymphocyte derived one and the myeloma derived one and were, therefore, like Sp6, of the HLGK-type. To obtain pure specific antibodies, sublines were selected which had lost the production of the non-specific G and K chains of the myeloma fusion partner (2, 42). It was observed that heavy chains are more easily lost than light chains. The sequence of chain loss as exemplified with Sp6 was Sp6/HLGK → Sp6/HLK → Sp6/HL and not Sp6/HLGK → Sp6/HLG → Sp6/HL. This was reminiscent of the results of Coffino and Scharff (22, 43), who, using the myeloma line MPCll, found that H-chain loss was observed at a rate of $1–2 \times 10^{-3}$ per cell and generation but L-chain loss was not observed even at a 100 times lower rate. After H chain loss, however, L chains were lost at a rate similar to H chains.

To test the generality of the observation a series of hybridoma lines were established making three, two and one antibody and their chain loss variants were monitored (42, Table 3). L chain loss was inhibited with increasing severity only in those combinations resulting in an increasing excess of H chain production. It is concluded that heavy chains not counterbalanced by enough light chains are toxic for the cells. Excess H chain producers have a growth disadvantage and cannot be cloned from the cell population. Additional support for the H chain toxicity hypothesis came from the analysis of Sp6 variants. A subline (Sp6/HLk) was derived producing a variant k chain, which combined less efficiently to the H chain (this property is symbolized by the lower k in Sp6/HLk). The Sp6/HLk line lost L at a 1 000 x lower frequency than the original Sp6/HLK line (34, 42). We concluded that the variant k could not substitute for the parental K chain to avoid toxic accumulation of free H chain. We have cloned the H and L chain gene of Sp6 and reintroduced them separately or physically linked into the Ig non-producing myeloma line X63Ag8.6.5.3 (Table 4). If production of free heavy chain is toxic for the cell, we would expect that μ-alone transformants are selected which produce lower amounts of μ when compared to (μ and ϰ)-transformants. A 5 to 13 fold difference was observed between these two groups of transformants. On average, about 100 times less μ RNA or protein is found in the μ-alone transformants. This low amount is comparable to the amount of μ alone found in pre-B cells (44) and may no longer be toxic for the cells. In proteins consisting of two

Table 4. Expression of Cloned Sp6μ and κ Genes in X63Ag8.6.5.3 B-Myeloma Cells

Gene	DNA Number of copies	RNA (% of Sp6 parent) μ	RNA (% of Sp6 parent) κ	Protein (% of Sp6 parent) μ	Protein (% of Sp6 parent) κ
κ	50 (41)	–	62 (10)	–	50 (4)
μ	7 (8)	1.3 (9)	–	1 (9)	–
μ + κ	5 (11)	6 (8)	4 (8)	13 (12)	13 (12)
Ratio $\frac{\mu+\kappa}{\mu}$	~1	5		13	

Data from J. McCubrey, unpublished. In parenthesis the number of stable transformants analysed.

different subunits such as H and L in immunoglobulins, often one chain is produced in excess (L chains in Ig producing cells). Potentially toxic effects of free light but not free heavy chains will therefore be selected against. In the multigene Ig-family with about 5–10.000 H and 250 L chains, elimination of cells producing H-L pairs with too low affinities to form antibody molecules may be an important control mechanism.

Somatic mutants of H and L chains dominate late immune responses (45, 46). It is conceivable that mutations occurring in the variable regions of low affinity H and L pairs enhance their pairing affinity. Such cells avoid elimination caused by free H chain toxicity and contribute to the somatic diversification of antibody molecules. Scaling up Ig production or switching Ig-classes may induce new rounds of selections (Fig. 2). Preferential association of origi-

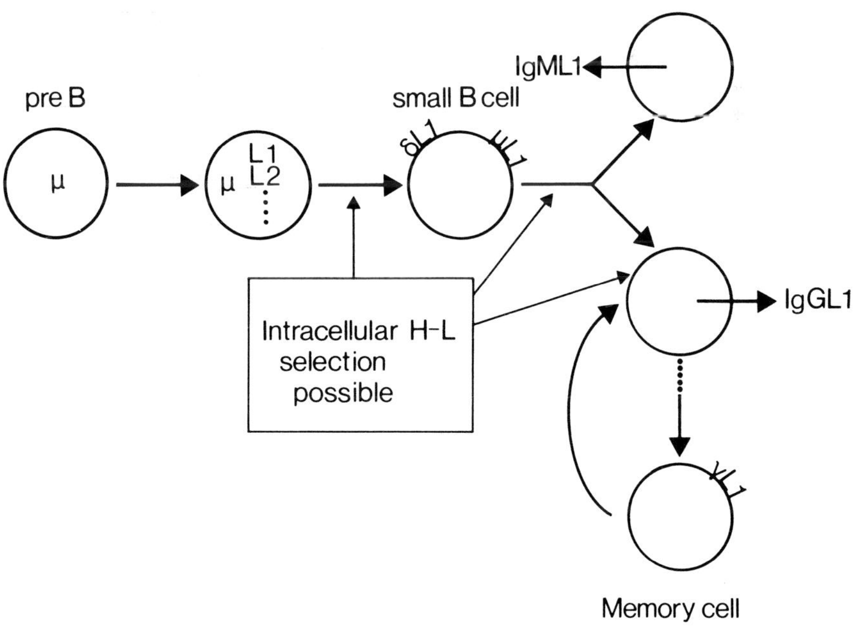

Fig. 2 – Schematic representation of B-cell development.
The points at which increased production of heavy chains or production of different heavy chain isotypes might become toxic for the cells are indicated. Variable region mutations increasing heavy (H) and light (L) chain pairing affinities may avoid toxicity of free heavy chains and be enriched.

nal H-L pairs in competition experiments (47, 48) are easily explained by this intracellular selection mechanism.

Arrays of antibodies with one specificity generated by secondary fusions

The Sp6/HK and Sp6/LK lines, (chain loss derivatives of Sp6/HLGK) are no longer producing antibodies with anti-TNP specificity (a property confined to the HL combination). Both lines were fused to unimmunized mouse spleen cells stimulated for four days with the mitogen lipopolysaccharide. The resulting hybrids were screened for restoration of the original anti-TNP specificity (49). In about every 100th hybridoma the anti-TNP activity of the Sp6/HK line was restored. In about every 2 000th hybridoma the anti-TNP activity of the Sp6/LK line was restored, which was indistinguishable from values obtained with the line Sp2/0, which does not contribute any Ig chain. The generality of this observation was confirmed by using three other pairs of hybridomas of the HK- or L-type, with similar results. The HK-lines could easily be complemented, the L-lines gave 'background' complementation. From these experiments the following conclusions were drawn (49):

1) Around 40 light chain variable region genes contribute to the light chain repertoire of the mouse.
2) Around 250 different light chains are found in unstimulated spleen cells, due to the combination of 40 V-region genes with 4 J (joining) segments and their early somatic diversification.
3) About 20–40 fold more heavy than light chains are expressed in early, unstimulated spleen lymphocytes.
4) A statistical analysis of the data leads to the generalisation, that, could one screen through all heavy chains with a given light chain one would find any given antibody specificity. Could one screen through all light chains with a given heavy chain a given antibody specificity is only found in every 20th case.

The above points have to be considered with caution, since they are based on limited numbers of hybridomas. Nevertheless, I think hybridomas have opened a new way to analyse the problem of antibody diversity, by analyzing H and L chains separately, thereby avoiding the enormous heterogeneity created by the combination of heavy and light chains in antibody molecules. The Sp6 line has been diversified into 10 different sub-lines, each of which makes a different light chain, which together with the Sp6 heavy chain gives rise to the original anti-TNP specificity. Such lines will provide insights of heavy and light chain variable region interactions necessary to create a given antigenic combining site. A more practical aspect of such secondary arrays of antibodies with one specificity would be the isolation of low affinity antibodies, which for certain antigen-purification procedures may be advantageous.

Diversification by mutant selection

We have studied mutants of the Sp6 and PC 700 (IgM anti phosphorylcholine (PC)) line. The selection method was simple. The hapten TNP or PC was covalently attached to the membranes of the cells. The cells were diluted and

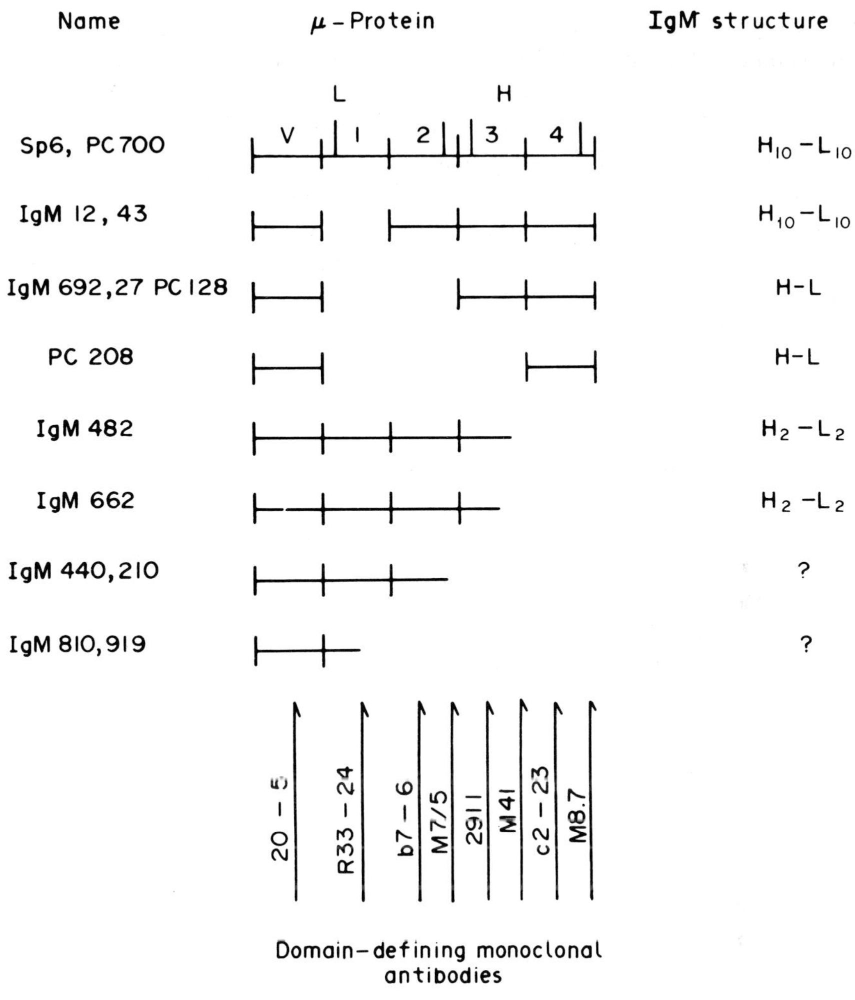

Fig. 3 – Deletions in mouse μ-chains.
The wild-type μ-protein (Sp6 and PC700) is drawn in its domain structure and disulphide bridges to the light (L) or heavy (H) chains are indicated. The light chain of IgM 12 and IgM 43 is not covalently bound to μ, whereas it is in all other mutants with known IgM structure. The monoclonal rat anti-mouse μ-antibodies were selected from about 20 antibodies and define independent binding sites. The line 20–5 makes a mouse anti-idiotypic antibody reacting with Sp6 but not PC700 IgM. Reprinted with permission from Pergamon Press Ltd, © 1984, Pergamon Press.

incubated in the presence of complement. Cells secreting wild type IgM preferentially bind their own antibody and are killed. Cells secreting less or less lytic IgM are enriched. Three to four selection rounds lead to an almost "pure" mutant-cell population. Table 5 summarizes the mutants so far characterized for the Sp6 line. The deletion variants were used to map a panel of monoclonal antibodies (12). Fig. 3 shows the IgM structure of some of the deletion variants and the mapping of monoclonal antibodies to each of the five μ-domains. The nonoclonal antibodies against 8 defined, non-crossreacting determinants of

Table 5. Mutations affecting Sp6 immunoglobulin M.

Defect	Number	Description	Reference
not determined, possibly point-mutations	2	10× lower affinity 200× reduced lytic activity	34
insertion	2	loss of L production or reduced L production	36
deletion	4	loss of Cμ1, Cμ1,2, Cμ1,2,3,*	35
frameshift	16	loss of C-terminal portions	50

* found in the PC 700 system (52) only.

mouse μ chains were used to define new IgM variants (50), to map Clq binding to the area of binding of C2–23 monoclonal antibody (38, unpublished) and to define the quaternary structure of some of the IgM-mutants (12).

Diversification by reverse genetics

The light chain gene of Sp6 has been cloned and sequenced (36). The Sp6 heavy chain variable region gene was placed into the V_{κ}-C_{κ} intron using the HindIII and XbaI restriction sites. After transfection into the X63Ag8.6.5.3 (X63/0) myeloma line (51) a V_{μ}-C_{κ} chimaeric protein was revealed (Fig. 4). Fusing this line to Sp6/L (IgM 10) resulted in hybrids producing covalently linked dimers containing the $V_{\varkappa}$-$C_{\varkappa}$ and the chimaeric V_{μ}-$C_{\varkappa}$ chain. Binding of the heavy-chain-dependent anti-idiotypic antibody 20–5 was restored in these dimers. We could not measure antigen binding activity, because of the low anti-TNP affinity of the Sp6-IgM (10^{-4} M, (34)). A similar chimaeric protein with antigen binding for arsonyl was, however, described (37). Small, antigen-binding molecules lacking the heavy chain constant region determinants might be useful when only one binding site is required (to avoid modulation of cell surface antigens), when a smaller molecule with less non-specific interactions (via Fc-receptors) and possibly a higher elimination rate is needed. Such molecules might also be less immunogenic. Since most therapeutically interesting antibodies are of mouse origin the construction of chimaeric antibodies using human constant regions and mouse variable regions could be a solution avoiding immunogenicity but keeping effector functions of the antibody. That this is possible has recently been shown with the Sp6 IgM (39). The mouse heavy and light chain variable regions were placed in front of human μ- and κ-constant regions, respectively. Mouse-human chimaeric IgM was recovered from supernatants of a plasmid carrying myeloma line. The IgM was pentameric and functional in terms of being able to lyse TNP coupled sheep erythrocytes.

The transgenic mouse model

So far three groups have introduced rearranged immunoglobulin genes into the germline of the mouse. A μ and a ϰ gene alone as well as a combination of both

Table 6. Immunoglobulin genes introduced into mouse germline cells

Gene (specificity)	Molecules injected	Zygotes implanted	Offspring (with trans-gene)	# of copies per genome	Gene in germline	References
μ (NP)	50	284	13 (5)	17–140	3	53
$\varkappa_{21}$ (?)	440	197	11 (6)	20–200	6	54, 55, 56
μ, ϰ (TNP)	50	13	5 (1)	4	1	57

genes derived from the Sp6 line have been used (Table 6). In the Sp6 case about 50 molecules of the heavy and light chain genes both cloned into a pBR322 plasmid vector ($pRHL_{TNP}$) were injected into fertilized eggs from Swiss albino mice. From 13 implanted zygotes 5 offspring mice were obtained, one of them carrying the gene. Analysis of the germline transmission pattern indicated integration of 4 copies of the $pRHL_{TNP}$ plasmid into one autosomal chromosome.

Expression of trans-immunoglobulins

In all cases with germline transmission of the immunoglobulin genes also expression of the respective chains was observed. Expression of $\varkappa_{21}$-chains was confined to the B cell lineage excluding Abelson virus transformed pre-B cell lines (55). Expression of μ_{NP}-chains was observed in B- and in T-cells (53). Other tissues like kidney, brain, heart, lung and liver did not transcribe the genes. By fluorescence analysis the Sp6 μ6 and ϰ6 chains were expressed on the membrane of B- but not T-cells, although the μ6 chain (ϰ6 was not testable) was expressed in around one quarter of Con A-stimulated, splenic T-cells. The Sp6 immunoglobulin was found as pentameric, functional IgM in the sera of the transgenic mice where it represented around one fifth to one third of the total IgM. It will be interesting to see whether such massive production of one specificity induces changes in idiotypic network interactions. Introducing antibodies with anti-self specificities into the germline of the mouse may give some insight on how self-tolerance is achieved at the B cell level. With the genes introduced so far another immunological question, the one of allelic exclusion, was analysed. How is it possible that B cells produce heavy and light chains from only one of the two homologous chromosomes, a prerequisite of effective clonal selection?

Influence of transgenic μ and ϰ chains on allelic exclusion

In order to create a functional heavy chain gene, three spatially separated DNA segments, the variable region —, the diversity — and the joining — (V, D, J_H) segments have to juxtapose. To create a functional light chain gene, a V_L and a J_L segment have to join. In roughly one quarter of the B cells allelic exclusion is achieved by having functionally rearranged VDJ_H^+ and VJ_L^+ units in one allele and non-functionally rearranged VDJ_H^- and VJ_L^- units in the other allele, which due to mistakes during the joining process are unable to produce the respective chains. In about 75 % of the B-cells either the heavy or the light

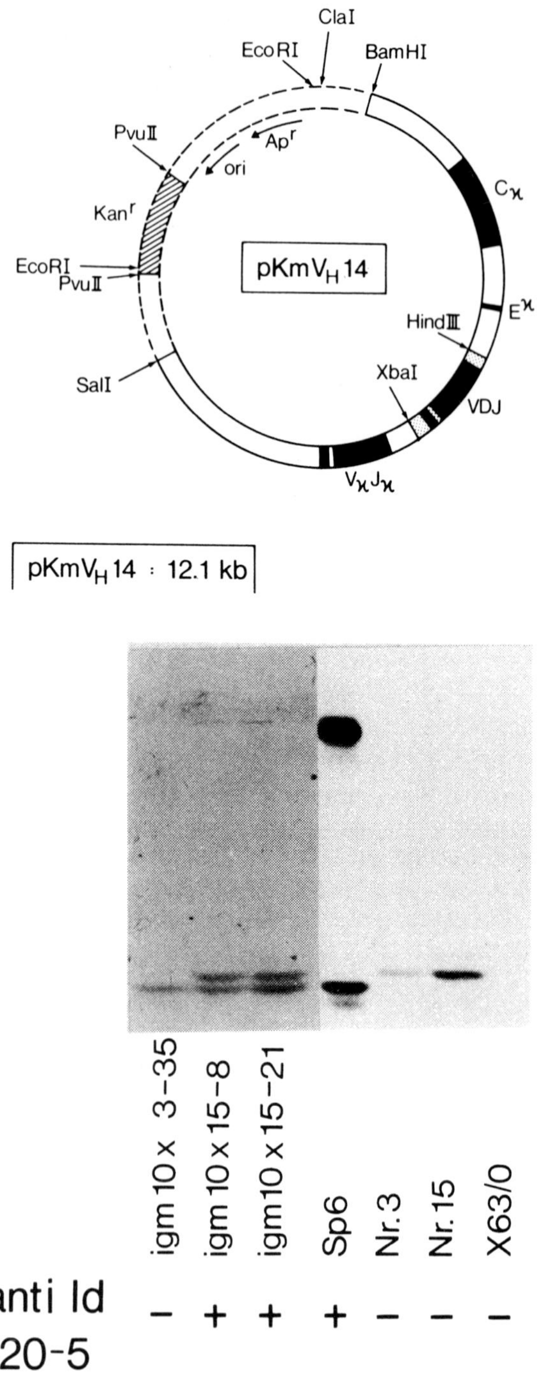

Fig. 4 – Production of chimaeric antibodies.

Top: The plasmid pKmV$_H$14 carries the kanamycin-resistance gene (shaded), pBR322 sequences (dashed) and an insert of the Sp6 kappa gene from the BamHI to the SalI restriction sites. The Sp6 variable heavy chain gene (VDJ) was inserted into the XbaI and the Hind3 restriction sites in front of the kappa enhancer element (E$^{\kappa}$). Bottom: SDS-gel electrophoreses under reducing conditions of products formed after transfection of pKmV$_H$14 plasmid into X63/0 myeloma cells. Stably trans-

Table 7. Allelic exclusion in trans hybridomas.

$\varkappa_{21}$:	11/24 Hybrids	H $\varkappa_{21}$ − (germline)		
ref 56	2	H − shift L		
	7	− $\varkappa_{21}$ L		
	2	− $\varkappa_{21}$ − (germline)		
	2	− $\varkappa_{21}$ − (rearranged)		
$\mu_6\varkappa_6$:	3/11 Hybrids	μ_6 $\varkappa_6$ H L		
ref (57)	8	μ_6 $\varkappa_6$ − L		
*5 lines still have both H-encoding chromosomes	→10 alleles:	VDJ^- 1	DJ 8	germline 1

H = endogeneous heavy chain, L = endogeneous light chain; in parenthesis light chain gene configuration.

* Heavy chain gene configuration of 5 of the 8 H-minus hybrids is shown.

chain variable region or both have not been completely rearranged. In order to explain this finding a regulatory feedback mechanism of the immunoglobulin chains on the rearrangement process was postulated (58).

To study the effect of rearranged and expressed immunoglobulin transgenes to the rearrangement process of endogeneous Ig genes, a series of B cell hybridomas were made and their endogeneous heavy and light chain expression monitored. In Table 7 our results and those obtained from transgenic mice carrying a rearranged kappa gene alone (56, 57) are summarized (results from the μ alone transgenic mice are not yet available). From 24 hybridomas derived from ϰ21 transgenic mice none expressed a heavy chain with two light chains. In seven cases where no heavy chain production is observed, two light chains, the endogeneous and the transgenic one, are expressed. These results are compatible with a negative feedback mechanism of H-L molecules on the light-chain gene rearranging process.

Expressing the transgenic μ_6- and κ_6-chains should, therefore, inhibit endogeneous light chain gene-rearrangement, which was not observed in any of the eleven hybridomas. However, the trans-κ_6-chain was expressed at one tenth the level of the endogeneous light chains whereas the μ_6-chain was made in excess. Thus, low amounts of μ_6-$\varkappa_6$ molecules in an excess of free μ_6-chains are not a feedback signal to stop light chain gene rearrangement. This observation is well in agreement with the exceptional light chain double producers found in myeloma cell lines such as S107 where one of the kappa light chains was unable to combine with the alpha heavy chain (59), or such as in the MOPC315 line where a truncated λ_1 chain unable to bind to its heavy chain is found together with a functional λ_2 light chain (60). A high proportion (8/11) of hybridomas

formed lines No. 3 and No. 15 show a band with slightly slower mobility than the Sp6 light chain. Light-chain and pKmV_H14-directed production of V_HC_K in (15 x igm 10) hybrids 8 and 21 is observed. Only in these hybrids is the heavy-chain-dependent idiotype 20−5 restored. Reprinted with permission from Pergamon Press Ltd, © 1984, Pergamon Press.

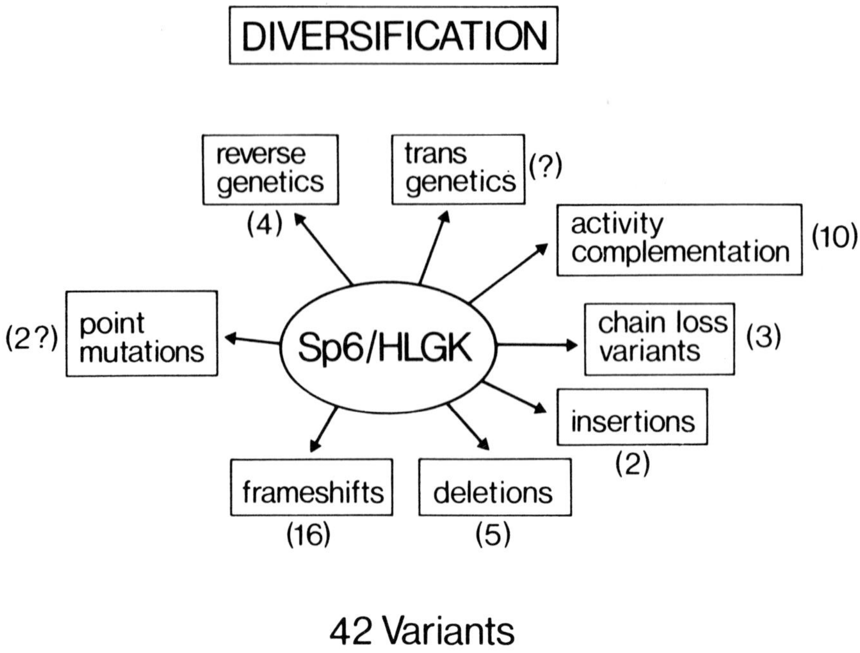

Fig. 5 – Summary of Sp6 variants.

derived from the $\mu_6\varkappa_6$ expressing mouse did not express a second endogeneous heavy chain. Five of these lines still retained both heavy chain encoding chromosomes. Only in one of the ten heavy chain alleles a complete (though inactive) VDJ_H^- unit was found. One allele was apparently frozen in the germline configuration, and eight stopped rearrangement at the immature DJ_H state. Thus again, the result is compatible with a (leaky) feedback regulation of the heavy chain gene rearrangement process by the mu chain itself. Whether also the light chains are involved in this process has to be clarified by analysing pre-B cell lines. However, other interpretations such as different cellular selection processes in transgenic versus normal mice are possible. Nevertheless, the power of the transgene approach to study immmunological phenomena is clearly demonstrated.

Let me now go back to the general theme of this paper and ask whether or not diversification of our Sp6 antibody could be achieved by having introduced its genes into the mouse. At the moment one can only speculate. A somatic mutation mechanism operates on immunoglobulin light and heavy chain genes at an approximate rate of 10^{-3} per base and generation (45,46). It seems possible that the introduced genes will profit from this mechanism and that we will generate a whole series of somatically mutated Sp6 IgM molecules. Figure 5 summarizes the variants obtained so far from the Sp6 line using different techniques and underlines my belief that a single monoclonal antibody will only be the starting point of a variety of man-made secondary antibodies, each manufactured to fulfill a special requirement.

REFERENCES

1. Köhler, G., and Milstein, C. Continuous cultures of fused cells secreting antibodies of predefined specificity. *Nature, 256*: 995, (1975).
2. Köhler, G., and Milstein, C. Derivation of specific antibodyproducing tissue culture and tumor lines by cell fusion. Eur. J. Immun. *6*: 511, (1976).
3. Polin, R.A. Monoclonal antibodies against streptococcal antigens, p 353–359, in *Monoclonal Antibodies* (eds. Kenneth, McKearn, Bechtol), Plenum Press, New York, (1980).
4. Flamand, A., Wiktor, T.J., and Koprowski, H. Use of hybridoma monoclonal antibodies in the detection of antigenic differences between rabies and rabies-related virus proteins. *J. Gen. Virol., 48*: 97, (1980).
5. Mitchell, G., Cruise, K.M., Garcia, E.G., and Anders, R.F. Hybridoma derived antibody with immunodiagnostic potential for schistosomiasis japonica. Proc. Nat. Acad. Sci. USA. *78*: 3165, (1981).
6. Howard, F.D., Ledbetter, J.A., Wong, J.,Bieber, C., Stinson, E.B., and Herzenberg, L.A. Human T lymphocyte differentiation marker defined by monoclonal antibodies that block E-rosette formation. J. Immunol., *126*: 2117, (1981).
7. Brodsky, F.M., Parham, P., Barnstable, C.J., Crumpton, M.J., and Bodmer, W.F. Monoclonal antibodies for analysis of the HLA system. *Immunol. Reviews, 47*: 3, (1979).
8. Koprowski, H., Steplewski, Z., Herlyn, D., and Herlyn, M. Production of monoclonal antibody against human melanoma by somatic cell hybrids. Proc. Nat. Acad. Sci. USA. *75*: 3405, (1978).
9. Secher, D.S., and Burke D.C. A monoclonal antibody for large-scale purification of human leucocyte interferon. *Nature, 285*: 446, (1980).
10. McMaster, W.R., and Williams, A.F. Monoclonal antibodies to Ia antigens from rat thymus: Crossreactions with mouse and human and use in purification of rat Ia glycoproteins.. *Immunol. Review, 47*: 117, (1979).
11. Miggiano, V., Stähli, C., Häring, P., Schmidt, J., Le Dain, M., Glatthaar, B., and Staehelin, T. Monoclonal antibodies to three tumor markers: human chorionic gonadotropin, prostatic acid phosphatase and carcinoembryonic antigen. p. 501–504 in *Protides of the Biological Fluids* (ed. Peters), Pergamon Press, Oxford, (1980).
12. Leptin, M., Potash, M.J., Grützmann, R., Heusser, C., Shulman, M., Köhler, G., and Melchers, F. Monoclonal antibodies specific for murine IgM. I. Characterization of antigenic determinants on the four constant domains of the μ chain. Eur. J. Immun. *14*: 554, (1984).
13. Yoshida, N., Nussenzweig, R.S., Potocnjak,, P., Nussenzweig, V., and Aikawa, M. Hybridoma produces protective antibodies directed against the sporozoite stage of malaria parasite. *Science, 207*, 71, (1980).
14. Mudgett-Hunter, M., Margolies, M.N., Smith, T.W., Nowotny, J., and Haber, E. Monoclonal antibodies to the cardiac glycoside digoxin, p 367–374 in *Monoclonal Antibodies and T cell Hybridomas* (eds. G. Hämmerling, and J. Kearney), Elsevier, Amsterdam. (1980).
15. Hollander, N., Pillemer, E., and Weissman, I.L. Blocking effect of Lyt-2 antibodies on T-cell functions. *J. exp. Med., 152*, 674, (1980).
16. Koskimies, S. Human lymphoblastoid cell line producing specific antibody against Rh-antigen D. *Scand. J. Immunol., 11*: 73, (1980).
17. Reth, M., Kelsoe, G., and Rajewsky, K. Idiotypic regulation by isologous monoclonal anti-idiotypic antibodies. *Nature, 290*: 257, (1981).
18. Holtkamp, B., Cramer, M., Lemke, H., and Rajewsky, K. Isolation of a cloned cell line expressing variant H2k using the fluorescence activated cell sorter. *Nature, 289*: 66, (1981).
19. Gerhard, W., and Webster, R.G. Antigenic drift in influenza A virus. I. Selection and characterization of antigenic variants of A/PR/8/34 (HONl) influenza virus with monoclonal antibodies. *J. exp. Med., 148*: 383, (1978).
20. Zipser, B., and McKay, R. Monoclonal antibodies distinguish identifiable neurones in the leech. *Nature, 289*: 549, (1981).
21. Mach, J.P., Buchegger, F., Forni, M., Ritchard, J., Berche, C., Lumbroso, J.-D., Schrejer, M., Girardet, Ch., Accolla, R.S., and Carrel, S. Use of radiolabelled monoclonal anti-CEA

antibodies for the detection of human carcinomas by external photoscanning and tomoscintography. *Immunology Today*, *2*: 239,(1981).

22. Coffino, P., and Scharff, M.D. Rate of somatic mutation in immunoglobulin production by mouse myeloma cells. Proc. Nat. Acad. Sci. USA. *68*: 219, (1971).
23. Morrison, S.L. Murine heavy chain disease. Eur. J. Immun. *8*: 194, (1978).
24. Morrison, S.L. Sequentially derived mutants of the constant region of the heavy chain of murine immunoglobulins. J. Immun. *123*: 793, (1979).
25. Mosmann, T.R., and Williamson, A.R. Structural mutations in a mouse immunoglobulin light chain resulting in failure to be secreted. Cell, *20*: 283, (1980).
26. Wu, G.E., Hozumi, N., and Murialdo, H. Secretion of a λ_z immunoglobulin chain is prevented by a single amino acid substitution in its variable region. Cell, *33*: 77, (1983).
27. Milstein, C., Adetugbo, K., Cowan, N. J., Köhler, G., Secher, D. S., and Wilde, C. D. Somatic cell genetics of antibody-secreting cells: studies of clonal diversification and analysis by cell fusion. *Cold Spring Harbor Symp. Quant. Biol. 41*: 793, (1977).
28. Neuberger, M.S., and Rajewsky, K. Switch from hapten specific immunoglobulin M to immunoglobulin D secretion in a hybrid mouse cell line. Proc. Nat. Acad. Sci. USA. *78*: 1138, (1981).
29. Radbruch, A., Liesegang, B., and Rajewsky, K. Isolation of variants of mouse myeloma X63 that express changed immunoglobulin class. Proc. Nat. Acad. Sci. USA. *77*: 2909, (1980).
30. Sablitzky, F., Radbruch, A., and Rajewsky, K. Spontaneous immunoglobulin class switching in myeloma and hybridoma cell lines differs from physiological class switching. *Immun. Rev. 67*: 59, (1982).
31. Krawinkel, V., Zoebelein, G., Brüggemann, M., Radbruch, A., and Rajewsky, K. Recombination between antibody heavy chain variable-region genes: evidence for gene conversion. Proc. Nat. Acad. Sci. USA. *80*: 4997, (1983).
32. Cook, W. D., and Scharff, M. D. Antigen-binding mutants of mouse myeloma cells. Proc. Nat. Acad. Sci. USA. *74*: 5687, (1977).
33. Rudikoff, S., Giusti, A.M., Cook, H.D., and Scharff, M.D. Single amino acid substitution altering antigen-binding specificity. Proc. Nat. Acad. Sci. USA. *79*: 1979, (1982).
34. Köhler, G., and Shulman, M. Immunoglobulin M mutants. Eur. J. *Immun. 10*: 467, (1980).
35. Köhler, G., Potash, M.J., Lehrach, H., and Shulman, M. Deletions in immunoglobulin mu chains. *EMBO 1*, 555, (1982).
36. Hawley, R.G., Shulman, M.J., Murialdo, H., Gibson, D.M., and Hozumi, N. Mutant immunoglobulin genes have repetitive DNA elements inserted into their intervening sequences. Proc. Nat. Acad. Sci. USA. *79*: 7425, (1982).
37. Sharon, J., Gefter, M.L., Mauser, T., Morrison, S.L., Oi, V.T., and Ptashne, M. Expression of a V_H-C_K chimaeric protein in mouse myeloma cells. Nature *309*: 364, (1984).
38. Köhler, G., Baumann, B., Iglesias, A., McCubrey, J., Potash, M.J., Traunecker, A., and Zhu, D. Different ways to modify monoclonal antibodies. Med. Oncol & Tumor Pharmacother, *1*: 227, (1984).
39. Boulianne, G.L., Hozumi, N., and Shulman, M. Production of functional chimaeric mouse/human antibody. Nature, *312*: 643, (1984).
40. Morrison, S.L., Johnson, M.J., Herzenberg, L.A., and Oi. V. Chimeric human antibody molecules: Mouse antigen binding domains with human constant region domains. Proc. Nat. Acad. Sci. USA. *81*: 6851, (1984).
41. Neuberger, M.S., Williams, G.T., and Fox, R.O. Recombinant antibodies possessing novel effector functions. Nature *312*: 604, (1984).

41a Cobbold, S. P., and Waldmann, H. Therapeutic potential of monovalent antibodies. *Nature 308*: 460 (1984).

41b Milstein, C., and Cuello, A. C. Hybrid hybridomas and their use in immunohistochemistry. *Nature 305*: 537 (1983).

42. Köhler, G. Immunoglobulin chain loss in hybridoma lines. Proc. Nat. Acad. Sci. USA. *77*: 2197, (1980).
43. Baumal, R., Birshtein, B.K., Coffino, P., and Scharff, M.D. Mutations in immunoglobulin-producing mouse myeloma cells. Science *182*: 164, (1973).

44. McHugh, Y., Yagi, M., and Koshland, M.E. The use of J and μ-chain analysis to assign lymphoid cell lines to various stages in B cell differentiation. In: B lymphocytes in the Immune Response: Functional, Developmental and Interactive Properties. Eds. N. Klinman, D. Mosier, I. Scher, and E. Vitetta. Elsevier/North Holland, N.Y. p 467.(1981).
45. McKearn, D., Huppi, K., Bell, M., Staudt, L., Gerhard, W., and Weigert, M. Generation of antibody diversity in the immune response of Balb/c mice to influenca virus hemagglutinin. Proc. Nat. Acad. Sci. USA. *81*: 3180, (1984).
46. Sablitzky, F., Wildner, G., and Rajewsky, K. Somatic mutation and clonal expansion of B cells in an antigen-driven immune response. EMBO J. in press, (1985).
47. de Preval, C., and Fougereau, M. Specific interaction between V_H and V_L regions of human monoclonal immunoglobulins. J. Mol. Biol. *102*: 657, (1976).
48. Klein, M., Kostan, C., Kells, D., and Dorrington, K. Biochemistry *18*: 1473, (1979).
49. Zhu, D., Lefkovits, I., and Köhler, G. Frequency of expressed immunoglobulin light chain genes in lipopolysaccharide-stimulated Balb/c spleen cells. J. Exp. Med. *160*: 971, (1984).
50. Baumann, B., Potash, M.-J., and Köhler, G. Consequences of frameshift mutations at the immunoglobulin heavy chain locus of the mouse. EMBO J. in press. (1984).
51. Kearney, J.F., Radbruch, A., Liesegang, B., and Rajewsky, K. A new mouse myeloma cell line that has lost immunoglobulin expression but permits the construction of antibody-secreting hybrid cell lines. J. Immun *123*: 1548, (1979).
52. Shulman, M.J., Heusser, C., Filkin C., and Köhler, G. Mutations affecting the structure and function of immunoglobulin M. Molec. Cell. Biol. *2*: 1033, (1982).
53. Grosschedl, R., Weaver, D., Baltimore, D., and Constantini, F. Introduction of a μ immunoglobulin gene into the mouse germ line. Specific expression in lymphoid cells and synthesis of functional antibody. Cell *38*: 647, (1984).
54. Brinster, R.L., Ritchie, K.A., Hammer, R.E., O'Brien, R.L., Arp, B., and Storb, U. Expression of a microinjected immunoglobulin gene in the spleen of transgenic mice. Nature. 306: 332, (1983).
55. Storb, U., O'Brien, R.L., McMullen, M.D., Gollahon, K.A., and Brinster, R.L. High expression of cloned immunoglobulin κ gene in transgenic mice is restricted to B-lmyphocytes. Nature *310*: 238, (1984).
56. Ritchie, K.A., Brinster, L.R., and Storb, U. Allelic exclusion and control of endogenous immunoglobulin gene rearrangement in transgenic mice. Nature *312*: 517, (1984).
57. Rusconi, S., and Köhler, G. A transgenic mouse line transmitting and expressing a specific pair of rearranged immunoglobulin μ and κ genes. Nature, in press.
58. Alt, W.F., Yancopoulos, G.D., Blackwell, K.T., Wood, C., Thomas, E., Boss, M., Coffmann, R., Rosenberg, N., Tonegawa, S., and Baltimore, D. Ordered rearrangement of immunoglobulin heavy chain variable region segments. EMBO *3*: 1209, (1984).
59. Kwan, S. ,Max, E., Seidman, J., Leder, P., and Scharff, M. Two kappa immunoglobulin genes are expressed in the myeloma s 107. Cell. *26*: 57, (1981).
60. Hozumi, N., Wu, G., Murialdo, H., Baumal, R., Mosmann, T., Winberry,, L., and Marks, A. Arrangement of λ light chain genes in mutant clones of the MOPC3lS mouse myeloma cells. J. Immunol. *129*: 260, (1982).
61. Shulman, M., Wilde, D.C., and Köhler, G. A better cell line for making hybridomas secreting specific antibodies. Nature, *276*: 269–270, (1978).
62. Szybalski, W., Szybalska, E.H., and Ragni, G. Genetic studies with human cell lines. Natn. Cancer Inst. Monogr. *7*: 75–89, (1962).
63. Littlefield, J.W. Selection of hybrids from matings of fibroblasts in vitro and their presumed recombinants. Science *145*: 709. (1964).

César Milstein

CÉSAR MILSTEIN

Born 8 october 1927, in Bahía Blanca, Argentina. Married in 1953, to Celia (née Prilleltensky). No children.

1939–1944	Colegio Nacional, Bahía Blanca (Bachiller)
1945–1952	Facultad de Ciencias, Universidad de Buenos Aires (Licenciado en Ciencias Químicas)
1950–1956	Part-time clinical analyst at Laboratorios Liebeschutz
1952–1957	Research Student at the Instituto de Química Biológica, Facultad de Ciencias Médicas, Universidad de Buenos Aires
1957	*Doctor en Química* (Universidad de Buenos Aires)
1957–1963	Staff of Instituto Nacional de Microbiología, Buenos Aires (Leave of absence 1958–1961)
1958–1960	British Council Fellowship at the Department of Biochemistry, University of Cambridge
1960	*Ph.D. degree* (University of Cambridge)
1960–1961	Scientific staff of Medical Research Council at the Department of Biochemistry, University of Cambridge
1961–1963	Head of División de Biología Molecular, Instituto Nacional de Microbiología, Buenos Aires
1963–	Scientific Staff of Medical Research Council Laboratory of Molecular Biology, Cambridge
1983	Head, Protein and Nucleic Acid Chemistry Division, Cambridge

Honorary member, Scandinavian Immunological Societies (1970); Member, European Molecular Biology Organization (1974); Fellow of the Royal Society (1975); Honorary member, American Association of Immunologists (1979); Fellow of Darwin College, Cambridge (1980); Honorary Fellow of Fitzwilliam College, Cambridge (1982); Foreign Associate, National Academy of Sciences, USA (1981); Honorary Fellow, Royal College of Physicians (1983); Foreign Honorary Member, American Academy of Art and Sciences (1983); Member of the Deutsche Akademie der Naturforscher Leopoldina (1983); Académico Correspondiente Extranjero of the Real Academia de Ciencias Exactas, Físicas y Naturales, Madrid (1984).

My father was a Jewish immigrant who settled in Argentina, and was left to his own devices at the age of 15. My mother was a teacher, herself the daughter of a poor immigrant family. For both my mother and my father, no sacrifice was too hard to make sure that their three sons (I was the middle one) would go to university. I wasn't a particularly brilliant student, but on the other hand I was very active in Student Union affairs and in student politics. It was in this way that I met my wife, Celia. After graduation, we married, and took a full year off in a most unusual and romantic honeymoon, hitch-hiking our way through most countries in Europe, including a couple of months working in Israel kibbutzim. As we returned to Argentina, I started seriously to work towards a doctoral degree under the direction of Professor Stoppani, the Professor of Biochemistry at the Medical School. My PhD thesis work was done with no economic support. Both Celia and I worked part-time doing clinical biochemistry, between us earning just enough to keep us going. My thesis was on kinetics studies with the enzyme aldehyde dehydrogenase. When that was finished, I was granted a British Council Fellowship to work under the supervision of Malcolm Dixon. There, in the Department of Biochemistry at the University of Cambridge, I started a project on the mechanism of metal activation of the enzyme phosphoglucomutase. It was through that enzyme that I started to collaborate with Fred Sanger. I have described this collaboration in some detail previously (Lynen Lecture; Miami Winter Symp. Proc., *In:* "From gene to protein: translation into biotechnology"; Ed. W. Whelan, Academic Press, 1982). It was after completing my PhD thesis that I took a short-term appointment with the Medical Research Council in Sanger's group, and then returned to Argentina for a period of two years. During that period I extended my studies of mechanisms of enzyme action to the enzymes phosphoglyceromutase and alkaline phosphatase. It was then that I had my first experience at directing other people's work, including my first research student. The political persecution of liberal intellectuals and scientists manifested itself as a vendetta against the director of the institute where I was working. This forced my resignation and return to Cambridge to rejoin Fred Sanger, who by then had been appointed Head of the Division of Protein Chemistry in the newly-formed Laboratory of Molecular Biology of the Medical Research Council. Following his suggestion, I shifted my interests from enzymology to immunology. The evolution of my research in this area is described in the Lynen Lecture as mentioned above and in the Nobel Lecture.

Prizes and Awards

Prize Herrero Doucloux of the Asociación Química Argentina (1957); CIBA Medal and Prize (1978); Lewis S. Rosenstiel Award, Brandeis University (1979); Avery-Landsteiner Prize, Society for Immunology (1979); V. D. Mattia Lectureship Award, Roche Institute (1979); Adolph Rosenberg Award, University of Miami (1980); Wolf Prize in Medicine, Wolf Foundation, Israel (1980); Louisa Gross Horwitz Prize, Columbia University (1980); Robert Koch Prize and Medal, Germany (1980); Royal Society Wellcome Foundation Prize (1980); Madonnina Award, Fondazione Carlo Erba, Milano (1981); William Bate Hardy Prize, Cambridge Philosophical Society (1981); Jiménez Díaz Memorial Award, Fundación Conchita Rabago de Jiménez Díaz, Spain (1981); General Motors Cancer Research Foundation Sloan Prize, USA (1981); The Gairdner Foundation Annual Award, Canada (1981); Krebs Medal, Federation of European Biochemical Societies (1981); Brown-Hazen Memorial Award, Albany, New York (1982); Lynen Medal, Miami Winter

Symposium (1982); Gerónimo Forteza Medal, Valencia, Spain (1982); David Pressman Memorial Award, U.S.A. (1982); Biochemical Analysis Prize 1982, German Society for Clinical Chemistry (1982); Karl Landsteiner Award, American Association of Blood Banks (1982); Royal Medal, Royal Society (1982); XI International Congress of Allergology and Clinical Immunology Award (1982); Rabbi Shai Shacknai Memorial Prize, Hebrew University, Jerusalem (1982); Philip Levine Award, American Society of Clinical Pathologists (1983); Franklin Medal, Franklin Institute, U.S.A. (1983); Mallinkrodt Award for Investigative Research, Clinical Ligand Assay Society, U.S.A. (1983); Carlos J. Finlay Prize for Meritorious Work in Microbiology, UNESCO (1983); Common Wealth Award in Science, Sigma XI Scientific Research Society, U.S.A. (1983); Dale Medal, Society for Endocrinology (1984); Albert Lasker Basic Medical Research Award, Albert and Mary Lasker Foundation (1984); John Scott Award, Board of Directors of City Trusts, Philadelphia, U.S.A. (1984).

FROM THE STRUCTURE OF ANTIBODIES TO THE DIVERSIFICATION OF THE IMMUNE RESPONSE

Nobel lecture, 8 December, 1984

by

CÉSAR MILSTEIN

Medical Research Council Laboratory of Molecular Biology, Hills Road, Cambridge, U.K.

> *Cuando se acerca el fin,* escribió Cartaphilus, *ya no quedan imágenes del recuerdo; sólo quedan palabras.* Palabras, palabras desplazadas y mutiladas, palabras de otros, fué la pobre limosna que le dejaron las horas y los siglos.
>
> *J. L. Borges*

When an animal is infected, either naturally or by experimental injection, with a bacterium, virus, or other foreign body, the animal recognises this as an invader and acts in such a way as to remove or destroy it. There are millions of different chemical structures that the animal has never seen and yet which it is able to recognise in a specific manner. How is this achieved? Scientists have been fascinated by this question for most of this century, and we continue to be fascinated by the intricacies and complexities that still need to be clarified. Even so, looking back over the years since I myself became involved in this problem, progress in the understanding of the process has been phenomenal. Suffice it to remind our younger colleagues that 20 years ago we were still trying to demonstrate that each antibody differed in its primary amino acid sequence.

What attracted me to immunology was that the whole thing seemed to revolve around a very simple experiment: take two different antibody molecules and compare their primary sequences. The secret of antibody diversity would emerge from that. Fortunately at the time I was sufficiently ignorant of the subject not to realise how naïve I was being.

Back in 1962, when I had by accident become the supervisor of Roberto Celis in Argentina, it occurred to me that antibody diversity might arise from the joining by disulphide bridges of a variety of small polypeptides in combinatorial patterns. I don't know whether anybody else had the same idea at that time, but of all the prevailing theories about antibody diversity that I am aware of, this is one that was widest of the mark. I hold it to my credit that I never put it into print. But it was of great value to me as it provided an intellectual justification to work on disulphide bonds of antibodies. By the time I joined the Laboratory of Molecular Biology in 1963, the model of two heavy and two light

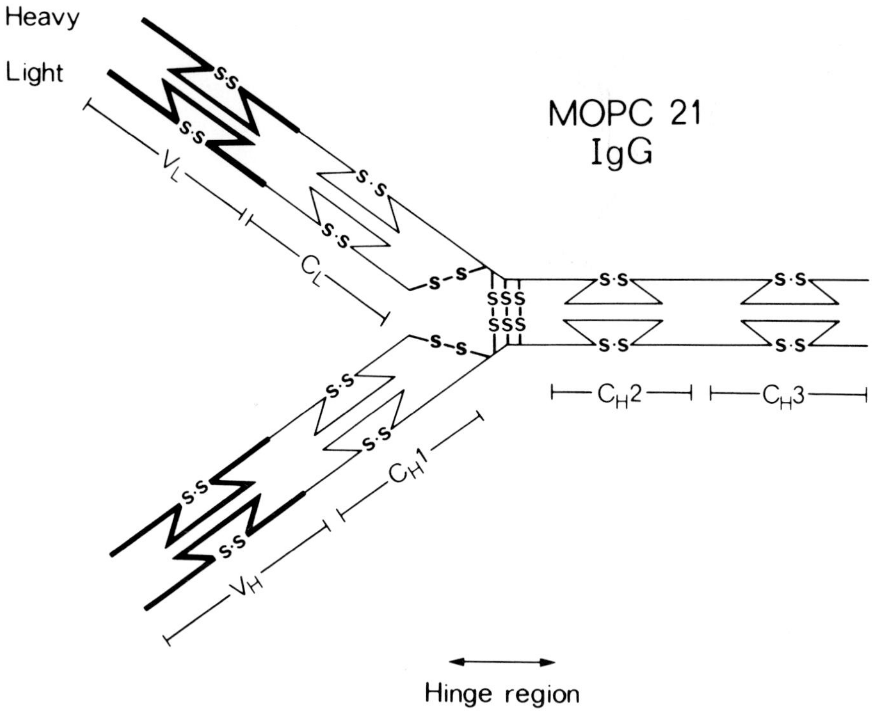

Fig. 1. Antibodies are made of two or more pairs of heavy and light chains joined by disulphide bonds. Each chain has two regions. The variable region differs in structure from one antibody to another and contains the combining site. The antibody combining site is located at the tips of a Y-shaped three-dimensional structure. The constant region is invariant within a given class or subclass, and is responsible for effector functions (complement binding, attachment to and transport across membranes etc). The number and position of the interchain disulphide bonds is characteristic for the different classes and subclasses. In this figure, the structure depicted is the mouse myeloma protein MOPC 21 which was the subject of much research in our laboratory.

chains joined by disulphide bonds (Fig. 1) had been established (1), and I was eager to accept Dr. Sanger's proposal that I should engage in studies of antibody combining sites.

The nature of antibody diversity

At first I looked for differences in fingerprints of digests of iodinated antibodies directed against different antigens. The pattern that emerged from those studies implied that purified antibodies were too complex and differed only in a subtle quantitative way from the totally unfractionated immunoglobulin. I never published those results, which only led me to the conviction that the protein chemistry of antibodies at that level was too difficult to tackle, and that a different approach was needed.

The study of the amino acid sequence around the disulphide bonds of the immunoglobulins was my own short-cut to the understanding of antibody diversity. I soon recognised the existence of what appeared to be a variable disulphide bridge and a common disulphide bridge (2,3), but the full meaning of that observation only became obvious when Hilschmann and Craig described the variable and constant halves of antibody light chains (4). The variable half contained one disulphide bond, and the constant half the other. This was followed, in later studies with Pink, Frangione, Svasti and others, by the observation of the repeating pattern of similar S-S loops as a distinctive common architectural feature of the different classes and subclasses of immunoglobulin chains. What distinguished them from each other was the diversity of interchain S-S bonds (5).

The period between 1965 and 1970 was full of excitement, both at the experimental and theoretical level. How were these variable and constant regions going to be explained? It was now not only a problem of millions of antibody structures, but that in addition those millions of structures were part of a polypeptide which otherwise had an invariant primary sequence encoded by only one or very few genes. How to solve the puzzle? Dreyer and Bennett (6) suggested that there were thousands of genes in the germline and that the paradox was easy to solve if we postulated a completely unprecedented scheme. This became known as the "two genes—one polypeptide" hypothesis. At the time we did not like that, and proposed a mechanism of hyper-mutation operating on selected segments of a gene (7). There were other ideas at the time to generate antibody diversity. One of them, widely discussed in a Cold Spring Harbor Symposium in 1967, was based on a mechanism of somatic cross-over between gene-pairs (8). It was very exciting for me when soon after the symposium I could show that in the human kappa chains at least three genes must be involved (9). The predicted thousands of V-regions could be grouped into a small number of families or subgroups. The fact that these families were encoded by non-allelic V-genes (10) – coupled to the genetics of the C-region, which indicated a single Mendelian C-gene – provided the experimental evidence that convinced me and many others that the "two genes—one polypeptide" hypothesis was inescapable.

After that, there was a period of consolidation and extension of the results. The concept of V-gene families or subgroups became firmly established, as was the existence of hypervariable residues within the variable segment (9,11). Crystallographic data showed that such hypervariable residues were near to each other, justifying the idea that they were part of the antibody combining site. This was directly shown with crystals of myeloma protein-antigen complexes (12). The work with myelomas was not only totally vindicated, but also generally accepted. The idea of separate pools of V- and C-genes that were under continuous expansion and contraction was the last element added to the picture. By 1970 we became convinced that "the section of the genome involved in the coding of immunoglobulin chains undergoes an expansion-contraction evolution: that the number of individual genes coding for basic sequences is not large, and that it varies in different species and even within species at different stages of its own history. The task of providing for the endless variety of individual chains is left to somatic processes" (13).

Light chain mRNA and the signal for secretion

I now began to feel a bit restless. It seemed that protein chemistry alone was not going to get us much further. Furthermore, there was a lot of excitement in the laboratory with the new methods for sequencing RNA being developed by Sanger and his group. Perhaps even more important, one of my closest friends at the laboratory, George Brownlee, was beginning to feel that the time was ripe to attack molecules more complicated than 5S or 6S RNA. So we joined forces in an attempt to isolate immunoglobulin mRNA. This was a difficult problem and when George's new research student, Tim Harrison, joined us we decided to move from solid tumours (14) to cell lines in culture which were kindly provided by colleagues from the Salk Institute (15). The first important breakthrough in the field was a paper reporting *in vitro* synthesis of immunoglobulin light chains (16). We immediately set to work to follow up that approach, and to our delight ran into the unexpected observation of the existence of a biosynthetic precursor of light chains. Further experiments led us to propose that the extra N-terminal sequence was a signal for vectorial transport across membranes during protein synthesis. That was the first evidence which indicated that the signal for secretion was an N-terminal segment, rapidly cleaved during protein synthesis (17,18).

However, our major concern remained the sequence of the messenger RNA for the light chains. In those days there was no DNA sequencing, only mRNA sequencing via elaborate fingerprints of radioactive mRNA. Every radioactive messenger preparation on which we could do sequence analysis involved the labelling of cells with inorganic ^{32}P-phosphate at levels of 50 mCi. So there we were, dressed up in our new-style laboratory coats (namely heavy lead aprons), behind a thick plastic screen, labelling cells and then frantically working up our messenger purification procedures and performing fingerprinting experiments, before the inexorable radioactive decay. Although we didn't go very far in our sequencing, we could isolate oligonucleotides that corresponded to the protein sequences (19). Among these were oligonucleotides spanning the V- and C-regions, demonstrating that the protein chain was made from a single messenger RNA and that, therefore, integration of the V- and C-genes did not take place during or after protein synthesis (20). At this stage the radioactive approach was stopped and we tested alternative methods for the sequencing of mRNA, using synthetic primers and cDNA synthesis. This approach went on in the background while our main efforts were moving in a different direction. Eventually however, it paid off (21). I will come back to that later, because it forms part of my story.

Spontaneous somatic mutants of a myeloma protein

The introduction of tissue culture methods to our laboratory had a major impact on the direction of our research. With my new research student, D. S. Secher, and soon after with R. G. H. Cotton, we decided to embark on an analysis of the rate and nature of somatic mutation of myeloma cells in culture. We were hoping that we might reveal a high rate of mutation of the hypervaria-

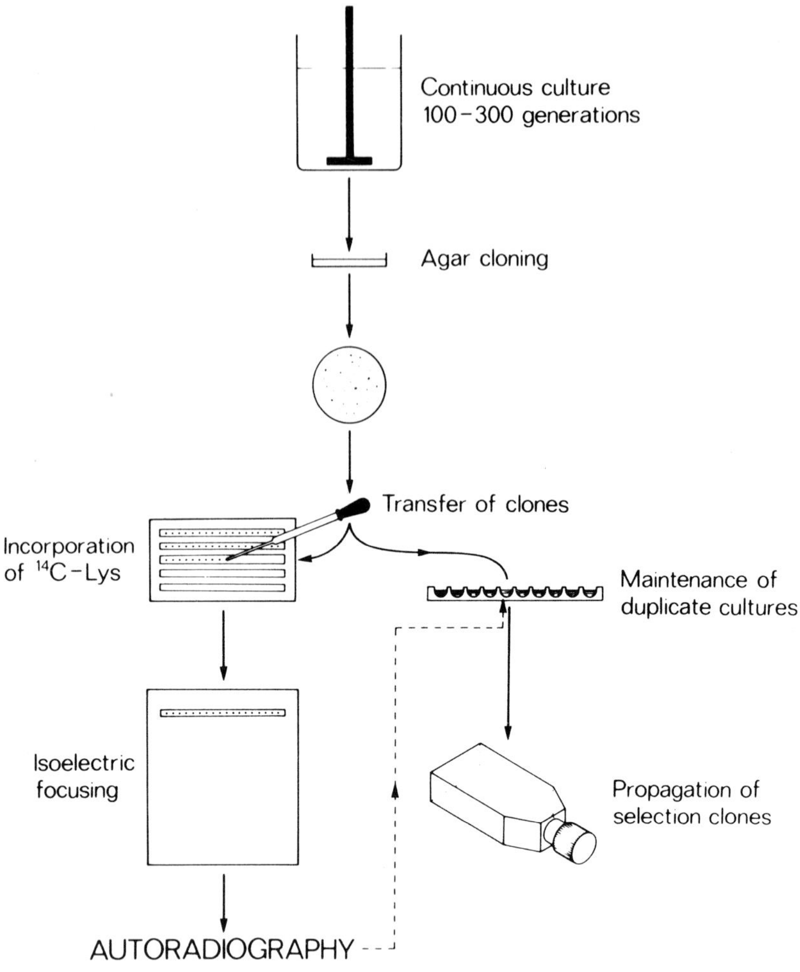

Fig. 2. Protocol used for the screening of the isoelectric focusing pattern of the immunoglobulin secreted by 7,000 clones of P3 myeloma cells. Mutants were detected and their primary defect analysed by amino acid and mRNA sequence analysis. The results are described in Table 1 (taken from Ref. 23).

ble segments. (The protocol is described in Fig. 2.) A continuous culture was grown for a minimum of three months to allow mutants to accumulate, and individual cells were taken and grown as colonies. These were incubated with labelled amino acids and the radioactive immunoglobulin analysed to detect mutants with altered electrophoretic properties. Our first structural mutant appeared after a few thousand clones (22), and the final analysis of 7,000 individual clones gave us a pool of mutants which are described in Table 1. We were relieved that this elaborate experiment provided the first evidence at the protein and nucleic acid levels of the existence of somatic mutations of mammalian cells (23). Furthermore, the rate at which these mutations occurred suggested an important role in the generation of diversity (24). But the mutations were not in the variable region, and we were forced to conclude that in the cells

Table 1. Spontaneous structural mutants of MOPC 21 heavy chains

Mutant	Protein defect	Genetic defect
IF1	Last 82 residues of CH3 missing; carbohydrate difference	Ser (387) → Ter small deletion?
IF2	Whole CHl deleted	5.5 K bases deleted including CHI exon. Aberrant switch?
IF3	Altered sequence of residues 367−380. Deletion of rest of CH3	Frameshift (−2). Premature "ochre" termination.
IF4	Asparagine 452 to Aspartic acid.	A to G transition. ("mis-sense").
NSII/1	Deletion of last 67 residues	Trp (406) → Ter G to A transition ("non-sense").

we were studying, there was no evidence for a hypermutable segment. So that in a sense we were back to square one.

Hybrid myelomas

While this work was going on, Cotton was preparing another type of experiment which turned out to be more important than we anticipated (25). This involved the fusion of two myeloma cells in culture (Fig. 3). That fusion demonstrated that the phenomenon of allelic exclusion was not dominant. On the contrary, fusion of two myeloma cells gave rise to a hybrid co-dominantly expressing the antibody chains of both parents. In addition, we proved that the expression of V- and C-regions was *cis*, probably because the V- and C-segments were already integrated at the DNA level by a translocation event in the precursors of plasma cells. This was in contrast to the assembly of heavy and light chains, which combined with each other to give rise to hybrid molecules.

Armed with these results, I went to Basel to give a seminar, and the important consequence was that Georges Köhler came to Cambridge. He joined in our main research project of looking at somatic mutants in immunoglobulin-producing cells, and in the other minor project concerning the phenotypic expression of somatic cell hybrids prepared between myelomas and myeloma mutants. It became increasingly clear that we could not go on looking for mutants by the procedure we had employed before, and the only way ahead was to use a culture of a myeloma cell line capable of expressing an antibody. Mutants from that cell could then be made based on the antibody activity. Although at that time there had been reports in the literature of myeloma cells

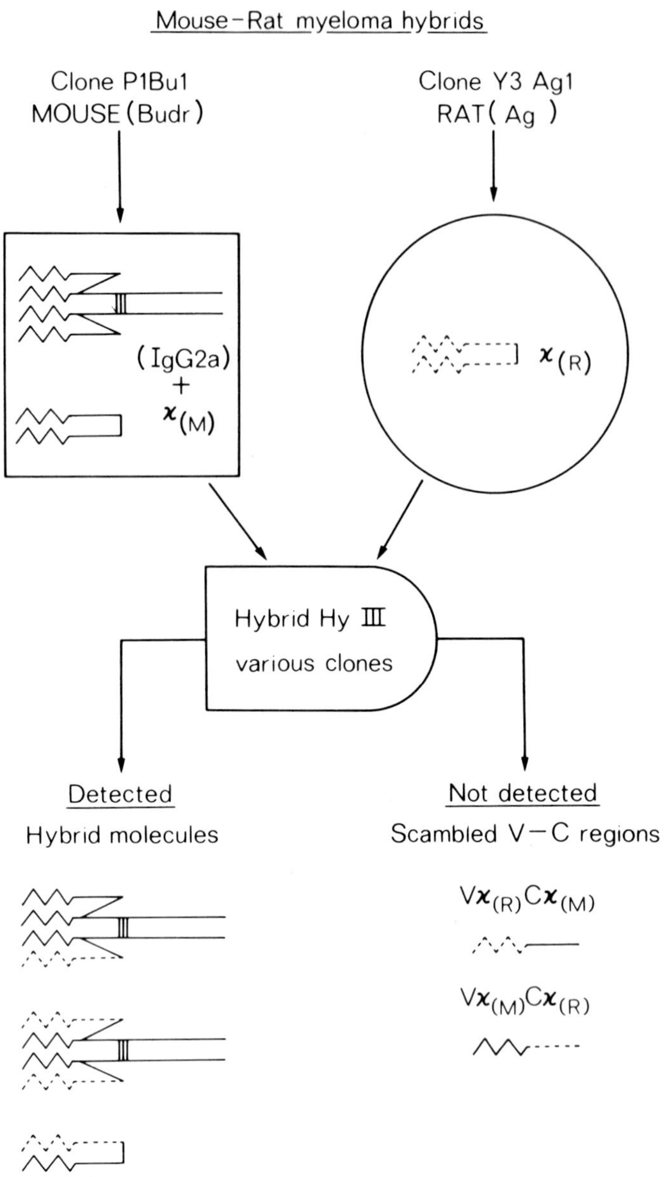

Fig. 3. Co-dominant *cis* expression of antibody genes in hybrids of myeloma cells. The diagram describes data taken from Ref. 25.

capable of fulfilling that role, none proved suitable in our hands. The myeloma cell line P3 (MOPC 21) would have been ideal from a chemical point of view, because at the time sequencing the protein was a major undertaking and we knew how to deal with MOPC 21. But we were unable to find a suitable antigenic binding activity to this myeloma protein. We failed, but others who were pursuing similar types of experiments succeeded. Scharff and his co-workers were the first to demonstrate that one can isolate somatic mutants of a variable region in that way (26).

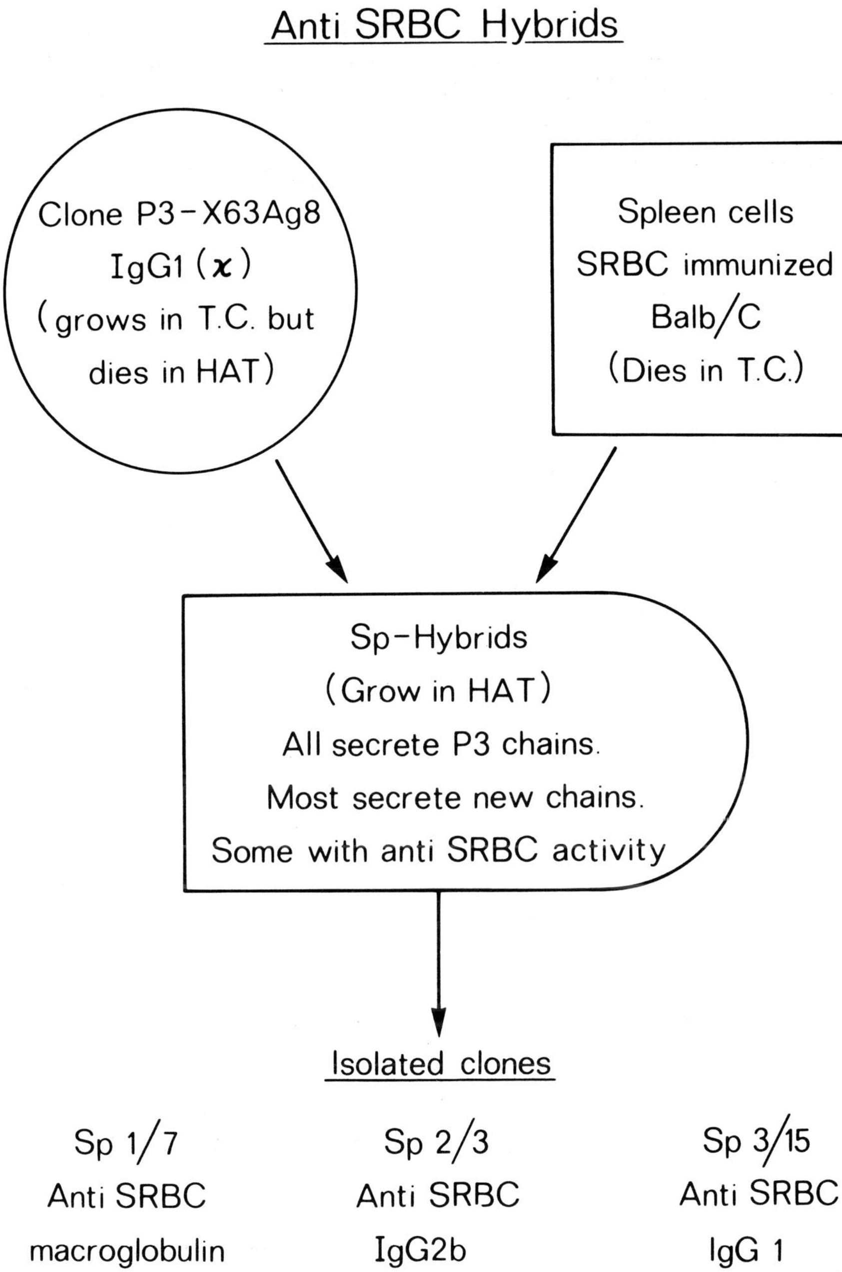

Fig. 4. The first successful hybridoma was prepared from cells from a mouse immunized with sheep red blood cells (SRBC) (56). These were fused to a myeloma cell line producing the IgG protein MOPC 21 (Fig. 1) growing in tissue culture and made resistant to azaguanine. Hybrids were selected by growth in HAT medium (57).

And yet in a funny way our lack of success led to our breakthrough; because, since we could not get a cell line off the shelf doing what we wanted, we were forced to construct it. And the little experiment being done in the background concerning hybridization between myeloma cells developed into a method for the production of hybridomas. Thus, as illustrated in Fig. 4, instead of hybridizing two myelomas, we hybridized a myeloma and an antibody producing cell. The resultant hybrid was an immortal cell capable of expressing the

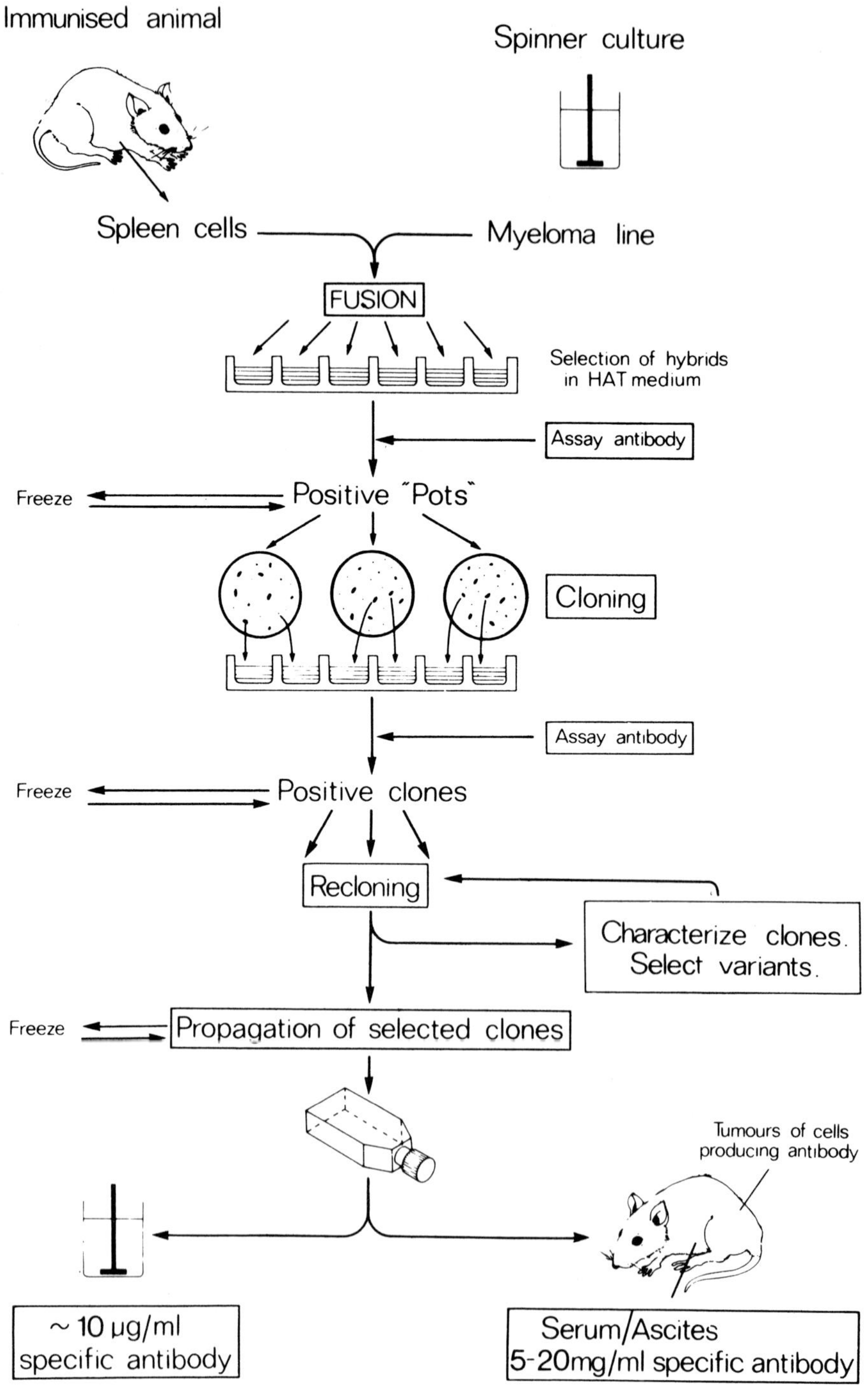

Fig. 5. Most generally used protocol for the derivation of hybridomas (taken from Ref. 58).

antibody activity of the parental antibody-producing cell, the immortality being acquired from the myeloma.

So finally, we were able to obtain a continuously-growing cell-line expressing a specific antibody and use it to search for mutants of the hypervariable region. This was undertaken by my research student, Deborah Wilde. While she got more and more discouraged by her lack of success in what she called "looking for a needle in a haystack", it dawned on me that it was up to us to demonstrate that the exploitation of our newly-acquired ability to produce monoclonal antibodies "à la carte" was of more importance than our original purpose. After our early success we ran into technical difficulties and could not get our

Table 2. Selected list of monoclonal antibodies derived in our laboratory

Hybridoma	Antigen	Purpose and use	Selected references
R3/13 R2/10P R2/10S	Rat MHC	Reagents for tissue typing Synergistic effects	Galfré *et al.*, 1977, (27) Howard *et al.*, 1979, (60)
W3/13 W3/25	Rat T cell markers	Analysis of cell surface antigens	Williams *et al.*, 1977, (29)
H6/31	Mouse IgD allotype	Standard allotype reagent	Pearson *et al.*, 1977, (61)
W6/32 W6/1 W6/34 and others	HLA-A,B,C Blood group A Controlled by chromosome 11	Tools for genetic analysis and biochemical studies	Barnstable *et al.*, 1978, (62)
M1/69, M1/70	Mac-1 and other mouse leukocyte surface antigens	Novel mouse leukocyte differentiation antigens.	Springer *et al.*, 1978, 1979, (63,64)
M1/22	Forssman	Embryonic development	Stern *et al.*, 1978, (30)
H9/25	Alloantigen on killer and plaque forming cells		Takei *et al.*, 1980, (65)
NA1/34	Subpopulation of human thymocytes (CD1)	Define subpopulations of human lymphoid cells	McMichael *et al.*, 1979, (66)
NC1/34	Substance P	Radioimmunoassay Immunocytochemical localization of neurotransmitters. Internally labelled antibodies.	Cuello *et al.*, 1979, (32,67)
YC5/45	Serotonin	Dual localization at the EM level	
6D4 NB1/19	Blood group A Blood group B	Standard blood group reagents	Voak *et al.*, 1980, (33,68)
NK2	Human anti-interferon	Large scale protein purification	Secher and Burke, 1980, (34)

fusion experiments to work for quite some time. Then Giovanni Galfré, who had recently joined us, got us out of the deadlock when he discovered that one of our stock solutions had become contaminated with a toxic substance. After this an improved reliable protocol was developed (Fig. 5) and quick progress made towards the first practical applications of the technology. For several years I shelved the antibody diversity problem to demonstrate the practical importance of monoclonal antibodies in other areas of basic research and in clinical diagnosis (Table 2). We were able to show that the hybrid myelomas were capable of being used for the production of standard reagents such as anti-histocompatibility antigens (27) and anti-Ig-allotypes (28). The procedure was ideally suited to the study of cell surface and tumour antigens and to providing reagents for cell fractionation (29–31). Monoclonal antibodies produced in this way were suitable for radioimmunoassays and for neuropharmacology (32), as blood group reagents (33) and for large scale purification of natural products (34). We also extended the hybrid myeloma technology to a second species—the rat (35) and to the production of bi-specific immunoglobulins (hybrid-hybridomas) (36).

Genetic origin of antibody diversity

In the period 1970–1975, a considerable effort was being made to measure the number of germline genes coding for the variable regions of immunoglobulin chains. Our own contributions started when we persuaded Terry Rabbitts to join us. After considerable effort and a lot more radioactivity we obtained results indicating that the number of germ line genes was not much higher than would be predicted from our understanding of subgroups, and this view was shared and reinforced by parallel work being conducted by others (37,38). By 1976 this view was gaining general support (39). But then the impact of the recombinant DNA revolution began to be very strongly felt. Within a few years, and largely through the work of Tonegawa, Leder, Rabbitts, Hood, Baltimore, and others, a coherent picture of the arrangement and rearrange-

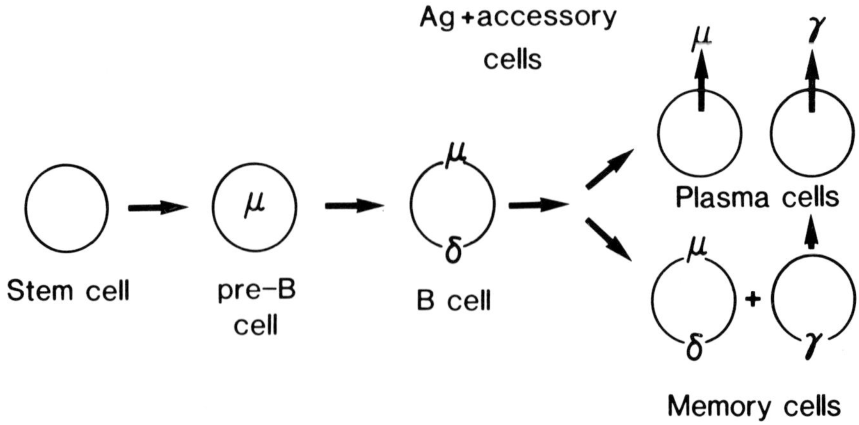

Fig. 6. Differentiation of B cells.

ment of immunoglobulin genes and their involvement in the generation of diversity began to emerge (40).

The precursors of the antibody producing cells do not express an immunoglobulin, but during their differentiation into pre-B cells and B cells, they express first the heavy chain and then the light chain (Fig. 6). The first antibody produced is membrane bound and functions as a receptor molecule, which receives antigenic signals. Triggered cells divide and differentiate to antibody producing cells and memory cells. These events at the cellular level are correlated with changes in the DNA structure (Fig. 7). The germline DNA contains the V- and C-genes on different DNA fragments, as predicted. But in addition, there are further fragmentations, and only some of them are shown in the figure. Light and heavy chains can only be transcribed and translated when certain fragments (one of the V and J in light chains, V, D and J in heavy chains) are integrated by a deletion mechanism. During this process of integration enormous diversity is generated.

To theorize about the genetic origin of antibody diversity was a "must" among molecular immunologists for quite a number of years. How do those theories contrast with the reality of today? The hard experimental facts made possible by the methodological advances in molecular biology show that, while none of them was right, most of them contained at least a grain of truth. There were two major currents of opinion. One consisted of germline theories whereby all the diversity was inherited as genes present in the germline. The other included somatic diversification theories, whereby somatic processes were responsible for the generation of diversity, starting from a small number of germline genes. As it turns out, the genetic mechanisms responsible for the

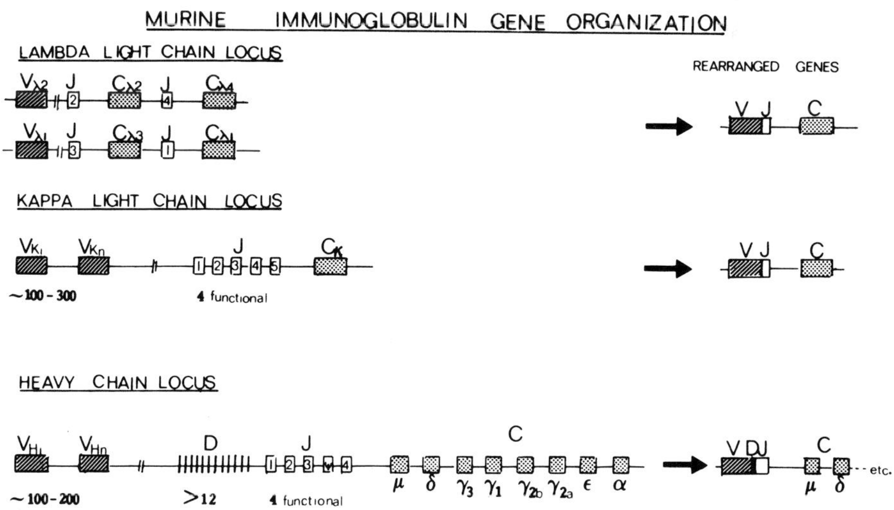

Fig. 7. Genetic arrangement of immunoglobulin genes in the germline. During differentiation into pre B cells and B cells large deletions of DNA lead to the integration of fragments (rearranged genes). Further proliferation leads to somatic mutation of the integrated gene and this is of major importance in the maturation of the response.

Table 3. Mechanisms that generate antibody diversity

1. GERMLINE: multiple V-gene segments	
2. COMBINATORIAL:	a) Different combinations of V-(D)-J
	b) Different combinations of V_H and V_L
3. JUNCTIONAL: variation at at V-J, V-D, and D-J boundaries	
4. SOMATIC POINT MUTATION: nucleotide substitutions throughout the V region	

generation of diversity include a little bit of everything (Table 3). There are between 50 and 300 gene fragments in the germline encoding the light or the heavy chains. The number varies from species to species. So there is a considerable germline contribution. Recombination and gene conversion are probably important genetic events in the evolution and maintenance of that germline gene pool. We still do not know whether these events are significant as somatic generators of diversity (41). As shown in Fig. 7, the V-region is encoded by V, D and J segments (heavy chain) and V and J segments (light chain). Their combinatorial integration into a single gene, although an important component of the generation of diversity, is not the critical mechanism predicted by the mini gene hypothesis (42). Also important is the diversity generated during the joining process, and this contains an element of the errors and aberrations during repair predicted by other theories (7,43). And then there are the somatic point mutations for which a mechanism remains to be elucidated. It may involve error-prone repair enzymes (7), genetic hot-spots (24), appropriate selection either by antigen (44) or by other network elements (45), or quite possibly by a mixture of all or some of these. The instructional theories were largely forgotten as soon as the chemical diversity of antibodies was established (46). Yet they also may contain a grain of truth. It has recently been proposed that peptide segments of the antigen which appear to be mobile are better immunogens, presumably because they adapt their structure to a predefined antibody structure (47,48). It is also possible that to some extent the antibody combining site itself has a certain degree of mobility, which has a limited capacity to accommodate its own structure to that of the antigen. Of course dynamic adaptation has a price to pay in terms of affinity. Adaptability should not be confused with the generation of specificity. As I discuss below, an improved fit of binding to the ligand is the result of somatic mutation and antigenic selection.

Molecular analysis of an immune response using monoclonal antibodies and mRNA sequencing

Let us return to an animal that is being immunized with a certain substance. The immune system recognizes the substance as foreign, and the B cells are triggered to produce antibody (Fig. 8). The different antibodies are secreted and mixed in the serum. The individual antibody molecules are extremely similar and once mixed cannot be separated from each other. For this reason, and until the advent of the hybridoma technology, it was impossible to study the diversity of the antibody response to a given immunogen. The derivation of immortal cell hybrids solved this problem, because it affords individual anti-

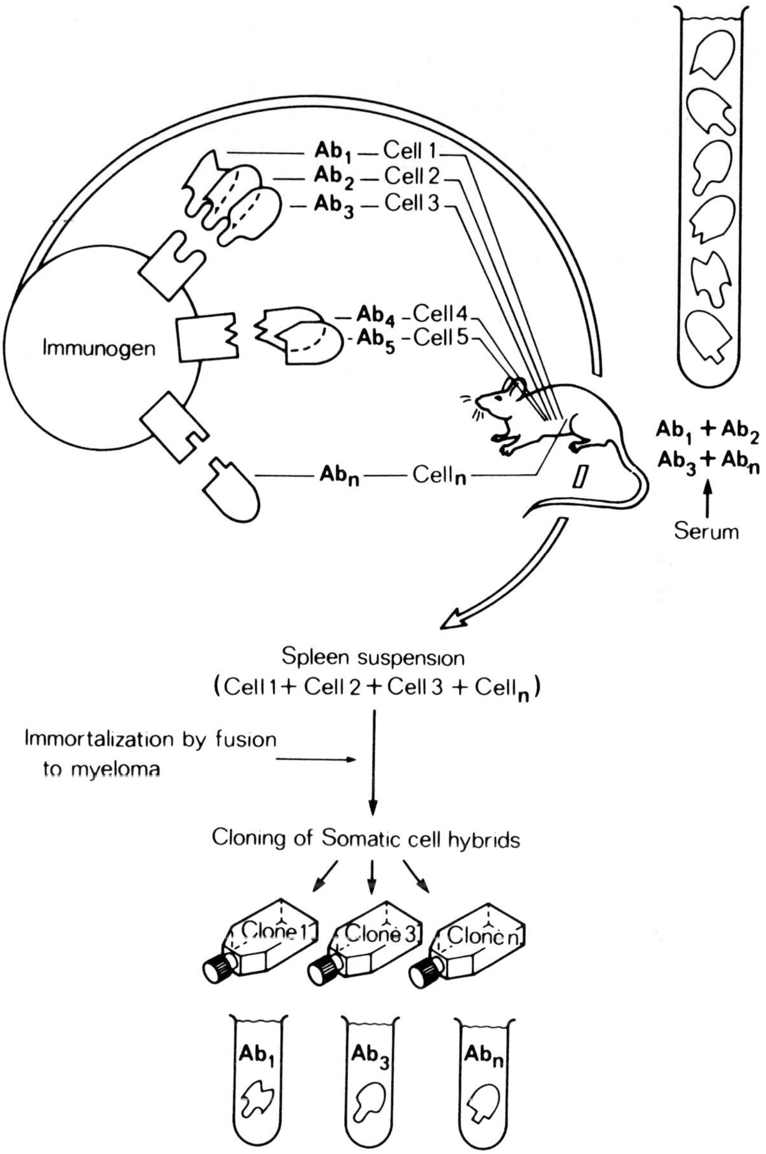

Fig. 8. The dissection of the immune response by the hybridoma technique. When an animal is injected with an immunogen the animal responds by producing an enormous diversity of antibody structures directed against different antigens, different determinants of a single antigen, and even different antibody structures directed against the same determinant. Once these are produced they are released into the circulation and it is next to impossible to separate all the individual components present in the serum. But each antibody is made by individual cells. The immortalization of specific antibody-producing cells by somatic cell fusion followed by cloning of the appropriate hybrid derivative allows permanent production of each of the antibodies in separate culture vessels. The cells can be injected into animals to develop myeloma-like tumours. The serum of the tumour-bearing animals contains large amounts of monoclonal antibody.

bodies separately produced, in culture vessels and as mouse myelomas. This permits dissection of the individual components of the antigen. Monoclonal antibodies prepared against hitherto undefined cellular components can themselves be used to identify the chemical nature of those components, to probe for their function, and later for use as reagents for diagnostic and therapeutic purposes. These are the fundamental properties behind the most important of the general applications of monoclonal antibodies. When we started to explore these applications, and until some years ago, it was possible to some extent to summarize the main results obtained (49). In recent years their application to basic research, clinical biochemistry, medical therapy, and in industry has been so widespread that I do not intend even to attempt to discuss it any further here.

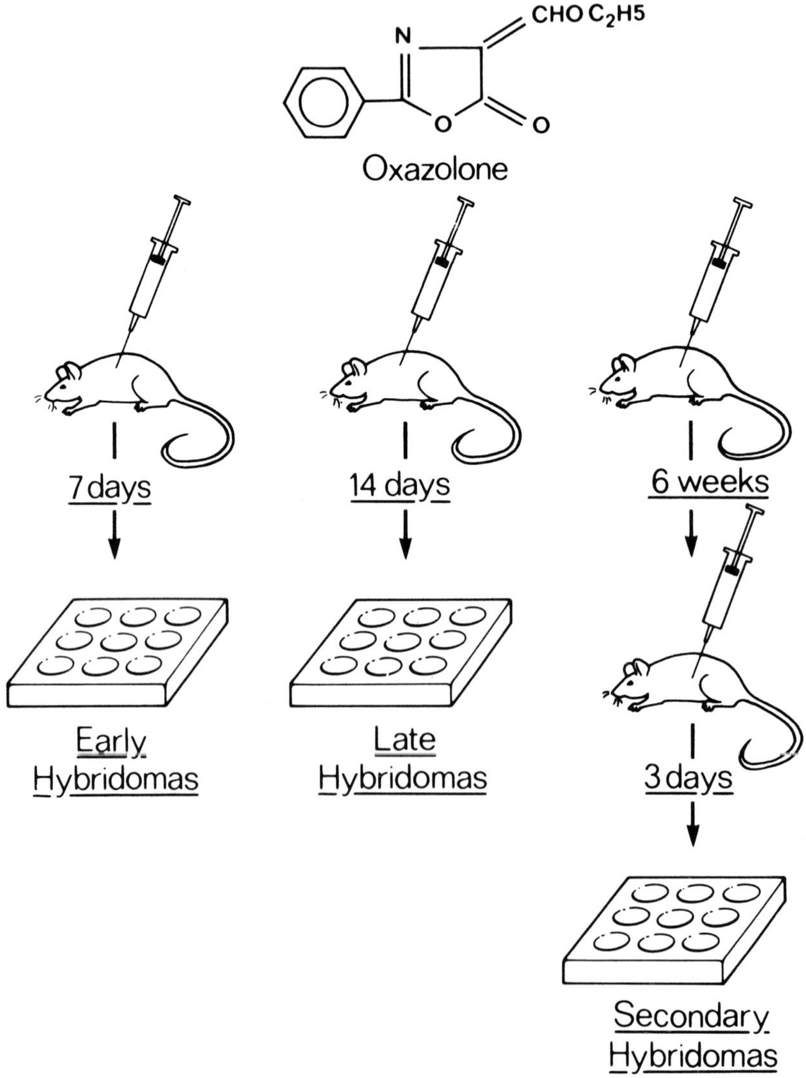

Fig. 9. Derivation of monoclonal antibodies at the onset and during the maturation of the response to oxazolone.

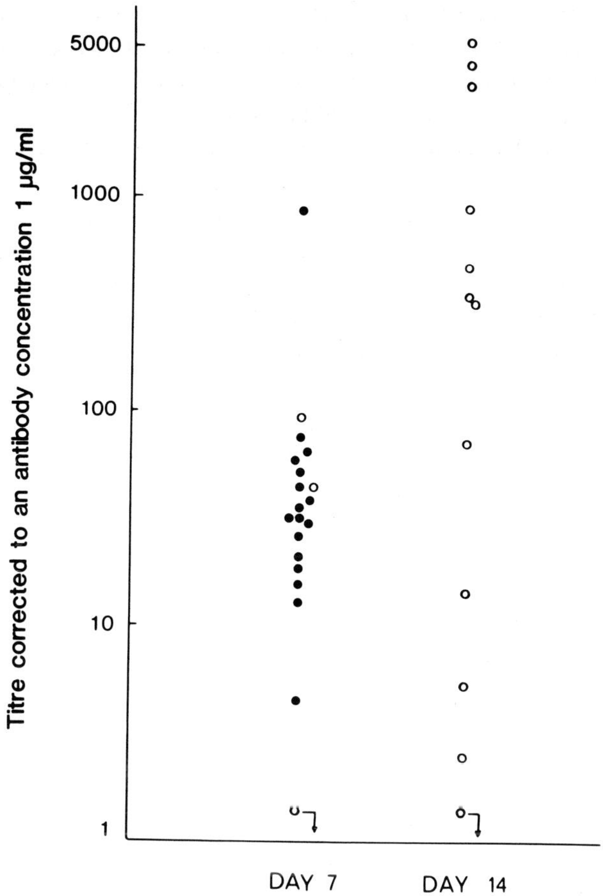

Fig. 10. Avidity of monoclonal antibodies 7 and 14 days after immunization. Haptenated phage inhibition (HPI) per μg of anti-phOx immunoglobulin from supernatants of IgG-secreting hybridomas. Those on the left were from day 7 and those on the right from day 14 fusions. Black circles represent oxazolone idiotype-positive IgG and open circles represent idiotype-negative IgG (taken from Ref. 50).

Different antibodies recognize different antigenic determinants of the immunogen, and the recognition of each determinant is complex in itself (Fig. 8). It has been known for a long time that even the simplest antigenic determinants are recognized by an unknown variety of antibody molecules. Monoclonal antibodies can be made pure and used to answer the old questions of how complex the collection of antibody molecules produced by the animal as a response to a particular antigen is, and how the individual molecules differ from each other. This brings me back to sequencing messenger RNA.

While in the late '70s the excitement about monoclonal antibodies and DNA recombinant methods was simmering, Pamela Hamlyn was quietly adapting Sanger's fast DNA sequencing methods to the sequencing of light chain mRNA. Her eventual success (21) added to our capacity to derive cell lines secreting monoclonal antibodies to a predefined antigen, and to our ability to sequence quickly the messenger RNA of the antibody molecule they produce. So, instead of asking the question "What is the nature of antibody diversity?", we were now in a position to ask the question "How do antibodies diversify during an immune response?" In other words, how, in real life in the animal, are all those genetic events capable of producing antibody diversity actually operate in response to an antigenic stimulus?

In collaboration with Matti Kaartinen, Gillian Griffiths and Claudia Berek, we have been conducting a study of the response to the hapten phenyl oxazolone (50,51). The essence of the experiment is described in Fig. 9. The hapten conjugated to chicken serum albumin as carrier is injected into mice, and 7 days and 14 days later animals are sacrificed, hybridomas are prepared and a number of random clones isolated in each case. Other animals are left for a couple of months, and hybridomas of the secondary response are prepared.

Hybridomas prepared 7 days and 14 days after primary immunization are compared in Fig. 10. Each point on the figure represents the avidity of each one of 32 monoclonal antibodies. The mixture of antibodies at each stage, as a first approximation, represents a cross-section of the complexity of a typical antiserum. The average titres of the antibodies at both stages are not very different, although the day 14 average is slightly higher. This is as expected. The antibody titre of an antiserum, as well as its average avidity, increases during the course of an immunization. It is what we refer to as the maturation of the response. What distinguishes the results of the day 7 and day 14 is that while the day 7 results cluster around the average, the scatter at day 14 is much wider.

Since each monoclonal antibody was the product of an immortal hybridoma, we could go one step further and study the total amino acid sequence of each one of these monoclonal antibodies. Better still, we could study the sequence of

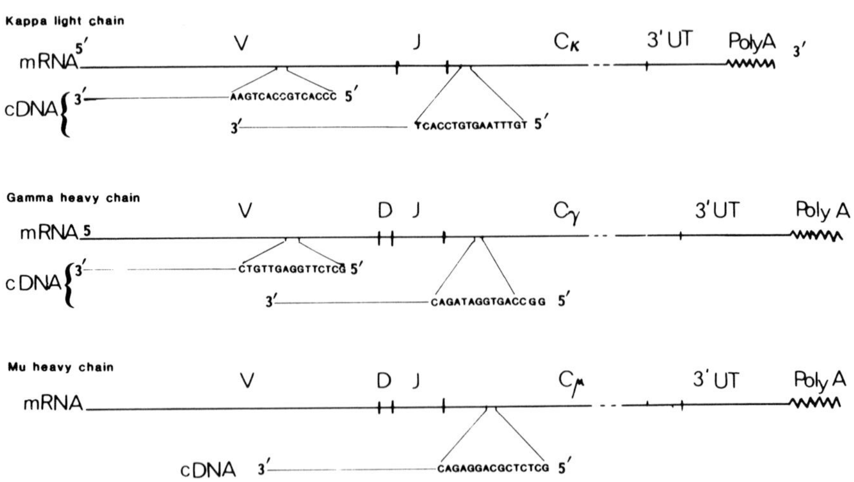

Fig. 11. mRNA sequencing strategy. Synthetic oligonucleotide primers designed to pair with defined bases within segments of mRNAs were used to initiate reverse transcription. Using dideoxynucleotides, specific stops in the cDNA can be generated and the nucleotide sequence determined by gel methods (taken from Ref. 59).

the mRNA coding for each amino acid sequence. This not only provided more information, but was also technically simpler. To do so, RNA was prepared from the hybridoma cells and direct sequencing done on the impure messenger preparations, as shown in Fig. 11. In this way, sequences of antibodies at different stages of the immune response could be compared.

What we have learned from this is that the majority of antioxazolone antibodies at day 7 express a single set of germline V-genes taken from the total pool of over 100 for each of the two chains (Fig. 12). This pair of germline genes (which we refer to as V_H-Ox1 and V_k-Ox1) are at this stage expressed in their unmutated form. The few differences between them arise by junctional diversity — that is the variations introduced during integration of the DNA fragments V, D and J which make up the variable region of the antibodies. At day 14 the same germ line genes V_H-Ox1 and V_k-Ox1 still seem to dominate the response.

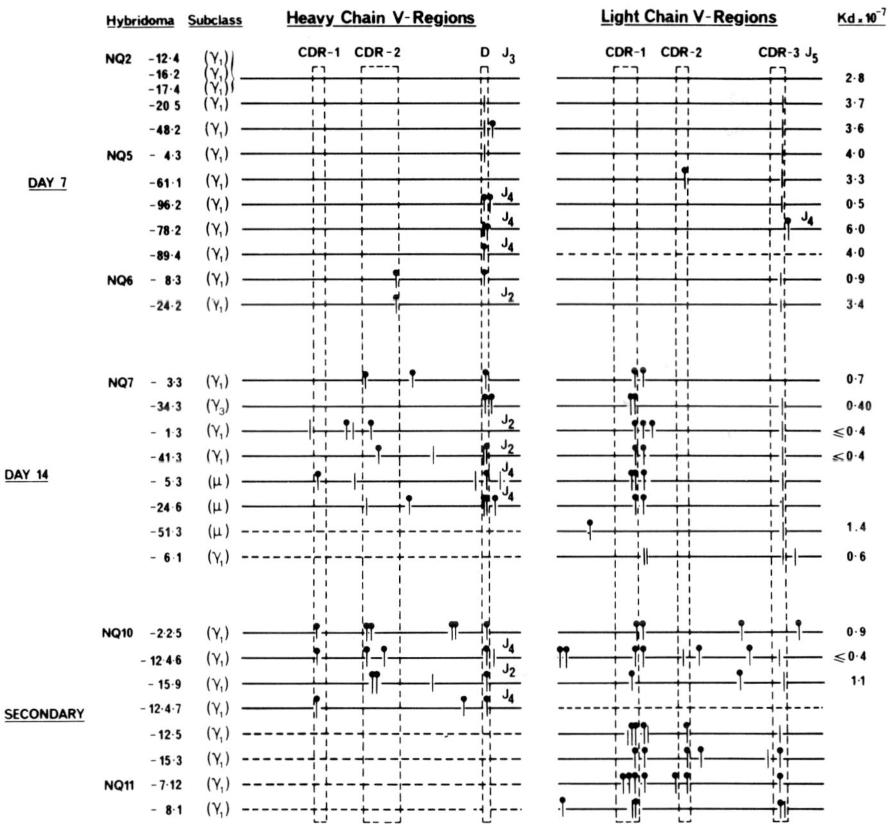

Fig. 12. Diagrammatic comparison of the mRNA sequences from anti-phOx-secreting hybridomas derived at different stages after immunization with Ox-CSA. Only sequences closely related to the prototype are shown. The variable region sequences of each hybridoma have been compared with the sequences of V_H-Ox1 and V_K-Ox1 respectively. Unbroken horizontal lines denote identical sequences, broken lines represent extensive sequence differences. A black circle indicates that these changes predict an amino acid difference at this position. Complementarity determining regions (CDR-1, −2, −3) have been marked as have the D and J regions. Where different J segments are observed these are represented accordingly. Dissociation constants determined by fluorescence quenching (Kd in moles/litre) are shown on the right (taken from Ref. 51).

However, in sharp contrast to day 7, the day 14 antibodies express a small number of point mutations which are responsible for a significant increase in affinity for the same hapten. In other words, as the response matures, new somatic mutants appear in a seemingly endless variety.

The antibodies obtained during the secondary response, expressing the germline gene combination characteristic of the primary response, show a further small increase in point mutations (Fig. 12). However, the most important feature of the secondary response is a shift towards other germline genes (see Table 4).

Table 4. Expression of germ line V genes in the maturation of the response to oxazolone

	V_HOX_1-V_kOX_1 combination	Other combinations	% V_HOX_1-V_kOX_1
DAY 7	11	4	73
DAY 14	6	5	54
SECONDARY	4	18	18

It appears therefore that the development and maturation of the immune response to oxazolone — which we take as a model system — proceeds basically in three stages. In the first the majority of the antibody reflects a very restricted choice from a vast repertoire of germline gene combinations, self-selected for their capacity to bind the antigen. In the second stage, cells expressing these combinations proliferate, and during this proliferation mutants arise which improve the affinity of the antibody for the antigen. In the third stage, as the first type of germline gene combinations and their mutants reach a certain limit of dissociation constants, new germline gene combinations and somatic mutants are selected for further improvements. Of course the three stages are not absolutely separate and all three processes overlap to a certain extent. In many ways, the system behaves as a Darwinian system, where adaptation is an improvement in antigen binding. It remains to be seen to what extent other regulatory constraints are critical to the process.

From monoclonal antibodies to antibody engineering

The immortalization of antibody-producing cells not only allows the permanent supply of an antibody of a constant chemical structure but, more important, affords all the advantages that can be derived from the techniques of cell culture and somatic cell genetics. The most obvious is cell cloning, and this has been at the root of the explosion in the use of this technology. And yet the derivation of cell lines producing specific antibodies cannot go beyond the immortalization of what already exists. We select hybrids producing monoclonal antibodies of desired properties, but if the immunized animal does not make it, there is no way of immortalizing it. Fortunately we can go further.

Hybridomas are established cell lines and are therefore capable of other "*in vitro*" manipulations using somatic cell genetic and molecular engineering techniques. We are at the beginning of a new era of immunochemistry, namely the production of "antibody based" molecules. The derivation of hybrid hybridomas is one example of the utilization of such methods for the biosynthesis of bi-specific antibodies (36). Another example is the derivation of class switch mutant antibodies (52).

Some years ago, I discussed the eventual use of recombinant DNA techniques to make more drastic changes (53). Recent developments have shown the feasibility and potential of the approach. Antibody genes have been put into suitable vectors, propagated, modified and re-introduced into myeloma cells which will then secrete recombinant antibodies possessing novel properties. For instance, in my laboratory Neuberger has developed a cell line which secretes a mouse-human antibody molecule with a mouse anti-nitrophenacetyl variable region and a human epsilon heavy chain constant region (54). In another example, the Fc portion of the mouse antibody was replaced by staphylococcal nuclease (55). A novel antibody was thus made which contains an antigen specific Fab portion joined to an enzymatic effector function replacing the normal Fc portion.

More elaborate modifications will be made possible by the fast-developing techniques of site-directed mutagenesis. These will allow well-planned specific modifications of antibody combining sites. In this way we will be able to test the contribution of individual point mutations to the generation of high affinity antibody during the process of the maturation of the response. This brings us back to the problems of the diversity of molecular recognition and the maturation of the immune response.

Exciting as these prospects are, they still require the basic starting genes taken from a hybridoma line. With them, we can introduce changes at the amino acid sequence level but with the exception of simple changes, the ultimate folding pattern and their effect on protein-ligand interaction cannot yet be reliably predicted. This will remain so for the time being. Total construction of antibody molecules to suit specific needs depends on a much better understanding of protein folding.

While selection is the strategy of the antibody response of an animal, the immunochemistry of the future will revert to an instructional approach where the antigen will tell us what antibody structure we should construct. Although this is not science fiction, we need to overcome the theoretical problems involved in the translation of one-dimensional reality into a valid three-dimensional prediction. Although the way ahead is full of pitfalls and difficulties, this is indeed an exhilarating prospect. There is no danger of a shortage of forthcoming excitement in the subject. Yet, as always, the highlights of tomorrow are the unpredictabilities of today.

Acknowledgements
The hybridoma technology was a by-product of basic research. Its success in practical applications is to a large extent the result of unexpected and unpredictable properties of the method. It thus represents another clear-cut example of the enormous practical impact of an investment in research which might not have been considered commercially worthwhile, or of immediate medical relevance. It resulted from esoteric speculations, for curiosity's sake, only motivat-

ed by a desire to understand nature. It is to the credit of the Medical Research Council in Britain to have fully appreciated the importance of basic research to advances in medicine. We are delighted to belong to the small, lucky group of those who are at the window-dressing end of the justification for the wisdom of that policy.

I learned what research was all about as a research student of Stoppani in Argentina, and then with Sanger in the Department of Biochemistry at Cambridge. I owe an enormous debt to the atmosphere of the Laboratory of Molecular Biology, where all the work I have described here was done, mostly under the Chairmanship of Max Perutz, and within the Division of Protein and Nucleic Acid Chemistry, of which Fred Sanger was the Head. From them, I always received an unspoken message which in my imagination I translated as "Do good experiments, and don't worry about the rest".

During my lecture I have tried to acknowledge those of my collaborators whose contributions were critical at specific stages of the work. In addition, so far unmentioned, is John Jarvis, my personal assistant for well over 20 years – only months less than my involvement with immunology. Since this prize mentions the discovery of the principles of the hybridoma technology, I would like to acknowledge specifically the importance of the contributions of Dick Cotton and David Secher, with whom preliminary work leading to that discovery was made, and my tissue culture assistant, Shirley Howe, who was directly involved not only in the preliminary work but also in some of the specific experiments conducted with Georges Köhler. I would also like to acknowledge other collaborators who were concerned in our own contributions to the demonstration of the practical potential of the hybridoma technology in a variety of fields, particularly G. Galfré, A. F. Williams and A. C. Cuello, and the technical assistance of Mr. B. W. Wright. The list of acknowledgements is certainly much longer, but I wouldn't like to end it without recording my indebtedness to my secretaries, Margaret Dowding, and Judith Firth. Their handling of the press in the immediate period after the announcement of this prize ranks high among my memories of that exciting moment.

REFERENCES

1. Edelman, G. M. and Porter, R. R. (1972) Nobel lectures.
2. Milstein, C. (1964) *J. Mol. Biol. 9*, 836.
3. Milstein, C. (1966) *Biochem. J. 101*, 352–368.
4. Hilschmann, N. and Craig, L. C. (1965) *Proc. Natl. Acad. Sci. USA 53*, 1403–1409.
5. Frangione, B., Milstein, C. and Pink, J. R. L. (1969) *Nature 221*, 145–148.
   Svasti, J. and Milstein, C. (1970) *Nature 228*, 932.
6. Dreyer, W. J. and Bennett, C. J. (1965) *Proc. Natl. Acad. Sci. USA 54*, 864–869.
7. Brenner, S. and Milstein, C. (1966) *Nature 211*, 242–243.
8. Smithies, O. (1967) Science *157*, 267.
9. Milstein, C. (1967) *Nature 216*, 330–332.
10. Milstein, C., Milstein, C. P. and Feinstein, A. (1969) *Nature 221*, 151–154.
11. Wu, T.T. and Kabat, E.A. (1970) *J. Exp. Med. 132*, 211–250.
12. Reviewed in: Amzel, L. M. and Poljak, R. S. (1979) *Ann. Rev. Biochem. 48*, 961–997.
13. Milstein, C. and Pink, J. R. L. (1970) In: "*Progress in Bioph. and Molec. Biol.*" Vol. *21*, pp. 209–263. Pergamon Press: Oxford and New York.
14. Potter, M. (1972) *Physiological Reviews 52*, 631–719.
15. Horibata, K. and Harris, A. W. (1970) *Exp. Cell Res. 60*, 61–70.
16. Stavnezer, J. and Huang, R.-C. C. (1971) *Nature New Biol. 230*, 172.
17. Milstein, C., Brownlee, G. G., Harrison, T. M. and Mathews, M. B. (1972) *Nature New Biol. 239*, 117–120.
18. Harrison, T. M., Brownlee, G. G. and Milstein, C. (1974) *Eur. J. Biochem. 47*, 613–620.
19. Brownlee, G. G., Cartwright, E. M., Cowan, N. J., Jarvis, J. M. and Milstein, C. (1973) *Nature 244*, 236–240.
20. Milstein, C., Brownlee, G. G., Cartwright, E. M., Jarvis, J. M. and Proudfoot, N. J. (1974) *Nature 252*, 354–359.
21. Hamlyn, P. H., Gait, M. J. and Milstein, C. (1981) *Nucleic Acids Res. 9*, 4485–4494.
22. Cotton, R. G. H., Secher, D. S. and Milstein, C. (1973) *Eur. J. Immunol. 3*, 136–140.
23. Adetugbo, K., Milstein, C. and Secher, D. S. (1977) *Nature 265*, 299–304.
24. Secher, D. S., Milstein, C. and Adetugbo, K. (1977) *Immunol. Rev. 36*, 51–72.
25. Cotton, R. G. H. and Milstein, C. (1973) *Nature 244*, 42–43.
26. Rudikoff, S., Giusti, A. M., Cook, W. D. and Scharff, M. D. (1982) *Proc. Natl. Acad. Sci. USA 79*, 1979–1983.
27. Galfré, G., Howe, S. C., Milstein, C., Butcher, G. W. and Howard, J. C. (1977) *Nature 266*, 550–552.
28. Pearson, T., Galfré, G., Ziegler, A. and Milstein, C. (1977) *Eur. J. Immunol. 7*, 684–690.
29. Williams, A. F., Galfré, G. and Milstein, C. (1977) *Cell 12*, 663–673.
30. Stern, P. L., Willison, K. R., Lennox, E. S., Galfré, G., Milstein, C., Secher, D. S. and Ziegler, A. (1978) *Cell 14*, 775–783.
31. Milstein, C., Galfré, G., Secher, D. S. and Springer, T. (1979) *Ciba Foundation Symp.* No. *66*, "Human Biology: Possibilities and realities." Excerpta Medica: Amsterdam, pp. 251–266.
32. Cuello, A. C., Galfré, G. and Milstein, C. (1979) *Proc. Natl. Acad. Sci. USA 76*, 3532–3536.
33. Voak, D., Sacks, S., Alderson, T., Takei, F., Lennox, E. S., Jarvis, J. M., Milstein, C. and Darnborough, J. (1980) *Vox Sang. 39*, 134–140.
34. Secher, D. S. and Burke, D. C. (1980) *Nature 285*, 446–450.
35. Galfré, G., Milstein, C. and Wright, B. W. (1979) *Nature 277*, 131–133.
36. Milstein, C. and Cuello, A. C. (1983) *Nature 305*, 537–540.
37. Rabbitts, T. H. and Milstein, C. (1977) In: "*Contemporary Topics in Molecular Immunology*" Vol. *6*, Eds. R. R. Porter and G. Ada. Plenum Press: New York and London, pp. 117–143.
38. Tonegawa, S., Hozumi, N., Matthyssens, G. and Schuller, R. (1976) *Cold Spring Harbor Symp. Quant. Biol. 41*, 877–889.
39. *Cold Spring Harbor Symp. Quant. Biol.* (1976) *41* (various papers).
40. Reviewed in: Tonegawa, S. (1983) *Nature 302*, 571–581.
41. Edelman, G. M. and Gally, J. A. (1967) *Proc. Natl. Acad. Sci. USA 57*, 353.
42. Kabat, E. A., Wu, T. T. and Bilofsky, H. (1978) *Proc. Natl. Acad. Sci. USA 75*, 2429–2433.

43. Alt, F. W. and Baltimore, D. (1982) *Proc. Natl. Acad. Sci. USA 79*, 4118–4122.
44. Cohn, M. (1971) *Ann. N.Y. Acad. Sci. 190*, 529.
45. Jerne, N. K. (1971) *Eur. J. Immunol. 1*, 1–9.
46. Haurowitz, F. (1967) *Symp. Quant. Biol. 32*, 559–567.
47. Westhof, E., Altschuh, D., Moras, D., Bloomer, A.C., Mondragon, A., Klug, A. and Van Regenmortel, M. H. V. (1984) *Nature 311*, 123–127.
48. Tainer, J. A., Getsoff, E. D., Alexander, H., Houghton, R. A., Olson, A. J., Lerner, R. A. and Hendrickson, W. A. (1984) *Nature 312*, 127–134.
49. For example: Waldmann, H. and Milstein, C. (1982) In: "*Clinical Aspects of Immunology*" 4th ed., Vol. 1. Eds. P. J. Lachmann and D. K. Peters, Blackwells Scientific Publications, Oxford, pp. 476–503.
50. Kaartinen, M., Griffiths, G. M., Hamlyn, P. H., Markham, A. F., Karjalainen, K., Pelkonen, J. L. T., Makela, O. and Milstein, C. (1983) *J. Immunol. 130*, 937–945.
51. Griffiths, G. M., Berek, C., Kaartinen, M. and Milstein, C. (1984) *Nature 312*, 271–275.
52. Preud'homme, J. L., Birshtein, B. K. and Scharff, M. D. (1975) *Proc. Natl. Acad. Sci. USA 72*, 1427–1430.
53. Milstein, C. Wellcome Foundation Lecture. (1980) *Proc. Roy. Soc. Lond. B* (1981) *211*, 393–412.
54. Neuberger, M. S., Williams, G. T., Mitchell, E. R., Jouhal, S. S., Flanagan, J. G. and Rabbitts, T. H. (1985) *Nature 314*, 268.
55. Neuberger, M. S., Williams, G. T. and Fox, R. O. (1984) *Nature, 312*, 604.
56. Köhler, G. and Milstein, C. (1975) *Nature 256*, 495–497.
57. Szybalski, W., Szybalska, E. H. and Ragni, G. (1962) *Natl. Cancer Inst. Monogr. 7*, 75–89.
58. Galfré, G. and Milstein, C. (1981) In: *Methods in Enzymol.* Vol. *73*. Eds. S. P. Colowick, N. O. Kaplan and J. J. Langone, H. Van Vunakis, Academic Press: New York and London, pp. 3–46.
59. Kaartinen, M., Griffiths, G. M., Markham, A. F. and Milstein, C. (1983) *Nature 304*, 320–324.
60. Howard, J. C., Butcher, G. W., Galfré, G., Milstein, C. and Milstein, C. P. (1979) In: *Immunological Reviews* Vol. *47*. Ed. G. Möller, Munksgaard: Copenhagen, pp. 139–174.
61. Pearson, T., Galfré, G., Ziegler, A. and Milstein, C. (1977) *Eur. J. Immunol. 7*, 684–690.
62. Barnstable, C. J., Bodmer, W. F., Brown, G., Galfré, G., Milstein, C., Williams, A. F. and Ziegler, A. (1978) *Cell 14*, 9–20.
63. Springer, T., Galfré, G., Secher, D.S. and Milstein, C. (1978) *Eur. J. Immunol. 8*, 539–551.
64. Springer, T., Galfré, G., Secher, D.S. and Milstein, C. (1979) *Eur. J. Immunol. 9*, 301–306.
65. Takei, F., Waldmann, H., Lennox, E.S. and Milstein, C. (1980) *Eur. J. Immunol. 10*, 503–509.
66. McMichael, A. J., Pilch, J. R., Galfré, G., Mason, D. Y., Fabre, J. W. and Milstein, C. (1979) *Eur. J. Immunol. 9*, 206–210.
67. Cuello, A. C., Priestley, J. V. and Milstein, C. (1982) *Proc. Natl. Acad. Sci. USA 79*, 665–669.
68. Voak, D., Lennox, E. S., Sacks, S., Milstein, C. and Darnborough, J. (1982) *Med. Lab. Sci. 39*, 109–122.

1985
Physiology or Medicine

MICHAEL S. BROWN
JOSEPH L. GOLDSTEIN

"for their discoveries concerning the regulation of cholesterol metabolism"

THE NOBEL PRIZE FOR PHYSIOLOGY OR MEDICINE

Speech by Professor VIKTOR MUTT of the Karolinska Institute. Translation from the Swedish text.

Your Majesties, Your Royal Highnesses, Ladies and Gentlemen,

At the meeting of the French Academy of Sciences on August 26, 1816, the chemist Michel Chevreul suggested that a substance, with fat-like properties, discovered some decades previously in gallstones by physicians in France and in Germany, should be named cholesterine, from the Greek: chole, bile, and stereos, solid.

Cholesterin, or cholesterol, as it later came to be called, proved not to be confined to gallstones but to occur also in all organs in humans as in all vertebrates and to be a substance of vital importance for them. It participates in the formation of various cellular membranes and is a substance necessary for the synthesis of bile acids (of importance for digestion) and of the vitally important stereoid hormones. For their elucidation of the complicated structures of cholesterol and of bile acids, Wieland and Windaus were awarded a Nobel Prize in 1928.

Cholesterol is of vital importance but may also be deleterious, and far more so than by causing gallstones. Since the middle of the nineteenth century it has been known that in atherosclerosis, cholesterol, or rather cholesterol esters, accumulate in high concentrations in the lesioned areas of the blood vessels, and since the late 1930s a specific inheritable disease, familial hypercholesterolemia, has been recognized, with greatly increased concentrations of cholesterol in the blood, and severe alterations in the normal structure of blood vessels.

Cholesterol is almost insoluble in water. Its solubility in blood plasma is — like other lipids – due to its being packaged into submicroscopic spherical particles with completely hydrophobic components inside, surrounded outside by a mosaic layer of less hydrophobic ones, such as phospholipids and protein. Such particles are called lipoproteins. Cholesterol occurs mainly in a type of lipoproteins called low density lipoproteins, LDL.

Not all organisms require cholesterol, and some which do so, such as insects, are incapable of producing it by themselves and are, therefore, entirely dependent on dietary sources for it. The mammalian cell, however, is capable of producing its own cholesterol, but it also obtains dietary cholesterol by way of the blood. Schoenheimer's investigations from the 1930s suggested that there was some kind of equilibrium between the amount of cholesterol which the cell itself synthesized and that which it obtained from the diet. How this equilibrium was maintained was, however, completely unknown, as was the cause of the highly increased blood cholesterol concentrations in familial hypercholesterolemia. The complicated mechanism for the cellular synthesis of cholesterol had, however, been elucidated, and investigations in this field by Bloch and by

Lynen had in 1964 been recognized by a Nobel Prize.

In elegant and systematic investigations — always in collaboration — the laureates of this year studied cholesterol metabolism in cultures of connective tissue cells from healthy persons and from patients with familial hypercholesterolemia, either *without* or *with* the addition of blood serum, and thereby cholesterol, to the culture medium. They made the surprising discovery that whereas cells from healthy persons had on their surfaces specific structures, receptors, for the binding of LDL, cells from the patients had either no such receptors or else decreased numbers of them, depending on whether the patient had acquired the disease from both parents or from only one. Equally surprising was the finding that the LDL, after being bound to the receptor moved together with the latter into the interior of the cell. There, the receptor was set free and returned to the cell surface, where it could again bind LDL. The LDL particle on the other hand disintegrated into its components, and the cholesterol thus released was found to have different functions: it contributed to meeting the requirements of the cell for cholesterol; it decreased the synthesis by the cell of endogenous cholesterol by suppressing the activity of a key enzyme, named HMG CoA reductase, for such synthesis; it decreased the number of LDL receptors and thereby the influx of more LDL; and it activated an enzyme in the cell which converts excessive cholesterol into a suitable storage form. The knowledge thus acquired concerning the normal intracellular metabolism of cholesterol, the "LDL pathway", has given not only insights into the causes of genetically determined rearrangements in cholesterol metabolism where, i.a. various defects in receptor structures have been revealed, but also insights into severe and common disease states where the amount of cholesterol in the diet may play a role. This suggests possibilities for the development of methods for treatment and prevention.

Research on cholesterol has been in continuous progress for two centuries and contains several fascinating chapters by eminent scientists. The chapter that this year's laureates have contributed is one of the most fascinating.

Professor Brown, Professor Goldstein,

In elegant and systematic studies you have discovered a physiological mechanism of great importance: the way in which mammalian cells strive to establish an equilibrium between their own synthesis of cholesterol and the cholesterol they obtain from the circulating blood influenced by diet. You have also elucidated important genetically determined aberrations from this mechanism.

This knowledge forms a rational basis for development of methods for the treatment and prevention of the widespread disabling diseases known to be a consequence of dearrangement in plasma cholesterol concentrations. You have also demonstrated something else: how successful co-operation can be a principle that should perhaps be more widely applied, both in science and in other areas of human endeavour.

As a representative of the Nobel Assembly of the Karolinska Institute, I convey to you the sincere congratulations of the Assembly and ask you now to receive your Prize from the hands of His Majesty the King.

Michael S. Brown

MICHAEL S. BROWN

Michael S. Brown was born on April 13, 1941, in Brooklyn, New York, the eldest child of Harvey Brown, a textile salesman, and Evelyn Brown, a housewife. His sister Susan was born three years later. When Brown was 11 years old the family moved to Elkins Park, Pennsylvania, a suburb of Philadelphia, where Brown attended Cheltenham High School. An amateur radio operating license obtained at age 13 led to a life-long fascination with science. A serious interest in journalism also developed early. These two passions, science and writing, have remained paramount ever since.

Brown graduated in 1962 from the College of Arts and Sciences of the University of Pennsylvania, with chemistry as his major subject. He spent most of his time at the headquarters of the student newspaper, the Daily Pennsylvanian, where he served as features editor and briefly as editor-in-chief. In 1966 Brown received his M.D. degree from the University of Pennsylvania School of Medicine. In 1964 he married Alice Lapin, a companion from childhood. The next two years were spent as intern and resident in Internal Medicine at the Massachusetts General Hospital in Boston. Here Brown met Joseph L. Goldstein, a fellow intern, and the two established the friendship and mutual respect that led to their long-term scientific collaboration.

The years 1968–1971 were spent at the National Institutes of Health where Brown served initially as Clinical Associate in gastroenterology and hereditary disease. He then joined the Laboratory of Biochemistry, headed by Earl R. Stadtman, a pioneer in the disclosure of the mechanisms by which enzymes are regulated. Here Brown learned the techniques of enzymology and the fundamental principles of metabolic regulation. Brown made an important contribution to the Stadtman effort when he and a colleague discovered that a regulatory enzyme in the glutamine synthetic pathway was controlled by covalent attachment of a nucleotide, uridine.

In 1971 Brown joined the division of Gastroenterology in the Department of Internal Medicine at the University of Texas Southwestern Medical School in Dallas. His selection of Dallas was strongly motivated by his friendship with Goldstein, who had graduated from the Southwestern Medical School. In Dallas, Brown came under the influence of Donald W. Seldin, Chairman of the Department of Internal Medicine, an inspirational figure whose passion for medical science shaped the lives of a generation of Texas students.

Soon after his arrival in Dallas, Brown succeeded in solubilizing and partially purifying 3-hydroxy-3-methylglutaryl coenzyme A reductase, a previously enigmatic enzyme that catalyzes the rate-controlling enzyme in cholesterol biosynthesis. He and Goldstein had developed the hypothesis that abnormalities in the regulation of this enzyme were the cause of familial hypercholesterolemia, a genetic disease in which excess cholesterol accumulates in blood and tissues. The formal scientific collaboration with Goldstein began one year later, in 1972, just after Goldstein returned to Dallas from a postdoctoral fellowship in Seattle. The two young physicians initially maintained separate laboratories, but by 1974 the laboratories had been formally joined.

Throughout the decade of the 1970's, when their scientific work was most intensive, Brown and Goldstein continued to function as academic physicians, each performing clinical attending rounds on the general medicine wards of Parkland Memorial Hospital for six to twelve weeks per year. They also conducted frequent teaching rounds in medical genetics. Their research efforts were aided by a number of talented senior collaborators and junior associates, as well as by frequent interchange with interested members of the Department of Internal Medicine.

In 1974, Brown was promoted to the rank of Associate Professor of Internal Medicine at the University of Texas Southwestern Medical School. He became a Professor in 1976. In 1977 he was appointed Paul J. Thomas Professor of Medicine and Genetics, and Director of the Center for Genetic Disease at the same institution. In 1985, Brown was appointed Regental Professor of the University of Texas.

Brown was elected to membership in the National Academy of Sciences of the United States in 1980. He is a member of the American Academy of Arts and Sciences, the American Society for Clinical Investigation, the Association of American Physicians, the American Society of Biological Chemists, and the American Society for Cell Biology. He is a Diplomate of the American Board of Internal Medicine and a Fellow of the American College of Physicians.

Brown received several student awards at the University of Pennsylvania, including a Proctor and Gamble Scholarship (1958–1962), the David L. Drabkin Prize in Biochemistry (1962), and the Frederick L. Packard Prize in Internal Medicine (1966). He was elected to Phi Beta Kappa and Alpha Omega Alpha. From 1974 to 1977 he was an Established Investigator of the American Heart Association. He has served on several review boards including the Molecular Cytology Study Section of the National Institutes of Health (1974–77) and the editorial boards of the *Journal of Lipid Research*, the *Journal of Cell Biology*, *Arteriosclerosis* and *Science*. He has been a member of the Board of Scientific Advisors of the Jane Coffin Childs Fund since 1980.

Brown received the honorary degree of Doctor of Science from the University of Chicago (1982) and Rensselaer Polytechnic Institute (1982). With his colleague, Goldstein, Brown has shared the following awards: Heinrich Wieland Prize for Research in Lipid Metabolism (1974); Pfizer Award for Enzyme Chemistry of the American Chemical Society (1976); Albion O. Bernstein Award of the New York State Medical Society (1977); Passano Award (1978); Lounsbery Award of the U.S. National Academy of Sciences (1979); Gairdner Foundation International Award (1981); New York Academy of Sciences Award in Biological and Medical Sciences (1981); Lita Annenberg Hazen Award (1982); V.D. Mattia Award of the Roche Institute of Molecular Biology

(1984); Distinguished Research Award of the Association of American Medical Colleges (1984); Research Achievement Award of the American Heart Association (1984); Louisa Gross Horwitz Award (1984); 3M Life Sciences Award of the Federation of American Societies for Experimental Biology (1985); William Allan Award of the American Society of Human Genetics (1985); and the Albert D. Lasker Award in Basic Medical Research (1985).

Brown and Goldstein jointly delivered the following lectures: Harvey Lecture (1977); Christian A. Herter Lectures at Johns Hopkins University (1979); Harry Steenboch Lectures at the University of Wisconsin at Madison (1980); Smith, Kline, and French Lectures at the University of California, Berkeley (1981); Duff Memorial Lecture of the American Heart Association (1981); Doisy Lectures at the University of Illinois at Urbana-Champaign (1983); the first Pfizer Lecture in Honor of Konrad Bloch at Harvard University (1985); and the Berzelius Lecture at the Karolinska Institutet, Stockholm (1985).

Brown and his wife, Alice, have two daughters: Elizabeth (born in 1973) and Sara (born in 1977).

Joseph L. Goldstein

JOSEPH L. GOLDSTEIN

Joseph L. Goldstein was born on April 18, 1940, in Sumter, South Carolina, the only son of Isadore E. and Fannie Alpert Goldstein. The family owned and operated a clothing store in Kingstree, South Carolina, a town of 5000 people After his education in the primary and secondary public schools of Kingstree, Goldstein attended Washington and Lee University in Lexington, Virginia, and received the B.S. degree in chemistry, *summa cum laude,* in 1962. He then attended Southwestern Medical School of the University of Texas Health Science Center in Dallas where he was inspired to pursue a career in academic medicine by Donald W. Seldin, then and now Chairman of the Department of Internal Medicine. During his last year in medical school, Seldin offered Goldstein a future faculty position if he would become trained in genetics and return to Dallas to establish a division of medical genetics in the Department of Internal Medicine. After receiving the M.D. degree in 1966, Goldstein moved to Boston where he was an Intern and Resident in Medicine at the Massachusetts General Hospital (1966–68). It was at the Massachusetts General Hospital that Goldstein first met and developed a friendship with Michael S. Brown, his long-term scientific collaborator.

After completion of his medical training, Goldstein spent two years (1968–70) at the National Institutes of Health, where he worked in the laboratory of Marshall W. Nirenberg and also served as a clinical associate at the National Heart Institute. The opportunity to work in a first-rate basic science laboratory while at the same time carrying a limited clinical responsibility proved highly influential in shaping Goldstein's career. In Nirenberg's laboratory, Goldstein and his colleague C. Thomas Caskey isolated, purified, and worked out the mechanism of action of several proteins required for termination of protein synthesis. Here he acquired scientific skills and taste, experienced the thrill of discovery and the excitement of science, and appreciated the power of a molecular biology approach to human disease. As a clinical associate, Goldstein served as physician to the patients of Donald S. Fredrickson, then Clinical Director of the National Heart Institute and an expert on disorders of lipid metabolism. His curiosity about hypercholesterolemia was aroused when he cared for patients with the striking clinical syndrome of homozygous familial hypercholesterolemia. These patients were intensively discussed with Brown. In view of his and Brown's common interest in metabolic disease, Goldstein convinced his colleague to join him as a faculty member at the University of Texas Health Science Center at Dallas, where they would work collaboratively on the genetic regulation of cholesterol metabolism. While at the National Institutes of Health, Goldstein and Brown became avid duplicate bridge players. Their successful bridge partnership proved to be a valid testing ground for their future scientific partnership.

Before returning to Dallas, Goldstein spent two years (1970–72) as a Special NIH Fellow in Medical Genetics with Arno G. Motulsky at the University of Washington in Seattle. Motulsky was one of the creators of human genetics as a medical specialty. In Seattle, Goldstein initiated and completed a population genetic study to determine the frequency of the various hereditary lipid disorders in an unselected population of heart attack survivors. He and his colleagues discovered that 20% of all heart attack survivors have one of three single-gene determined types of hereditary hyperlipidemia. One of these disorders was the heterozygous form of familial hypercholesterolemia, which was found to affect 1 out of every 500 persons in the general population and 1 out of every 25 heart attack victims. During his fellowship in Seattle, he became conversant with tissue culture techniques, which proved to be invaluable in the subsequent studies with Brown.

In 1972, Goldstein returned to the University of Texas Health Science Center at Dallas, where he was appointed Assistant Professor in Seldin's Department of Internal Medicine and head of the medical school's first Division of Medical Genetics. He became Associate Professor of Internal Medicine in 1974 and Professor in 1976. In 1977, he was made Chairman of the Department of Molecular Genetics at the University of Texas Health Science Center at Dallas and Paul J. Thomas Professor of Medicine and Genetics, a position that he currently holds. In 1985, he was named Regental Professor of the University of Texas.

Goldstein was elected to membership in the National Academy of Sciences in 1980. He is also a member of the American Academy of Arts and Sciences, Association of American Physicians, American Society for Clinical Investigation (President, 1985–86), American Society of Biological Chemists, American Society of Human Genetics, American Society for Cell Biology, and the American Federation for Clinical Research (National Council, 1979–82). He is also a Fellow of the American College of Physicians and is a Diplomate of the American Board of Internal Medicine. Goldstein has served on study sections for the American Heart Association (1975–78) and the National Institutes of Health (1975–78). He served on the Scientific Review Board of the Howard Hughes Medical Research Institute (1978–84) and is presently a member of its Medical Advisory Board (1985-present). In 1983 he became a Non-resident Fellow of The Salk Institute for Biological Sciences. He is, or has been, a member of the Editorial Board of the following journals: *Annual Review of Genetics* (1979–84), *Arteriosclerosis* (1981-present), *Cell* (1982-present), *Journal of Biological Chemistry* (1980–85), *Journal of Clinical Investigation* (1977–82), and *Science* (1985-present).

He has received honorary Doctor of Science degrees from the University of Chicago (1982) and Rensselaer Polytechnic Institute (1982). His other academic honors include membership in Phi Beta Kappa and Alpha Omega Alpha. He was also the recipient of the Ho Din Award for Outstanding Medical School Graduate of the University of Texas Southwestern Medical School (1966) and of a Research Career Development Award from the National Institutes of Health (1972–77).

In addition to the 1985 Nobel Prize for Physiology or Medicine, Goldstein and his colleague Brown have been jointly honored for their research with the following awards: Heinrich Wieland Prize for Research in Lipid Metabolism (1974); Pfizer Award for Enzyme Chemistry of the American Chemical Society (1976); Albion O. Bernstein Award of the New York State Medical Society (1977); Passano Award (1978); Lounsbery Award of the U.S. National Academy of Sciences (1979); Gairdner Foundation International Award (1981); New York Academy of Sciences Award in Biological and Medical Sciences (1981); Lita Annenberg Hazen Award (1982); V.D. Mattia Award of the Roche Institute of Molecular Biology (1984); Distinguished Research Award of the Association of American Medical Colleges (1984); Research Achievement Award of the American Heart Association (1984); Louisa Gross Horwitz Award (1984); 3M Life Sciences Award of the Federation of American Societies for Experimental Biology (1985); William Allan Award of the American Society for Human Genetics (1985); and the Albert D. Lasker Award in Basic Medical Research (1985).

Goldstein and his colleague Brown have shared the podium for a number of distinguished lectureships, including the Harvey Lecture (1977), Christian A. Herter Lectures at Johns Hopkins University (1979), Harry Steenboch Lectures at the University of Wisconsin at Madison (1980); Smith, Kline, and French Lectures at the University of California, Berkeley (1981); Duff Memorial Lecture of the American Heart Association (1981); Doisy Lectures at the University of Illinois at Urbana-Champaign (1983); the first Pfizer Lecture in Honor of Konrad Bloch at Harvard University (1985); and the Berzelius Lecture at the Karolinska Institute, Stockholm (1985).

A RECEPTOR-MEDIATED PATHWAY FOR CHOLESTEROL HOMEOSTASIS

Nobel lecture, 9 December, 1985

by

MICHAEL S. BROWN AND JOSEPH L. GOLDSTEIN

Department of Molecular Genetics, University of Texas Health Science Center, Southwestern Medical School, 5323 Harry Hines Blvd. Dallas, Texas, U.S.A.

In 1901 a physician, Archibald Garrod, observed a patient with black urine. He used this simple observation to demonstrate that a single mutant gene can produce a discrete block in a biochemical pathway, which he called an "inborn error of metabolism". Garrod's brilliant insight anticipated by 40 years the one gene-one enzyme concept of Beadle and Tatum. In similar fashion the chemist Linus Pauling and the biochemist Vernon Ingram, through study of patients with sickle cell anemia, showed that mutant genes alter the amino acid sequences of proteins. Clearly, many fundamental advances in biology were spawned by perceptive studies of human genetic diseases (1).

We began our work in 1972 in an attempt to understand a human genetic disease, familial hypercholesterolemia or FH. In these patients the concentration of cholesterol in blood is elevated many fold above normal and heart attacks occur early in life. We postulated that this dominantly inherited disease results from a failure of end-product repression of cholesterol synthesis. The possibility fascinated us because genetic defects in feedback regulation had not been observed previously in humans or animals, and we hoped that study of this disease might throw light on fundamental regulatory mechanisms.

Our approach was to apply the techniques of cell culture to unravel the postulated regulatory defect in FH. These studies led to the discovery of a cell surface receptor for a plasma cholesterol transport protein called low density lipoprotein (LDL) and to the elucidation of the mechanism by which this receptor mediates feedback control of cholesterol synthesis (2,3). FH was shown to be caused by inherited defects in the gene encoding the LDL receptor, which disrupt the normal control of cholesterol metabolism. Study of the LDL receptor in turn led to the understanding of receptor-mediated endocytosis, a general process by which cells communicate with each other through internalization of regulatory and nutritional molecules (4). Receptor-mediated endocytosis differs from previously described biochemical pathways because it depends upon the continuous and highly controlled movement of membrane-embedded proteins from one cell organelle to another in a process termed receptor recycling (4). Many of the mutations in the LDL receptor that occur in FH patients disrupt the movement of the receptor between organelles. These mutations define a new type of cellular defect that has broad implications for normal and deranged human physiology.

In this lecture we first discuss the peculiar problem of plasma cholesterol transport. We then present some historical aspects of FH and the origin of the LDL receptor concept. Next, we summarize current knowledge of this receptor and the mechanism by which it functions in cells. Finally, we relate these findings to the pathogenesis of FH and to the common clinical problem of high blood cholesterol levels and atherosclerosis in human subjects.

THE PROBLEM OF CHOLESTEROL TRANSPORT

Cholesterol is the most highly decorated small molecule in biology. Thirteen Nobel Prizes have been awarded to scientists who devoted major parts of their careers to cholesterol (5). Ever since it was first isolated from gallstones in 1784, almost exactly 200 years ago, cholesterol has exerted a hypnotic fascination for scientists from the most diverse domains of science and medicine. Organic chemists have been fascinated with cholesterol because of its complex four-ring structure. Biochemists have been fascinated because cholesterol is synthesized from a simple two-carbon substrate, acetate, through the action of at least 30 enzymes, many of which are coordinately regulated. Physiologists and cell biologists have been fascinated with cholesterol because of its essential function in membranes of animal cells, where it modulates fluidity and maintains the barrier between cell and environment, and because cholesterol is the raw material for the manufacture of steroid hormones and bile acids. And finally, physicians have been fascinated because elevated levels of blood cholesterol accelerate the formation of atherosclerotic plaques leading to heart attacks and strokes. The studies of cholesterol therefore embrace almost all disciplines of modern biology. If the role of cholesterol in biomedicine is to be elucidated, all of these disciplines must be employed.

Cholesterol is a Janus-faced molecule. The very property that makes it useful in cell membranes, namely its absolute insolubility in water, also makes it lethal. For when cholesterol accumulates in the wrong place, for example within the wall of an artery, it cannot be readily mobilized, and its presence eventually leads to the development of an atherosclerotic plaque. The potential for errant cholesterol deposition is aggravated by its dangerous tendency to exchange passively between blood lipoproteins and cell membranes. If cholesterol is to be transported safely in blood, its concentration must be kept low, and its tendency to escape from the bloodstream must be controlled.

Multicellular organisms solve the problem of cholesterol transport by esterifying the sterol with long-chain fatty acids and packaging these esters within the hydrophobic cores of plasma lipoproteins (Fig. 1). With its polar hydroxyl group esterified, cholesterol remains sequestered within this core, which is essentially an oil droplet composed of cholesteryl esters and triglycerides, solubilized by a surface monolayer of phospholipid and unesterified cholesterol and stabilized by protein. The small amounts of unesterified cholesterol on the

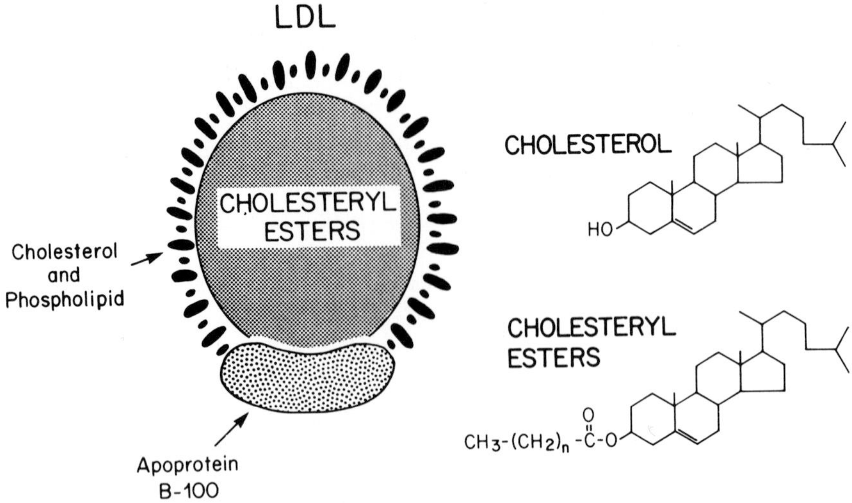

Fig. 1. Structure of plasma LDL (*left*) and its cholesterol and cholesteryl ester components (right). LDL is a spherical particle with a mass of 3×10^6 daltons and a diameter of 22 nanometers. Each LDL particle contains about 1500 molecules of cholesteryl ester in an oily core that is shielded from the aqueous plasma by a hydrophilic coat composed of 800 molecules of phospholipid, 500 molecules of unesterified cholesterol, and 1 molecule of a 387,000-dalton protein called apoprotein B-100 (129). Elevations in blood cholesterol are usually attributable to an increase in the number of LDL particles.

surface of the particle are maintained in equilibrium exchange with the cholesterol of cell membranes, but the larger amounts of cholesteryl esters remain firmly trapped in the core of the particle and leave the particle only as the result of highly controlled processes.

The major classes of plasma lipoproteins were delineated in the 1950's and 1960's through work in many laboratories, most notably those of Oncley (6), Gofman (7), and Fredrickson (8). The four major classes are very low density lipoprotein (VLDL), intermediate density lipoprotein (IDL), low density lipoprotein (LDL), and high density lipoprotein (HDL). A schematic representation of LDL, the most abundant cholesterol-carrying lipoprotein in human plasma, is shown in Fig. 1.

Packaging of cholesteryl esters in lipoproteins solves the problem of nonspecific partitioning of cholesterol into cell membranes, but it creates another problem, namely one of delivery. Cholesteryl esters are too hydrophobic to pass through membranes. How then can esterified cholesterol be delivered to cells? The delivery problem is solved by lipoprotein receptors, of which the prototype is the LDL receptor (9). Strategically located on the surfaces of cells, these receptors bind LDL and carry it into the cell by receptor-mediated endocytosis. The internalized lipoprotein is delivered to lysosomes where its cholesteryl esters are hydrolyzed. The liberated cholesterol is used by the cell for the synthesis of plasma membranes, bile acids, and steroid hormones, or stored in the form of cytoplasmic cholesteryl ester droplets. Two properties of the receptor – its high affinity for LDL and its ability to cycle multiple times in and out of the cell – allow large amounts of cholesterol to be delivered to body tissues, while at the same time keeping the concentration of LDL in blood low enough to avoid the buildup of atherosclerotic plaques. When LDL receptor function is inappropriately diminished as a result of genetic defects or in response to regulatory signals, the protective mechanism is lost, cholesterol builds up in plasma, and atherosclerosis ensues (10).

FAMILIAL HYPERCHOLESTEROLEMIA: ORIGIN OF THE LDL RECEPTOR CONCEPT

As a disease, FH has a rich clinical history. It was first described in 1938 by Carl Müller, a clinician at the Oslo Community Hospital in Norway, as an "inborn error of metabolism" that produced high blood cholesterol levels and myocardial infarctions in young people (11). Müller astutely concluded that FH is transmitted as a single gene-determined autosomal dominant trait. In the mid 1960's and early 1970's, Khachadurian (12) at the American University in Beirut, Lebanon, and Fredrickson and Levy (13) at the National Institutes of Health showed that FH exists clinically in two forms: the less severe heterozygous form and the more severe homozygous form.

FH heterozygotes, who carry a single copy of a mutant LDL receptor gene, are quite common, accounting for 1 out of every 500 persons among most ethnic groups throughout the world (14). These individuals have a two-fold increase in the number of LDL particles in plasma from the time of birth. They begin to have heart attacks at 30 to 40 years of age. Among people under age 60 who suffer myocardial infarctions, about 5% have the heterozygous form of FH, a 25-fold enrichment over the incidence in the general population (15–17).

The attractiveness of FH as an experimental model stems from the existence of homozygotes. These rare individuals, who number about 1 in 1 million persons, inherit two mutant genes at the LDL receptor locus, one from each parent. Their disease is much more severe than that of heterozygotes. They have six to ten-fold elevations in plasma LDL levels from the time of birth, and they often have heart attacks in childhood (12–14). The severe atherosclerosis that develops in these patients without any other risk factors is formal proof that high levels of plasma cholesterol can produce atherosclerosis in humans. Experimentally, the availability of FH homozygotes permits study of the manifestations of the mutant gene without any confounding effects from the normal gene.

At the time that our studies began in 1972, it was generally felt that all important events in cholesterol metabolism take place in the liver or intestine (18). It was obviously impossible to perform meaningful studies in livers of humans with FH. Our only chance to explain its mysteries depended on the mutant phenotype being faithfully manifest in long-term cultured cells such as skin fibroblasts. Techniques for growing such cells had been established over the preceding two decades. Moreover, inherited enzyme defects were known to be expressed in cultured fibroblasts from patients with rare recessive diseases such as galactosemia, the Lesch Nyhan syndrome, and Refsum's syndrome. By

1970, Neufeld's classic studies of the mucopolysaccharidoses, a form of lysosomal storage disease, were beginning to establish the value of cultured skin fibroblasts in elucidating complex cellular pathways (19).

There was some reason to believe that the FH derangement might be manifest in cultured skin fibroblasts. Studies in the 1960's by Bailey (20) and Rothblat (21) had demonstrated that several types of cultured animal cells synthesize cholesterol and that this synthesis is subject to negative feedback regulation. When serum was present in the medium, cultured cells produced little cholesterol from radioactive acetate. When serum lipoproteins were removed from the culture medium, cholesterol synthesis increased.

Regulation of HMG CoA Reductase by LDL in Fibroblasts

We began our work by setting up a micro-assay for 3-hydroxy-3-methylglutaryl coenzyme A reductase (HMG CoA reductase), the rate-determining enzyme of cholesterol biosynthesis. This assay was used to measure HMG CoA reductase activity in extracts of cultured fibroblasts (2,22). Earlier studies in rat livers by Bucher and Lynen (23) and by Siperstein (24) had shown that the activity of this enzyme was reduced when rats ingested cholesterol and that this reduction limited the rate of cholesterol synthesis. We soon found that the activity of HMG CoA reductase was subject to negative regulation in fibroblasts (2,22). As shown in Fig. 2A, when normal human fibroblasts were grown in the presence of serum, HMG CoA reductase activity was low. When the lipoproteins were removed from the culture medium, the activity of HMG CoA reductase rose by at least 50-fold over 24 hr period. The induced enzyme was rapidly suppressed when lipoproteins were added back to the medium (Fig. 2B).

Not all lipoproteins could suppress HMG CoA reductase activity. Of the two major cholesterol-carrying lipoproteins in human plasma, LDL and HDL, only LDL was effective (22,25). This specificity was the first clue that a receptor might be involved. The second clue was the concentration of LDL that was required. The lipoprotein was active at concentrations as low as 5 μg of protein per ml, which is less than 10^{-8} molar (22,25). A high affinity receptor mechanism must be responsible for enzyme suppression.

The key to this mechanism emerged from studies of cells from patients with homozygous FH (2,25). When grown in serum containing lipoproteins, the homozygous FH cells had HMG CoA reductase activities that were 50 to 100-fold above normal (Fig. 2A). This activity did not increase significantly when the lipoproteins were removed from the serum, and there was no suppression when LDL was added back. Clearly, the genetic defect was expressed in cell culture (Figs. 2A and 2B).

The simplest interpretation of these results was that FH homozygotes had a defect in the gene encoding HMG CoA reductase that rendered the enzyme resistant to feedback regulation by LDL-derived cholesterol. This working hypothesis was immediately disproved by the next experiment. Cholesterol, dissolved in ethanol, was added to normal and FH homozygote cells. When

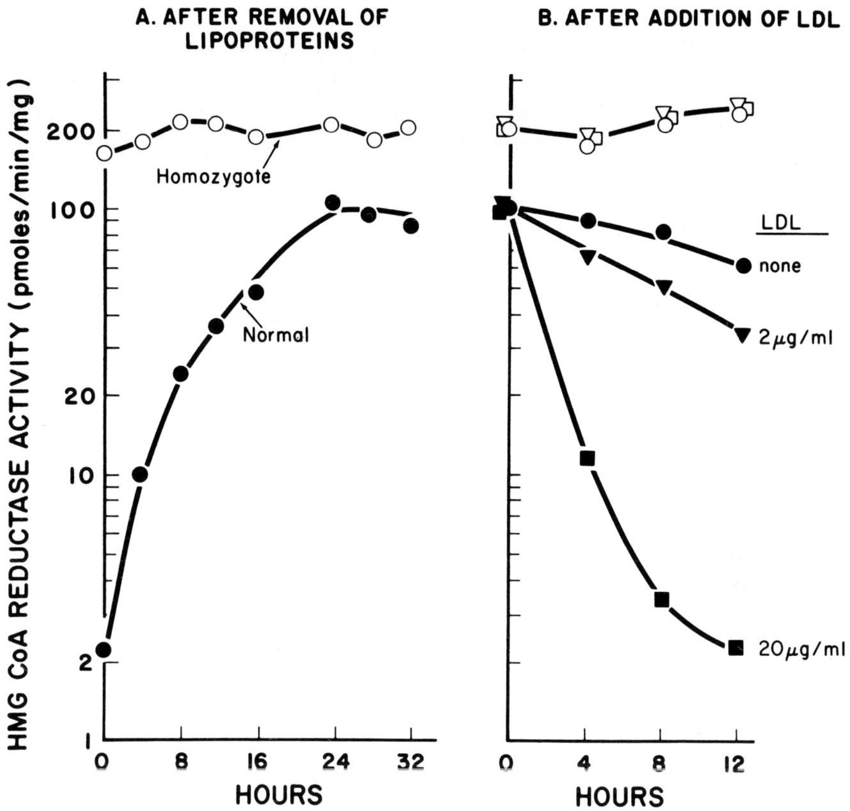

Fig. 2. Regulation of HMG CoA reductase activity in fibroblasts from a normal subject (○) and from an FH homozygote (●). *Panel A:* Monolayers of cells were grown in dishes containing 10% fetal calf serum. On day 6 of cell growth (zero time), the medium was replaced with fresh medium containing 5% human serum from which the lipoproteins had been removed. At the indicated time, extracts were prepared and HMG CoA reductase activity was measured. *Panel B:* 24 hours after addition of 5% human lipoprotein-deficient serum, human LDL was added to give the indicated cholesterol concentration. HMG CoA reductase activity was measured in cell free extracts at the indicated time. (Reprinted from ref. 2.)

mixed with albumin-containing solutions, cholesterol forms a quasi-soluble emulsion that enters cells passively, apparently by diffusion through the plasma membrane. When cholesterol was added in this form, the HMG CoA reductase activities of normal and FH homozygote fibroblasts were suppressed at the same rate and to the same extent (25).

Clearly, the defect in the FH homozygote cells must reside in their ability to extract cholesterol from the lipoprotein, and not in the ability of the cholesterol, once extracted by the cells, to act. But how do normal cells extract the cholesterol of LDL? The high affinity of the process suggested that a cell surface receptor was involved. The existence of cell surface receptors for protein hormones and other chemical messengers had been known for many years. It was generally thought that these receptors acted by binding the ligand at the surface and then generating a "second messenger" on the intracellular side of the plasma membrane. The classic second messenger was cyclic adenosine monophosphate (cyclic AMP) (26). Perhaps LDL was binding to a receptor and generating some second messenger that suppressed HMG CoA reductase.

Delineation of the LDL Receptor Pathway

The existence of an LDL receptor was confirmed when LDL was radiolabeled with ^{125}Iodine and incubated with normal and FH homozygote fibroblasts. These studies showed that normal cells had high affinity binding sites for ^{125}I-LDL, whereas FH homozygote cells lacked high affinity receptors (3,27). This seemed to explain the genetic defect in FH, but it did not reveal how LDL generated the signal that suppressed HMG CoA reductase. The answer came from studies of the fate of the surface-bound ^{125}I-LDL. Techniques were developed to distinguish surface-bound from intracellular ^{125}I-LDL (28), and these revealed that the receptor-bound LDL remained on the surface for less than 10 min on average (*Fig. 3A*). Within this time most of the surface-bound LDL particles entered the cell; within another 60 min the protein component of ^{125}I-LDL was digested completely to amino acids and the ^{125}I, which had been attached to tyrosine residues on LDL, was released into the culture medium as ^{125}I-monoiodotyrosine (27,28). Meanwhile, the cholesteryl esters of LDL were hydrolyzed, generating unesterified cholesterol which remained within the cell (29).

The only cellular organelle in which LDL could have been degraded so completely and rapidly was the lysosome. Originally described by de Duve (30), lysosomes were known to contain a large number of acid hydrolases that could easily digest all of the components of LDL. The hypothesis of lysosomal digestion of LDL was confirmed through the use of inhibitors such as chloroquine (31), which raises the pH of lysosomes and inhibits lysosomal enzymes (32), and through studies of cultured fibroblasts from patients with a genetic deficiency of lysosomal acid lipase (29). Cells from the latter patients bound and internalized LDL but failed to hydrolyze its cholesteryl esters, even though they were able to degrade its protein component.

The cholesterol that was generated from LDL within the lysosome proved to be the second messenger responsible for suppressing HMG CoA reductase activity. We now know that cholesterol (or an oxygenated derivative that is formed within the cell) acts at several levels, including suppression of transcription of the HMG CoA reductase gene (33) and acceleration of the degradation of the enzyme protein (34). The LDL-derived cholesterol also regulates two other cellular processes in a coordinated action that stabilizes the cell's cholesterol content. It activates a cholesterol-esterifying enzyme, acyl CoA: cholesterol acyltransferase (ACAT), so that excess cholesterol can be stored in the cytoplasm as cholesteryl ester droplets (35). It also suppresses synthesis of LDL receptors by lowering the concentration of receptor mRNA (36,37). The latter action allows cells to adjust the number of LDL receptors to provide sufficient cholesterol for metabolic needs without causing cholesterol overaccumulation (9). Through these regulatory mechanisms, cells keep their level of

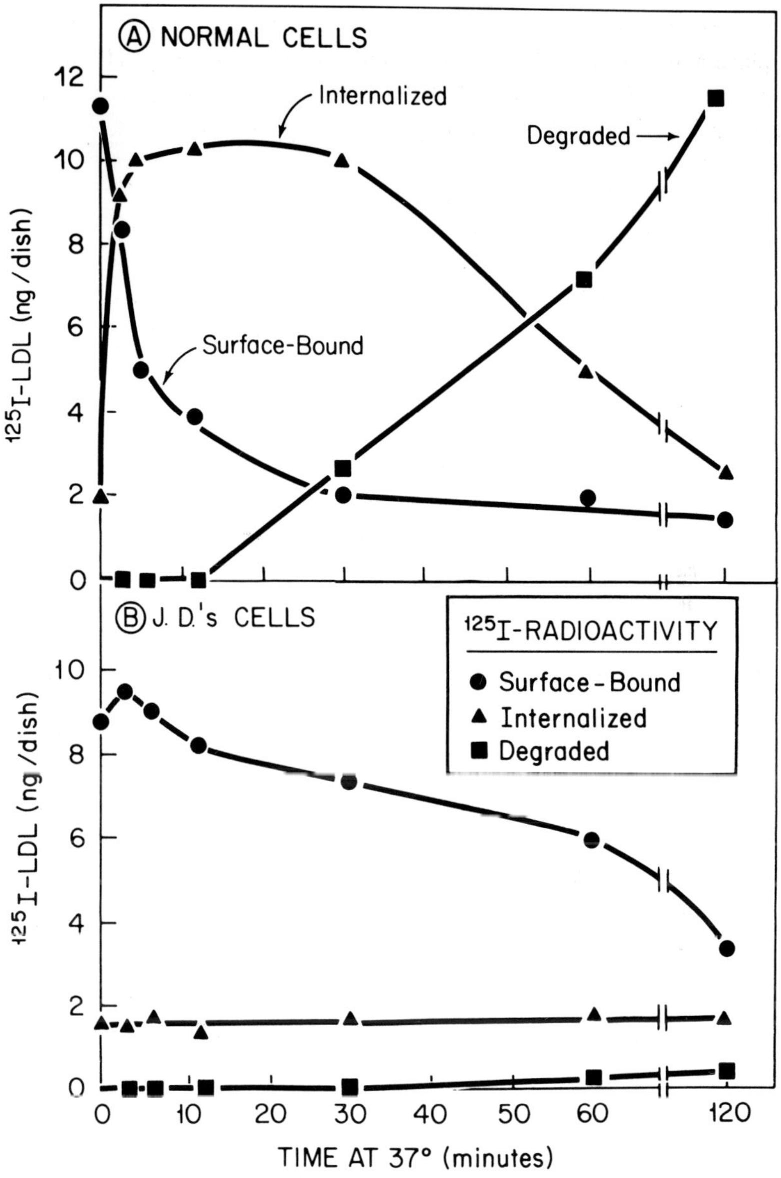

Fig. 3. Internalization and degradation at 37°C of ^{125}I-LDL previously bound to the LDL receptor at 4°C in fibroblasts from a normal subject (*Panel A*) and from J.D., a patient with the internalization-defective form of FH (*Panel B*). Each cell monolayer was allowed to bind ^{125}I-LDL (10 ug protein/ml) at 4°C for 2 hr, after which the cells were washed extensively. In one set of dishes, the amount of ^{125}I-LDL was determined by measuring the amount of ^{125}I-LDL that could be released from the surface by treatment with heparin. Each of the other dishes then received warm medium, after which they were incubated at 37°C. After the indicated interval, the dishes were rapidly chilled to 4°C, and the amounts of surface-bound (heparin-releasable) ^{125}I-LDL (●), internalized (heparin-resistant) ^{125}I-LDL (▲), and degraded (trichloroacetic acid-soluble) ^{125}I-LDL (■) were measured. (Reprinted with permission from ref. 41.)

unesterified cholesterol remarkably constant despite wide fluctuations in cholesterol requirements and exogenous supply.

Receptor-Mediated Endocytosis of LDL: Binding Coupled to Internalization in Coated Pits

The rapidity of internalization of receptor-bound LDL and the completeness with which the protein of LDL was hydrolyzed implied that fibroblasts have a special mechanism for transport of the lipoprotein from the cell surface to the lysosome. The likely mechanism was endocytosis, the process by which surface membranes pouch inward and pinch off to form vesicles that eventually fuse with lysosomes. Endocytosis was first demonstrated by cinematography of phagocytic cells in the 1930's, and its universal occurrence in all cells was established in the 1950's by the electron microscopic studies of Palade (38). Endocytosis was felt to be a nonspecific process that transported bulk fluid and its contents into cells. There was no precedent for entry of specific receptors into cells by this route.

To determine whether endocytosis was involved in LDL uptake, we began in 1975 a collaboration with Richard G.W. Anderson, a cell biologist at our medical school in Dallas. Through the use of LDL coupled to electron-dense ferritin, we found that receptor-bound LDL was internalized by endocytosis. More important, however, these morphological studies explained the efficiency of internalization: efficiency was contingent upon the clustering of the LDL receptors in small pockets on the surface called coated pits (39). Coated pits had been described in detail by Roth and Porter (40) in 1964 during electron microscopic studies of the uptake of yolk proteins by mosquito oocytes. These investigators showed that coated pits pinch off from the surface to form coated endocytic vesicles that carry extracellular fluid and its contents into the cell.

The finding that LDL receptors were clustered in coated pits raised the general possibility that these structures serve as gathering-places for cell surface receptors that are destined for endocytosis (4). Other cell surface proteins, being excluded from coated pits, could not rapidly enter the cell.

This interpretation of coated pit function was strengthened by study of fibroblasts from a unique FH homozygote. Cells from most of these subjects simply failed to bind LDL. But cells from one FH patient, whose initials are J.D., bound LDL, but failed to internalize it (Fig. 3B and refs. 41 and 42). In collaboration with Anderson, we showed that the receptors in these mutant cells were excluded from coated pits (43). This was an important finding, for it established the essential role of coated pits in the high efficiency uptake of receptor-bound molecules (4).

Figure 4 summarizes the sequential steps in the LDL receptor pathway as deduced from the biochemical, genetic, and ultrastructural studies performed between 1972 and 1976. Figure 5 shows the striking "all-or-none" biochemical differences in the metabolism of LDL and its regulatory actions in fibroblasts derived from a normal subject and from an FH homozygote with a complete deficiency of LDL receptors.

Soon after the initial studies of the LDL receptor pathway, Pearse (44)

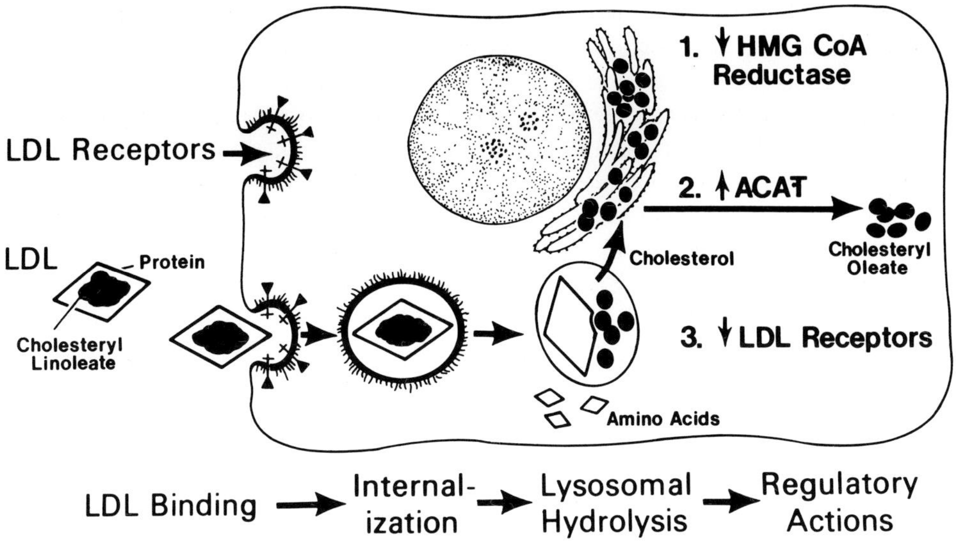

Fig. 4. Sequential steps in the LDL receptor pathway of mammalian cells. HMG CoA reductase denotes 3-hydroxy-3-methylglutaryl CoA reductase; ACAT denotes acyl-CoA: cholesterol acyl-transferase. Vertical arrows indicate the directions of regulatory effects. (Reprinted from ref. 130.)

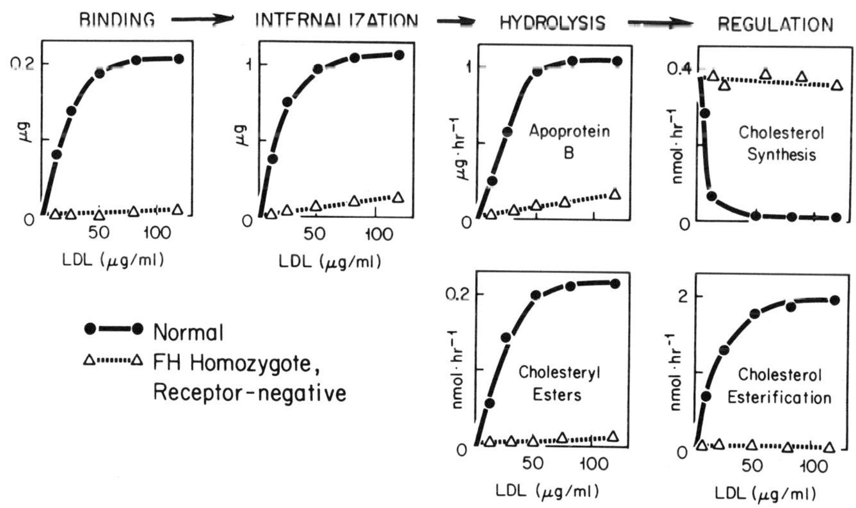

Fig. 5. Actions attributable to the LDL receptor in fibroblasts from a normal subject (●) and from a homozygote with the receptor-negative form of FH (△) incubated with varying concentrations of ^{125}I-LDL or unlabeled LDL at 37°C for 5 hr. Assays were performed in growing cells in monolayers as previously described (128). All data are normalized to 1 mg of total cell protein. The units for each assay are as follows: *Binding*, μg of ^{125}I-LDL bound to cell surface; *Internalization*, μg of ^{125}I-LDL contained within the cell; *Hydrolysis of apoprotein B-100*, μg of ^{125}I-LDL degraded to ^{125}I-monoiodotyrosine per hr; *Hydrolysis of cholesteryl esters*, nmol of [^{3}H]cholesterol formed per hr from the hydrolysis of LDL labeled with [^{3}H]cholesteryl linoleate; *Cholesterol Synthesis*, nmol of [^{14}C]acetate incorporated into [^{14}C]cholesterol per hr by intact cells; *Cholesterol esterification*, nmol of [^{14}C]oleate incorporated into cholesteryl [^{14}C]oleate per hr by intact cells. (Reprinted from ref. 130.)

purified coated vesicles and found that the cytoplasmic coat was composed predominantly of a single protein that she named clathrin. At the same time, Cohen and his collaborators performed their elegant studies of the action of epidermal growth factor (EGF) on cultured fibroblasts (45). They found that this peptide hormone was internalized by cells in a manner that was indistinguishable from that of LDL. Similar observations were made by Terris and Steiner (46) with insulin in hepatocytes, by Neufeld and coworkers (47) and by Sly and coworkers (48) with lysosomal enzymes in fibroblasts, and by Ashwell, Morell, and coworkers (49) with asialoglycoproteins in hepatocytes. Moreover, Helenius, Simons, and their coworkers (50) showed that several lipid-enveloped viruses enter cells by this route. Clearly, receptor-mediated endocytosis did not exist solely for cholesterol delivery: it was a general process by which cells internalized and degraded many extracellular molecules (4,51). In all instances in which adequate morphologic studies were performed, this internalization was attributable to clustering of receptors in coated pits. Indeed, Pastan and Willingham (51) and Carpentier, *et al.* (52) showed that receptors for several different ligands co-localize in the same coated pit.

The early LDL receptor studies also exposed another feature of receptor-mediated endocytosis – namely, that receptors can be recycled (4,28). After internalization the receptors dissociate from their ligands. From the work of Maxfield (53) and of Helenius and coworkers (54), we now know that such dissociation is triggered by a drop in pH within a special class of endocytic vesicles called endosomes (discussed below). After dissociation the receptors find their way back to the cell surface. The LDL receptor makes one round trip into and out of the cell every 10 min for a total of several hundred trips in its 20-hr lifespan (4,28).

THE LDL RECEPTOR: STRUCTURE ADAPTED TO FUNCTION

The LDL receptor is a cell surface glycoprotein that contains approximately two asparagine-linked (*N*-linked) oligosaccharide chains of the complex type and approximately 18 serine/threonine-linked (*O*-linked) oligosaccharide chains (55,56). About two-thirds of the *O*-linked sugars are clustered in one region of the molecule (57). The LDL receptor binds two proteins: 1) apo B-100, the 387,000-dalton glycoprotein that is the sole protein of LDL (27); and 2) apo E, a 34,000-dalton protein that is found in multiple copies in intermediate density lipoprotein (IDL) and a subclass of HDL (58,59). Innerarity and Mahley (59) demonstrated that lipoproteins which contain multiple copies of apo E bind to LDL receptors with up to 20-fold higher affinity than LDL, which contains only one copy of apo B.

Figure 6 illustrates the circuitous itinerary followed by the LDL receptor from its site of synthesis to its site of internalization in coated pits and its site of recycling in endosomes. The receptor is synthesized in the rough endoplasmic reticulum (ER) as a precursor (60) that contains high mannose *N*-linked carbohydrate chains and the core sugar (N-acetylgalactosamine) of the *O*-linked chains (56). The *O*-linked core sugars are added before the mannose

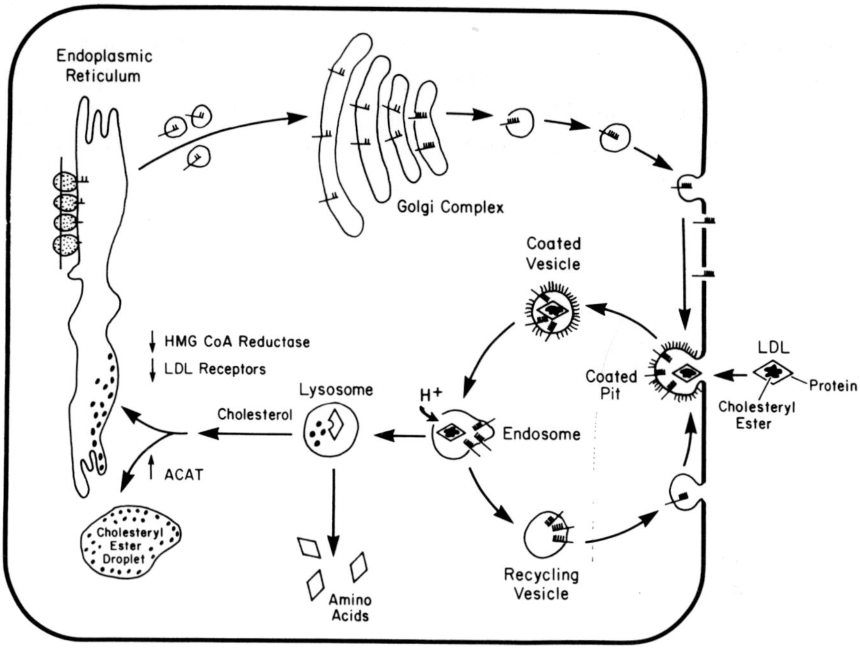

Fig. 6. Itinerary of the LDL receptor in mammalian cells. The receptor begins life in the endoplasmic reticulum from which it travels to the Golgi complex, cell surface, coated pit, endosome, and back to the surface. HMG CoA reductase denotes 3-hydroxy-3-methylglutaryl CoA reductase; ACAT denotes acyl-CoA: cholesterol acyltransferase. Vertical arrows indicate the direction of regulatory effects. (Reprinted from ref. 131 with permission.)

residues of the *N*-linked chains are trimmed, i.e., while the receptor is still in the endoglycosidase H-sensitive stage. Thus, the *O*-linked sugars must be added either in the ER or in a transitional zone between the ER and the Golgi apparatus. The receptor precursor migrates on sodium dodecyl sulfate (SDS) polyacrylamide gel electrophoresis as a single band corresponding to an apparent molecular weight of 120,000 (60).

Within 30 min after its synthesis, the LDL receptor decreases in mobility on SDS gels. The apparent molecular weight increases from 120,000 to 160,000 (60). This change is coincident with the conversion of the high-mannose *N*-linked oligosaccharide chains to the complex endoglycosidase H-resistant form (56). At the same time, each *O*-linked chain is elongated by the addition of one galactose and one or two sialic acid residues (56). The amount of carbohydrate is not sufficient to account for an increase in molecular mass of 40,000 daltons. Rather, the decrease in electrophoretic mobility is primarily caused by a change in conformation of the protein that results from the elongation of the clustered *O*-linked sugars (56,57).

About 45 min after synthesis, LDL receptors appear on the cell surface, where they gather in coated pits. Within 3 to 5 min of their formation, the coated pits invaginate to form coated endocytic vesicles. Very quickly, the clathrin coat dissociates. Multiple endocytic vesicles then fuse to create larger sacs of irregular contour called endosomes or receptosomes (4,61). The pH of the endosomes falls below 6.5, owing to the operation of ATP-driven proton pumps in the membrane (53,54,61). At this acid pH, the LDL dissociates from the receptor. The latter returns to the surface, apparently by clustering with other receptors in a segment of the endosomal membrane that pinches off to form a recycling vesicle. Once it reaches the surface, the receptor binds another lipoprotein particle and initiates another cycle of endocytosis (4). Each LDL receptor makes one round-trip every 10 min in continuous fashion whether or not it is occupied with LDL (4,62). The LDL that dissociates from the receptor is delivered to a lysosome when the membranes of the endosome and lysosome fuse. There the protein component of LDL is hydrolyzed to amino acids and the cholesteryl esters are hydrolyzed by an acid lipase, liberating cholesterol (discussed above).

The striking feature of this pathway is that it requires the continuous movement of a membrane embedded protein from one organelle to another in a highly ordered fashion. Each time it moves, the receptor must be segregated from neighboring membrane proteins that do not follow the same route. This raises a crucial question: What are the signals that direct the highly selective movement of receptors from one membrane organelle to another? Clearly, the signals must lie in the structures of the receptors. What do we know about the structure of the LDL receptor?

The LDL Receptor: A Multi-Domain Protein

The LDL receptor was purified from bovine adrenal cortex by Wolfgang Schneider in our laboratory (55). A partial amino acid sequence was obtained, and this sequence was used by David Russell and Tokuo Yamamoto to isolate a full-length cDNA for the human LDL receptor (37,63). Biochemical studies of the receptor protein, coupled with the amino acid sequence that was deduced from the nucleotide sequence of the cDNA, have provided insight into the structural domains of the LDL receptor (Fig. 7 and refs. 63–65).

At the extreme NH_2-terminus of the LDL receptor, there is a hydrophobic sequence of 21 amino acids that is cleaved from the receptor immediately after it is translated. This segment functions as a classic signal sequence to direct the receptor-synthesizing ribosomes to the ER membrane. Because it does not appear in the mature receptor, the signal sequence is omitted from the structural domain that is described below. The mature receptor (without the signal sequence) consists of 839 amino acids (63).

The *first domain* of the LDL receptor consists of the NH_2-terminal 292 amino acids, which is composed of a sequence of 40 amino acids that is repeated with some variation seven times (65,66). Studies of anti-peptide antibody binding to intact cells revealed that this domain is located on the external surface of the plasma membrane (67). Each of the seven 40-amino acid repeats contains six cysteine residues, which are in register for all of the repeats. The receptor cannot be labeled with [^{3}H]iodoacetamide without prior reduction, suggesting that all of these cysteines are disulfide-bonded (65). This region of the receptor

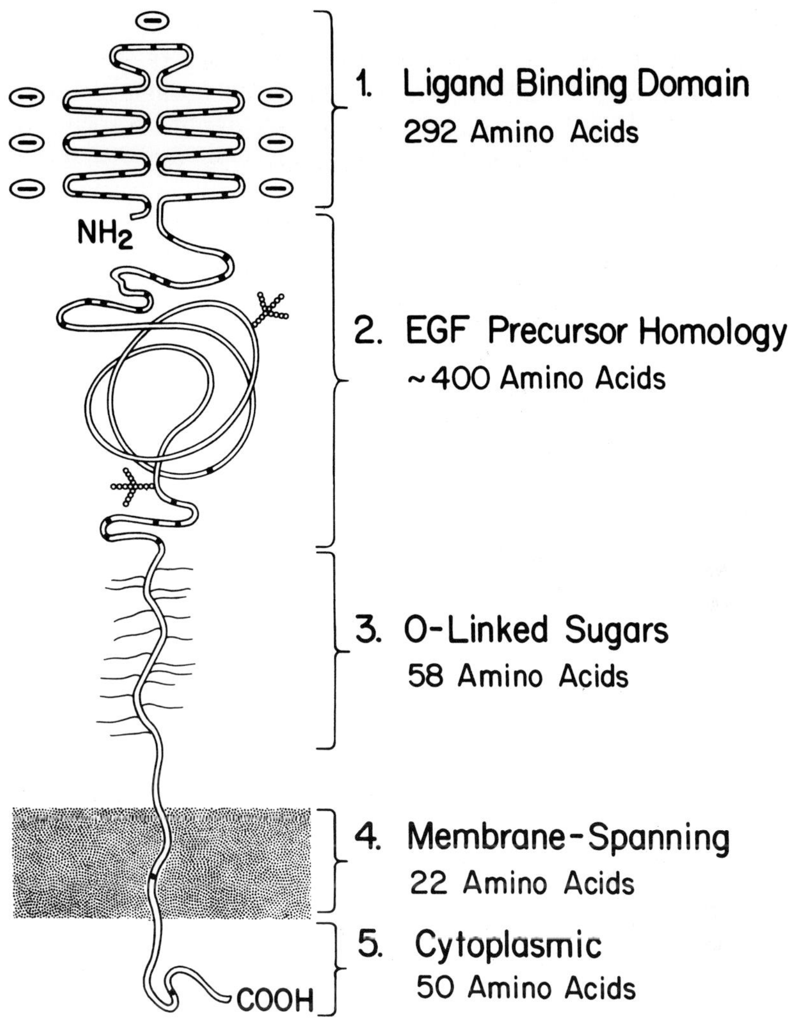

Fig. 7. The LDL receptor: a single protein with five domains. The significance of each domain is discussed in the text. Each black dot in the protein structure denotes the position of a cysteine residue.

must therefore exist in a tightly cross-linked, convoluted state. This explains the extreme stability of the binding domain of the receptor; the receptor can be boiled in strong denaturants and still retain its binding activity as long as the disulfide bonds are intact (65).

A striking feature of each cysteine-rich repeat sequence is a cluster of negatively-charged amino acids near the COOH-terminus of each repeat (65,66). The charges on these sequences are complementary to a cluster of positively-charged residues that are believed to occupy one face of a single α-helix in apo E, the best studied ligand for the LDL receptor (68). Elegant studies by Mahley and Innerarity (68) with mutant and proteolyzed forms of apo E and with monoclonal antibodies against different regions of apo E showed that the positively-charged region contains the site whereby this protein binds to the LDL receptor. It is therefore tempting to speculate that the negatively-charged clusters of amino acids within the cysteine-rich repeat sequence of the LDL receptor constitute multiple binding sites, each of which binds a single apo E molecule by attaching to its positively-charged α-helix (65).

The *second domain* of the LDL receptor, consisting of ~ 400 amino acids, is 35% homologous to a portion of the extracellular domain of the precursor for epidermal growth factor (EGF) (63,64,69). The EGF precursor is a molecule of 1217 amino acids that appears to span the plasma membrane one time like the LDL receptor (69–72). Analysis of the amino acid sequence of the EGF precursor, as revealed from the sequence of the cloned cDNA (70,71), suggests that EGF, a peptide of 53 amino acids, is liberated from the EGF precursor by proteolysis. The sequence of EGF is not homologous to the LDL receptor. Rather, the homology involves a part of the EGF precursor that is on the NH_2-terminal side of EGF itself. The function of this region in either the LDL receptor or the EGF precursor is unknown.

The *third domain* of the LDL receptor lies immediately external to the membrane-spanning domain and consists of a stretch of 58 amino acids that contains 18 serine or threonine residues (63,66). This domain is encoded within a single exon (see below). Proteolysis studies reveal that this region contains the clustered *O*-linked sugar chains (64).

The *fourth domain* consists of a stretch of 22 hydrophobic amino acids that span the plasma membrane, as demonstrated by proteolysis experiments (63,64). Comparison of the amino acid sequences of the human and bovine LDL receptors reveals that the membrane-spanning region is relatively poorly conserved (65). Of the 22 amino acids in this region, 7 differ between human and cow, but all of the substitutions retain a hydrophobic character.

The *fifth domain* is the cytoplasmic tail. The human and bovine LDL receptors each contain a COOH-terminal segment of 50 amino acids that projects into the cytoplasm (63,64). Localization of this domain to the cytoplasmic side of the membrane was determined through use of an anti-peptide antibody directed against the COOH-terminal sequence (64). When inside-out membrane vesicles containing receptor were digested with pronase, the antibody-reactive material was removed, and the molecular weight of the receptor was reduced by approximately 5000. The cytoplasmic sequence is strongly conserved among species. Of the 50 amino acids in this region, only 4 differ between human and cow, and each of these substitutions is conservative with respect to charge (65).

The cytoplasmic domain of the LDL receptor plays an important role in clustering in coated pits, either through interaction with clathrin itself or with some protein that is associated with clathrin on the cytoplasmic side of the membrane (4). This conclusion is based on a molecular analysis of three naturally-occurring mutations at the LDL receptor locus that produce receptors that bind LDL normally but fail to cluster in clathrin-coated pits. All three of these mutations produce defects in the cytoplasmic tail (discussed below and refs. 65 and 73).

The LDL Receptor: A Mosaic Gene

The haploid human genome contains a single copy of the LDL receptor gene (66) on chromosome 19 (74). Sequences representing almost the entire gene were isolated from bacteriophage lambda and cosmid libraries by Thomas Südhof and David Russell (66). The position of each intron within the gene was mapped, and the sequence of each exon-intron junction was determined.

The LDL receptor gene spans ~ 45 kilobases and is made up of 18 exons separated by 17 introns (66). There is a striking correlation between the exons in the gene and the functional domains of the protein (Fig. 8). The first intron is located just at the end of the DNA encoding the cleaved signal sequence. The binding domain is encoded by exons 2 to 6. Within this domain (which contains the seven cysteine-rich repeats), introns occur precisely at the ends of repeats I, II, V, VI, and VII (Fig. 8). Repeats III, IV, and V are included in one exon. The binding domain is terminated by an intron at amino acid 292, the last residue in the seventh repeat. Thus, the binding domain is composed of a single exon that has been duplicated multiple times to produce seven repeats of a single 40 amino acid sequence. Each of these seven repeats in the LDL receptor is strongly homologous with a stretch of 40 amino acids that occurs in the middle of the C9 component of complement, a plasma protein of 537 amino acids that participates in the complement cascade (66,75).

The next eight exons in the LDL receptor gene (exons 7 to 14) encode the region that is homologous with the EGF precursor (Fig. 8). The gene for the EGF precursor contains the same eight exons (69). These exons form a cassette that has been lifted out of some ancestral gene during evolution and placed in the middle of the EGF precursor gene and the LDL receptor gene. Three of these exons have also been used by another class of genes. These exons encode a cysteine-rich sequence of 40 amino acids (labeled A, B, and C in Fig. 8) that is repeated three times in the LDL receptor and occurs once in several proteins of the blood clotting system, including factor IX, factor X, and protein C (69,76). Thus, these exons have been used by members of at least three different gene families.

The *O*-linked sugar domain is also encoded by a single exon (exon 15). However, not all domains of the protein are encoded by single exons. Thus, the membrane-spanning region is encoded by parts of two exons (exons 16 and 17). The cytoplasmic tail is also encoded by two exons (exons 17 and 18) (Fig. 8).

The sharing of exons between the LDL receptor gene and other genes provides strong evidence to support Gilbert's hypothesis concerning the nature and function of introns (77). As originally proposed by Gilbert, introns permit functional domains encoded by discrete exons to shuffle between different proteins, thus allowing proteins to evolve as mosaic combinations of preexisting functional units. The LDL receptor is a vivid example of such a mosaic protein (66,78). It seems likely that other cell surface receptors will also be found to be mosaic structures assembled from exons shared with other genes.

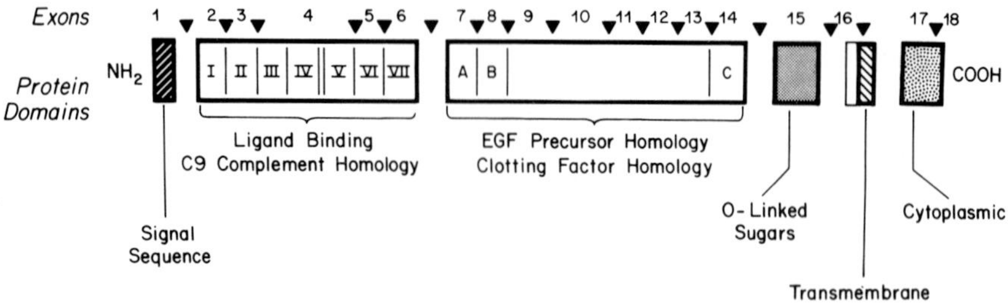

Fig. 8. Correlation of exon organization with protein domains in the human LDL receptor. The domains of the protein are delimited by thick black lines and are labeled in the lower portion. The seven cysteine-rich, 40-amino acid repeats in the LDL binding domain (Fig. 7) are assigned roman numerals I to VII. Repeats IV and V are separated by eight amino acids. The three cysteine-rich repeats in the domain that is homologous with the EGF precursor are lettered A to C. The positions at which introns interrupt the coding region are indicated by arrowheads. Exon numbers are shown between the arrowheads. (Reprinted with permission from ref. 66.)

GENETIC DEFECTS IN THE LDL RECEPTOR

The mutations in the LDL receptor gene in FH patients have helped to delineate the crucial steps of receptor-mediated endocytosis. At the time of this writing, we have studied fibroblasts from 110 patients with the clinical phenotype of homozygous FH. All of them show evidence of defects in the LDL receptor, but not all defects are the same. At least 10 different mutations can be distinguished by structural criteria (65). These fall into four classes as shown in Fig. 9. Many of the apparent FH homozygotes are actually compound heterozygotes who inherit different mutant alleles from each parent.

Class I Mutations: No Receptors Synthesized

This is the most common class of mutant alleles, accounting for approximately half of the mutations so far analyzed. These genes produce either no LDL receptor protein or only trace amounts as determined by reaction with polyclonal or monoclonal antibodies. One of these alleles has been analyzed by molecular cloning; the gene bears a large deletion that extends from exon 13 to an *Alu* repetitive element in intron 15 (79). This deletion is easily recognized on Southern blots of genomic DNA. We have not found evidence of a similar deletion in any other individual with the receptor-negative phenotype, so this particular deletion must be rare.

Class II Mutations: Receptor Synthesized, But Transported Slowly From ER to Golgi

This is the second most common class of mutations. These alleles produce receptors that are synthesized as precursors whose apparent molecular weights vary from 100,000 to 135,000. Most have an apparent molecular weight similar to that of the normal precursor (120,000). These receptors contain high mannose *N*-linked sugars and the core N-acetylgalactosamine of the *O*-linked sugars (56,80). However, the *N*-linked sugars are not converted to the complex endoglycosidase H-resistant form nor are the *O*-linked sugar chains elongated.

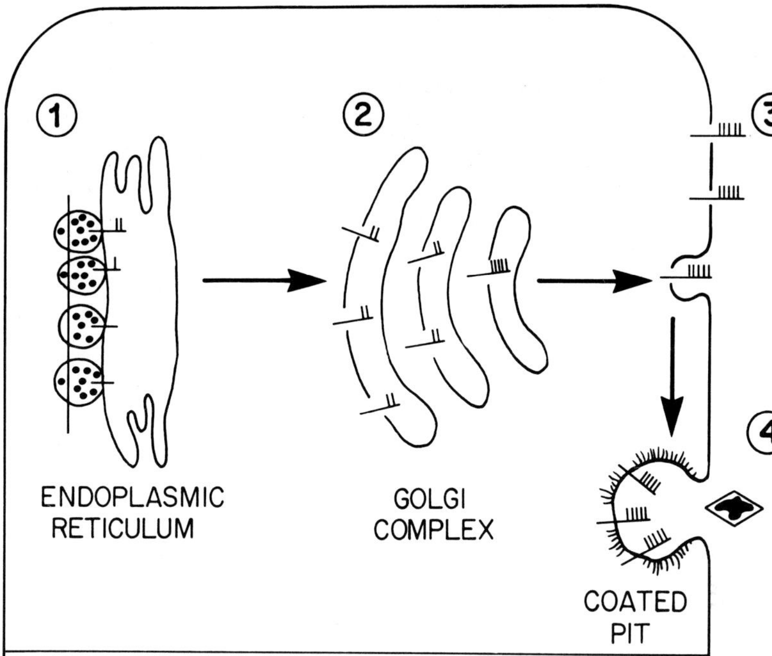

Class of Mutation	Synthesis	Transport from ER to Golgi	Binding of LDL	Clustering in Coated Pits
1	X			
2	→	X		
3	→	→	X	
4	→	→	→	X

Fig. 9. Four classes of mutations that disrupt the structure and function of the LDL receptor and cause FH. Each class of mutation affects a different region in the gene and thus interferes with a different step in the process by which the receptor is synthesized, processed in the Golgi complex, and transported to coated pits. Each class of mutation can be further subdivided into different mutant alleles that are described in detail in ref. 65. (Reprinted with permission from ref. 132.)

These mutant receptors do not appear on the surface of the cell; rather, they seem to remain in the ER until they are eventually degraded. Some mutations in this class are complete, i.e., there is no detectable processing of carbohydrate. Others are partial, i.e., some of the receptors are processed and move to the surface at a rate that is one-tenth of normal (80,81). The molecular defect in this class of mutations has not been determined.

Class III Mutations: Receptors Processed and Reach Cell Surface, But Fail to Bind LDL Normally

In the mature form, these mutant receptors can have a normal apparent molecular weight of 160,000 or aberrant apparent molecular weights of 140,000 or 210,000 (65). They are all synthesized as precursors that appear to be 40,000 daltons smaller than the mature form. They all undergo normal carbohydrate processing and reach the cell surface, and they bind a variety of antibodies directed against the LDL receptor. However, they have a markedly reduced ability to bind LDL. We suspect that these mutations may involve amino acid substitutions, deletions, or duplications in the cysteine-rich LDL binding domain or the EGF precursor region, but none has yet been fully elucidated at the molecular level.

Class IV Mutations: Receptors Reach Cell Surface and Bind LDL, But Fail to Cluster in Coated Pits

Study of these internalization-defective mutations at the cellular level originally revealed the importance of coated pits in receptor-mediated endocytosis (42,43). Three of the mutations have now been elucidated in molecular detail. All involve alterations in the cytoplasmic tail of the receptor, i.e., the 50 amino acids at the COOH-terminus that project into the cytoplasm (Fig. 10). The mutations have been unraveled through the preparation of genomic DNA libraries and the subsequent isolation and sequencing of exons 17 and 18, which encode the cytoplasmic domain. In the most drastic case a tryptophan codon has been converted to a nonsense (stop) codon at a position that is 2 residues distal to the membrane-spanning region (73). This produces a recep-

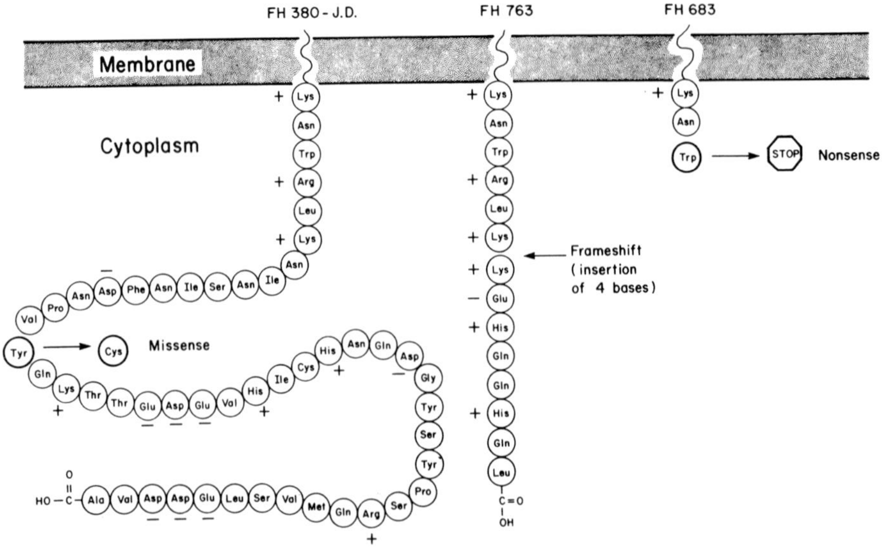

Fig. 10. Mutations affecting the cytoplasmic domain of the LDL receptor in three FH homozygotes with the internalization-defective form of FH.

tor with only 2 amino acids in the cytoplasmic tail. Another mutation involves a duplication of 4 nucleotides following the codon for the 6th amino acid of the cytoplasmic tail (73). This duplication alters the reading frame and leads to a sequence of 8 random amino acids followed by a stop codon. This receptor has only 6 of the normal amino acids in the cytoplasmic domain. Protein chemistry studies have confirmed that these two proteins lack the normal COOH-terminus (73).

The third mutation is the most informative. In this patient, who was the original internalization-defective subject to be described (J.D., Fig. 3), a single base change leads to the substitution of a cysteine for a tyrosine residue at position 807, which is in the middle of the cytoplasmic tail domain (*Fig. 10*). We have recently reproduced this amino acid substitution in the normal LDL receptor cDNA by oligonucleotide-directed mutagenesis. When the altered cDNA was introduced into Chinese hamster ovary cells by gene transfer techniques, it produced a receptor that bound LDL but did not cluster in coated pits, confirming that the single base change is responsible for the internalization defect in J.D.'s cells (82).

Inasmuch as all three internalization-defective mutations involve the cytoplasmic tail, this region must normally play a crucial role in the clustering of LDL receptors in coated pits. It is likely that the cytoplasmic tail binds to clathrin or some other protein that is itself linked to clathrin. The puzzling feature at the moment is that other cell surface receptors that cluster in coated pits do not show obvious homology with the LDL receptor in the amino acid sequences of their cytoplasmic tails (65). Thus, the precise structure that links receptors to coated pits remains a mystery.

We have identified several interesting variants of the Class IV mutations in which the mutant genes produce LDL receptors that are secreted into the culture medium. In two mutants of this class (each from an unrelated family), the responsible mutation is a large deletion that results from a recombination between two repetitive *Alu* sequences, one in intron 15 and the other in the 3' untranslated region of exon 18. The deletion joints in the two mutants are similar but not identical, indicating that the two mutations arose by independent events (83 and unpublished observations). In each mutant, the deletion removes the exons encoding the membrane-spanning region as well as the cytoplasmic tail. Presumably these prematurely terminated proteins have a short random sequence of amino acids at the COOH-terminus, owing to read-through of an unspliced mRNA. The receptors are transported to the surface, where some of them remain bound to the membrane. The vast majority, however, are released into the culture medium (83 and unpublished observations). The few receptors that remain on the surface bind LDL, but do not migrate to coated pits, thus giving rise to an internalization-defective phenotype. These findings emphasize the importance of the membrane-spanning region in anchoring the LDL receptor to the plasma membrane.

Figure 11 shows the location of nine mutations in the LDL receptor gene that have been analyzed at the molecular level. Each FH family examined to date has had a different mutation, and multiple types of mutational events have

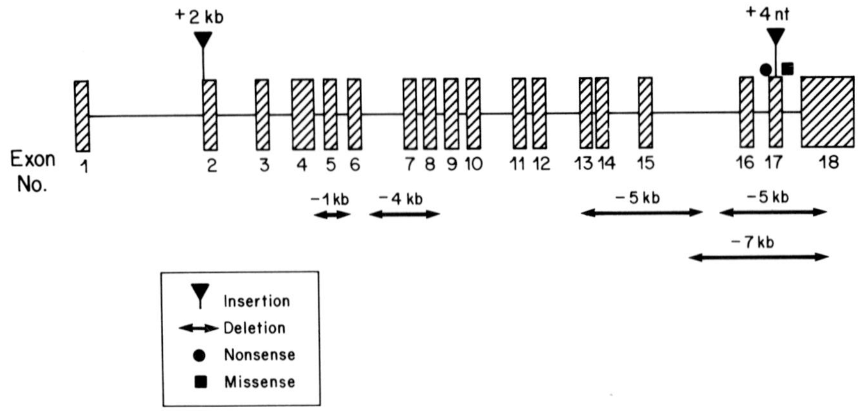

Fig. 11. Location of mutations in the LDL receptor gene. To date, nine mutations have been identified by molecular cloning and DNA sequence analysis or by restriction endonuclease analysis of genomic DNA. Five of the nine mutations are described in detail in refs. 73, 79, 82, and 83. kb denotes kilobases; nt denotes nucleotides.

occurred. Of the nine mutations, two involve single base substitutions, two involve insertions (one small and one large), and five involve large deletions. Many of the deletion joints occur in *Alu* repetitive elements.

FUNCTIONS OF THE LDL RECEPTOR IN THE BODY

The LDL receptor was elucidated by an investigative route that is opposite to the one usually employed to uncover metabolic pathways in animals. These pathways are usually observed first in intact animals or tissues and then they are studied in isolated cells. The LDL receptor was first observed in a totally artificial environment – namely, tissue culture. The question immediately arose: What tissues express LDL receptors in the body, and how do they work? We knew at the outset that the receptor must play some role in the body as evidenced by the devastating consequences of LDL receptor deficiency in FH homozygotes and the proportionately less severe abnormalities in FH heterozygotes. Clearly, the receptor must be functioning somewhere. But where?

Detection of LDL Receptor Expression In Vivo

The first cells that were demonstrated to have LDL receptor activity *in vivo* were circulating blood lymphocytes. In the initial studies carried out with Y.K. Ho in 1975, we isolated lymphocytes from the bloodstream and incubated them for 67 hr *in vitro* in the absence of exogenous cholesterol so as to "derepress" receptor synthesis (84). Under these conditions the lymphocytes expressed abundant LDL receptors as determined by measurements of the high affinity uptake and degradation of ^{125}I-LDL (Fig. 12A). Lymphocytes from FH homozygotes did not express detectable LDL receptor activity, and lymphocytes from FH heterozygotes had an intermediate level consistent with the presence of only a single functional gene (85). LDL receptors were also detectable on lymphocytes immediately after their isolation from the bloodstream, although the level of activity was lower than it was after derepression for 67 hr (85). Thus, LDL receptors were expressed in at least one cell type *in vivo*.

Another early clue to the function of LDL receptors *in vivo* came from studies of the rate of disappearance of intravenously-injected ^{125}I-LDL from plasma (Fig. 12B). Such LDL is removed from the circulation more slowly in FH heterozygotes than it is in normals (86,87). In FH homozygotes the removal defect is even more profound (87–89). The sluggishness of LDL catabolism *in vivo* correlates with the relative deficiency of LDL receptors as determined in isolated lymphocytes.

More detailed demonstrations of LDL receptor function *in vivo* have been obtained in experimental animals. Together with Sandip K. Basu, we established an assay for the binding of ^{125}I-LDL to membranes from homogenates of cultured cells and various tissues of the cow and other animals (90). Using this assay, Petri Kovanen found that most tissues of the cow had detectable high affinity ^{125}I-LDL binding; the adrenal gland and ovarian corpus luteum had the highest activity on a per gram basis (91). When the weight of the organ was taken into consideration, the liver was found to produce by far the largest number of LDL receptors. Similar results were obtained in studies of human fetal tissues (91). In collaboration with Havel's laboratory, we showed that ^{125}I-LDL was taken up by perfused rat livers by a high affinity receptor-mediated process that could be markedly accelerated by administration of the estrogenic hormone, 17α-ethinyl estradiol (92).

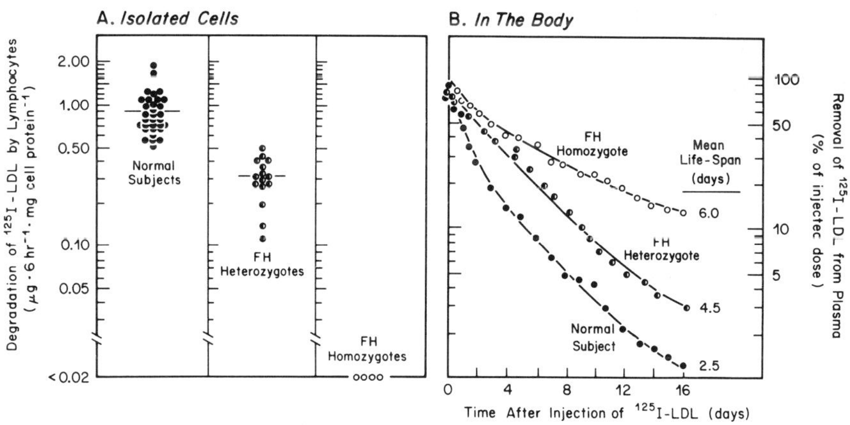

Fig. 12. Measurement of the number of LDL receptors in blood lymphocytes (Panel A) and in living subjects (*Panel B*). *Panel A:* Lymphocytes were isolated from venous blood of 32 normal subjects (●), 15 FH heterozygotes (◐), and 4 FH homozygotes (0). After incubation for 67 hr at 37°C in medium containing 10% lipoprotein-deficient serum, LDL receptor activity was assessed by measurement of the high affinity degradation of ^{125}I-LDL at 37°C. (Data replotted from ref. 85). *Panel B:* In the whole-body assay, a tracer amount of ^{125}I-LDL was injected intravenously, and the radioactivity remaining in the circulation over the next 16 days was measured in samples of venous blood (87,89). The higher the number of LDL receptors on body cells (*Panel A*), the faster the removal of ^{125}I-LDL from the blood (*Panel B*).

High levels of hepatic LDL receptors were also observed when radiolabeled LDL was injected into the circulation of experimental animals and its uptake into various tissues was compared. Steinberg and coworkers (93) and Dietschy and coworkers (94) showed that approximately 70% of the total-body uptake of radiolabeled LDL took place in the liver by LDL receptor-dependent pathways, but that the highest rates of uptake on a weight basis were seen in the adrenal gland. Various other tissues also showed receptor-mediated uptake of LDL in excess of that seen with nonspecific bulk phase markers such as radiolabeled albumin.

Measurements of receptor-mediated LDL uptake by tissues of animals were made more practical as a result of two developments: 1) Steinberg and coworkers developed a method to label LDL with radioactive sucrose and later with tyramine-cellobiose (95). In contrast to ^{125}I-labeling of tyrosines, the latter methods produced a radioactive marker that remained trapped in lysosomes after uptake and degradation, thus allowing slow rates of uptake to be quantified cumulatively over long periods. 2) Shepherd and Packard (96) showed that LDL whose arginine residues were modified by reaction with cyclohexanedione was cleared from the human circulation much more slowly than was native LDL. The rationale for these latter studies lay in previous work from our laboratory (97) and from Mahley's laboratory (98), which showed that modification of arginine or lysine residues on LDL abolished its ability to bind to the LDL receptor. These observations provided a crude estimate of the fraction of LDL clearance that was attributable to LDL receptors.

We had earlier estimated the fraction of total LDL clearance that was receptor-dependent by comparing the rate of catabolism of intravenously injected ^{125}I-LDL in normal individuals and in FH homozygotes (99). The fractional catabolic rate for LDL, i.e., the fraction of the total plasma pool of LDL removed per unit time, was 3-fold higher in normal subjects than in FH homozygotes (87). From this observation we reasoned that approximately two-thirds of LDL clearance is normally mediated through the LDL receptor (99). This conclusion has generally been borne out by a number of studies comparing the degradation rates for native versus lysine-modified or arginine-modified LDL both in normal human subjects and in a wide variety of experimental animals (100).

The WHHL Rabbit and the Role of the LDL Receptor in Clearance of IDL

One of the most important functions of LDL receptors *in vivo* was appreciated only in the past few years as a result of studies performed in Watanabe Heritable-Hyperlipidemic (WHHL) rabbits. This strain of mutant rabbits was discovered in the late 1970's by Yosio Watanabe, a veterinarian in Kobe, Japan (101). These rabbits have a mutation in the LDL receptor gene that is similar to the Class II mutations in human FH (81,102). When present in the homozygous form, this mutation gives rise to extremely high LDL-cholesterol levels; the rabbits develop atherosclerosis early in life (101,102).

The WHHL rabbits proved invaluable in explaining a previously puzzling feature of homozygous FH. Kinetic studies of ^{125}I-LDL metabolism by Myant and coworkers (88), and by Bilheimer and Grundy (87,89) indicated that FH subjects have a dual defect. In addition to degrading LDL more slowly, FH homozygotes and heterozygotes also appeared to overproduce LDL. How does a genetic defect in the LDL receptor lead simultaneously to overproduction and reduced degradation of LDL? The answer lies in the complex biosynthetic pathway for LDL.

Early studies by Gitlin (103) and later those of Bilheimer, Levy, and Eisenberg (104) suggested that LDL is not secreted directly from the liver, but rather produced in the circulation from a blood-borne precursor, very low density lipoprotein (VLDL) (Fig. 13A). VLDL is a large, triglyceride-rich lipoprotein that is secreted by the liver; it transports triglyceride to adipose tissue and muscle. The triglycerides in VLDL are removed in capillaries by the enzyme lipoprotein lipase, and the VLDL returns to the circulation as a smaller particle with a new name, intermediate-density lipoprotein (IDL). The IDL particles have lost most of their triglyceride, but they retain cholesteryl esters. Some of the IDL particles are rapidly taken up by the liver; others remain in the circulation where they undergo further triglyceride hydrolysis and are converted to LDL. A distinguishing feature of the IDL particles is their content of multiple copies of apo E in addition to a single copy of apo B-100. The multiple copies of apo E allow IDL to bind to the LDL receptor with very high affinity. When IDL is converted to LDL, the apo E leaves the particle and only apo B-100 remains. Thereafter, the affinity for the LDL receptor is much reduced (102).

With Toru Kita, we showed that the apparent overproduction of LDL in WHHL rabbits is due to the failure of IDL to be removed from the plasma (102,105) (Fig. 13B). Thus, when ^{125}I-VLDL was administered to WHHL rabbits, the resultant IDL was not taken up by the liver, as it was in normal rabbits (105). Rather it remained in the circulation and was converted in increased amounts to LDL. These findings strongly suggest that IDL is normally cleared from plasma by binding to LDL receptors in the liver. Although experiments of similar detail cannot be carried out in humans, the observations of Soutar, Myant, and Thompson (106) are consistent with the notion that enhanced conversion of IDL to LDL also occurs in FH homozygotes, thus accounting for much of the apparent overproduction of LDL.

Figure 13A illustrates the dual role of the LDL receptor in LDL metabolism as determined from the studies of WHHL rabbits. First, the receptor limits LDL production by enhancing the removal of the precursor, IDL, from the circulation. Second, it enhances LDL degradation by mediating cellular uptake of LDL. A deficiency of LDL receptors causes LDL to accumulate as a result both of overproduction and of delayed removal (*Fig. 13B*). By this quirk of dual functionality, LDL receptors become crucially important modulators of plasma LDL levels in humans and animals.

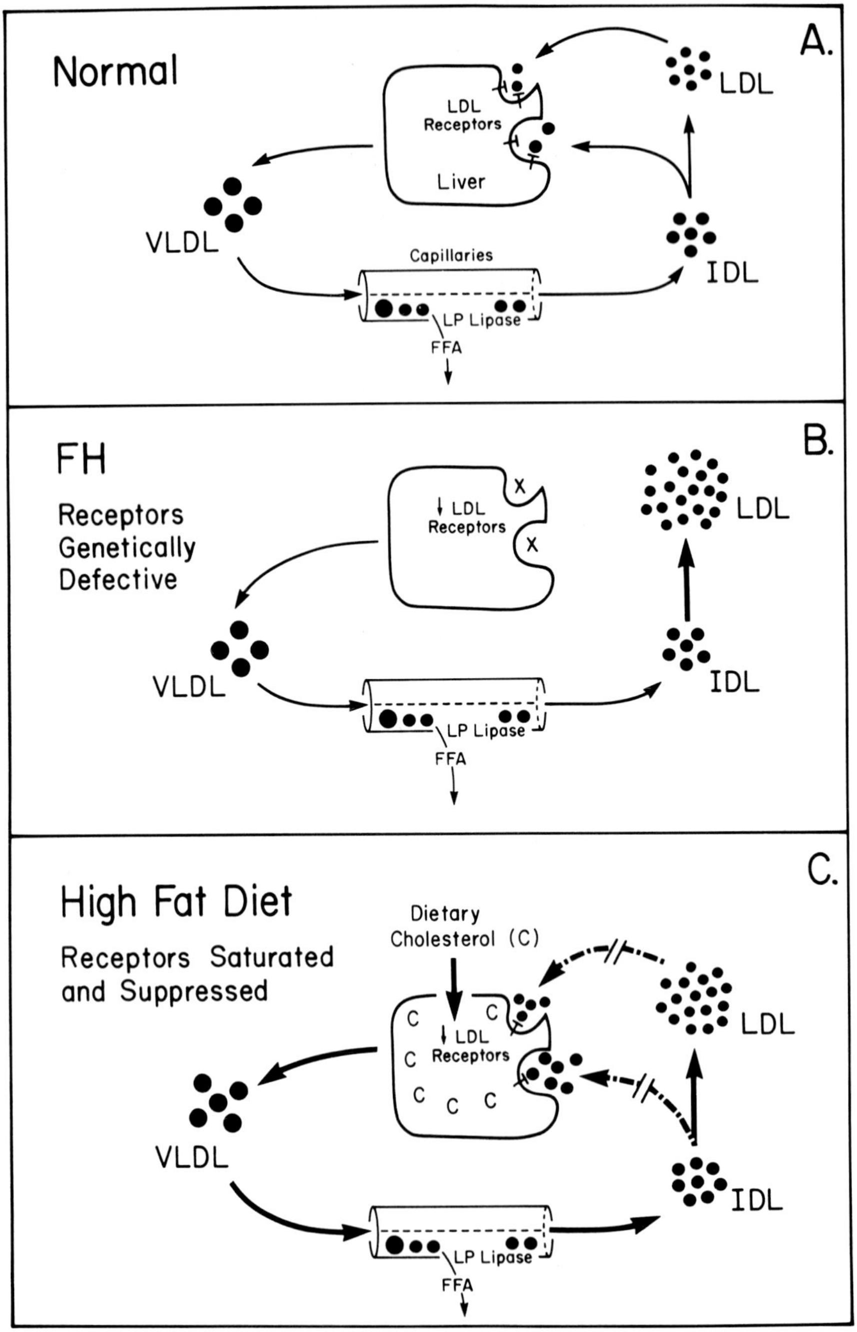

Fig. 13. Schematic model of the mechanism by which LDL receptors in the liver control both the production and catabolism of plasma LDL in normal human subjects (*Panel A*), in individuals with FH (*Panel B*), and in individuals consuming a diet rich in saturated fats and cholesterol (*Panel C*). VLDL denotes very low density lipoprotein; IDL denotes intermediate density lipoprotein. (Modified from ref. 132 with permission.)

PERSPECTIVES

Receptor Regulation: Therapeutic Implications

Knowledge of the fundamental properties of the LDL receptor has important implications for the therapy of FH and other hypercholesterolemic states. This knowledge also provides fuel for certain speculations about the role of the LDL receptor as a protective factor against atherosclerosis in human beings.

The therapeutic implications of the LDL receptor studies center on strategies for increasing the production of LDL receptors in the liver, thereby lowering plasma LDL-cholesterol levels. In FH heterozygotes this goal can be attained by stimulating the normal gene to produce more than its usual number of LDL receptors, thus compensating for the defective allele (107). The rationale for such therapy emerged from studies of cultured fibroblasts, which showed that the production of LDL receptors is driven by the cell's demand for cholesterol (9,36). When demands for cholesterol are high, the cells have high levels of mRNA for the LDL receptor. Conversely, when demands for cholesterol are reduced, excess cholesterol accumulates in cells, and the amount of receptor mRNA falls (36,37).

Inasmuch as the liver is the major site of expression of LDL receptors, the therapeutic problem is reduced to the development of methods to increase hepatic demands for cholesterol. This can be achieved by two techniques: 1) inhibition of the intestinal reabsorption of bile acids; and 2) inhibition of cholesterol synthesis. These techniques can be used alone or in combination, as illustrated in Fig. 14.

The liver requires cholesterol for conversion into bile acids, which constitute the major route by which cholesterol is excreted from the body (18). However, only a fraction of the bile acids secreted by the liver actually leaves the body. The vast bulk of bile acids are reabsorbed in the terminal ileum and returned to the liver for reutilization. As a result, the liver converts only a minimal amount of cholesterol into bile acids (Fig. 14, *left*). The liver's demand for cholesterol can be enhanced by the ingestion of resins that bind bile acids in the intestine and prevent their reabsorption. Since the liver can no longer re-use old bile acids, it must continually make new bile acids and the liver's demand for cholesterol increases. In order to obtain this cholesterol, the liver makes a dual response: 1) it synthesizes increased amounts of cholesterol through an increase in the activity of HMG CoA reductase; and 2) it attempts to take up additional plasma cholesterol by increasing the production of LDL receptors. The increased LDL receptor activity causes plasma LDL levels to fall (Fig. 14, *center*). The problem with bile acid resin therapy (and the physiologically equivalent procedure of ileal bypass surgery) is that the effects are not profound. The increase in cholesterol production partially offsets the hepatic demand for cholesterol and so there is only a 15 to 20% increase in the synthesis of LDL receptors and only a 15 to 20% drop in plasma LDL-cholesterol levels.

The second method for increasing LDL receptor production, namely, inhibition of hepatic cholesterol synthesis, is much more powerful than bile acid depletion. The technique emerged from the discovery in 1976 of a class of

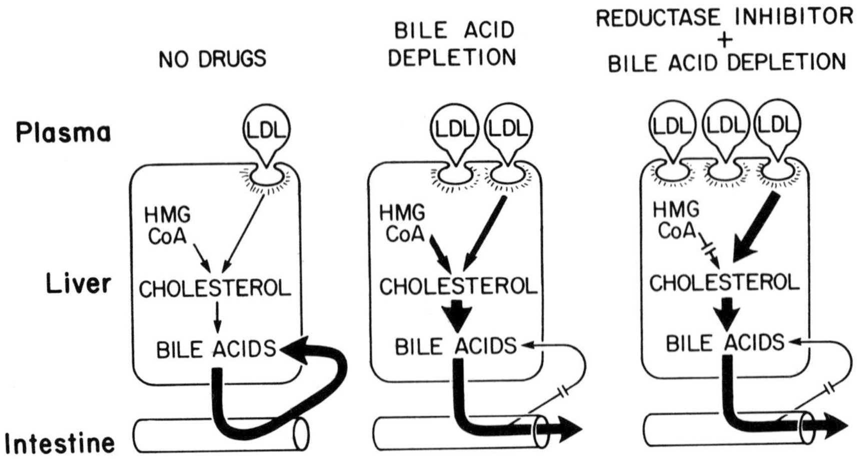

Fig. 14. Rationale for the use of a bile acid binding resin and an inhibitor of 3-hydroxy-3-methylglutaryl Coenzyme A reductase in the treatment of FH heterozygotes. A detailed discussion of this figure is presented in the text.

fungal metabolites that inhibit HMG CoA reductase. The original compound, discovered by Akira Endo at the Sankyo Drug Company in Japan, is called compactin (108). A more recent version, developed by the Merck, Sharp and Dohme Research Laboratories in the United States, is called mevinolin (109). These two agents are potent competitive inhibitors of HMG CoA reductase; the inhibitory constant is approximately 10^{-9} molar (108).

When given to experimental animals, compactin or mevinolin initially inhibit cholesterol synthesis in the liver, and this triggers a complex regulatory mechanism that lowers the plasma LDL-cholesterol level. With Kovanen and Kita, we showed that the inhibition of cholesterol synthesis elicits a dual compensatory response: 1) hepatocytes synthesize increased amounts of HMG CoA reductase; and 2) they synthesize increased numbers of LDL receptors (110). When a new steady state is attained, the increase in HMG CoA reductase is almost sufficient to overcome the inhibitory effects of compactin. Total body cholesterol synthesis is only slightly reduced (111). Meanwhile, the plasma LDL level has fallen as a result of the increase in LDL receptors. The fall in plasma LDL levels is balanced by the increase in LDL receptors, and so the absolute amount of cholesterol entering the liver through the receptor pathway is the same as it was earlier. The difference, however, is that this delivery is now occurring at a lower plasma LDL level (107).

When given as a single agent to FH heterozygotes, mevinolin routinely produces a 30% fall in plasma LDL-cholesterol levels. When given together with cholestyramine, mevinolin blocks the compensatory increase in cholesterol synthesis, and the increase in LDL receptors is even more profound (Fig. 14, *right*). Plasma LDL cholesterol levels fall by 50 to 60% (112).

The important principle to emerge from these studies is that stimulation of LDL receptor activity lowers the plasma LDL-cholesterol level without grossly distorting cholesterol delivery (107,111). At present mevinolin and related compounds are in the early stages of clinical testing. Their efficacy in lowering plasma LDL-cholesterol levels has been well established, but there is no information regarding long-term toxicity in man. If these drugs turn out to be non-toxic, they will have an important role in the therapy of FH heterozygotes and probably of other hypercholesterolemic individuals as well.

The principles applied to treatment of FH heterozygotes cannot, unfortunately, be applied to homozygotes, especially those who have totally defective LDL receptor genes. These individuals do not respond to the above-mentioned drugs because they cannot synthesize LDL receptors (113). Current therapy for these individuals involves removal of LDL from plasma extracorporeally through repeated plasmapheresis (114). Such procedures, which must be repeated every two to three weeks, are technically difficult and are very demanding of patient and physician.

Recently, a more direct therapeutic approach was taken in an FH homozygote, whose initials are S.J. and who has two mutant genes at the LDL receptor locus. This six-year-old girl, who is a patient of our colleague David Bilheimer in Dallas, had a total plasma cholesterol level over 1,000 mg/dl (greater than 6 times above normal limits), and she sustained repeated episodes of myocardial infarction. After she failed to respond to two coronary bypass procedures plus a mitral valve replacement, she was subjected to combined heart-liver transplantation by a team of surgeons led by Thomas E. Starzl at the University of Pittsburgh (115). The liver transplant was designed to provide a source of LDL receptors. The heart transplantation was necessitated because of the poor condition of her own heart as a result of the atherosclerotic process.

Immediately after the operation, S.J.'s total plasma cholesterol level fell from 1100 mg/dl to the range of 200 to 300 mg/dl, and it remained in that range for the succeeding 13 months (Fig. 15A). Thereafter she was started on the HMG

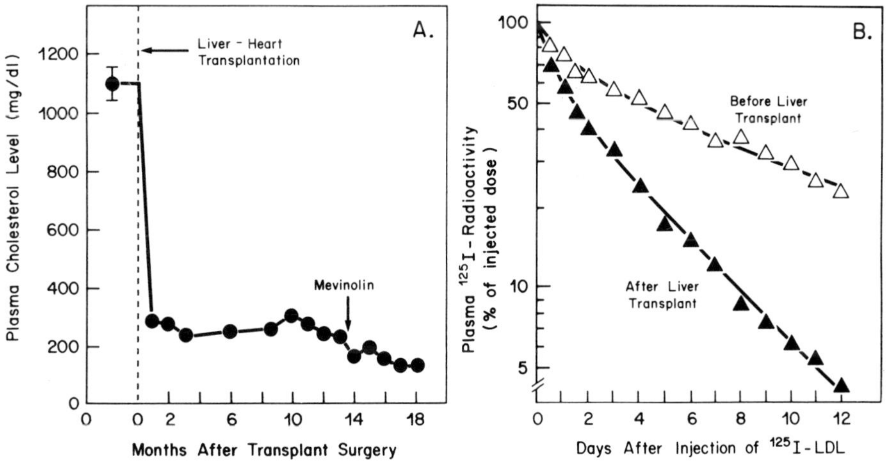

Fig. 15. LDL metabolism in S.J., a patient with homozygous FH, before and after liver-heart transplantation. *Panel A:* Total cholesterol levels in plasma. *Panel B:* Plasma decay curves of ^{125}I-LDL after intravenous injection of tracer amounts of ^{125}I-LDL before (△) and after (▲) liver-heart transplantation. (Data in *Panel B* reprinted with permission from ref. 115).

CoA reductase inhibitor mevinolin, and her cholesterol fell further to the range of 150–200 mg/dl (Fig. 15A). Liver transplantation not only lowered the plasma cholesterol level but it also restored responsiveness to mevinolin, which requires a normal LDL receptor gene in order to act. Lipoprotein turnover studies performed six months after surgery confirmed that the new LDL receptors furnished by the transplanted liver were responsible for the dramatic drop in plasma cholesterol level (Fig. 15B). S.J. remains asymptomatic at the time of this writing, and her cutaneous xanthomas have resolved. However, she requires continuous therapy with cyclosporin to prevent rejection of the transplanted organs, and her long-term prognosis is uncertain.

The response to liver transplantation in S.J. underscores the importance of hepatic LDL receptors *in vivo* and raises the possibility that other FH homozygotes may respond to similar transplantation procedures. Hopefully, in appropriate cases liver transplantation can be performed before heart transplantation becomes necessary.

Speculations: LDL Receptors and the General Problem of Atherosclerosis

We now leave the realm of solidly established scientific fact and enter the much more controversial realm of speculation about the relation between cholesterol levels, LDL receptors, and atherosclerosis in the general population. After all, FH heterozygotes account for only 5% of myocardial infarctions in patients under the age of 60. What causes the other 95% of heart attacks?

Extensive epidemiologic studies performed in many populations in many countries over the past three decades have pointed strongly to a general association of high blood cholesterol levels with heart attacks. Among the most striking examples is the seven-country study of coronary artery disease directed by Ansel Keys (116). A similar correlation has been observed within a single population in the extensive studies in Framingham, Massachusetts (117).

These studies have all shown that the incidence of myocardial infarction rises in proportion to the plasma cholesterol level, more specifically the plasma level of LDL-cholesterol. When LDL-cholesterol levels are below 100 mg/dl (equivalent to a total plasma cholesterol level of ~ 170 mg/dl), heart attacks are rare. When LDL-cholesterol levels are above 200 mg/dl (equivalent to a total plasma cholesterol level of ~ 280 mg/dl), heart attacks are frequent. Controversy arises over the middle ground, i.e., individuals with plasma LDL-cholesterol levels between 100 and 200 mg/dl (total plasma cholesterol of 170 to 280 mg/dl). This is the range in which the vast bulk of heart attacks occur. Somewhere within this range there is a threshold value of cholesterol at which heart attacks begin to become more frequent. In this middle ground how much of the heart attack burden is attributable to plasma cholesterol? There is no definitive answer. In addition to cholesterol, heart attacks in this group are aggravated by smoking, hypertension, stress, diabetes mellitus, and poorly understood genetic factors. However, it seems reasonable to propose that plasma cholesterol does have something to do with heart attacks in these subjects, and that the incidence of heart attacks would be reduced if plasma cholesterol could be lowered (10).

The LDL receptor studies lend experimental support to the epidemiologists' suggestion that the levels of plasma cholesterol usually seen in Western industrialized societies are inappropriately high (9). This support derives from knowledge of the affinity of the LDL receptor for LDL. The receptor binds LDL optimally when the lipoprotein is present at a cholesterol concentration of 2.5 mg/dl (28). In view of the 10 to 1 gradient between concentrations of LDL in plasma and interstitial fluid, a level of LDL-cholesterol in plasma of 25 mg/dl would be sufficient to nourish body cells with cholesterol (118). This is roughly one-fifth of the level usually seen in Western societies (Fig. 16 and ref. 119). Several lines of evidence suggest that plasma levels of LDL-cholesterol in the range of 25–60 mg/dl (total plasma cholesterol of 110 to 150 mg/dl) might indeed be physiologic for human beings. First, in other mammalian species that do not develop atherosclerosis, the plasma LDL-cholesterol level is generally less than 80 mg/dl (Fig. 16 and ref. 120). In these animals the affinity of the LDL receptor for their own LDL is roughly the same as the affinity of the human LDL receptor for human LDL, implying that these species are designed by evolution to have similar plasma LDL levels (9,119). Second, the LDL level in newborn humans is approximately 30 mg/dl (121), well within the range that seems to be appropriate for receptor binding (Fig. 16). Third, when humans are raised on a low fat diet, the plasma LDL-cholesterol tends to stay in the range of 50 to 80 mg/dl. It only reaches levels above 100 mg/dl in individuals who consume a diet rich in saturated animal fats and cholesterol that is customarily ingested in Western societies (116,122).

What is the mechanism for the high levels of plasma LDL that are so frequent in Western industrialized societies? Extensive evidence implicates two major factors: diet and heredity. When people habitually consume diets low in animal fats, their plasma LDL-cholesterol levels generally tend to remain low. When even moderate amounts of animal fat are introduced into the diet, the plasma cholesterol level rises (116,122). However, the level does not rise equally in every person. Clearly, genetic as well as dietary factors play a role.

How might a diet rich in animal fats and cholesterol elevate the plasma LDL-cholesterol level? Here we believe that two properties of the LDL receptor play a role — saturation and suppression. As the plasma LDL level rises, the receptors become saturated. This saturation of receptors sets an upper limit on the rate at which LDL can be removed efficiently from plasma (123). Each receptor can handle only one particle of LDL at a time. Once the receptors become saturated, the rate of removal of LDL can be accelerated only by an increase in clearance by non-receptor pathways that operate at low efficiency. In order to drive these alternate pathways, the LDL level must be quite high (99). At ordinary levels of LDL, the major factor that limits the removal of LDL from plasma is saturation of the LDL receptor (123).

Once LDL receptors become saturated, the removal rate of LDL is proportional to the number of receptors. Whenever the number of receptors is reduced, plasma LDL levels must rise. Experiments in animals indicate that the consumption of a high fat diet decreases the number of LDL receptors in the liver (123, 124). We believe that this mechanism operates through feedback suppression as described above. That is, when excess dietary cholesterol accumulates in the liver, the liver responds by decreasing the production of LDL receptors (Fig. 13C). The entry of dietary cholesterol into the liver is mediated by a receptor, termed the chylomicron remnant receptor, whose activity is genetically distinct from the LDL receptor (125). The chylomicron remnant receptor is unaffected by cholesterol accumulation (126), and it causes cholesterol to accumulate to high levels in liver when the diet contains excess fat.

The combination of saturation and suppression of hepatic LDL receptors contributes in a major way to the buildup of LDL in plasma when a diet rich in saturated fats and cholesterol is ingested. Insofar as such a diet also may increase production of LDL in the face of a fixed or declining removal capacity, the LDL level would rise even higher.

If the LDL receptor does limit the removal of LDL from plasma, then maneuvers that increase LDL receptor activity might be effective in individuals who have high plasma LDL-cholesterol levels, but who do not have defective LDL receptor genes. Such therapy seems feasible with the development of HMG CoA reductase inhibitors. However, it is still too early to tell whether such therapy would decrease the incidence of myocardial infarctions in individuals with moderately elevated plasma LDL-cholesterol levels in the range of 100 to 200 mg/dl. There is much circumstantial evidence to expect such improvement (127), but unequivocal data are simply not there. Hopefully, with the availability of powerful receptor-stimulating drugs, the hypothesis should be susceptible to testing in the near future.

In considering the role of diet and drugs in treatment of high cholesterol levels, physicians and public health authorities must bear in mind the genetic variability between individuals. This variability exists at three levels: 1) The degree of increase in plasma cholesterol upon ingestion of a high cholesterol diet is variable. Not all people develop hypercholesterolemia. Some people, such as the Pima Indians, maintain low plasma cholesterol levels despite ingestion of a high fat diet (10). 2) Even when the plasma cholesterol level becomes elevated, the propensity for atherosclerosis varies. For example, a substantial proportion of FH heterozygotes (10 to 20%) escape myocardial infarction until the 8th or 9th decade despite pronounced hypercholesterolemia from birth (14). 3) Genetic susceptibility to contributory risk factors is variable. Some people can withstand hypertension and cigarette smoking for decades without developing atherosclerotic complications; others are highly sensitive.

In view of this genetic variability in susceptibility, dietary and drug recommendations must be individualized. The family history of the individual must be constantly borne in mind, particularly the familial incidence of premature heart attacks or strokes. An important goal of future research will be to dissect this genetic variability. Hopefully, it will become possible to identify the genes that determine such predispositions and to analyze them in each individual. For example, are there alleles that produce subtle defects in the LDL receptor that predispose to "garden-variety", diet-responsive hypercholesterolemia? Are there subtle abnormal alleles at other loci such as those governing cholesterol absorption, cholesterol synthesis, or bile acid synthesis? It seems likely that variability at such loci exists; the scientific challenge is to expose it.

Receptor Recycling: A Novel Cellular Pathway

The studies of the LDL receptor have revealed a new process by which membrane-embedded receptors cycle continuously into and out of cells. The receptors move from one organelle to another as a result of two sequential events: 1) segregation from other proteins by lateral movement in the plane of the membrane, and 2) pinching off of receptor-enriched membranes to form vesicles that eventually fuse with a different organelle. These receptors have been designated as "migrant" membrane proteins to distinguish them from "resident" membrane proteins that do not move in this fashion (4). The purpose of such intracellular traffic is to integrate the behavior of multiple organelles to form coherent biochemical pathways. Thus, the movement of the LDL receptor links the cell surface to the endosome and to the lysosome. The cholesterol liberated from LDL in lysosomes exerts regulatory effects in two other organelles, the endoplasmic reticulum and the nucleus. Selective movement of membrane proteins from one organelle to another allows such multi-organelle regulation to occur.

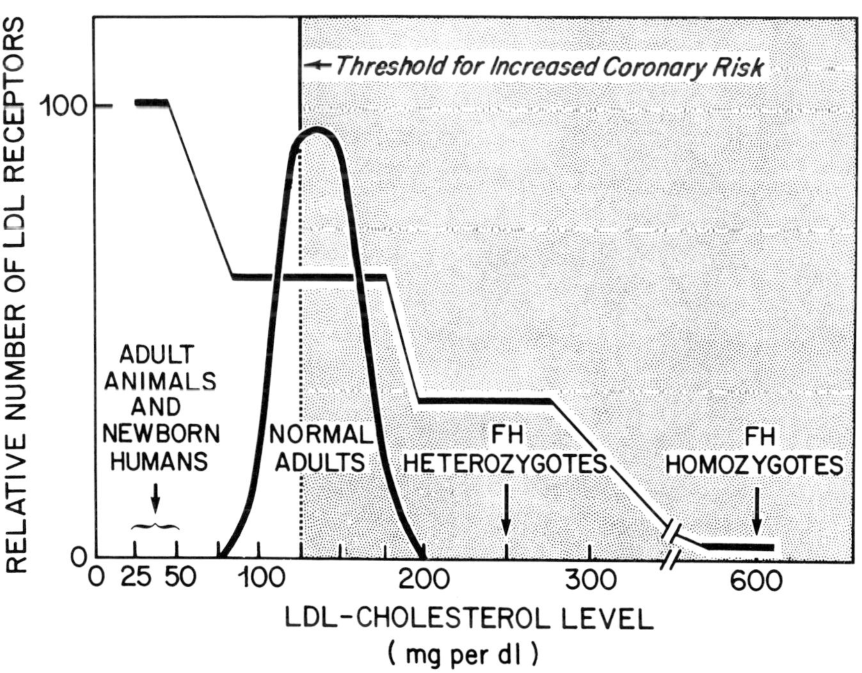

Fig. 16. Range of LDL levels in "normal" adults in Western industrial societies, indicated by the bell-shaped curve (127), is compared with the range in adult animals (120) and human infants (121) and with the levels seen in FH patients (14). Levels in the shaded region of the chart are above the threshold associated with accelerated atherosclerosis; more than half of the adults have LDL levels above this threshold. The LDL level is inversely associated with the number of LDL receptors. (Modified from ref. 10 with permission.)

What are the signals that dictate the path that each migrant membrane protein must follow? We are beginning to obtain some insight into the signals necessary for LDL receptors to be incorporated into one sorting structure, the coated pit. However, there is still no information with regard to signals that cause proteins to leave other organelles such as the endoplasmic reticulum and move to different organelles such as the Golgi complex. Delineation of these sorting signals is a major challenge facing the field of cell biology.

CONCLUDING REMARKS

In his Nobel Lecture of 1964, Konrad Bloch summarized his brilliant studies on the biological synthesis of cholesterol. At the end of his talk, Bloch predicted that the next era in cholesterol research would involve the elucidation of homeostatic control mechanisms (128). A decade later, in 1973, the LDL receptor concept was advanced to explain the homeostasis of plasma cholesterol and to account for regulatory abnormalities in cholesterol metabolism that were observed in patients with familial hypercholesterolemia. During the next 12 years, the LDL receptor was transformed from a genetic abstraction to a well characterized protein whose structural domains have been defined. Studies of this receptor taught us about receptor-mediated endocytosis and the novel route by which receptors cycle in and out of cells. We have learned that receptors contain multiple functional domains that direct each step in this movement and that these domains are encoded on exons that can be shared among many proteins. We have learned that genetic defects in the receptor can cause cholesterol to accumulate in plasma, producing premature atherosclerosis. Together with others, we have also learned that the liver is the most important site of action of LDL receptors and that liver replacement can offer a successful form of therapy for children with homozygous FH. Finally, we have learned that regulation of this receptor through drugs and diet can profoundly change the LDL-cholesterol level and that saturation and suppression of receptors may contribute to the high incidence of hypercholesterolemia in industrialized society. It is hoped that these insights will lead to a deeper understanding of the biology of cells and thereby to more effective forms of treatment for diseases such as familial hypercholesterolemia.

We wish to express our deepest appreciation to Dr. Donald W. Seldin, chairman of the Department of Internal Medicine at the University of Texas Southwestern Medical School for more than 30 years and creator of the intellectual environment that made our work possible. We are grateful for the contributions of many students, postdoctoral fellows, and faculty members who have advanced this effort over the past 12 years. Finally, we acknowledge the National Heart, Lung, and Blood Institute of the National Institutes of Health, the Moss Heart Foundation, and the Leland Fikes Foundation for their long-term financial support.

REFERENCES AND NOTES

1. Garrod, A.E. (1923) *Inborn Errors of Metabolism.* Oxford University Press, London, 2nd edition. pp. 1–216; Beadle, G.W. (1959) Genes and chemical reactions in neurospora. *Science 129:* 1715–1719; Tatum, E.L. (1959) A case history in biological research. *Science 129*: 1711–1714; Pauling L., Itano H.A., Singer, S.J., and Wells, I.C. (1949) Sickle cell anemia: A molecular disease. *Science 110:* 543–548; Ingram, V.M. (1957) Gene mutations in human haemoglobin: The chemical difference between normal and sickle cell haemoglobin. *Nature 180:* 326–328.
2. Goldstein, J.L., and M.S. Brown. (1973) Familial hypercholesterolemia: Identification of a defect in the regulation of 3-hydroxy-3-methylglutaryl coenzyme A reductase activity associated with overproduction of cholesterol. *Proc. Natl. Acad. Sci. USA 70:* 2804–2808.
3. Brown, M.S., and J.L. Goldstein. (1974) Familial hypercholesterolemia: Defective binding of lipoproteins to cultured fibroblasts associated with impaired regulation of 3-hydroxy-3-methylglutaryl coenzyme A reductase activity. *Proc. Natl. Acad. Sci. USA 71:* 788–792.
4. Goldstein, J.L., R.G.W. Anderson, and M.S. Brown. (1979) Coated pits, coated vesicles, and receptor-mediated endocytosis. *Nature 279:* 679–685; Brown, M.S., R.G.W. Anderson, and J.L. Goldstein. (1983) Recycling receptors: The round-trip itinerary of migrant membrane proteins. *Cell 32:* 663–667.
5. Heinrich O. Wieland (1928), Adolf O.R. Windaus (1928), Leopold Ruzicka (1939), Robert Robinson (1947), and Otto P.H. Diels (1950) were awarded the Nobel Prize in Chemistry in part for work that led to the elucidation of the structure of cholesterol, a brilliant chapter in the history of organic chemistry. Konrad Bloch and Feodor Lynen were awarded the Nobel Prize in Medicine or Physiology in 1964 for their landmark studies of the cholesterol biosynthetic pathway, a complex sequence involving at least 30 steps. Robert B. Woodward, who pioneered the stereochemical synthesis of cholesterol, received the Nobel Prize in Chemistry in 1965 "for his outstanding achievement in the art of organic synthesis." Derek H.R. Barton and Odd Hassel were awarded the Nobel Prize in Chemistry in 1969 "for developing and applying the principles of conformation in chemistry," which included establishing the *all chair* conformation of cholesterol. John W. Cornforth, who collaborated with George Popjak to establish the orientation of all of the hydrogen atoms in the cholesterol molecule, received the Nobel Prize in Chemistry in 1975 "for his work on the stereochemistry of enzyme-catalyzed reactions."
6. Oncley, J.L. (1956) Lipoproteins of human plasma. *Harvey Lect 50:* 71–91.
7. Gofman, J.W., O. Delalla, F. Glazier, N.K. Freeman, F.T. Lindgren, A.V. Nichols, B. Strisower, A.R. Tamplin. (1954) The serum lipoprotein transport system in health, metabolic disorders, atherosclerosis and coronary heart disease. *Plasma 2:* 413–484.
8. Fredrickson, D.S. (1974) Plasma lipoproteins and apolipoproteins. *Harvey Lect. 68:* 185–237.
9. Goldstein, J.L., and M.S. Brown. (1977) The low-density lipoprotein pathway and its relation to atherosclerosis. *Ann. Rev. Biochem. 46:* 897–930.
10. Brown, M.S., and J.L. Goldstein. (1984) How LDL receptors influence cholesterol and atherosclerosis. *Sci. Am. 251:* 58–66.
11. Müller, C. (1938) Xanthomata, hypercholesterolemia, angina pectoris. *Acta Med. Scand. 89: (suppl)* 75–84.
12. Khachadurian, A.K. (1964) The inheritance of essential familial hypercholesterolemia. *Am. J. Med. 37:* 402–407.
13. Fredrickson, D.S., and R.I. Levy. (1972) Familial hyperlipoproteinemia. Chapter 28. In *The Metabolic Basis of Inherited Disease.* 3rd edition. J.B. Stanbury, J.B. Wyngaarden, and D.S. Fredrickson, editors. McGraw-Hill Book Co., New York. pp. 545–614.
14. Goldstein, J.L., and M.S. Brown. (1983) Familial hypercholesterolemia. Chapter 33. In *The Metabolic Basis of Inherited Disease.* J.B. Stanbury, J.B. Wyngaarden, D.S. Fredrickson, J.L. Goldstein, and M.S. Brown, editors. 5th edition. McGraw-Hill Book Co., New York. pp. 672–712.
15. Goldstein, J.L., H.G. Schrott, W.R. Hazzard, E.L. Bierman, and A.G. Motulsky. (1973) Hyperlipidemia in coronary heart disease. II. Genetic analysis of lipid levels in 176 families

and delineation of a new inherited disorder, combined hyperlipidemia. *J. Clin. Invest. 52:* 1544–1568.

16. Patterson, D., and J. Slack. (1972) Lipid abnormalities in male and female survivors of myocardial infarction and their first-degree relatives. *Lancet i:* 393–399.
17. Nikkila, E.A., and A. Aro. (1973) Family study of serum lipids and lipoproteins in coronary heart-disease. *Lancet i:* 954–958.
18. Dietschy, J.M., and J.D. Wilson. (1970) Regulation of cholesterol metabolism. *N. Engl. J. Med. 282:* 1128–1138, 1179–1183, 1241–1249.
19. Neufeld, E.F., and J.C. Fratantoni. (1970) Inborn errors of mucopolysaccharide metabolism. *Science 169:* 141–146.
20. Bailey, J.M. (1973) Regulation of cell cholesterol content. In *Atherogenesis: Initiating Factors,* Ciba Found. Symp. Edited by R. Porter and J. Knight, Vol. 12. Elsevier, Amsterdam. pp. 63–92.
21. Rothblat, G.H. (1969) Lipid metabolism in tissue culture cells. *Adv. Lipid Res. 7:* 135–162.
22. Brown, M.S., S.E. Dana, and J.L. Goldstein. (1973) Regulation of 3-hydroxy-3-methylglutaryl coenzyme A reductase activity in human fibroblasts by lipoproteins. *Proc. Natl. Acad. Sci. USA 70:* 2162–2166.
23. Bucher, N.L.R., K. McGarrahan, E. Gould and A.V. Loud. (1959) Cholesterol biosynthesis in preparations of liver from normal, fasting, x-irradiated, cholesterol-fed, triton, or Δ^4-cholesten-3-one-treated rats. *J. Biol. Chem.* 234: 262–267; Bucher, N.L.R., P. Overath, and F. Lynen. (1960) β-hydroxy-β-methylglutaryl coenzyme A reductase, cleavage and condensing enzymes in relation to cholesterol formation in rat liver. *Biochim. Biophys. Acta 40:* 491–501.
24. Siperstein, M.D. (1970) Regulation of cholesterol biosynthesis in normal and malignant tissues. *Curr. Topics Cell. Reg. 2:* 65–100.
25. Brown, M.S., S.E. Dana, and J.L. Goldstein. (1974) Regulation of 3-hydroxy-3-methylglutaryl coenzyme A reductase activity in cultured human fibroblasts: Comparison of cells from a normal subject and from a patient with homozygous familial hypercholesterolemia. *J. Biol. Chem. 249:* 789–796.
26. Sutherland, E.W. (1972) Studies on the mechanism of hormone action. *Science 177:* 401–408.
27. Goldstein, J.L., and M.S. Brown. (1974) Binding and degradation of low density lipoproteins by cultured human fibroblasts: Comparison of cells from a normal subject and from a patient with homozygous familial hypercholesterolemia. *J. Biol. Chem. 249:* 5153–5162.
28. Goldstein, J.L., S.K. Basu, G.Y. Brunschede, and M.S. Brown. (1976) Release of low density lipoprotein from its cell surface receptor by sulfated glycosaminoglycans. *Cell 7:* 85–95.
29. Goldstein, J.L., S.E. Dana, J.R. Faust, A.L. Beaudet, and M.S. Brown. (1975) Role of lysosomal acid lipase in the metabolism of plasma low density lipoprotein: Observations in cultured fibroblasts from a patient with cholesteryl ester storage disease. *J. Biol. Chem. 250:* 8487–8495.
30. de Duve, C. (1983) Lysosomes revisited. *Eur. J. Biochem. 137:* 391–397.
31. Goldstein, J.L., G.Y. Brunschede, and M.S. Brown. (1975) Inhibition of the proteolytic degradation of low density lipoprotein in human fibroblasts by chloroquine, concanavalin A, and Triton WR 1339. *J. Biol. Chem. 250:* 7854–7862.
32. de Duve, C., T. DeBarsy, B. Poole, A. Trouet, P. Tulkens, and F. Van Hoof. (1974) Lysosomotropic agents. *Biochem. Pharmacol. 23:* 2495–2534.
33. Luskey, K.L., J.R. Faust, D.J. Chin, M.S. Brown, and J.L. Goldstein. (1983) Amplification of the gene for 3-hydroxy-3-methylglutaryl coenzyme A reductase, but not for the 53-kDa protein, in UT-1 cells. *J. Biol. Chem. 258:* 8462–8469.
34. Gil, G., J.R. Faust, D.J. Chin, J.L. Goldstein, and M.S. Brown. (1985) Membrane-bound domain of HMG CoA reductase is required for sterol-enhanced degradation of the enzyme. *Cell 41:* 249–258.
35. Goldstein, J.L., S.E. Dana, and M.S. Brown. (1974) Esterification of low density lipoprotein cholesterol in human fibroblasts and its absence in homozygous familial hypercholesterolemia. *Proc. Natl. Acad. Sci. USA 71:* 4288–4292.
36. Brown, M.S., and J.L. Goldstein. (1975) Regulation of the activity of the low density lipoprotein receptor in human fibroblasts. *Cell 6:* 307–316.

37. Russell, D.W., T. Yamamoto, W.J. Schneider, C.J. Slaughter, M.S. Brown, and J.L. Goldstein. (1983) cDNA cloning of the bovine low density lipoprotein receptor: Feedback regulation of a receptor mRNA. *Proc. Natl. Acad. Sci. USA 80:* 7501–7505.
38. Palade, G.E. (1953) Fine structure of blood capillaries. *J. Applied Physics 24*: 1424.
39. Anderson, R.G.W., J.L. Goldstein, and M.S. Brown. (1976) Localization of low density lipoprotein receptors on plasma membrane of normal human fibroblasts and their absence in cells from a familial hypercholesterolemia homozygote. *Proc. Natl. Acad. Sci. USA 73:* 2434–2438; Anderson, R.G.W., M.S. Brown, and J.L. Goldstein. (1977) Role of the coated endocytic vesicle in the uptake of receptor-bound low density lipoprotein in human fibroblasts. *Cell 10*: 351–364.
40. Roth, T.F., and K.R. Porter. (1964) Yolk protein uptake in the oocyte of the mosquito *Aedes aegypti. L. J. Cell. Biol. 20:* 313–332.
41. Brown, M.S., and J.L. Goldstein. (1976) Analysis of a mutant strain of human fibroblasts with a defect in the internalization of receptor-bound low density lipoprotein. *Cell 9:* 663–674.
42. Goldstein, J.L., M.S. Brown, and N.J. Stone. (1977) Genetics of the LDL receptor: Evidence that the mutations affecting binding and internalization are allelic. *Cell 12:* 629–641.
43. Anderson, R.G.W., J.L. Goldstein, and M.S. Brown. (1977) A mutation that impairs the ability of lipoprotein receptors to localise in coated pits on the cell surface of human fibroblasts. *Nature 270*: 695-699.
44. Pearse, B.M.F. (1976) Clathrin: A unique protein associated with intracellular transfer of membrane by coated vesicles. *Proc. Natl. Acad. Sci. USA 73*: 1255–1259.
45. Carpenter, G., and S. Cohen. (1976) ^{125}I-labeled human epidermal growth factor: Binding, internalization, and degradation in human fibroblasts. *J. Cell Biol. 71:* 159–171.
46. Terris, S., and D.F. Steiner. (1975) Binding and degradation of ^{125}I-insulin by rat hepatocytes. *J. Biol. Chem. 250:* 8389–8398.
47. Neufeld, E.F., G.N. Sando, A.J. Garvin, and L.H. Rome. (1977) The transport of lysosomal enzymes. *J. Supramol. Struct. 6:* 95–101.
48. Gonzalez-Noriega, A., J.H. Grubb, V. Talkad, and W.S. Sly. (1980) Chloroquine inhibits lysosomal enzyme pinocytosis and enhances lysosomal enzyme secretion by impairing receptor recycling. *J. Cell Biol. 85:* 839–852.
49. Ashwell, G., and J. Harford. (1982) Carbohydrate-specific receptors of the liver. *Ann. Rev. Biochem. 51:* 531–554; Stockert, R.J., D.J., Howard, A.G. Morell, and I.H. Scheinberg. (1980) Functional segregation of hepatic receptors for asialoglycoproteins during endocytosis. *J. Biol. Chem. 255:* 9028–9029.
50. Helenius, A., J. Kartenbeck, K. Simons, and E. Fries. (1980) On the entry of Semliki Forest virus into BHK-21 cells. *J. Cell. Biol. 84:* 404–420.
51. Pastan, I.H., and M.C. Willingham. (1981) Journey to the center of the cell: Role of the receptosome. *Science 214:* 504–509.
52. Carpentier, J-L., P. Gorden, R.G.W. Anderson, J.L. Goldstein, M.S. Brown, S. Cohen, and L. Orci. (1982) Co-localization of ^{125}I-epidermal growth factor and ferritin-low density lipoprotein in coated pits: A quantitative electron microscopic study in normal and mutant fibroblasts. *J. Cell Biol. 95:* 73–77.
53. Maxfield, F.R. (1982) Weak bases and ionophores rapidly and reversibly raise the pH of endocytic vesicles in cultured mouse fibroblasts. *J. Cell. Biol. 95:* 676–681.
54. Marsh, M., E. Bolzau, and A. Helenius. (1983) Penetration of Semliki forest virus from acidic prelysosomal vacuoles. *Cell 32:* 931–940.
55. Schneider, W.J., U. Beisiegel, J.L. Goldstein, and M.S. Brown. (1982) Purification of the low density lipoprotein receptor, an acidic glycoprotein of 164,000 molecular weight. *J. Biol. Chem. 257:* 2664–2673.
56. Cummings, R.D., S. Kornfeld, W.J. Schneider, K.K. Hobgood, H. Tolleshaug, M.S. Brown, and J.L. Goldstein. (1983) Biosynthesis of the N- and O-linked oligosaccharides of the low density lipoprotein receptor. *J. Biol. Chem. 258:* 15261–15273.
57. Davis, C.G., A. Elhammer, D.W. Russell, W.J. Schneider, S. Kornfeld, M.S. Brown, and J.L. Goldstein. (1986) Deletion of clustered O-linked carbohydrates does not impair function of low density lipoprotein receptor in transfected fibroblasts. *J. Biol. Chem. 261:* 2828–2838.

58. Bersot, T.P., R.W. Mahley, M.S. Brown, and J.L. Goldstein. (1976) Interaction of swine lipoproteins with the low density lipoprotein receptor in human fibroblasts. *J. Biol. Chem. 251:* 2395–2398.
59. Innerarity, T.L., and R.W. Mahley. (1978) Enhanced binding by cultured human fibroblasts of apo-E-containing lipoproteins as compared with low density lipoproteins. *Biochemistry 17:* 1440–1447.
60. Tolleshaug, H., J.L. Goldstein, W.J. Schneider, and M.S. Brown. (1982) Posttranslational processing of the LDL receptor and its genetic disruption in familial hypercholesterolemia. *Cell 30:* 715–724.
61. Helenius, A., I. Mellman, D. Wall, and A. Hubbard. (1983) Endosomes. *Trends Biochem. Sci. 8:* 245–250; Pastan, I., and M.C. Willingham. (1983) Receptor-mediated endocytosis: Coated pits, receptosomes and the Golgi. *Trends Biochem. Sci. 8:* 250–254.
62. Brown, M.S., R.G.W. Anderson, S.K. Basu, and J.L. Goldstein. (1981) Recycling of cell surface receptors: Observations from the LDL receptor system. *Cold Spring Harbor Symp. Quant. Biol. 46:* 713–721; Basu, S.K., J.L. Goldstein, R.G.W. Anderson, and M.S. Brown. (1981) Monensin interrupts the recycling of low density lipoprotein receptors in human fibroblasts. *Cell 24*: 493–502.
63. Yamamoto, T., C.G. Davis, M.S. Brown, W.J. Schneider, M.L. Casey, J.L. Goldstein, and D.W. Russell. (1984) The human LDL receptor: A cysteine-rich protein with multiple Alu sequences in its mRNA. *Cell 39:* 27–38.
64. Russell, D.W., W.J. Schneider, T. Yamamoto, K.L. Luskey, M.S. Brown, and J.L. Goldstein. (1984) Domain map of the LDL receptor: Sequence homology with the epidermal growth factor precursor. *Cell 37:* 577–585.
65. Goldstein, J.L., M.S. Brown, R.G.W. Anderson, D.W. Russell, and W.J. Schneider. (1985) Receptor-mediated endocytosis: Concepts emerging from the LDL receptor system. *Ann. Rev. Cell Biol. 1:* 1–39.
66. Südhof, T.C., J.L. Goldstein, M.S. Brown, and D.W. Russell. (1985) The LDL receptor gene: A mosaic of exons shared with different proteins. *Science 228:* 815–822.
67. Schneider, W.J., C.J. Slaughter, J.L. Goldstein, R.G.W. Anderson, D.J. Capra, and M.S. Brown. (1983) Use of anti-peptide antibodies to demonstrate external orientation of NH_2-terminus of the LDL receptor in the plasma membrane of fibroblasts. *J. Cell Biol. 97:* 1635–1640.
68. Mahley, R.W., and T.L. Innerarity. (1983) Lipoprotein receptors and cholesterol homeostasis. *Biochim. Biophys. Acta 737:* 197–222; Innerarity, T.L., K.H. Weisgraber, K.S. Arnold, S.C. Rall, Jr., and R.W. Mahley. (1984) Normalization of receptor binding of apolipoprotein E2: Evidence for modulation of the binding site conformation. *J. Biol. Chem. 259:* 7261–7267.
69. Südhof, T.C., D.W. Russell, J.L. Goldstein, M.S. Brown, R. Sanchez-Pescador, and G.I. Bell. (1985) Cassette of eight exons shared by genes for LDL receptor and EGF precursor. *Science 228:* 893–895.
70. Scott, J., M. Urdea, M. Quiroga, R. Sanchez-Pescador, N. Fong, M. Selby, W.J. Rutter, and G.I. Bell. (1983) Structure of a mouse submaxillary messenger RNA encoding epidermal growth factor and seven related proteins. *Science 221:* 236–240.
71. Gray, A., T.J. Dull, and A. Ullrich. (1983) Nucleotide sequence of epidermal growth factor cDNA predicts a 128,000-molecular weight protein precursor. *Nature 303:* 722–725.
72. Doolittle, R.F., D.-F. Feng, and M.S. Johnson. (1984) Computer-based characterization of epidermal growth factor precursor. *Nature 307:* 558–566.
73. Lehrman, M.A., J.L. Goldstein, M.S. Brown, D.W. Russell, and W.J. Schneider. (1985) Internalization-defective LDL receptors produced by genes with nonsense and frameshift mutations that truncate the cytoplasmic domain. *Cell 41*: 735-753.
74. Francke, U., M.S. Brown, and J.L. Goldstein. (1984) Assignment of the human gene for the low density lipoprotein receptor to chromosome 19: Synteny of a receptor, a ligand, and a genetic disease. *Proc. Natl. Acad. Sci. USA 81:* 2826–2830.
75. Stanley, K.K., H.-P. Kocher, J.P. Luzio, P. Jackson, and J. Tschopp. (1985) The sequence and topology of human complement component C9. *EMBO J. 4:* 375–382.

76. Doolittle, R.F. (1985) The genealogy of some recently evolved vertebrate proteins. *Trends Biochem. Sci. 10:* 233–237.
77. Gilbert, W. (1978) Why genes in pieces? *Nature 271:* 501.
78. Gilbert, W. (1985) Genes-in-pieces revisited. *Science 228:* 823–824.
79. Lehrman, M.A., D.W. Russell, J.L. Goldstein, and M.S. Brown. (1986) Exon-Alu recombination deletes 5 kilobases from low density lipoprotein receptor gene, producing null phenotype in familial hypercholesterolemia. *Proc. Natl. Acad. Sci. USA* In Press.
80. Tolleshaug, H., K.K. Hobgood, M.S. Brown, and J.L. Goldstein. (1983) The LDL receptor locus in familial hypercholesterolemia: Multiple mutations disrupting the transport and processing of a membrane receptor. *Cell 32:* 941–951.
81. Schneider, W.J., M.S. Brown, and J.L. Goldstein. (1983) Kinetic defects in the processing of the LDL receptor in fibroblasts from WHHL rabbits and a family with familial hypercholesterolemia. *Mol. Biol. Med. 1:* 353–367.
82. Davis, C.G., M.A. Lehrman, D.W. Russell, R.G.W. Anderson, M.S. Brown, and J.L. Goldstein. (1986) The J.D. mutation in familial hypercholesterolemia: Substitution of cysteine for tyrosine in cytoplasmic domain impedes internalization of LDL receptors. *Cell 45:* 15–24.
83. Lehrman, M.A., W.J. Schneider, T.C. Südhof, M.S. Brown, J.L. Goldstein, and D.W. Russell. (1985) Mutation in LDL receptor: Alu-Alu recombination deletes exons encoding transmembrane and cytoplasmic domains. *Science 227:* 140–146.
84. Ho, Y.K., M.S. Brown, D.W. Bilheimer, and J.L. Goldstein. (1976) Regulation of low density lipoprotein receptor activity in freshly isolated human lymphocytes. *J. Clin. Invest. 58:* 1465–1474.
85. Bilheimer, D.W., Y.K. Ho, M.S. Brown, R.G.W. Anderson, and J.L. Goldstein. (1978) Genetics of the low density lipoprotein receptor: Diminished receptor activity in lymphocytes from heterozygotes with familial hypercholesterolemia. *J. Clin. Invest. 61:* 678–696.
86. Langer, T., W. Strober, and R.I. Levy. (1972) The metabolism of low density lipoprotein in familial type II hyperlipoproteinemia. *J. Clin. Invest. 51:* 1528–1536.
87. Bilheimer, D.W., N.J. Stone, and S.M. Grundy. (1979) Metabolic studies in familial hypercholesterolemia: Evidence for a gene-dosage effect *in vivo*. *J. Clin. Invest. 64:* 524–533.
88. Simons, L.A., D. Reichl, N.B. Myant, and M. Mancini. (1975) The metabolism of the apoprotein of plasma low density lipoprotein in familial hyperbetalipoproteinaemia in the homozygous form. *Atherosclerosis 21:* 283–298.
89. Bilheimer, D.W., J.L. Goldstein, S.M. Grundy, and M.S. Brown. (1975) Reduction in cholesterol and low density lipoprotein synthesis after portacaval shunt surgery in a patient with homozygous familial hypercholesterolemia. *J. Clin. Invest. 56:* 1420–1430.
90. Basu, S.K., J.L. Goldstein, and M.S. Brown. (1978) Characterization of the low density lipoprotein receptor in membranes prepared from human fibroblasts. *J. Biol. Chem. 253:* 3852–3856.
91. Kovanen, P.T., S.K. Basu, J.L. Goldstein, and M.S. Brown. (1979) Low density lipoprotein receptors in bovine adrenal cortex. II. Low density lipoprotein binding to membranes prepared from fresh tissue. *Endocrinology 104:* 610–616; Brown, M.S., P.T. Kovanen, and J.L. Goldstein. (1979) Receptor-mediated uptake of lipoprotein-cholesterol and its utilization for steroid synthesis in the adrenal cortex. *Recent Prog. Hormone Res. 35:* 215–257.
92. Chao, Y.S., E.E. Windler, G.C.Chen, and R.J. Havel. (1979) Hepatic catabolism of rat and human lipoproteins in rats treated with 17α-ethinyl estradiol. *J. Biol. Chem. 254:* 11360–11366; Kovanen, P.T., M.S. Brown, and J.L. Goldstein. (1979) Increased binding of low density lipoprotein to liver membranes from rats treated with 17α-ethinyl estradiol. *J. Biol. Chem. 254:* 11367–11373.
93. Pittman, R.C., T.E. Carew, A.D. Attie, J.L. Witztum, Y. Watanabe, and D. Steinberg. (1982) Receptor-dependent and receptor-independent degradation of low density lipoprotein in normal rabbits and in receptor-deficient mutant rabbits. *J. Biol. Chem. 257:* 7994–8000.
94. Spady, D.K., D.W. Bilheimer, and J.M. Dietschy. (1983) Rates of receptor dependent and independent low density lipoprotein uptake in the hamster. *Proc. Natl. Acad. Sci. USA. 80:* 3499–3503.

95. Steinberg, D. (1983) Lipoproteins and atherosclerosis: A look back and a look ahead. *Arteriosclerosis 3:* 283–301.
96. Shepherd, J., S. Bicker, A.R. Lorimer, and C.J. Packard. (1979) Receptor mediated low density lipoprotein catabolism in man. *J. Lipid Res. 20:* 999–1006.
97. Basu, S.K., J.L. Goldstein, R.G.W. Anderson, and M.S. Brown. (1976) Degradation of cautionized low density lipoprotein and regulation of cholesterol metabolism in homozygous familial hypercholesterolemia fibroblasts. *Proc. Natl. Acad. Sci. USA 73:* 3178–3182.
98. Mahley, R.W., T.L. Innerarity, R.E. Pitas K.H. Weisgraber J H Brown and E. Gross. (1977) Inhibition of lipoprotein binding to cell surface receptors of fibroblasts following selective modification of arginyl residues in arginine-rich and B apoproteins. *J. Biol. Chem. 252:* 7279–7287; Weisgraber, K.H., T.L. Innerarity, and R.W. Mahley. (1978) Role of the lysine residues of plasma lipoproteins in high affinity binding to cell surface receptors on human fibroblasts. *J. Biol. Chem. 253:* 9053–9062.
99. Goldstein, J.L., and M.S. Brown. (1977) Atherosclerosis: The low-density lipoprotein receptor hypothesis. *Metabolism 26:* 1257–1275.
100. Brown, M.S. and J.L. Goldstein. (1983) Lipoprotein receptors in the liver: Control signals for plasma cholesterol traffic. *J. Clin. Invest. 72:* 743–747.
101. Watanabe, Y. (1980) Serial inbreeding of rabbits with hereditary hyperlipidemia (WHHL-rabbit). Incidence and development of atherosclerosis and xanthoma. *Atherosclerosis 36:* 261–268.
102. Goldstein, J.L., T. Kita, and M.S. Brown. (1983) Defective lipoprotein receptors and atherosclerosis: Lessons from an animal counterpart of familial hypercholesterolemia. *N. Engl. J. Med. 309:* 288–295.
103. Gitlin, D., D.G. Cornwell, D. Nakasato, J.L. Oncley, W.L. Hughes, Jr., and C.A. Janeway. (1958) Studies on the metabolism of plasma proteins in the nephrotic syndrome. II. The lipoproteins. *J. Clin. Invest. 37:* 172–184.
104. Bilheimer, D.W., S. Eisenberg, and R. I. Levy. (1972) The metabolism of very low density lipoprotein proteins. I. Preliminary *in vitro* and *in vivo* observations. *Biochim. Biophys. Acta 260:* 212–221.
105. Kita, T., M.S. Brown, D.W. Bilheimer, and J.L. Goldstein. (1982) Delayed clearance of very low density and intermediate density lipoproteins with enhanced conversion to low density lipoprotein in WHHL rabbits. *Proc. Natl. Acad. Sci. USA 79:* 5693–5697.
106. Soutar, A.K., N.B. Myant, and G.R. Thompson. (1982) The metabolism of very low density and intermediate density lipoproteins in patients with familial hypercholesterolaemia. *Atherosclerosis. 43:* 217–231.
107. Brown, M.S., and J.L. Goldstein. (1981) Lowering plasma cholesterol by raising LDL receptors (editorial). *N. Engl. J. Med. 305:* 515–517.
108. Endo, A., M. Kuroda, and K. Tanzawa. (1976) Competitive inhibition of 3-hydroxy-3-methylglutaryl coenzyme A reductase by ML-236A and ML-236B fungal metabolites, having hypocholesterolemic activity. *F.E.B.S. Lett. 72:* 323–326; Endo, A. (1985) Compactin (ML-236B) and related compounds as potential cholesterol-lowering agents that inhibit HMG-CoA reductase. *J. Medicinal Chem. 28:* 401–405.
109. Alberts, A.W., J. Chen, G. Kuron, V. Hunt, J. Huff, C. Hoffman, J. Rothrock, M. Lopez, H. Joshua, E. Harris, A. Patchett, R. Monaghan, S. Currie, E. Stapley, G. Albers-Schonberg, O. Hensens, J. Hirshfield, K. Hoogsteen, J. Liesch, and J. Springer. (1980) Mevinolin, a highly potent competitive inhibitor of HMG-CoA reductase and cholesterol lowering agent. *Proc. Natl. Acad. Sci. USA 77:* 3957–3961.
110. Kita, T., M.S. Brown, and J.L. Goldstein. (1980) Feedback regulation of 3-hydroxy-3-methylglutaryl coenzyme A reductase in livers of mice treated with mevinolin, a competitive inhibitor of the reductase. *J. Clin. Invest.* 66: 1094–1100; Kovanen, P.T., D.W. Bilheimer, J.L. Goldstein, J.J. Jaramillo, and M.S. Brown. (1981) Regulatory role for hepatic low density lipoprotein receptors *in vivo* in the dog. *Proc. Natl. Acad. Sci. USA 78:* 1194–1198.
111. Grundy, S.M., and D.W. Bilheimer. (1984) Inhibition of 3-hydroxy-3-methylglutaryl-CoA reductase by mevinolin in familial hypercholesterolemia heterozygotes: Effects on cholesterol balance. *Proc. Natl. Acad. Sci. USA 81:* 2538–2542.

112. Mabuchi, H. Sakai, T., Sakai, Y., Yoshimura, A., Watanabe, A., Wakasugi, T., Koizumi, J., and Takeda R. (1983) Reduction of serum cholesterol in heterozygous patients with familial hypercholesterolemia: additive effects of compactin and cholestyramine. *N. Engl. J. Med. 308:* 609–613; Bilheimer, D.W., S.M. Grundy, M.S. Brown, and J.L. Goldstein. (1983) Mevinolin stimulates receptor-mediated clearance of low density lipoprotein from plasma in familial hypercholesterolemia heterozygotes. *Proc. Natl. Acad. Sci. USA. 80*: 4124-4128; Illingworth, D.R. (1984) Mevinolin plus colestipol in therapy for severe heterozygous familial hypercholesterolemia. *Ann. Intern. Med. 101:* 598–604.

113. Yamamoto, A., H. Sudo, and A. Endo. (1980) Therapeutic effects of ML-236B in primary hypercholesterolemia. *Atherosclerosis 35:* 259–266.

114. Thompson, G.R., N.B. Myant, D. Kilpatrick, C.M. Oakley, M.J. Raphael, and R.E. Steiner. (1980) Assessment of long-term plasma exchange for familial hypercholesterolaemia. *Brit. Heart J. 43:* 680–688.

115. Bilheimer, D.W., Goldstein, J.L., Grundy, S.C., Starzl, T.E., and Brown, M.S. (1984) Liver transplantation provides low density lipoprotein receptors and lowers plasma cholesterol in a child with homozygous familial hypercholesterolemia. N. *Engl. J. Med. 311:* 1658–1664.

116. Keys, A. (1980) *Seven countries: A multivariate analysis of death and coronary heart disease.* Harvard University Press, Cambridge, MA. pp. 1–381.

117. Kannel, W.B., W.P. Castelli, T. Gordon, and P.M. McNamara. (1971) Serum cholesterol, lipoproteins, and the risk of coronary heart disease: The Framingham Study. *Ann. Int. Med. 74:* 1–12.

118. Reichl, D., N.B. Myant, M.S. Brown, and J.L. Goldstein. (1978) Biologically active low density lipoprotein in human peripheral lymph. *J. Clin. Invest. 61*: 64–71.

119. Goldstein, J.L., and M.S. Brown. (1982) Lipoprotein receptors: Genetic defense against atherosclerosis. *Clin. Res. 30:* 417–426.

120. Mills, G.L., and C.E. Taylaur. (1971) The distribution and composition of serum lipoproteins in eighteen animals. *Comp. Biochem. Physiol. 40B:* 489–501; Calvert, G.D. (1976) Mammalian low density lipoproteins. In: *Low Density Lipoproteins*, edited by C.E. Day and R.S. Levy, pp. 281–319. Plenum Press, New York.

121. Kwiterovich, P.O., Jr., R.I. Levy, and D.S. Fredrickson. (1973) Neonatal diagnosis of familial type-II hyperlipoproteinaemia. *Lancet i:* 118–122.

122. Connor, S.J., and W.E. Connor. (1984) The interactions of genetic and nutritional factors in hyperlipidemia. In: *Genetic Factors in Nutrition*, edited by A. Velazquez and H. Bourges. Academic Press, Orlando, FL. pp. 137–155; Applebaum-Bowden, D., S.M. Haffner, E. Hartsook, K.H. Luk, J.J. Albers, and W.R. Hazzard. (1984) Down-regulation of the low-density lipoprotein receptor by dietary cholesterol. *Amer. J. Clin. Nutrition 39:* 360–367.

123. Kovanen, P.T., M.S. Brown, S.K. Basu, D.W. Bilheimer, and J.L. Goldstein. (1981) Saturation and suppression of hepatic lipoprotein receptors: A mechanism for the hypercholesterolemia of cholesterol-fed rabbits. *Proc. Matl. Acad. Sci. USA 78:* 1396–1400.

124. Hui, D.Y., T.L, Innerarity, and R.W. Mahley. (1981) Lipoprotein binding to canine hepatic membranes: Metabolically distinct apo-E and apo B,E receptors. *J. Biol. Chem. 256:* 5646–5655; Spady, D.K., S.D. Turley, and J.M. Dietschy. (1985) Rates of low density lipoprotein uptake and cholesterol synthesis are regulated independently in the liver. *J. Lipid Res. 26:* 465–472.

125. Kita, T., J.L. Goldstein, M.S. Brown, Y. Watanabe, C.A. Hornick, and R.J. Havel. (1982) Hepatic uptake of chylomicron remnants in WHHL rabbits: A mechanism genetically distinct from the low density lipoprotein receptor. *Proc. Natl. Acad. Sci. USA 79:* 3623–3627.

126. Sherrill, B.C., and J.M. Dietschy. (1978) Characterization of the sinusoidal transport process responsible for uptake of chylomicrons by the liver. *J. Biol. Chem. 253:* 1859–1867.

127. Lipid Research Clinics Program. (1984) The lipid research clinics coronary primary prevention trial results: I. Reduction in incidence of coronary heart disease. *J. Amer. Med. Assn. 251:* 351–364; Lipid Research Clinics Program. (1984) The lipid research clinics coronary primary prevention trial results: II. The relationship of reduction in incidence of coronary heart disease to cholesterol lowering. *J. Amer. Med. Assn. 251:* 365–374.

128. Bloch, K. (1965) The biological synthesis of cholesterol. *Science 150:* 19–28.

129. Deckelbaum, R.J., G.G. Shipley, and D.M. Small. (1977) Structure and interactions of lipids in human plasma low density lipoproteins. *J. Biol. Chem. 252:* 744–754; Elovson, J., J.C. Jacobs, V.N. Schumaker, and D.L. Puppione. (1985) Molecular weights of apoprotein B obtained from human low-density lipoprotein (apoprotein B-PI) and from rat very low density lipoprotein (apoprotein B-PIII). *Biochemistry 24:* 1569–1578.
130. Brown, M.S., and J.L. Goldstein. (1979) Receptor-mediated endocytosis: Insights from the lipoprotein receptor system. *Proc. Natl. Acad. Sci. USA 76:* 3330–3337.
131. Brown, M.S., and J.L. Goldstein. (1985) The LDL receptor and HMG-CoA reductase – Two membrane molecules that regulate cholesterol homeostasis. *Curr. Topics Cell. Reg. 26:* 3–15.
132. Goldstein, J.L., and M.S. Brown. (1984) Progress in understanding the LDL receptor and HMG CoA reductase, two membrane proteins that regulate the plasma cholesterol. *J. Lipid Res. 25:* 1450–1461.

1986
Physiology or Medicine

STANLEY COHEN
RITA LEVI-MONTALCINI

"for their discoveries of growth factors"

THE NOBEL PRIZE FOR PHYSIOLOGY OR MEDICINE

Speech by Professor KERSTIN HALL of the Karolinska Institute.
Translation from the Swedish text

Your Majesties, Your Royal Highnesses, Ladies and Gentlemen,

We have all been small infants who have grown tall. It is growth hormone, released from the pituitary gland, which regulates growth after birth. Lack of this hormone during infancy results in growth retardation, and a growth hormone deficient man is a minicopy of his potential self. The pituitary growth hormone, however, has no direct growth promoting effect on cells, and growth before birth occurs independently of growth hormone.

We are all derived from one single cell which carries the genetic material coding for all the different characteristics expressed in the thousands of billions of cells present in adult man. This first cell divides into two identical daughter cells. These daughter cells grow and then they too divide. During the cell divisions which follow, the cells begin to express specific characteristics, in other words, they differentiate. The newborn infant already has all the different types of cells found in the adult.

The pattern of growth and differentiation has long been established, but the mechanisms regulating prenatal development remained unknown — growth hormone does not control these events. The discovery of growth factors in tissues other than the pituitary led to a new understanding — growth and differentiation are regulated by signal substances released from cells and acting on neighbouring cells. The first such signal substances to be identified were nerve growth factor (NGF) and epidermal growth factor (EGF). The discovery of NGF by Rita Levi-Montalcini and EGF by Stanley Cohen initiated a new era in the research area of growth and differentiation and was followed by the identification of several other growth factors released by different types of cells.

It all began when the Italian developmental biologist Rita Levi-Montalcini was invited to Viktor Hamburgers laboratory in St. Louis, Missouri. There she repeated a previously performed study, but the conclusion she reached was different. When transplanting mouse tumour to chick embryos she found an outgrowth of certain nerve fibres in the chicken. The nerve outgrowth was similar when the transplantation was performed without direct contact between the tumour and the chick embryo. Rita Levi-Montalcini concluded that the tumour released a substance which promoted nerve growth. She developed a bioassay using cultured nerve cells for identification of the factor.

The biochemist Stanley Cohen joined the research group in the early 1950s. He observed that saliva and salivary gland from the male mouse contained far more NGF than the mouse tumours. He purified NGF from salivary glands and raised antibodies against NGF.

The discovery, identification and isolation of NGF created a breakthrough in the research field of developmental neurobiology. For the first time a chemically well-characterized substance became available for use in studies of nerve growth. Rita Levi-Montalcini showed, in a series of brilliantly performed studies, that NGF is not only necessary for the survival of certain nerves but also regulates the directional growth of the nerve fibres. The nerve cells die when NGF is blocked by antibodies. NGF is produced by the target cells which lure the nerve fibres to grow in the direction. Injections of NGF into the brain cause the outgrowth of specific nerve fibres. This neurotropic effect of NGF offers an explanation of how nerve fibres can find their way through the tangle of nerves in the brain.

Stanley Cohen, who purified NGF, is also the discoverer of epidermal growth factor or EGF. While investigating the effects of NGF he observed that injection of salivary gland extracts to newborn mice accelerated their development. They displayed precocious opening of their eyelids and early eruption of their teeth. Stanley Cohen realized that the salivary gland extracts contained some additional growth factor apart from NGF. He isolated, characterized and established the amino acid sequence of this factor and showed that it accelerated the healing of corneal wounds.

EGF has proven to be a general growth factor with action not only on epithelial cells but also on a large variety of other cells. A prerequisite for its action is the presence of specific binding sites, termed receptors, on the surface of the target cells. Stanley Cohen isolated and characterized the EGF-receptor. He discovered that the receptor consisted of one part on the outside of the cell membrane, which captures EGF, and the other part on the inside of the cell which displays enzyme activity. When EGF binds to the receptor on the outside of the cell it activates this internal enzyme activity. Gradually a new concept has emerged — this type of enzyme activity is a general pathway by which the action of growth factors is initiated. Furthermore, some viral oncogenes cause tumour growth code for proteins with the same kind of enzyme activity as the EGF receptor.

NGF and EGF were discovered in mice, but since then one has moved from mouse to man. The chemical structures of human NGF and EGF are established today, and recombinant human NGF and EGF are produced by DNA-technology. This has opened the way for the use of NGF and EGF in clinical medicine. Deficiency or overproduction of these growth factors may be of importance in the pathogenesis of malformations and errors of development, degenerative changes with regeneration defects, delayed wound healing and tumour diseases. The role of NGF in diseases of the central nervous system, such as senile dementia, and the possibility of using NGF after damage to peripheral nerves are currently being explored. Application of EGF has already been shown to enhance the healing of wounds of cornea, skin and intestine. Autotransplantation of skin rapidly cultivated outside the body with the help of EGF can be used to cover burns.

Rita Levi-Montalcini and Stanley Cohen were the first to discover and isolate growth factors. Their pioneering contributions stimulated the search for other growth factors and several such substances have been characterized today. Their work has opened up a research field of potential importance to future medicine. Rita Levi-Montalcini and Stanley Cohen have advanced our knowledge from a stage when growth and differentiation could only be described as phenomena and growth factors were unknown, to a situation today when the role of growth factors in cell proliferation, organ differentiation, and tumour transformation is generally recognized. Rita Levi-Montalcini is the great developmental biologist who showed how the outgrowth of the nerves was regulated. Stanley Cohen is the brilliant biochemist who purified the first growth factors and improved our understanding about how a growth signal from the outside is relayed into the cells.

As a representative of the Nobel Assembly at the Karolinska Institute, I convey to you the sincere congratulations of the Assembly and ask you now to receive your Prize from the hands of His Majesty the King.

STANLEY COHEN

I was born in Brooklyn in 1922. Both my mother and father were Russian Jewish emigrants who came to America in the early 1900's. My father was a tailor and my mother, a housewife. Though of limited education themselves, they instilled in me the values of intellectual achievement and the use of whatever talents I possessed.

I was educated in the public school system of New York City and was bright enough to be accepted at Brooklyn College. Fortunately for me, my college education was most thorough (I majored in both Biology and Chemistry). Perhaps equally important was the fact that Brooklyn College was a city school and had a policy of no tuition; the cost of an education would have been prohibitive for my parents.

My scientific interests throughout my undergraduate days were directed to cell biology and especially the mysteries of embryonic development. I think my one insight into these problems was the recognition that much could be learned by the application of chemistry to biology.

After working for a short period as a bacteriologist in a milk processing plant to save enough money to go to graduate school, fellowships enabled me to continue my education, first at Oberlin College, where I received an M.A. in Zoology in 1945, and then in the Biochemistry Department at the University of Michigan where I received a Ph.D. in 1948. My Ph.D. thesis concerned the metabolic mechanism by which the end product of nitrogen metabolism in the earthworm is switched from ammonia to urea during starvation. I remember spending my nights collecting over 5,000 worms from the University campus green.

I believe it was my ability to stomach-tube earthworms that convinced Dr. Harry Gordon to offer me my first job in the Pediatrics and Biochemistry Departments of the University of Colorado, where I was involved in metabolic studies of premature infants.

Feeling the need to gain experience with the then emerging application of radioisotope methodology to biological research, I left Colorado and went to Washington University in 1952 to work with Martin Kamen in the Department of Radiology at Washington University as a postdoctoral fellow of the American Cancer Society. I learned isotope methodology while studying carbon dioxide fixation in frog eggs and embryos, and also derived a priceless education participating in the journal club administered by Dr. Arthur Kornberg who had just arrived at Washington University.

In 1953 I became associated with the Department of Zoology under the leadership of Viktor Hamburger at Washington University with a two-fold purpose in mind. I joined with Rita Levi-Montalcini to isolate a Nerve Growth Factor (NGF) that Dr. Levi-Montalcini had discovered in certain mouse tumors and to become educated in the field of experimental embryology. I leave it to Dr. Levi-Montalcini, with whom I am honored to share this Nobel Award, to recount the results of our early collaboration.

I came to Vanderbilt University in 1959 as an Assistant Professor in the Biochemistry Department where I have been ever since, exploring the chemistry and biology of epidermal growth factor (EGF) that is the subject of this lecture.

In 1976 I was appointed an American Cancer Society Research Professor and in 1986 Distinguished Professor. The works recognized by this Nobel Prize are clearly a group effort of achievement as may be seen from the names associated with our publications on EGF. They share in this honor. I have received much recognition during my research career and I am most grateful.

HONORS

Research Career Development Award, National Institutes of Health (1959–1969).
National Paraplegia Foundation's Second Annual William Thomson Wakeman Award (1974).
American Cancer Society Research Professor of Biochemistry (1976).
Earl Sutherland Prize for Achievement in Research, Vanderbilt University (1977).
Albion O. Bernstein, M.D. Award (Medical Society of the State of New York) (1978).
National Academy of Science (1980).
H.P. Robertson Memorial Award, National Academy of Science (1981).
Lewis S. Rosenstiel Award, Brandeis University(1982).
General Motors Cancer Research Foundation, Alfred P. Sloan Award (1982).
Louisa Gross Horwitz Prize, Columbia University (1983).
Distinguished Achievement Award of the UCLA Laboratory of Biomedical and Environmental Sciences (1983).
Lila Gruber Memorial Cancer Research Award, American Academy of Dermatology (1983).
Bertner Award of M.D. Anderson Hospital, University of Texas (1983).
American Academy of Arts and Sciences (1984).
Charles B. Smith Visiting Research Professorship, Sloan Kettering (1984).
Honorary Doctor of Science, University of Chicago (1985).
Gairdner Foundation International Award (1985).
Feodor Lynen Lecturer, University of Miami 18th Miami Winter Symposium (1986).
National Medal of Science (1986).
Steenbock Lecturer, University of Wisconsin (1986).
Fred Conrad Koch Award, The Endocrine Society (1986).
Albert Lasker Basic Medical Research Award, Albert and Mary Lasker Foundation (1986).
Nobel Prize in Physiology and Medicine (1986).

EPIDERMAL GROWTH FACTOR

Nobel lecture, 8 December, 1986

by

STANLEY COHEN

Vanderbilt University, School of Medicine, Nashville, Tennessee, U.S.A.

Introduction

Upon the foundations provided by experimental embryology, endocrinology, cell biology, biochemistry, and molecular biology, the intricacies of the regulatory processes that occur during embryonic development are slowly coming to light. While the importance of "classical" hormones in the control of growth and development has long been recognized, we now know that many more intercellular signals are involved in these highly complex processes. The recent advances in this area have, somewhat unexpectedly, also provided mechanisms that may lead to a more detailed understanding of important biomedical questions, such as the growth behavior of malignant cells.

My own efforts in this area of research over the past thirty years have been directed toward the understanding, on a biochemical level, of two biological observations, both initially made in the Department of Zoology, directed by Dr. Viktor Hamburger, at Washington University.

The first observation was that of Dr. Rita Levi-Montalcini who noted that certain mouse tumors, when implanted into chick embryos, released a factor that stimulated the growth of specific embryonic neurons. The second biological observation was made during my study of the nerve growth factor detected in male mouse submaxillary glands. It was noted [1] that when crude submaxillary gland preparations were injected into newborn mice, unexpected "side effects" not related to the activities of nerve growth factor, were produced. These effects included precocious eyelid opening (6–7 days, compared with 12–14 days for controls) and precocious tooth eruption (5–6 days, compared with 8–10 days for controls).

After I transferred to the Biochemistry Department at Vanderbilt University in 1959, these "side effects" were to become the focus of my research. From my training in embryology, I felt that any substance that altered the timing of specific developmental processes would be of biological significance. I, of course, did not foresee that the biochemical mechanism by which these extracts induced precocious eyelid opening would be related to those involved in oncogenic transformation by one class of retroviruses. This lecture summarizes briefly some of the thoughts and key experiments that have led to our present understanding of epidermal growth factor (EGF).

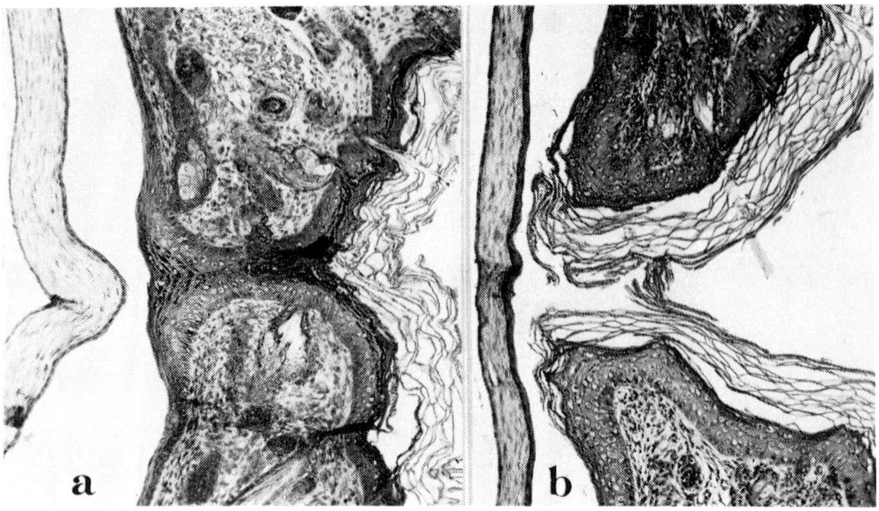

Figure 1a and b. Cross sections of the eyelid area from control, 1a, and experimental, 1b, 8-day-old rats. The experimental animal had received daily injections (1 μg per 1 gm body weight) of the active protein. × 100. (reprinted from *Journal of Investigative Dermatology* (1963) 40, 1–5.)

The First Decade. By employing precocious eyelid opening as an assay, the factor, a small protein, responsible for these effects was isolated from murine submaxillary glands in the early 1960's [2]. Histological examination (Fig. 1) of control and EGF-treated newborn animals (mouse, rat, rabbit) revealed that the observed precocious eyelid separation was the consequence of a more generalized biological effect, namely, an enhancement of epidermal growth and keratinization [3]. The apparent precocious incisor eruption induced by EGF was, in reality, caused by an enhanced differentiation of the lids of the treated animals.

Since these were whole animal experiments, we were faced with the problem of whether the factor operated directly on epidermal cells or whether growth was induced indirectly, possibly by the increased production of a more "classical" hormone.

The techniques of tissue and organ culture seemed ideally suited for resolving this problem. A preliminary organ culture study was carried out during a sabbatical at the Istituto Superiore di Sanità in Rome in collaboration with Drs. Rita Levi-Montalcini and Domenica Attardi, and the study subsequently was continued at Vanderbilt University. The name epidermal growth factor, or EGF, was first used in the initial reports of these studies [4]. The results demonstrated that EGF directly stimulated the proliferation of epidermal cells in organ cultures of chick embryo skin; this mitogenic action of EGF, therefore, did not necessarily depend on other systemic or hormonal influences. During these experiments, the range of responsive animals was widened to include birds as well as mammals, suggesting that knowledge of the evolutionary origins of EGF would contribute to our understanding.

By 1970 we had accumulated a spectrum of information regarding many aspects of the physiology of EGF:

1. We described a series of metabolic alterations (enhancement of polysome formation, induction of ornithine decarboxylase, etc.) that accompany the growth stimulating effects of EGF on epidermal cells. Many of these changes are now known to take place in a variety of cells when a growth stimulus is applied.
2. We identified the tubular cells of the submaxillary gland, which in the mouse exhibit sexual dimorphism, as the major site of synthesis of EGF in this species and noted, with the aid of a radioimmunoassay, that the synthesis of EGF, especially in female mice, was markedly enhanced by the administration of testosterone.
3. We demonstrated that, in crude homogenates of the mouse submaxillary gland, EGF existed as a high molecular weight noncovalent complex (~ 75,000 daltons) consisting of two molecules of EGF and two molecules of an EGF-binding protein that possessed arginyl esterase activity.
4. On a more practical level, we and others found that the topical application of EGF accelerated corneal re-epithelialization in rabbits with wounded corneas.

The reader is referred to a number of early review articles wherein this information is detailed and references provided [5, 6]. By the end of the first decade, I was convinced that EGF plays a normal physiological role in many species, either during embryonic development or in homeostasis; what this role was at the whole animal level and how EGF interacted with cells at the molecular level were problems for the future.

The Second Decade. The development of a rapid, essentially two-step procedure for isolation of milligram quantities of EGF from murine submaxillary glands in the early 1970's [7] permitted the purification of sufficient quantities of mEGF (mouse-derived EGF) for a thorough characterization. This single technical advance opened the door to the application of many biochemical methodologies and insights. Amino acids analysis revealed that mEGF is a 53-residue polypeptide, entirely devoid of alanyl, phenylalanyl, or lysyl residues [8]. The primary sequence of mEGF [9] and the position of the three internal disulphide bonds [10] were determined and are depicted in Fig. 2. Though mEGF has yet to be crystallized and subjected to X-ray diffraction analysis, considerable spectroscopic data have been accumulated suggesting that the hormone has little periodic secondary structure; the presence of β-sheet structures have been detected (reviewed in (11)).

At about this time (1973) a new facet of the biology of EGF was uncovered. Armelin [12] and Hollenberg and Cuatrecasas [13] were the first to report that fibroblasts in culture responded to EGF with enhanced DNA synthesis. These findings were corroborated in our laboratory with human fibroblasts [14, 15].

The finding that mouse-derived EGF was a potent mitogen for human cells indicated that receptors for EGF were present on human cells and, therefore, a polypeptide similar to EGF might be found in human tissue. We took two approaches in an attempt to isolate EGF-like molecules from human urine.

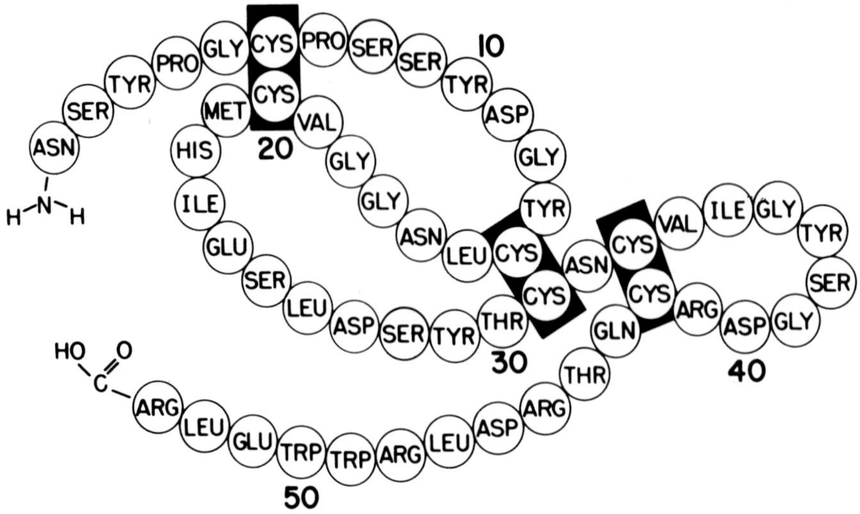

Figure 2. The amino acid sequence of EGF with placement of disulfide bonds. (Reprinted from *Journal of Biological Chemistry* (1973) 248, 7669–7672.)

First, an immuno-affinity column procedure (using anti-mouse EGF antibodies) was used to purify partially a substance from human urine that was similar to the mouse hormone in its biological activity [16]. In another approach, we developed a sensitive and specific radioreceptor competitive binding assay for EGF-related polypeptides, using cultured human fibroblasts and ^{125}I-labelled mouse EGF, that permitted the isolation of microgram quantities of pure growth factor from protein concentrates of human urine [17]. The biological effects of the purified human polypeptide were qualitatively identical to those previously described for the mouse growth factor. These included the stimulation of the proliferation *in vitro* of fibroblasts and corneal epithelial cells, as well as the induction of precocious eyelid opening in the newborn mouse, which still remains the most specific biological assay for EGF. The amino acid compositions of the human and mouse polypeptides differed, but clear similarities were noted. Both polypeptides apparently competed for the same site on the cell membrane and antibodies to the mouse polypeptide crossreacted with the human hormone. We concluded that we had isolated the human counterpart of murine EGF.

As is usual in science, an unexpected and completely new aspect of the biology of EGF emerged with the report by Gregory [18] that urogastrone, a gastric antisecretory hormone isolated from human urine, appeared to be identical to human EGF and closely related to murine EGF. Human EGF (urogastrone) and murine EGF are now believed to invoke identical response in all target cells. The relationship between human EGF and urogastrone could only have been detected from a structural comparison of these molecules; even today, no rationale is available to connect inhibition of acid secretion and stimulation of cell growth.

Given a cell culture system (human fibroblasts) in which EGF acted as a "growth factor" we were faced in 1975 with a rather formidable task: how does EGF stimulate cell growth? Although neuronal uptake and retrograde transport of nerve growth factor had been demonstrated in 1974 [19], almost all endocrinologists were of the opinion that peptide hormones, after binding to their receptors on the plasma membrane, were released into the extracellular environment.

Our initial experiments [20] utilizing ^{125}I-EGF and human fibroblasts confirmed the presence of plasma membrane receptors for EGF. Two additional and significant observations were made. First, the binding of ^{125}I-EGF to the cell surface of intact fibroblasts was rapidly followed by proteolytic degradation of the growth factor by a cell-mediated process. Secondly, it was noted that NRK cells lost their ability to bind ^{125}I-EGF following transformation by the Kirsten virus. The former observation directed us to an examination of the possibility that cell-bound EGF was internalized prior to degradation [21]. The latter observation was later generalized to include a variety of cells transformed by retroviruses [22] and eventually led George Todaro and others to isolate the EGF-related polypeptide, α-transforming growth factor, and to propose the autocrine hypothesis [62].

As a step in defining the biochemical events that occur during and subsequent to the interaction of EGF with the cell surface, we examined the metabolic fate of the bound hormone. We came to the conclusion (21) that subsequent to the initial binding of ^{125}I EGF to specific plasma membrane receptors, the EGF:receptor complex is internalized and the hormone is ultimately degraded in lysosomes. These conclusions, drawn from studies of the interaction of ^{125}I-EGF with human fibroblasts, were based on the following series of observations:

1. Cell-bound ^{125}I-EGF was rapidly degraded to [^{125}I]monoiodotyrosine at 37°. At 0°, however, cell-bound ^{125}I-EGF was not degraded, but slowly dissociated from the cell surface.
2. When the binding of ^{125}I-EGF was first carried out at 37° and the cells then incubated at 0°, almost no release of cell-bound radioactivity was detected.
3. Degradation of ^{125}I-EGF, but not binding, required metabolic energy.
4. The degradation was blocked by drugs that inhibit lysosomal function, such as chloroquine and ammonium chloride.
5. When ^{125}I-EGF was bound to cells at 0°, the hormone was readily accessible to surface reactive agents, such as trypsin and antibodies to EGF. However, when the hormone was bound to cells at 37° it was much less accessible to either of these reagents.
7. Exposure of fibroblasts to EGF resulted in an apparent loss of plasma membrane receptors for EGF.

Taken together, these observations, which have subsequently been extented by others to a number of polypeptide hormones, provided quantitative biochemical evidence for a complex mechanism through which cells interact with extracellular regulatory signals.

The challenge of direct visualization of the internalization of EGF was

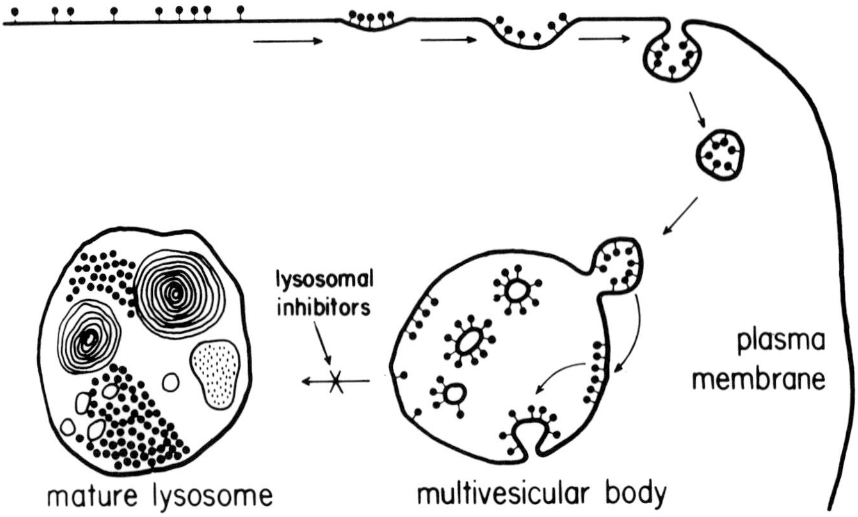

Figure 3. Diagram of F-EGF interaction with A-431 cells. F-EGF-receptor complexes, identified by the characteristic spatial relationship of particles and membrane (4- to 6-nm separation), are apparent at initial binding and are preserved through the processes of clustering, pinocytosis, and incorporation into MVBs. Further incubation at 37°C allows disruption of the F-EGF-receptor complex (attested by pools of free ferritin), a process blocked by the presence of amines. (reprinted from *Proceeding of National Academy of Science USA* (1979) 76, 5689–5693.)

approached using three general procedures: the preparation and tracing of fluorescent derivatives of EGF [23, 24], the tracing of ^{125}I-EGF by electron microscopic autoradiography [25], and the preparation and tracing of EGF-ferritin conjugates by electron microscopy [26, 27].

Although all three morphological approaches confirmed the original biochemical studies with ^{125}I-EGF [21], the use of the biologically active EGF-ferritin conjugate (F-EGF) provided the clearest and most direct picture of the metabolic fate of EGF. At 4° the EGF-ferritin conjugate specifically bound to the plasma membrane of cells and appeared to be randomly distributed on specific receptor sites. When the cells were warmed to 37°, the EGF-ferritin redistributed on the surface of the plasma membrane within one minute to form many small clusters. The clusters of receptor-bound EGF-ferritin molecules were then rapidly internalized into endocytic vesicles. Within 30 min approximately 84% of the ferritin was seen in multivesicular bodies that were considered to be lysosome-related. These data also provided morphological evidence for the hypothesis that "down-regulation" of surface receptors for EGF involves internalization of intact hormone-receptor complexes. A diagram that illustrates our conclusions is presented in Fig. 3. It was subsequently demonstrated, by metabolic labelling and immunoprecipitation with anti-receptor antibodies, that EGF-mediated internalization of the EGF:receptor complex is associated not only with the degradation of EGF but also with enhanced degradation of the receptor [28].

A critical question in this area of hormone research is whether the intracellular processing of hormones and their receptors is related to, or necessary for, the generation of biological responses to the hormone. My opinion is that no clear experimental evidence exists to answer this very important question.

In view of our inability to define the relevance of receptor-mediated internalization to the growth factor's biological activity and our belief that cellular alterations induced by EGF result from the amplification and propagation of a series of "signals" generated during the binding and internalization of the hormone, we sought, in the late 1970's, to obtain a cell-free system that responded *in vitro* to the addition of EGF. Since the A-431 human epidermoid carcinoma cell line had been shown to have an extraordinarily high concentration of EGF receptors, $2-3 \times 10^6$ receptors/cell [29, 24], we utilized a membrane preparation from these cells to look for an EGF-dependent alteration of membrane structure and/or function. Like the technical turning point that the rapid purification of milligram quantities of EGF provided (see above), the identification of the A-431 cell line as an enriched source of EGF receptors was instrumental for both biochemical and molecular biological studies of the mechanism of action of EGF.

As expected, membranes from these cells were able to specifically bind relatively large quantities of ^{125}I-EGF. Since phosphorylation and dephosphorylation reactions participate in the control of many metabolic pathways and membranes contain endogenous protein kinases and phosphatases, a study [30] was initiated to assess the possible role of EGF as a modulator of these regulatory processes. Aliquots of the A-431 membrane preparation were examined for their ability to phosphorylate endogenous membrane components and to determine whether the binding of EGF resulted in a perturbation of this biochemical system. The incubation of A-431 membranes at 0° with [γ-^{32}P]ATP in the presence of Mg^{++} or Mn^{++}, resulted in the incorporation of radioactivity into trichloroacetic acid-insoluble material. Of key importance was the discovery that the prior addition of EGF to the reaction mixture resulted in a 3-fold enhancement of the phosphorylation of endogenous membrane-associated protein (Fig. 4).

The enchanced incorporation of ^{32}P into the membrane preparations was specific for EGF; the major phosphorylated membrane components detected were proteins having molecular weights of ~170,000 and 150,000. The addition of EGF to A-431 membrane preparations stimulated the phosphorylation of not only endogenous membrane proteins, but also a number of exogenously added protein substrates.

It was suggested at that time (1978) that the phosphorylation of membrane or membrane-associated components might be an initial event in the generation of intracellular signals that regulate cell growth. The reader is referred to a review (31) that summarized our knowledge as of 1979.

By the end of this second decade, I was encouraged and excited by the prospect that we had made a significant inroad into the understanding of the mechanism of action of EGF at the cellular and biochemical level.

The Third Decade. The detection of a direct effect of EGF on a chemical

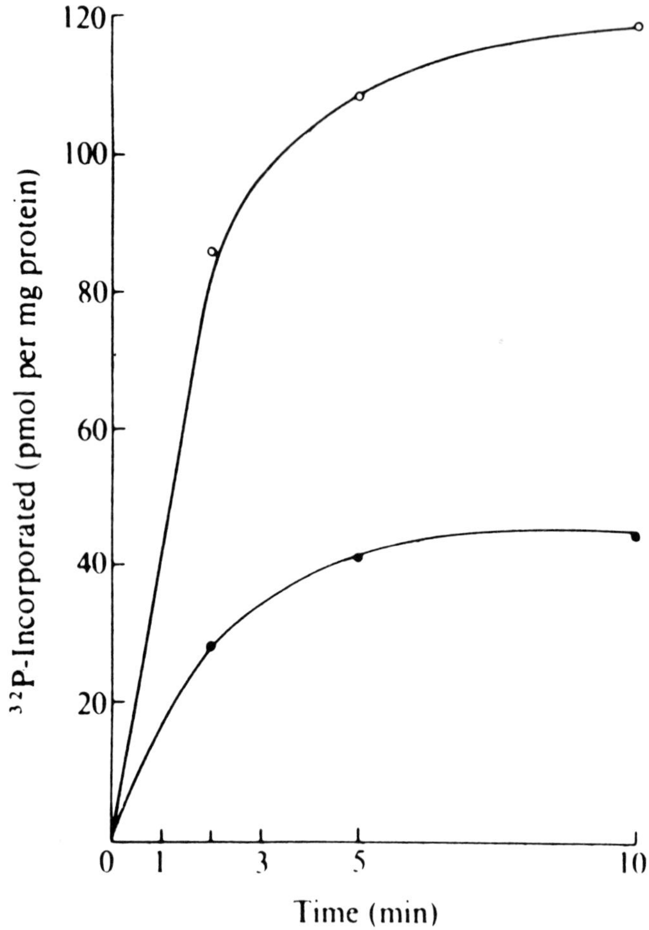

Figure 4. Stimulation of EGF of the incorporation of ^{32}P-hosphate from [γ-^{32}P] ATP into cell membranes. The reaction mixtures contained A-431 membrane (27 μg protein), HEPES buffer (20 mM, pH 7.4), $MnCl_2$ (1 mM), [γ-^{32}P]ATP (15 μM, 8×10^5 c.p.m.), EGF (40 ng) and BSA (7.5 μg) in a final volume of 60 μl. The reaction tubes were placed on ice and preincubated for 10 min in the presence (0) or absence (●) of EGF. The reaction was initiated by the addition of labelled ATP and incubation at 0°C was continued for the indicated times. The reaction was terminated by pipetting 50-μl aliquots on to squares (2 cm) of Whatman 3 MM filter paper and immediately dropping the paper into a beaker of cold 10% TCA containing 0.01 M sodium pyrophosphate. The filter papers were washed extensively with pyrophosphate-containing 10% TCA at room temperature, extracted with alcohol and ether, dried, and the radioactivity measured in a Nuclear Chicago gas-flow counter. (reprinted from *Nature* (1978) 276, 408–410.)

reaction in a cell-free system led to a more detailed biochemical characterization of the reaction in A-431 membranes and to its extension to membrane preparations from normal human placenta and cultured human fibroblasts [32, 33, 34]. The EGF-stimulated kinase activity of A-431 membranes was not removed by extraction of the membranes with a variety of solutions, such as high salt or urea, suggesting that the kinase, receptor, and substrates were integral membrane proteins. We, at this time, were aware of reports from several laboratories [35, 36, 37] that the molecular weight of the putative receptor for EGF, as detected by crosslinking with ^{125}I-EGF, was in the range of our major phosphorylated membrane glycoproteins, i.e. 150–170 kDa.

Our studies provided the following data concerning the mechanism by which EGF regulated protein phosphorylation in the cell-free A-431 membrane system:

1. Activation of the membrane-associated kinase activity by EGF was a rapid process, even at 0°.
2. Dephosphorylation reactions in the membrane also occurred with great rapidity, but were not affected by the presence of EGF.
3. EGF does not cause the release from the membrane of either a soluble protein kinase or modulator of the kinase.
4. The EGF-induced activation of the membrane kinase could be reversed by removal of the hormone from the membrane by anti-EGF IgG, indicating that proteolytic activation was not involved.

We originally assumed, based on data that clearly indicated different heat sensitivities of the receptor and the kinase [32], that at least two separate entities were involved. However, the possibility that the receptor and the kinase activities were present in the same molecule was raised by two unexpected observations. First, A-431 membranes could be solubilized by detergents with retention of EGF-enhanced phosphorylation activity as well as ^{125}I-EGF binding activity. Second, the EGF-receptor, the EGF-dependent kinase activity, as well as the substrate, were co-purified by EGF-affinity chromatography as a major 150 kDa protein.

We originally had reported [32] that the EGF-stimulated protein kinase phosphorylated mainly threonine residues, and in this and other regards it resembled the kinase activity of the transforming protein of Rous sarcoma virus (RSV). Soon thereafter, however, it was reported that the RSV-associated protein kinase and tumor virus-associated protein kinases [38, 39] actually phosphorylated tyrosine residues, originally mistakenly identified as threonine due to co-migration of the two phosphorylated amino acids in the electrophoretic system employed. Since we had employed a similar electrophoretic system, we reinvestigated the nature of the EGF-stimulated protein kinase reaction and discovered that the affinity purified EGF-activated receptor-kinase also phosphorylated tyrosine residues [40].

To determine whether the EGF receptor associated kinase activity might be due to trace contamination with pp60^{src} we looked for an interaction of the EGF receptor kinase preparations with pp60^{src} antisera. All though the receptor kinase was able specifically to phosphorylate these anti-*src* antibodies, the receptor-kinase was not precipitated by such antisera [41, 42]. We interpreted these results to mean that while the EGF receptor-kinase might be related to pp60^{src}, the two kinases were not identical.

We then attempted to purify further the EGF receptor [43]. When the previous affinity purification procedure [34] was applied to A-431 membrane vesicles in the absence of Ca^{2+}, the receptor-kinase was isolated as a 170 kDa protein. The 150 kDa receptor protein initially observed in preparations from scraped membranes has been shown to be a proteolytic fragment of the 170 kDa native species, produced by the action of a Ca^{2+}-dependent neutral protease [44, 45].

The receptor properties of both the 170 kDa and 150 kDa preparations were demonstrated not only by their capacities to bind ^{125}I-EGF, but also by covalent crosslinking to ^{125}I-EGF. The major functional difference between the 170 and 150 kDa preparations appeared to be the ability of the former to "autophosphorylate" at a rate of 5 to 10 times greater than the latter. This observation is understandable since it is now known that the major autophosphorylated tyrosine residues are located near the carboxy-terminus of the 170 kDa receptor and are not present in the 150 kDa proteolytic fragment.

We addressed the question of whether the three domains present in our receptor preparation (binding, kinase, and substrate) reside in one or more than one molecule by applying more stringent purification procedures. The three domains remained associated not only following EGF affinity chromatography but also lentil lectin-Sepharose chromatography, indicating that both the receptor and the kinase were associated with lectin-reactive carbohydrate groups. More importantly, the three detectable domains remained associated following electrophoresis in nondenaturing gels and immunoprecipitation using antisera to the purified receptor. Although these results were not considered definitive, they encouraged speculation that all of the domains were present in the same molecule.

The question was resolved by a series of experiments designed to identify the EGF-stimulated kinase by affinity labelling [46, 47]. When A-431 membrane vesicles were treated with 5'-p-FSO_2BzAdo, a reagent previously shown to affinity label ATP or ADP binding sites in a variety of enzymes, the EGF-stimulated kinase was irreversibly inhibited. When A-431 membrane vesicles were labelled with 5'-p-FSO_2Bz[^{14}C]Ado and then subjected to SDS-polyacrylamide gel electrophoresis and autoradiography, most of the covalently attached affinity label migrated with the 170 kDa receptor and its 150 kDa proteolytic fragment. The labelling observed was at an ATP binding site, as it was inhibited by AMP-PNP, a hydrolysis-resistant ATP analog, but not by adenosine, AMP, ADP, GTP, or NAD. Furthermore, after labelling A-431 vesicles or scraped membranes with 5'-p-FSO_2Bz[^{14}C]Ado, the receptor was affinity purified under conditions previously shown to co-purify the receptor and the kinase. The receptor was the only component of the purified preparation which contained detectable affinity label. Thus, we concluded that the receptor and the kinase are two domains of the same polypeptide. Lastly, if the EGF-sensitive kinase activity was inactivated, by mild heating or exposure to n-ethylmaleimide, the 150 kDa and 170 kDa receptor species could not be labelled with the ATP affinity reagent. The mechanism by which binding of EGF to the external domain of the receptor activates the cytoplasmic catalytic domain is not yet known [11, 48].

That the EGF receptor is a glycoprotein was first suggested by observations that a variety of lectins inhibit the binding of ^{125}I-EGF to cultured human fibroblasts (49) or to human placental membranes [50] and that the receptor may be purified by lectin chromatography [51]. The biosynthesis and glycosylation of the receptor in A-431 cells have recently been addressed in several studies in which cells have been metabolically labelled and the receptor species identified by immunoprecipitation [48, 52].

It is not possible to consider here the thousands of reports regarding EGF and its receptor in biology and medicine. The reader is referred to several recent reviews that summarize various aspects of this ever burgeoning area [11, 48, 53, 54].

I now, very briefly, indicate some of the major advances made in other laboratories throughout the world that I believe will lead to a more complete understanding of the role of EGF and its receptor/kinase in growth regulation.

(1) The elucidation of the amino acid sequence of the EGF receptor, deduced from the nucleotide sequence of cDNA clones, and the discovery that the *erb*-B transforming gene of avian erythroblastosis virus probably is derived from the avian EGF receptor [55].

(2) The elucidation of the nucleotide sequence of the cDNA for prepro EGF which predicts a 128,000 molecular weight protein precursor [56, 57]. The EGF precursor may be a membrane-spanning protein, conceivably a receptor for an as yet unknown ligand. Of great interest in this regard has been the detection of preproEGF in the kidney [58], the detection of two EGF-related loci (in *Drosophila* and in *Caenorhabditis*) that regulate development [59, 60], and the detection of an EGF-related sequence in the genome of the vaccinia virus (61). These findings suggest that EGF is of ancient origin and may have been used for a variety of functional roles.

(3) The discovery that both fetal and malignant cells produce an EGF-related protein (α-TGF) that appears to interact with the EGF receptor and mimics the biological activity of EGF [62]

(4) The discovery that the receptors for insulin as well as a number of other growth factors are ligand activated tyrosine kinases [48, 54].

Although our current working hypothesis is that the initial functional signal transmitted by EGF is related to the tyrosine kinase activity of its receptor, the exact pathway of growth activation, expecially between the receptor and cell nucleus, remains elusive. This is true not only for EGF, but also for the other growth factors whose receptors are tyrosine kinases as well as those oncogenes whose products are tyrosine kinases.

Where do we go from here? — Do we look for specific cellular proteins whose functions are altered by tyrosine phosphorylation? Is the intracellular translocation of tyrosine kinases of physiological significance? Is it possible that autophosphorylated receptors or related oncogene proteins serve some still unidentified regulatory role? What are the mechanisms for sending stimulatory or inhibitory signals to the nucleus? What is the normal physiological role of EGF during development and homeostasis? The answers to these and a host of other questions must be found before we can fully comprehend this important regulatory system.

ACKNOWLEDGEMENTS

To the many colleagues, students and technical assistants who have contributed to these investigations, I am most grateful. I also wish to acknowledge, with gratitude, the support of the National Institutes of Health and the American Cancer Society.

REFERENCES

1. Cohen, S. (1960) *Proc. Natl. Acad. Sci. USA* 46, 302–311.
2. Cohen, S. (1962) *J. Biol. Chem.* 237, 1555–1562.
3. Cohen, S. and Elliott, G. A. (1963) *J. Invest. Dermatol.* 40, 1–5.
4. Cohen, S. (1965) *Dev. Biol.* 12, 394–407.
5. Cohen, S. and Taylor, J. M. (1974) in *Recent Progress in Hormone Research,* Vol. 30 pp. 533–550 (R. O. Greep, ed.) Academic Press.
6. Cohen, S. and Savage, C. R., Jr. (1974) in *Recent Progress in Hormone Research,* Vol. 30, pp. 551–574 (R. O. Greep, ed.) Academic Press.
7. Savage, C. R., Jr., and Cohen, S. (1972) *J. Biol. Chem.* 247, 7609–7611.
8. Taylor, J. M., Mitchell, W. M. and Cohen, S. (1972) *J. Biol. Chem.* 247, 5928–5934.
9. Savage, C. R., Jr., Inagami, T. and Cohen, S. (1972) *J. Biol. Chem.* 247, 7612–7621.
10. Savage, C. R., Jr., Hash, J. H. and Cohen, S. (1973) *J. Biol. Chem.* 248, 7669–7672.
11. Staros, J. V., Cohen, S., and Russo, M. W. (1985) in *Molecular Mechanisms of Membrane Signalling* pp. 253–277 (Cohen and Houslay, eds.) Elsevier.
12. Armelin, H. (1973) *Proc. Natl. Acad. Sci. USA* 70, 2702–2706.
13. Hollenberg, M. D. and Cuatrecasas, P. (1973) *Proc. Natl. Acad. Sci. USA* 70, 2964–2968.
14. Cohen, S., Carpenter, G. and Lembach, K. J., (1975) *Adv. Metab. Disord.* 8, 265–284.
15. Carpenter, G. and Cohen, S. (1976) *J. Cell. Physiol.* 88, 227–237.
16. Starkey, R. H., Cohen, S., and Orth, D. N. (1975) *Science* 189, 800–802.
17. Cohen, S. and Carpenter, G. (1975) *Proc. Natl. Acad. Sci. USA* 72, 1317–1321.
18. Gregory H. (1975) *Nature* (London) 257, 325–327.
19. Hendry, I. A., Stoeckel, K., Thoenen, H. and Iverson, L. L. (1974) *Brain Res.* 68, 103–121.
20. Carpenter, G., Lembach, K. J., Morrison, M. M., and Cohen, S. (1975) *J. Biol. Chem.* 250, 4297–4304.
21. Carpenter, G. and Cohen, S. (1976) *J. Cell Biol.* 71, 159–171.
22. Todaro, G. J., DeLarco, J. E. and Cohen, S. (1976) *Nature* 264, 26–31.
23. Schlessinger, J., Shecter, Y., Willingham, M. C. and Pastan, I., (1978) *Proc. Natl. Acad. Sci. USA* 75, 2659–2663.
24. Haigler, H. T., Ash, J. F., Singer, S. J. and Cohen, S. (1978) *Proc. Natl. Acad. Sci. USA* 75, 3317–3321.
25. Gordon, P., Carpentier, J.-L., Cohen, S. and Orci, L. (1978) *Proc. Natl. Acad. Sci. USA* 75, 5025–5029.
26. Haigler, H. T., McKanna, J. A. and Cohen, S. (1979) *J. Cell Biol.* 81, 382–395.
27. McKanna, J. A., Haigler, H. T. and Cohen, S. (1979) *Proc. Natl. Acad. Sci. USA* 76, 5689–5693.
28. Stoscheck, C. M. and Carpenter, G. (1984) *J. Cell Biol.* 98, 1048–1053.
29. Fabricant, R. N., DeLarco, J. E. and Todaro, G. J. (1977) *Proc. Natl. Acad. Sci. USA* 74, 565–569.
30. Carpenter, G., King, L., Jr. and Cohen, S. (1978) *Nature (London)* 276, 408–410.
31. Carpenter, G. and Cohen, S. (1979) *Ann. Rev. Biochem.* 48, 193–216.
32. Carpenter, G., King, L., Jr., and Cohen, S. (1979) *J. Biol. Chem.* 254, 4884–4891.

33. King, L., Jr., Carpenter, G., and Cohen, S. (1980) *Biochemistry* 19, 1524–1528.
34. Cohen, S., Carpenter, G., and King, L., Jr. (1980) *J. Biol. Chem.* 255, 4834–4842.
35. Das, M., Miyakawa, T., Fox, C. F., Pruss, R. M., Aharonov, A., and Herschman, H. R. (1977) *Proc. Natl. Acad. Sci.USA* 74, 2790–2794.
36. Wrann, M. M. and Fox, C. F. (1979) *J. Biol. Chem.* 254, 8083–8086.
37. Hock, R. A., Nexo, E., and Hollenberg, M. D. (1979) *Nature* 277, 403–405.
38. Hunter, T., and Sefton, B. M. (1980) *Proc. Natl. Acad. Sci. USA* 77, 1311–1315.
39. Witte, O. N., Dasgupta, A., and Baltimore, D. (1980) *Nature* 283, 826–831.
40. Ushiro, H. and Cohen, S. (1980) *J. Biol. Chem.* 255, 8363–8365.
41. Chinkers, M. and Cohen, S. (1981) *Nature (London)* 290, 516–519.
42. Kudlow, J. E., Buss, J. E. and Gill, G. N. (1981) *Nature (London)* 290, 519–521.
43. Cohen, S., Ushiro, H., Stoscheck, C. M. and Chinkers, M. (1982) *J. Biol. Chem.* 257, 1523–1531.
44. Gates, R. E. and King, L. E., Jr. (1982) *Mol. Cell. Endocrinol.* 27, 263–276.
45. Cassel, D. and Glaser, L. (1982) *J. Biol. Chem.* 257, 9845–9848.
46. Buhrow, S. A., Cohen, S. and Staros, J. V. (1982) *J. Biol. Chem.* 257, 4019–4022.
47. Buhrow, S. A., Cohen, S., Garbers, D. L. and Staros, J. V. (1983) *J. Biol. Chem.* 258, 7824–7827.
48. Carpenter, G. (1987) *Ann. Rev. Biochem.* 56, (in press).
49. Carpenter, G. and Cohen, S. (1977) *Biochem. Biophys. Res. Com.* 79, 545–552.
50. Hock, R. A., Nexo, E. and Hollenberg, M. D. (1979) *Nature* 277, 403–405.
51. Cohen, S., Fava, R. A. and Sawyer, S. T. (1982) *Proc. Natl. Acad. Sci. USA* 79, 6237–6241.
52. Soderquist, A. M. and Carpenter, G. (1986) *J. Membrane Biol.* 90, 97–105.
53. Carpenter, G. and Zendegui, J. G. (1986) *Exptl. Cell. Res.* 164, 1–10.
54. Hunter, T. and Cooper, J. A. (1985) *Ann. Rev. Biochem.* 54, 897–930.
55. Ullrich, A., Coussens, L., Hayflick, J. S., Dull, T. J., Gray, A., Tam, A. W., Lee, J., Yarden, Y., Liberman, T. A., Schlessinger, J., Downward, J., Mayes, E. L. V., Whittle, N., Waterfield, M. D. and Seeburg, P. H. (1984) *Nature* 309, 418–425.
56. Scott, J., Urdea, M., Quiroga, M., Sanchez-Pescador, R., Fong, N., Shelby, M., Rutter, W. J. and Bell, G. I. (1983) *Science* 221, 236–240.
57. Gray, A., Dull, T. J. and Ulllrich, A. (1983) *Nature* 303, 722–725.
58. Rall, L. B., Scott, J., Bell, G. I., Crawford, R. J., Penschow, J. D., Niall, H. D. and Coghlaw, J. P. (1985) *Nature* 313, 228–231.
59. Wharton, K. A., Johansen, K. M., Xu, T. and Artavanis-Tsakonas, S. (1985) *Cell* 43, 567–581.
60. Greenwald, D. (1985) *Cell* 43, 583–590.
61. Twardzik, D. R., Brown, J. P., Rauchalis, J. E., Todaro, G. J. and Moss, B. (1985) *Proc. Natl. Acad. Sci. USA* 82, 5300–5304.
62. Sporn, M. B. and Roberts, A. B. (1985) *Nature* 313, 745–747.

Rita Levi Montalcini

RITA LEVI-MONTALCINI

My twin sister Paola and I were born in Turin on April 22, 1909, the youngest of four children. Our parents were Adamo Levi, an electrical engineer and gifted mathematician, and Adele Montalcini, a talented painter and an exquisite human being. Our older brother Gino, who died twelve years ago of a heart attack, was one of the most well known Italian architects and a professor at the University of Turin. Our sister Anna, five years older than Paola and myself, lives in Turin with her children and grandchildren. Ever since adolescence, she has been an enthusiastic admirer of the great Swedish writer, the Nobel Laureate Selma Lagerlöf, and she infected me so much with her enthusiasm that I decided to become a writer and describe Italian saga "à la Lagerlöf". But things were to take a different turn.

The four of us enjoyed a most wonderful family atmosphere, filled with love and reciprocal devotion. Both parents were highly cultured and instilled in us their high appreciation of intellectual pursuit. It was, however, a typical Victorian style of life, all decisions being taken by the head of the family, the husband and father. He loved us dearly and had a great respect for women, but he believed that a professional career would interfere with the duties of a wife and mother. He therefore decided that the three of us—Anna, Paola and I—would not engage in studies which open the way to a professional career and that we would not enroll in the University.

Ever since childhood, Paola had shown an extraordinary artistic talent and father's decision did not prevent her full-time dedication to painting. She became one of the most outstanding women painters in Italy and is at present still in full activity. I had a more difficult time. At twenty, I realized that I could not possibly adjust to a feminine role as conceived by my father, and asked him permission to engage in a professional career. In eight months I filled my gaps in Latin, Greek and mathematics, graduated from high school, and entered medical school in Turin. Two of my university colleagues and close friends, Salvador Luria and Renato Dulbecco, were to receive the Nobel Prize in Physiology or Medicine, respectively, seventeen and eleven years before I would receive the same most prestigious award. All three of us were students of the famous Italian histologist, Giuseppe Levi. We are indebted to him for a superb training in biological science, and for having learned to approach scientific problems in a most rigorous way at a time when such an approach was still unusual.

In 1936 I graduated from medical school with a summa cum laude degree in Medicine and Surgery, and enrolled in the three year specialization in neurology and psychiatry, still uncertain whether I should devote myself fully to the medical profession or pursue at the same time basic research in neurology. My perplexity was not to last too long.

In 1936 Mussolini issued the "Manifesto per la Difesa della Razza", signed by ten Italian 'scientists'. The manifesto was soon followed by the promulgation of laws barring academic and professional careers to non-Aryan Italian citizens. After a short period spent in Brussels as a guest of a neurological institute, I returned to Turin on the verge of the invasion of Belgium by the German army, Spring 1940, to join my family. The two alternatives left then to us were either to emigrate to the United States, or to pursue some activity that needed neither support nor connection with the outside Aryan world where we lived. My family chose this second alternative. I then decided to build a small research unit at home and installed it in my bedroom. My inspiration was a 1934 article by Viktor Hamburger reporting on the effects of limb extirpation in chick embryos. My project had barely started when Giuseppe Levi, who had escaped from Belgium invaded by Nazis, returned to Turin and joined me, thus becoming, to my great pride, my first and only assistant.

The heavy bombing of Turin by Anglo-American air forces in 1941 made it imperative to abandon Turin and move to a country cottage where I rebuilt my mini-laboratory and resumed my experiments. In the Fall of 1943, the invasion of Italy by the German army forced us to abandon our now dangerous refuge in Piemonte and flee to Florence, where we lived underground until the end of the war.

In Florence I was in daily contact with many close, dear friends and courageous partisans of the "Partito di Azione". In August of 1944, the advancing Anglo-American armies forced the German invaders to leave Florence. At the Anglo-American Headquarters, I was hired as a medical doctor and assigned to a camp of war refugees who were brought to Florence by the hundreds from the North where the war was still raging. Epidemics of infectious diseases and of abdominal typhus spread death among the refugees, where I was in charge as nurse and medical doctor, sharing with them their suffering and the daily danger of death.

The war in Italy ended in May 1945. I returned with my family to Turin where I resumed my academic positions at the University. In the Fall of 1947, an invitation from Professor Viktor Hamburger to join him and repeat the experiments which we had performed many years earlier in the chick embryo, was to change the course of my life.

Although I had planned to remain in St. Louis for only ten to twelve months, the excellent results of our research made it imperative for me to postpone my return to Italy. In 1956 I was offered the position of Associate Professor and in 1958 that of Full Professor, a position which I held until retirement in 1977. In 1962 I established a research unit in Rome, dividing my time between this city and St. Louis. From 1969 to 1978 I also held the position of Director of the Institute of Cell Biology of the Italian National Council of Research, in Rome. Upon retirement in 1979, I became Guest Professor of this same institute.

THE NERVE GROWTH FACTOR: THIRTY-FIVE YEARS LATER

Nobel lecture, December 8, 1986

by

RITA LEVI-MONTALCINI

Istituto di Biologia Cellulare, via G. Romagnesi 18/A, ROMA, Italy

1) Neurogenesis and its early experimental approach
2) Experimental neuroembryology in the forties
3) The unexpected break: a gift from malignant tissues
4) NGF at its early in-vitro and in-vivo debuts
5) The vital role of NGF in the life of its target cells
6) NGF as a retrograde trophic messenger and tropic factor
7) Neuronal and non-neuronal target cells
8) The I.D. card of NGF
9) NGF, growth factors and protoncogenes
10) NGF in exocrine glands: a fortuitous presence or a biological function?
11) Foreseeable approaches and predictions of the unpredictable

Neurogenesis and its early experimental approach

"Embryogenesis is in some way a model system. It has always been distinguished by the exactitude, even punctitio, of its anatomical descriptions. An experiment by one of the great masters of embryology could be made the text of a discourse on scientific method. But something is wrong, or has been wrong. There is no *theory* of development in the sense in which Mendelism is a theory that accounts for the results of breeding experiments. There has therefore been little sense of progression or timeliness about embryological research. Of many papers delivered at embryological meetings, however good they may be in themselves, one too often feels that they might have been delivered five years beforehand without making anyone much the wiser, or deferred for five years without making anyone conscious of a great loss" [1].

This feeling of frustration, so incisively conveyed by these considerations by P. Medawar, pervaded in the forties the field of experimental embryology which had been enthusiastically acclaimed in the mid-thirties, when the upper lip of the amphibian blastopore brought this area of research to the forefront of the biological stage. The side branch of experimental neuroembryology, which had stemmed out from the common tree and was entirely devoted to the study of the trophic interrelations between neuronal cell populations and between these and the innervated organs and tissues, was then in its initial vigorous growth phase. It in turn suffered from a sharp decrease in the enthusiasm that had inflamed the pioneers in this field, ever since R. G. Harrison delivered his celebrated lecture on this topic at the Royal Society in London in 1935 [2]. Although the alternate 'wax and wane' cycles are the rule rather than the exception in all fields of human endeavor, in that of biological sciences the 'wane' is all too often indicative of a justified loss of faith in the rational and methodical approach that had at first raised so much hope.

A brief account of the state-of-the-art of experimental neuroembryology in the forties, when interest in this approach to the study of the developing nervous system was waning, is a prerequisite for understanding the sudden unforseeable turn of events which resulted in the discovery of the Nerve Growth Factor.

Experimental neuroembryology in the early forties

The replacement, in 1934 by Viktor Hamburger, of the chick embryo with that of the amphibian larva as object of choice for the analysis of the effects of limb bud estirpation on spinal motor neurons and sensory nerve cells innervating the limbs [3], signed the beginning of a long series of investigations centered on the analysis of this and related experimental systems in avian embryos. Here I shall only list the major advantages offered by the chick embryo over amphibian larvae as object of neurological investigations.

The avian nervous system is built according to a more elaborate design than that of amphibians, and it lends itself to a more rigorous analysis of its nerve centers than that of lower vertebrates. Extensive fundamental studies on the nervous system of the chick embryo, with use of the invaluable silver specific techniques by Ramon y Cajal and coworkers, extended recently by myself and other investigators [4, 5], provided a very accurate blueprint of most nerve centers and their developmental history during neurogenesis. This allowed the detection of even small infractions to normal developmental rules in experimentally manipulated embryos. At variance with ontogenetic processes in amphibians, the same processes in chick embryos unfold according to a rigidly scheduled time sequence which never departs from the anticipated. It is therefore possible to compare the central and peripheral nerve centers of experimental and control specimens in embryos incubated at the same temperature and environmental conditions. The analysis, in amphibian larvae, was extended to the brain, spinal cord and peripheral nervous system under various experimental situations. In the chick embryo, it was mainly confined to the study of the effects called forth by estirpation of limb primordia or implantation of additional wing or leg buds on their innervating motor and sensory nerve centers. In 1934, Viktor Hamburger published an article [3] on the effects of wing bud estirpation on the development of the brachial spinal motor segment and sensory dorsal root ganglia innervating the wing. He came to the conclusion that the hypoplasia of motor nerve cells of the ventral horn and of other nerve cells of the same hemisection of the spinal cord resulted from lack of stimuli centripetally transmitted by nerve fibers of the first differentiated neurons. These normally exert a regulatory effect on proliferation and differentiation of neighboring nerve cells. A reinvestigation of the effects produced by limb bud estirpation prospected a different control mechanism of the developing nerve centers by peripheral tissues. Through serial studies of silver stained embryos, the conclusion was reached that the severe hypoplasia of nerve centers deprived of their fields of innervation, resulted from death of differentiated neurons and not from failure of recruitment of neurons from a pool of still uncommitted nerve cell precursors [6, 7]. In 1947, Hamburger invited me to join him for the purpose of reinvestigating this problem. This invitation marked the beginning of a thirty year period that I spent at Washington University and of my life-long friendship with Viktor. Our 1949 article [7] confirmed the hypothesis previously submitted by G. Levi and myself [6]. The satisfaction of this confirmation of an important theoretical issue, and the successful analysis of other neuroembryological problems [8, 9] was, however, perturbed by the awareness of the low resolution power of the techniques in our possession for in depth exploration of the tremendously complex neurogenetic processes. The temptation to abandon the experimental analysis of the developing nervous system and move into the phage field, in full blossom in the forties, did not take hold, however, thanks to unpredictable and most fortunate events which occurred at the same time and opened a new era in developmental neurobiology.

The unexpected break: a gift from malignant tissues

In a 1948 article, a former student of Viktor Hamburger, Elmer Bueker, reported the results of a bold and imaginative experiment consisting in grafting fragments of mouse sarcoma 180 into the body wall of three-day chick embryos. The histological study of the embryos fixed 3–5 days later, showed that sensory nerve fibers emerging from adjacent dorsal root ganglia had gained access into the neoplastic tissue while no motor nerve fibers entered into the tumor [10]. The author concluded that histochemical properties of the fast growing mouse sarcoma offered a favorable field for growth of sensory fibers. This condition, in turn, resulted in a slight but consistent volume increase of these ganglia as compared to that of homologous ganglia innervating the wing of the contralateral side. Viktor and I reinvestigated this remarkable phenomenon adopting the method I had developed during my first neuroembryological studies, i. e., that of a daily inspection of control and experimental embryos serially sectioned and impregnated with a specific silver technique. Our results confirmed those reported by Bueker, but at the same time uncovered other effects elicited by the grafts of the mouse tumor, which hardly fit in with the hypothesis that they were in the same range and of the same nature as those called forth by transplants of normal embryonic tissues. They differed from the latter in the following, most significant, respects: sympathetic and not only sensory fibers gained access into the neoplastic tissues where they built a network of extraordinarily high density; nerve fibers branched at random between tumor cells without, however, establishing synaptic connections with them; sensory and sympathetic ganglia innervating the tumor underwent a progressive increase in volume, attaining, in the case of sympathetic ganglia, a size about six times larger than that of same control ganglia [11]. Subsequent experiments uncovered another astonishing deviation from the norm in embryos bearing transplants of mouse sarcoma 180 or of another tumor of identical origin, known as sarcoma 37. It was found that embryonic viscera which in normal specimens are devoid of innervation, such as the mesonephroi, or which become scarcely innervated only in late developmental stages, such as the sex glands, the thyroid, the parathyroid and the spleen, were loaded with sympathetic nerve fibers during early embryonic stages (12). A patent infraction of all developmental rules came to light with the finding of thick sympathetic fiber bundles inside the veins of the host where they protruded in the form of large neuromas obstructing blood circulation (Fig. 1). All sympathetic chain ganglia, and not only ganglia adjacent to or in direct connection with neoplastic tissues, were enormously enlarged. The hypothesis that these anomalous effects could result from the release by neoplastic cells of a soluble, diffusible agent which altered the differentiative and growth properties of its target cells, received full

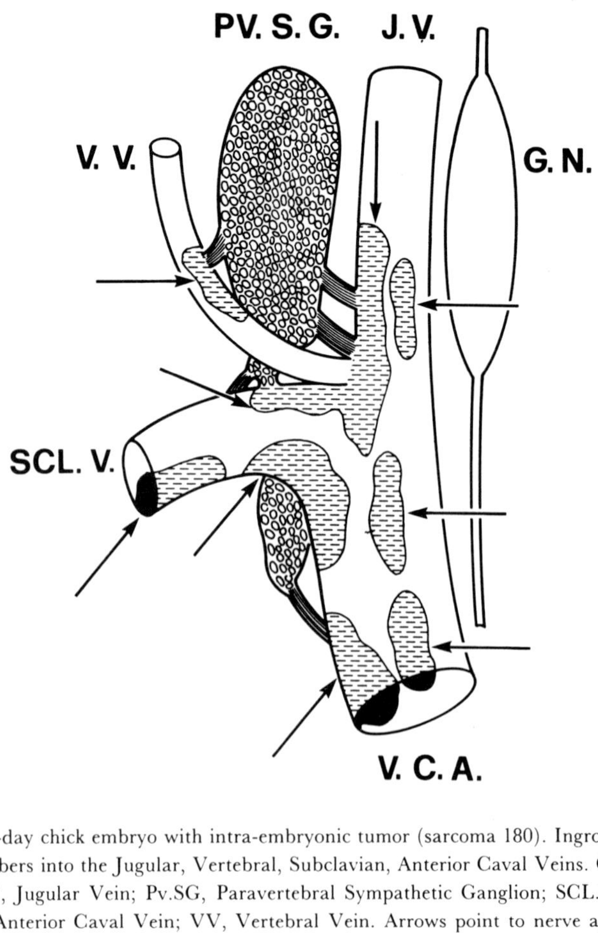

Fig 1. Sixteen-day chick embryo with intra-embryonic tumor (sarcoma 180). Ingrowth of sympathetic nerve fibers into the Jugular, Vertebral, Subclavian, Anterior Caval Veins. GN, Ganglium Nodosum; JV, Jugular Vein; Pv.SG, Paravertebral Sympathetic Ganglion; SCL.V, Subclavian Vein; VCA, Anterior Caval Vein; VV, Vertebral Vein. Arrows point to nerve agglomerations. (from Ref. 12)

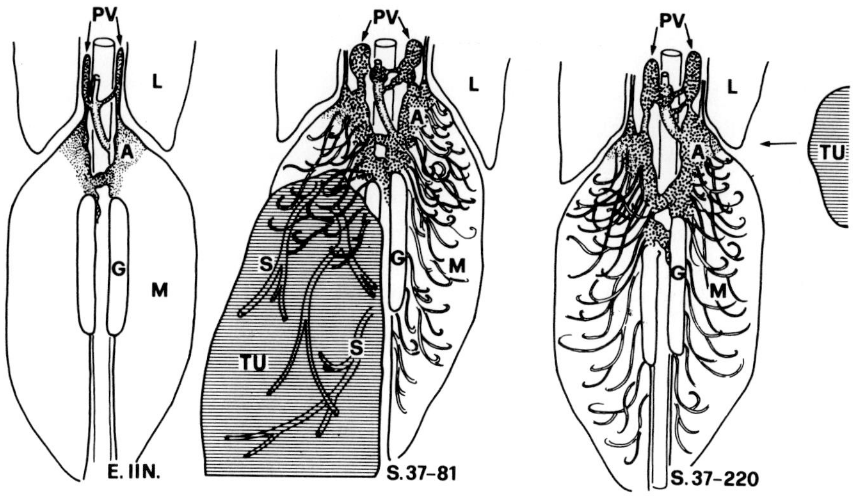

Fig 2. Semi-diagrammatic reconstruction of a normal 11-day chick embryo (E 11n), of an 11-day embryo carrying an intra-embryonic transplant of mouse sarcoma (S 37–81) and of an 11-day embryo with transplant of sarcoma 37 on the chorio-allantoic membrane (S 37–220). Note the hyperplastic growth of the prevertebral chain ganglia in embryos carrying tumor transplants. Visceral nerve fibres from these ganglia invade the nearby mesonephroi. A, adrenal; G, gonad; L, lung; M, mesonephros; PV, prevertebral ganglia; S, sensory nerves; Tu, tumor. (from Ref. 12)

confirmation from experiments transplanting one or the other mouse sarcoma onto the chorio-allantoic membrane of 4 to 6-day chick embryos, in such a position as to prevent direct contact between embryonic and neoplastic tissues (Fig. 2). Embryonic and tumor tissues were, however, in reciprocal connection through the circulatory system. The finding that these extra-embryonic transplants elicited the same effects as intraembryonic grafts gave definite evidence for the diffusible nature of the tumoral nerve growth promoting factor [12, 13].

Attempts to replicate these effects by implanting dried tumor pellets or by injecting extract of either sarcoma were unsuccessful. I then thought of resorting to the tissue culture technique, which I had practiced with G. Levi at the University of Turin. Lack of facilities in this field in the Department of Zoology at Washington University, prompted me to ask hospitality from Professor Carlos Chagas, Director of the Biophysics Institute of the University of Brasil in Rio de Janeiro. There, a friend of mine, Hertha Meyer, had built and was director of a most efficient tissue culture unit. Upon approval and invitation by Professor Chagas, I boarded a plane for Rio de Janeiro, carrying in my handbag two mice bearing transplants of mouse sarcomas 180 and 37.

The Nerve Growth Factor at its early in-vitro and in-vivo debut

"The tumor had given a first hint of its existence in St. Louis, but it was in Rio de Janeiro that it revealed itself, and it did so in a theatrical and grand way, as if spurred by the bright atmosphere of that explosive and exuberant manifestation of life that is the Carnival in Rio" [14].

The discovery of the growth response elicited by a soluble tumoral agent revealed the receptivity of developing nerve cells to hitherto unknown humoral factors and in this way opened a new area of investigation. The in vitro bioassay offered a practical and invaluable tool for uncovering the identity card of this factor and paved the way for the study of its mechanism of action. Ink drawings, which I enclosed in several letters mailed from Rio to Viktor, give an eloquent account of the spectacular way in which this still unknown agent revealed itself. Sensory and sympathetic ganglia explanted from 8-day chick embryos in a semi-solid medium in proximity to, but not in contact with, fragments of mouse sarcoma 180 or 37 produced, in a 24 hour period, a halo of nerve fibers of maximal density on the side facing the tumor [15] (Fig. 3). The euphoric state elicited by this discovery was, however, soon dampened by the discovery that normal mouse tissues, at variance with those of chick embryos, induce a milder, but not substantially different effect from that of mouse sarcomas. In retrospect, this should have alerted us to a novel and even more significant aspect of these in vitro experiments; namely, the widespread presence of the factor endowed with nerve growth promoting activity in normal and neoplastic tissues. The failure to realize the significance of this 'mouse effect' was beneficial rather than detrimental, since for the next two years our attention was entirely focussed on the study of the chemical nature of the factor released by the two mouse sarcomas, in much larger quantities than from normal mouse tissues.

A young biochemist, Stanley Cohen, who joined our Group shortly before my return from Rio, isolated from the two tumors a nucleoprotein fraction endowed with the in vitro nerve growth promoting activity [16]. Chance, rather than calculated search, signed a new, most fortunate turn of events. In order to degrade the nucleic acids present in this active fraction, Stan made use of snake venom which contains, among other enzymes, also the nucleic acid degrading enzyme, phosphodiesterase. Its addition in minute amounts to the nucleoprotein tumor fraction, was expected to suppress the formation of the fibrillar halo if nucleic acids rather than the protein were responsible for the nerve growth promoting effect elicited by this fraction. The startling result was a marked increase in the density of the fibrillar halo around the ganglia incubated in the presence of the tumoral fraction treated with snake venom. Since a dense fibrillar halo was produced also around ganglia cultured in the presence of minute amounts of snake venom alone, it became apparent that the venom itself was a most potent source of nerve growth promoting activity. On the basis of biochemical studies, Cohen was in fact able to show that equivalent growth stimulation effects were obtained by 15,000 μg of a sarcoma 180 homogenate and 6 μg of the moccasin snake venom. From the latter he isolated, after several purification steps, a non-dialyzable, heat-labile substance endowed with nerve growth promoting activity, identified as a protein molecule with a molecular weight in the order of 20,000 [17, 18]. Microgram quantities of the purified snake venom fraction endowed with nerve growth promoting property, injected daily into the yolk of 6 to 8-day chick embryos for a 3 to 5-day period, resulted in the overgrowth of sensory and sympathetic ganglia and excessive

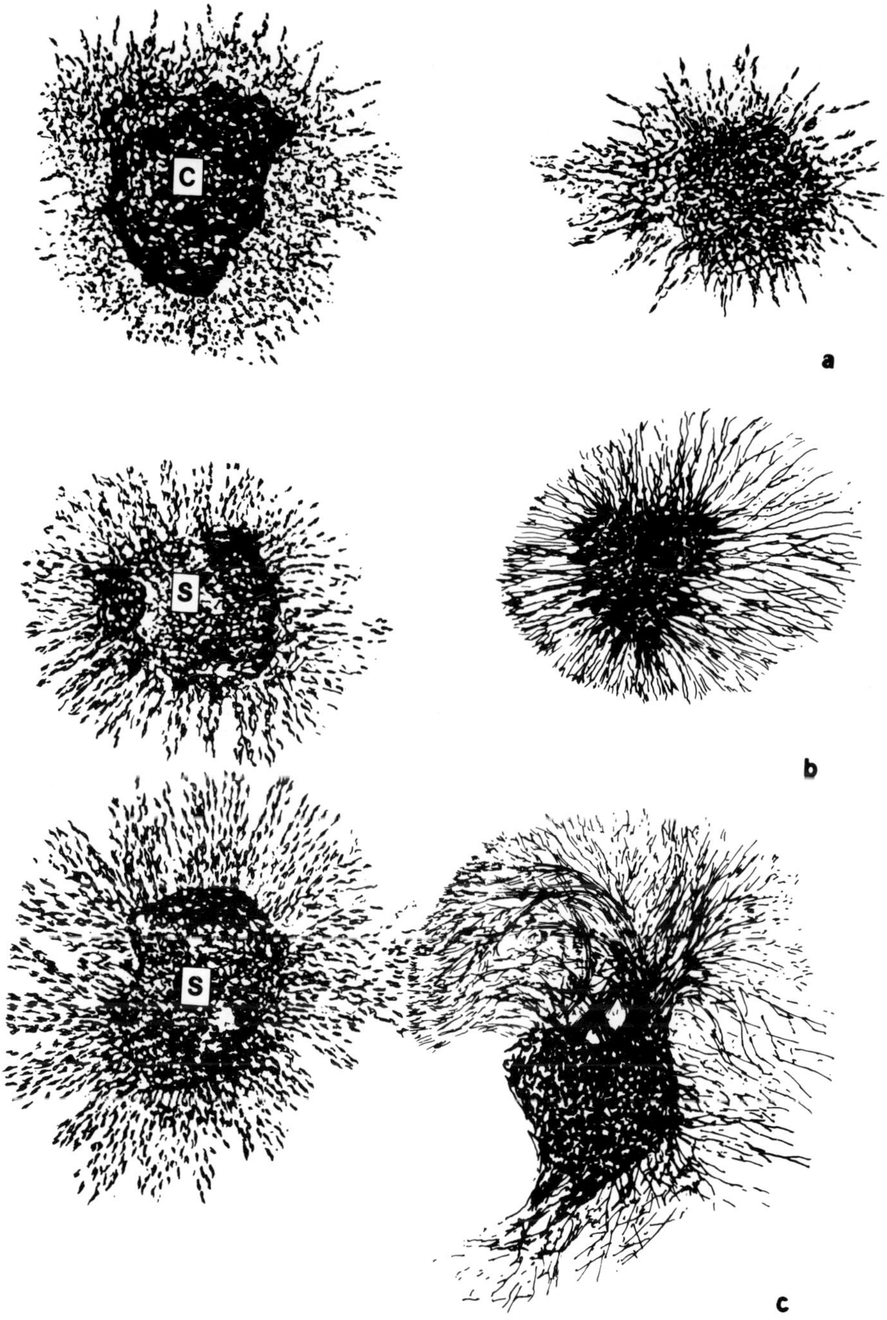

Figure 3. Drawings illustrating the in vitro "halo" effect on 8-day chick embryo sensory ganglia cultured in the presence of fragments of mouse sarcoma 180 for 24 hours (b) or 48 hours (c). In (a), the ganglion, which faces a fragment of chick embryonic tissue, shows fibroblasts but few nerve fibers. In (b) and (c), the ganglia, facing fragments of sarcoma 180, show the typical "halo" effect elicited by the growth factor released from the sarcoma. Note in (c) the first evidence of a neurotropic effect of the growth factor.

production of their fibers. Sympathetic nerve bundles branched profusely into the viscera and protruded into the cavity of the veins, mimicking in all details the effects elicited by grafts of mouse sarcomas [19].

If chance brought to our attention the unforeseeable presence of two nerve growth promoting sources, mouse sarcomas and snake venom, the sub-

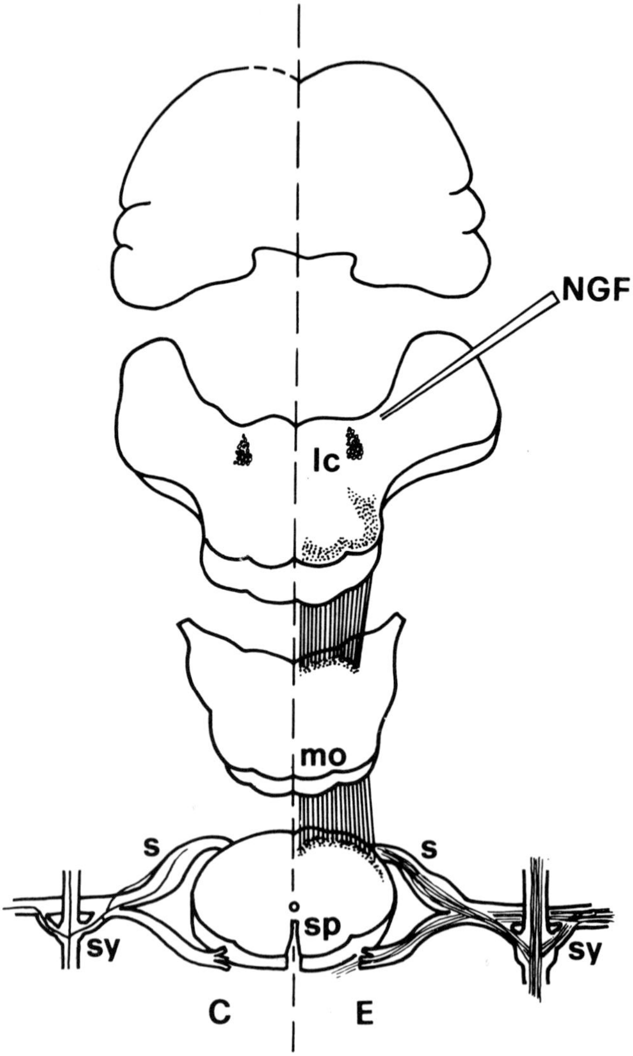

Figure 4. Diagrammatic representation of sympathetic fiber bundles which enter the spinal cord and medulla oblongata from adjacent sympathetic ganglia in intracerebrally NGF injected neonatal rats. Left half: control (C). Right half: experimental (E) embryo. NGF, site of injection of NGF into the floor of the fourth ventricle, lc, locus coeruleus; mo, medulla oblongata; sp, spinal cord; s, sensory ganglia; sy, sympathetic ganglia. Sympathetic fibers run across the sensory ganglion and enter into the neural tube with the dorsal roots. (from Ref. 40)

sequent finding that mouse submandibular salivary gland extract added in a minute quantity to the culture medium elicits an even denser and more compact fibrillar halo, was the result of a calculated search. These glands, as the homologue of the snake venom glands, were chosen by Stanley Cohen [20] as more likely that other organs screened with the in vitro bioassay, to store the nerve growth factor (NGF). These results were soon followed by purification and identification by Cohen of the salivary factor as a protein molecule with a molecular weight of 44,000 [20]. Its availability in larger quantities than the venom NGF, and its moderate toxicity when injected in a highly purified form, made possible the exploration of its biological activity in neonatal, young and adult mammals [21]. The results of these investigations signed the beginning of an ever more extensive and systematic in vivo and in vitro analysis of the salivary NGF, its chemical structure, as well as its mechanism and spectrum of action. Only the most significant findings reported from several laboratories in original and review articles will be considered in the following pages.

The vital role of NGF in the life of its target cells

In spite of, or perhaps because of its most unusual and almost extravagant deeds in living organisms and in-vitro systrems, NGF did not at first find enthusiastic reception by the scientific community, as also indicated by the reluctance of other investigators to engage in this line of research. The finding that a protein molecule from such diverse and unrelated sources as mouse sarcomas, snake venom and mouse salivary glands, elicited such a potent and disrupting action on normal neurogenetic processes, did not fit into any conceptual pre-existing schemes, nor did it seem to bear any relationship to normal control mechanisms at work during ontogenesis. It was in this skeptical atmosphere that NGF asserted, in a most forceful way, its vital role in the life of its target cells. Previous in vitro experiments had shown that incubation of snake venom with its antiserum inhibited the fiber outgrowth induced by the venom NGF. A specific antiserum to salivary NGF likewise abolished the formation of the in vitro fibrillar halo. These results suggested testing the effect of daily injections of small amounts of this antiserum (AS-NGF) in neonatal rodents. The inspection of treated mice, performed at the end of the first month with stereo and optic microscopes, revealed the near total disappearance of sympathetic para- and prevertebral chain ganglia [22–24]. This dramatic effect, which deprives newborn rodents and other neonatal mammals injected with antiserum to salivary NGF of the sympathetic system, without interfering with their normal development and vitality, became known as immunosympathectomy [25, 26]. The same treatment produces much less damaging effects in young and adult animals.

Two alternative hypotheses were submitted to explain the mechanism underlying the destructive effects of the antiserum: 1) a complement-mediated cytotoxic effect, or 2) inactivation of NGF or of an NGF-like protein essential for differentiation and survival of sympathetic nerve cells. Although the first hypothesis was favored in early articles, the second progressively gained support and is now generally accepted on the basis of this and an in-vitro experimental approach which provided additional, unequivocable evidence of the essential role NGF plays in the early differentiation stages of its target cells. The in vitro experiments consisted of the dissociation of sensory and sympathetic nerve cells from ganglia of 8–11 day chick embryos and their incubation in minimum essential media. Nerve cells failed to survive unless nanogram quantities of NGF were added daily to the culture medium [27]. The in vitro evidence for the role of NGF in the early phases of development of sensory nerve cells, received confirmation from subsequent experiments which proved that administration of NGF antiserum to rodent fetuses [28, 29] and autoimmunization of pregnant rodents against endogenous NGF, [30] result in failure of sensory ganglia to undergo normal development.

NGF as a retrograde trophic messenger and tropic factor

The evidence in favor of the hypothesis that immunosympathectomy results from removal, by AS-NGF [31], of circulating endogenous NGF, raised the questions of how NGF reached its target cells and what were its sources of production. Subsequent experimental pharmacological and surgical approaches provided satisfactory answers to both questions, and in view of the interest in these problems, techniques and main findings will be briefly reported.

Administration to neonatal rodents of drugs such as 6-hydroxydopamine, which destroys adrenergic nerve endings [32], or of vinblastine, which blocks axonal transport [33], results in death of the large majority of sympathetic nerve cells in their most active phase of differentiation and growth. The degenerative effects produced by these drugs are of the same magnitude as those produced by administration of AS-NGF and result in the destruction of para- and prevertebral sympathetic ganglia through a process which became known as chemical sympathectomy [32, 33]. A third experimental manipulation, consisting of the surgical transectomy of postganglionic axons of superior cervical ganglion performed in neonatal rodents, results in death of about 90% of immature sympathetic cells in this ganglion [34]. The experimental evidence that in all instances nerve cell death is prevented by an exogenous supply of NGF [30, 35–37] demonstrates the vital role played by this molecule in the life and differentiation of these cells. The subsequent demonstration that labelled NGF is taken up by the nerve endings of sympathetic [38] or sensory fibers [39] and is retrogradely transported to the cell perikarya, lent strong support to the concept of NGF as a strophic messenger, conveyed through nerve fibers from peripheral cells to the innervating neurons. Disconnection of the partners by chemical or surgical axotomy results in death of differentiating nerve cells deprived of this essential molecule.

At the same time as the vital role of NGF in developing sympathetic and sensory nerve cells was assessed and its regrograde transport from peripheral tissues was well documented, another important property of NGF—its ability to direct growing or regenerating axons of sensory and sympathetic fibers along its concentration gradient (neutrotropism)—was definitely established through different in vivo and in vitro experimental approaches [40–45].

The first strong evidence for a NGF neurotropic effect was obtained from experiments of daily micro-injections of NGF into the floor of the fourth ventricle. A 7-day treatment resulted in the penetration of fiber bundles originating from sympathetic ganglia inside the neural tube and in their ending at the level of experimentally produced NGF pools [40–41] (Fig. 4). In vitro experimental approaches gave more rigorous proof that neurites of NGF target cells grow along a NGF concentration gradient and deflect their route according to the changed position of the NGF releasing pipette [42]. While these studies unequivocably establish the NGF neurotropic effect as independent from its trophic action, they leave unanswered the question of whether this effect is exerted via a local control of growth cone motility [43], altered adhesion of this locomotor organelle to the substratum [44–45], or other mechanism(s) [46–47].

Neuronal and non-neuronal target cells

As indicated in Table I, targets of NGF action that have been well characterized up to now, can be classified under three main categories: 1) neural-crest derivatives, 2) central nervous system (CNS) neurons; and 3) cells of non-neuronal origin. For a thorough analysis of the many and diversified effects exerted by NGF on each one of these cells, the reader is referred to review articles on this specific matter [47–52]. In this context I only wish to make some general considerations.

A generally valid rule is that all cells are maximally responsive to NGF action during their early differentiation; the response undergoes progressive restriction in the adult without, however, being totally effaced. Long sympathetic neurons and sensory neurons, with particular reference to those of the dorsomedial quadrant of spinal ganglia in chick embryo [12], provided a most valuable system for demonstrating the three main activities of NGF, i.e., 1) its vital trophic role during the early developmental stages, 2) its property of enhancing differentiative processes such as neurite outgrowth, and 3) of guiding the growing or regenerating neurites along its own concentration gradient [43–44]. These same cells offered an in vivo model system to study the induction of enzymes involved in neurotransmitter synthesis [53] and were also instrumental in providing the first demonstration of the retrograde transport of NGF [54] and its role as a trophic messenger [55]. If sensory and sympathetic cells played a key role in revealing these properties of NGF, chromaffiin cells and their neoplastic counterpart, the clonal cell line PC12, became the model of choice for studying the capacity of NGF to modulate the phenotypic expression and molecular mechanism subserving this process [56]. The phenotypic shift induced by NGF both in chromaffin [57–58] and PC12 cells [56, 59], resulting in their neuronal differentiation accompanied by a pletora of chemical, ultrastructural and morphological changes characteristic of the neuronal rather than glandular phenotype, is too well known to warrant a detailed description [50]. These cells, moreover, uncovered the startling capacity of NGF to act both as a mitogenic [60] and a antimitotic agent [56], even within the context of the same clonal cell line PC12 and of a mutated version of it [61].

This, in turn, clearly pointed to the 'versatility' of NGF receptors and of their transduction machinery, whose message is evidently read and interpreted in different ways according to the cell type and previous cell history. The 'priming model' prospected to give a molecular account for the very fast and very slow onset of neurite outgrowth occurring, respectively, in sensory and sympathetic cells [24] on the one hand and PC12 cells on the other [50], is an excellent example of the contribution of these latter cells to studies on the mode of action of NGF.

Other examples of the wide and at the same time diversified NGF effects are illustrated by other sympatho-adrenal derivatives such as paraganglia, small intensely fluorescent cells (SIF) and carotid body cells [62–64]. A particularly impressive evidence of the capacity of NGF to modulate phenotypic expression is the case of SIF cells which have been hypothesized as immediate precursors of both sympathetic and chromaffin cells. When these cells are cultured under appropriate conditions, they can be channelled towards the first or the second phenotype in media supplied with NGF or with dexamethasone [63, 64]. Such an interplay, even in fully differentiated cells, between NGF and steroid hormones, is also indirectly suggested by in vivo studies on the short adrenergic neurons which innervate the genito-urinary system in both sexes [65].

In more recent years, two new populations came to the forefront of studies on NGF target cells: CNS neurons and cells originating from the hematopoietic system.

Small and large neuronal populations located in different brain areas have been shown to exhibit all properties and responses typical of sensory and sympathetic cells, such as: 1) the presence of specific receptors [66], 2) retrograde transport of NGF [67], 3) increased neurotransmitter synthesis with special reference to acetylcholine [68–70], and 4) trophic response manifested as protection by exogenous NGF administration to selective noxious treatments or surgical transection otherwise leading to cell death [71, 72]. A role for NGF in the development of as yet unidentified hypothalamic brain centers has been suggested by the finding that injections of affinity purified polyclonal antibodies against NGF in rat fetuses induce a severe postnatal neuroendocrine syndrome [29]. The loop of an unquestionable NGF role in brain is completed by the demonstration that other nerve cells, especially those located in the hippocampus and cortical areas, manufacture large quantities of NGF mRNA and NGF protein, thus closing the functional link between NGF-producing and NGF-responding cells [73–75]. As prospected in the last section of this article, although the range of NGF action in the CNS is qualitatively comparable to that previously observed in peripheral neurons, the actual extent of the NGF role in brain is far from foreseeable due to the vast repertoire of possible responses from nerve cells in the CNS.

An analogous general consideration holds for the effect exerted by NGF on mast cells and possibly on other cells of the immune system. The increased in-vivo and in vitro number of mast cells following NGF treatment [76, 77], as well as the effect of this growth factor on histamine release [78–80], point to an unquestionable role in the physiology of these cells. It is not yet clear, however, whether such an effect is exerted through a generalized action on all mast cell precursors or through a sort of clonal selection mechanism. The more recent report of an NGF effect on other spleen cells, such as mononuclear cells [81], and the existence of NGF receptors on thymocytes [82], clearly suggests that the NGF action extends also to cells belonging to a network of enormous functional significance. The role played by histamine as an immunomodulator and the obvious involvement of spleen cells in the immune response of the organism prospect new scenarios in which NGF may gain access, not through a back door, but through the main entrance.

The I.D. card of NGF

Sequencing of mouse submandibular gland NGF, achieved in 1971 [83], provided invaluable information not only on its primary structure but was recently instrumental in the preparation of synthetic oligonucleotides which resulted in the identification of NGF cDNA [84]. The cloning which followed in rapid succession of mouse [84], human [85], bovine [86] and chick [87] genes, demonstrated their high degree of homology. The NGF gene, located in the human species on the proximal short arm of chromosome 1 [88], codes for a large polypeptide of 307 amino acid residues which, upon cleavage(s), gives rise to the 118 amino acid mature NGF subunit protein and, possibly, to other peptides of unknown function and with no sequence homology with presently identified proteins [84]. NGF is a dimer composed of two identical subunits held together by non-covalent bonds. The dimer can be isolated as such [89] or under the form of a complex also consisting of two other proteins, one with an esteropeptidase activity, probably involved in the processing of an NGF precursor, and the other with an as yet unknown function [90–92]. While it remains to be established whether each NGF subunit is biologically active, it has been demonstrated that a covalently cross-linked form of the dimer maintains full activity [91, 92]. Between the two well-indentified molecular entities of NGF and of its coding gene, which can be visualized as the summit and the base of an iceberg, are several other possible intermediate forms of unknown nature and biological properties. Their identification would answer important questions such as: Are other biologically active peptides coded for by the NGF gene? What is the significance of different splicing in different cells of NGF mRNA [93]? Is the processing of pre-pro-NGF identical in all neuronal and non-neuronal cells or, as in other peptides [94], do alternate processing pathways result in the production of peptides endowed with different biological functions? Since the same peptides may undergo post-transcriptional or post-translational modification, the submerged areas of the NGF iceberg loom very large.

Studies on the immunological and biological relatedness of NGFs purified from different species strongly support the hypothesis that the site(s) of interaction with their receptors has remained structurally more constant than is the case for other epitopes, probably free to mutate in view of their less fundamental biological functions [95].

NGF, growth factor and oncogenes

The discovery of NGF, soon followed by that of Epidermal Growth Factor (EGF), led to the biological identification of an ever-growing list of polypeptide growth factors [48]. In the seventies, another apparently unrelated area of biology came to the forefront of research with the discovery of single gene products (oncogenes) causing transformation. Polypeptide Growth Factors (PGF) and oncogene research, pursued at first independently of each other, converged when homology between some oncogenes and growth factors or their receptors was shown by sequence analysis. Evidence is steadily increasing that excessive synthesis, or an altered version of PGFs or of their receptors, may result in transformation of recipient cells [95–98]. More recently, the demonstration that the opposite is also true, namely, that certain oncogene products may induce differentiation of recipient cells, called attention to another facet of this intricate interplay between differentiative and transforming processes. The case of H-ras and that of v-src, whose expression into PC12 cells [99, 100] result in mitotic arrest and neuronal differentiation comparable to those elicited by NGF, provide instances of a list most likely to grow. The obvious conclusion is that a given polypeptide growth factor, or intracellular proteins playing essential roles in the cell cycle or in differentiation of some cells, may exert markedly different actions in distinct cell types. In the case of NGF, one wonders if and how other actions are elicited by this versatile molecule. For instance, is an altered version of NGF or of its receptors capable of causing transformation of some recipient cells, as has been shown for other PGFs? If this is the case, could NGF in a modified version or its receptors be implicated in neoplasia in the central and peripheral nervous systems?

NGF in exocrine glands: a fortuitous presence or a biological function?

The early discovery that mouse submandibular glands synthesize and release large quantities of NGF into the saliva, that the synthesis of this protein molecule is under the control of testosterone and of thyroxine [101, 102] and that the NGF protein content is about ten-fold higher in male than in female mice, remained for about three decades a puzzling and unexplained finding. The conflicting but altogether negative attempts to reveal the presence of this molecule in the circulating blood [49, 51], and the lack of any adverse effects on sympathetic and sensory cells by removal of these glands, which deprived these rodents of such a large NGF source, militated against the hypothesis that salivary NGF gains access to their target cells. An alternative biological function for salivary NGF was first hypothesized by our group [103], and recently proved by us [104] and another investigator [105]. We demonstrated that intraspecific fighting, experimentally induced in adult male mice by 6–9 weeks of social isolation, results in massive NGF release into the blood stream, an event prevented by previous sialoadenectomy. Since injections of NGF induce weight and size increase of the adrenal glands [105] and stimulate the synthesis of the catecholamine key enzyme, TH [106], we suggested that such a massive discharge into the blood circulation of endogenous salivary NGF may be instrumental in the defence and/or offense mechanisms of vital significance for male mice that engage in intraspecific fighting among individuals of the same sex. In favor of this hypothesis, is a recent report that aggressive behavior results in the release into the blood of another biologically active protein, renin, synthesized in the same tubular portions of these glands [107]. The mechanism triggering this NGF release is not yet understood, nor is it known whether other stations are activated and play a role in this specific stress syndrome.

As for the presence of large NGF sources in snake venom [18] and male genital organs [108, 109], they may be conceived as instances of bizarre evolutionary gene expression. Alternatively, in these cases NGF may subserve other functions which may somehow be linked with the poisonous action of snake venom, or the reproductive activity of the genital apparatus. In the case of snake venom, one can envisage the possibility that a highly specific neurotropic molecule such as NGF is utilized by reptiles as a carrier of other neurotoxins devoid of specific receptors in the central and peripheral nervous systems. For instance, enzymes such as phospholipases, phosphodiesterases and proteases of various nature, which by themselves may lack specific recognition sites in target cells, may exploit NGF as a carrier to gain access inside cells wherever there are specific NGF receptors. The widespread distribution of these specific molecules also in several non-neuronal cells could offer some toxins or enzymes a better access to their target organs.

In the reproductive tract, NGF could participte in fertilization mechanisms by cytoskeletal mediated activation of spermatozoa locomotion much in the same way as in neurite outgrowth, or by favoring egg implantation, via inhibition of rejection through the immune system. This latter hypothesis is presently under investigation (Geraci, Cocchiara and Calissano) by assessing the effect of NGF on uterine mast cells which, through histamine release, are postulated to prevent the local immune reaction [110].

Foreseeable approaches and predictions of the unpredictable

The most obvious among the foreseeable approaches is the search for other NGF target cells, using the ever more sophisticated in vivo and in vitro techniques which became available in these last decades. It was this multimodal approach which in recent years led to the discovery of NGF target cells in the CNS of lower and higher vertebrates and in cell lineages playing a role in the immune system. This list is likely to increase, as the search extends to other neuronal and non-neuronal cell populations. Furthermore, one should take into account the fact that some of these populations are receptive to NGF mainly during developmental stages in prenatal life. This was already demonstrated in sensory cells of avian and mammalian species [49, 51, 52], and in cells lining the third ventricle in amphibian tadpoles [111] and prenatal and neonatal rodents (Aloe and Levi-Montalcini, unpublished observations). Likewise, the systematic screening of neuroendocrine and hematopoietic cell lines in in vitro and in vivo systems may reveal other as yet uncovered roles of this growth factor.

Another approach now in progress in many laboratories is the search NGF-like factors active on other neuronal populations. These factors may be subdivided into two major classes: 1) those coded by the NGF gene itself but processed through alternate post-transcriptional or post-translational pathways leading to PGFs with a somewhat different structure and function: 2) other proteins or peptides having the trophic, chemotactic and/or differentiative activity of NGF, but coded by other genes.

The search for factors belonging to the first group and their identification will take advantage of the techniques of molecular biology and immunology. These should provide valuable information on some of the still unexplored, submerged areas of the NGF iceberg, dealing with the processes of the NGF gene transcription or translation. Of particular importance would be the identification of the NGF sequence responsible for its binding to receptors which may presumably trigger a given cellular response [47]. As previously surmised, [95], this portion has possibly been better conserved than other parts of the molecule. Once identified, it will be feasible to introduce, in its synthetic counterpart, amino acid substitutions and/or chemical modifications and explore the biological potency of the newly manufactured peptide. This approach should not only provide invaluable information on the nature and properties of the NGF active center, but, hopefully, will result in the synthesis of peptides endowed with an activity even higher than that of NGF itself, so brilliantly achieved in the field of other biologically active peptides [112, 113].

Within this category of studies on NGF and its coding gene, one can conceive a strategy aimed at exploiting the property of non-neuronal cells in peripheral tissues and of neurons and satellites in the CNS to manufacture and release NGF by resorting to pharmacological agents that modify NGF gene expression or processing. The well-established findings that NGF synthesis is increased following transection of nerve fibers connecting NGF receptive nerve cells to their targets [114] or via hormonal action [101, 102], are an additional indication of the remarkable plasticity of the mechanisms controlling the expression of the NGF gene. It is conceivable that this property might be modulated by pharmacological agents acting on the same path as those involved in the regulation of the synthesis and release of NGF.

The search for neurotrophic factors coded by genes other than the NGF gene could take advantage, at least in its main lines, of the classical approach so successfully applied in the isolation and identification of NGF. Two main causes may explain why extensive work invested in this attempt has not been so successful in providing evidence for the existence of other PGFs activating different neuronal cell lines: 1) the lack of fast and reliable bioassays such as those developed for NGF and 2) the failure to detect large sources of these factors comparable to those fortuitously discovered in early NGF studies. The availability of rapid, highly reliable bioassays can, however, now be achieved by resorting to the use of most stringent, chemically defined media, permitting survival and differentiation of only given cell types, upon addition to the medium of putative growth factors extracted from different sources and screened with the in vitro bioassay for their potential specific growth enhancing activity. The problem of finding by sheer chance large sources of NGF-like peptides, such as those which played a key role in the discovery of NGF, can be solved by resorting to techniques of protein chemistry and recombinant DNA technology. A few micrograms of purified protein are sufficient to decipher the sequence, prepare the corresponding cDNAs, identify the gene of the PGF in question, and express it in bacteria, thus replacing a search once guided by unpredictable strokes of luck, with a rational and systematic strategy.

Predictions of the unpredictable are encouraged by the same history of NGF which may be defined as a long sequence of unanticipated events which each time resulted in a new turn in the NGF unchartered route, and opened new vistas on an ever-changing panorama. This trend, which became manifest from the very beginning and in fact alerted me to the existence of NGF, is perhaps the most attractive, even though elusive trait of this thirty-five year long adventure. One can at present only predict where future developments are most likely to occur. The main causes of unpredictability of the findings, reside in the intricacy of the new surroundings where NGF is moving—the CNS and the immune system—rather than in NGF itself. The enormous complexity of these two networks, which on the basis of recent findings are closely interrelated and influence each other through bidirectional signals [115, 116], opens endless possibilities for NGF activation of distinct repertoires of cells belonging to one or the other system. How many indirect effects can be elicited by direct NGF action on cholinergic, adrenergic and peptidergic neurons interlinked via fiber pathways and humoral channels or through short-distance diffusion? Likewise, how many effects could follow the simple histamine release by NGF activated mast cells, considering the well-established role of this amine as an immunomodulator or an immunosuppressor? These considerations hold also for the potential utilization of NGF in brain and immunosystem disorders. For instance, whenever cell death of specific neuronal populations may be linked to a decreased local availability of neurotrophic factors, such as NGF, its exogenous supply or stimulation of its endogenous production via pharmacological agents may offer a promising approach to presently incurable diseases.

I shall end this account of the unfolding of the NGF story with a remark made more than a decade ago by Viktor Hamburger: "— — — the fact that this discovery, which grew out of a seemingly peripheral problem (peripheral in every sense of the word), has blazed so many new trails is its greatest contribution in neuroembryology" [117]. Studies in this last decade have not only provided new strong evidence of the most important contributions of NGF in the field of neuroembryology, but brought to the fore its significance in the more general field of neuroscience and also prospect its role in that of the immune system.

I dedicate this article to Viktor Hamburger, who promoted and took part in this search, and to whom I am forever indebted for invaluable suggestions and generosity. Without him, the Nerve Growth Factor would never have come to our attention.

To my dear friends, Pietro Calissano and Luigi Aloe, I wish to express my deepest gratitude for their fundamental contributions. In this thirty-five year long investigation, a large number of colleagues, research associates and graduate students took part in this scientific adventure. I am particularly indebted and I very gratefully acknowledge the most important work performed by two of them: Drs. Piero Angeletti and Vincenzo Bocchini. To Professor Carlos Chagas, for his generous hospitality in the Biophysic Institute of the University of Brasil, and to Dr. Hertha Meyer who helped me in devising the tissue culture bioassay of NGF, my warmest thanks.

Table 1. *NGF TARGET CELLS*

NEURAL CREST DERIVATIVES	*Sympathoadrenal*	*Long sympathetic neurons*	
		Short sympathetic neurons	
		Cells of *paraganglia* (carotid & abdominal paraganglia)	
		SIF (small, intensely fluorescent) cells	
		Chromaffin cells	normal
			neoplastic (PC12)
	Sensory neurons		
CENTRAL NERVOUS SYSTEM	*Cholinergic neurons:*	corpus striatum, basal forebrain, septum,	
	Adrenergic, indoleaminergic,	nucleus diagonal band of Broca	
	Peptidergic neurons	Xenopus laevis tadpoles	
NON NEURONAL ORIGIN	*Mast cells*		

REFERENCES

1. Medawar P. B., in The Art of the Soluble, (Methuen & Co., LTD, 1967) pp. 106–107.
2. Harrison R. G., Proc. Roy. Soc. London Series B 118, 155–196 (1935).
3. Hamburger V., J. Exp. Zool., 68, 449–494 (1934).
4. Tello J. F., Trabajos Lab. Invest. Biol. Univ. Madrid 21, 1–93 (1922).
5. Levi-Montalcini R., Prog. Brain Res. 4, 1–29 (1964).
6. Levi-Montalcini R., Levi G., Arch. Biol. Liège 54, 183–206 (1943).
7. Hamburger V., Levi-Montalcini R., J. Exp. Zool. 111, 457–502 (1949).
8. Levi-Montalcini R., J. Comp. Neurol. 91, 209–242 (1949).
9. Levi-Montalcini R., J. Morphol. 86, 256–283 (1950).
10. Bueker E. D., Anat. Rec. 102, 369–390 (1948).
11. Levi-Montalcini R., Hamburger V., J. Exp. Zool. 116, 321–362 (1951).
12. Levi-Montalcini R., Ann. N.Y. Acad. Sci. 55, 330–343 (1952).
13. Levi-Montalcini R., Hamburger V., J. Exp. Zool. 123, 233–288 (1953).
14. Levi-Montalcini R., in The Neurosciences: Paths of Discovery, F. G. Worden, J. P. Swayzey, G. Adelman, Eds. (MIT Press, 1975), pp. 245–265.
15. Levi-Montalcini R., Meyer H., Hamburger V., Cancer Res. 14, 49–57 (1954).
16. Cohen S., Levi-Montalcini R., Hamburger V., Proc. Natl. Acad. Sci. USA 40, 1014–1018 (1954).
17. Cohen S., Levi-Montalcini R., Proc. Natl. Acad. Sci. USA 42, 571–574 (1956).
18. Cohen S., J. Biol. Chem. 234, 1129–1137 (1959).
19. Levi-Montalcini R., Cohen S., Proc. Natl. Acad. Sci. USA 42, 695–699 (1956).
20. Cohen S., Proc. Natl. Acad. Sci. USA 46, 302–311 (1960).
21. Levi-Montalcini R., Booker B., Proc. Natl. Acad. Sci. USA 46, 373–384 (1960).
22. Levi-Montalcini R., Booker B., Proc. Natl. Acad. Sci. USA 46, 384–391 (1960).
23. Levi-Montalcini R., Science 143, 105–110 (1964).
24. Levi-Montalcini R., The Harvey Lectures 60, 217–219 (1966).
25. Levi-Montalcini R., Angeletti P. U., Pharmacol. Rev. 18, 819–828 (1966).
26. Steiner G., Schönbaum E. Eds., Immunosympathectomy (Elsevier Publishing Co., Amsterdam, 1972).
27. Levi-Montalcini R., Angeletti P. U., Dev. Biol. 7, 653–659 (1963).
28. Levi-Montalcini R., Aloe L., Calissano P., Cozzari C., in 1st Meeting of the Internatl. Society for Dev. Neuroscience, Strasbourg, vol. 1, pp. 5 (1980).
29. Aloe L., Cozzari C., Calissano P., Levi-Montalcini R., Nature 291, 413–415 (1981).
30. Johnson E. M., Gorin P. D., Brandeis L. D., Pearson J., Science 210, 916–918 (1980).
31. Goedert M., Otten U., Schaefer T., Schwab M., Thoenen H., Brain Res. 201, 399–409 (1980).
32. Angeletti P. U., Levi-Montalcini R., Proc., Natl. Acad. Sci. USA 65, 114–121 (1970).
33. Calissano P., Monaco G., Menesini-Chen M. G., Chen J. S., Levi-Montalcini R., in Contractile Systems in Non-muscle Tissue, S. W. Perry, A. Margret, R. S. Adelstein, Eds. (Elsevier, Amsterdam, 1976a), pp. 201–211.
34. Hendry I. A., Brain Res. 90, 235–244 (1975).
35. Levi-Montalcini R., Aloe L., Mugnaini E., Oesch F., Thoenen H., Proc. Natl. Acad. Sci. USA 72, 595–599 (1975).
36. Hendry I. A., and Campbell, J. J. of Neurocytol. 5, 351–360 (1976).
37. Aloe L., Mugnaini E., Levi-Montalcini R., Arch. Ital. Biol. 113, 326–353 (1975).
38. Stöckel K., Paravicini U., Thoenen H., Brain Res. 76, 413–421 (1974).
39. Hamburger V., Brunso-Bechtold V. J. K., Yip J. W., J. Neurosci. 1, 60–71 (1981).
40. Levi-Montalcini R., Prog. Brain Res. 45, 235–258 (1976).
41. Menesini-Chen M. L., Chen J. S., Levi-Montalcini R., Arch. Ital. Biol. 116, 53–84 (1978).

42. Gundersen R. W., Barrett J. N., Science 206,1079–1080 (1979).
43. Gundersen R. W., Barrett J. N., J. Cell Biol. 87, 546–554 (1980).
44. Campenot R. B., Dev. Biol. 93, 1–12 (1982).
45. Campenot R. B., Dev. Biol. 93, 13–41 (1982).
46. Pfenninger K. H., Johnson M. P., Proc. Natl. Acad. Sci. USA 78, 7797–7800 (1981).
47. Calissano P., Cattaneo A., Aloe L., Levi-Montalcini R., in Hormonal Proteins and Peptides, C. H. Li, Ed. (Academic Press, 1984), Vol. XII, pp. 1–56.
48. Bradshaw R. A., Annu. Rev. Biochem. 47, 191–216 (1978).
49. Thoenen H., Barde Y. A., Physiol. Rev. 60, 1284–1335 (1980).
50. Greene L. A., Shooter E. M., Annu. Rev. Neurosci. 3, 353–402 (1980).
51. Calissano P., Cattaneo A., Biocca S., Aloe L., Mercanti D., Levi-Montalcini R., Exp. Cell Res. 154, 1–9 (1984).
52. Levi-Montalcini R., Calissano P., Trends Neurosci. 9, 473–476 (1986).
53. Thoenen H., Angeletti P. U., Levi-Montalcini R., Kettler R., Proc. Natl. Acad. Sci. USA 68, 1598–1602 (1971).
54. Paravicini U., Stoeckel K., Thoenen H., Brain Res. 84, 279–291 (1975).
55. Johnson E. M., Yip H. K., Nature 314, 751–753 (1985).
56. Greene L. A., Tischler A. S., Proc. Natl. Acad. Sci. USA 73, 2424–2428 (1976).
57. Unsicker K., Krisch B., Otten U., Thoenen H., Proc. Natl. Acad. Sci. USA 75, 3498–3502 (1978).
58. Aloe L., Levi-Montalcini R., Proc. Natl. Acad.Sci. USA 76, 1246–1250 (1979).
59. Greene L. A., Liem R. K. H., Shelanski M. L., J. Cell Biol. 96, 76–83 (1983).
60. Lillien L. E., Claude P., Nature 317, 632–634 (1985).
61. Burnstein, D. E., Greene L. A., Dev. Biol. 94, 477–482 (1982).
62. Levi-Montalcini R., Aloe L., Adv. Biochem. Psychopharmacol. 25, 3–16 (1980).
63. Doupe A. J., Patterson P. H., Landis S. C., J. Neurosci. 5, 2143–2160 (1985).
64. Doupe A. J., Landis S. C., Patterson P. H., J. Neurosci. 5, 2119–2142 (1985).
65. Owman C., Sjoberg N. O., in Proceedings of the 5th International Congress of Endocrinology, V. H. T. James, Ed. (Excerpta Medica, Amsterdam, 1977), vol. I, pp. 205–209.
66. Szutowicz A., Frazier W. A., Bradshaw R. A., J. Biol. Chem. 251, 1516–1523 (1976).
67. Seiler M., Schwab M., Brain Res. 30, 33–39 (1984).
68. Gnahn H., Hefti F., Heumann R., Schwab M. E., Thoenen H., Dev. Brain Res. 9, 45–52 (1983).
69. Hefti F., Dravid A., Hartikka J., Brain Res. 293, 305–311 (1984).
70. Mobley W. C., Rutkowski J. L., Tennekoon G. I., Buchanan K., Johnston M. V., Science 229, 284–287 (1985).
71. Williams L. P., Varon S., Peterson G. M., Wictorin K., Fischer W., Bjorklund A., Gage F. H., Proc. Natl. Acad. Sci. USA 83, 9231–9235 (1986).
72. Kromer L. F., Science 235, 214–216 (1987).
73. Korsching S., Auburger G., Heumann R., Scott J., Thoenen H., EMBO J. 4, 1389–1393 (1985).
74. Shelton D. L., Reichardt L. F., Proc. Natl. Acad. Sci. USA 83, 2714–2718 (1986).
75. Whittemore S. R., Ebendal T., Lärkfors L., Olson L., Seiger A., Strömberg I., Persson H., Proc. Natl. Acad. Sci. USA 83, 817–821 (1986).
76. Aloe L., Levi-Montalcini R., Brain Res. 133, 358–366 (1977).
77. Böhm A., Aloe L., Acad. Naz. dei Lincei 80, 1–6 (1986).
78. Bruni A., Bigon E., Boarato E., Leon A., Toffano G., FEBS Lett. 138, 190–192 (1982).
79. Sugiyama K., Suzuki Y., Furuta H., Arch. Oral. Biol. 30, 93–95 (1985).
80. Mazurek N., Weskamp G., Erne P., Otten U., FEBS Lett. 198, 315–320 (1986).
81. Thorpe L. W., Werrbach-Perez K., Perez-Polo J. R., 2° Internatl. Workshop on Neuroimmunomodulation, Dubrovnik, Abstract, pp. 151 (1986).

82. Cattaneo A., Secher D. S., Exp. Cell Res. (in press).
83. Hogue-Angeletti R., Bradshaw R. A., Proc. Natl. Acad. Sci. USA 68, 2417–2420 (1971).
84. Scott J., Selby M., Urdea M., Quiroga M., Bell G., Rutter W.J., Nature 302, 538–540 (1983).
85. Ullrich, A., Gray A., Berman C., Dull T.J., Nature 303, 821–823 (1983).
86. Meier R., Becker-Andre M., Gotz R., Heumann R., Shaw A., Thoenen H., EMBO J. 5, 1489–1493 (1986).
87. Ebendal T., Larhammar D., Persson H., EMBO J. 5, 1483–1487 (1986).
88. Francke V., De Martinville B., Coussens L., Ullrich A., Science 222, 1248–1250 (1983).
89. Bocchini V., Angeletti P. U., Proc. Natl. Acad. Sci. USA 64, 787–792 (1969).
90. Varon S., Nomura J., Shooter E. M., Biochemistry 6, 2202–2210 (1967).
91. Varon S., Nomura J., Shooter E. M., Biochemistry 7, 1296–1303 (1968).
92. Stach R. W., Shooter E. M., J. Biol. Chem. 249, 6668–6674 (1974).
93. Edwards R. H., Selby M.J., Rutter W.J., Nature 319, 784–787 (1986).
94. Eipper B. A., Mains R. E., Herbert E., Trends Neurosci. 100, 463–467 (1986).
95. Doolittle R., Hunkapiller M. W., Hood L. E., DeVare S. G., Robbins K. L., Aaronson S. A., Antoniades H. N., Science 221, 275–276 (1983).
96. Downward J., Yraden Y., Mayes E.,Scrace G., Totty N., Stockwell P., Ullrich A., Schlessinger J., Waterfield M. D., Nature 307, 521–527 (1984).
97. Sherr C. J., Retternmier C. W., Sacca R., Roussel M. F., Look T. A., Stanley E. R., Cell 41, 665–676 (1985).
98. Weinberger C., Hollenberg S. M., Rosenfeld M. G., Evans R. M., Nature 318, 670–673 (1985).
99. Bar-Sagi D., Feramisco J. R., Cell 42, 841–848 (1985).
100. Alemà S., Casalbore P., Agostini E., Tatò F., Nature 316, 557–559 (1985).
101. Levi-Montalcini R., Angeletti P. U., in Salivary Glands and their Secretions, L. M. Sreebny, J. Meyer, Eds. (Pergamon, Oxford, 1964), pp. 129–141.
102. Aloe L., Levi-Montalcini R., Exp. Cell Res. 125, 15–22 (1980a).
103. Aloe L., Cozzari C., Levi-Montalcini R., Brain Res. 332, 259–265 (1985).
104. Aloe L., Alleva E., Böhm A., Levi-Montalcini R., Proc. Natl. Acad. Sci. USA 83, 6184–6187 (1986).
105. Lakshmanan J., Am. J. Physiol. 250, E 386–391 (1986).
106. Otten U., Schwab M., Gagnon C., Thoenen H., Brain Res. 133, 291–303 (1977).
107. Bing J., Poulsen K., Hackenthal E., Rix E., Taugner R., J. Histochem. Cytochem. 28, 874–880 (1980).
108. Harper G. P., Barde Y. A., Burnstock G., Carstairs J. R., Dennison M. E., Suda K., Vernon C. A., Nature 279, 160–162 (1979)
109. Harper G. P., Thoenen H., J. Neurochem. 34, 893–903 (1980).
110. Beer D.J., Matloff S. M., Rocklin R. E., Adv. Immunol. 35, 209–215 (1984).
111. Levi-Montalcini R., Aloe L., Proc. Natl. Acad. Sci. USA 82, 7111–7115 (1985).
112. Kaiser E. T., Lawrence D. S., Science 226, 505–511 (1984).
113. Rajashekhar B., Kaiser E. T., J. Biol. Chem. 261, 13617–13623 (1986).
114. Ebendal T., Olson L., Seiger A., Hedlund K. O., Nature 286, 25–28 (1980).
115. Roszman T. L., Jackson J. C., Cross R.J., Titus M.J.., Markesbery W. R., Brooks W. H., J. Immunol. 135, 769s–772 (1985).
116. Hall N. R., McGillis U. P., Spangelo B. L., Goldstein A. L., J. Immunol. 135, 806–811.
117. Hamburger V., Perspectives in Biol. and Med. 18, 162–178 (1975).

1987

Physiology or Medicine

SUSUMU TONEGAWA

"for his discovery of the 'genetic principle for generation of antibody diversity' "

THE NOBEL PRIZE FOR PHYSIOLOGY OR MEDICINE

Speech by Professor HANS WIGZELL of the Karolinska Institute.
Translation from the Swedish text.

Your Majesties, Your Royal Highnesses, Ladies and Gentlemen,

The defence of our body against infections is carried out by the immune system, a talented cellular society with a capacity to distinguish between self and non-self and with a memory capable of remembering a previous contact for decades. The system is managing this through the inbuilt capacity in a single human being to produce billions of different forms of protective molecules, antibodies. The Nobel Prize of this year is given for the elucidation of the unique capacity of the immune system to produce this enormous diversity of specific antibodies.

Susumu Tonegawa is the great molecular biologist in immunology. In a series of ingenious experiments carried out in the middle of the 1970's he solved the problem how our limited genetic material is capable of generating the diversity required to create protection against established as well as future disease provoking microorganisms. When Tonegawa did his experiments at the Basel Institute of Immunology in Switzerland other scientists had already generated a considerable amount of knowledge regarding the features and functions of antibodies. But this knowledge had also led to uncertainty and even confusion. Antibodies are proteins and their structure is strictly ruled by genes, by the DNA in our chromosomes. When Tonegawa carried out his experiments it was commonly believed that each protein, each polypetide chain, was governed by its gene in a relation one to one. But at the same time calculations on the number of genes in the chromosomes in man determining proteins gave a number probably below one hundred thousand genes. They should suffice to all the proteins in the body, to the hemoglobin in the red blood cells, to the pigment in our eyes and so on. Only a minor part, maybe one percent, could probably be used for the creation of antibodies. Around one thousand genes being able to create billions of different forms of antibodies? The equation seemed impossible to solve.

Our antibodies are made up of two sorts of polypeptide chains, short and long ones. Tonegawa did first acquire a toolbox, filling it with the best precision tools there were of hybrid-DNA nature, developed new methods and started to study the actual construction of the genes determining the short chains of antibody molecules. He discovered something entirely new and revolutionary in genetics. On the chromosome where the gene for the short chain was expected to be located, there was not one single, but a string, of pearls of genes. One special gene resided at one position whereas two other sets of variable genes create two gene families, in all maybe around one hundred genes. When a cell should start to make antibodies — this was preceded by a gene-lottery.

One member of the largest gene family selected at random was cut out from the chromosome and moved close to a member of the second gene family, whereafter they created a functional gene for the short chain together with the solitary gene. Three and not one gene participate in the creation of the short chain of antibody molecules. Each member in one family can probably be linked to any one of the members of the second gene family, increasing variability by multiplication. The results showed beyond doubt that our body has the capacity to carry out advanced recombinant DNA processes. The intelligence of Nature can also be seen as the studies went on. The recombination of genes and their coupling together do not occur in exactly the correct manner. While such relative misfits should in other systems be bad, here they constituted yet another mechanism of increasing the diversity of antibodies. Experiments by Tonegawa as well as other scientists also revealed that the same genetic lottery principle did apply to the generation of the long chain although here the number of variants were even larger. Four different genes could be shown to create these chains together. The number of variant short chains should then be multiplied by the combinatorial possibilities of the heavy chain to give the variation at the antibody level, a fact which will also drastically enhance the diversity of antibodies.

The equation was in essence solved. A few hundred genes are used by the body in a new, revolutionary way and can thus generate billions of different antibodies. Through this genetic lottery the immune system is always prepared to react against known as well as unknown microorganisms. The economic usage of precious DNA is compensated by wasting more dispensable material. Every minute our body produces several millions of white blood cells — lymphocytes. Each one of these has undergone the hybrid-DNA procedure and is prepared with its own, unique antibodies. If not called upon to react they will rapidly die. If, however, they make contact with the fitting foreign structures they receive a reward, i.e., they are allowed to proliferate and live longer. After the great randomized gene lottery natural selection will pick the winners, thereby generating specific immunity, the cheapest and most efficient protection there is against infections.

Dr. Tonegawa,

On behalf of the Nobel Assembly of the Karolinska Institute I would like to congratulate you on your outstanding accomplishments and ask you to receive the Nobel Prize in Physiology or Medicine from the hands of His Majesty the King.

利根川 進

SUSUMU TONEGAWA

I was born in Nagoya, Japan on September 6th, 1939, the second of three sons. I have also a younger sister. My father was an engineer working for a textile company that had several factories scattered in rural towns in the southern part of Japan. The company policy made it necessary for my father to move from one factory to another every few years. I and my brothers and sister spent most of our childhood in these small provincial towns, enjoying the space and freedom of the countryside. As my elder brother and I reached adolescence, however, my parents decided to send us to Tokyo so that we could receive a better education.

I commuted to the prestigious Hibiya high school from my Uncle's home in Tokyo. During the high school years I developed an interest in chemistry, so upon graduation, I chose to take an entrance examination for the Department of Chemistry of the University of Kyoto, the old capital of Japan. After having failed once, I was admitted to this University in 1959. This happened to be one year before the first ten-year term of the defence treaty between Japan and the United States expired and the governments of both countries were preparing for a second ten-year term.

The nation was deeply divided between the pragmatic pro-American conservatives and the idealistic anti-military leftists. Being the home of the most radical leftist student groups, classes at Kyoto University were often cancelled due to frequent political discussions and demonstrations on the streets. I was only a passive participant, withdrawn from the turmoil, but could not help having a feeling of defeat shared with many of my classmates when the treaty was finally extended for the next ten-year term. I believe that this experience might have been a major factor in making me give up the original goal of becoming a chemical engineer to pursue the academic life.

I became fascinated by the then blossoming science of molecular biology when in my senior year I happened to read the papers by Francois Jacob and Jacque Monod on the operon theory. I decided to pursue graduate study in molecular biology and was accepted by Professor Itaru Watanabe's laboratory at the Institute for Virus Research at the University of Kyoto, one of a few laboratories in Japan where U.S.-trained molecular biologists were actively engaged in research. However, only two months after I started my work in his laboratory, Professor Watanabe called me into his office and suggested that I carry out my graduate study in the United States. He explained how inadequate the graduate training programs in molecular biology laboratories were in Japan, including his own, and offered to help in my application to some major universities in the United States, if I would seriously consider studying abroad.

At that time, it was a common career development for a Japanese molecular biologist to go to the United States for a few years of postdoctoral study after obtaining the Ph.D. in Japan. I already had a vague wish to follow that pattern. Professor Watanabe's advice to enroll in an American graduate school therefore came to me as a bit of a surprise, but I was excited by the idea and accepted his help immediately. I cannot thank Professor Watanabe enough for this critical suggestion in the early phase of my scientific career.

With the additional help of Dr. Takashi Yura, then an assistant professor in Watanabe's laboratory, I was admitted to the graduate school of the Department of Biology of the University of California at San Diego that had recently been established by Professor David Bonner in La Jolla, the beautiful southern Californian town near the Mexican border.

At UCDS I studied in the laboratory of Professor Masaki Hayashi, carrying out a thesis project on the transcriptional control of phage λ and received my Ph.D. in molecular biology in 1978. I remained in Professor Hayashi's laboratory as a postdoctoral fellow working on the morphogenesis of a phage, ØX174, until early 1979. Then I moved, also as a postdoctoral fellow, across the street to the laboratory of Dr. Renato Dulbecco at the Salk Institute.

Like many others, I believed that the golden age of prokaryotic molecular biology was coming to an end and that the great excitement would be in higher organisms. However, the complexity of high organisms was baffling and the necessary tools seemed hopelessly insufficient. Small tumor viruses like polyoma and simian virus 40, the biological material primarily dealt with in Dulbecco's laboratory, seemed to offer a bridge for the gap between prokaryotes and eukaryotes. Indeed Dulbecco's laboratory was filled with first-class postdoctoral fellows from around the world, who were trained in prokaryotic molecular biology and who came there intending to expand their research into eukaryotic molecular biology.

My project was to define the transcripts of SV40 during lytic infection and in transformed cells. Since this was the pre-restriction enzyme and pre-recombinant DNA age, the information I could obtain was very limited. However, being a member of the best laboratory in the field I glimpsed the excitement of the cutting edge of scientific research. Furthermore, I very much enjoyed the free and stimulating atmosphere of the laboratory. Unfortunately, as an awardee of a Fulbright travel grant, my U.S. visa was to expire by the end of 1970 and I had to leave the country for at least two years before I was eligible for another U.S. visa.

I had two or three job possibilities outside of the United States, but none were particularly interesting. In the autumn of 1970, only a few months before my visa was to expire I received a letter from Renato Dulbecco who was travelling in Europe. Renato mentioned the newly established Basel Institute for Immunology in Basel, Switzerland, and suggested that the time might be ripe for a molecular biologist to tackle immunological problems. I had very little knowledge of immunology, but decided to take Dr. Dulbecco's advice and sent an application letter to the Director of the Institute, Professor Niels Kaj Jerne, who offered me a two-year contract.

In the winter of 1971, I thus found myself surrounded by immunologists in this small town located in the middle of Europe. I must admit that the first year in the Institute was not easy for me. I had a continuing interest in work on SV40, but I was also keenly aware that I would not be able to take much advantage of the my circumstances if I isolated myself by pursuing that subject. I therefore decided to study immunology in the hope of finding an interesting project.

An immunologist, Dr. Ita Askonas, and a geneticist, Charles Steinberg, were very helpful to me on my entering the new field. By the end of 1971, I was introduced to the great debate on the genetic origins of antibody diversity. I felt from the beginning that I could contribute to resolving this question by applying the recently invented techniques of molecular biology, namely, restriction enzymes and recombinant DNA. Initially I worked only with my skillful technicians, Monica Shöld and Rita Schuller, but was soon joined by Drs. Nobumichi Hozumi, Minoru Hirama, and Christine Brack. Later, as my research group expanded, I had the good fortune to work with many capable postdoctoral fellows and devoted technical assistants. In addition, Charles Steinberg was a very important collaborator and consultant, particularly in the initial phase of the research.

Looking back, the research progressed with amazing speed from 1974 to 1981, the year I left Basel. We all worked hard and had had a great deal of fun. Our work resolved the long held debate on the genetic origin of antibody diversity. It turned out that this diversity is generated by somatic recombination of the inherited gene segments and by somatic mutation. To our very good fortune, Director Niels Jerne was quick to understand the importance of our approach and became a staunch supporter of the research in its early phase.

In the beginning of the 1980's I began to feel that the great mystery of antibody diversity had been solved, at least in its outlines. I thought that it might be good to change my environment to launch into a new project. I also recalled that I had initially come to Switzwerland with the intention of staying for two years and then returning to the United States. Fortunately, I received a few offers from the United States and decided in 1981 to take a professorship at the Center for Cancer Research at M.I.T. Professor Salvadore Luria, Director of the Cancer Center, was extremely helpful, not only in bringing me to M.I.T., but also in providing me with a beautiful laboratory.

The research projects on which I had decided concerned two major problems. One was to investigate the role of somatic rearrangement in the activation of the rearranged antibody gene, and the second was to extend the research in Basel to "the other half" of the immune system, namely, to the antigen receptor of T cells. Fortunately, we could contribute to the understanding of both problems by discovering a tissue-specific transcriptional enhancer in the immunoglobulin heavy chain gene and by identifying, cloning, and sequencing genes coding for the polypeptide subunits of the T cell receptor. A particularly intriguing development made during the latter study was the identification of a gene that led to the discovery of a new T cell receptor, $\gamma\delta$. While the function of the T cells bearing this receptor is currently unknown, data accumulated during the past year in our laboratory as well as many other laboratories suggest that these T cells may be involved in an entirely new aspect of immunity.

When I look back on my scientific career to-date, I am amazed at my good fortune. At every major turn, I met scientists who were not only at the very top in their own fields, but who also gave me insightful advice and generous help. I am most grateful to Professors Itaru Watanabe, Renato Dulbecco, Niels Kaj Jerne, Charles Steinberg, and Salvadore Luria. I also wish to extend my unending gratitude to many colleagues and technical assistants.

My parents were firm believers that education is the best asset that parents can give to their children. I am deeply grateful to them for their outstanding support of my study and professional career. I am extremely grateful to my wife, Mayumi, whom I married in September 1985 for her devotion, interest, encouragement and criticism. I also wish to express my sincere thanks to my first wife, Kyoko, for her limitless devotion during my days in La Jolla and Basel.

I have been fortunate enough to receive many professional honors which include: The Cloëtta Prize of Foundation Professor Dr. Max Cloëtta, Switzerland (1978), Warren Triennial Prize of the Massachusetts General Hospital, U.S.A. (1980), Genetics Grand Prize of Genetics Promotion Foundation, Japan (1981), Avery Landsteiner Prize of the Gesselshat für Immunologie, West Germany (1981), Asahi Prize of Asahi – Shimbun (Asahi Press), Tokyo, Japan (1982), Louisa Gross Horwitz Prize of Columbia University, New York, U.S.A. (1982), The V.D. Mattia Award of the Roch Institute of Molecular Biology, Nutley, U.S.A. (1983), Gairdner Foundation International Awards of the Gairdner Foundation, Toronto, Canada (1983), Person of Cultural Merit "Bunkakorosha" of the Japanese Government (1983), Order of Culture "Bunkakunsho" from the Emperor of Japan (1984), Bristol-Myers Award for Distinguished Achievement in Cancer Research (1986), Robert Koch Prize of the Robert Koch Foundation, West Germany (1986). Albert and Mary Lasker Award, New York City (1987) and NOBEL PRIZE in Physiology or Medicine, Stockholm, Sweden (1987).

SOMATIC GENERATION OF IMMUNE DIVERSITY

Nobel lecture, December 8, 1987

by

SUSUMU TONEGAWA

Center for Cancer Research and Department of Biology, Massachusetts Institute of Technology, Cambridge, Massachusetts, U.S.A.

One day in the fall of 1970, I received an airmail letter from Renato Dulbecco who was travelling in Europe. At that time I was a postdoctoral fellow in his laboratory at the Salk Institute. The letter, written on stationery of the Hotel Hassler in Rome said:

"Dear Susumu,

I don't know what arrangements you have made for after your departure from La Jolla at the end of the year but I would like to mention to you another possibility. The Institute of Immunology in Basel, Switzerland will start operating in a month. They already have an excellent collection of immunologists, but have not yet built an adequate background in molecular biology. I talked about you to Niels Jerne, the Director, and they are interested in having you there ... There are many immunologically interesting phenomena obtained with crude RNA preparations but they are unreliable because RNA is not characterized. In general, it seems the best system for understanding development at a molecular level and you may like to get into such a field. If you are interested, write to Niels K. Jerne, Basel Institute for Immunology, 487 Grenzacherstrasse..."

Thanks partly to this remarkably prophetic letter and partly to the U.S. immigration law that prevented me from remaining in the U.S.A., in February 1971 I found myself in this cozy Swiss town almost completely surrounded by immunologists. For someone who had had no formal training in immunology whatsoever and had never even visited Switzerland, it was a rather drastic change. Indeed, the first twelve months at the Basel Institute were not easy.

After I arrived in Basel I initially attempted to continue the project of my days in Dulbecco's laboratory, namely, the transcriptional control of the simian virus 40 genes. However, I soon realized that this was not a subject that aroused great interest in an Institute almost entirely staffed by immunologists nor one that allowed me to take full advantage of my many talented colleagues. I therefore decided to learn immunology by talking to them, reading papers, and asking questions. An immunologist, Ita Askonas, and a geneticist, Charlie

Steinberg, became my tutors and were most helpful to me in getting into a new field. It was during this process that I was introduced to the problem of the origin of antibody diversity.

The problem

Immunologists agreed that an individual vertebrate synthesizes many millions of structurally different forms of antibody molecules even before it encounters an antigen. Moreover, Gerald Edelman and Rodney Porter had shown that a typical antibody molecule is composed of two identical light chains and two identical heavy chains (1,2). It had also been found that each of these two types of chain exhibits great sequence variability in the amino terminal region between one antibody molecule and the next and little sequence variability in the carboxyl terminal regions (3). These two regions were then referred to as the variable, or V, and the constant, or C, regions. However, immunologists and geneticists were divided for many years into two schools of thought with respect to the issue of whether the genetic diversity required for the synthesis of these proteins is generated during evolution and is carried in the germline or during development in which case it would be present in somatic but not germline cells. One school of thought held that the germline must include a separate gene for every polypeptide that ultimately appears in an antibody molecule (4). In this germline theory, antibody or immunoglobulin genes are expressed in exactly the same way as those for any other protein, and no special gene-processing mechanisms are needed. On the other hand, the model requires an enormous number of immunoglobulin genes inherited from the parents. While the four chain structure of an immunoglobulin molecule allows diversity to be generated by chain paring, the number of genes required for both light and heavy chains is still very large. One major difficulty for germline theories of antibody diversity was the observation that all antibody polypeptide chains of a given type share a common genetic marker (allotype) that segregates as a single Mendelian gene. If there were many thousands of light and heavy chain genes, how could the same genetic marker in all of these genes have been maintained?

The second theory supposed that there are only a limited number of antibody genes in the germline, and that these genes somehow diversify as the antibody-forming B lymphocytes emerge from their stem cells. In other words, the diversification of antibody gene sequences takes place in specialized somatic, or body, cells rather than being carried from generation to generation by the germ cells (5–7). One attraction of this latter theory is that it relieves the host of the need to commit a disproportionately large fraction of the inherited genes to code for antibodies, but the theory demands an unprecedented mechanism for diversifying the inherited genes somatically.

Arguments for and against these contrasting ideas were made both vocally and in written form for many years. However, all of these arguments were based on the interpretation of amino acid sequences of immunoglobulin polypeptide chains or on the generally accepted principles of evolution and genetics. No direct evidence for either view had been obtained. This was because no technique was available that would allow an analysis of the fine structure of specific genes from higher organisms.

Gene counting

In the early seventies the technology for purifying a specific eukaryotic mRNA was just becoming available. Furthermore, a method to determine the number of copies of a specific gene by kinetic analysis of nucleic acid hybridization had already been established (8,9). These technical developments led some scientists, including myself, to think that one can experimentally determine the number of immunoglobulin genes contained in a germline genome and thereby decide which of the two major theories of antibody diversity is correct. The validity of this approach is based in part on the fact that the V region of a given chain type, while being different, exhibits a high degree of amino acid sequence homology. It was therefore thought that mRNA coding for a specific immunoglobulin polypeptide chain would hybridize not only with its own gene but also with many other immunoglobulin genes, if they existed in a germline genome.

I thus obtained mouse myeloma cells and put my effort to purifying immunoglobulin mRNA and carrying out the hybridization studies. However, the initial studies focusing on the mouse κ light chain and heavy chain genes gave ambiguous results. The difficulty was primarily due to uncertainty about the purity of the mRNA used as the hybridization probe as well as a lack of knowledge on the extent to which a probe will hybridize with the related but not identical genes, and the precise effect of sequence differences on hybridization kinetics. Thus, it turned out to be nearly impossible to make a convincing interpretation of the data obtained in these early studies in relation to the issue of the evolutionary versus somatic generation of antibody diversity.

One subsequent series of experiments which I carried out on genes coding for the mouse λ light chains, however, was very encouraging (10). Using a mRNA preparation that was more than 95% pure, I could show that the mouse λ light chain gene is reiterated no more than the β globin gene. The latter gene had been shown to be essentially unique. Fortunately, Weigert, Cohn and their coworkers had identified at least eight different V_λ region sequences among BALB/c-derived myelomas (11). Since these V regions were highly homologous, differing by only one, two or three amino acid residues, it was very likely that the corresponding genes would crosshybridize extensively if they existed separately in the germline genome. Furthermore, statistical analysis of λ light chain-secreting myelomas strongly suggested that a BALB/c mouse has the capacity to synthesize many more than the eight different V_λ regions identified. Thus, the number of mouse λ genes determined experimentally (no more than a few) was far smaller than the number of different V_λ regions (at least eight, most probably many more) detected in proteins. On the basis of these results I was convinced that a somatic diversification occurs in this gene system.

Rearrangement

In the meantime I became aware that some immunologists had been speculating that immunoglobulin polypeptide chains may be encoded by two separate DNA segments, one each for the V and C regions. Drawing an analogy from the elegant Campbell model (12) on the integration and excision of a phage λ genome, Dreyer and Bennett had further suggested that one of many "V genes" may be excised out from the original chromosomal position and joined with the single "C gene" in an immunoglobulin-producing B cell (13). This model successfully explained the maintenance of the common genetic marker in all immunoglobulin polypeptide chains of a given type by postulating a single C gene for that chain type. Although a somatic recombination between the "V and C genes" is an inherent aspect of the model it is clearly a version of the germline theory of antibody diversity because the model assumed that the germline genome carries many "V genes", one for every V region that an organism can synthesize.

When the Dreyer and Bennett model was published in 1965, it was not widely accepted by biologists. This is understable because the model was built on two hypotheses, both of which violated the then current dogmas of biology. These were the principles of one gene encoding one polypeptide chain and of the constancy of the genome during ontogeny and cell differentiation. My personal reaction to the model when I learned of it in the early nineteen seventies was also that of skepticism. However, at the same time I thought that the model might be testable if one were to use restriction enzymes. While in Dulbecco's laboratory I had heard of Daniel Nathan's breakthrough in the analysis of the SV4O genome by an application of the then newly discovered restriction enzymes (14). As one who used to struggle to define the transcriptional units of this DNA virus I was keenly aware of the power of these enzymes for the analysis of DNA structure. However, an extension of the restriction enzyme analysis from a viral genome of 5×10^3 base pairs to the 2×10^9 base pair genome of an eucaryote as complex as a mouse, required the use of an additional trick for the detection of a specific DNA fragment in a vast array of irrelevant fragments. An obvious solution seemed to lie in the combination of an electrophoretic separation of enzyme-digested DNA and the sensitive technique of nucleic acid hybridization. I discussed with Charlie Steinberg the need for developing a method that allows an *in situ* detection of a specific DNA sequence among the electrophoretically fractionated DNA fragments, but we really could not come up with a good idea worthy of exploring. As we all now know, a very simple and elegant method ideal for this purpose was later developed by Edward Southern (15).

A few weeks passed by before I accidentally saw in one of the Institute's cold rooms a huge plexiglass tray in which someone was fractionating serum proteins by starch gel electrophoresis. I thought one may be able to fractionate a sufficient amount of digested DNA in a gel of such dimensions, so that the DNA eluted from gel slices could be used for liquid phase hybridization. A quick calculation indicated that the experiment was feasible. Nobumichi Hozumi, a postdoctoral fellow in my laboratory, and I therefore decided to give it a try although we were keenly aware of the intense labor required by this type of experiment. As hybridization probes we used purified ϰ or λ light chain mRNA (V+C-probe) and its 3'-half fragment (C-probe) that had been iodinated to a high specific activity. The rationale of the experiment was as follows: First, if an immunoglobulin polypeptide chain is encoded by two "genes" V and C in the germline genome, it is highly probable that treatment with a restriction enzyme will separate these DNA sequences into fragments of distinct size, germline genome, it is highly probable that treatment with a restriction enzyme will separate these DNA sequences into fragments of distinct size, thus allowing their electrophoretic separation. Second, if a somatic rearrangement joins the V and C "genes" it is also highly probable that the myeloma DNA digested with the same restriction enzyme will contain a DNA fragment carrying both V and C "genes".

The results obtained were clear cut: To our pleasant surprise the patterns of hybridization of the embryo (a substitute of germline) DNA and a ϰ-myeloma DNA were not only drastically different but also consistent with the occurrence of separate V and C "genes" and a joined V plus C gene, respectively (16). We were of course aware of the alternative interpretations of the results, such as a fortuitious modification of the enzyme cleavage sites in one of the two types of DNA. However, we considered these alternative explanations of the results unlikely because they all required multiple fortuitious events. Our confidence was fortified soon afterwards as the development of Southern blot techniques allowed us to carry out more extensive analyses using a variety of restriction enzymes and myeloma cells.

Joining of gene segments

While the experiments with restriction enzymes were informative, details of the rearrangement were difficult to come by with this approach. Forunately, recombinant DNA technology was just becoming available and was the ideal means for this purpose. Debates on the possible hazards of this type of research were flaring initially in the United States and shortly afterwards in European countries. In order to make sure that our research would not become a target of controversy, Charlie and I got in touch with Werner Arber at the University of Basel who was coordinating recombinant DNA research activities in Switzerland. A small informal working group was set up by the local researchers interested in this technique. The consensus of the group, which was supported by most of the other Swiss researchers, was that we should all follow the practices and guidelines being adopted in the United States. We met about once a month and exchanged information regarding both ethical and practical aspects of the technology.

On the basis of the previous experiments attempting to count immunoglobulin genes, I thought that it would be wise to start with the mouse λ light chain system, the simplest of all chain types that had been studied. Our goal was to clone the V_λ and C_λ "genes" in the germline state from embryonic cells as well as the rearranged V plus C "genes" from a λ myeloma, and to determine the relationship between these genomic DNA clones by electronmicroscopy and

DNA sequencing. No precedent existed at that time for cloning "unique" eucaryotic genes. We therefore had to devise a few tricks as we attempted to clone the first immunoglobulin gene. For instance, our available probe at that time was again 95 % pure mRNA rather than a cDNA clone. This situation made the screening of a large number of DNA clones difficult because of the high background. To avoid this problem we pre-enriched the λ gene-containing genomic DNA fragments as much as possible using preparative R-loop formation (17,18), so that the DNA library constructed would have the clone of interest at a high frequency.

Starting with the embryonic DNA we could isolate a clone that clearly hybridized specifically with the λ mRNA (18). When an electronmicroscopist, Christine Brack, who had just joined us from the Biozentrum of the University of Basel, examined the mixture of this clone and λ mRNA that had been annealed under an appropriate condition, she found a beautiful R-loop from which about a half of the mRNA strand protruded. This and additional analysis convinced us that we had cloned a V_{λ} "gene" to which no C "gene" was contiguously attached, thus confirming at the DNA clone level that the V and C "genes" are indeed separate in the germline genome. A subsequent DNA sequencing study carried out in collaboration with Allan Maxam and Walter Gilbert of Harvard University revealed that this DNA clone corresponded to the V "gene" for the λ_2 subtype (19).

In the meantime Minoru Hirama, another postdoctoral fellow, succeded in preparing λ and ϰ cDNA clones. Once these probes became available isolation of the genomic clones became much easier. My assistant Rita Schuller and I isolated a number of genomic DNA clones from λ and ϰ chain-synthesizing myelomas as well as from embryos (20,21). Analysis of these DNA clones by electromicroscopy, by restriction enzyme mapping and by DNA sequencing, not only confirmed the somatic rearrangement of immunoglobulin genes but also revealed some striking features of their arrangement and rearrangement (Fig. 1). These can be summarized as follows:

1. Although the V and C "genes" are rearranged and are much closer to each other in myeloma cells than in embryo cells, they are not contiguous and are separated by a few kilobases of DNA sequence that does not participate in coding of the polypeptide chain. This untranslated DNA sequence present within the rearranged, complete immunoglobulin gene was unanticipated and was also among the first demonstrations of an intron in eucaryotic genes (22).
2. The V "gene" found in the germline genome is about 13 codons short when it is compared to the length of the conventionally defined V region. The missing codons were found in a short stretch of DNA referred to as a *J* (or joining) gene segment that is located many kilobases away from the incomplete V "gene" (referred to as a *V* gene segment) and a few kilobases upstream of the C "gene" (also referred to as a *C* gene segment). In myeloma cells the rearrangement event attaches the *J* gene segment to the *V* gene segment and thereby creates a complete V region "gene" (20,23).
3. The signal peptide is encoded in yet another DNA segment referred to as the

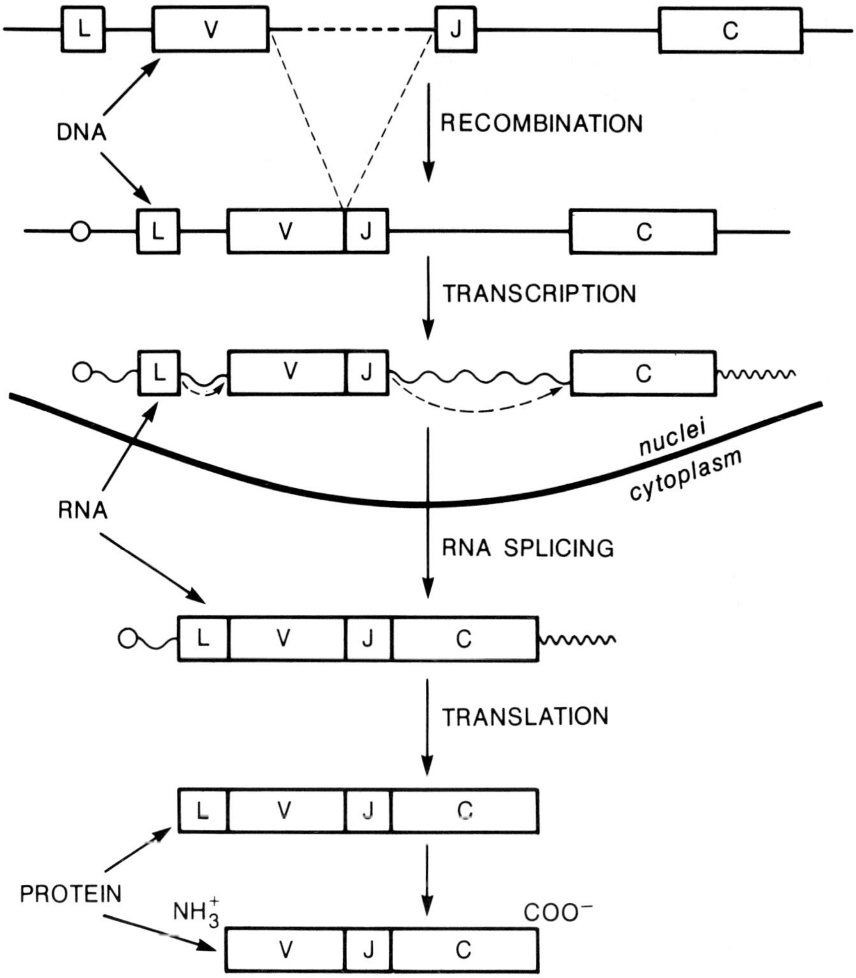

Figure 1. The basic scheme for rearrangement and expression of an immunoglobulin light chain gene. At top is an arrangement of the gene segments on a germline genome. Somatic rearrangement links the *V* and *J* gene segment and generates a complete light chain gene shown just below the germline genome. The entire gene containing the leader exon (L), the V region exon (V and J), the C region exon (C), and the introns present between these exons are transcribed into a premRNA in the nuclei of the B cell. The premRNA is processed by RNA splicing as it is transported from the nuclei to the cytoplasm. The resulting mRNA, devoid of introns, is translated in the endoplasmic reticulum into a nascent polypeptide chain from which a mature λ light chain is generated after cleavage of the signal peptide.

L (or leader) exon that is separated from the *V* gene segment by a short intron (19,23).

Finding that the V_λ "gene" was split into two gene segments, V_λ and J_λ, in the germline genome was completely unexpected. But as soon as this discovery was made its implication for the somatic generation of antibody diversity was obvious. If the germline genome carries multiple copies of different *V* and *J* gene segments the number of complete V "genes" that can be generated by random joinings between these two types of gene segments would be much greater than the total number of the inherited gene segments. Thus, contrary to the Dreyer and Bennett original concept, DNA rearrangement could provide a major means for the somatic diversification of antibody molecules. The amino acid sequence data of the κ light and heavy chains were consistent with the view that the germline genome carries multiple different *V* and *J* gene segments (24,25). Indeed, the nucleotide sequence analysis of the mouse κ chain gene complex carried out both in my laboratory and in Phillip Leder's laboratory at The United States National Institutes of Health confirmed that a germline genome contains multiple *V* and *J* gene segments and that these gene segments are joined in different combinations in each myeloma cell (20,26). Four different J_κ gene segments were found several kilobases upstream of the C_κ gene segment. The exact number of V_κ gene segments is unknown even today, but it is estimated to be two to three hundred (27).

Heavy chain genes

Inasmuch as an immunoglobulin heavy chain is also composed of V and C regions, it was reasonable to expect that its gene also would undergo the type of DNA rearrangement discribed for the light chain genes. This supposition was confirmed by Leroy Hood and his coworkers at California Institute of Technol-

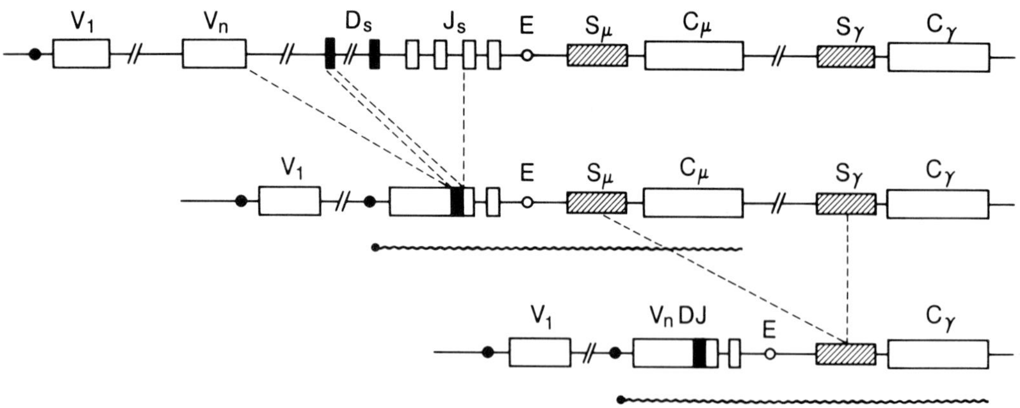

Figure 2. Organization of the immunoglobulin heavy chain gene family. At top, middle, and bottom are organization in a germline genome, in a genome of B cells synthesizing a μ class heavy chain, and in a genome of a plasma cell synthesizing a γ class heavy chain, respectively. A mouse haploid genome carries several hundred different *V* gene segments, about a dozen *D* gene segments, four *J* gene segments, and one copy of *C* gene segment for each of the eight different classes or subclasses of immunoglobulin heavy chains. In a virgin B cell one copy each of the *V*, *D*, and *J* gene segment pools have been linked up and the joined *VDJ* DNA sequence is transcribed into a premRNA together with the C_μ gene segment. In different B cells of the same organism a different set of *V*, *D*, and *J* gene segments are usually hooked up and expressed. As the virgin B cell differentiates either to a plasma cell or to a memory B cell (see Fig. 5) the second type of somatic recombination called "switch recombination" often occurs between a region (S_μ) located upstream of the C_μ gene segment and another region (S_γ) located upstream of the C_γ gene segment. As shown at the bottom, the switch recombination replaces the C_μ gene segment with the C_γ gene segment without changing the *VDJ* exon. Filled circles designate transcriptional promotors present at the upstream of every *V* gene segment. The open circle designates the transcriptional enhancer (102, 103) which together with the promotor activate the rearranged heavy chain gene for a high level expression.

ogy and by ourselves (Fig. 2) (28,29). As in κ genes four *J* gene segments were found several kilobases upstream of the *C* gene segments coding for the C region of the μ class heavy chain. Multiple *V* gene segments were also identified.

While these features of the organization of heavy chain genes are essentially the same as those of the light chain genes, one observation made during these studies suggested that the somatic assembly of gene segments plays an even more prominent role in the diversification of heavy chains than of light chains. It was found that from one or two to a dozen amino acid codons that are present in the V-J junction region of the assembled gene are not found in either of the apparently corresponding germline *V* or *J* gene segments (30,31). This suggested that a third type of short gene segment referred to as D (or diversity) might participate in the somatic assembly of a heavy chain gene. Indeed, Hitoshi Sakano and Yoshi Kurosawa, two postdoctoral fellows in my laboratory, soon discovered about a dozen *D* gene segments (32,33) which were subsequently mapped in a region upstream of the *J* cluster in the germline genome (34,35). Thus, the construction of a complete heavy chain V "gene" requires two DNA recombinational events, one joining a *V* with a *D* gene segment and the other the same *D* with a *J* gene segment.

Recombination Rule

The joining of V-J or V-D-J involves a site-specific recombination. It might therefore be expected that these gene segments would carry sequences in the vicinity of the joining ends that are recognized by a putative site-specific recombinase. Furthermore, such recognition sequences are likely to be common for all gene segments of a given type (e.g. V_κ's), because they all seem to be capable of joining with the common set of gene segments of the appropriate type (e.g. J_κ's). There are indeed a heptamer and a nonamer that are conserved in the region immediately downstream of each V_κ gene segment (Fig. 3) (36,37). Sequences complementary to the V_κ heptamer and nonamer were also found in the region immediately upstream of each of the four J_κ gene segments. The same sets of sequences were also found in the corresponding regions of the V_λ and J_λ gene segments (36). When the heavy chain V and J gene segments were analyzed subsequently they too had the common conserved sequences (30,31). Furthermore, *D* gene segments carry the heptamer and nonamer sequences both upstream and downstream (32,33). Another interesting feature of these putative recognition sequences is the fact that the length of the spacer between the heptamer and nonamer is either about 12 or 23 base pairs (30,31). Furthermore, a gene segment carrying a recognition sequence with one type of spacer is able to join only with a gene segment with the spacer of the other type. This 12/23 base pair spacer rule seems to be adhered to strictly. Little is currently known about the recombinase, but proteins with an affinity to the heptamer or nonamer have been identified in the extract of Abelson virus transformed pre B cell lines in which the rearrangement occurs *in vitro* at a relatively high frequency (38,39).

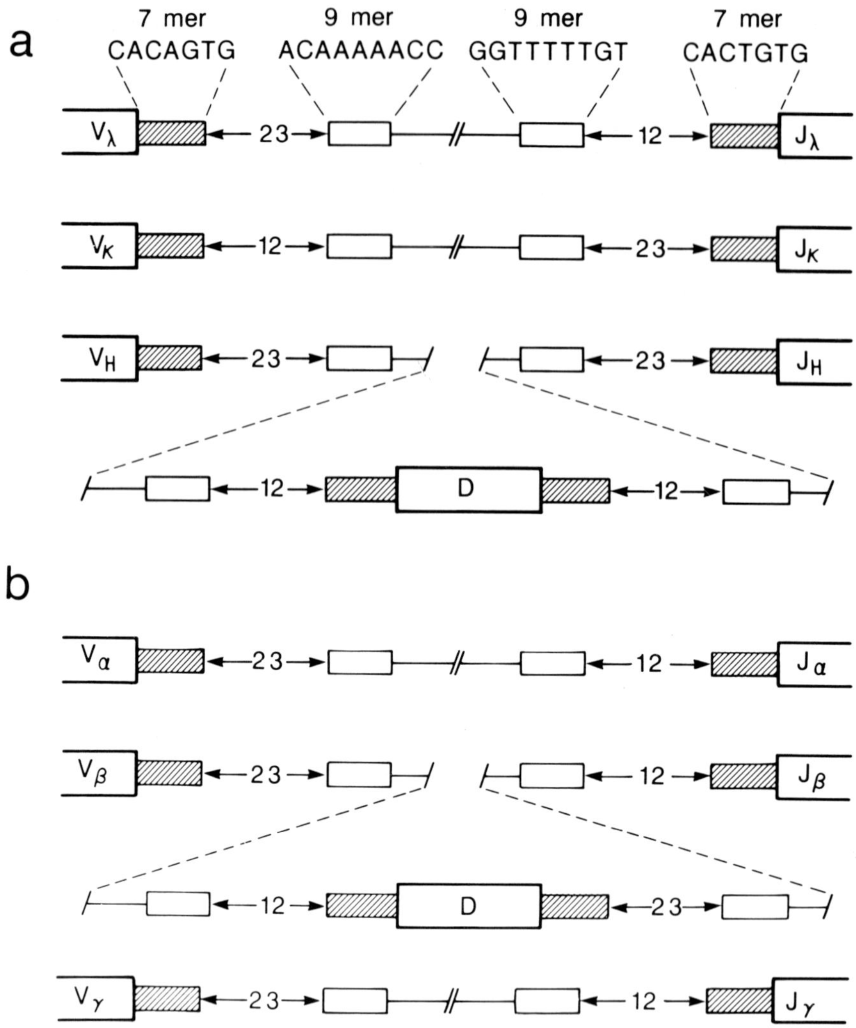

Figure 3. Putative recognition sequences for the rearrangement of immunoglobulin and T cell receptor genes. The conserved heptamer and nonamer sequences and the length of the spacer between these sequences are schematically illustrated for immunoglobulin (Panel a) and for T cell receptor (Panel b) gene families. The sequences shown on top are consensus sequences. Individual sequences may deviate from these consensus sequences by a few nucleotides.

Diversity generated at the joins

When the deduced amino acid sequence of a germline J_κ gene segment was compared with the determined amino acid sequences of those κ chains that are encoded in part by that J_κ gene segment, it was noticed that the 5' end of the J_κ gene segment is not prefixed but rather shifts toward upstream or downstream by several base pairs in different joining events (36,37). This flexibility in the precise site of the joining was subsequently found to be a characteristic of the joining ends of other gene segments rather than of just J_κ -gene segments (31). It applies even when the same pair of gene segments were joined in different B cell precursors, such that the completed V "genes" are likely to have slightly different codons in the junction regions.

The V-D and D-J junctions exhibit diversity of yet another type. We found that up to a dozen base pairs of essentially random sequence are inserted in these junctions apparently without a template during the breakage and reunion of the recombining gene segments (32,33). While the precise mechanism is yet unknown, the terminal deoxynucleotide transferase which is found in early B lymphatic nuclei or an enzyme with similar characteristics is thought to play a role in this phenomenon (40).

The part of the V region affected by the above two diversification mechanisms is limited. But this does not mean that they do not play a significant role in the determination of antibody specificity. On the contrary, the junctions encode the most variable two of the six loops of polypeptides that make up the antigen binding region of the antibody molecule (Fig. 4). Furthermore, specific cases are known where the affinity of an antibody to a defined antigen is drastically altered by a slight change in one junctional sequence (41). Thus, the junctional variation is also a potent somatic generator of antibody diversity.

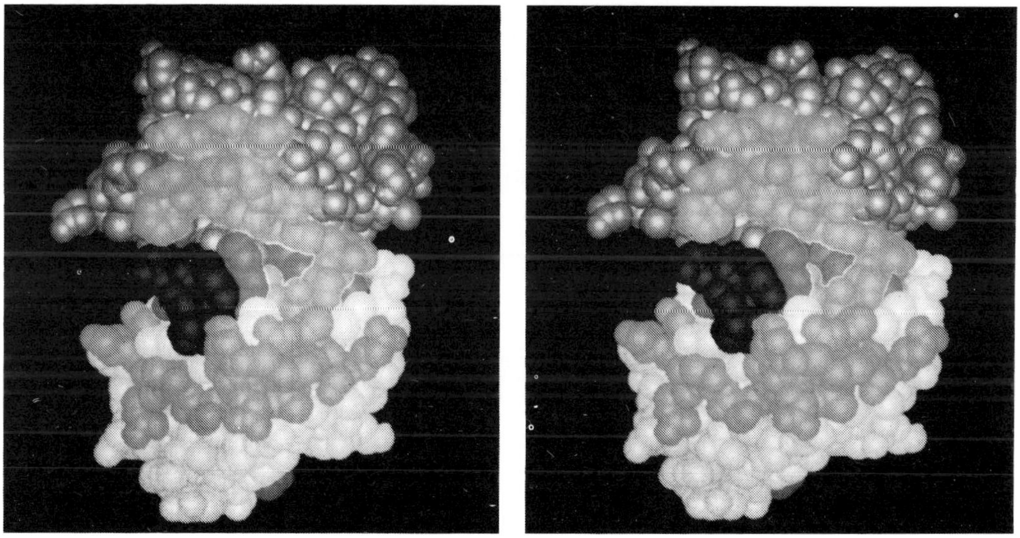

Figure 4. Space-filling, stereo image of an antibody combining site. Atomic coordinates of mouse immunoglobulin MOPC 603 (104) were used to produce the picture. The heavy chain variable domain is color-coded dark grey, the light chain variable domain light grey. The hypervariable regions (except the VH third hypervariable region) are blue, the heavy chain segment coded for by the D gene is red, and the heavy and light chain segments coded for by the J genes are yellow. The D segment corresponds virtually exactly to the third heavy chain hypervariable region; hypervariable regions were defined as in Novotney *et al.* (105) except for the heavy chain second hypervariable region, which is marked as defined by Kabat *et* al. (25). The antigen of this particular immunoglobulin, phosphoryl choline, binds into the cavity in the middle of the picture in between the VH and VL domains, making contacts to amino acid residues belonging to the VH and J segments of the heavy chain and the VL segment of the light chain. Importance of the D segment is well illustrated in the two crystallographic structures of antibodies which bind the protein antigen lysozyme (106, 107). There, the contact area contributed by the D segment amount to 50% and 24%, respectively, of the total heavy chain contact area. This image was computer-generated by Jiri Novotny using the program SPHERE of Robert Bruccoleri.

Somatic mutation

When F. Macfarlane Burnet proposed the clonal selection theory he recognized the need for some kind of random genetic process in order to generate antibodies able to bind specifically to the vast variety of antigens (42). He considered somatic mutations as the most plausible mechanism. Subsequently, this idea was adopted and forcefully presented by many including Joshua Lederberg, Niels Kaj Jerne and Melvin Cohn (5,6,7).

The amino acid sequence data accumulated by Martin Weigert in Melvin Cohn's laboratory at the Salk Institute provided an excellent opportunity to examine directly the role of somatic mutations in antibody diversity (7,11). They had analyzed the λ_1 light chains derived from eighteen myelomas. All the mice were of an inbred strain BALB/c and should thus have been genetically identical. They found that twelve of the $V_{\lambda 1}$ regions were identical but that the other six differed both from the majority sequence and from one another by only one, two, or three amino acid residues. They proposed that BALB/c mice may carry only one germline $V_{\lambda 1}$ "gene" which codes for the majority sequence, and that all the other $V_{\lambda 1}$ regions observed are encoded by somatic mutants of this single $V_{\lambda 1}$ "gene" that arose in B cell development. As I already mentioned in an earlier section our gene-counting experiment by hybridization kinetics suggested that the germline BALB/c genome carries no more than a few $V_{\lambda 1}$ "genes". This number was reduced to one when we reevaluated the copy number by the more reliable Southern blotting method (20). The final proof of somatic mutation in $V_{\lambda 1}$ came when we cloned and sequenced the sole germline $V_{\lambda 1}$ gene segment and the rearranged λ_1 genes expressed in a myeloma (23). As Weigert and Cohn guessed the nucleotide sequence of the germline $V_{\lambda 1}$ gene segment corresponded to the major amino acid sequence, while the λ_1 gene expressed in the myeloma had been altered by single base changes.

Since this work several subsets of κ light and heavy chains and their germline V gene segments have been analyzed by cloning and sequencing (43–46). These results have all confirmed that somatic mutations further amplify the diversity encoded in the germline genome. Particularly revealing was the analysis carried out by Patricia J. Gearhart, Leroy Hood and their coworkers for the V_H regions associated with the binding of phosphorylclorine (PC). They demonstrated the single base changes can be extensive and yet are restricted to the joined VDJ sequences and the immediately adjacent regions (47,48).

Developmental control of rearrangement and hypermutation

Why have two extraordinary somatic genetic mechanisms, recombination and hypermutation, evolved in the immune system in order to carry out what appears to be one task, namely to diversify antibodies?

I believe that the answer may be the differential roles of these two genetic mechanisms. Thanks to the efforts of several independent groups of cellular and molecular immunologists a general picture is emerging that describes the relationship between the stages of B cell development and the occurance of somatic recombination or mutation (Fig. 5) (49–55). Somatic recombinations contributing to diversity are initiated first for the heavy chain and then for

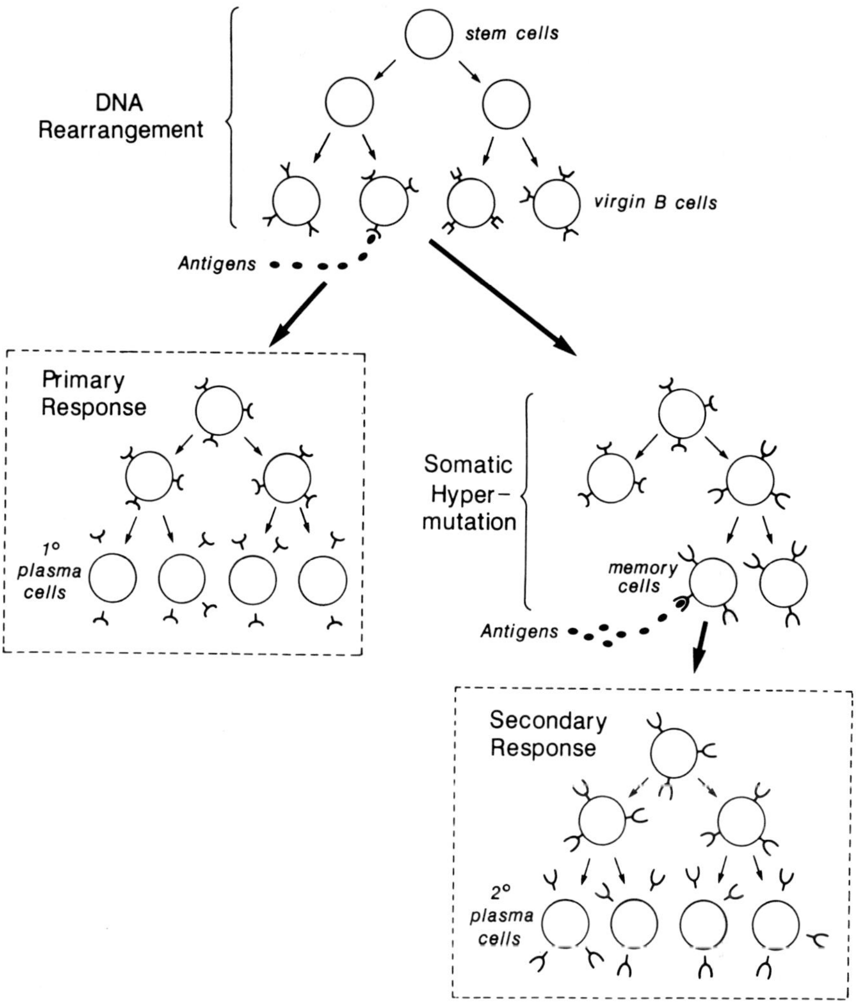

Figure 5. Differentiation of B cells. Note that the receptors present on the memory cells and the antibody molecules secreted by the plasma cells of the secondary response have a tighter fit to the antigen than the receptors on the ancestral virgin B cells or the antibodies secreted by the plasma cells of the primary response. See text for the full explanation.

the light chain during the differentiation of progenitor cells, and the completion of somatic recombination is accompanied by the appearance of virgin B cells (56–58). These B cells form clones each of which is composed of cells bearing homogeneous IgM molecules as surface receptors. Thus, somatic recombination is completed *prior to* any possible interaction of a B cell with antigens.

When an antigen enters the lymphatic system for the first time, it will be screened by these virgin B cells. The small fraction of these B cells that happen to have sufficient affinity for the antigenic determinants in question will respond and follow either of two pathways: they will produce the primary antibody response, or they will contribute to the generation of memory B cells.

In the former pathway, the selected B cells will proliferate and differentiate into antibody-secreting plasma cells. During this process, the C region of the heavy chain can switch from μ to another class, but mutation is rare in either the heavy or the light chain V region. Consequently, the antibodies secreted by plasma cells in the primary response would largely have the same V regions as the immunoglobulin receptors on the virgin B cells from which they derive.

By contrast, immunoglobulin remains in the cell surface receptor form during the other pathway taken by the antigen-activated virgin B cells, namely the generation of memory B cells. 'During this process the hypermutation apparatus appears to be most active, and the rate of the mutation approaches 10^{-3} base substitution per cell per generation. Antigen selects, in a stepwise fashion, better and better fitting mutants so that the immunoglobulins on the surface of memory B cells achieve a substantially higher affinity than the immunoglobulins on the ancestral virgin B cells. Switch recombination also occurs frequently during this process too. When the same antigen as the one that elicited the primary response reenters the body, the memory B cells are selectively propagated and differentiate into plasma cells. This is the so called secondary antibody response, which, therefore, consists of high affinity antibodies of "mature" isotype, and these antibodies show extensive somatic mutation in their V regions. Somatic mutations appear to cease after memory cells are generated, and little or no further mutation takes place during the secondary antibody response.

This scheme of B cell differentiation can be rephrased as follows. An organism is prepared for infection with pathogens bearing virtually any antigens with a large variety of resting B cells. These B cells bear unique immunoglobulin receptors encoded by one copy each of complete light and heavy chain genes that have been constructed by a random or quasi-random assembly of the inherited gene segments. Since the assembly occurs independent of antigens, and since the inherited gene segments are not usually selected during evolution for precise fit to most antigens, the antibody secreted by the plasma cells derived directly from the selected resting virgin B cells during a primary antibody response usually have a relatively low affinity. By contrast, the frequent single base changes that occur during the generation of memory B cells provide the organism with a great variety of finely altered immunoglobulin receptors from which only those with the best fit to the antigen in question will be selected. Since the plasma cells generated during the secondary antibody response are mostly direct descendants of these memory B cells, having no further alterations in the antigen-combining sites, these antibodies usually exhibit a much higher affinity for the antigen than do the primary antibodies. This explains the long known phenomenon of affinity maturation of antibodies during the course of repeated immunizations (59).

Thus, somatic creation of antibody genes can be viewed as a two step process. In the first step, blocks of gene segments are employed to build, in an antigen-independent fashion, a set of genes coding for antibodies of great diversity but with low affinity. In the second step, once the antigen is defined, a small selected set of B cells bearing low affinity antibodies as cell surface receptors undergo somatic mutations with the result that a fraction of them develop a higher affinity to that antigen and can be selected for further expansion. This process improves the ability of the immune system to detect a low concentration of antigens. One wonders what happens to those cells in which mutation did not improve affinity. A recent study suggests that at least some of these cells may be set aside for selection by different antigens (54). Thus, somatic mutation may also contribute to the repertoire of receptors specific for antigens not previously introduced into an immune system.

T cell receptors

As the mystery of the genetic origin of antibody diversity was unravelled at least in its basic outlines, it seemed natural to extend our research to "the other half" of the lymphoid system, namely T cells. Although we often discussed the idea of research on the nature of antigen recognition by T cells in the laboratory in the late seventies while I was still in Basel, the real work did not start until the early eighties in my new laboratory at M.I.T. Although T cells were known to recognize and distinguish antigens as precisely as B cells, nothing was known about the biochemical nature of the molecules responsible for this task, namely T cell receptors (TCR). This lack of information was in stark contrast to the wealth of information about antibodies. Much debate took place among cellular immunologists on the nature of these molecules. Some argued that T cell receptors are just another class of immunoglobulins. Others thought T cell receptors would have to be quite different from immunoglobulins. Indeed studies carried out in the late seventies had shown that the way in which a T cell recognizes an antigen is quite different from the way a B cell does: The T cell reacts to antigens on a cell surface, and the T cell receptor simultaneously recognizes both an antigen and a determinant present on a glycoprotein encoded by a gene in the major histocompatibility complex (MHC) (60–62). This discovery raised another issue: Does a T cell recognize two determinants with one receptor or does it have two receptors, one for antigen and the other for a MHC product?

The receptor protein was first detected in 1983 in experiments carried out by three independent groups of scientists headed by James P. Allison, Ellis L. Reinherz, and Phillipa Marrack and John Kappler (63–65). They prepared antibodies that bind to a protein on the T cell surface. Since these proteins were similar but exhibited clonally distributed structural diversity, they were thought to be a good candidate for the receptor. Furthermore, the antibodies they prepared were T cell clone-specific, and they could show that these antibodies blocked activation of the T cell clone in a clone-specific fashion. The receptor identified by these experiments was composed of two polypeptide subunits, designated α and β, that are held together by a disulfide bond. These studies were critical in the sense that the receptor was finally identified, its overall structure defined, and its predicted structural variability confirmed. However, the paucity of the protein on the T cell surface and the absence of the secreted form of the receptor made it very difficult to obtain further information about the structure of this molecule, especially its amino acid sequence.

α and β genes

In the meantime, molecular biologists were attempting to identify the genes coding for the T cell receptor. This turned out to be a much more difficult endeavor than the cloning of immunoglobulin genes. Although T cell lines and hybridomas expressing a homogeneous receptor were becoming available, these cells were more difficult to grow than myelomas, and the amount of the receptor made was at least two orders of magnitude less than the amount of immunoglobulin produced by a myeloma cell. In 1984, Mark Davis and his coworkers at Stanford University and Tak Mak and his coworkers at the University of Toronto independently made a breakthrough (66–68). Their experimental strategy depended on two assumptions. First that mRNAs coding for the α and β polypeptide chains are present in a T cell hybridoma or T cell tumor but are absent in B cell tumors. And second that the α and β chain genes are rearranged in T cells in a manner similar to the immunoglobulin genes in B cells. Thus, they made a library from the fraction of T cell-cDNA that did not hybridize with B cell-derived mRNA and tested each T cell-specific cDNA clone for rearrangement of the corresponding gene in T cells. As the source of T cells, Davis' group used a hybridoma obtained by fusing a mouse helper T cell recognizing antigen plus self class II MHC molecules with a T cell tumor while Mak's group used a human T cell tumor. The two groups came up with one common class of cDNA clones that satisfied the above criteria. The nucleotide sequence showed that the corresponding polypeptide chain is significantly (30–35%) homologous to immunoglobulin chains. Furthermore, the cDNA clones contained sequences homologous to V and C regions in the correct orientation. Thus, it seemed certain that the gene represented by this class of cDNA clones codes for one of the two subunits of the T cell receptor. That this gene encodes the β subunit was soon confirmed by determination of the partial amino acid sequence of the human β chain (69).

In my laboratory at M.I.T. Haruo Saito and I collaborated with David Kranz and Herman Eisen to isolate both α and β cDNA clones from another type of T cell, namely a cytotoxic T cell clone specific for class I MHC molecule. In 1984, using a modified subtractive cDNA library method, we identified two classes of cDNA clones that also satisfied the criteria for a T cell receptor gene (70). One class of these clones represented the β subunit. Taken together with the earlier finding by Davis' group this demonstrated that the two major classes of T cells, helper T cells and cytotoxic T cells, employ the same set of genes at least for the β subunit. The same conclusion was drawn subsequently for the α subunit. This point is significant because the two types of T cell are specific for two distinct subclasses of MHC gene products. Thus, the same T cell receptor genes mediate recognition of both class I and class II MHC.

The polypeptide chains encoded by the other class of rearranging T cell-specific cDNAs isolated by Saito and myself was also homologous to immunoglobulin chains by 30–35%. These cDNA clones were, however, clearly distinct from β cDNA clones because the polypeptide chains encoded by the two sets of cDNAs were homologous by only 30–35% not only in the V regions but also in the C regions. Since only two subunits, α and β, were known for the T cell receptor, we initially proposed that this second class of cDNAs represented the α gene (70). However, even before the work was published a question arose about the assignment of this gene as encoding the α chain. The putative α cDNAs do not carry codons for N-linked glycosylation sites, while unpublished studies from Charlie Janeway's laboratory at Yale University and from Jim Allison's at the University of California, Berkeley, indicated that both α and β subunits of at least some T cells carry N-linked carbohydrates. While it was still possible that the apparent discrepancy in glycosylation could be explained by differences in the type of T cell or in mouse strains used, continued screening of our subtracted cDNA library yielded within a few weeks a third class of clones whose genes also rearrange specifically in T cells (71). This gene not only was as homologous to immunoglobulin genes as the first two classes of T cell-specific genes, but also had two potential sites for N-linked glycosylation, and therefore was a better candidate for the α gene. This proposition was soon confirmed by comparing its nucleotide sequence with the partial amino acid sequence of the human α subunit (72). Furthermore, the α gene was also cloned at about the same time from a helper hybridoma (73).

Once cDNA encoding the α and β chains were identified, it was straightforward to determine the organization of the corresponding genes in genomic DNA. These studies demonstrated that both α and β genes are organized in the germline genome and rearranged in T cells in a way remarkably similar to the immunoglobulin genes (74–77). Thus, the organism inherits the genetic information for these polypeptide chains as V_α and J_α gene segments or V_β , D_β. and J_β gene segments, and a random assembly of these gene segments occurs exclusively during T cell development to generate a diversity comparable to that of immunoglobulins for receptors expressed on the surface of mature T cells. Even the presumed recognition sequences for the site specific recombinase, the so called heptamers and nonamers with a 12 or 23 base pair spacer, seem to be common for both the immunoglobulin and T cell receptor genes (Fig. 3).

The complete primary structure of a T cell receptor can be deduced from the nucleotide sequences of the α and β cDNA clones. Its comparison with the primary structure of an immunoglobulin molecule suggests that the external part of the receptor is composed of four compact, immunoglobulin-like globular domains associated in two non-covalently bound pairs V_α V_β and C_α -C_β, and further stabilized by an interchain disulfide bond between the C domain and the transmembrane region. This extracellular part of the receptor is anchored on the membrane lipid bilayer through two transmembrane peptides, one each from the α and β chains (Fig. 6) (70).

Determining the structure and organization of genes encoding the T cell receptors settled the issue of their relationship with immunoglobulins and accounted for the genetic origin of their diversity. However, these studies did not illuminate the mechanism by which these receptors can accomplish the dual recognition of an antigen and a MHC determinant. This last issue is particularly tantalizing because recent studies using a technique for injecting T

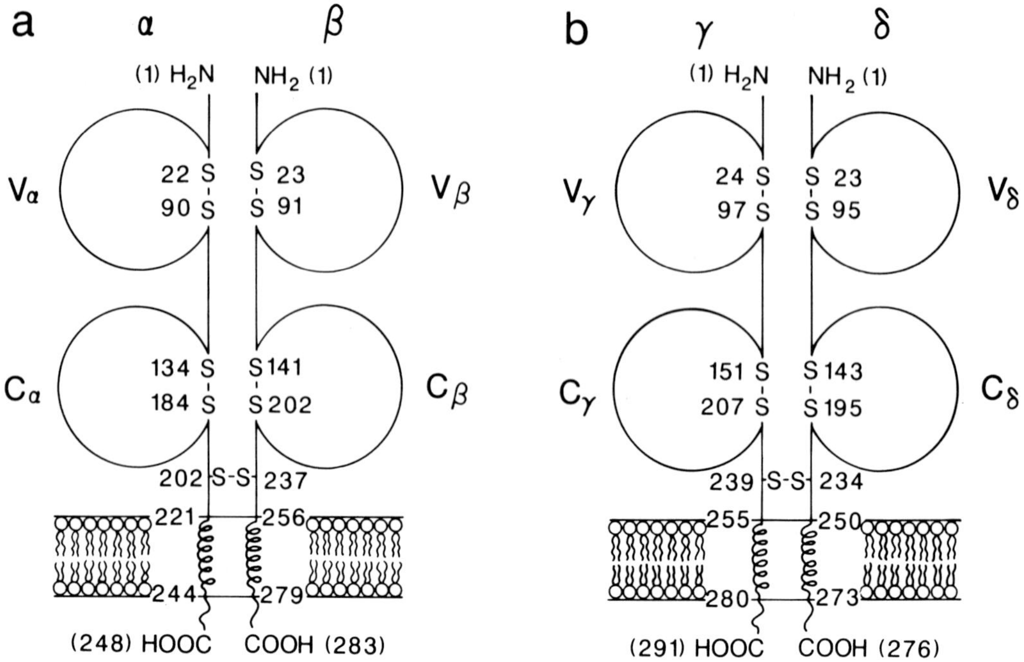

Figure 6. Diagram showing the subunit structures of T cell receptors αβ (panel a) and γδ (panel b) as deduced from the nucleotide sequences of cDNA clones. The αβ receptor is from an alloreactive cytotoxic mouse T cell clone, 2C and the γδ receptor from a mouse thymocyte hybridoma, KN6, prepared by Osami Kanagawa of Lilly Research Laboratories, La Jolla, California. Intra- and inter-chain disulfide bonds are indicated. The receptors are thought to be anchored on the membrane lipid bilayer by transmembrane peptides. The invariant CD3 complex associated with the heterodimers are not shown.

cell receptor genes into cloned, functional T cells confirmed that the αβ heterodimer alone is sufficient to mediate this dual specificity (78). In order to understand how the heterodimer simultaneously recognizes the two determinants much more information is needed as to the structure of the receptor and its compound ligand. It seems that the ultimate solution has to come from X-ray crystallographic analysis of the receptor protein.

A new T cell receptor, γδ

Since it was established that the third T cell-specific rearranging gene discovered was for the α subunit, the second one became an orphan. This gene is so closely related to the other two genes, however, that it seems certain that it must have some role in recognition by T cells. Nevertheless, previous immunochemical studies did not reveal any polypeptide chain that might be considered as a candidate for the protein product of this gene. The γ gene is also assembled somatically from *V*, *J*, and *C* gene segments and shares a number of characteristics with the α and β genes as well as with immunoglobulin genes (79,80).

A number of possibilities were considered initially as firsthand roles for the putative γ chain. For instance, it was thought that the γ chain may be a subunit for a second T cell receptor coexpressed with the αβ heterodimer. This hypothesis is in line with the two receptor model of dual recognition of antigen and MHC by T cells. Another possibility proposed was that there may be a switch in the subunits of the T cell receptor during T cell development. A model was proposed in which a T cell receptor composed initially of a γβ heterodimer switches to an αβ heterodimer as T cells differentiate in the thymus (81,82). This model, which seemed to be supported by the time course kinetics of appearance of α-, β-, and γ-specific RNA in developing hymocytes, was an attempt to explain an apparent dilemma in the intrathymic selection of the T cell repertoire (for instance see ref. 83).

Subsequent studies carried out in my and several other laboratories, however, revealed a number of features of the γ gene and its expression which are not necessarily consistent with these hypotheses. First, the γ gene is not rearranged in some T cell clones or hybridomas. Furthermore, even in many of those T cells in which this gene is rearranged, the joining of the *V* and *J* gene segments does not allow the *J* region codons to be translated in phase with the *V* region codons (84-86). Thus, the γ gene product does not seem to be universally expressed in conventional, αβ receptor-positive cytotoxic and helper T cells. Second, the γ polypeptide chain is expressed on the surface of a small (less than 0.5 %) subset of peripheral T cells as a component of a heterodimer referred to as γδ (87-89). The majority of these T cells exhibit none of the CD4 or CD8 glycoproteins of conventional αβ receptor T cells on their surface, and therefore belong to a distinct cell population. Third, like the αβ heterodimer, the γδ heterodimer is associated relatively tightly with another glycoprotein, CD3 (87). The invariant CD3 protein complex contains a subunit that seems to play a critical role in the transmission of the signal received by the variable heterodimer into the cell (90). Thus, the similarly between the γδ- and αβ-receptor heterodimer includes both their structure and transmembrane signal transmission. Fourth, γδ-bearing cells are relatively abundant in the $CD4^-$, $CD8^-$ fraction of fetal and adult thymocytes (91-94). For instance, thymocytes of 16-day old fetal mice, which are mostly double negative cells (i.e. $CD4^-$ and $CD8^-$), are a relatively rich source of γδ-bearing cells. Since double negative thymocyte populations contain precursors for mature, functional, αβ-bearing T cells (95), a natural question that arises is whether γδ-bearing thymocytes are precursors for αβ-bearing T cells. Fifth, another major site of expression of the γδ-receptor is epidermal tissues. It was recently found by two groups that this tissue contains $Thy\text{-}1^+$ (another cell surface marker shared by all types of T cells), $CD3^+$, $CD4^-$ and $CD8^-$ cells bearing γδ-heterodimers (96,97). Unlike conventional T cells, these cells look more like dendritic cells and are therefore referred to as dendritic epidermal cells (DEC). Finally, the δ gene also undergoes rearrangement. *D*, *J*, and *C* gene segments for δ polypeptide chains have recently been mapped within the α gene family between V_α and J_α gene segments (98). The nested configuration of α and δ gene segments is intriguing and provokes curiosity about the possible relationship of gene organization with the regulation of the rearrangement and expression of the two types of genes. Another intriguing question is whether and to what extent the pool of $V\alpha$ and V_δi gene segments overlap.

Despite the rapid progress made in the characterization of the γ and δ genes and their products, the most intriguing problem, namely the physiological role of the γδ receptor-bearing cells is currently unknown. One can only speculate on this matter using the currently available information. As to the ligand of this new receptor, it is probably correct to emphasize the fact that the receptor shares with the immunoglobulin B cell and the αβ T cell receptors the same genetic basis for somatic diversification. It therefore is likely that the ligand in question will also exhibit structural diversity. In this respect it is interesting that recent studies by us and others suggest that at least part of the ligand is encoded in MHC (99,100). The effector function of the cells bearing γδ receptors has not yet been defined with certainty, but recent studies using human and mouse γδ cell clones suggest that many of these cells have cytotoxic capacity (89,100,101). The finding of a new type of T cell with an apparently distinct set of T cell receptors in epidermal tissues stimulates one's imagination. It may be that occurrence of this type of T cell is not restricted to the outer epithelial tissues but extends to all epithelial layers including the lining of various internal organs (C. Janeway, pers. commun.). If so, these cells may have evolved to protect the part of the body that is most vulnerable to infection, namely its external and internal epithelial surface that are in direct contact with the environment. However, the prominence of γδ cells in the thymus suggests an intrathymic role of these cells as well. An exciting possibility that has not been ruled out is a role for these cells in the intrathymic selection of appropriate αβ T cells.

Concluding remarks

Use of restriction enzymes and recombinant DNA methods allowed resolution of a long standing and central issue in immunology, the genetic origin of antibody diversity. It turned out that an organism does not inherit even a single complete gene for antibody polypeptide chains. Rather, the genetic information is transmitted in germline as no more than several hundred gene segments. Through a series of specialized somatic recombinations occuring specifically during the differentiation of B lymphocytes, these gene segments are assembled into tens of thousands of complete genes. Somatic hypermutation occurring in these assembled genes further diversifies antibody polypeptide chains, so that B cells displaying immunoglobulin receptors having a better fit to a given antigen can be selected in a later phase of B cell differentiation. Thus, in the immune system, organisms have exploited two major processes for modification of DNA, recombination and mutation, as means to diversify somatically the limited amount of inherited genetic information in order to cope with the vastly diverse antigen universe.

Why has somatic diversification been necessary in the evolution of the immune system? Microorganisms and substances produced by them are the primary source of biologically relevent antigens against which vertebrates need to produce antibodies for survival. Since the generation time of microorganisms is several orders of magnitude shorter than that of vertebrates, the former can produce genetic variants much faster than the latter. Thus, if genetic alterations in the germline genome were to be the only source of antibody diversity, vertebrates would be unable to deal effectively with the rapidly changing world of antigens. Somatic diversification allows the individual organism to generate a virtually limitless number of lymphocyte variants. Like organisms in an ecosystem, these lymphocytes are subject to selection by antigens and the fittest will survive. Thus, as Jerne and Burnet were aware, the individual immune system can be conceived of as a kind of Darwinian microcosm.

The molecular biological approach played an even more fundamental role in the analysis of the T cell receptor in that very little structural information existed prior to the cloning of the receptor genes. It was demonstrated that the polypeptide chains composing the receptor protein are encoded by genes that share a common ancestor with the immunoglobulin genes. Like immunoglobulins, T cell receptors are diversified by somatic recombination, but unlike immunoglobulins, these receptor proteins have not been observed to undergo further diversification by somatic mutation. The reason for this difference is unknown, but it seems likely that the explanation will be as follows. First, unlike immunoglobulins, T cell receptors function exclusively as cell surface receptors which are specialized for interacting with cellbound antigens. Since both ligand and receptor are distributed in two dimensional space when a T cell interacts with an antigen-presenting cell, and as T cells have mechanisms for transiently adhering to other cells, these receptor-ligand interactions occur under conditions of high local concentration. Thus, improvement of the affinity beyond the one accomplished by somatic rearrangement may not be necessary in T cell recognition.

Second, the ligand consists in part of an essentially invariant component, self MHC. As the T cell receptor is selected for self MHC recognition, both during development and during immunization, the extreme variability availabe via somatic mutation may not only be unnecessary but even disadvantageous. Third, and probably more important, T cells appear to be selected early in development for self tolerance, the inability to recognize self antigen. Somatic mutation during antigenic stimulation, as occurs in B lymphocytes, could lead to the development of autoreactivity. While autoreactive B cells depend upon the additional presence of autoreactive helper T cells to generate autoimmunity, this is not true of autoreactive T cells, which can directly produce disease. Thus, Ehrlich's famous concept of "honor autotoxins", originally developed for antibodies, is probably critical only for T cells. It will be interesting to examine autoreactive T cell receptors for any evidence of post-thymic somatic diversification.

Finally, it is interesting to notice that during the fifteen years in which I have studied the immune system, the role of molecular genetics in immunological research has altered radically. When I started investigating the problem of antibody diversity, there was abundant information about the structure and function of antibody molecules, while virtually nothing was known about their genes. By contrast, in the most recent study on T cell recognition no gene product was known at all when the rearranging gene, γ was discovered. From the structure of the gene and its rearranging behavior, it was deduced to be a receptor gene, and this discovery has led directly to new insights into T cell development and T cell biology. This short history of research in one area, lymphocyte receptors, is yet another witness to the power of DNA technology, and to the ability of this approach not only to explain known biological phenomena, but also to contribute to the discovery of new biological systems.

Acknowledgement
The work summarized in this article is the result of collaboration with many colleagues, students and technical assistants. I wish to extend my sincere thanks to everyone of them. I also extend my thanks to Hoffman LaRoche Company which so generously supported my work in Basel. My special thanks are extended to Charles A. Janeway Jr., Nancy Hopkins and Yohtaroh Takagaki for many useful comments on the manuscript, to Jiri Novotny for preparing Figure 4, and to my secretary, Eleanor Lahey Basel, for her tireless devotion.

REFERENCES

1. Porter, R.R., Science *180*:713–716 (1973).
2. Edelman, G.M., Science *180*:830–840 (1973).
3. Hilschmann, N. and Craig, L.C., Proc. Natl. Acad. Sci. USA *53*:1403–1409 (1965).
4. Hood, L., and Talmage, D.W. Science *168*:325–334 (1970).
5. Lederberg, J. Science *129*:1649–1653 (1959).
6. Jerne, N.K. Eur. J. Immuno. *1*:1–14 (1971).
7. Cohn, M., Blomberg, B., Geckeler, W., Raschka, W., Riblet, R. and Weigert, M. in "The Immune System: Genes, Receptors, Signals" (E.E. Sercarz, A.R. Williamson and C.F. Fox, eds, PP 89 Academic Press, New York and London (1974).
8. Gelderman, A.H., Rake, A.V., and Britten, R.J. Proc. Natl. Acad. Sci. USA *68*:172–176 (1971).
9. Bishop, J.O. Biochem. J. *126*:171–185 (1971).
10. Tonegawa, S., Proc. Natl. Acad. Sci. USA, *73*:203–207, (1976).
11. Weigert, M., Cesari, M., Yonkovich, S.J. and Cohn, M. Nature *228*: 1045–1047 (1970). (1970).
12. Campbell, A., Advan. Genet. *11*: 101 (1962).
13. Dryer, W.J., and Bennett, J., Proc. Natl. Acad. Sci. USA *54*, 864–869 (1965).
14. Danna, J.J., Sack Jr., G.H., and Nathans, D. J. Mol. Biol. *78*:363 (1973).
15. Southern, E., J. Mol. Biol. *98*:503 (1975).
16. Hozumi, N., and Tonegawa, S., Proc. Natl. Acad. Sci., USA, *73*:3628–3632, (1976).
17. Thomas, M., White, R.L. and Davis, R.N. Proc. Natl. Acad. Sci. USA *73*:2294–2298 (1976).
18. Tonegawa, S., Brack, C., Hozumi, N. and Schuller, R., Proc. Natl. Acad. Sci. USA, *71*:3518–3522 (1977).
19. Tonegawa, S., Maxam, A.M. Tizard, R., Bernard, O., and Gilbert, W., Proc. Natl. Acad. Sci. USA, *75*:1485–1489 (1978).
20. Brack, C. Hirama, M., Lenhard-Schuller, R., and Tonegawa, S. Cell, *15*:1–14 (1978).
21. Lenhard-Schuller, R., Hohn, B., Brack, C., Hirama, M., and Tonegawa, S., Proc. Natl. Acad. Sci. USA, *75*:4709–4713, (1978).

22. Brack, C. and Tonegawa, S., Proc. Natl. Acad. Sci. USA, *74*:5652–5656, (1977).
23. Bernard, O., Hozumi, N., and Tonegawa, S., Cell, *15*:1133–1144 (1978).
24. Weigert, M., Gatmaitan, L., Loh, E., Schilling, J., and Hood, L., Nature *276*, 785–789 (1978).
25. Kabat, E.A., WU, T.T., Bilofsky, H., Reid-Miller, M., and Perry, H., U.S. Dept. of Health and Human Services Publication (1983).
26. Seidman, J.G., Leder, A., Nau, M., Norman, B., and Leder, P., Science *202*:11–16 (1978).
27. Cory, S., Tyler, B.M., Adams, J.M. J. Mol. Appl. Genet. *1*:103–116 (1981).
28. Davis, M., Calame, K., Early, P., Livant, D., Joho, R., Weisman, I., and Hood, L. Nature *283*:733–738 (1980).
29. Maki, Richard; Traunecker, André; Sakano, Hitoshi; Roeder, William; and Tonegawa, S., Proc. Natl. Acad. Sci. USA, *77*:2138–2142 (1980).
30. Early, P., Huang, H., Davis, M., Calame, K., and Hood, L., Cell *19*:981–992 (1980).
31. Sakano, H., Maki, R., Kurasawa, Y., Roeder, W., and Tonegawa, S., Nature *286*:676–683 (1980).
32. Sakano, H., Kurosawa, Y., Weigert, M., and Tonegawa, S., Nature *290*: 562–565 (1981).
33. Kurosawa, Y., von Boehmer, H., Haas, W., Sakano, H., Traunecker, A., and Tonegawa, S., Nature *290*:565–570 (1981).
34. Kurosawa, Y., and Tonegawa, S., J. Exp. Med. *155*:201–218 (1982).
35. Wood, C., and Tonegawa, S., Proc. Natl. Acad. Sci. USA *80*:3030–3034 (1983).
36. Sakano, H., Huppi, K., Heinrich, G., and Tonegawa, S., Nature *280*, 288–294 (1979).
37. Max, E.E., Seidman, J.G., and Leder, P., Proc. Natl. Acad. Sci. USA *76*:3450–3454 (1979).
38. Halligan, B.D., and Desiderio, S.V. Proc. Natl. Acad. Sci. USA *81*:7019–7023 (1987).
39. Aguitera, R.J., Akira, S., Okagaki, K. and Sakano, H. Cell (in press)
40. Alt, F., and Baltimore, D., Proc. Natl. Acad. Sci. USA *79*:4118–4122 (1982).
41. Azuma, T., Igras, V., Reilly, E., and Elsen, H.N., Proc. Natl. Acad. Sci. USA *81*:6139–6143 (1984).
42. Burnet, F.M. The Clonal Selection Theory of Acquired Immunity. London: Cambridge Univ. (1959).
43. Bothwell, A.L.M., Paskind, M., Reth, M., Imanishi-Kari, T., Rawesky, K., and Baltimore, D., Cell *24*:624 (1981).
44. Crews, S , Griffin, J., Huang, H., Calame, K., and Hood, L., Cell *25*:59–66 (1981)
45. Givol, D., Zakut, R., Effron, K., Rechavi, G., Ram, D., and Cohen, J.B., Nature *292*:426–430 (1981).
46. Selsing, E., and Storb, U., Nucl. Acids Res. *9*:5725–5735 (1981).
47. Gearhart, P.J., Johnson, N.D., Douglas, R., and Hood, L. Nature *291*:29–33 (1981).
48. Kim, S., Davis, M., Sinn, E., Patten, P., and Hood, L., Cell *27*:573–580 (1981).
49. Askonas, B.A. and Williamson, A.R. Eur. J. Immunol. *2*:487–493 (1972).
50. Berek, C., Griffiths, G.M., and Milstein, C. Nature *316*:412–418 (1985).
51. McKean, D., Hüppi, K., Bell, M., Standt, L., Gerhard, W., and Weigert, M. Proc. Natl. Acad. Sci. USA *81*:3180–3184 (1984).
52. Okumura, K., Julius, M.H., Tsu, T., Herzenberg, L.A., and Herzenberg, L.A. Eur. J. Immunol. *6*:467–472 (1976).
53. Sablitzberg, M., Kocks, C., and Rajewsky, K. EMBO J. *4*:345–350 (1985).
54. Sickevitz, M., Kocks, C., Rajewsky, K., and Dildrop, R. Cell *48*:757–770 (1987).
55. Wysocki, L.J., Manser, T., and Gefter, M.L. Proc. Natl. Acad. Sci. USA *83*: 1847–1851 (1986).

56. Maki, R., Kearney, J., Paige, C., and Tonegawa, S., Science, *209*:1366–1369 (1980).
57. Perry, R.P., Kelley, D.E., Coleclough, C., and Kearney, J.F., Proc. Natl. Acad. Sci. USA *78*:247–251 (1981).
58. Reth, M.G., Ammirati, P.A., Jackson, S.J., and Alt. F.W., Nature *317*:353–354 (1985).
59. Eisen, H.N. and Siskind, G.W., Biochemistry *3*:996–1000 (1964).
60. Rosenthal, A.S. and Shevach, E.M., J. Exp. Med. *138*:1194–1212 (1973).
61. Katz, D.H., Hamaoka, T., Dorf, M.E., and Benacerraf, B. Proc. Natl. Acad. Sci. USA *70*:2624–2628 (1973).
62. Zinkernagel, R.M. and Doherty, P.C. J. Exp. Med. *141*:1427–1436 (1975).
63. Allison, J.P., McIntyre, B.W., and Bloch, D., J. Immunol. *129*:2293–2300 (1982).
64. Meuer, S.C., Acuto, O., Hussey, R.E., Hodgdon, J.C., Fitzgerald, K.A., Schlossman, S.F., and Reinherz, E.L. Nature *303*:808–810 (1983).
65. Haskins, K., Kubo, R., White, J., Pigeon, M., Kappler, J., and Marrack, P., J. Exp. Med. *157*:1149–1169 (1983).
66. Hedrick, S.M., Cohen, D.I., Nielsen, E.A., and Davis, M.M., Nature *308*:149–153 (1984).
67. Hedrick, S.M., Nielsen, E.A., Kavaler, J., Cohen, D.I., and Davis, M.M., Nature *308*:153–158 (1984).
68. Yanagi, Y., Yoshikai, Y., Leggett, K., Clark, S.P., Aleksander, I., and Mak. T.W., Nature *308*:145–149 (1984).
69. Acuto, O., Fabbi, M., Smart, J., Poole, C., Protentis, J., Royer, H.D., Schlossman, S., and Reinherz, E.L., Proc. Natl. Acad. Sci. USA *81*:3851–3855 (1984).
70. Saito, H., Kranz, D.M., Takagaki, Y., Hayday, A.C., Eisen, H.N. and Tonegawa, S., Nature *309*:757–762 (1984).
71. Saito, H., Kranz, D.M., Takagaki, Y., Hayday, A.C., Eisen, H., and Tonegawa, S., Nature *312*:36–39 (1984).
72. Hannum, T.H., Kappler, J.W., Trowbridge, I.S., Marrack, P., and Freed, J.H., Nature *312*:65–67 (1984).
73. Chien, Y-h., Becker, D.M., Lindsten, T., Okamura, M., Cohen, D.J., and Davis, M.M., Nature *312*:31–35 (1984).
74. Chien, Y-h, Gascoigne, N., Kavaler, J., Lee, N.E., and Davis, M.M., Nature *309*:322–326 (1984).
75. Siu, G., Kronenberg, M., Strauss, E., Haars, R., Mak, T.W., and Hood, L., Nature *311*:344–350 (1984).
76. Hayday, A.C., Diamond, D.J., Tanigawa, G., Heilig, J.S., Folsom, V., Saito, H., Tonegawa, S., Nature *316*:828–832 (1985).
77. Winoto, A., Mjolsness, S., and Hood, L., *Nature 316*:832–836 (1985).
78. Dembič, Z., Haas, W., Weiss, S., McCubray, J., Kiefer, H., von Boehmer, H., and Steinmetz, M. Nature *320*:232–238 (1986).
79. Kranz, D.M., Saito, H., Heller, M., Takagaki, Y., Haas, W., Eisen, H.N., and Tonegawa, S., Nature *313*:752–755 (1985).
80. Hayday, A.C., Saito, H., Gillies, S.D., Kranz, D.M., Tanigawa, G., Eisen, H.N., Tonegawa, S., Cell *40*:259–269 (1985).
81. Raulet, D.H., Garman, R.D., Saito, H. and Tonegawa, S., Nature *314*:103–107 (1985).
82. Pernis, B. and Axel, R., Cell *41*:13–16 (1985).
83. Schwartz R. in Fundamental Immunology (ed.) William Paul, Raven Press, NY (1984) pp. 379–438.
84. Reilly, E.B., Kranz, D.M., Tonegawa, S., and Eisen, H.N., *Nature 321*:878–880 (1986).
85. Rupp, F., Frech, G., Hengartner, H., Zinkernagel, R.M. and Joho, R., Nature *321*:876–878 (1986).
86. Heilig, J.S. and Tonegawa, S. Proc. Natl. Acad. Sci. USA *84*:8070–8074 (1987).

87. Brenner, M.B., McLean, J., Dialynas, D.P., Strominger, J.L., Smith, J.A., Owen, F.L., Seidman, J.G., Ip, S., Rosen, F. and Krangel, M.S. Nature *322*:145–149 (1986).
88. Lanier, L.L., Federspiel, N.A., Ruitenberg, J.J., Phillips, J.H., Allison, J.P., Littman, D., and Weiss, A. J. Exp. Med. *165*:1076–1094 (1987).
89. Borst, J., Van de Griend, R.J., van Oostveen, J.W., Ang, S-L., Melief, C.J., Seidman, J.G., Bolhuis, R.L.H., Nature *325*:683–688 (1987).
90. Wauve, J.P. van, Mey, J.R. de, and Goosens, J.G. J. Immunol. *124*:2708–2812 (1980).
91. Bank, I., DePinho, R.A., Brenner, M.B., Cassimeris, J., Alt, F.W., and Chess, L. Nature *322*:179–81 (1986).
92. Lew, A.M., Pardoll, D.M., Maloy, W.L., Fowlkes, B.J., Kruisbeck, A., Cheng, S-F, Germain, R.N., Bluestone, J.A., Schwartz, R.H., and Coligen, J.E. Science *234*:1401–1405 (1986)
93. Nakanishi, N., Maeda, K., Ito, K., Heller, M., and Tonegawa, S., Nature *325*:720–723 (1987).
94. Pardoll, D.M., Fowlkes, B.J., Bluestone, J.A., Kruisbeck, A., Maloy, W.L., Coligan, J.E. and Schwartz, R.H. Nature *326*:79–81 (1987).
95. Fowlkes, B.J., Thesis, George Washington University Graduate School of Arts and Sciences, Washington DC (1984).
96. Koning, F., Stingl, G., Yokoyama, W.M., Maloy, W.L., Tschachler, E., Shevach, E.M., Coligan, J.E. Science *236*:834–837 (1987).
97. Kuziel, W.A., Takashima, A., Bonyhadi, M., Bergstresser, P.R., Allison, J.P., Tigelaar, R.E., and Tucker, P.W. Nature *328*:263–266 (1987).
98. Chien, Y-h., Iwashima, M., Kaplan, K.B., Elliott, J.F. and Davis, M.M. Nature *327*:677–682 (1987).
99. Maeda, K., Nakanishi, N., Rogers, B.L., Haser, W.G., Shitara, K., Yoshida, H., Takagaki, Y., Augustin, A.A. and Tonegawa, S. Proc. Natl. Acad. Sci. USA *84*:6536–6540 (1987).
100. Matis, L.A., Cron, R., and Bluestone, J.A. Nature *330*:262–264 (1987).
101. Moingeon, P., Jitsukawa, S., Faure, F., Troalen, F., Triebel, F., Graziani, M., Forestier, F., Bellet, D., Bohuon, C., and Hercend, T. Nature *325*:723–726 (1987).
102. Gillies, S.D., Morrison, S.L., Oi, V.T., and Tonegawa, S., Cell *33*:717–728 (1983).
103. Banerji, J., Olson, L., and Schaffner, W., Cell *33*:729–739 (1983).
104. Satow, Y., Cohen, G.H., Padlan, E.A., Davies, D.R. Mol. Biol. *190*:593–604 (1986).
105. Novotny, J., Bruccoleri, R.E., Newell, J., Murphy, D., Haber, E., Karplus, M. J. Biol. Chem. *258*:14433–14437 (1983).
106. Amit, A.G., Mariuzza, R.A., Phillips, S.E.V., Poljak, R. Science *233*:747–752 (1986).
107. Sheriff, S., Silverton, E.W., Padlan, E.A., Cohen, G.H., Smith-Gill, S.J., Finzel, B.C., Davies, D.R. Proc. Natl. Acad. Sci. USA *84*:8075–8079

1988
Physiology or Medicine

SIR JAMES W. BLACK
GERTRUDE B. ELION and
GEORGE H. HITCHINGS

"for their discoveries of 'important principles for drug treament' "

THE NOBEL PRIZE IN PHYSIOLOGY OR MEDICINE

Speech by Professor FOLKE SJÖQVIST of the Karolinska Institute.
Translation from the Swedish text.

This year's Nobel Prize in Physiology or Medicine has been awarded for the discoveries of important principles for drug treatment which have been applied successfully to the treatment of a wide variety of serious illnesses. Sir James Black's findings have made possible the development of new and improved agents for the treatment of angina pectoris, myocardial infarction, high blood pressure and peptic ulcer. In Sweden alone, approximately half a million people were treated in 1987 with the types of drugs which resulted from his research. Gertrude Elion's and George Hitchings' research has led to drugs effective against such widely differing conditions as leukemia, rejection of transplanted organs, gout, malaria and bacterial and viral infections. The drugs which have resulted from the discoveries of this year's prizewinners are now well-proven medications which have stood the test of time over the past 15–35 years, and which remain today front-line agents for the treatment of a wide spectrum of illnesses. They also appear in the World Health Organizations's list of so-called "Essential Drugs", which denotes those medicines which should be available worldwide in order that the goal of "Health for All by the Year 2000" can be realized.

In a letter to a close friend, written a few months before his death in 1896, Alfred Nobel wrote, "My heart trouble will keep me here in Paris for another few days at least ... Isn't it the irony of fate that I have been prescribed nitroglycerin to be taken internally! They call it Trinitrin, so as not to scare the chemist and the public". By the latter remark, Nobel was referring to the use of nitroglycerin as an ingredient of dynamite. Nitroglycerin reduces the pain of angina pectoris by dilating the cardiac blood vessels and thereby increasing the supply of oxygen to the heart. Sir James Black was the first to recognize that an alternative therapeutic strategy for angina would be to use a drug which decreases the heart's requirement for oxygen. He therefore focused his attention on the specialized binding sites, the so-called β-receptors, on heart-muscle cells to which the stress hormones adrenaline and noradrenaline bind and thereby increase the workload and oxygen demand of the heart. The stress hormones are the keys, and the receptors are the "locks" which have to be opened in order for this biological process to occur. In 1962, Black was successful in developing clinically useful compounds which blocked specifically the effect of stress hormones on the heart by inhibiting their binding to receptors. These compounds functioned simply as false keys in the locks. Patients treated with such agents, the so-called β-receptor blocking drugs, were able to increase their physical effort without over-exerting heart function and experiencing chest pain as a result. Subsequently, it was found that these same compounds could be used in the treatment of high blood pressure and in reducing fatalities resulting from myocardial infarction.

Sir James Black also developed a fundamentally new approach to the treatment of peptic ulcer. A formerly widely used diagnostic test of a patient's susceptibility to develop peptic ulcer was to administer histamine, which is a powerful stimulant of gastric acid secretion. Black raised the question of why the so-called antihistamines failed to inhibit this production of gastric acid, despite the fact that they successfully blocked the allergic reactions mediated by histamine. The explanation proved to be that histamine present in the stomach acts upon a quite distinct family of receptors from those blocked by the antihistamines. Through altering the chemical structure of the histamine molecule, Black succeeded in developing false keys for this new type of receptor, and produced compounds which almost immediately stopped the production of gastric acid (1972). These so-called histamine-2 receptor blockers proved to be highly effective in the treatment of peptic ulcers, and decreased the need for surgical intervention in patients suffering from this disease.

While Sir James Black worked with the structure of the exterior of cells, Gertrude Elion and George Hitchings studied the nucleus and its genetic content, the nucleic acids. During the early 1950's they published the hypothesis that, with the aid of drugs, it should be possible to, selectively inhibit the synthesis of nucleic acids used by e.g. cancer cells and bacteria, without simultaneously impeding the growth of normal cells. In these days our knowledge of how nucleic acids were synthesized within the cell was very limited. However, it was known that cells used certain simple building blocks in the manufacture of their nucleic acids. Elion and Hitchings studied how false building blocks, the so-called antimetabolites, could be used to interrupt cellular growth. As early as 1951, they discovered a compound, 6-mercaptopurine, which was used successfully in a hitherto incurable form of leukemia. Through a simple chemical alteration of the 6-mercaptopurine structure, they developed another drug, azathioprine (1957), which inhibited the property of white cells to reject transplanted organs. Twenty years later, one of the world's leading transplant surgeons asserted that, as a result of the discovery of this immunosuppressant drug, 20 000 individuals had been able to receive a new kidney. Furthermore, a new strategy for the treatment of gout with allopurinol was developed by Elion and Hitchings in 1963.

Hitchings and co-workers also developed the anti-malarial drug pyrimethamine (1950) and the anti-bacterial agent trimethoprim (1956). An important observation of theirs was that the effects of both of these drugs were enhanced by sulfonamides, which led to the use of the trimethoprim-sulfa combination in the treatment of severe bacterial infections, such as those encountered in AIDS patients, and also in an important antimalarial preparation. A subsequent outcome of Elion's and Hitchings' research programme was the successful development of the first effective antiviral drug, acyclovir, which is used against infections caused by e.g. the herpes virus. The virus-infected cell is tricked into transforming acyclovir into a compound which inhibits cell growth and thereby suppresses the ability of the virus particle to reproduce. Only cells infected by the virus are attacked.

Through their research efforts, Black, Elion and Hitchings succeeded in developing a rational approach to the discovery of new drugs, based upon basic scientific studies of biochemical and physiological processes. As a result, a new era in drug research was born which offers promise for the development of new therapeutic strategies for the treatment of illnesses against which existing drugs are either unsatisfactory or simply do not exist.

Dr. Black, Dr. Elion and Dr. Hitchings,

On behalf of the Nobel Assembly of the Karolinska Institute I would like to congratulate you on your outstanding accomplishments and ask you to receive the Nobel Prize in Physiology or Medicine from the hands of His Majesty the King.

James Black

JAMES BLACK

I have never wanted to check out the family folklore that we could be traced back to a dominie at the hamlet of Balquhidder in the Scottish highlands. The romantic notion that I might have tenuous roots with two great traditions — with the political rebelliousness of Rob Roy McGregor and with the Scottish tradition of rural education, arguably one of the best anywhere — was too enjoyable to be seriously tested. The outcome, the fourth in an issue of five boys born into a staunch Baptist home, meant that from the beginning I was taught to be respectful of others no less than myself, influencing ever since both my political and administrative attitudes. My father, a mining engineer and colliery manager, gave his brood many advantages not least of which, for me, was his love of singing which gave music a central place in our lives.

Apart from two periods of intense study, of music between the ages of 12 and 14 and of mathematics between the ages of 14 and 16, I coasted, day-dreaming, through most of my school years. The imprinting mathematical influence was Dr Waterson at Beath High School, a brilliant and rumbustious teacher, who more or less man-handled me into sitting the competitive entrance examination for St Andrews University. This led to an interview with the Vice-Chancellor, the redoubtable Sir James Irvine, flanked by elderly academic worthies, all poking into the mind of a nervous 15 year old boy. I was awarded the Patrick Hamilton Residential Scholarship, mercifully unaware at the time that the family budget couldn't otherwise have stretched to yet another university student.

As a condition of the Patrick Hamilton Residential Scholarship I spent my undergraduate years in St Salvator's Hall, a fine new building modelled on the Oxbridge colleges. The young aficionados of St Salvator's Hall in my day were culled from every imaginable class and state from the United Kingdom and overseas. The few intimate years I spent in this company were an extraordinarily mind-broadening experience for the country boy from the coalfields of Fife, no doubt much as the scholarship architects had intended.

I chose to study Medicine mainly under the influence of an elder brother, William, a graduate in Medicine at St Andrews some years earlier. In the cold, forbidding, greyness of St Andrews — with its dedication to "causes purely spiritual and intellectual, to religion and learning" (Andrew Lang) — I learned, for the first time, the joys of substituting hard, disciplined study for the indulgence of day-dreaming. Undergraduate prizes seemed to confirm that I was working harder than my colleagues in a new-found love affair with knowledge. An important catalyst in my conversion to scholarship was my first year encounter with Professor D'Arcy Wentworth Thompson, last of the great Victorian polymaths, author inter alia of the classic allometric study "On Growth and Form", and an intellectual giant if ever there was one.

I met Hilary Vaughan at a Student Ball in 1944 and we married in the summer of 1946, as soon as I graduated. I joined the Physiology Department under Professor R.C. Garry in October 1946 and Hilary, completing her degree in Biochemistry, was the best student I ever had. Had she chosen a sectarian approach to study she would have become a visible star but her eclectic pursuit of knowledge and her unwavering support for her family led her to study law and choose poetry as a distillate of her wisdom. Intellectually she was the most exciting person I have ever known and, quite simply, the mainspring of my life until she died in 1986.

My first year of research in Garry's laboratory introduced me to some simple ideas which, in a variety of ways, have dominated my thinking ever since. Garry was trying to find out how the intestine was able to absorb sugars selectively. Na iodoacetate treatment eliminated selective absorption and Verzar had deduced that the selectivity was based on phosphorylation. Learning that Garry's research student was showing that iodoacetate destroyed the intestinal epithelium, I wondered if iodoacetate was a general poison. What did it do to blood pressure, for example? When I developed the technology to show that, in rats, iodoacetate rapidly and irreversibly reduced the blood pressure to about 40 mm Hg, I was faced with the question which has influenced my thinking ever since: when and to what extent does local blood flow act as a metabolic throttle?

We went to Singapore at the end of 1947 – an inevitable result of marriage, debts accumulated to pay for the completion of my medical studies, and pitiful academic prospects. As a Lecturer at the King Edward VII College of Medicine I experimented with learning how to teach Physiology; and I learned that experimenting in Physiology was too difficult if the inspiration was no more than wishful thinking. Nevertheless, I made some progress in relating mucosal blood flow to rates of intestinal absorption to use in my carpet-bagging efforts later in London.

We paid off our debts, we learned some, made friends and returned in 1950 with a larger view of life. I had, however, no home, no income of any kind and no prospects whatsoever. I knocked on the doors of Physiology Departments all over London and met more sympathy than I expected; then a chance encounter with Professor Garry in Oxford Street led me to William Weipers, subsequently knighted, Director of the newly "nationalised" University of Glasgow Veterinary School. He gave me the opportunity to start a new Physiology Department, and during the next eight years I built a state-of-the-art physiology teaching laboratory based on my enduring belief that our brains work best when doing focuses our thinking. We had a daughter, Stephanie, born in 1951; I built a workshop-coupled research laboratory providing the most advanced cardiovascular technology I knew;

and persuaded George Smith and Adam Smith, academic surgeons, to join me.

As I slowly learned, like a primitive painter, how to be an effective experimenter, ideas began to ferment. Work with Adam Smith on the effects of 5-hydroxytryptamine on gastric acid secretion was to surface again later on in my interest in the pharmacology of histamine-stimulated acid secretion. Work with George Smith, concerned with finding ways of increasing the supply of oxygen to the heart in patients with narrowed coronary arteries, led me to propose that reducing myocardial demand for oxygen by annulling cardiac sympathetic drive might be equally effective. By 1956, I had clearly formulated the aim, based on Ahlqvist's dual adrenoceptor hypothesis, of finding a specific adrenaline β-receptor antagonist. Egged on by their local representative, I successfully approached I.C.I. Pharmaceuticals Division for help and ended up being employed by them at their exciting new laboratories at Alderley Park, Cheshire. During my six years with them Dr Garnet Davey (subsequently Research Director) constantly supported me and, I have no doubt, fought many battles on my behalf to keep the initially controversial programme going. All I ever promised was that I was sure I could develop a new pharmacological agent which might answer a physiological question. Any utility would be implicit in that answer.

My years at I.C.I., between 1958–1964, were some of the most exciting of my life. I was assigned a brilliant chemist, John Stephenson. He taught me about modern deductive organic chemistry; how to be more than merely curious about a molecule with an interesting biological effect: how to ask questions about it. He converted me to pharmacology. Indeed, my whole experience at I.C.I. was an educational tour de force. I had to learn how to collaborate across disciplines, how to change gears when changing from research to development, how to make industry work — in short, how to be both effective and productive.

Among the numerous people who were involved in bringing the first β-receptor antagonist to the marketplace, three played crucial roles. Bert Crowther masterminded the medicinal chemistry development. Genial, enthusiastic and highly experienced he was a splendid colleague. Bill Duncan, biochemist, brilliantly controlled the linchpin between research and development. He illuminated the black box between drug delivery and effect, developing analytical methods for estimating the levels and tissue distribution of a drug and its metabolites which allowed us to monitor and control toxicity tests, human pharmacology and clinical trials. Duncan brings brio and bravura to everything he does; and he is reliably my severest critic. Without him I would have made many more mistakes than I did. Brian Pritchard, clinical pharmacologist at University College, London, spearheaded the clinical development of the β-adrenoceptor antagonists and crusaded on their behalf — as well as revolutionising their use by his discovery of their antihypertensive effect.

By 1963, I faced opposing pressures. I saw that the success of the β-receptor antagonist programme would suck me more and more into the role of giving the young propranolol technical support and promotion — just as I was itching to start a new programme. I was convinced that the histamine antagonists of the day were analogous to the α-receptor antagonists and that the equivalent of a β-receptor antagonist was needed to block, for example, histamine-stimulated acid secretion. Then Edward Paget, Head of Pathology at I.C.I., who had accepted the Research Directorship at Smith, Kline & French Laboratories asked my advice about finding a pharmacologist to run the biological research there. Half-jokingly, I asked what was wrong with me. So we made a deal: I would run his biological research provided I had a free hand to run my new project. Bill Duncan joined me to run the Biochemistry Department, so maintaining a tremendously successful partnership which lasted 15 years.

The histamine project, modelled by analogy with the β-adrenoceptor project, was also somewhat controversial at the beginning. It succeeded because of the faith of my managers and the scientific skill and devotion of my colleagues. When I was struggling at the front, Bill Duncan was defending the rear. Mike Parsons adopted the new pharmacology with rare enthusiasm and commitment and became one of the doughtiest colleagues I have ever had. I think we made a good team. Graham Durant made the initial breakthrough with a partial agonist, and Robin Ganellin exploited that lead by brilliant, deductive, medicinal chemistry. The years I spent working with Ganellin were the most sustained, intellectually exciting and productive period of medicinal chemistry I have ever experienced. John Wyllie, surgeon from University College London, contributed the last critical piece in a sucessful mission.

By 1972, the H_2-receptor antagonist programme was launched, cimetidine was in development and I was looking for a new project. I was now totally committed to arranging marriages between bioassay and medicinal chemistry. Obvious candidates existed, such as 5-hydroxytryptamine, but other shadowy ideas were lurking about in my imagination.

The potential freedom from commercial constraints in academia was looking more and more attractive. Yet, when I was eventually offered the Chair in Pharmacology at University College, London, I was apprehensive about my ability to achieve my new goals. I had developed two ambitions. In research, I wanted to establish the medicinal chemistry/bioassay conjugation as an academic pursuit, as exciting to the imagination as astrophysics or molecular biology. In teaching, I wanted to offer a general pharmacology course based on chemical principles, biochemical classification and mathematical modelling. In the event I achieved neither of my ambitions. I failed to raise support for my medicinal chemistry project — by academic peer-review standards my proposals were altogether too wispy and expensive. My ideas about teaching based on a catechismal approach to drugs in general, rather than cataloguing drugs in particular, turned out to have too many curricular difficulties. I did help to set up an undergraduate course in medicinal chemistry and made progress in modelling and analysing pharmacological activity at the tissue level, my new passion. But after four years, I was suffering from withdrawal symptoms from lack of a chemical collaboration. Thus, I eagerly accepted John Vane's invitation to join the Wellcome Foundation.

My years at the Foundation (from 1977 to 1984) were an emotional roller-coaster. I wanted to make use of ideas I had been chiselling out, over the years, about the differences between successful and failed industrial projects. The division I took over at Wellcome, however, was remarkable for its traditional, conservative, ways and feudal structures. Entrenched attitudes can absorb reformist efforts like a punch bag. Yet despite disappointment in my managerial role, I made great progress in my own research. Working with brilliant young investigators such as Paul Leff, I began to see analytical pharmacology as a viable discipline. I had found myself a new mission – and once more my recurring dilemma between corporate commercial needs and personal scientific ambitions was solved unexpectedly. The Wellcome Foundation offered me the chance to establish a small academic research unit, modestly funded, but with total independence. The real opportunity, however, came from King's College, London. The College and Medical School between them have not only solved problems and smoothed difficulties, they have positively welcomed and supported my small unit. In intellectual terms the last five years at King's have been the most productive in my life. Surrounded by talented researchers and PhD students, I feel I have found my niche at last.

DRUGS FROM EMASCULATED HORMONES: THE PRINCIPLES OF SYNTOPIC ANTAGONISM

Nobel Lecture, December 8, 1988

by

JAMES BLACK

King's College School of Medicine & Dentistry, Analytical Pharmacology
Rayne Institute, London, England.

In this lecture I want to give an outline of the early stages in the discovery of adrenaline β-receptor antagonists and of the histamine H_2-receptor antagonists. I will end with a brief personal view about future research.

Adrenaline β-receptor antagonists

The work that is the theme of this lecture began in the early summer of 1958 when I joined Imperial Chemical Industries' Pharmaceuticals Division. I had gone there to pursue a very clear project that had been developing in my mind for several years. The idea had clinical, therapeutic, physiological and pharmacological elements.

Clinically, angina pectoris was known to be precipitated by anxiety and emotion just as well as by exercise. Indeed, the initiation of pain by an injection of adrenaline had been used as a diagnostic test. Partial thyroidectomy had been found to relieve severe angina pectoris whether or not associated with hyperthyroidism. At that time, tachycardia seemed to me to be the connecting link in these disorders.

Therapeutically, nitroglycerine could quickly relieve an attack of angina. Nitroglycerine also produced facial flush and headache. The relief of angina was attributed to similar vasodilatation in the coronary arteries. However, the newer, synthetic, selective coronary vasodilators, such as dipyridamole, were clinically ineffective despite the enhanced coronary artery dilatation that they provided. Here was a question mark against the widely practiced industrial strategy of seeking better drugs to increase coronary blood flow for angina.

Physiologically, Smith and Lawson (1958) had found that hyperbaric oxygen, at two atmospheres pressure, reduced the incidence of ventricular fibrillation associated with occlusion of a coronary artery even although the oxygen carrying capacity of the blood had increased by a maximum of only 25%. Might not an equivalently small *decrease* in the myocardial demand for oxygen be just as effective? That was my question.

Myocardial oxygen consumption is determined by the work of the heart and is a function of arterial blood pressure and heart rate. Lowering blood pressure by systemic vasodilatation might dangerously reduce the perfusion pressure and blood flow through disease-narrowed coronary arteries. Indeed, hypotension was known to be able to induce a heart attack. Heart rate, on the other hand, is largely determined by the cardiac autonomic nervous system. Heart rate would thus be reduced by cardiac sympathetic blockade. In addition, there was much discussion in those days about a postulated "anoxiating" action of adrenaline, proposing that the price of rapidly increasing cardiac power was a decrease in cardiac metabolic efficiency.

These clinical, therapeutic and physiological features of hearts coping with coronary artery disease all seemed to point to the potential advantage of annulling the actions of the sympathetic hormones, noradrenaline and adrenaline, on the heart.

Pharmacologically, the anti-adrenaline drugs were a well-recognised class in 1958. All of them showed a pattern of actions similar to those seen by Dale (1906) with the ergot alkaloids. Characteristically, they reversed the blood pressure rise produced by adrenaline to a fall in pressure, but they did not suppress the associated tachycardia. Konzett (1940) had shown that isoprenaline, the purely synthetic isopropyl derivative of noradrenaline, produced only the actions such as tachycardia, vasodilatation and bronchodilatation which the antiadrenaline drugs were not able to suppress. These were the actions of isoprenaline that Ahlquist (1948) could not explain on the basis of Cannon and Rosenblueth's (1939) prevailing hypothesis involving sympathins E and I. Ahlquist went on to propose that the widespread physiological effects of adrenaline were mediated by two classes of receptors, α and β. In this new classification, the antiadrenaline drugs of the day were α-receptor antagonists, and isoprenaline was a selective stimulant of β-receptors.

So I started at I.C.I. with a clear goal — I wanted to find a β-receptor antagonist. I expected this to reduce pulse rates at rest and during exercise and hoped that it would decrease the susceptibility of patients to angina pectoris. The unknown factor for me at that time was the significance of adrenaline's "anoxiating" activity.

John Stephenson was the medicinal chemist assigned to work with me. As no compounds were known to annul the actions of adrenaline on the heart, the programme had to be cold-started. The structure of isoprenaline (Fig. 1), the selective β-receptor stimulant, was our only clue. We thought that if N-substitution of adrenaline with isopropyl produced a selective *agonist,* then perhaps substitution with a different, larger, group might produce a selective *antagonist.* We thought that symmetrical, doubled-up, analogues of isoprenaline and dibenzylethylamines, might be interesting targets.

We were making compounds and testing them, admittedly without success when, early in 1959, we read Powell and Slater's (1958) report about the properties of DCI, an analogue of isoprenaline in which the ring hydroxyl groups were replaced by chlorine atoms (Fig. 1). In trying to

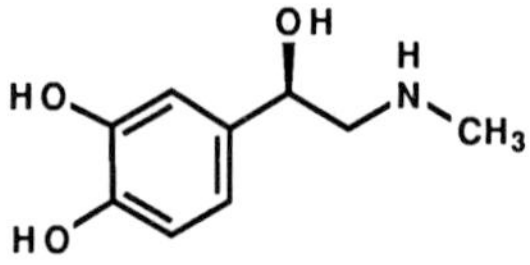

ADRENALINE

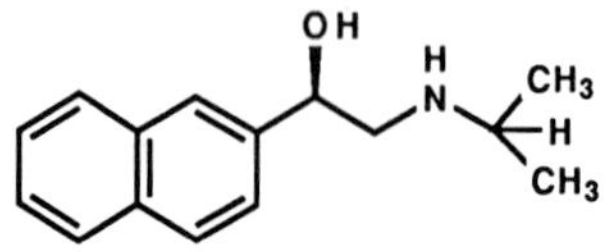

PRONETHALOL

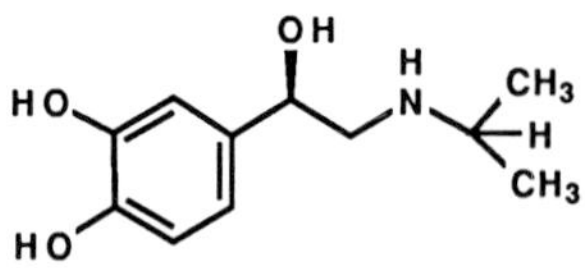

ISOPRENALINE

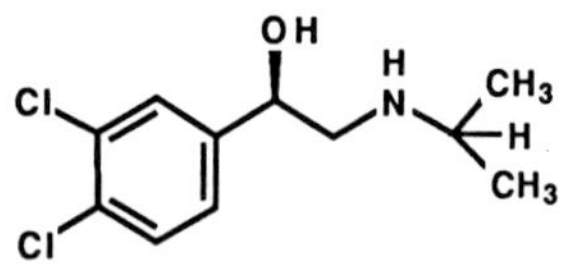

DICHLOROISOPRENALINE

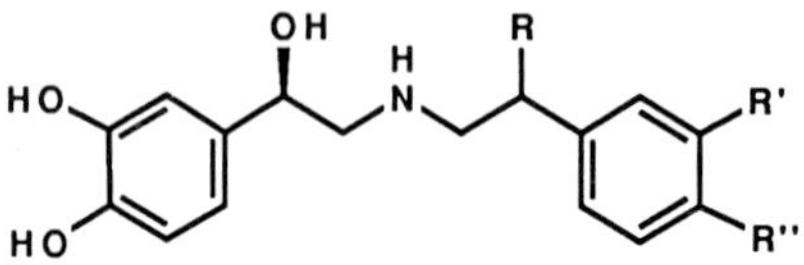

DIBENZYLETHYLAMINES

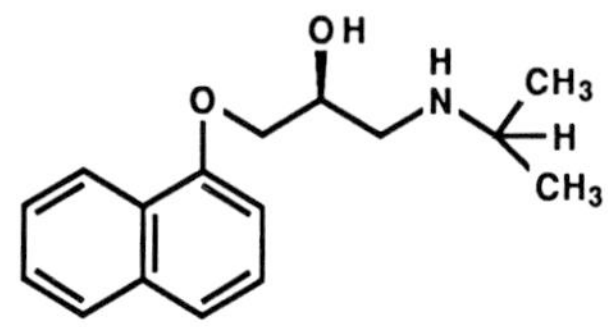

PROPRANOLOL

Figure 1. Chemical structures of adrenaline-related compounds.

exploit the bronchodilator properties of isoprenaline by making a long-acting variant, the Lilly group had discovered a compound that intrigued them by displaying – instead of isoprenaline's bronchodilator activity – the opposite property, namely antagonism. Soon afterwards, Moran and Perkins (1958) reported that DCI could annul the inotropic effects of adrenaline on the heart and classified DCI as a β-receptor antagonist. Stephenson immediately made some DCI for us to test.

We had started our bioassays using the classical Langendorff preparation, the isolated, spontaneously beating guinea-pig heart. Isoprenaline is a powerful stimulant of both the rate and force of beating in this preparation, but we measured the amplitude of contractions which compounded both changes. In this system DCI turned out to be as powerful a stimulant as isoprenaline and so was not at all what we were looking for. We had also developed a technique for simultaneously recording blood pressure and heart rate, in analogue form, in anaesthetised animals. Potential antagonists could now be given economically by slow intravenous infusion, and this allowed the effect on a wide range of systems to be monitored. Here, too, the powerful stimulant effects of DCI on heart rate were clearly seen, although there was less hypotension due to vasodilatation than we had expected (Fig. 2). On the basis of these experiments, we decided that DCI was not the lead we needed.

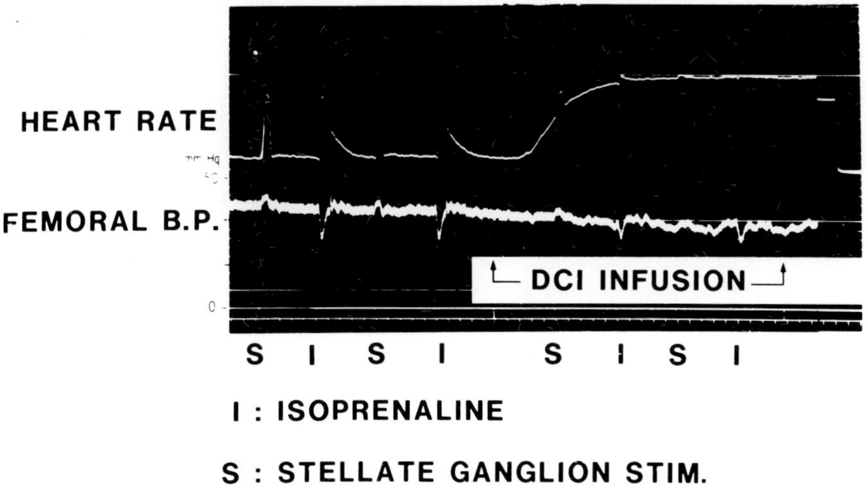

Figure 2. Effects of DCI (100μg/kg/min) (above) and pronethalol (100μg/kg/min) (below) infusion upon heart rate and blood pressure responses to injections of isoprenaline (0.4μg/kg i.v.) and to sympathetic stimulation (square wave pulses, 10 ms duration, 2.5 volts, 15 pulses/s for 30s) in anaesthetised cats. Lower axis shows time markers in one minute intervals.

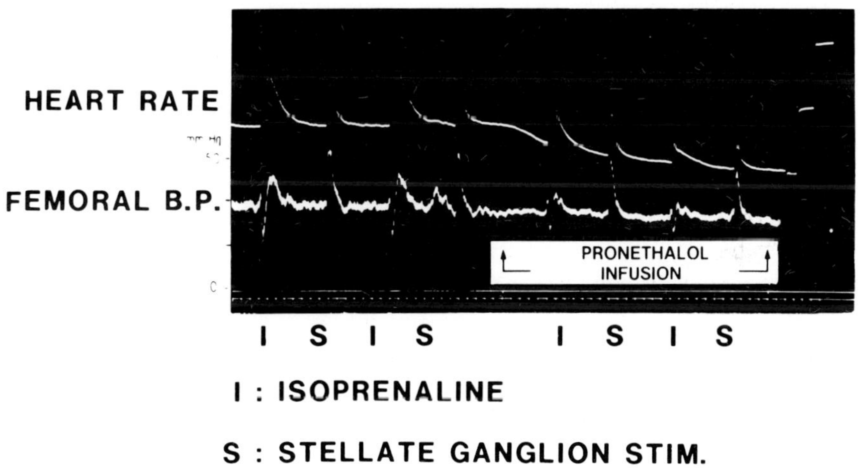

As analytical pharmacologists, what we are allowed to see of a new molecule's properties is totally dependent on the techniques of bioassay we use. The prismatic qualities of an assay distort our view in obscure ways and degrees. Our only defence lies in restless improvement in technique and experimental design, in the hope that collimation of several techniques will improve the reliability of our vision. We would make the change self-consciously today, but then it was intuitive.

We developed a new in vitro assay based on guinea-pig cardiac papillary muscles as a way of measuring the contractile effects of isoprenaline independently of rate changes. Then, we reassessed many early compounds, including DCI. On the new preparation, DCI had no stimulant activities

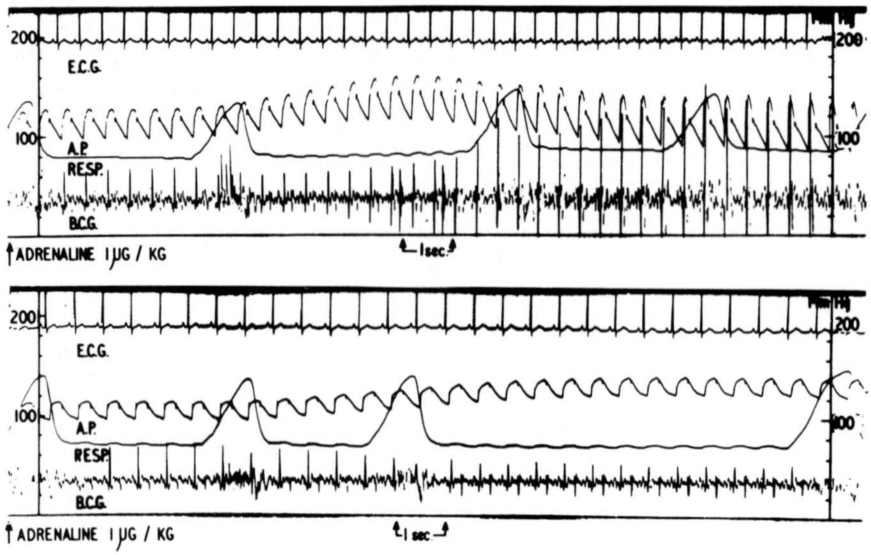

Figure 3. Effects of 38,174 (pronethalol, 5mg/Kg i.v.) on cardiac and respiratory responses to adrenaline (lµg/Kg i.v.) in anaesthetised dogs. E.C.G.: electrocardiogram; A.P.: aortic blood pressure; RESP: respiration; B.C.G.: ballistocardiogram.

itself but simply antagonised the effects of adrenaline and isoprenaline, although the stimulant activity on pacemaker tissue could be clearly seen in the atrial preparation. We were astonished. Today, we classify DCI as a partial agonist. Ariens (1954) and R.P. Stephenson (1956) had introduced the concept of partial agonists a few years earlier but nothing in their writing, as far as I recall, alerted us to expect that the agonist activity of these compounds could be so tissue dependent.

I shall never forget John Stephenson's reaction to this discovery: "We'll make the naphthyl analogue of isoprenaline" (Fig. 1). He had realised immediately that while a fused benzene ring would have similar steric and electronic properties to the two chlorine atoms, there was also the possible advantage of extended π-bonding. Compound ICI 38174 – nicknamed nethalide for a time, but finally christened pronethalol – was conceived in excitement and thrilled us at its birth. Pronethalol was an antagonist without any sign of agonist activity in both atrial and ventricular tissues. In anaesthetized animals, pronethalol reduced the resting heart rate and depressed the increments from isoprenaline or stimulation of cardiac sympathetic nerves (Fig. 2).

Having got over the first hurdle much more easily than I had dared to imagine, I was impatient to tackle the next one. How would someone, restricted by β-receptor blockade, cope with a surge of adrenaline or a burst of exercise? I had always imagined that the combination of Starling's "Law of the Heart" and the buffering capacity of the arterio-venous oxygen difference ought to be able to take up the slack of a reduction in cardiac output. We had developed the non-invasive technique of acceleration ballistocardiography to estimate the force of cardiac contractions in anaesthetised dogs (Fig. 3). Adrenaline increased heart rate, aortic blood pressure and force of contractions. After pronethalol, basal heart rate and force were reduced and the effects of adrenaline were abolished. However, the vasodilator effects of adrenaline were also blocked, thus exposing the heart to a vasoconstrictor load mediated by the unblocked α-receptors. The heart was able to maintain its output and produced an enhanced rise in blood pressure. This was the experiment that convinced me that the new compound might be more than a laboratory curiosity. In fact, I did notice in these early experiments, that the cardiac ballistic action was reduced under load. I noticed also that the time taken from ventricular excitation to the opening of the aortic valves — that is from the R-wave to the upstroke of aortic pressure — was increased under load. These were tell-tale signs that the cardiac reserve was reduced, but I persuaded myself at the time that this was a reasonable price to pay for the possibility of increasing the work capacity of a heart with restricted coronary flow.

The early clinical studies seemed to confirm that judgement. Dornhorst and Robinson (1962) studied the interaction between pronethalol and isoprenaline in healthy volunteers. Isoprenaline, infused into the brachial artery, produced a large increase in forearm blood flow. However, when repeated after an intra-arterial infusion of pronethalol, the first route of administration into man, the vasodilator effect was abolished. Isoprenaline given by slow intravenous infusion increased heart rate, respiratory amplitude, arterial pulse pressures and forearm blood flow. The subjects in these studies often seemed to get a fear of impending doom and became visibly restless. After pronethalol, all of these effects of isoprenaline were suppressed (Fig. 4). By chance, an athlete and a loafer were the first pair to

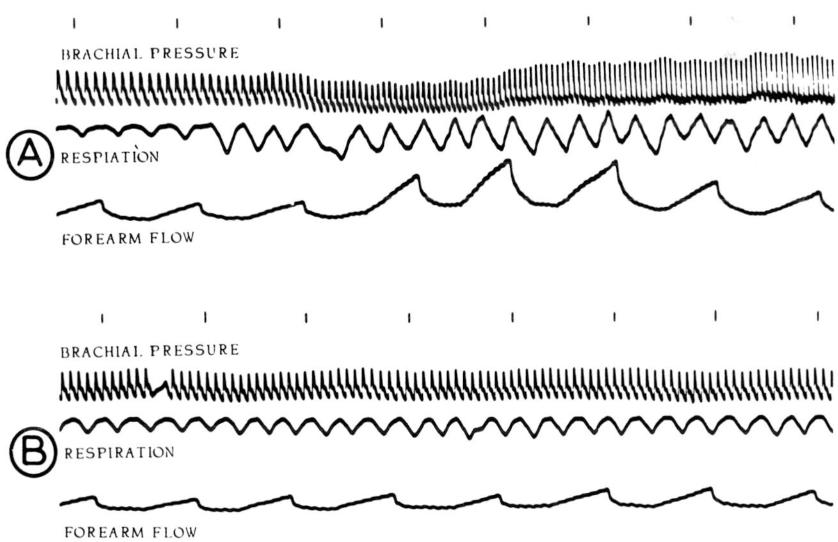

Figure 4. Effects of intravenous infusion of isoprenaline (10μg/min), (A) before and (B) after intravenous infusion of pronethalol (110mg) upon arterial blood pressure, respiration and forearm blood flow in a healthy volunteer (reproduced by kind permission from the Lancet). Time intervals: 10 seconds.

do maximal exercise after pronethalol. Compared to the control run, the athlete's heart rate at rest and exercise were little changed and his capacity to work was reduced. The loafer's heart rate was substantially reduced by pronethalol at rest and during exercise. He was less distressed by his lower heart rate. The potential benefits of β-adrenoceptor blockade for people with embarrassed hearts was also seen in the first patient with angina of effort. After pronethalol he was able to do more work before the onset of pain forced him to stop when his heart rate had eventually reached the same level as in the control run (Fig. 5).

Pronethalol always seemed to us to be a prototype drug, good enough to answer questions of principle, but not good enough to be marketable. So a large chemical group, directed by Crowther (Black et al. 1964), was assembled to try to find a more active, safer replacement for pronethalol. The discovery of ICI 45520, propranolol, a naphthyloxy propanolamine derivative, was the result (Fig. 1).

Our bioassays which had been developed as qualitative screens had now to be adapted for comparative quantitative bioassay. The isolated spontaneously-beating guinea-pig right atrial preparation proved to be excellent for

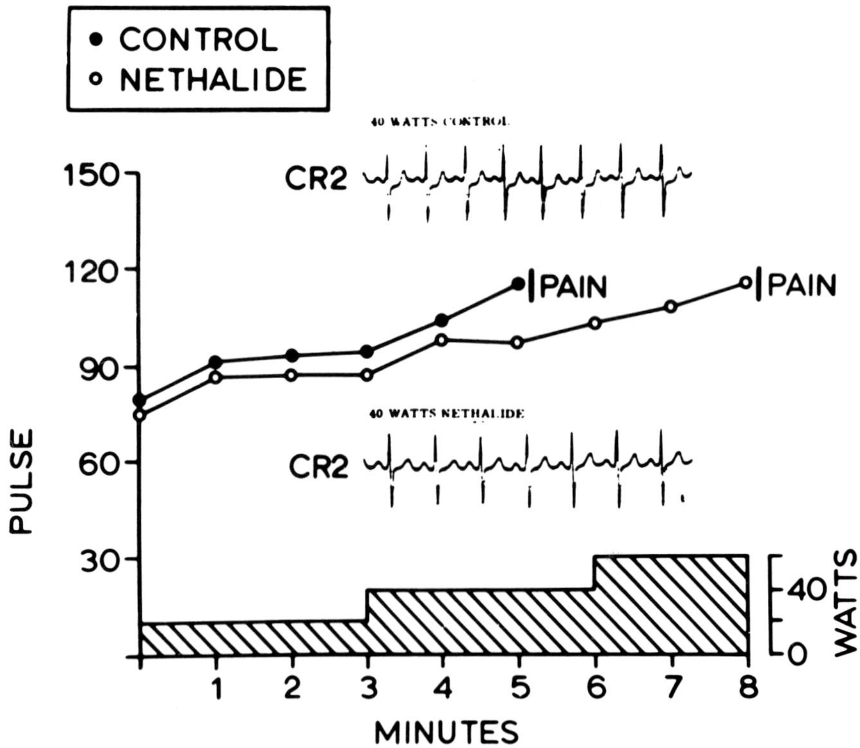

Figure 5. Effect of nethalide (pronethalol, 250 mg p.o.) upon time taken and work exerted before the onset of chest pain in a patient with angina pectoris performing graded exercise on a bicycle. Inserts show the ECG before and after nethalide (reproduced by kind permission from the Lancet).

these assays. The nature of the surmountable antagonism by propranolol was analysed by relating the rightward displacement of cumulatively derived dose-response curves to the concentration of antagonist. The linearity and slope of the Schild plot relating these variables indicated the likelihood that adrenaline and propranolol were competing for the same sites. This is the evidence that propranolol can be classified as a *syntopic* antagonist to the native hormones which activate β-adrenoceptors. Note here that I follow A. J. Clark (1933) in using the term 'hormone' very broadly: when one cell secretes a chemical to which another responds physiologically, I define that chemical as a hormone.

Histamine H_2-receptor antagonists

In 1964, I went to Smith, Kline & French Laboratories Ltd. to pursue another project that I had been thinking about for some time. Again, the idea had clinical, therapeutic, physiological and pharmacological implications.

The clinical problem was gastric and duodenal ulcers. I had thought a lot about the problem when I worked with Adam Smith (1953) on the effects of 5-hydroxytryptamine on gastric secretion. The immediate cause of ulceration was recognised to be hyperscretion of acid but the nature of the driving stimulus was unknown. The one clear fact was that patients with duodenal ulcers gave an exaggerated secretory response to histamine, the basis of a diagnostic test.

The therapeutic problem was that only surgical intervention, partial gastrectomy in those days, was recognized to be effective. The potential value of anticholinergic drugs, like atropine, was obscured by unacceptable side-effects. Antacids could be shown to promote ulcer healing, but only with clinically-unacceptable regimens.

The physiological problem was the relationship between gastrin and histamine, both of them powerful stimulants of acid secretion and both synthesized in the mucous membrane of the stomach. MacIntosh (1938) had proposed that histamine was the final stimulant of secretion when the vagus was stimulated and both Code (1965) and Kahlson (Rosengren and Kahlson, 1972) had extended that idea to gastrin as well, making histamine the final *common* chemostimulant. Mainstream thinking in gastroenterology, however, regarded gastrin as the *direct* hormone of secretion in its own right, thus, the question of the function of histamine in the stomach was unsettled (Grossman, 1974).

The pharmacological problem was the selective blocking properties of the antihistamines (Loew, 1947). The available antihistamines were a diverse group, chemically unrelated to histamine and reminiscent in this respect of the class of adrenaline α-receptor antagonists. They were powerful inhibitors of histamine-induced visceral muscle contractions but had no effect at all against histamine-induced acid secretion, uterine relaxation or cardiac stimulation. Other effects of histamine, such as vasodilatation were well known to be insensitive to the antihistamines. The parallels with the spectrum of activity of the anti-adrenalines seemed obvious.

In 1964, I had no doubts that histamine had its "β-receptors" and that a new type of selective histamine antagonist could be found. Ambiguity about the physiological role of histamine in acid secretion left me unsure about the clinical value of such drugs. At the very least, however, I expected to answer the physiological question of the gastrin-histamine connection.

The bioassay systems were easily selected. For in vitro assays, guinea-pig ileal muscle was the classical system for studying antihistamines such as mepyramine. Guinea-pig atrial tissue looked like a good assay for mepyramine-refractory histamine responses. The assay for acid secretion was harder to choose. No in vitro assays were available at that time. We chose the Ghosh and Schild (1958) method of lumen perfusion of the stomach of anaesthetized rats, but the method worked reliably only after it had been substantially modified by Parsons (1969), my new colleague.

The chemical programme, from the start, concentrated on making analogues and derivatives of histamine. As the whole project was conceived by

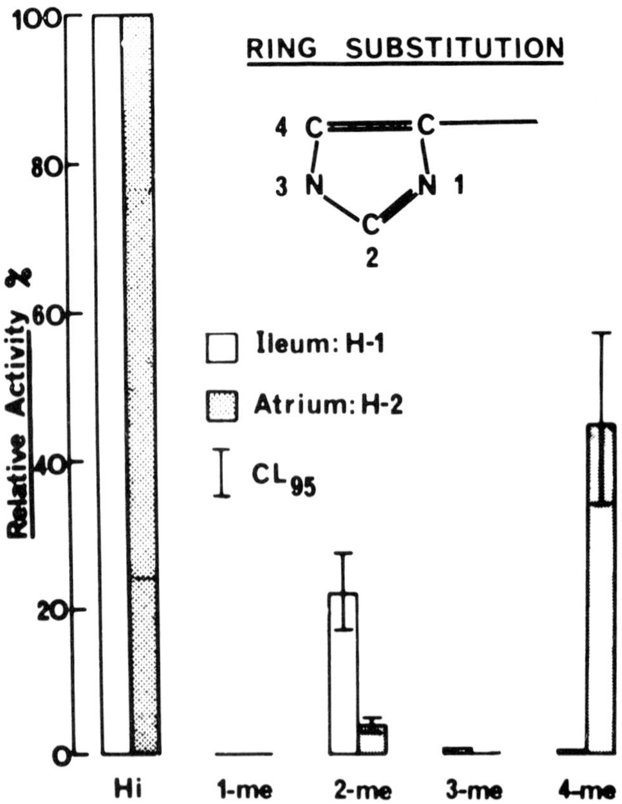

Figure 6. Selectivity of histamine and several methyl-substituted analogues in the guinea-pig ileum (histamine H_1 receptor assay) and right atrium (histamine H_2 receptor assay). Activity values, relative to histamine were calculated from parallel line assays. Error bars show 95% confidence limits.

analogy with the adrenaline β-receptor story, I was particularly anxious to concentrate on varying the imidazole ring end of histamine. Some of the early ring-substituted compounds turned out to be very important (Black et al. 1972) (Fig. 6). Methyl substitution on either of the ring nitrogens produced inactive compounds. Methyl substitution on the 2-position was found to be less active than histamine itself, but nevertheless showed a clear preference for the ileal assay. However, 4-methylhistamine was exciting for

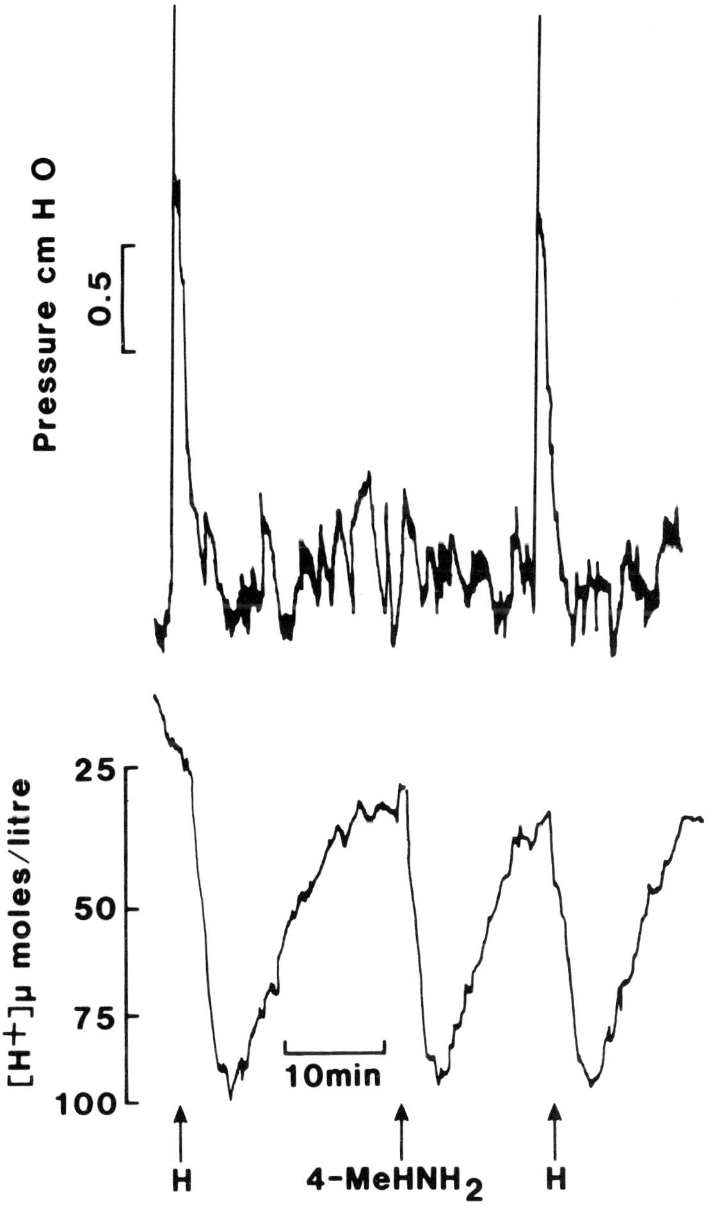

Figure 7. Effects of i.v. bolus injections of histamine (H, 2ug/kg) and 4-methyl histamine (4-MeHNH$_2$, 5ug/kg) upon secretion of gastric acid and stomach wall contraction in an anaesthetised rat.

me. Nearly half as active as histamine on the atrial assay, 4-methylhistamine was practically inactive on the ileal assay. We were able to confirm this selectivity in vivo. In the anaesthetized rat preparation, an injection of histamine produced a fast, spasmodic contraction of the stomach wall followed by a phasic burst of acid secretion (Fig. 7); 4-methylhistamine produced an equivalent output of acid without any muscle contractions. Thus, 4-methylhistamine was a selective agonist analogous to isoprenaline at adrenergic receptors. This result was the compelling clue which kept us going through several lean years of negative screening.

This observation assumed even greater importance when we compared 4-methylhistamine and 2-methylhistamine with the 1,2,4-triazole analogue of histamine. Using a number of additional assays, both in vitro and in vivo, we found that the triazole analogue was clearly non-selective, whereas 2-methylhistamine was selective for the mepryamine-sensitive responses and 4-methylhistamine was selective for the refractory responses (Fig. 8). When Ash and Schild (1966) proposed that the mepyramine-sensitive histamine receptors should be classified as H_1, we used this pattern of bioassay results to argue for the homogeneity of a non-H_1 class of histamine receptor.

A very large number of compounds were made, predominantly ring substitutions and fused ring heterocycles, all of them inactive. I vividly remember wondering suddenly if the strategy was all wrong. Perhaps we should have spent more time exploring the role of the side-chain amino

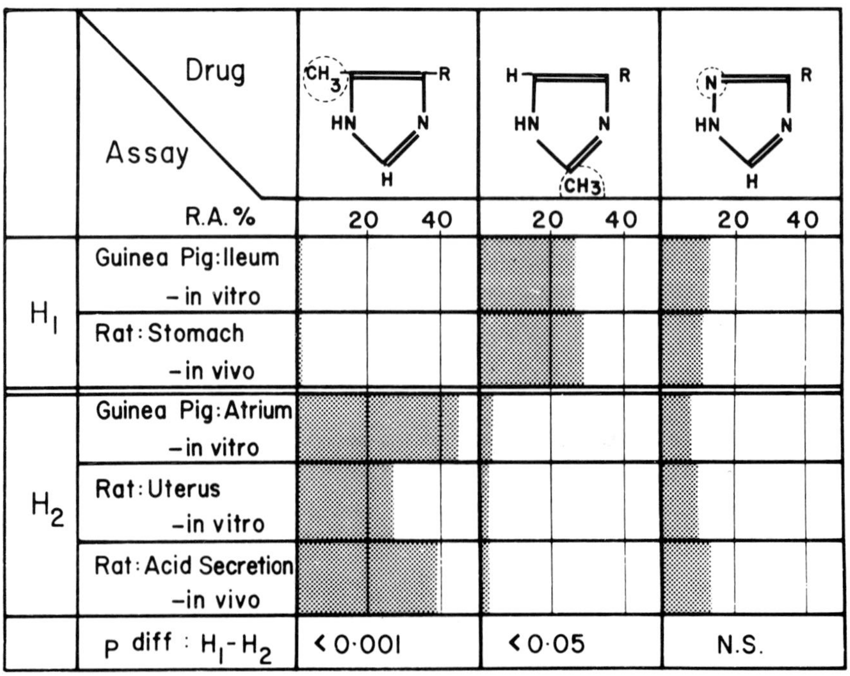

Figure 8. Selectivity of 2-methylhistamine, 4-methylhistamine and 1,2,4-triazole analogue of histamine in several in vitro and in vivo, H_1- and H_2- histamine receptor assays. R.A. %: relative activity to histamine was calculated from parallel line assays.

group. On this suggestion, Parsons quickly scanned through the earlier compounds looking for examples of side-chain variations. He came up with N^{α}-guanylhistamine (Fig. 11). This compound was one of the earliest we had tested and had proved to be quite a potent agonist when injected, like histamine, intravenously. However, over the years we had changed the design of the screening assay. A continuous intravenous infusion of histamine was used to produce a stable background of near-maximal acid secretion. A new compound could now be quickly screened by giving a succession of increasing doses intravenously. Even antagonism of very short duration would be detectable. In the new experimental design, the guanidino- analogue of histamine now exhibited a small degree of inhibition, about 5% reduction. There it was — a partial agonist! Guanylhistamine on histamine receptors was the analogue of DCI on β-adrenoceptors. However, unlike DCI, the efficacy of histamine had been reduced by modifying the side-chain rather than the ring system.

This was the lead that Ganellin and his colleagues in chemistry had been waiting for. One of the early analogues simply increased the length of the side chain from ethyl to propyl. In the rat stomach assay this compound showed good antagonism of histamine without much agonist activity of its own. However, in other species, particularly cat and dog, the compound was nearly a full agonist. Similarly, in the isolated guinea-pig atrial preparation, although much less potent than histamine, the compound achieved a maximum response about 80% of the histamine maximum. The true nature of this partial agonism could be seen by repeating the dose-response curve in the presence of a nearly maximal concentration of histamine. Only

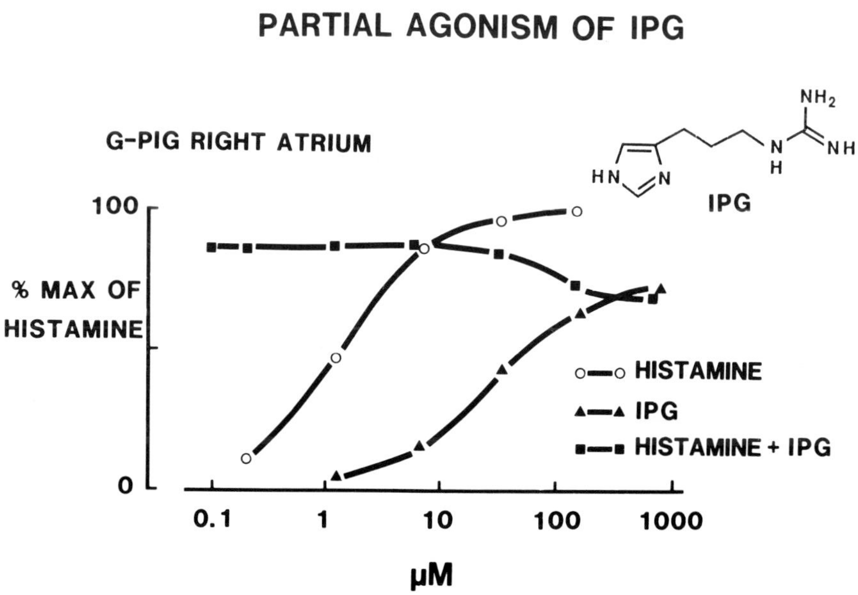

Figure 9. Effects of cumulative additions of histamine, imidazole propylguanidine (IPG) and IPG in the presence of 10μM histamine on the pacemaker frequency of the isolated guinea-pig right atrium. Error bars have been removed for clarity (n=7).

antagonism was now seen with the maximum inhibition, about 20%, being equal to the agonist maximum (Fig. 9).

Lengthening the side-chain to four carbon atoms and replacing the strongly basic guanidino group by the neutral methyl thiourea group produced burimamide, the first antagonist of moderate activity which had low enough efficacy to avoid being an agonist in any of our assays (Fig. 11). Burimamide, having relatively low potency and poor oral bioavailability, was clearly only a prototype. Ganellin saw the way forward. The non-basic, electron-releasing side-chain in burimamide, compared to the basic, electron-withdrawing, side-chain in histamine, raised the pK_a of the ring and favoured the opposite tautomer. Inserting the electro-negative thioether linkage in place of a methylene and introducing the 4-methyl group to favour H_2-receptor selectivity produced metiamide, which was much more potent and better absorbed than burimamide (Fig. 11). Toxicity associated with thiourea was then eliminated by replacing the thiourea sulphur with a cyano-imino group to produce cimetidine.

On the atrial assay in vitro, burimamide produced surmountable antagonism, shown by rightward parallel displacement of the histamine dose-response curves. When analysed by the Schild method, burimamide behaved like a syntopic antagonist to histamine. The estimated dissociation constants (K_B) were found to be independent of the potency of the titrating agonist and also independent of the tissue (atrium or uterus) used for the assay, substantially confirming the syntopic classification (Fig. 10). The high value on the ileum, an H_1 system, disclosed the compound's selectivity. As Ash and Schild (1966) had proposed the notation H_1, we proposed that burimamide should be classed as an H_2-receptor antagonist (Black et al. 1972).

BURIMAMIDE K_B (μM)

	ATRIUM	UTERUS	ILEUM
HISTAMINE	7.8	6.6	288
4-Me HISTAMINE	7.2		
2-Me HISTAMINE	6.9		

Figure 10. Equilibrium dissociation constants (K_B) for burimamide in guinea-pig right atrium and rat uterus (H_2-histamine receptor assays) and in guinea-pig ileum (H_1-histamine receptor assay) against histamine and its analogues.

The analytical capability to distinguish an antagonist which acts at the same site as the native hormone from one which does not act syntopically — that is a functional antagonist — seems to me to be important in drug research for two reasons. For a defined homogeneous population of receptors, widely disseminated across tissues, the properties of the syntopic antagonist can be generalised. As the mechanism of action of the functional antagonist is unknown, however, its properties have to be identified on a tissue-by-tissue basis. Again, the analytical power of a syntopic antagonist — that is its ability to prove hormonal involvement in physiological processes — is likely to be greater than for a functional antagonist. Of course, a compound that is a syntopic antagonist at one receptor system can also be a functional antagonist at a different receptor system. However, a possible combination of syntopic and functional properties would be expected to vary between different molecules, and the confusion can be eliminated by building up a class of syntopic antagonists that are chemically distinct but pharmacologically homogeneous. Syntopic antagonists are the best tools that analytical pharmacologists possess.

This problem of the resolving power of a receptor antagonist was seen from the beginning with metiamide. The histamine-induced acid secretion in the rat assay, having reached a plateau, was promptly inhibited when metiamide was given intravenously. Metiamide was found to be equally effective at inhibiting pentagastrin-induced secretion but much less effective against carbachol-induced secretion. Failure to inhibit cholinergically-stimulated secretion showed that metiamide was not a non-specific inhibitor of acid secretion. The ability of metiamide to inhibit the effects of gastrin

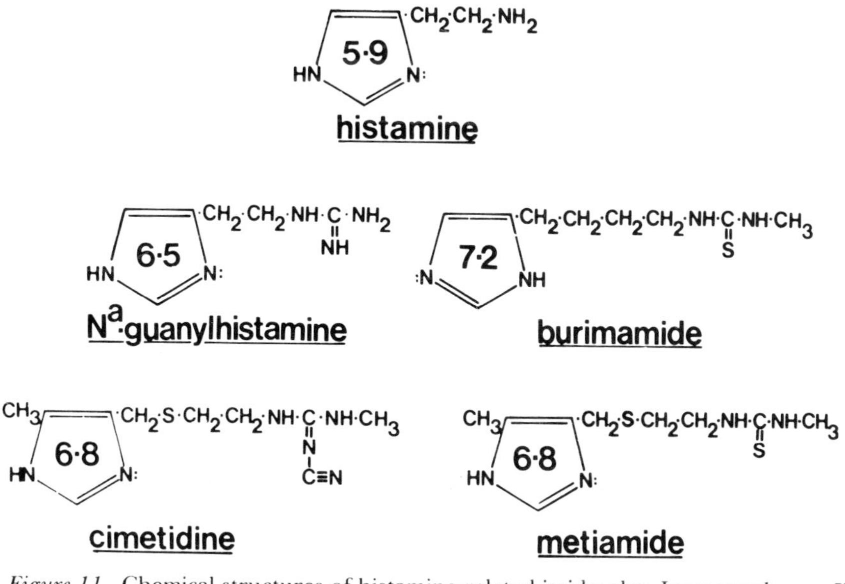

Figure 11. Chemical structures of histamine-related imidazoles. Inset numbers: pK_a values.

pointed to potential clinical utility, but it was not at all clear that this result might contribute to a resolution of the gastrin-histamine controversy.

The problem became clear when the interactions between metiamide and the various stimulants were studied quantitatively in dogs with Heidenhain pouches (Black, 1973). Metiamide displaced the histamine dose-response curves in parallel to the right, as would be expected for a surmountable, syntopic antagonist (Fig. 12). Carbachol's steep dose-response curves were relatively refractory to inhibition. However, the flatter dose-response curves to pentagastrin were depressed downwards as well as being displaced to the right. At that time I did not know what to expect if H_2-receptor blockade inhibited gastrin only because histamine was the final common chemostimulant. I could not rule out the possibility that these drugs were functional antagonists of gastrin. Subsequently, when other workers had confirmed the different patterns of inhibition in other species, and with different compounds, an unspecific inhibitory action seemed unlikely. The pattern also became understandable when I was able to model indirect competitive antagonism and applied the model to tyramine, a well-characterised indirectly acting agonist, to show that it was inhibited insurmountably by propranolol, just like the gastrin-metiamide interaction (Black et al. 1980).

When we took the H_2-receptor antagonists into human volunteers, there were no surprises in the patterns of secretory inhibition. However, we did get a surprise, right at the start, with burimamide. We followed the standard clinical practice of giving the volunteers mepyramine before giving them histamine intravenously. Even so, the subjects showed marked skin and conjunctival vasodilatation. The surprise was that treatment with burimamide completely blocked this vasodilatation. In the laboratory, burimamide

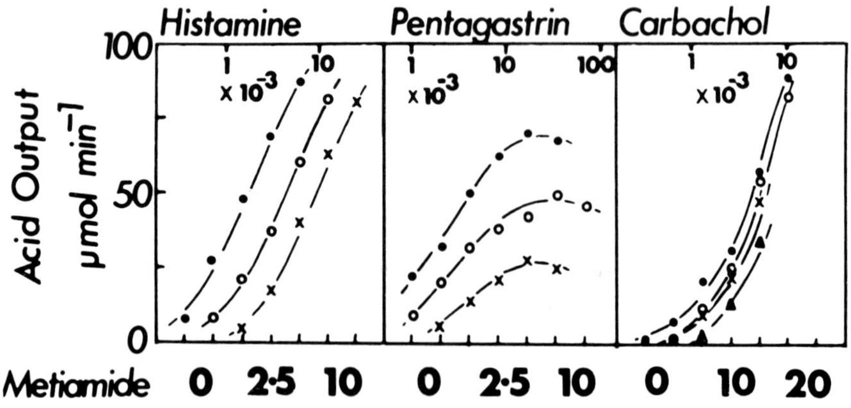

Figure 12. Effects of i.v. infusion of metiamide against histamine-, pentagastrin- and carbachol-induced gastric acid secretion in dogs with Heidenhain pouches. Doses shown are μmol/kg/min for the agonists (upper abscissae); μmol/kg/h for metiamide (closed circles 0, open circles 2.5, crosses 10, triangles 20). Reproduced by kind permission from: International Symposium of H_2receptor blockade, 1973, Eds. Wood and Simpkins, Publ. SK&F Laboratories.

alone had had no effect on histamine-induced vasodilatation. As both H_1- and H_2-receptors were involved, both antagonists were needed. This finding thus explained the results of Folkow et al. (1942) of 30 years earlier.

Hormone-receptor antagonists in the future

The histamine project was started by analogy with my experience of the adrenaline project. In retrospect, I think they have some features in common which helped them to succeed. Both started from well-recognised clinical problems at a time when they could be illuminated by specific hypothetical modelling at the laboratory level. The laboratory modelling defined the chemical starting points and the types of bioassay. The clinical problem defined how the newly classified drug should be tested in volunteers and patients. If the intimate coupling of clinical experience and pharmacological modelling has the effect of helping to eliminate wishful thinking in drug research, then the limiting step in the future will be the development and improvement of these models.

Models in analytical pharmacology are not meant to be descriptions, pathetic descriptions, of nature; they are designed to be accurate descriptions of our pathetic thinking about nature. They are meant to expose assumptions, define expectations and help us to devise new tests.

Traditionally, pharmacological modelling of hormone-receptor systems has been based on the application of the Law of Mass Action to reversible interactions. Therefore, they are all chemical, molar models characterised by thermodynamic parameters. The discovery, often in a homologous series of compounds, that not all agonists could produce the same maximum response (now defined as partial agonists) led to models which had both binding, or affinity, parameters *and* efficacy, or response-generating, parameters. In both of the studies that I have sketched for you today, chemical modification of a native hormone produced, first of all, selective agonists, then quite separate chemical changes produced partial agonists and finally pure antagonists. The assumption is that partial agonists and antagonists are associated with a relative loss of efficacy — emasculated hormones.

The discovery of partial agonists was, in both reports, crucial to the development of syntopic antagonists. Yet I very nearly, and could quite easily, have failed to discover them. The choice of tissue for the assay was vital. So, why is the expression of efficacy so tissue dependent? How can we try to choose tissues which are most likely to allow us to detect partial agonists?

To illuminate these questions, Leff and I developed an operational model of agonism which defined three mutually-connected surfaces such that knowing the shape of the function on any two allowed us, syllogistically, to deduce the necessary shape of the function on the third space (Black and Leff, 1983) (Fig. 13a). On the right-hand side of the graphical display of the model is the measured function that relates agonist concentration, on a logarithmic scale (log A), to the tissue effect which it produces. The pharmacological assumption is that the agonist initiates an effect by binding to a

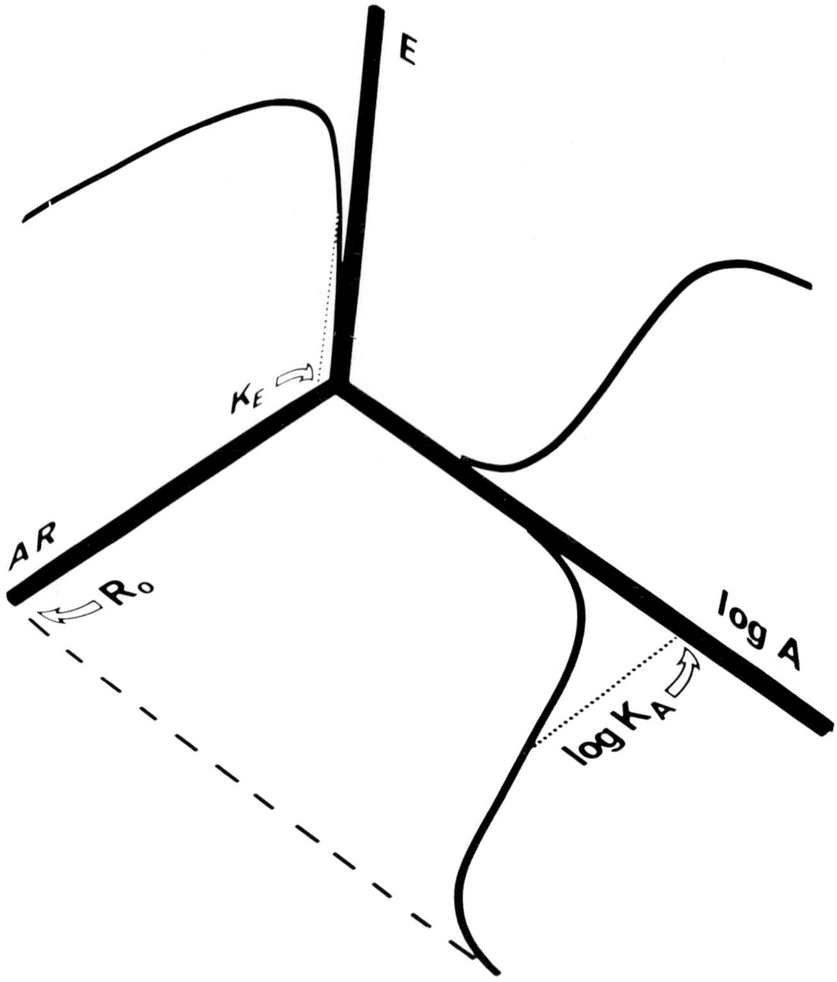

Figure 13a. Three-dimensional display of the Operational Model of Agonism (Black and Leff, 1983). E: pharmacological effect, log [A]: logarithmic concentration of agonist A, log K_A: log equilibrium agonist dissociation constant, [R_O]: operational receptor concentration, [AR]: concentration of receptors occupied by A, K_E: concentration of AR required for half-maximal tissue response. The three planes of the figure represent pharmacological effect (right panel), binding or affinity (base panel) and efficacy or transduction (left panel).

receptor (R); then the bound receptor (AR) activates a messenger system that produces the effect. Therefore, the base of the display shows the assumed relationship between agonist concentration and the concentration of bound receptors — the *affinity* relation. Then, the left-hand panel shows the deduced relation between bound receptor concentration and effect — the *efficacy* relation. The behaviour of this model is critically determined by the ratio of R_0, the total receptor concentration, to K_E, the concentration of bound receptor needed to produce a half-maximal effect. For example, when R_0 is equal to K_E the agonist can only produce a half-maximal response, thus defining a 50% partial agonist (Fig. 13b).

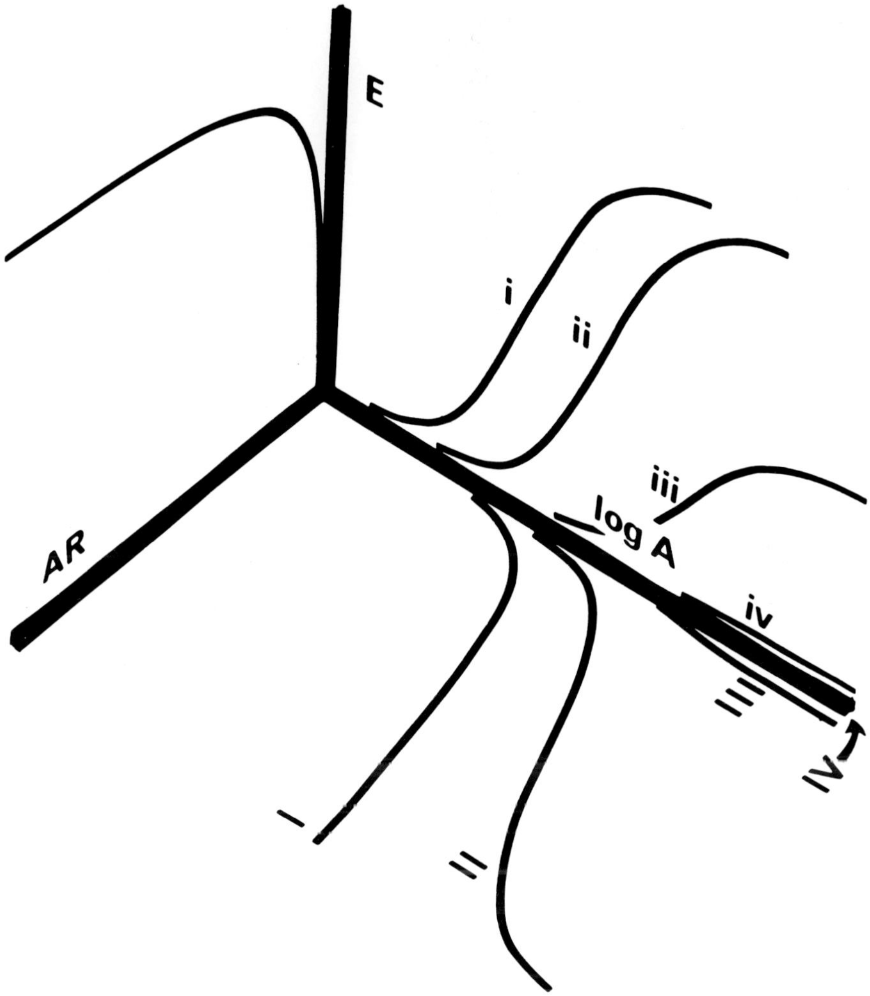

Figure 13b. Predictions of the influence of changes in $[R_o]$ upon pharmacological effect using the Operational Model of Agonism. Curves I to IV show R_o concentration decreasing successively by ten-fold giving rise to effect curves i to iv, respectively.

One of the first uses of the model was to fit simultaneously all of the data which Kenakin and Beek (1980) had got from comparing isoprenaline and prenalterol, a partial agonist at β-receptors, on six different tissues (Fig. 13c). The model of agonism allowed all of the data to be fitted by the theoretical curves shown when only one parameter, total receptor density Ro, was allowed to vary. The concentration of receptors is now known to vary between tissues, so that this seems to me to be an attractive way of accounting for the tissue dependence of a partial agonist's efficacy. Superimposing all the dose-response curves clearly demonstrates that the tissues most sensitive to isoprenaline support the greatest maximum responses to prenalterol and vice versa (Fig. 13d). Practically, therefore, the best way to avoid missing a partial agonist is to measure the potency of the native hormone or full agonist on as many tissues as possible and select assays

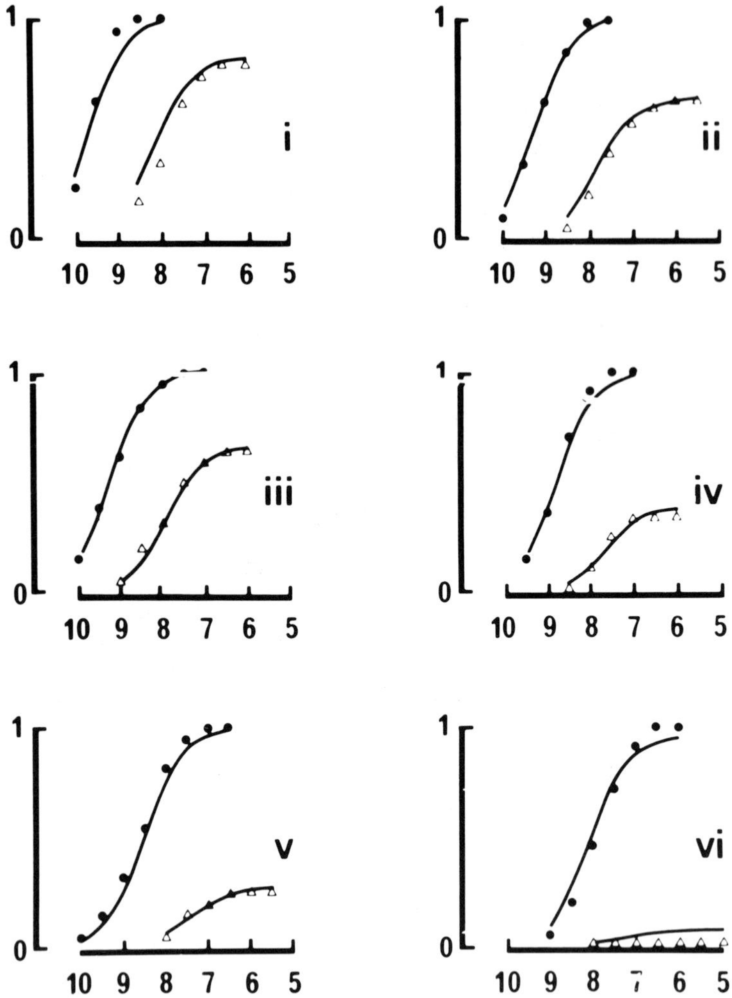

Figure 13c. Behaviour of isoprenaline (circles) and prenalterol (triangles) in (i) guinea-pig tracheal muscle, (ii) cat left atria, (iii) rat left atria, (iv) cat papillary, (v) guinea-pig left atria and (vi) guinea-pig extensor digitorum longus muscle; (Kenakin and Beek, 1980). The data has been regressed to the Operational Model of Agonism allowing only $[R_O]$ to vary between tissues. Abscissa: log molar agonist concentration. Ordinate: Fractional response to isoprenaline.

expressing both high and low efficacy. This seems to be a robust test, relatively insensitive to the mechanisms underlying the differences in sensitivity.

Partial agonists, as empirical facts, have been recognised for many years. Pharmacological modelling of partial agonism and the related concept of efficacy has, however, developed more slowly. Fundamental problems about the nature of efficacy, either as a molar, thermodynamic, concept or as a molecular problem in wave mechanics have still to be tackled. However, there seems no doubt about the pragmatic utility in drug research of

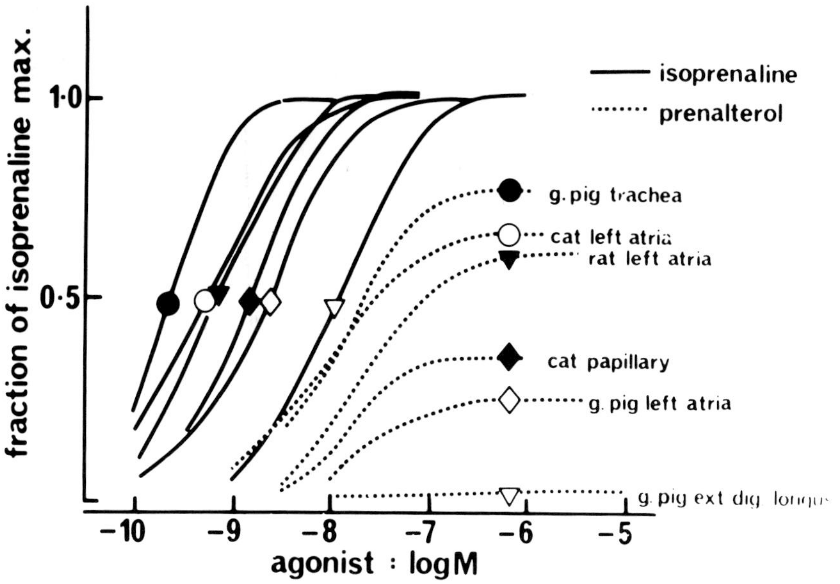

(after Kenakin & Beek, 1980)

Figure 13d. Representation of the model-simulated curves from Fig. 13c superimposed on a single pair of axes.

distinguishing potency changes from efficacy changes in the bioassay of hormone analogues. The discovery of a partial agonist is the vital clue in developing useful syntopic antagonists.

We have gone on to use the model to study the effects of the slope of dose-response curves (Black et al. 1985a), functional antagonism (Leff et al. 1985), indirect competitive antagonism (Black et al. 1985b), dual receptor systems and receptor distribution when there is a relatively high concentration of transducer molecules.

The dose-response relation can, of course, be broken down into any two necessarily-connected steps. Thus, the gastrin dose-response curve of acid secretion can be broken down into the gastrin/histamine-released relation followed by the histamine-released/acid secretion response relation. Two of the relations are known from measurement and the third, the gastrin/histamine-released relation can be deduced. Using this model, we were able to make a correct estimate of the K_B of an H_2-receptor antagonist from the family of unsurmountable curves produced by its interaction with pentagastrin, an important piece of evidence linking local histamine release to the physiological action of gastrin. (Black et al. 1985b.)

The approach I have outlined so far has regarded hormones and their conjugate receptors as simple, linear, "command-control" systems. The approach has undoubtedly had some success. Nevertheless, I think if we want to continue to try to develop new drugs by mimicking and manipulating physiological chemical control systems, our ideas will have to become more sophisticated. There is plenty of evidence now that hormone receptors and their dependent messengers are not insulated from each other. Mutually-enhancing interactions between any two receptor-messenger systems can occur at many different points leading to different kinds of physiological advantage.

When one hormone can interact with two allosterically-linked receptors on the same cell, the continuous gearing can change the relatively-flat concentration-response curve characteristic of the Mass-Action Law behaviour of a one-receptor system into steep curves. This greatly reduces the change in concentration needed to sweep the cellular response through its full range. This could be an advantage for fast-responding cells. When there are two hormones and two receptors, mutual potentiation can lead to threshold changes, pulsing signals and, more importantly, by making the activity of one hormone depend on the other, the convergence changes the type of behaviour from obligatory responses to conditional responses, like nerve cells, based on summation.

The rich possibilities of hormonal convergence plus inter-receptor amplification are now being discovered in the area of neuroendocrine secretion. While co-existence of multiple hormones in a single nerve ending does not necessitate co-transmission, there seems as yet no need to doubt it. Hökfelt, in a recent review (Hökfelt et al 1987), pointed out that the distribution of these hormones was not random. For example, neurones classified as 5-HT-, noradrenaline- or dopamine-transmitting had each got different groups of peptides co-existing in their terminals. Neurobiologists have plenty of ideas about the significance of the very large and rapidly growing number of pharmacologically active substances which have been identified in nervous tissue. However, as an outsider looking in on all their excitement, I sense that my colleagues have problems with the Principle of Parsimony. Neurotransmission involving discrete, microscopic events is unlikely to generate problems with chemical crosstalk in the brain at large. So, is there not now an embarrassing number of potential neurotransmitters?

On the other hand, biologists concerned with brain development probably do need an abundance of specific cell markers. Sperry's Chemoaffinity Hypothesis (1963), one of the earliest attempts to account for the details of pattern development in the embryonic brain, required that cells have individual chemical identification markers almost down to the level of single cells. Edelman's modulation hypothesis (1984) — how the composition and density of nerve cell adhesion molecules can be locally regulated by the cells themselves — is chemically much more economical. These molecules subserving cell-to-cell interactions can provide a framework for guiding neurite growth cones. Diffusible growth factors, such as the specific Nerve Growth Factor, can provide a general engine for neural growth into a supporting network. The question that intrigues me, however, is whether the framework and the engine are enough to account for the exquisite fine-tuning of synaptic connections which occurs during brain maturation and for the control of synaptic plasticity now known to be a feature of the mature brain.

I like the idea that these synaptic connections are determined chemotactically. An effective chemotactic address might then involve the co-operative signalling of two or more chemicals. The possibilities for chemotactic signatures are factorial. If an effective signal involved just three chemicals, then 100 hormones could provide over a hundred thousand different signatures in any one compartment.

As our ideas about hormone-receptor systems become progressively more complicated in terms of multiplex pathways, hierarchy due to cellular conjunctions and biochemical cascades, the reductionist methods of molecular biology would seem to offer modern drug research a way out: simplify the systems by receptor isolation and expression. Molecular biology undoubtedly holds out the promise of the most direct and productive route ever to screening chemicals as hormone receptor reagents. However, once classified at the molecular level, a new reagent will have to be evaluated at the level of tissue complexity to confirm its classification and define its selectivity. These tissue bioassays, such as I've discussed today, may seem old-fashioned but, properly designed, they are arguably the best methods we have for making reliable predictions about clinical outcomes. They have served us well but they need to be continually improved, both technically and in their related operational models, to match our changing ideas. Molecular biology will continue to provide drug research with extraordinary analytical methods and lend a richer texture to our imagination.

These reflections suggest that there will be both great opportunities, and potential dangers, for the development of specific hormone receptor reagents in the future. The limiting factors, however, are likely to be the verisimilitude of our models and the complexity of our bioassays. Analytical pharmacology has got an important and exciting future

REFERENCES

Ahlquist, R.P. (1948) Am. J. Physiol., *153*, 586.
Ariens, E.J. (1954) Archiv. Int. Pharmacodyn. Ther., *99*, 32.
Ash, A.S.F. and Schild, H.O. (1966) Br. J. Pharmacol., *27*, 427.
Black, J.W. (1973) In Int. Symp. on Histamine H_2-Receptor Antagonists, (eds). Wood, C.J. and Simpkins, M.A., SK&F, Welwyn Garden City, UK, 219
Black, J.W., Crowther, A.F., Shanks, R.G. Smith, L.H. and Dornhorst, A.C. (1964) Lancet, *i*, 1080.
Black, J.W., Duncan, W.A.M., Durant, G.J., Ganellin, C.R. and Parsons, M.E. (1972) Nature (London), *23*, 385.
Black, J.W., Duncan, W.A.M. and Shanks R.G. (1965) Br. J. Pharmacol., *2*, 577.
Black, J.W., Fisher, E.W. and Smith, A.N. (1958) J. Physiol., Lond, *141*, 27.
Black, J.W., Jenkinson, D.H. and Kenakin, T.P. (1980) Eur. J. Pharmacol., *65*, 1.
Black, J.W. and Leff, P. (1983) Proc. Roy. Soc. B, *220*, 141.
Black, J.W., Leff, P., Shankley, N.P. and Wood, J. (1985a) Br. J. Pharmacol., *84*, 561.
Black, J.W., Leff, P. and Shankley, N.P. (1985b) Br. J. Pharmacol., *86*, 589.
Black, J.W. and Stephenson, J.S. (1962) Lancet, *ii*, 311.
Cannon, W.B. and Rosenbleuth, A. (1939) In *Autonomic Neuroeffector Mechanisms.* The Macmillan Co., New York.
Clark, A.J. (1933) In *The Mode of Action of Drugs in Cells,* Edward Arnold Co., London.
Code, C.F. (1965) Fed. Proc., *24*, 1311.
Dale, H.H. (1906) J. Physiol., Lond, *34*, 163.
Dornhorst, A.C. and Robinson, B.F. (1962) Lancet, *ii*, 314.
Edelman, G.M. (1984) Ann. Rev. Neurosci., *7*, 339.
Folkow, B., Maeger, K. and Kahlson, G. (1948) Acta Physiol. Scand., *15*, 264.
Ghosh, M.N. and Schild, H.O. (1958) Br. J. Pharmacol. Chemother., *13*, 54.
Grossman, S.P. (1967) In Handbook of Physiology, *1*, Washington, USA, Sect. 6, 287.
Hökfelt, T., Millhorn, D., Seroogy, K., Tsuruo, Y., Ceccatelli, S., Lindh, B., Meister, D., Melander, T., Schalling, M., Bartfai, T. and Terenius, L. (1987) Experientia, *43*, 768.
Kahlson, G. and Rosengren, E. (1972) Experientia, *28*, 993.
Kenakin, T.P. and Beek, D. (1980) J. Pharmacol. Exp. Ther., *213*, 406.
Konzett, H. (1940) N. Arch. Exp. Path. Pharmak., *197*, 27.
Leff, P., Martin, G.R. and Morse, J.M. (1985) Br. J. Pharmacol., *85*, 655.
Loew, E.R. (1947) Physiol. Rev., *27*, 542.
MacIntosh, F.C. (1938) Quart. J. Exp. Physiol., *28*, 87.
Moran, N.C. and Perkins, M.E. (1958) J. Pharmacol. Exp. Ther., *124*, 223.
Parsons, M.E. (1969) *Quantitative Studies of Drug-Induced Acid Gastric Secretion.* PhD Thesis, University of London.
Powell, C.E. and Slater, I.H. (1958) J. Pharmac. Exp. Ther., *122*, 480.
Smith, G. and Lawson, D.D. (1958) Scot. Med. J., *3*, 346.
Sperry, R.W. (1963) Proc. Natl. Acad. Sci. (USA) *50*, 703.
Stephenson, R.P. (1956) Br. J. Pharmacol., *11*, 379.

Gertrude B. Elion

GERTRUDE B. ELION

I was born in New York City on a cold January night when the water pipes in our apartment froze and burst. Fortunately, my mother was in the hospital rather than at home at the time. My father emigrated from Lithuania to the United States at the age of 12. He received his higher education in New York City and graduated in 1914 from the New York University School of Dentistry. My mother came at the age of 14 from a part of Russia which, after the war, became Poland; she was only 19 when she was married to my father. My first seven years were spent in a large apartment in Manhattan where my father had his dental office, with our living quarters adjoining it.

My brother was born about six years after I was, and shortly thereafter we moved to the Bronx, which was then considered a suburb of New York City. There were still many open lots where children could play and large parks, including the Bronx Zoo, to which I was very much devoted. My brother and I had a happy childhood. We went to a public school within walking distance of our house. Our classrooms were generally quite crowded, but we received a good basic education.

I was a child with an insatiable thirst for knowledge and remember enjoying all of my courses almost equally. When it came time at the end of my high school career to choose a major in which to specialize I was in a quandary. One of the deciding factors may have been that my grandfather, whom I loved dearly, died of cancer when I was 15. I was highly motivated to do something that might eventually lead to a cure for this terrible disease. When I entered Hunter College in 1933, I decided to major in science and, in particular, chemistry.

By this time my father was not financially well-off since he, like many others, had invested heavily in the stock market, and in the crash of 1929 had gone into bankruptcy. Fortunately, he still had his profession and his loyal patients. Had it not been that Hunter College was a free college, and that my grades were good enough for me to enter it, I suspect I might never have received a higher education. My brother also was able to take advantage of a free higher education, going to the College of the City of New York where he studied physics and engineering.

I remember my school days as being very challenging and full of good comradery among the students. It was an all-girls school and I think many of our teachers were uncertain whether most of us would really go on with our careers. As a matter of fact, many of the girls went on to become teachers and some went into scientific research. Because of the depression, it was not possible for me to go on to graduate school, although I did apply to a number of universities with the hope of getting an assistantship or fellowship.

Jobs were scarce and the few positions that existed in laboratories were not available to women. I did get a three-month job teaching biochemistry to nurses in the New York Hospital School of Nursing. Unfortunately, because of the trimester system, the same job would not have been available again for nine months. By chance, I met a chemist who was looking for a laboratory assistant. Although he was unable to pay me any salary at that time, I decided that the experience would be worthwhile. I stayed there for a year and a half and was finally making the magnificient sum of $20 a week. By then I had saved some money and, with help from my parents, entered graduate school at New York University in the fall of 1939. I was the only female in my graduate chemistry class but no one seemed to mind, and I did not consider it at all strange.

After a year of graduate studies I had finished all the required courses but now needed to do the research work for my Master's degree. During this period, I took a job as a teacher-in-training and then as a substitute teacher in the New York City secondary schools, teaching chemistry, physics and general science for two years. In the meantime, I did my research work at night and on week-ends at New York University, and obtained my Master of Science degree in chemistry in 1941.

By this time, World War II had begun and there was a shortage of chemists in industrial laboratories. Although I was finally able to get a job in a laboratory, it was not in research. I did analytical quality control work for a major food company. After a year and a half, during which I learned a good deal about instrumentation, I became restless because the work was so repetitive and I was no longer learning anything. I applied to employment agencies for a research job, and was chosen to go to a laboratory at Johnson and Johnson in New Jersey. Unfortunately, that laboratory was disbanded after about six months. At that time I was offered a number of positions in research laboratories but the one which intrigued me most was a position as assistant to George Hitchings. My thirst for knowledge stood me in good stead in that laboratory, because Dr. Hitchings permitted me to learn as rapidly as I could and to take on more and more responsibility when I was ready for it. From being solely an organic chemist, I soon became very much involved in microbiology and in the biological activities of the compounds I was synthesizing. I never felt constrained to remain strictly in chemistry, but was able to broaden my horizons into biochemistry, pharmacology, immunology, and eventually virology.

At the same time, I was eager to get my doctorate degree and began to go to school at night at Brooklyn Polytechnic Institute. After several years of long range commuting, I was informed that I would no longer be able to continue my doctorate on a part-time basis, but would need to give up my job and go to school full-time. I made what was then a critical decision in my life, to stay with my job and give up the pursuit of a doctorate. Years later, when I received three honorary doctorate degrees from George Washington University, Brown University and the University of Michigan, I decided that perhaps that decision had been the right one after all. Unfortunately, neither of my parents lived to see this recognition.

The work became fascinating almost from the very beginning. We were exploring new frontiers, since very little was known about nucleic acid biosynthesis or the enzymes involved with it. I had been assigned quite early to work on the purines and, with the exception of a few deviations into the pteridines and into some other condensed pyrimidine systems, the remainder of my work concentrated almost completely on the purines. Each series of studies was like a mystery story in that we were constantly trying to deduce what the microbiological results meant, with little biochemical information to help us. Then, in the mid-1950's came the work of Greenberg, Buchanan, Kornberg and others which elucidated the pathways for the biosynthesis and utilization of purines, and many of our findings began to fall into place. When we began to see the results of our efforts in the form of new drugs which filled real medical needs and benefited patients in very visible ways, our feeling of reward was immeasurable.

Over the years, my work became both my vocation and avocation. Since I enjoyed it so much, I never felt a great need to go outside for relaxation. Nevertheless, I became an avid photographer and traveler. Possibly my love for travel stems from the early years when my family seldom went away on vacation. Thus, my curiosity about the rest of the world did not begin to be satisfied until I began to travel. I have traveled fairly widely over the world, but there still remain many places for me to explore. Another major interest is music, not because I am musically talented, but because I love to listen to it. I am an opera lover and have been a subscriber to the Metropolitan Opera for over 40 years. I also enjoy concerts, ballet and theater.

Although I never married, my brother fortunately did, and I have had the pleasure of watching his three sons and daughter grow up. Several of them now have children of their own. We have been a close-knit family, although often separated by distance, and have shared each other's happiness, sorrows, and aspirations.

In my professional career I was promoted frequently, and in 1967 I was appointed Head of the Department of Experimental Therapy, a position which I held until I retired in 1983. This department was sometimes termed by some of my colleagues a "mini-institute" since it contained sections of chemistry, enzymology, pharmacology, immunology and virology, as well as a tissue culture laboratory. This made it possible to coordinate our work and cooperate in a manner that was extremely useful for development of new drugs.

I have been associated with the National Cancer Institute in many capacities, from 1960 when I served on one of its study sections, to serving later on a number of its advisory committees and the Board of Scientific Counselors for the Division of Cancer Treatment, and most recently as a member of the National Cancer Advisory Board. I have taken an active part in the American Association for Cancer Research, serving on its Board of Directors, its program committees, and in 1983–84 as its President. In addition, I have served on Advisory Committees for the American Cancer Society, the Leukemia Society of America, and a number of committees for the Tropical Disease Research division of the World Health Organization, currently serving as Chairman of the Steering Committee on the Chemotherapy of Malaria. I am a member of the American Chemical Society, the Royal Society of Chemistry, the Transplantation Society, the American Society of Biological Chemists, the American Society of Pharmacology and Experimental Therapeutics, the American Association for Cancer Research, the American Society of Hematology, the American Association for the Advancement of Science, the American Association of Pharmaceutical Scientists, and am a Fellow of the New York Academy of Sciences.

After my official retirement as Department Head from Burroughs Wellcome, I have remained there as a Scientist Emeritus and Consultant, and have tried to take an active part in the discussions, seminars and staff meetings relating to research. In addition, I have become a Research Professor of Medicine and Pharmacology at Duke University and each year work with one third-year medical student who wishes to do research in the areas of tumor biochemistry and pharmacology. This has been a very stimulating experience and one that I hope to continue for some time to come. I serve on a number of editorial boards and continue to lecture and write. In a sense, my career appears to have come full circle from my early days of being a teacher to now sharing my experience in research with the new generations of scientists.

THE PURINE PATH TO CHEMOTHERAPY

Nobel Lecture, December 8, 1988

by

GERTRUDE B. ELION

Wellcome Research Laboratories, Burroughs Wellcome Co.,
Research Triangle Park, U.S.A.

When I joined the Wellcome Research Laboratories in 1944, World War II was in progress. Since the work I am about to discuss covers a period of some 40 years, it may be pertinent to consider the "state of the art" at that time. Our ultraviolet absorption spectra were measured with a Bausch and Lomb spectrograph which had a carbon arc as the light source and photographic plates for recording the amount of light absorbed at each wavelength. There was no paper or ion-exchange chromatography, and purines were isolated and separated as copper and silver salts, or picrates, by fractional crystallization. Tritium and ^{32}P were available, but no ^{14}C or ^{35}S. Geiger counters were used for counting radioactivity; scintillation counters came much later. Some heavy isotopes, e.g., ^{15}N and ^{13}C were obtainable but required the use of a mass spectrometer, which few laboratories had. The state of knowledge of nucleic acids was rather rudimentary. We knew they contained purines and pyrimidines, but the sequences were not known. The prevailing theory was that there were two purines and two pyrimidines in each tetranucleotide and that these tetranucleotides were strung together in some fashion. However, the nature of the internucleotide linkage had not been established, and the helical structure of DNA had not yet been proposed.

In 1940, Woods and Fildes (1,2) had put forth the antimetabolite theory to explain the action of sulfonamides on bacteria, suggesting that the sulfonamides interfered with the utilization of a necessary nutrient, para-aminobenzoic acid. Hitchings theorized that, since all cells required nucleic acids, it might be possible to stop the growth of rapidly dividing cells (e.g., bacteria, tumors, protozoa) with antagonists of the nucleic acid bases. One might hope to take advantage of the faster rate of multiplication of these cells compared with normal mammalian cells and eventually sort out the biochemical differences between various types of cells by the way they responded to these antimetabolites (3,4). It was my assignment to work on purines, pteridines, and some other condensed pyrimidine systems.

It was, of course, necessary to have some biological systems to determine the potential activities of the new compounds. Essentially nothing was known at that time about the anabolic pathways leading to the utilization of purines for nucleic acid synthesis. A number of catabolic enzymes were known: nucleases, nucleotidases, nucleosidases, deaminases (for guanine, adenine, adenosine and adenylic acid), xanthine oxidase and uricase. In 1947, Kalckar described the reversibility of nucleoside phosphorylase (5). The enzymes guanase and xanthine oxidase were useful in our laboratory to examine the purines as substrates or inhibitors of these enzymes (6,7). However, it was the microorganism *Lactobacillus casei* upon which we mainly relied. It could grow on adenine, guanine, hypoxanthine or xanthine, provided the pyrimidine thymine was added. It could also synthesize purines and thymine if given a source of folic acid in the form of liver powder. (The structure of folic acid was not elucidated until 1946 by the Lederle group (8)). Hitchings and Falco had devised a screening test in which it was possible to determine whether a compound could substitute for thymine (9) or a natural purine (4,10) or inhibit its utilization, and could also determine whether a compound was a folic acid antagonist (11,12).

Few chemists were interested in the synthesis of purines in those days and I relied mainly on methods in the old German literature. The transformation reactions were carried out mainly by the methods of Emil Fisher and the syntheses from pyrimidine intermediates by the methods of Traube. The direct replacement of oxygen by sulfur by the method of Carrington (13) also proved to be exceedingly useful for synthesizing the mercaptopurines (14,15).

In 1948, we found that 2,6-diaminopurine inhibited the growth of *L. casei* very strongly and that the inhibition was reversed specifically by adenine but not by the other natural purines (4, 16). However, low concentrations of diaminopurine could also be reversed by folic acid, an attribute which diaminopurine had in common with other diaminopyrimidines and diamino-pyrimidine condensed systems (10). Studies on a diaminopurine-resistant strain of *L. casei* revealed that it grew poorly on adenine as a source of purine. We deduced that adenine and 2,6-diaminopurine must be anabolized by the same enzyme, and that the product of diaminopurine anabolism interfered with purine interconversion (17). That enzyme was reported by Kornberg in 1955 to be adenylate pyrophosphorylase (adenine phosphoribosyltransferase) (18). When tested on mouse tumors and the AKR mouse leukemia (19) or tumor cells in tissue culture (20) diaminopurine proved to be strongly inhibitory. It produced two good clinical remissions in chronic granulocytic leukemia in adults but produced severe nausea and vomiting as well as severe bone marrow depression in two other patients (21). Interestingly, diaminopurine showed activity against vaccinia virus, a DNA virus, *in vitro* (22), but its toxicity in animals led us to abandon that possible utility.

Antileukemic Drugs

By 1951, we had made and tested over 100 purines in the *L. casei* screen (23) and discovered that the substitution of oxygen by sulfur at the 6-position of guanine and hypoxanthine produced inhibitors of purine utilization. 6-Mercaptopurine (6-MP) and 6-thioguanine (TG) were tested at the Sloan-Kettering Institute, with whom we had estabished a collaboration, and were found to be active against a wide spectrum of rodent tumors and leukemias. Of special interest was the finding by Clarke (24) that 6-MP-treated tumors, although they had not regressed completely in the host mouse, were not transplantable into other mice. After some animal toxicology studies by Philips et al. (25), Burchenal proceeded rapidly to clinical trial with 6-mercaptopurine (6-MP) in children with acute leukemia (26). At that time the only drugs available for the treatment of these terminally ill children were methotrexate and steroids, and the median life expectancy was between 3 and 4 months; only 30% lived for as long as one year. The findings that 6-MP could produce complete remissions of acute leukemia in these children, although most of them relapsed at various intervals thereafter, led the Food and Drug Administration to approve the drug for this use in 1953, a little more than two years after its synthesis and microbiological investigation. A symposium on 6-MP was held at the New York Academy of Sciences in 1954 (27). The addition of 6-MP to the antileukemia armamentarium increased the median survival time to 12 months in these children, and a few remained in remission for years with 6-MP and steroids. This convinced us, as well as many other investigators in the cancer field, that antimetabolites of nucleic acid bases were fruitful leads to follow. Today 6-MP remains one of the dozen or more drugs found useful in the treatment of acute leukemia. With the use of combination chemotherapy with three or four drugs to produce and consolidate remission, plus several years of maintenance therapy with 6-MP and methotrexate, almost 80% of children with acute leukemia can now be cured.

Although we felt we were on the right track in 1952, there were still many unanswered questions. How did 6-MP work? What was the reason for its differential effect on neoplastic cells? How could one improve this differential effect? Reversal studies with 6-MP in *L. casei* did not pinpoint antagonism for any single purine. The inhibition was reversed by hypoxanthine, adenine, guanine and xanthine (28). However, studies with a 6-mercaptopurine-resistant strain of *L. casei* revealed that 6-MP was unable to utilize hypoxanthine for growth (29). Again, as with the earlier studies with 2,6-diaminopurine, we concluded that 6-MP and hypoxanthine were anabolized by the same enzyme and that interference with purine inter-conversions at the nucleotide level were involved (30). In 1955, (two years after the introduction of 6-MP into clinical use) the enzyme which converts hypoxanthine and 6-MP to their respective nucleotides was identified as hypoxanthine phosphoribosyltransferase (HGPRT) (18). Also in the mid-fifties the pioneering work of Greenberg (31) and of Buchanan (32,33) revealed the pathways of the biosynthesis of purines and the importance of hypoxanth-

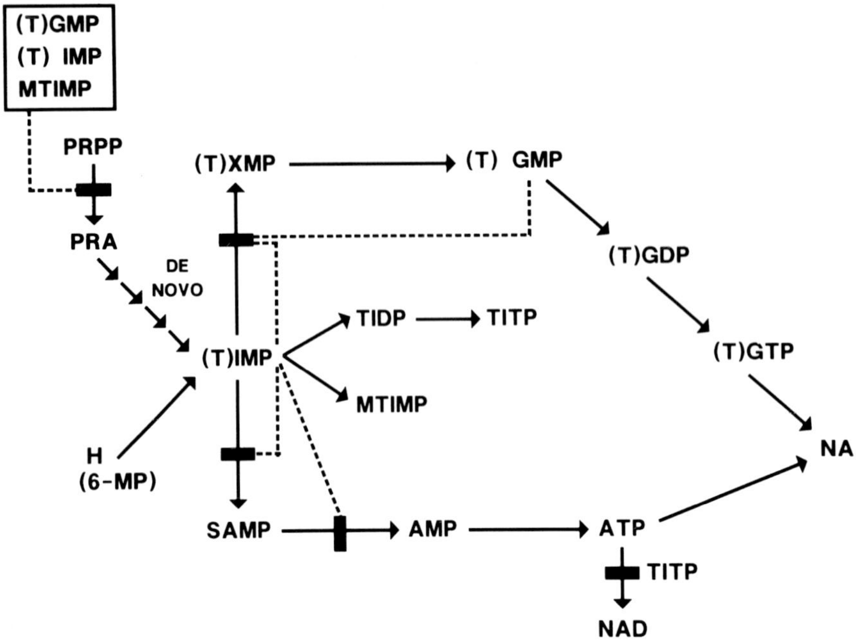

Figure 1. Pathways in the anabolism of 6-MP and loci of action of the nucleotides derived from 6-MP. In addition, TGDP is converted to dTGDP and dTGTP for incorporation into DNA.

ine ribonucleotide (inosinic acid, IMP) as the first purine nucleotide formed in this biosynthetic pathway. It took a number of investigators and a period of years to unravel all of the pathways in which the nucleotide of 6-MP, thioionsinic acid (TIMP), and nucleotides derived from TIMP participated as substrates and inhibitors (34). These are summarized in Fig. 1. Based on our current knowledge of enzyme inhibition constants and the concentrations of the various nucleotides achieved with the therapeutic regimens, the principal sites of action appear to be feedback inhibition of *de novo* purine synthesis (particularly by methylthioinosinic acid), inhibition of inosinate dehydrogenase, and incorporation into DNA in the form of thioguanine (35). Selectivity for neoplastic cells probably depends on the levels of the individual anabolic and catabolic enzymes. Catabolic enzyme levels are generally much lower in tumor cells than in normal cells. In addition, mitotic rate, drug transport and metabolite pool sizes can be responsible for selective toxicity.

While these biochemical studies were going on, we studied the metabolic fate of 6-MP, first in mice (36) and then in humans (37). Pharmacokinetic and metabolic studies were then in their infancy, perhaps because the methodology for separation of metabolites and the counting of radioactive samples as thin films in a flow Geiger counter were so tedious and time-consuming. Nevertheless, using Dowex-1 and Dowex-50 ion-exchange columns and paper chromatography, we investigated the fate of 6-MP *in vivo* and attempted to discover whether it was possible to modify this metabolism and thus improve the efficacy of 6-MP. We continued to synthesize deriva-

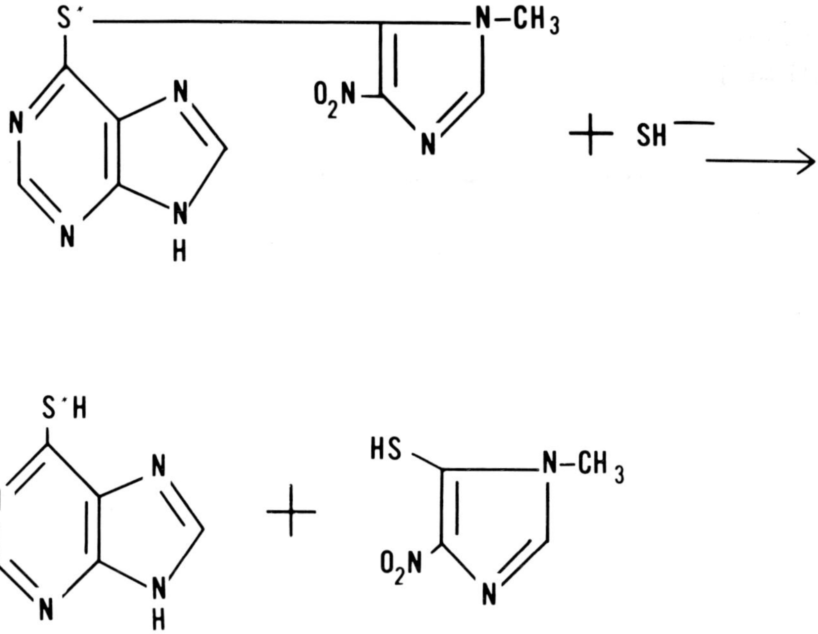

Figure 2. Reaction of sulfhydryl ion upon azathioprine to release 6-MP. Reproduced from Ref. 46 with permission.

tives of 6-MP and thioguanine and investigate structure-activity relationships (38 – 40). Thioguanine, which we had synthesized earlier than 6-MP, was more active but also more toxic (25). It was also more difficult to synthesize and, since its mechanism of action appeared to be similar to that of 6-MP, its metabolic fate and clinical activity were explored somewhat later (41 – 43). Thioguanine later found its main utility in the treatment of acute myelocytic leukemia in adults, in combination with cytosine arabinoside.

Studies of the urinary metabolites of 6-MP revealed that extensive metabolic transformations occurred *in vivo* (36, 37, 44–46). The single product present in highest amount was 6-thiouric acid, formed by the action of xanthine oxidase on 6-MP. In addition, there were various substances in which the sulfur had been methylated, and the methylthio derivative had been oxidized on the sulfur or on the purine ring. A considerable amount of the sulfur had been removed and converted by oxidation to inorganic sulfate, which also appeared as ethereal sulfates. Very little 6-MP was excreted unchanged.

In an attempt to modify the metabolism of 6-MP, we introduced substitutents at the 2 and/or 8 positions of the purine ring and on the ring nitrogens. This led to the loss of antitumor activity, with the exception of the 2-amino-6-mercapto derivative, thioguanine, which we had previously found to have strong antimetabolic activity. Attempts were then made to protect the sulfur from oxidation and hydrolysis by blocking groups which might be removed intracellularly to release 6-MP, hopefully by some tumor-specific enzyme. The most successful compound to emerge from this approach was the 1-methyl-4-nitro-5-imidazolyl derivative, the compound now known as azathioprine (Imuran®). This compound acts as a pro-drug for 6-MP which, due to the proximity of the ortho-nitro group, is subject to attack by sulfhydryl groups and other nucleophiles (45–49) (Fig. 2). In particular, the glutathione present in red cells reacts with azathioprine, releasing 6-MP back into the plasma (47). This compound had a better therapeutic index in mice bearing adenocarcinoma 755, being as active as 6-MP but less toxic (50). In patients with leukemia, however, the chemotherapeutic index of 6-MP and azathioprine were similar (51).

Immunosuppression and Transplantation

In 1958, a new horizon appeared. Robert Schwartz, working with William Dameshek in Boston, investigated the effect of 6-MP on the immune response, based on the rationale that the immunoblastic lymphocyte formed during an immune response closely resembled leukemic lymphocytes. Schwartz showed that when 6-MP was administered to rabbits for several days, beginning with the time of injection of a foreign antigen, e.g., bovine serum albumin, they were unable to mount an antibody response to that antigen (52). He worked out the importance of drug dose and timing and showed that 6-MP was most effective when treatment was started at the time of antigen administration (52,53). He also demonstrated that animals could be made tolerant to a particular antigen while still retaining immunological reactivity to other antigens (54). At Schwartz's instigation, we set up an immunological screening test that consisted of measuring the antibody response of mice to sheep red cells (55,56). It enabled us to identify new active agents, synergistic combinations of drugs, and to show that immunosuppression was greater at higher doses of antigen and of drug.

Roy Calne, a young British surgeon, stimulated by Schwartz's papers, decided to examine the effect of 6-MP on kidney transplant rejection in dogs. He obtained a 44-day survival of a kidney from an unrelated donor in a dog given daily doses of 6-MP (57). This was considerably longer than the expected 9 to 10 day graft survival in control animals. When Calne asked us for compounds related to 6-MP which he might investigate, we suggested that azathioprine might have some advantages. The studies which followed showed azathioprine to be superior to 6-MP for preventing rejection of canine kidney homografts (58,59). Successful transplantation of kidneys to unrelated recipients became a reality in man in 1962, with regimens of immunosuppression consisting of azathioprine and prednisone (60). By 1977, the Kidney Registry had records of 25,000 kidney transplants done between 1965 and 1972 (61), and the numbers have continued to increase yearly (62, 63). Today the procedure is considered therapeutic rather than experimental, and the importance of histocompatibility matching is recognized (63). Other organ transplants, e.g., liver, heart and lung, have likewise become possible. Other immunosuppressive drugs, e.g., Cyclosporin, have come into use in recent years, but azathioprine remains a mainstay in kidney transplantation.

The immunosuppressive effects of azathioprine have been studied in a wide variety of immunological systems. The earlier work is reviewed in (64) and more recent studies in (65). Immunosuppressive drugs have also shown utility in the treatment of autoimmune disease. Remissions with 6-MP, thioguanine and azathioprine have been reported in autoimmune hemolytic anemia, systemic lupus, and chronic active hepatitis (65). Azathioprine is now an approved drug for the treatment of severe rheumatoid arthritis (66).

Gout and Hyperuricemia

It was time now to try a new approach to the potentiation of 6-MP activity. Since we knew from metabolic studies that 6-thiouric acid was one of the principal products of 6-MP catabolism, it seemed possible that we could interfere with this oxidation by inhibiting the enzyme responsible for it, xanthine oxidase. In the early days of seeking antimetabolites for the natural purines in our laboratory, xanthine oxidase had been one of test enzymes. Doris Lorz had identified many substrates as well as inhibitors of this enzyme (7). These compounds had also been tested on *L. casei* and on animal tumors. To test for xanthine oxidase inhibition *in vivo,* we chose a compound that had no inhibitory effects on bacteria or tumors, was non-toxic, but which was a potent inhibitor of xanthine oxidase. This compound was the hypoxanthine analog, 4-hydroxypyrazolo (3,4-d) pyrimidine (allopurinol). When allopurinol was given to mice together with 6-MP, it did indeed inhibit the oxidation of 6-MP and potentiated the antitumor and immunosuppressive properties of 6-MP three to fourfold (45,67,68). Moreover, the toxicity of 6-MP to mice appeared to be potentiated only two fold, so that the chemotherapeutic index of 6-MP had been increased. Was it possible that the same phenomenon would occur in man? With the collaboration of Wayne Rundles, we explored this possibility in patients with chronic granulocytic leukemia in whom the efficacy and metabolism of 6-MP could be investigated. As had happened in mice, the oxidation of 6-MP to thiouric acid was inhibited in a dose-related manner, and the antileukemic activity of 6-MP was potentiated proportionally (45, 69–71). Figure 3 illustrates the four fold reduction in thiouric acid and increase in 6-MP when 300 mg allopurinol was given together with 110 mg 6-MP. Later investigations (72) showed that the increased activity of 6-MP was accompanied by a proportional increase in toxicity. Thus, although less 6-MP was required to produce an antileukemic effect, the therapeutic index of 6-MP for leukemia remained unchanged.

Xanthine oxidase is responsible not only for the oxidation of 6-MP, but also for the formation of uric acid from hypoxanthine and xanthine (Fig. 4). Consequently, treatment with allopurinol produces a marked decrease in both serum and urinary uric acid (69,71,73,74). This presented the possibility of a unique approach to the treatment of gout and other forms of hyperuricemia.

It was recognized at the outset that the inhibition of an enzyme like xanthine oxidase *in vivo* might present some difficulties. First, there was the

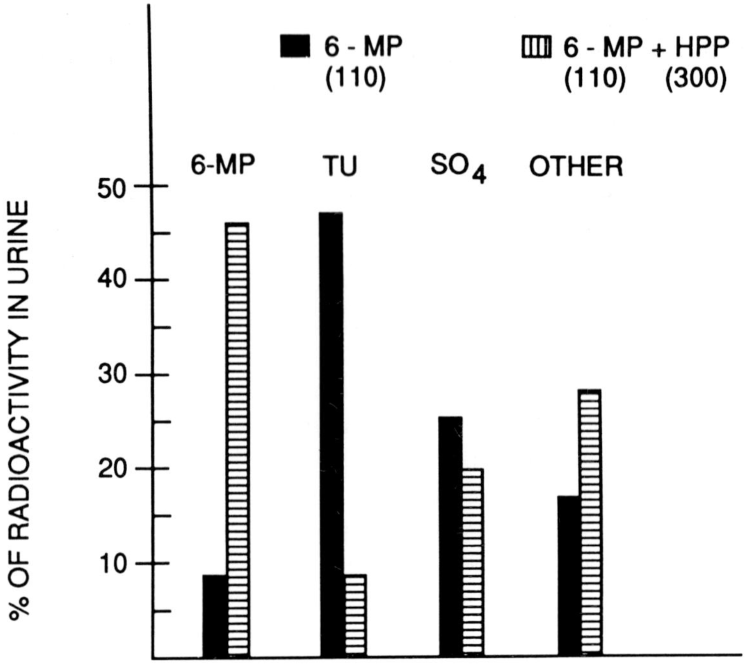

Figure 3. Radioactive metabolites in the urine following administration of ^{35}S-6-MP (110 mg) to a patient with and without 300 mg of allopurinol (HPP). TU = thiouric acid. Reproduced from Ref. 71 with permission.

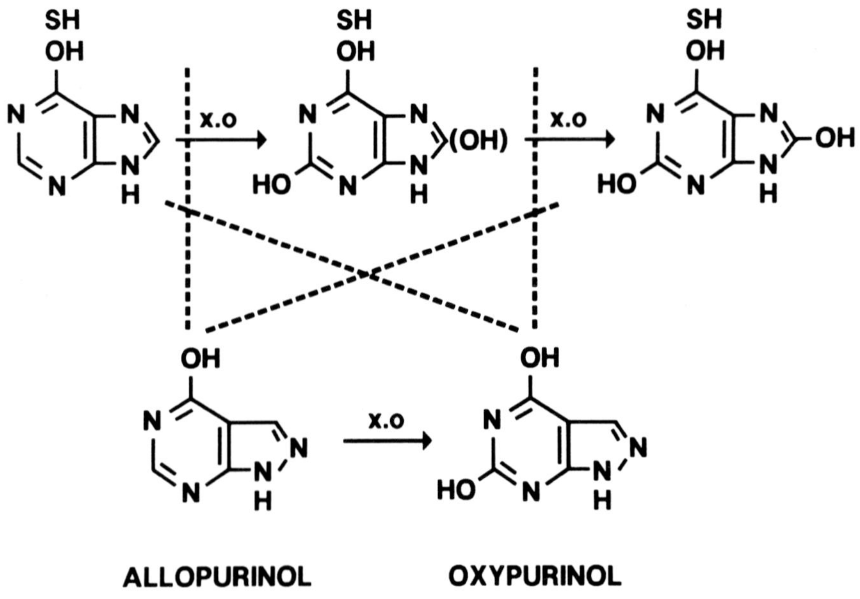

Figure 4. Pathways in the oxidation of hypoxanthine and 6-MP by xanthine oxidase (x.o) inhibited by allopurinol and oxypurinol. For 6-MP the intermediate is 8-hydroxy-6-MP; for hypoxanthine the intermediate is 2, 6-dihydroxypurine (xanthine).

question as to whether the inhibitor had a sufficiently long half-life in the body to produce a persistent reduction in uric acid production. There was also the possibility that the inhibition would result in an unacceptable accumulation of the intermediates, hypoxanthine and xanthine, the latter compound being about as insoluble as uric acid. Moreoever, it was possible that more enzyme would be induced by the presence of excess amounts of the intermediates, leading to the need for ever-increasing amounts of inhibitor. Finally, one had to consider carefully the long-term effects of this inhibitor, since gout patients would probably continue to take the drug for life. All of these possibilities were thoroughly examined, first in animals and then in man.

Allopurinol is not only a potent competitive inhibitor of xanthine oxidase, but it is also a substrate (75), the oxidation resulting in the corresponding xanthine analog, oxypurinol (called alloxantine, oxoallopurinol, or oxipurinol in early papers) which is also a xanthine oxidase inhibitor (Fig. 4). Oxypurinol also has the unusual property of binding very tightly to the reduced form of the enzyme, thereby inactivating it (75,76). The enzyme activity can be restored by oxidation, which takes place slowly in the presence of air (76,77). Although allopurinol itself has a short half-life in plasma (about 90 to 120 minutes) oxypurinol has a very long half-life, 18 to 30 hours (75,78,79). This is due to the fact that oxypurinol is reabsorbed in the proximal tubule of the kidney (80). Consequently, steady-state levels of oxypurinol are achieved in a few days, and uric acid concentrations can be maintained at the desired level by proper dose-adjustment (80). Because allopurinol is completely absorbed orally, whereas oxypurinol is not, allopurinol remains the ideal pro-drug for oxypurinol.

The fate of the intermediate oxypurines, hypoxanthine and xanthine, turned out to be a fascinating one. These oxypurines do not accumulate in the serum. In fact, their serum levels rise very little during allopurinol treatment (79,81). There are two reasons for this. One is that both hypoxanthine and xanthine can be reutilized for nucleic acid synthesis via the enzyme hypoxanthine-guanine phosphoribosyltransferase (HGPRT) (82, 83). The nucleotides formed, IMP and XMP, are the normal intermediates for adenine and guanine nucleotides (AMP, GMP). Through a process of feedback inhibition IMP, AMP and GMP can reduce the *de novo* synthesis of purines by inhibiting PRPP-amidotransferase (84). Thus, the salvage of hypoxanthine and xanthine serves to regulate purine biosynthesis, reducing it when it is excessive. When the oxypurines are not reutilized, they are excreted by the kidney by glomerular filtration, since they are not reabsorbed by the kidney tubule to any significant degree (85,86).

Long-term studies with allopurinol in animals and in man have shown that new enzyme is not induced and that allopurinol is a safe and effective drug for long-term treatment (87). A few percent of patients develop a rash when taking allopurinol. Patients with poor kidney function require lower doses of drug because of the pharmacological properties of oxypurinol and its long half-life (88).

Allopurinol helps to alleviate several of the clinical problems associated with gout. The hyperuricemia of gout produces deposits of small crystals of uric acid in joints which results in extreme pain, or large deposits, tophi, which result in gouty arthiritis and restricted joint movement. In patients who excrete excessive amounts of uric acid in the urine, urate stones often form in the kidney. With allopurinol it has been possible to prevent and reduce the tophaceous deposits and to prevent urate stone formation (74,81,87,89,90). Likewise, the secondary hyperuricemia which is associated with the therapy of malignancy can be reduced with allopurinol treatment (91).

Antiprotozoal Effects of Allopurinol

I would like to digress at this point, to describe another use for allopurinol which materialized about 10 years later, because it typifies the kind of chemotherapeutic selectivity which can be achieved with purine analogs as the result of differences in the specificities of parasitic and mammalian enzymes. The finding by Joseph Marr (92) that allopurinol inhibits the replication of *Leishmania donovani* led us into an extended collaboration with Marr to discover the biochemical basis for this unexpected activity (93–101). Leishmaniae and trypanosomes, like many other unicellular parasites, lack the ability to synthesize purines *de novo*. They do, however, have ample quantities of salvage enzymes which enable them to utilize preformed purines and nucleosides present in the mammalian blood stream. Leishmaniae and trypanosomes have large amounts of the enzyme HPRT capable of utilizing allopurinol as a substrate to a much greater extent than does the human enzyme (97). Moreover, the enzyme adenylosuccinate synthetase which has as its function the conversion of IMP to SAMP, the intermediate for AMP, has a broader substrate specificity than does the mammalian enzyme (96,101). The protozoal enzymes consequently convert allopurinol ribonucleotide to the adenylate analog, 4-aminopyrazolopyrimidine ribonucleotide, which is then converted to a di- and tri-phosphate and incorporated in RNA (93,94,98,100). In the mammalian host this amination does not occur, and allopurinol is not incorporated into nucleic acids (102). The differences in the metabolism of allopurinol by the mammalian host and the trypanosomes is illustrated in Fig. 5. Allopurinol has shown activity in the treatment of leshmaniasis (103) and of Chagas' disease (caused by *Tryeanosoma cruzi*) (104). Allopurinol riboside, a minor urinary metabolite of allopurinol in man, can likewise be converted by leishmaniae in this way (94), whereas allopurinol riboside is not further metabolized in man.

Antiviral Drugs

In 1968, we decided to return to a path which had intrigued us as early as 1948, the path to antivirals. The antiviral activity of 2,6-diaminopurine had been provocative (22), although its toxicity had been discouraging. Meanwhile, the pursuit of 6-MP, thioguanine, azathioprine and allopurinol had

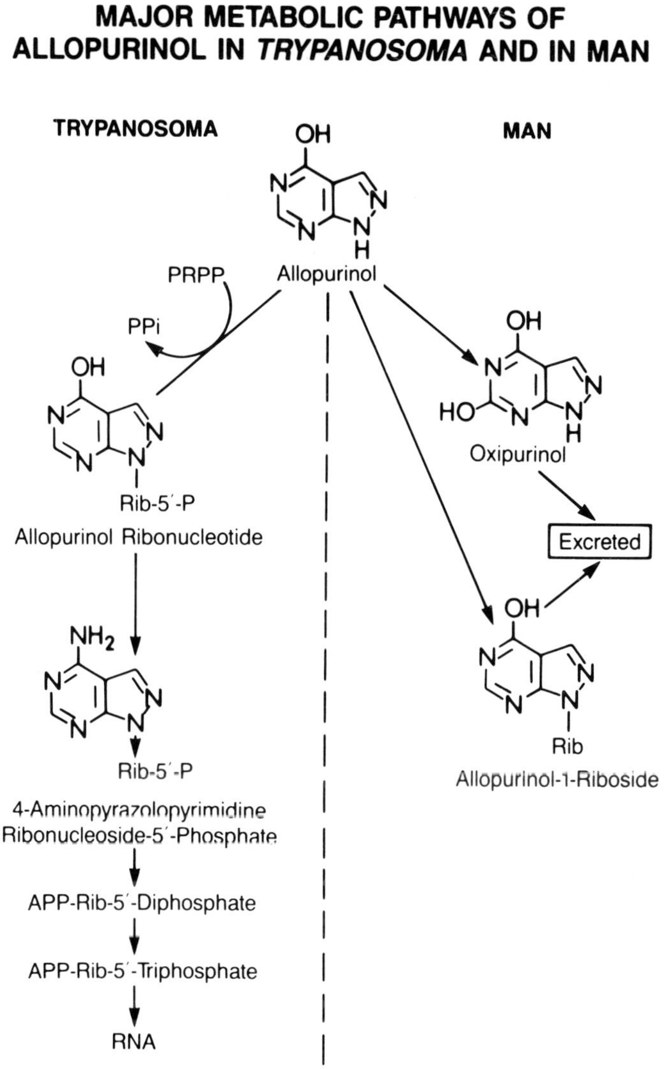

Figure 5. Metabolic pathways of allopurinol in trypanosoma and in man.

occupied 20 years. When it was discovered that adenine arabinoside (ara-A) inhibited the growth of both DNA and RNA viruses (105), the information started a train of thought. Would the arabinoside of 2,6-diaminopurine be equally active? After all, diaminopurine had mimicked adenine in many reactions and could be converted to a nucleoside and nucleotide by adenine-metabolizing enzymes. Moreover, diaminopurine riboside was a poorer substrate for adenosine deaminase than was adenosine. It therefore seemed possible that the arabinoside might persist longer than ara-A, a compound which is rapidly deaminated *in vivo*.

Diaminopurine arabinoside was synthesized by Janet Rideout, and, since we did not have a virus laboratory on site at the time, we sent the compound to our colleague, John Bauer, in the Wellcome Research Laboratories in the U.K. for antiviral screening. In a few weeks he informed us that this compound was highly active against both herpes simplex virus and vaccinia virus. Moreover, the compound was less cytotoxic to mammalian cells than ara-A. Thus began our antiviral odyssey. For several years my group pursued studies on the purine arabinosides, exploring structure-activity relationships, seeking better synthetic methods, and doing metabolic studies in mice (106). Bauer and Collins studied the activity of these compounds in rabbits and mice (106). We found that diaminopurine arabinoside (ara-DAP) was deaminated to guanine arabinoside in mice and that the guanine derivative (ara-G) was as active an antiviral as ara-DAP. In this respect, ara-G had an advantage over the deamination product of ara-A, hypoxanthine arabinoside, which had very poor antiviral activity. We were not certain whether this advantage was sufficient to warrant the full-scale development of ara-DAP.

In 1970 our laboratories moved to North Carolina, and Howard Schaeffer joined us as head of the Organic Chemistry Department. He had been studying analogs of adenosine as substrates and inhibitors of the enzyme adenosine deaminase, and had examined a variety of acyclic side chains on the 9-position of adenine to determine what changes the enzyme would tolerate in a substrate (107). He found that 9-(2-hydroxyethoxymethyl)-adenine could still serve as a substrate for adenosine deaminase. This suggested that other enzymes might also recognize such a side chain as a pentose and that nucleoside analogs of this kind might have antimetabolite properties.

When the acyclic adenosine analog was tested in the antiviral screen, it showed antiherpetic activity *in vitro* at about twice the concentration of ara-A. The antiviral program now concentrated on the acyclic nucleoside analogs, with the syntheses conducted by Schaeffer and Beauchamp, the antiviral testing by Bauer and Collins, and the mechanisms of action, enzymology and *in vivo* metabolism by my group (108,109). As was the case with the purine arabinosides, the 2,6-diaminopurine analog proved highly active on herpes simplex virus *in vivo* as well as *in vitro*. However, unexpectedly, the guanine analog, acyclovir (acycloguanosine), was over 100 times as active as the diamino compound (109). Acyclovir is a metabolic product of the diaminopurine derivative, formed by the action of adenosine deaminase and is undoubtedly responsible for the antiviral activity of the diamino compound observed *in vivo* (110 – 112).

Acyclovir

One of the most intriguing aspects of the antiviral activity of acyclovir (ACV) is not only its high potency but its unusual degree of selectivity (Fig. 6) (108, 109, 113–115). It is highly active against herpes simplex virus, types 1 and 2, and varicella zoster virus. It has activity against several other herpes type viruses, e.g., Epstein-Barr virus, pseudorabies, but only slight activity against the human cytomegalovirus (HCMV). It is not cytotoxic to the mammalian cells, in which these viruses are grown, at concentrations hundreds of times greater than the concentrations required for

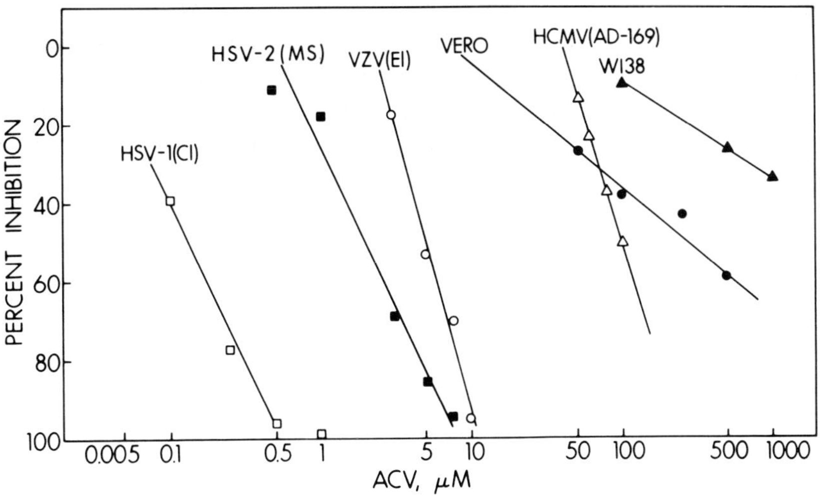

Figure 6. Dose response curves of various viruses and cells to ACV. Reproduced from Ref. 113 with permission.

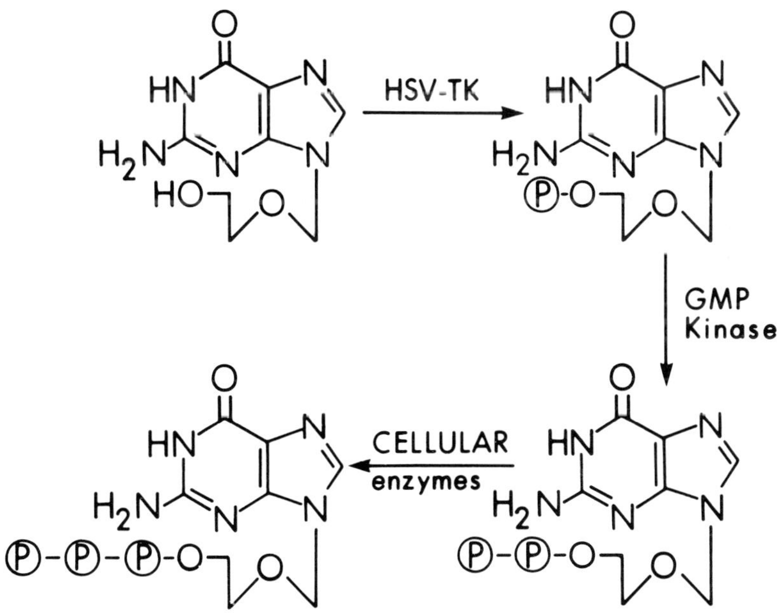

Figure 7. Pathways for the formation of ACV mono-, di- and triphosphates. Reproduced from Ref. 113 with permission.

antiviral activity. Moreover, the compound is inactive against other DNA viruses, e.g., vaccinia, as well as RNA viruses (109). We decided that it was highly important to determine the reason for this unusual selectivity, since this would undoubtedly offer exploitable information about the herpes viruses. In order to do these biochemical studies effectively, we set up a virus laboratory in-house with Phillip Furman as its head. Radioactive acyclovir labeled in the 8-position of the guanine with ^{14}C or with ^{3}H in the side chain was synthesized. Vero cells, uninfected or infected with herpes simplex virus type (HSV-1), were incubated with both types of radioactive acyclovir for 7 hours. Extracts of these cells were then examined by high pressure liquid chromatography. The extracts of the uninfected cells showed only the presence of unchanged acyclovir. In the HSV-infected cells three new radioactive compounds were formed (108). These products were identified by enzymatic methods as the mono-, di and triphosphates (ACV-MP, ACV-DP, ACV-TP) of acyclovir, and this was later confirmed by comparison with authentic synthetic samples. The enzyme responsible for the conversion of acyclovir to its monophosphate was laboriously isolated, purified and identified by James Fyfe as a herpes virus-specified thymidine kinase (108, 116). While this enzyme had been reported to be formed in herpes virus-infected cells, it was unexpected that an acyclic nucleoside of guanine could serve as its substrate. Nevertheless, this proved to be the case. A similar enzyme is specified by the varicella zoster virus. Once the first phosphate has been added, the second phosphate is added by cellular guanylate kinase (117), while several other cellular kinases can add the third phosphate (118). Since the cellular thymidine kinase cannot use acyclovir as a substrate, very little ACV-TP is formed in uninfected cells (108,119). The small amount of phosphorylation which occurs in normal cells is due to a 5'-nucleotidase (120). The pathways for the formation of ACV-TP are shown in Fig. 7.

When it was apparent that the active antiviral compound was ACV-TP, the interaction of this compound with viral and cellular DNA polymerases was investigated. ACV-TP proved to be a more potent inhibitor of the herpes virus DNA polymerase than of cellular DNA polymerase-α (121,122). The quantitative aspects of these differences in the amounts of ACV-TP formation and DNA polymerase inhibition (K_i value) by ACV-TP in virus-infected cells and uninfected cells are illustrated in Fig. 8 (123). Moreover, ACV-TP serves as a substrate for the herpes virus DNA polymerase, but chain termination occurs when it is incorporated, because of the absence of the 3'-hydroxyl group needed for chain elongation (113). Thus, only very small fragments of viral DNA are formed (124). In addition, ACV-TP serves not only to inhibit, but also to inactivate, the viral DNA polymerase following the formation of the enzyme-template-acyclovir monophosphate complex (125). This inactivation does not occur with cellular DNA polymerase.

The high selectivity of acyclovir for those herpes viruses which induce a herpes-specified thymidine kinase can thus be explained. This enzyme has

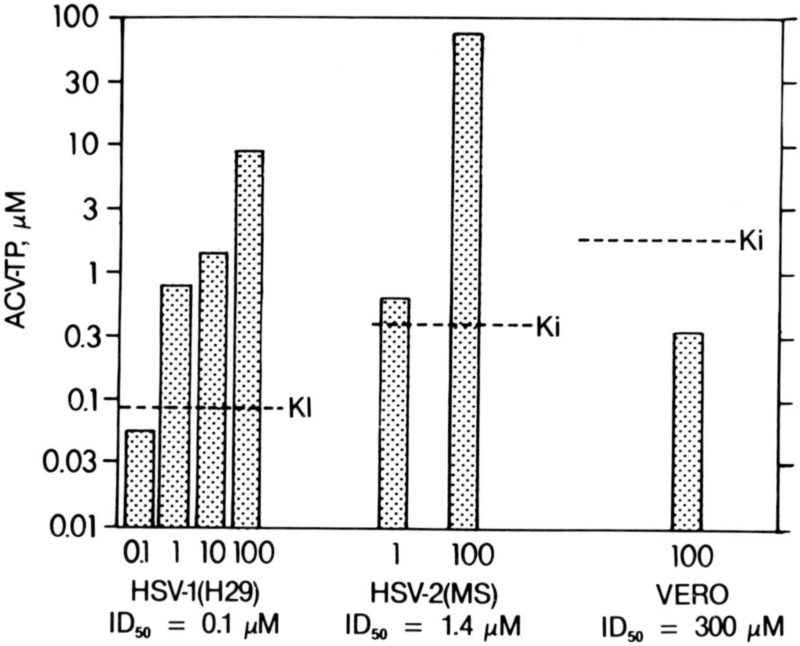

Figure 8. Amounts of ACV-triphosphate formed in uninfected, HSV-1 and HSV-2 infected Vero cells in the presence of varying concentrations (μM) of ACV. Dotted lines represent the K_i values for the individual DNA polymerases. ID_i values are concentrations for 50% inhibition of plague formation. Reproduced from Ref. 123 with permission.

proved to be a very useful tool for determining the structural requirements for other potential substrates of this enzyme (126). Its absence or alteration also explains the resistance of some herpes virus isolates to acyclovir (127,128,129). On the other hand, transfection of cells with a portion of the herpes virus genome containing the HSV thymidine kinase gene transforms normally resistant cells to ACV-sensitive cells (130). Resistance to ACV may also result from mutations in the viral DNA polymerase (127,128,129).

Epstein-Barr virus and human cytomegalovirus infection do not induce a specific kinase capable of phosphorylating acyclovir. However, the DNA polymerase of Epstein-Barr virus is exquisitely sensitive to the small amount of ACV-TP formed in EBV-infected cells (131). Interestingly, although HCMV-infected cells do not phosphorylate acyclovir to any extent, they do phosphorylate the closely related acyclovir derivative, ganciclovir, (formerly called BW B759U, DHPG, and 2′NDG) which has an extra hydroxymethyl group on the side chain. Consequently, ganciclovir has proven to be a much better inhibitor of HCMV replication than acyclovir (132–134).

Investigation of the pharmacokinetics and metabolism of acyclovir in several animal species and then in man revealed that it was a remarkably stable compound. Only two metabolites were found, the 2-carboxymethoxymethylguanine (CMMG) and 8-hydroxyacyclovir (135). In man, the carboxy derivative CMMG accounted for 9–14% of an intravenous dose of acyclovir, while essentially none of the 8-hydroxy metabolite appeared in the urine

(136,137). In other species there were somewhat increased amounts of both metabolites (135,137,138). Acyclovir has a plasma half-life of approximately three hours (139) so that an intravenous infusion is generally given every eight hours. Because acyclovir is excreted at about twice glomerular filtration rate and has a limited solubility in water (2.5 mg/ml at 37°C) it is administered as a one-hour intravenous infusion rather than as a bolus injection. The compound is also active topically and orally. Although the oral bioavailability is limited, the blood levels attained by oral administration of acyclovir are adequate for therapeutic efficacy against herpes simplex and varicella zoster viruses (140).

The compound is distributed in all body tissues and crosses the blood-brain barrier (109,135,141). Toxicology studies in mice, rats and dogs showed the drug to be non-toxic at doses well above those required for therapeutic efficacy (142).

Acyclovir has now been in clinical use for about eight years and it seems appropriate to mention some of the areas in which it has decreased suffering and saved lives. First episodes of genital herpes infections are characterized by pain, prolonged viral excretion, and slow healing of lesions. Acyclovir given by either the intravenous or oral routes produces a significant alleviation of symptoms and decreases the time of viral shedding and time to healing (Table 1) (143). Recurrent episodes of genital herpes are generally much shorter and less severe than the initial one. Therefore, the benefits of acyclovir treatment appear to be less dramatic. However, in a study in which patients initiated oral therapy, there was a significant difference in new lesion formation, 23% on placebo, 6.5% on ACV (147). For patients with frequently recurring genital herpes, e.g., once a month, of several years duration, it has been possible to decrease the frequency of recurrence to a mean of 1.8 episodes during the first year and 1.4 episodes during the second year by oral prophylaxis with acyclovir (148). The percentage of patients having no recurrences in a year of oral prophylactic therapy was 45–50% (149).

Herpes zoster, commonly known as shingles, is caused by a reactivation of latent varicella zoster virus. It produces severe pain during the acute two to three week episode, as well as a post-herpetic neuralgia months later in about 10 percent of patients. Acyclovir causes a significant decrease in the duration of acute pain but has little effect on the post-herpetic neuralgia (150). In immunocompromised individuals, herpes zoster can produce seri-

Table 1. Effect of ACV on first episodes on genital herpes.

Route	Median duration in days (ACV/placebo)			Ref.
	Virus Shedding	Pain	Healing	
i.v.	2/13	3/7	9/21	(144)
p.o.	1/13	4/8	6/11	(145)
p.o.[a]	3.9/13.4	2.8/3.4	9.5/13.7	(146)

[a] Women only. Mean duration

ous sequelae in the form of progressive skin dissemination and visceral disease. The latter can be fatal in a small percentage of patients. Intravenous treatment with acyclovir has effectively prevented this dissemination (151).

Herpes simplex infections in immunosuppressed individuals is a serious medical problem. Virus continues to be shed for a long time and healing is slow. In two studies which involved 97 immunocompromised patients (152) with a variety of diseases and 34 bone marrow transplant recipients (153) the effects of intravenous acyclovir treatment on viral shedding, pain and time to healing were highly significant (Table 2). In bone marrow transplant patients and in leukemic patients, prophylaxis with acyclovir has made it possible to prevent the reactivation of latent virus during therapy so that the patients can remain free of herpetic episodes during the period of maximum immunosuppression (154,155). The treatment of herpes encephalitis, a frequently fatal disease, with intravenous acyclovir has been successful in a large percentage of patients if begun before the patient is comatose (156,157).

In addition to the clinical utility of acyclovir, the lessons learned from its development have proven to be extremely valuable for future antiviral research. In depth studies of mechanisms of action have led to a better understanding of the enzymatic differences between normal and virus-infected cells. It has given impetus to the search for other viral-specific enzymes which are capable of therapeutic application.

In my attempt to cover 40 years of research on purines and purine analogs, I have been able to give you only a bird's eye view. However, I hope that I have successfully conveyed our philosophy that chemotherapeutic agents are not only ends in themselves but also serve as tools for unlocking doors and probing Nature's mysteries. This approach has served us well and has led into many new areas of medical research. Selectivity remains our aim and understanding its basis our guide to the future.

Table 2. Effect of intravenous ACV in HSV infections in immunocompromised patients.

Patients	Median duration in days (ACV/placebo).		
	Virus Shedding	Pain	Healing
all (97)	2.8/16.8	8.9/13.1	13.7/20.1
Bone marrow transplants (34)	3/17	10/16	14/28

REFERENCES

1. D.D. Woods, *Brit. J. Exp. Pathol.* **21,** 74 (1940).
2. P. Fildes, *Lancet* **1,** 955 (1940).
3. G.H. Hitchings, G.B. Elion, E.A. Falco, P.B. Russell, M.B. Sherwood, H. VanderWerff, *J. Biol. Chem.* **183,** 1 (1950).
4. G.H. Hitchings, G.B. Elion, E.A. Falco, P.B. Russell, H. VanderWerff, *Ann. N.Y. Acad. Sci.* **52,** 1318 (1950).
5. H.M. Kalckar, *J. Biol. Chem.* **167,** 461 (1947).
6. G.H. Hitchings and E.A. Falco, *Proc. Natl. Acad. Sci. U.S.A.* **30,** 294 (1944).
7. D.C. Lorz and G.H. Hitchings, *Fed. Proc.* 9, 197 (1950).
8. R.B. Angier, J.H. Boothe, B.L. Hutchings, et al., *Science* **103,** 667 (1946).
9. G.H. Hitchings, E.A. Falco, M.B. Sherwood, *Science* **102,** 251 (1945).
10. G.B. Elion and G.H. Hitchings, *J. Biol. Chem.* **185,** 651 (1950).
11. G.H. Hitchings, G.B. Elion, H. VanderWerff, E.A.Falco, *J. Biol. Chem.* **174,** 765 (1948).
12. G.H. Hitchings, G.B. Elion, H. VanderWerff, *J. Biol. Chem.* **174,** 1037 (1948).
13. H.C. Carrington, *J. Chem. Soc.* **124** (1944).
14. G.B. Elion, E. Burgi, G.H. Hitchings, *J. Am. Chem. Soc.* **74,** 411 (1952).
15. G.B. Elion and G.H. Hitchings, *J. Am. Chem. Soc.* **77,** 1676 (1955).
16. G.B. Elion and G.H. Hitchings, *J. Biol. Chem.* **187,** 511 (1950).
17. G.B. Elion, H. VanderWerff, G.H. Hitchings, M.E. Balis, D.H. Levin, J.B. Brown, *J. Biol Chem.* **200,** 7 (1953).
18. A. Kornberg, I. Lieberman and E.S. Simms, *J. Biol. Chem.* **215,** 417, (1955).
19. J.H. Burchenal, A. Bendich, G.B. Brown, G.B. Elion, G.H.Hitchings, C.P. Rhoads, et al., *Cancer* **2,** 119 (1949).
20. J.J. Biesele R.E. Berger, A.Y. Wilson, G.H. Hitchings, G.B. Elion, *Cancer* **4,** 186 (1951).
21. J.H. Burchenal, D.A. Karnofsky, E.M. Kingsley-Pillers et al., *Cancer* **4,** 549 (1951).

22a. R.L. Thompson, M.L. Wilkin, G.H. Hitchings, G.B. Elion, E.A. Falco, P.B. Russell, *Science* **110,** 454 (1949).

22b. R.L. Thompson, M.L. Price, S.A. Menton, Jr., G.B. Elion, G.H. Hitchings, *J. Immunol.* **65,** 529 (1950).

23. G.B. Elion, G.H. Hitchings, H. VanderWerff, *J. Biol. Chem.* **192,** 505 (1951).
24. D.A. Clarke, F.S. Phillips, S.S. Sternberg, C.C. Stock, G.B. Elion, G.H. Hitchings, *Cancer Res.* **13,** 593 (1953).
25. F.S. Philips, S.S. Sternberg, L. Hamilton, D.A. Clarke, *Ann. N.Y. Acad. Sci.* **60,** 283 (1954).
26. J.H. Burchenal, M.L. Murphy, R.R. Ellison et al., *Blood,* **8,** 965 (1953).
27. 6-Mercaptopurine, Ed. C.P. Rhoads, *Ann. N.Y. Acad. Sci.* **60,** 183 (1954).
28. G.B. Elion, S. Singer, G.H. Hitchings, *Ann. N.Y. Acad. Sci.* **60,** 200 (1954).
29. G.B. Elion, S. Singer, G.H. Hitchings, *J. Biol. Chem.* **204,** 35 (1953).
30. M.E. Balis, D.H. Levin, G.B. Brown, G.B. Elion, H.C. Nathan, G.H. Hitchings, *Arch. Biochem. Biophys.* **71,** 358 (1957).
31. G.R. Greenberg and L. Jaenicke, in: *The Chemistry and Biology of Purines,* Ed. G.E.W. Wolstenholme and C.M. O'Connor (Churchill, London, 1957) pp. 204–232.
32. J.M. Buchanan, J.G. Flaks, S.C. Hartman, B. Levenberg, L.N. Lukens, L. Warren, in: *The Chemistry and Biology of Purines,* Ed. G.E.W. Wolstenholme and C.M. O'Connor (Churchill, London, 1957) pp. 233–252.
33. J.B. Buchanan and S.C. Hartman, *Adm. Enzymol.* **21,** 199 (1959).
34. G.H. Hitchings, G.B. Elion, in: *Cancer Chemotherapy II Twenty-Second Hahnemann Symposium,* Ed. I. Brodsky, S.B. Kahn, J.H. Moyer (Grune and Stratton, New York, 1972) pp. 23–32.

35. G.B. Elion, in: *Pharmacologdal Basis of Cancer Chemotherapy* (Williams and Wilkins, Baltimore, 1975) pp 547–564.
36. G.B. Elion, S. Bieber, G.H. Hitchings, *Ann. N.Y. Acad. Sci.* **60,** 297 (1954).
37. L. Hamilton and G.B. Elion, *Ann. N.Y. Acad. Sci.* **60,** 304 (1954).
38. G.B. Elion, *Proc. Royal Soc. Med.* **50,** 7 (1957).
39. D.A. Clarke, G.B. Elion, G.H. Hitchings, C.C. Stock, *Cancer Res.* **18,** 445 (1958).
40. G.B. Elion, I. Goodman, W. Lange, G.H. Hitchings, *J.Am. Chem. Soc.* **81,** 1898 (1959).
41. G.B. Elion, S. Bieber, G.H. Hitchings, *Cancer Chemother. Rep.* **8,** 36 (1960).
42. G.B. Elion, S.W. Callahan, G.H. Hitchings, R.W. Rundles, *Cancer Chemother. Rep.* **8,** 47 (1960).
43. G.B. Elion, S.W. Callahan, G.H. Hitchings, R.W.Rundles, in: *Proc. VIIIth International Congress Hematology,* Vol I (Pan-Pacific Press, Tokyo, 1961) pp. 642–645.
44. G.B. Elion, S. Mueller, G.H.Hitchings, *J. Am. Chem. Soc.* **81,** 3042 (1959).
45. G.B. Elion, S. Callahan, R.W. Rundles, G.H.Hitchings, *Cancer Res.* **23,** 1207 (1963).
46. G.B. Elion, *Fed. Proc.* **26,** 898 (1967).
47. P. deMiranda, L.M. Beacham III, T.H. Creagh, G.B. Elion, *J. Pharmacol. Exp. Ther.* **187,** 588 (1973).
48. G.B. Elion and G.H. Hitchings, in: *Antineoplastic and Immunosuppressive Agents, Handbook of Experimental Pharmacology,* Vol. **38/2,** Ed. A.C. Sartorelli and D.G. Jones (Springer-Verlag, Berlin, 1975) pp. 404–425.
49. P. de Miranda, L.M. Beacham III, T.H. Creagh, G.B. Elion, *J. Pharmacol. Exp. Ther.* **195,** 50 (1975).
50. G.B. Elion, S. Callahan, S. Bieber, R.W. Rundles, *Cancer Chemother. Rep.* **14,** 93 (1961).
51. R.W. Rundles, J. Laszlo, T. Itoga, G.B. Hobson and F.E. Garrison, Jr., *Cancer Chemother. Rep.* **14,** 99 (1961).
52. R. Schwartz, J. Stack, W. Dameshek, *Proc. Soc. Exp. Biol. Med.* **99,** 164 (1958).
53. D. Chanmougan and R.S. Schwartz, *J. Exp. Med.* **124,** 363 (1966).
54. R. Schwartz and W. Dameshek, *Nature* **183,** 1682 (1959).
55. H.C. Nathan, S. Bieber, G.B. Elion, G.H. Hitchings, *Proc. Soc. Exp. Biol. Med.,* **107,** 796 (1961).
56. S. Bieber, G.B. Elion, G.H. Hitchings, D.C. Hooper, H.C. Nathan, *Proc. Soc. Exp. Biol. Med.* **111,** 334 (1962).
57. R.Y. Calne, *Lancet* I, 417 (1960).
58. R.Y. Calne, *Transplant Bull.* **28,** 65 (1961).
59. R.Y. Calne, G.P.J. Alexandre, J.E. Murray, *Ann. N.Y. Acad. Sci.* **99,** 743 (1962).
60. J.E. Murray, J.P. Merrill, J.H. Harrison, R.E. Wilson, G.J. Dammin, *N. Engl. J. Med.* **268,** 1315 (1963).
61. The Twelfth Report of the Human Renal Transplant Registry, Advisory Committee to the Renal Transplant Registry, *J. Am. Med. Assoc.* **233,** 787 (1975).
62. S.T. Perdue, P.I. Terasaki, S. Cats, M.R. Mickey, *Transplantation* **36,** 658 (1983).
63. H. Takiff, D.J. Cook, N.S. Himaya, M.R. Mickey, P.I. Terasaki, *Transplantation* **45,** 410 (1988).
64. G.H. Hitchings and G.B. Elion, *Pharmacol. Rev.* **15,** 365 (1963).
65. G. Wolberg, in: *Pharmacology of Lymphocytes Handbook of Experimental Pharmacology,* Vol. **85,** Ed. M.A. Bray and J. Morley (Springer-Verlag, Berlin, 1988) pp. 517–533.
66. T. Hunter, M.B. Urowitz, D.A. Gordon, H.A. Smythe, M.A. Ogryzlo, *Arthritis Rheum.* **18,** 15 (1975).

67. G.B. Elion, S.W. Callahan, G.H. Hitchings, R.W. Rundles, J. Laszlo, *Cancer Chemother, Rep.* **16,** 197 (1962).
68. G.B. Elion, S. Callahan, H. Nathan, S. Bieber, R.W. Rundles, G.H. Hitchings, *Biochem. Pharmacol.* **12,** 85 (1963).
69. R.W. Rundles, J.B. Wyngaarden, G.H. Hitchings, G.B. Elion, H.R. Silberman, *Trans. Assoc. Am. Physicians* **76,** 126 (1963).
70. R.W. Rundles, *Ann. Rheum. Dis.* **25,** 615 (1966).
71. G.H. Hitchings and G.B. Elion, *Cancer Res.* **45,** 2415 (1985).
72. W.R. Vogler, J.A. Bain, C.M. Huguley, Jr., H.G. Palmer, Jr., M.E. Lowrey, *Am. J. Med.* **40,** 548 (1966).
73. G.H. Hitchings, *Ann. Rheum. Dis.* **25,** 601 (1966).
74. Symposium on Allopurinol, Ed. J.T. Scott, *Ann. Rheum. Dis.* **25,** 599 (1966).
75. G.B. Elion, *Ann. Rheum. Dis.* **25,** 608 (1966).
76. V. Massey, H. Komai, G. Palmer, G.B. Elion, *J. Biol. Chem.* **246,** 2837 (1970).
77. V. Massey, H. Komai, G. Palmer, G.B. Elion, *Vitam. Horm.* **28,** 505 (1970).
78. G.B. Elion, A. Kovensky, G.H. Hitchings, E. Metz, R.W.Rundles, *Biochem. Pharmacol.* **15,** 863 (1966).
79. G.B. Elion in: *Uric Acid, Handbook of Experimental Pharmacology,* Vol. **51,** Ed. W.N. Kelley and I.M. Wiener (Springer-Verlag, Berlin, 1978) pp. 485–514.
80. G.B. Elion, T.-F.Yü, A.B. Gutman, G.H. Hitchings, *Am. J. Med.* **45,** 69 (1968).
81. T.F. Yü and A.B. Gutman, *Am. J. Med.* **37,** 886 (1964).
82. R. Pomales, S. Bieber, R. Friedman, G.H. Hitchings, *Biochim. Biophys. Acta.* **72,** 119 (1963).
83. R. Pomales, G.B. Elion, G.H. Hitchings, *Biochim. Biophys. Acta.* **95,** 505 (1965).
84. J.B. Wydgaarden and D.M. Ashton, *J.Biol. Chem.* **234,** 1492 (1959).
85. S. Goldfinger, J.R. Klinenberg and J.E. Seegmiller, *J. Clin. Invest.* **44,** 623 (1965).
86. J.R. Klinenberg, S.E. Goldfinger, J.E. Seegmiller, *Ann. Intern. Med.* **62,** 639 (1965).
87. R.W. Rundles, *Arch. Intern. Med.* **145,** 1492 (1985).
88. G.B. Elion, F.M. Benezra, T.D. Beardmore, W.N. Kelley, in: *Purine Metabolism in Man III A,* Ed. A. Rapado, R.W.E. Watts, C.H.M.M. DeBruyn (Plenum, New York, 1980) pp. 263–267.
89. R.W. Rundles, E. Metz, H.R. Silberman, *Ann. Intern. Med.* **64,** 229 (1966).
90. A. deVries, M. Frank, U.A. Liberman, O. Sperling, *Ann. Rheum. Dis.* **25,** 691 (1966).
91. I.H. Krakoff, *Arthritis Rheum.* **8,** 896 (1965).
92. M.A. Pfaller and J.J. Marr, *Antimicrob. Agents Chemother.* **6,** 469 (1974)
93. D.J. Nelson, C.J.I. Bugge, G.B. Elion, R.L. Berens, J.J. Marr, *J. Biol. Chem.* **254,** 3959 (1979).
94. D.J. Nelson, S.W. LaFon, J.V. Tuttle, et al., *J. Biol. Chem.* **254,** 11544 (1979).
95. J. Marr, R. Berens, D. Nelson, *Science* **201,** 1018 (1978).
96. T. Spector, T.E. Jones, G.B. Elion, *J. Biol. Chem.* **254,** 8422 (1979).
97. T.A. Krenitsky, G.W. Koszalka, J.V. Tuttle, D.L. Adamczyk, G.B. Elion, J.J. Marr, in: *Purine Metabolism in Man III B,* Ed. A. Rapado, R.E. Watts, C.H.M.M. DeBruyn (Plenum, New York, 1980) pp 271–276.
98. R.L. Berens, J.J. Marr, D.J. Nelson, S.W. LaFon, *Biochem. Pharmacol.* **29,** 2397 (1980).
99. S.W. LaFon, D.J. Nelson, R.L. Berens, J.J. Marr, *Biochem. Pharmacol.* **31,** 231 (1982).
100. R.L. Berens, J.J. Marr, F. Steele De Cruz, D.J. Nelson, *Antimicrob. Agents. Chemother.* **22,** 657 (1982).
101. T. Spector, R.L. Berens, J.J. Marr, *Biochem. Pharmacol.* **31,** 225 (1982).
102. D.J. Nelson and G.B. Elion, *Biochem. Pharmacol.* **24,** 1235 (1975).

103. P.A. Kager, P.H. Rees, B.T. Wellde et al., *Trans. R. Soc. Trop. Med. Hyg.* **75,** 556 (1981).
104. C.I. Meirovich, H.L. Montrull, R.H. Gallerano, R.R. Sosa, *Arq. Bras. Cardiol.* **45,** 217 (1985).
105. F.M. Schabel, Jr., *Chemotherapy* **13,** 321 (1968).
106. G.B. Elion, J.L. Rideout, P. de Miranda, P. Collins, D.J. Bauer, *Ann. N.Y. Acad. Sci.* **255,** 468 (1975).
107. H.J. Schaeffer, S. Gurwara, R. Vince S. Bittner, *J. Med. Chem.* **14,** 367 (1971).
108. G.B. Elion, P.A. Furman, J.A. Fyfe, P. deMiranda, L. Beauchamp, H.J. Schaeffer, *Proc. Natl. Acad. Sci. U.S.A.* **74,** 5716 (1977).
109. H.J. Schaeffer, L. Beauchamp, P. deMiranda, G.B. Elion, D.J. Bauer, P. Collins, *Nature* **272,** 583 (1978).
110. S.S. Good and P. de Miranda, *Fed. Proc.* **41,** 1733 (1982).
111. H.C. Krasny, S.H.T. Liao, S.S. Good, B. Petty, P.S Lietman, *Clin. Pharmacol. Ther.* **33,** 256 (1983).
112. T. Spector, T.E. Jones, L.M. Beacham III, *Biochem. Pharmacol.* **32,** 2505 (1983).
113. G.B. Elion, *Am. J. Med.* **73** (1A), 7 (1982).
114. P. Collins, *J. Antimicrob. Chemother.* **12** (Suppl. B), 19 (1983).
115. G.B. Elion, *J. Antimicrob. Chemother.* **12** (Suppl. B), 9 (1983).
116. J.A. Fyfe, P.M. Keller, P.A.Furman, R.L. Miller, G.B. Elion, *J. Biol. Chem.* **253,** 8721 (1978).
117. W.H. Miller and R.L. Miller, *J. Biol. Chem.* **255,** 7204 (1980).
118. W.H. Miller and R.L. Miller, *Biochem. Pharmacol.*, **31,** 3879 (1982).
119. P.A. Furman, P. de Miranda, M.H. St. Clair, G.B. Elion, *Antimicrob. Agents Chemother.* **20,** 518 (1981).
120. P.M. Keller, S.A. McKee and J.A. Fyfe, *J. Biol. Chem.* **260,** 8664 (1985).
121. P.A. Furman, M.H. St. Clair, J.A. Fyfe, J.L. Rideout, P.M. Keller, G.B. Elion, *J. Virol.* **32,** 72 (1979).
122. M.H. St.Clair, P.A. Furman, C.M. Lubbers, G.B. Elion, *Antimicrob. Agents Chemother.* **18,** 741 (1980).
123. G.B. Elion, in: *Antiviral Chemotherapy: New Directions for Clinical Application and Research,* Ed. J. Mills and L. Corey (Elsevier, New York, 1986) pp 118–137.
124. P.V. McGuirt, J.E. Shaw, G.B. Elion, P.A. Furman, *Antimicrob. Agents Chemother.* **25,** 507 (1984).
125. P.A. Furman, M.H. St. Clair, T. Spector, *J. Biol. Chem.* **259,** 9576 (1984).
126. P.M. Keller, J.A. Fyfe, L. Beauchamp, et al., *Biochem. Pharmacol.* **30,** 3071 (1981).
127. L.E. Schnipper and C.S. Crumpacker, *Proc. Natl. Acad. Sci. U.S.A.* **77,** 2270 (1980).
128. P.A. Furman, D.M. Coen, M.H. St. Clair, P.A. Schaffer, *J. Virol.* **40,** 936 (1981).
129. K.K. Biron, J.A. Fyfe, J.E. Noblin, G.B. Elion, *Am. J. Med.* **73**(1A), 383 (1982).
130. P.A. Furman, P.V. McGuirt, P.M. Keller, J.A. Fyfe, G.B. Elion, *Virology* **102,** 420 (1980).
131. J.S. Pagano, J.W. Sixbey, J.-C. Lin, *J. Antimicrob. Chemother.* **12** (Suppl. B) 113 (1983).
132. W.T. Ashton, J.D. Karkas, A.K. Field, R.L. Tolman, *Biochem. Biophys. Res. Commun.* **108,** 1716 (1982).
133. Y.C. Cheng, S.P. Grill, G.E. Dutschman, K. Narayama, K.F. Bastow, *J. Biol. Chem.* **258,** 12460 (1983).
134. K.K. Biron, S.C. Stanat, J.B. Sorrell, et al., *Proc. Natl. Acad. Sci. U.S.A.* **82,** 2473 (1985).
135. P. de Miranda, H.C. Krasny, D.A. Page, G.B. Elion, *J. Pharmacol. Exp. Ther.* **219,** 309 (1981).

136. P. de Miranda, S.S. Good, O.L. Laskin, H.C. Krasny, J.D. Connor, P.S. Lietman, *Clin. Pharmacol Ther.* **30,** 662 (1981).
137. P. de Miranda, S.S. Good, H.C. Krasny, J.D. Connor, O.L. Laskin, P.S. Lietman, *Am. J. Med.* **73**(1A), 215 (1982).
138. S.S. Good and P. de Miranda, *Am. J. Med.* **73** (1A), 91 (1982).
139. P. de Miranda, R.J. Whitley, M.R. Blum et al., *Clin. Pharmacol. Ther.* **26,** 718 (1979).
140. P. de Miranda and M.R. Blum, *J. Antimicrob. Chemother.* **12** (Suppl. B) 29 (1983).
141. M.R. Blum, S.H.T. Liao, P. de Miranda, *Am. J. Med.* 73 (1A), 186 (1982).
142. W.E. Tucker, *Am. J. Med.* **73** (1A), 27 (1982).
143. G.B. Elion, *Cancer Res.* **45,** 2943 (1985).
144. L. Corey, J. Benedetti, C. Critchlow, et al., *J. Antimicrob. Chemother.* **12** (Suppl. B), 79 (1983).
145. A.E. Nilsen, T. Aasen, A.M. Halsos et al., *Lancet* **2,** 571 (1982).
146. Y.J. Bryson, M. Dillon, M. Lovett et al., *N. Engl. J. Med.* **308,** 916 (1983).
147. R.C. Reichman, G.J. Badger, D.C. Mertz et al., *J. Am. Med. Assoc.* **251,** 2103 (1984).
148. G.J. Mertz, L. Eron, R. Kaufman et al. *Am. J. Med.* 85(2A), 14 (1988).
149. H.R. Mattison, R.C. Reichman, J. Benedetti et al., *Am. J. Med.* **86** (2A), 20 (1988).
150. B. Bean and C. Braun, *Lancet* 2, 118 (1982).
151. H.H. Balfour, K.A. McMonigal, B. Bean, *J. Antimicrob. Chemother.* **12** (Suppl. B), 169 (1983).
152. J.D. Meyers, J.C. Wade, C.D. Mitchell et al., *Am. J. Med.* **73** (1A), 229 (1982).
153. J.C. Wade, B. Newton, C. McLaren et al., *Ann. Intern. Med.* **96** 265 (1982).
154. R. Saral, W.H. Burns, O.L. Laskin et al., *N. Engl. J. Med.* **305,** 63 (1981).
155. R. Saral, R.F. Ambinder, W.H. Burns et al., *Ann. Intern. Med.* **99,** 773 (1983).
156. B. Sköldenberg, M. Forsgren, K. Alestig et al., *Lancet* **2,** 707 (1984).
157. R.J. Whitley, C.A. Alford, M.S. Hirsch et al., *N. Engl. J. Med.* **314,** 144 (1986).

GEORGE H. HITCHINGS, JR.

My forebears all came from the United Kingdom. On my father's side, they migrated from London and County Derry in Northern Ireland to Londonderry, New Hampshire. When the American Revolution came, they, as loyalists, moved on to Canada. My father, grandfather and great-grandfather were born in St. Andrews, New Brunswick. In 1865, my grandfather, Andrew Hitchings, moved his family to Eureka, California. Andrew was a skilled craftsman in the building of wooden ships, and my father, George Herbert Hitchings, Sr., followed in his footsteps, eventually becoming a marine architect and master builder.

On my mother's side, Scottish and English prevailed. The first American was one Thomas Littlejohn from near Edinburgh, who came to the New World about 1735. His descendants, including Shaws, Eldridges and Thomases, moved about in the Maritime Provinces and New England. My maternal grandfather and great-grandfather were descendants of the Matthews family that emigrated twenty-four strong from near Glasgow to Prince Edward Island about 1800. My grandfather, Peter Matthews, married Sara Elizabeth Eldridge, and my mother, Lillian Matthews, was born in Maine. In 1875, my grandfather moved his family across the United States. He, too, was a shipbuilder and settled in Eureka.

My mother and father met and were married in Eureka, and my two sisters were born there. About 1897, Peter Matthews established a shipyard in Hoquiam, Washington, to build lumber carriers for the E.K. Wood Lumber Company. The company built several schooners a year. When Peter Matthews died, my father succeeded to the management, which then became Hitchings and Joyce. Later, my father was master builder and supervisor in Bellingham, Washington, and Coos Bay, Oregon, and between times he engaged in marine architecture. He worked in the period between sail and steam and was especially noted for the design of the transition vessel, the steam schooner, which had a wooden hull and was steam propelled.

I was born in Hoquiam in 1905. Family wanderings put me in grade school in Berkeley and San Diego, California, as well as in Bellingham and Seattle, Washington. I enjoyed a warm and loving home environment. A high standard of ethics prevailed in our family, together with a thirst for knowledge and an urge to teach. In their schooling, my mother and father were limited to what was available in Eureka, but they were avid readers, especially my father. It is clear to me in retrospect that he would have been a scientist had opportunities been more easily attainable.

My father died after a prolonged illness when I was twelve years old. The deep impression made by this event turned my thoughts toward medicine. This objective shaped my selection of courses in high school and expressed itself when I was salutatorian at my class graduation. I chose the life of Pasteur as the subject for my oration. The blending of Pasteur's basic research and practical results remained a goal throughout my career.

My experiences at Franklin High School in Seattle were notable for another reason. We had a most heterogeneous population, one that blended upper class and minorities including blacks, Filipinos, Japanese, Chinese and first generation Catanians. As a result I lost any self-consciousness I felt in dealing with people from different cultures and backgrounds.

I entered the University of Washington as a premedical student in 1923. The enthusiasm of faculty and students in the Chemistry Department was very infectious, however, and by the end of the first year I had become a chemistry major. I earned top grades, election to Phi Beta Kappa in my junior year, and a degree *cum laude* in 1927.

I stayed on to earn a master's degree in 1928 with a thesis based on work carried out during the summer of 1927 at the Puget Sound Biological Station at Friday Harbor, Washington. This institution later became a branch of the Oceanographic Laboratories of the University of Washington, largely created and directed by Thomas C. Thompson, who had been my mentor for my master's thesis. Thompson taught analytical chemistry and was notable for the keen wit and humorous twists that made his teaching memorable. Perhaps the most useful lessons I learned from him have to do with the mathematics of the precision of measurement.

For further graduate work I was offered fellowships at the Mayo Foundation and at Harvard. I chose Harvard, and after one year as a Teaching Fellow in the Department of Chemistry at Cambridge, I was accepted as a Teaching Fellow in the Department of Biological Chemistry at Harvard Medical School. I had intended to work with Otto Folin, but it was his habit to assign first-year Fellows to Cyrus Fiske for a year. By the end of the year, I was caught up in the Fiske-Subbarow program, and Folin very generously allowed me to continue there. After the discovery of phosphocreatine, this group had detected and isolated adenosine triphosphate. My assignment was to prepare for physiological studies by developing analytic methods (on a scale then viewed as 'micro' – 1 mg or less) for the purine bases. These methods constituted my dissertation and several early publications.

I earned my Ph.D. in 1933, a year memorable for another great event in my life – my marriage to Beverly Reimer. Her forebears were German, Austrian (Pennsylvania Dutch), Scottish and English. Her father, Azariah Reimer, was a Methodist-Episcopal minister who was pastor of a number of parishes in the Greater Boston area and superintendent of the city missionary society. Beverly had experience with many races and cultures and grew up having friends among all.

Beverly was highly artistic and intelligent. She was very accurate in her intuitive appraisals of people, almost always empathic, almost never disparaging. As she said, "The same quality is often exhibited in a person's most liked and most disliked behavior." Beverly expressed her talents in painting, jewelry making, writing and teaching. As my research career progressed, we traveled together and raised two children—Laramie Ruth and Thomas Eldridge.

Our marriage and my career began in the middle of the Great Depression. I experienced a nine-year period of impermanence, both financial and intellectual. I held temporary appointments at the C.P. Huntington Laboratories of Harvard in cancer research, at The Harvard School of Public Health in nutrition research, and at Western Reserve University, Department of Medicine, in electrolyte research.

My career really began in 1942 when I joined the Wellcome Research Laboratories in Tuckahoe, New York, as head and sole member of the Biochemistry Department. Support was limited, but I was free to develop my own program of research.

Elvira Falco was the first permanent member of my staff; Gertrude Elion joined in 1944, and Peter Russell in 1947. Additional help was added here and there, but our numbers were always small. Russell came from Alex Todd's laboratory at Cambridge University and brought not only competence in organic chemistry but a sense of the workings of medicinal chemistry as well. Elion took part in most of the projects dealing with purines, and Falco participated in everything from bacteriology and animal feeding to organic chemistry. For several years our group was housed in one large laboratory. Under the leadership of Falco, a constant flow of banter developed covering a wide range of subjects and degrees of seriousness. We never had any obstacles to interpersonal communication.

In the mid-1940s we began a project that seemed like a digression at times, but one that had a notable reprise some 40 years later. This was the antiviral work carried out in collaboration with Randall L. Thompson, then at Western Reserve. It focused principally on vaccinia virus, and it produced some active compounds. It also convinced us that effective curative chemotherapy of viral infections would have to be applied early in the multiplication cycle.

In 1947, we began to send compounds to the Sloan Kettering Institute to be screened for activity. Among the first few compounds we submitted was 2,6-diaminopurine, which proved active and later produced several notable remissions in acute leukemia.

The association with Sloan Kettering was a major impetus for our growth. The director, C.P. Rhoads, offered us financial support to enable us to increase our search for antitumor agents. This rather unusual circumstance resulted from Rhoads' realization that our compounds were of special interest, both intrinsically and because they were accompanied by a package of biological information. The external financial help allowed us to expand to about 15 persons. The arrangement continued for a number of years. By that time Burroughs Wellcome Co. was able to furnish our support completely. The arrangement with Sloan Kettering was productive and very satisfying for the contacts it provided, especially with C.P. Rhoads, C.C. Stock, J.H. Burchenal, F. Philips, D. Hutchison and others.

In 1948, we began to divide responsibilities with respect to developing purine and pyrimidine analogs. In 1947, Falco had synthesized *p*-chlorophenoxy-2,4-diaminopyrimidine, and it was apparent at once that we had a new kind of antifolic acid in hand. This line was pursued vigorously by Falco and Russell, and within a short time yielded a very exciting line of investigation — the end of which is not yet in sight.

The decision to refer the "thiation" of hypoxanthine to Elion was based on her developing expertise in the field of purine metabolism. Elion participated in much of the subsequent work with the compound and the agents that followed, including azathioprine and allopurinol.

It was always stimulating to work with Elion. She is intelligent, hard working and ambitious. She became my first assistant, and as I was promoted she succeeded to the position just left. She became head of the Department of Experimental Therapy, a large segment of the Chemotherapy Division. There she elucidated the mode of action of acyclovir, a study which is described as a major part of her Nobel address.

In 1967, I was offered the position of Vice President in Charge of Research of Burroughs Wellcome Co. It was not a post I had sought, but my experience had suggested that a scientist was much better able to support the interests of working scientists than were administrators who got science second hand. I owe much to D.W. Adamson, Wellcome's Group Research and Development Director, for his support and encouragement.

By 1968, Burroughs Wellcome Co. had outgrown its facilities at Tuckahoe, and we were plunged into a new set of administrative problems by the decision to move the company to a new site. In the end we chose North Carolina and, acceding to my strong representations, selected a site in the new Research Triangle Park. The move provided *Lebensraum*, a good environment and excellent relationships with the three local universities — Duke, the University of North Carolina at Chapel Hill, and North Carolina State. The move to North Carolina was a monumental undertaking, but the company soon took root and today has grown fourfold.

I left my position as Vice President with its heavy administrative duties to become Scientist Emeritus in 1976. This allowed more time for my own research and for travel. By 1971, Beverly was handicapped by strange afflictions classed as "collagen disease" that required close monitoring and constant medication. She exhibited remarkable courage and continued to be a joy as a companion. During our last 10 years together we traveled nearly 400,000 miles, much of it on lecture tours. Beverly's disease ended in her death in December 1985.

For the past 20 years I have pursued my growing interests in philanthropy. I became Director of The Burroughs Wellcome Fund in 1968 and its President in 1971. The Fund is a nonprofit foundation dedicated to the support of biomedical research. The Fund is supported solely by Burroughs Wellcome Co. and is a relatively small foundation. We have focused The

Fund's resources on underfunded areas of medical research with competitive grants in fields including clinical pharmacology and innovative methods in drug design. It has been very rewarding to guide this enterprise and see it grow.

In 1983, I founded what is now the Greater Triangle Community Foundation. It has been remarkably successful in fulfilling needs in the Triangle area. I am designated as Founder and Director for Life.

I have been involved also with a number of volunteer civic activities. These were undertaken partly to provide for activity in the retirement that has not yet come. They include United Way (Director and Vice President), American Red Cross (Director and Committee Chairman), Foundation for Better Health of Durham (Director, President and Chairman), N.C. Board of Science and Technology, Carolina Consulting Scientists and Engineers (Director); Royal Society of Medicine Foundation (Director); The Life Sciences Research Foundation.

Somehow these activities have found a place in my scientific career with no more difficulty than my former administrative duties. Today I devote one-third of my time to philanthropy and two-thirds to science.

I am vitally interested in current developments in innovative methods in drug design, and I look back with pride at our contributions to this field. Our research was untargeted, and the line of inquiry we had begun in the 1940s yielded new drug therapies for malaria (pyrimethamine), leukemia (6-mercaptopurine and thioguanine), gout (allopurinol), organ transplantation (azathioprine) and bacterial infections (cotrimoxazole (trimethoprimÅ). The new knowledge contributed by our studies pointed the way for investigations that led to major antiviral drugs for herpes infections (acyclovir) and AIDS (zidovudine).

My greatest satisfaction has come from knowing that our efforts helped to save lives and relieve suffering. When I was baptised, my father held me up and dedicated my life to the service of mankind. I am very proud that, in some measure, I have been able to fulfill his hopes.

SELECTIVE INHIBITORS OF DIHYDROFOLATE REDUCTASE

Nobel Lecture, December 8, 1988

by

GEORGE H. HITCHINGS, JR.

President of The Burroughs Wellcome Fund, Research Triangle Park, North Carolina, U.S.A.

My interest in nucleic acids, their constitutents and metabolism can be traced to the discovery of adenosine triphosphate in muscle by Fiske and Subbarow (1). I had entered Harvard University as a candidate for a Ph.D. degree in 1928 and transferred to Harvard Medical School in 1929. Otto Folin, then head of the Department of Biological Chemistry, designated my first predoctoral year to be spent with Cyrus J. Fiske, then involved in momentous discoveries of phosphocreatine and the labile phosphorus compounds of muscle and other tissues. I was soon immersed in the discovery of analytical tools with which to follow the metabolism of adenosine triphosphate (2).

Other lines of thought coalesced with this interest when I joined Burroughs Wellcome Co. in 1942 as the sole member of the Biochemistry Department. Meantime, the antimetabolite principle had been expressed by Woods (3) and Fildes (4). I saw the opportunity to explore nucleic acid biosynthesis in a new and revealing way by employing synthetic analogues of the purine and pyrimidine bases in a system utilizing these heterocyclic compounds for biosynthesis.

I was able to interest Elvira Falco who was then an assistant in the company's Bacteriology Department. Together we worked out a system using *Lactobacillus casei*, which would grow either with a then-unknown "L-casei" growth factor or with a mixture of thymine and a purine (Fig. 1) (5). This system quickly gave us encouraging results. In a simple screening test for antibacterial activity, analogues were found to inhibit strongly not only the *L. casei* system but pathogenic bacteria as well. We had added toxicity testing in growing rats and other biological screening procedures and were becoming more and more excited by the results.

By 1947, six or seven of us were pursuing this work, and the feeling in the group was, "Now we have the chemotherapeutic agents; we need only to find the diseases in which they will be active." At that point I made two arrangements for collaborative studies, one with Sloan Kettering Institute for antitumor testing using sarcoma 180 in mice, and another with outside

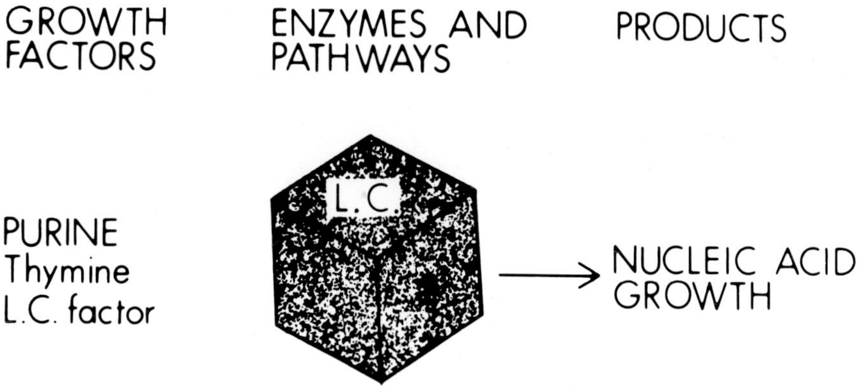

Figure 1. The use of *Lactobacillus casei* as a black box of enzymes and metabolic pathways concerned with nucleic acid metabolism (9).

laboratories for expansion of antibacterial and antimalarial testing. The antimalarial testing was included through the insight of Peter B. Russell, also a member of our research group. Russell noted the resemblance of a 5-phenyl-2,4-diaminopyrimidine to a hypothetical conformation of the anti-malarial proguanil. It turned out later that this theory was prescient; the dihydrotriazine is the active metabolite of proguanil (6).

The next year, two leads developed almost simultaneously. Falco began a series of selective inhibitors of dihydrofolate reductase with the synthesis of 2,4-diamino-5-phenoxypyrimidine, and Gertrude Elion synthesized 2,6-diaminopurine (7). The latter was among the first four compounds we submitted to Sloan Kettering Institute. It was found to be active in the S-180 test, was taken into clinical trial by Joseph H. Burchenal and gave at least one spectacular remission (8). This was sufficient to establish cancer chemotherapy as a continuing primary goal of our group. The purine analogue story, which has been a major theme in Gertrude Elion's career, is the topic of her Nobel address.

The main theme of this essay is the continuing topic of selective inhibitors of dihydrofolate reductase (DHFR). Biochemical knowledge of the role of folates was developing concomitantly. Figure 2 shows that dihydrofolate is synthesized *de novo* in prokaryotes (microorganisms) while the higher species of eukaryotes (host) must have the vitamin preformed. I will pick up the story of selective inhibitors of dihydrofolate reductase *in medias res* and carry it forward to current exciting developments.

A short review of our line of research was presented in a symposium honoring Sir Henry Wellcome (9). Papers in *Advances in Enzymology* (10), *Advances in Enzyme Regulation* (11), and *Enzyme Inhibitors as Drugs* (12) tell the story of selectivity among 2,4-diaminopyrimidines from its first recognition before 1950 to its confirmation. Proof was developed through inhibitor analysis sequencing, conformations, induced mitogenesis, computer assisted conformation studies, and new syntheses based on this type of information.

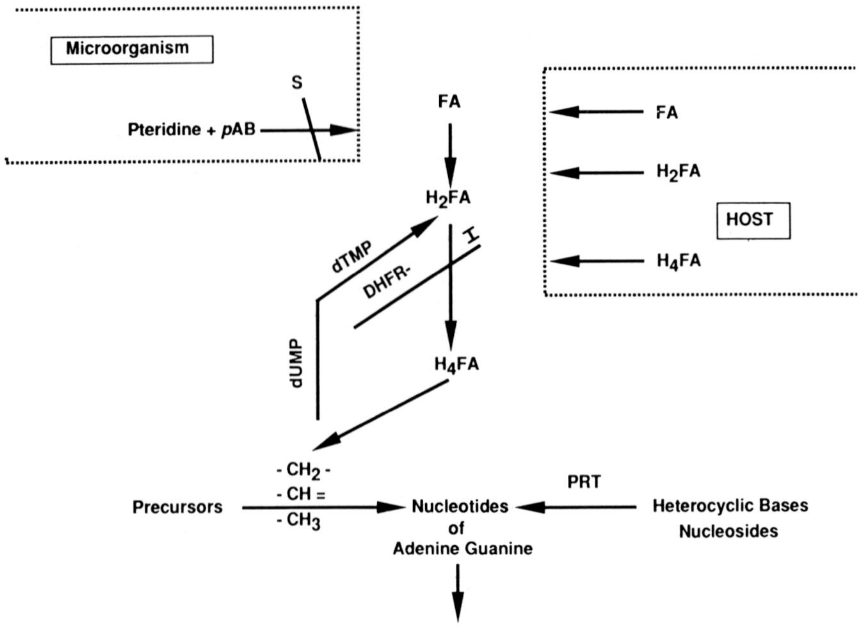

Figure 2. Folate metabolism. Upper left shows biosynthesis in prokaryotes; upper right shows the uptake of preformed vitamin in mammals. The subsequent utilization for the biosynthesis of nucleic acid components is shown below. The importance of dihydrofolate reductase (DHFR) and the selectivity of its inhibitors takes a central place. Folic acid (FA); dihydrofolic acid (H_2FA); tetrahydrofolate (H_4FA); thymidylate (dTMP); deoxyuridylate (dUMP); p-aminobenzoic acid (pAB); phosphoribosyl transferase (PRT) (10).

The earliest proof of the mechanism of action of these compounds is illustrated in Fig. 3 (13). The growth of *Streptococcus faecium* was easily inhibited by a diaminopyrimidine when the growth was induced by folic acid, but it took 500 to 1000 times as much inhibitor when folinic acid (tetrahydrofolate) was supplied. This was correctly interpreted as inhibition of the yet-unknown enzyme responsible for the reduction of folate to tetrahydrofolate.

By 1950, we had concluded that we were dealing with selective inhibitors of dihydrofolate reductase. From this work it appeared probable that details of fine structure in dihydrofolic reductase vary from species to species and that a given inhibitor may exhibit considerable selectivity as a result of looser or tighter binding to the corresponding enzyme of host or parasite, respectively (14). The full structural analogue of MTX shows little selectivity, while pyrimethamine is highly active against the malarial enzyme and trimethoprim against a bacterial enzyme. Neither has notable toxicity against the rat liver enzyme.

Isolation and characterization of this enzyme led first to inhibitor analysis (Table 1). This documented unequivocally the different structures of representative dihydrofolate reductases from different sources. This was soon followed by amino acid analyses and sequence determinations.

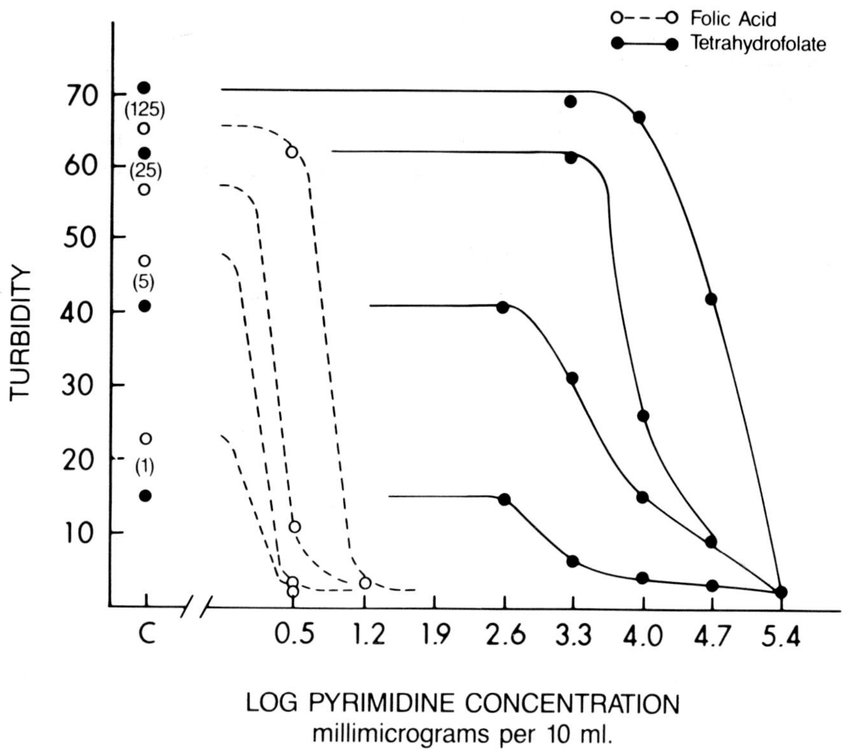

Figure 3. The concentration of pyrimethamine required to inhibit growth of *Streptococcus faecium* depends on whether folic acid (○) or a form of tetrahydrofolate (●) is used in the medium (13).

Table 1. The concentration (I.C.$_{50}$) of methotrexate (MTX) pyrimethamine (Pyr) or trimethoprim (Tmp) required to inhibit the DHFR derived from *E. coli*, rat liver and *P. berghei* (9).

INHIBITOR ANALYSIS
I.C.$_{50}$ × 10^8

Enzyme	*E. Coli*	Rat Liver	*P. Berghei*
Compound			
MTX	0.1	0.2	0.07
Pyr	250	70	0.05
Tmp	0.5	3 × 10^3	7.0

Concomitantly developing knowledge of the roles of folate derivations in biosyntheses and metabolism gave a fuller appreciation of the significance of inhibition of dihydrofolate reductase (Figs. 4, 5). The activity of the enzyme is necessary to produce tetrahydrofolate initially and to recycle it after its reoxidation, molecule for molecule, in the formation of thymidylate from deoxyuridylate.

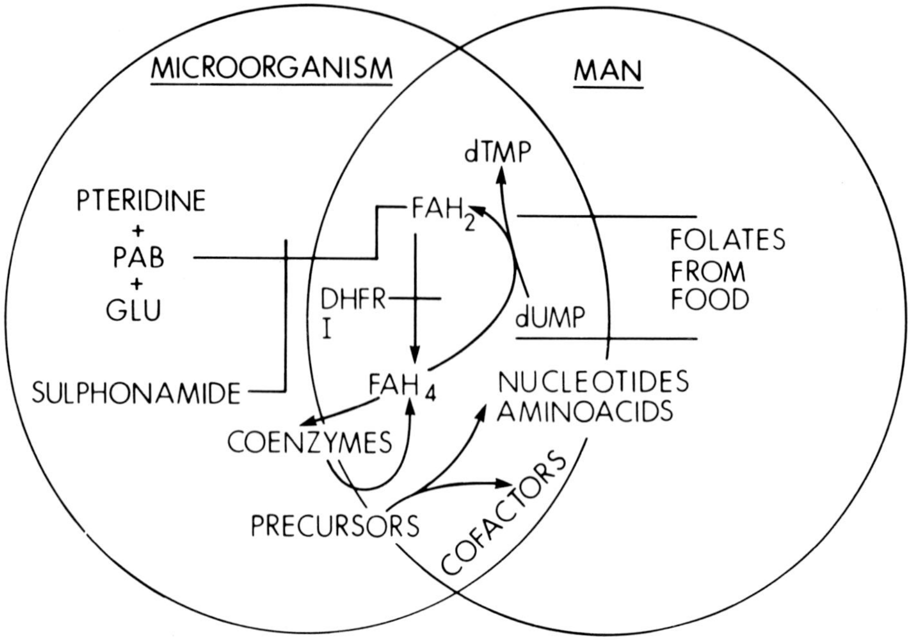

Figure 4. The metabolic reactions catalyzed by derivatives of tetrahydrofolate (FAH4): thymidylate (dTMP); deoxyuridylate (dUMP); *p*-aminobenzoic acid (PAB); glutamate (GLU) (12).

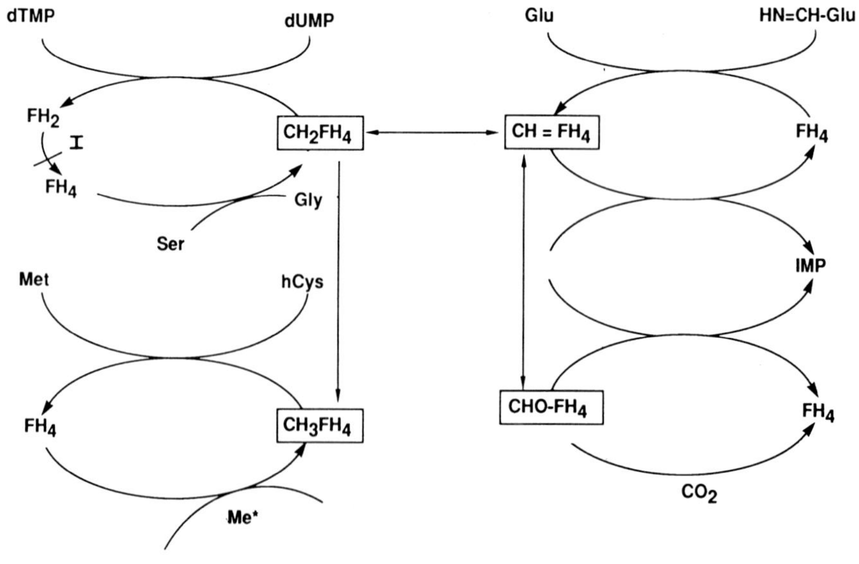

Figure 5. Details of reactions of specific folate-containing cofactors. Counter clockwise, formyl, methanyl, methylene, and methyl tetrahydrofolates (FH4). Serine (Ser); methionine (Met); inosinate (IMP). See Table 1 and Fig. 4 for other abbreviations (12).

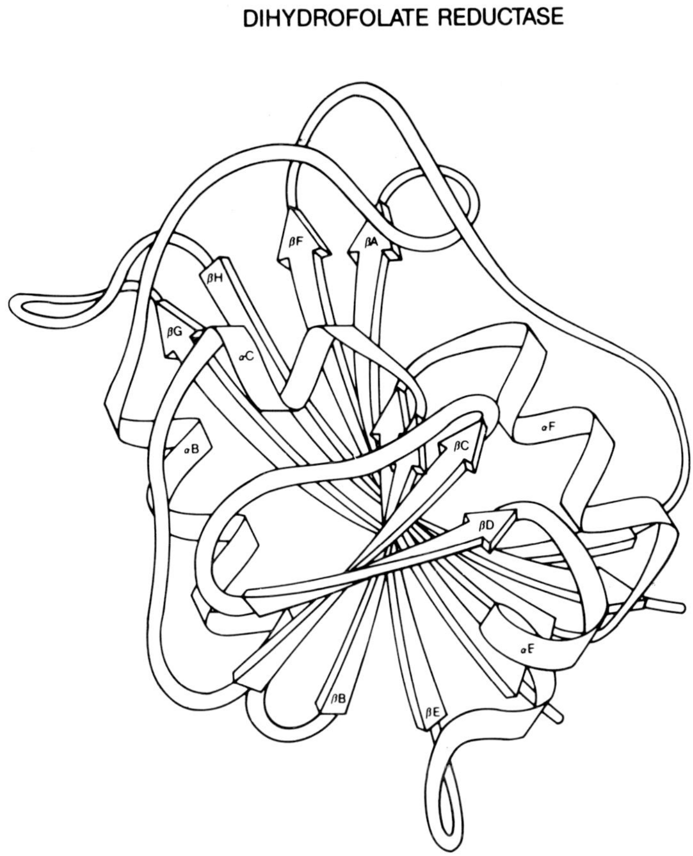

Figure 6. A graphic illustration of the configuration of the dihydrofolate reductase molecule (16).

Amino acid sequence determinations brought out a full appreciation of the high variability of the basic constituents of this enzyme in a range of species (15). Between the enzymes of two bacterial species there is only 30 percent homology, and between prokaryotes and eukaryotes only 30 percent. It is only the higher species that exhibit as much as 90 percent homology. Hitchings and Roth found 16 identities between the enzymes from *Escherichia coli* and those from the mouse tumor L1210 (12). They predicted correctly that study of a wider range of enzymes would reduce the number of identities. If one takes into account enzymes not in the mainstream, e.g. those from protozoa plasmids, the number is even smaller.

The mainstream enzymes (from bacterial and mammalian sources) have similar conformations. That published by Richardson (16) may be regarded as the type (Fig. 6). Although mammalian enzymes are larger than this bacterial enzyme by some 3000 daltons, the extra residues exist as loops that do not greatly alter the main conformation. On the other hand, the enzymes from other types of organism can be so different as to raise doubts about whether they are intrinsically dihydrofolate reductase, or whether their activity in that field is secondary to some other unknown function, e.g., Matthews (17). Figure 7 (18) shows some possible origins of Type II DHFR.

If its normal function is:

To reduce DHFR

1. Other intestinal coliforms
2. T4 phage
3. Anaerobes
4. Fungi (yeasts)
5. Protozoa

To reduce/oxidize other compounds

1. Pteridine/flavin
2. Non-specific dehydrogenases

Figure 7. Possible origins of untypical dihydrofolate reductases (18).

It is pertinent at this point to refer to some of the applications of selective inhibitors of dihydrofolate reductase. A major application is, of course, the establishment of co-trimoxazole as an antibacterial of major importance. Its creation derived from the knowledge that in combining trimethoprim with sulfamethoxazole, one was creating a sequential blockade of a major biosynthetic pathway in a bacterium or in other prokaryotes. The minimal effects on the host reflect the absence in eukaryotes of the reactions leading to the biosynthesis of dihydrofolate such as occurs in prokaryotic cells.

The strong potentiation that occurs is indicated by the data of Table 2 (19), which show that minimum inhibitory concentrations of one component may be reduced as much as 10- or 20-fold when the second component is also present. Moreover, the combination may be effective against organisms that would not be inhibited by the individual drugs, e.g., *Klebsiella sp.* and *Streptococcus* Group C, where the individual inhibitory concentrations would be borderline or unattainable.

One further illustration of the possible utility of biochemically related inhibitors is given in Fig. 8 (20), where the addition of 8-azaguanine to an already potentiative combination of diaveridine (B.W. 210U49) with sulfadiazine enhances the potency. Such biochemically orientated triple combinations have not been used in a major way, but their potential remains exploitable.

I should like to focus now on the uses of individual inhibitors of the enzyme dihydrofolate reductase.

Chronologically first, and perhaps still first in importance in cancer chemotherapy, is methotrexate from the Lederle Laboratories. Methotrexate had assumed a role in the therapy of acute leukemia as early as 1948 (21) and is still of major interest today. As shown in Table 1, however, it is

Table 2. Effect on minimal inhibitory concentration (MIC) on combining one part of trimethoprim with 20 parts sulfamethoxazole.

	MIC (μg/ml)			
	Sulfamethoxazole		Trimethoprim	
Organism	Alone	Mixture	Alone	Mixture
Streptococcus pyogenes	>100	1	1	0.05
Streptococcus pneumoniae	30	2	2	0.1
Staphylococcus aureus	3	0.3	1	0.015
Haemophilus influenzae	10	0.3	1	0.015
Bordetella pertussis	50	4	3	0.2
Klebsiella pneumoniae	>100	4	1	0.2
Klebsiella aerogenes	>100	4	1	0.2
Escherichia coli	3	1	0.3	0.05
Salmonella typhimurium	10	1	0.3	0.05
Shigella sonnei	10	1	0.3	0.05
Proteus vulgaris	30	3	3	0.15

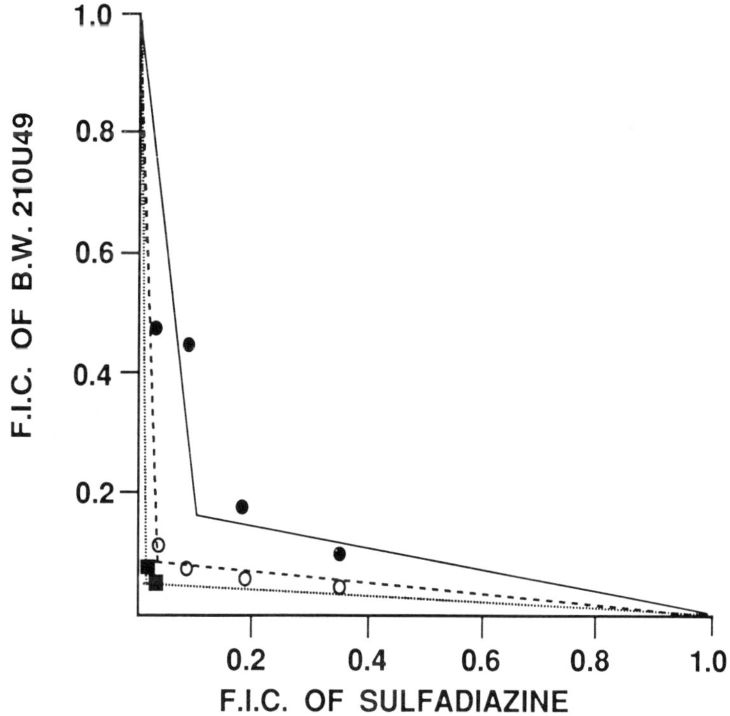

Figure 8. Fractional inhibitory concentrations of B.W. 210U49 and sulfadiazine required for 50 percent inhibition of *P. vulgaris* in the presence and absence of 8-azaguanine: ●—●, no 8-azaguanine; 0—0, 1 μg/ml.; ■—■, 10 μg/ml. (20).

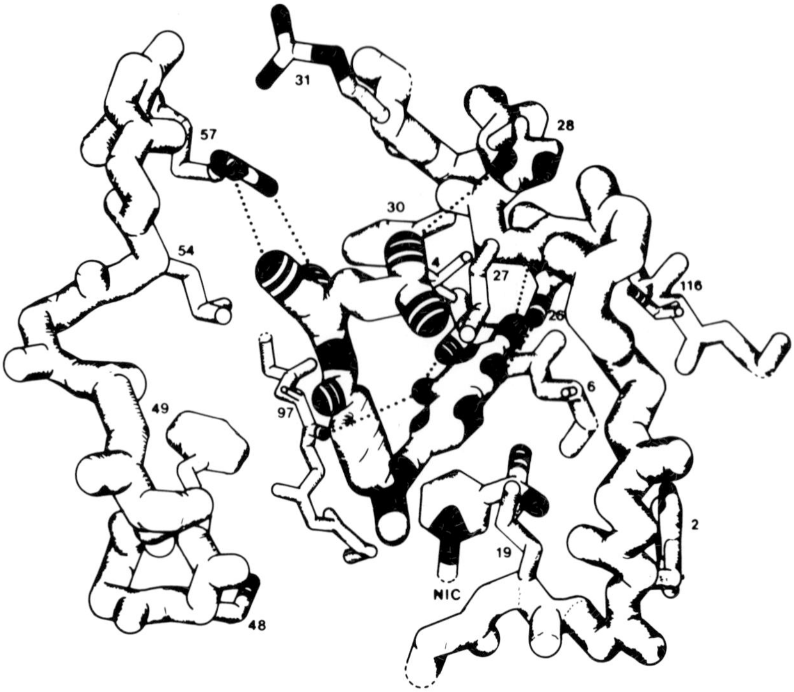

Figure 9. Schematic illustration of the active site of *L. casei* dihydrofolate reductase showing the binding of methotrexate and nicotinamide adenine dinucleotide phosphonate (NADPH) (22).

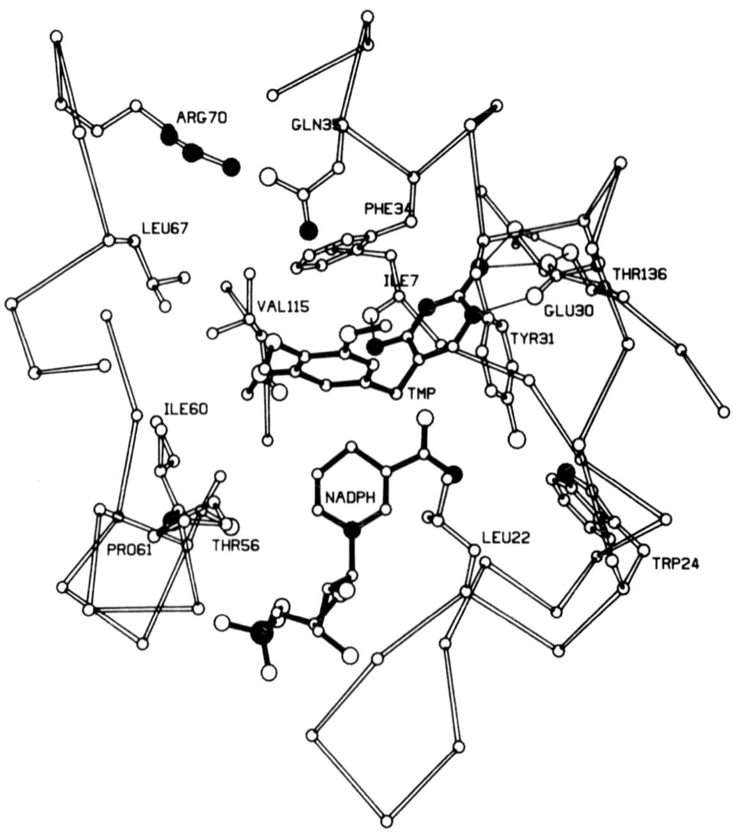

Figure 10. Schematic illustration of the cofactor in chicken liver DHFR showing the binding of trimethoprim (TMP) and NADPH (23).

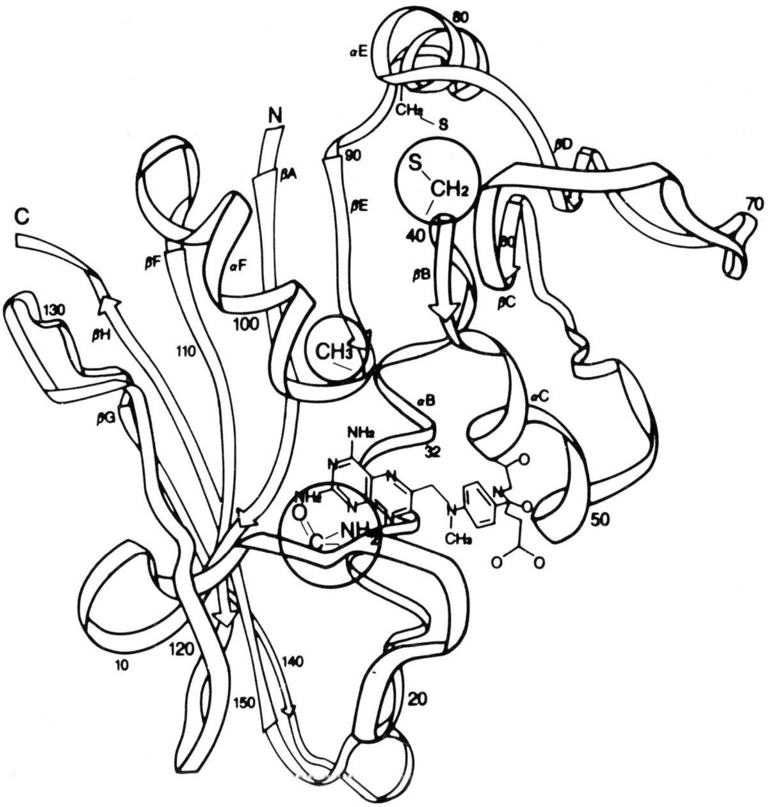

Figure 11. E. coli dihydrofolate reductase with induced specific mutations (24).

relatively unselective. Its therapeutic utility depends mainly on differences in uptake, glutamylation, cellular excretion, cellular metabolic balance, and other factors.

Whereas methotrexate is a close structural analogue of folate, our own antifols were significantly different and highly selective. These facts will be brought out by examination of details of structure and active centers of several enzymes.

For example, it may be illustrated by contrasting the fine structure of the complex of methotrexate and cofactor in the *L. casei* DHFR (Fig. 9) (22) with that of trimethoprim and cofactor in the chicken liver enzyme (Fig. 10) (23). It is obvious that the methotrexate molecule is much more space-filling than trimethoprim. Such contrasts are observable in all situations where the small molecule inhibitors are observed, and the contrasts in the active centers of the enzymes are consistent with the selectivities of the small molecule inhibitors.

There have been a number of studies of the functions of specific amino acid residues. One of the most cogent was reported by Kraut and coworkers (24) in a paper that combined several experiments. In this study of the *E. coli* enzyme, the replacement of aspartate by asparagine reduced activity to

Table 3. Comparison of binding constants and kinetic values of polyglutamates in enzymatic reactions in folate metabolism. Thymidylate synthase (T.S.); AICAR transformylase (AICAR TF); GAR transformylase (GAR TF) (25). Number of glutamate residues (n).

Enzyme	n	n/1 K_m	n/1 V_{max}
DHFR	5	1	—
serine $HOCH_2$- transferase	5	0.0025	—
5, 10-CH_2-H_4F dehydrogenase	3	<1	1.7
10-CHO-H_4F synthetase	3	0.01	0.5
10-CHO-H_4F synthetase	5	0.001	0.25
5, 10 CH-H_4F cyclohydrolase	5	0.5	1.5
T.S.	3	0.2	—
T.S.	5	0.014	3
AICAR-TF	3	0.01	1
GAR-TF	3	0.8	5

0.1% of that of the unaltered enzyme, showing the importance of the ionizable carboxyl group (Fig. 11). The substitution of alanyl for glycyl, next up, completely inactivated the enzyme, probably by distorting its conformation. However, substitution of SH for OH (cysteine for serine) had little effect. This type of experiment may provide the background for the synthesis of new and useful inhibitors.

An aspect of folate metabolism that appeared relatively late is the identification of the high molecular weight derivatives of folic acid as polyglutamates. Since the chain length of most of these exceeds that of the longest form that can be transported into the cell efficiently, most cells contain a pteroylpolyglutamate synthetase. The number of glutamyl residues varies with time and with cell type. The immense complexities of this situation are only beginning to appear. Nevertheless, it is interesting and instructive to compare the binding constants and kinetic values of polyglutamates in some of the enzymatic reactions involved in folate metabolism (Table 3) (25).

Table 4. The effect of pentaglutamylation on the binding of methotrexate (MTX) to specific tetrahydrofolate-utilizing synthetic enzymes (26).

Enzyme	Folate Cofactor	Ki (μM) MTX	Ki (μM) MTX-Glu_5	MTX / MTX-Glu_5
Thymidylate Synthase	5,10-CH_2-$H_4PteGlu_1$	13.0	0.047	277
	5,10-CH_2-$H_4PteGlu_5$	14.3	0.056	254
AICAR Transformylase	10-CHO-$H_4PteGlu_1$	143.9	0.057	2508
	10-CHO-$H_4PteGlu_5$	40.0	5.89	6.8
GAR Transformylase	10-CHO-$H_4PteGlu_1$	80	2.5	32
	10-CHO-$H_4PteGlu_5$	84	22.0	4

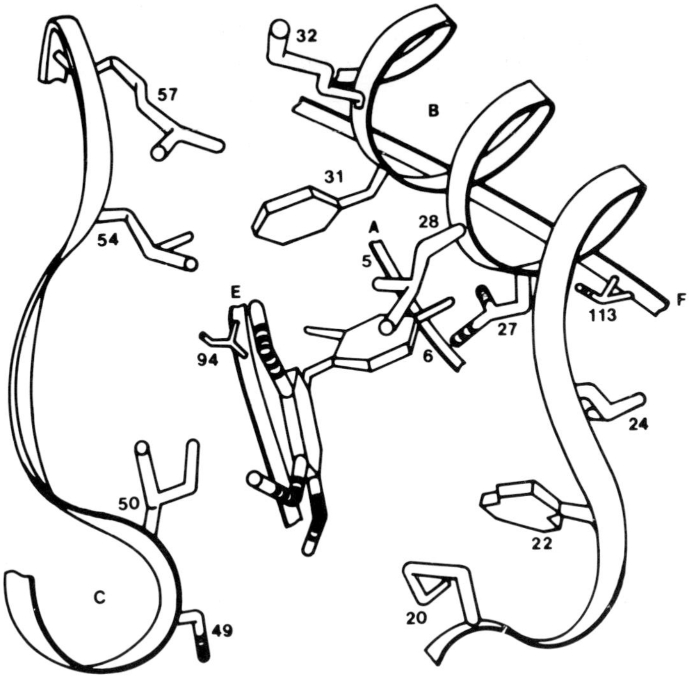

Figure 12. Fit of trimethoprim (center) in the DHFR of *E. coli* (27).

The inhibitor methotrexate is subject to polyglutamylation as well as the tetrahydrofolate. In many cases the ratio of folate polyglutamate to inhibitor polyglutamate is not markedly changed, but it is predictable that among the vast number of possibilities some of the anomalies and some of the therapeutic indices will find solid explanations. A few of the effects of glutamylation on specific biosynthetic enzymes are shown in Table 4 (26).

I wish to turn now to the intimate details of the structure of specific enzymes and their exploitation in the synthesis of more active inhibitors. Figure 12 (27) shows the fit of trimethoprim in the dihydrofolate reductase of *E. coli*. In the upper left (residue 57), one can see the guanidino group of arginine. This was exploited by Lee Kuyper who replaced the *m*-methoxyl of trimethoprim with a series of carboxyalkoxy radicals. The optimum length proved to be 5 carbon atoms, and the product, the carboxyl amyloxy derivation, bound the enzyme some 50 times more tightly than did trimethoprim (28). In the same vein, Barbara Roth synthesized bromoacetoxyphenoxy analogues of trimethoprim and found that the derivative bound to the histidine-22 residue of the enzyme with a covalent linkage (29). These examples may be regarded as probes of a vast and exciting future.

Resistance to antifols in multiple aspects appeared soon after work on these inhibitors began. Among various expressions of resistance (Fig. 13) was a loss of the cellular transport mechanism, which has stimulated the investigation of lipophilic inhibitors that do not depend on this mechanism

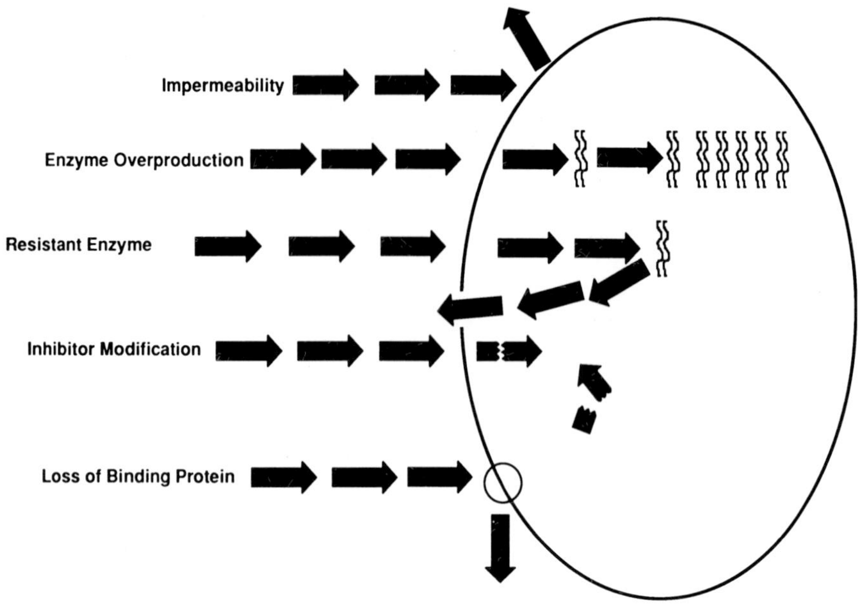

Figure 13. Mechanisms of resistance to inhibitors of dihydrofolate reductase (18).

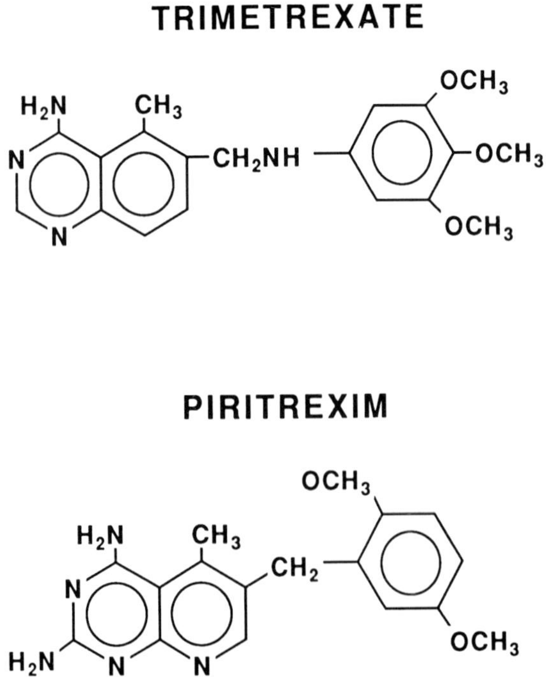

Figure 14. Chemical structures of trimetrexate and piritrexim, two potentially useful drugs.

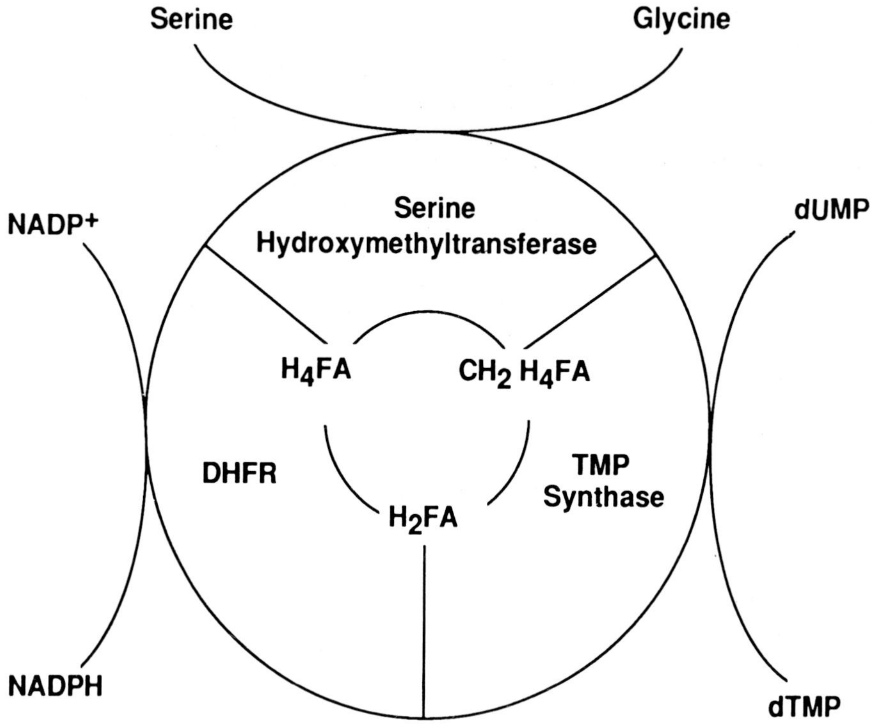

Figure 15. The thymidylate synthetic cycle. Methylene tetrahydrofolate (CH2 H4FA); Nicotinamide adenine dinucleotide phosphate (NADPH).

for transport into the cell. At present the two potentially useful drugs are trimetrexate and piritrexim (Fig. 14). Exploitation of these inhibitors is in its infancy, but each has exhibited interesting properties of its own. The effects of trimetrexate and methotrexate are compared in Table 5 (30).

Table 5. Efficacy of trimetrexate and methotrexate against transplantable rodent tumors and in xenografts (30).

	Tumor System	Trimetrexate	MTX
Murine	L1210 leukemia	+ +	+ +
	P388 leukemia	+ +	+ +
	B16 melanoma	+ +	– –
	Lewis lung carcinoma	– –	– –
	Colon 26	+ +	– –
	Colon 38	+	– –
	CD8F1 mammary	+ +	+
Human xenograft	Mammary MX-1	– –	– –
	Lung LX-1	– –	– –
	Colon CX-1	– –	– –

Table 6. Effect of the piritrexim-sulfadiazine combination in treatment of murine toxoplasmosis. Sulfadiazine (S) was administered orally by gavage at concentrations of 4, 40, and 400 mg/kg per day, and piritrexim (P) was administered ip at a concentration of 20 mg/kg per day (32).

	Percent cumulative mortality: day						
Treatment (concentration)*	**3**	**6**	**9**	**12**	**15**	**18**	**20**
None	0	0	100	—	—	—	—
S (4)	0	20	60	100	—	—	—
P + S (4)	0	20	40	60	80	80	80
S (40)	0	20	40	100	—	—	—
P + S (40)	20	20	20	40	80	80	80
S (400)	0	0	80	100	—	—	—
P + S (400)	0	0	0	0	0	0	0

Both trimetrexate and piritrexim have shown interesting activities against infectious prokaryotic organisms. The combination of trimetrexate and leucovorin has been used successfully to treat pneumocystic pneumonia in AIDS patients (31). Table 6 shows the effect of combining piritrexim and sulfadiazine to treat murine toxoplasmosis (32).

Finally, to be mentioned is a selective inhibitor of thymidylate synthase (Fig. 15). The first of these is 5,8-dideazaisopteroylglutamate (IAHQ) (Fig. 16). IAHQ and specific inhibitors of other biosynthetic reactions involving tetrahydrofolate represent the beginning of a biochemically orientated improved cancer chemotherapy (33).

I am incredibly blessed to have been involved for well over four decades in a field that continues to become more exciting with every passing year. Figure 17 (34) charts the history of increasing knowledge concerning dihydrofolate reductase, beginning with our work in the 1940s. Those early, untargeted studies led to the development of useful drugs for a wide variety of diseases (Fig. 18) and has justified our belief that this approach to drug discovery is more fruitful than narrow targeting. In 1988, this line of research continues to generate useful new compounds.

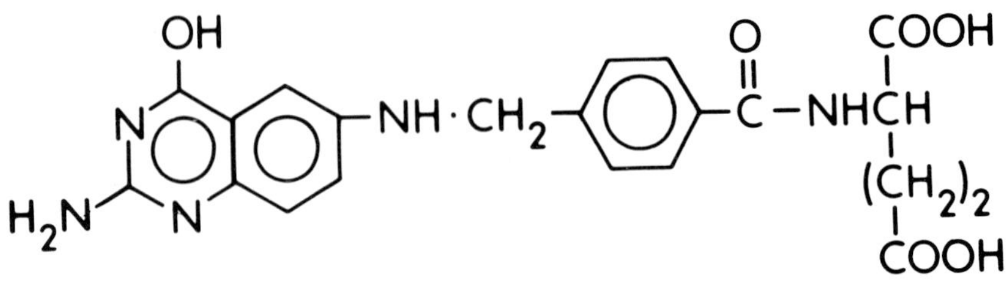

IAHQ

Figure 16. Chemical structure of IAHQ (5,8-dideazaisopteroylglutamate), a selective inhibitor of thymidylate synthase.

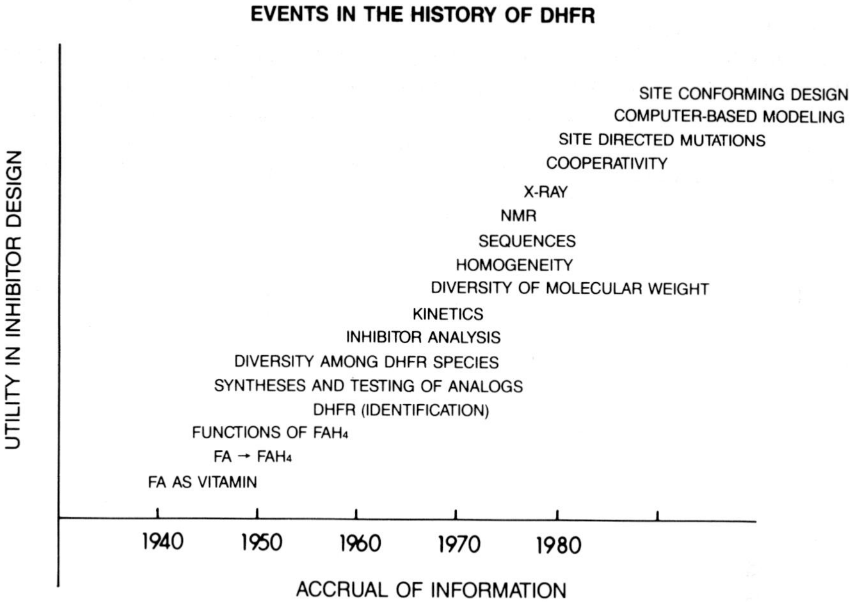

Figure 17. Accrual of information about dihydrofolate reductase, beginning with untargeted studies in the 1940s and continuing in the 1980s with discovery of new useful compounds and methods of drugs design (34).

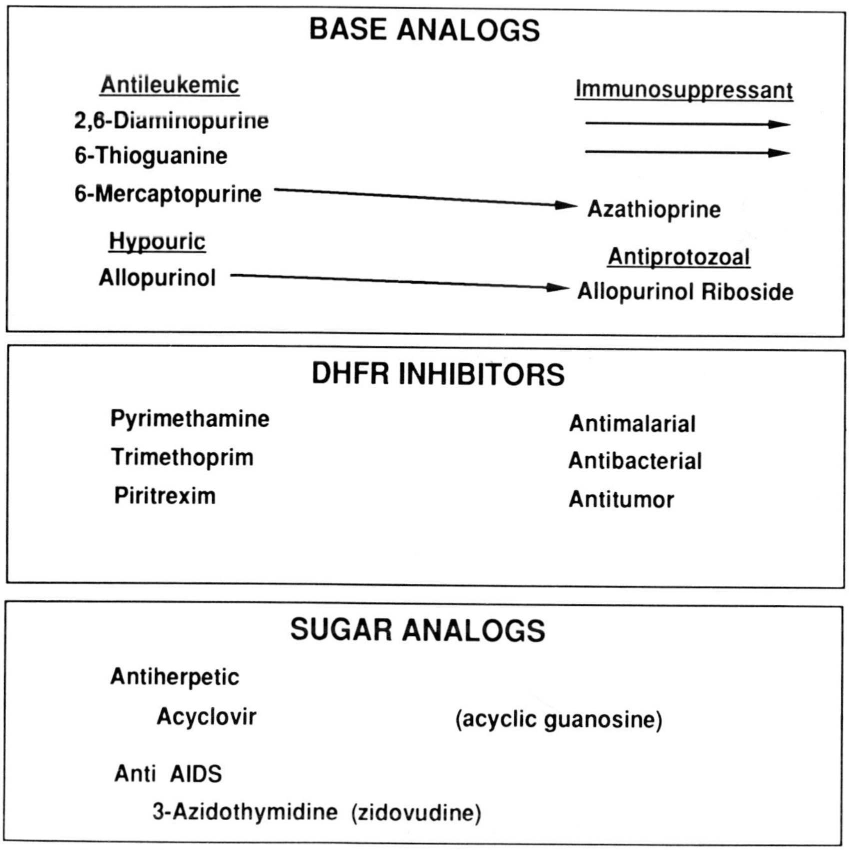

Figure 18. Drug therapies that have emerged from dihydrofolate reductase inhibitors.

I should like to close with a quotation from one of my own papers:

"To the biochemical chemotherapist, it is not only a matter of faith, but an obvious fact, that every cell type must have a characteristic biochemical pattern, and therefore be susceptible to attack at some *locus* or *loci* critical for its survival and replication." (35).

REFERENCES

1. C.H. Fiske and Y. Subbarow, *Science,* **70,** 381 (1929).
2. G.H. Hitchings, Dissertation, Harvard University (1933).
3. D.D. Woods, *Brit. J. Exptl. Path.,* **21,** 74 (1940).
4. P. Fildes, *Lancet I,* 955 (1940).
5. G.H. Hitchings, E.A. Falco and M.B. Sherwood, *Science,* **102**, 251 (1945).
6. E.A. Falco, G.H. Hitchings, P.B. Russell and H. VanderWerff, *Nature,* **164,** 107 (1949).
7. G.H. Hitchings, G.B. Elion, H. VanderWerff and E.A. Falco, *J. Biol. Chem.,* **174,** 765 (1948).
8. J.H. Burchenal, D.A. Karnofsky, E.M. Kingsley-Pillers, C.M. Southam, W.P. Laird Meyers, G.C. Escher, L.F. Craver, H.W. Dargeon and C.P. Rhoads, *Cancer,* **4,** 549 (1951).
9. G.H. Hitchings, *Drug Intell. Clin. Pharm.,* **16,** 843 (1982).
10. G.H. Hitchings and J.J. Burchall, *Adv. Enzymol.,* **27,** 417 (1965).
11. G.H. Hitchings and S.L. Smith, *Adv. Enzyme Reg.,* **18,** 349 (1980).
12. G.H. Hitchings and B. Roth, in *Enzyme Inhibitors as Drugs,* Ed. by M. Sandler, Baltimore, University Park Press, pp. 263 – 280 (1980).
13. G.H. Hitchings, E.A. Falco, G.B. Elion, S. Singer, G.B. Waring, D.J. Hutchison and J.H. Burchenal, *Arch. Biochem. Biophys,* **40,** 479 (1952).
14. G.H. Hitchings, in *Drugs, Parasites and Hosts,* Ed. by L.G. Goodwin and R.H. Nimmo-Smith, Boston, Little-Brown, pp. 196 – 210 (1962).
15. A.C.Y. Chang, J.H. Nunberg, R.J. Kaufman, H.A. Erlich, R.T. Schimke and S.N. Cohen, *Nature,* **275,** 617 (1978).
16. S. Richardson, *Adv. Protein Chem.,* **34,** 167 (1981).
17. D.A. Matthews, S.L. Smith, D.P. Baccanari J.J. Burchall S.J Oatley and J. Kraut, *Biochemistry,* **25,** 4194 (1986).
18. J. Burchall, Wellcome Scientist Lecture (1983).
19. S.R.M. Bushby, in *Trimethoprim-sulfamethoxazole, Microbiological, Pharmacological and Clinical Considerations,* Ed. by M. Finland and E.H. Kass, pp. 10 – 30 (1973).
20. G.B. Elion, S. Singer and G.H. Hitchings, *Antibiot. Chemother.,* **10,** 556 (1960).
21. S. Farber et al., *N. Engl. J. Med.,* **238,** 787 (1948).
22. B. Roth, E. Bliss, and C.R. Beddell, in *Molecular Aspects of Anti-Cancer Drug Action,* Ed. by S. Neidle and M.J. Waring, London, Macmillan, pp. 363 – 393 (1983).
23. D.A. Matthews, J.T. Bolin, J.M. Burridge, D.J. Filman, K.W. Volz, B.T. Kaufman, C.R. Beddell, J.N. Champness, D.K. Stammers and J. Kraut, *J. Biol. Chem.,* **260,** 381 (1985).
24. J.E. Villafranca, E.H. Howell, D.H. Voet, M.S. Strobel, R.C. Ogden, J.N. Abelson and J. Kraut, *Science,* **222,** 782 (1983).
25. J.J. McGuire and J.K. Coward, in *Folates and Pterins,* Vol 1, Ed. by R.L. Blakley and S.J. Benkovic, New York, Wiley, pp. 135 – 190 (1984).
26. C.J. Allegra, J.C. Drake, J. Jolivet and B.A. Chabner, in *Proceedings of the Second Workshop on Folyl and Antifolyl Polyglutamates,* Ed. by I.D. Goldman, New York, Praeger, pp. 348 – 359 (1985).

27. C.R. Beddell, in *X-ray Crystallography and Drug Action,* Ed. by A.S. Horn and C.J. DeRanter, Oxford, Oxford Univ. Press, 169–193 (1984).
28. L.F. Kuyper, B. Roth, D.P. Baccanari, R. Ferone, C.R. Beddell, J.N. Champness, D.K. Stammers, J.G. Dann, F.E. Norrington and D.J. Baker, *J. Med. Chem.,* **28,** 303 (1985).
29. J.H. Chan, L.F. Kuyper and B. Roth, *189th Am. Chem. Soc. Meeting Abstracts* MEDI 87 (1985).
30. J.T. Lin and J.R. Bertino, *J. Clin. Oncol.,* **5,** 2032 (1987).
31. C.J. Allegra, B.A. Chabner, C.U. Tuazon, D. Ogata-Arakaki, B. Baird, J.C. Drake, J.T. Simmons, E.E. Lack, J.H. Shelhamer, F. Balis, R. Walker, J.A. Kovacs, H.C. Lane and H. Masur, *N. Engl. J. Med.,* **317,** 978 (1987).
32. F. Aruajo, D.R. Guptill and J.S. Remington, *J. Infect. Dis.,* **156,** 828 (1987).
33. D.J. Fernandes, J.R. Bertino and J.B. Hynes, *Cancer Res.,* **43,** 1117 (1983).
34. G.H. Hitchings, in *Discoveries,* Ed. by A. Burgen, Cambridge, Cambridge University Press, in press.
35. G.H. Hitchings, *Cancer Res.,* **29,** 1895 (1969).

Acknowledgment: I am greatly indebted to Mara Gabriel and Coyla Barry for skillful and energetic work that made completion of this paper possible.

1989
Physiology or Medicine

HAROLD E. VARMUS
J. MICHAEL BISHOP

"for their discovery of the cellular origin of retroviral oncogenes"

THE NOBEL PRIZE IN PHYSIOLOGY OR MEDICINE

Speech by Professor ERLING NORRBY of the Karolinska Institute.
Translation from the Swedish text.

Your Majesties, Your Royal Highnesses, Ladies and Gentlemen.

Our body is composed of independent living entities, which we call cells. The number of cells in a single individual is about 1 000 times larger than the number of all individuals on this planet. Still, all these cells can interact in a remarkably controlled fashion. The orchestration of this interaction is one of the great wonders of biology.

When we cut one of our fingers, wound-healing is initiated. By a marvellously controlled process of cell division, the skin and neighboring tissues at the wound are restored to their original condition. The discovery which is being honored with this year's Nobel Prize in Physiology or Medicine has given us completely new insights into the mechanisms that control cell growth and division. However, this discovery was not made by studying the balanced growth of normal cells, but through investigations of a virus which causes tumors in chickens.

In 1966, Peyton Rous was awarded the Nobel prize in Physiology or Medicine for his discovery 55 years earlier of a tumor-inducing virus, which was later named after him. It was known in the mid-1970s that the Rous virus has a separate part of its genetic material – a gene – which determines whether tumors occur. This gene is not required for virus replication. This year's prize-winners, Michael Bishop and Harold Varmus, and their collaborators managed to develop a molecular probe, which could selectively identify the tumor-inducing gene in the Rous virus. By use of this probe it was demonstrated that the critical gene was present in normal cells from all species. To their own great surprise and that of the scientific community, they had to draw the conclusion that the tumor-inducing gene in the Rous virus was of cellular origin. Does this mean that we are carrying cancer genes in our cells? Obviously not. However, in our cells there is a family of probably several hundred genes which are old in evolutionary terms and which control the normal growth and division of cells. Disruptions in the functioning of one or more of these genes can cause one cell to slip out of the network of growth control. The cell runs amok and a tumor may be the result.

As so often in medical research, we gained new insights into the normal functioning of a biological system through studies of the unbalanced state. The abnormal is the mirror image of the normal, or to quote the 19th century Swedish poet Erik Johan Stagnelius, "chaos is the neighbor of God". Since the family of growth-controlling genes, of which we have now identified more than 60, was demonstrated in tumor cells, they were given the somewhat illogical name *oncogenes.* This name derives from the Greek term onkos meaning tumor. Our knowledge of the proteins synthesized under the direction of oncogenes has given us broad insights into the complicated growth-controlling signal systems in cells. These signal chains include growth factors, receptors for these factors on the cell surface, substances which transmit signals from the cell surface into the genetic material in the cell nucleus, and finally, substances which directly influence the genetic material.

Cancer originates in disturbances in the genetic material of cells. Yet in most cases a single disturbance is not sufficient, but instead an accumulation of several critical injuries is required. This is the reason why cancers usually occur relatively late in life. Abnormally functioning oncogenes have now been demonstrated in many types of tumors in man. For the first time we are beginning to comprehend the complicated mechanisms behind the development of this group of diseases. New opportunities for diagnosis and treatment of various forms of cancer are now becoming available.

Michael Bishop and Harold Varmus,

Through your discovery of the cellular origin of retroviral oncogenes you set in motion an avalanche of research on factors that govern the normal growth of cells. This research has given us a new perspective on one of the most fundamental phenomena in biology and as a consequence also new insights into the complex group of diseases that we call cancer. On behalf of the Nobel Assembly of the Karolinska Institute I would like to convey to you our warmest congratulations and I now ask you to step forward to receive your Nobel Prizes from the hands of His Majesty the King.

Harold E Varmus

HAROLD E. VARMUS

I was born in the shadow of World War II, on December 18, 1939, on the south shore of Long Island, a product of the early twentieth century emigration of Eastern European Jewry to New York City and its environs. My father's father, Jacob Varmus, left a village of uncertain name near Warsaw just after the turn of the century to become a farmer in Newburgh, New York, and later a hatter in Newark, New Jersey. His wife, Eleanor, was a victim of the influenza epidemic of 1918, when my father was eleven. My mother's parents, Harry and Regina Barasch, came from farming villages around Linz, Austria, to found a children's clothing store, still in existence, in Freeport, New York. As children of immigrants, my parents both had notable educations, my father (Frank) at Harvard College (until financial considerations required him to withdraw after two years) and at Tufts Medical School, and my mother (Beatrice) at Wellesley College and the New York School of Social Work.

Three years before my birth, my parents settled in Freeport, my mother's home town, where my father established a general medical practice, while my mother commuted to a social services job in New York City. With the entry of the United States into the War, however, my father was assigned to an Air Force Hospital near Winter Park, Florida, and my first memories were to be of long beaches, and bass fishing on a lake with alligators. We remained in Florida, spared the pain of war, until early in 1946. In the interim, my only sibling, Ellen Jane, was born; she is now a genetic counselor and mother of three in Berkeley, California.

My growing-up in Freeport was undemanding and in many ways privileged. The public schools I attended were dominated by athletics and rarely inspiring intellectually, but I enjoyed a small circle of interesting friends, despite my ineptitude at team sports and my preference for reading. Life was enriched by frequent outings to Jones Beach State Park (where my father was the medical officer for many years), family skiing vacations to New England, and many outdoor adventures with the Boy Scouts and later the Putney Summer Work Camp.

The most decisive turn in my intellectual history came in the fall of 1957, when I entered Amherst College intending to prepare for medical school. The evident intensity and pleasure of academic life there challenged my presumptions about my future as a physician, and my course of study drifted from science to philosophy and finally to English literature. At the same time, I became active in politics and journalism, ultimately serving as the editor of the college newspaper. Following graduation from Amherst, a

Woodrow Wilson Fellowship enabled me to test the depth of my interest in literary scholarship by beginning graduate studies at Harvard University. Within the year, I again felt the lure of medicine and entered Columbia College of Physicians and Surgeons. Although I began medical school with strong interests in psychiatry and international health, I was influenced towards basic medical sciences by the lectures of (among others) Elvin Kabat, Harry Rose, Herbert Rosenkrantz, Erwin Chargaff and Paul Marks. My desires to practice medicine abroad were also tempered by an apprenticeship in a mission hospital in Bareilly, India.

In preparation for a career in academic medicine, I worked as a medical house officer at Columbia-Presbyterian Hospital from 1966 to 1968, and then joined Ira Pastan's laboratory at the National Institutes of Health as a Clinical Associate. This provided me with my first serious exposure to laboratory science and to the excitement of experimental success. Our studies of bacterial gene regulation by cyclic AMP (in collaboration with Bob Perlman and Benoit de Crombrugge) and the evening courses offered to incipient physician-scientists at NIH stimulated me to seek further post-doctoral training in molecular biology, specifically in tumor virology. This decision, combined with an interest in trying life in the San Francisco area, led me to Mike Bishop's door in 1969. I joined him as a post-doctoral fellow at UC San Francisco in 1970, was appointed Lecturer shortly thereafter, and in 1972 became a regular member of the faculty in the Department of Microbiology and Immunology (led initially by Ernest Jawetz, later by Leon Levintow), ascending to the rank of Professor by 1979.

Throughout the nearly two decades I have been associated with UCSF, most of my research interests have been focused upon the behavior of retroviruses: various aspects of their unusual life cycle, the nature and origin of their transforming genes, and their potential to cause genetic change. Much of this work has been performed in collaboration with Mike Bishop, particularly in the years before 1984 when we shared facilities, personnel, and funds. Other faculty interactions during the 1970's stimulated work on hemoglobinopathies (with Y.W. Kan) and on glucocorticoid action (with Gordon Tomkins and Keith Yamamoto). During the 1980's, I also worked extensively on hepatitis B viruses in collaboration with Don Ganem (who was initially a post-doctoral fellow and later a faculty colleague). My career at UCSF has been greatly enhanced by the extraordinary collegiality of the faculty, the excellence of our graduate and medical students, an unremitting stream of first-rate post-doctoral fellows, and the loyalty of our staff research associates, especially Suzanne Ortiz, Nancy Quintrell, and Jean Jackson.

In 1969 I married Constance Louise Casey, then a reporter for *Congressional Quarterly* in Washington, D.C., her home town, and now the Book Critic for the *San Jose Mercury News*. Shortly after we moved to California, my parents died, my mother of breast cancer in 1971, my father of coronary artery disease in 1972. Our lives have been made more interesting by the births of Jacob Carey in 1973 and Christopher Isaac in 1978; the boys attend public schools in San Francisco, root for the Giants, and are musically-inclined (Jacob, especially, is a talented trumpeter). California weather has promoted my love of outdoor sports, particularly bicycling, running, backpacking, skiing, and fishing, but I also maintain strong interests in the arts – literature, theatre, music, and film. We have lived almost continuously since 1971 in a Victorian house in the Haight-Ashbury district of San Francisco, with the exception of 1978–79, when I was a sabbatic visitor in Mike Fried's laboratory at the Imperial Cancer Research Fund in London, and 1988–89, when the award of a Nieman Fellowship to Connie brought her to Harvard and me to the laboratories of Bob Weinberg and David Baltimore at the Whitehead Institute.

Most of the significant honors I have received have been awarded jointly to Mike Bishop, with whom I also share the Nobel Prize. The earlier awards include California Scientist of the Year (1982), the Albert Lasker Basic Medical Research Award (1982), the Passano Foundation Award (1983), the Armand Hammer Cancer Prize (1984), the Alfred P. Sloan Prize from the General Motors Cancer Foundation (1984), the Gairdner Foundation International Award (1984), and the American College of Physicians Award (1987). In addition, I was elected to the National Academy of Science (1984) and the American Academy of Arts and Sciences (1988). I received an honorary degree from Amherst College (1985) and the Alumni Gold Medal from the College of Physicians and Surgeons (1989), and I have been the American Cancer Society Professor of Molecular Virology since 1984.

RETROVIRUSES AND ONCOGENES I

Nobel Lecture, December 8, 1989

by

HAROLD E. VARMUS

Departments of Microbiology and Immunology, and Biochemistry and Biophysics, University of California at San Francisco, San Francisco, U.S.A.

INTRODUCTION

The story that Mike Bishop and I will tell in these two lectures is one in which retroviruses, oncogenes, our personal histories, and the history of tumor virology are closely interwoven. It begins with some simple questions about the origin and behavior of viral genes and takes us to a vantage point from which we can survey many aspects of retroviruses and animal cells, including some of the aberrations that lead to cancer. We now know that retroviruses capture normal cellular genes and convert them to cancer-causing genes, called oncogenes. Such transductions are rare, but depend upon the normal events of an intricate virus life cycle. Retroviruses have introduced us to more than forty cellular genes with the potential to become oncogenes, some discovered as components of viral genomes, others as genetic targets for viral insertion mutations.

It has been our privilege to participate in a generous share of the experiments that established these principles. But we have required the help of many talented people in our laboratories at UCSF, as well as the collaboration and friendly competition of others elsewhere. (I mention as many names as the narrative can bear, but inevitably I must apologize to valued colleagues who remain anonymous here.) Several viruses also figure in our tale, but Rous sarcoma virus again has the leading role, yet another of many tributes to the pioneering work of Peyton Rous and to the principle of delayed gratification in science. The product of his diligence in pursuing a single chicken tumor nearly eighty years ago (1), Rous' virus remains the only retrovirus that could have satisfied the genetic and biochemical criteria for the work we accomplished in the era that preceded molecular cloning.

FIRST TASTE OF MOLECUALR BIOLOGY: HYBRIDIZATIONS WITH THE LAC OPERON

My commitment to experimental science occurred, by today's standards, dangerously late in a prolonged adolescence. As an undergraduate at Amherst College, I was devoted to Dickensian novels and anti-Establishment journalism, while marginally fulfilling premedical requirements. I then indulged myself with a year of Anglo-Saxon and metaphysical poetry at Harvard graduate school, before beginning medical studies at Columbia

University, with a primary interest in psychiatry. But my ambitions soon turned towards an academic career in internal medicine. So just after graduation in 1966, like many of my contemporaries, I applied for research training at the National Institutes of Health. Perhaps because his wife was a poet, Ira Pastan agreed to take me into his laboratory, despite my lack of scientific credentials.

At the time, Ira was studying the biochemical effects of thyroid stimulating hormone on tissue slices, a subject close enough to clinical endocrinology not to be intimidating. But one day, while still an intern at Columbia-Presbyterian Hospital, I received a telephone call from Ira, telling me that a lecture by Earl Sutherland had inspired him to begin work on the effects of cyclic AMP on regulation of the *lac* operon in *E.coli.* Late that night, alone in the house staff library, I peered for the first time into the *Journal of Molecular Biology* – it is no small tribute to Columbia that this journal was there – and attempted to read the seminal papers on the *lac* operon by Jacob and Monod (2). I knew then that, one way or another, my life was about to change.

Science is largely the making of measurements, and I soon learned from Ira how much more important a new measurement could be than an old theory. He and Bob Perlman had just discovered that cyclic AMP reversed catabolite repression of the *lac* operon (3). They suggested that I use the relatively new technique of molecular hybridization to ask whether regulation by cyclic AMP occurs at the transcriptional level. Apart from the pleasure of just getting results (as Gunther Stent has said, results are wonderful because they give us something to talk about (4), these measurements had enormous intellectual appeal, because they very simply resolved the ambiguity of hypothesis, demonstrating unequivocal changes in synthesis of *lac* messenger RNA (5). Furthermore, they were carried out with technical subtleties that ultimately shaped the way I later thought about the problems of detecting single genes in more complex, eukaryotic cells. We annealed radiolabeled *E.coli* RNA to filter-bound DNA from a pair of bacteriophages that differed only by the presence or absence of the *lac* operon; and we minimized irrelevant hybridization by including, as competitor, unlabeled RNA from an *E.coli* mutant from which the lac operon was deleted. An aesthetic merger of genetics with molecular biology, itself as pleasing as the results!

INTRODUCTION TO THE PROVIRUS AND THE VIROGENE-ONCOGENE HYPOTHESES

A major feature of life at the NIH in late 1960's was the extraordinary offering of evening courses for physicians attempting to become sciantists as they neared thirty. Two classes had direct and specific effects on my subsequent work because they introduced me to important problems I believed approachable with the methods I had acquired in my brief apprenticeship.

Like many of my peers, I was excited by the prospect of applying reductionist methods to eukaryotic organisms, particularly in a way that might be informative about human disease. From some dilatory reading in the early 1960's, I knew enough about viruses and their association with tumors in animals to understand that they might provide a relatively simple entry into a problem as complex as cancer. In fact, for anyone interested in the genetic basis of cancer, viruses seemed to be the only game in town. What surprised and beckoned me were two rather simple but heretical hypotheses that described curious ways the genes of RNA tumor viruses might mingle with the chromosomes of host cells.

The more daring of these two hypotheses was the provirus hypothesis, first enunciated by Howard Temin (6). (John Bader, one of our NIH lecturers, was among the few others to espouse it in public (7).) The provirus hypothesis stated that the genes of RNA tumor viruses were copied into DNA, which became stably associated with the host cell; the proviral DNA then provided the information for production of new virus particles. With its existence supported principally — some said feebly — by studies with inhibitors of DNA and RNA synthesis, and its plausibility doubted in the absence of any precedent for information transfer from RNA to DNA, the provirus seemed to be a provocative target for a definitive decision with molecular hybridization.

The other hypothesis, the virogene-oncogene hypothesis, was more complex (8). George Todaro and Robert Huebner proposed that normal cells must contain genes related to those found in RNA tumor viruses, since viral proteins could often be found in cells of apparantly uninfected animals, especially chickens and mice. Such genes, known as virogenes, were believed to be transmitted vertically as components of chromosomes, expressed in response to a variety of agents, and acquired by infection of germ cells at some time in the past. Since some RNA tumor viruses were known to be highly oncogenic, it was also proposed that tumor-inducing genes of such viruses (viral oncogenes) might also be transmitted through the germ line as a consequence of ancient infection. Activation of these endogenous viral oncogenes by substances we recognize as carcinogens — chemicals, radiation, other viruses — could serve to initiate a neoplastic process.

TRANSITION TO RNA TUMOR VIROLOGY IN REVOLUTIONARY TIMES

During the summer of 1969, I combined a backpacking vacation in California with a search for a suitable place to study tumor viruses. Acting on a tip from Harry Rubin in Berkeley, I sought out a small group, composed of Mike Bishop, Leon Levintow, and Warren Levinson, that was beginning to work with Rous sarcoma virus at UC San Francisco. (Rubin, it should be said, was more eager for me to meet Peter Duesberg, but Duesberg was out of town.) A brief conversation with Mike was sufficient to convince me of our intellectual compatibility (happily, one of the few convictions to have survived twenty years in this field), and I made plans to join the UCSF group as a post-doctoral fellow the following summer.

Before the intervening year had passed, however, two major discoveries changed the landscape of tumor virology. Satoshi Mizutani and Temin (9) and David Baltimore (10) found the predicted enzyme, reverse transcriptase, in virus particles, thereby erasing most of the skepticism about the provirus hypothesis by providing a means to synthesize the heretical DNA copy of an RNA genome. And Steve Martin isolated a crucial mutant of Rous sarcoma virus (11), one that lost its ability to transform cells at elevated temperature and regained it when the temperature was reduced. Martin's mutant offered the first clear definition of the gene we later called *src,* and it showed that the gene – and, by implication, a protein the gene encoded – was required to instigate and sustain the transformed state. Since the mutant virus grew normally at the temperature that blocked transformation, oncogenic and replicative functions could be dissociated, a facet of the story that will soon resurface.

FIRST FORAYS WITH RSV: SEEKING PROVIRAL DNA

Reverse transcriptase was properly greeted as strong evidence for the provirus hypothesis, and defused the urgency of challenging it. Yet it still seemed important to detect the provirus directly, most obviously by molecular hybridization, and to follow the pathway of its synthesis, especially in infected cells, not just in vitro. Reverse transcriptase obligingly offered a means to simplify the job, through the synthesis of potentially powerful probes, radioactive virus-specific DNA copied from a template of viral RNA.

Some of my initial efforts to proceed with these problems look, in retrospect, frustrating, if not quixotic – though the results were published in prominent journals. We chose at first to use double-stranded products of the RSV reverse transcriptase as hybridization probes, in order to measure gene copies through the accelerating effects of cellular DNA on reassociation kinetics (12). However, the RNA template was unevenly copied by reverse transcriptase in vitro (13), so that the products had complicated reannealing kinetics and did not uniformly represent the viral genome (whose genetic composition was in any case still unknown.) When I attempted to measure RSV-related DNA in the most obvious settings – uninfected and RSV-infected chicken cells – I found multiple copies of virus-related DNA in the normal cells, in apparent confirmation of at least some aspects of the virogene-oncogene hypothesis (14). But I was unable to detect the anticipated increment of RSV DNA in infected chicken cells, until I switched to the use of single-stranded DNA as probes (15). But by then Paul Neiman had already measured the increment by hybridization with radiolabeled RNA from virions (16). And, in the meantime, Hill and Hillova had provided more dramatic support for the provirus hypothesis in an entirely different way, by DNA transfection: addition of DNA from RSV-infected cells to new cells allowed the recovery of the original virus (17). So the entire viral genome must have been present in the DNA of infected cells.

Our approach to the provirus was eased when I abandoned chicken cells, the traditional hosts for RSV in culture, in favor of cells from other birds — ducks and quail — and from mammals (18). Because we could detect little virus-related DNA in these cells prior to infection, it was relatively simple to measure new copies of RSV DNA following infection, to follow the time course of DNA synthesis, to show that reverse transcription occurred in the cytoplasm, and to define linear, circular, and integrated forms of viral DNA (19).

MAKING A PROBE TO TEST THE VIROGENE-ONCOGENE HYPOTHESIS

The varied abilities of normal avian DNAs to anneal to RSV-derived probes helped to focus our attention upon the sorts of virus-related sequence we could detect in chicken DNA. Did these sequences constitute genes for viral structural proteins? More importantly, did they include the viral transforming gene, as predicted by the oncogene-virogene hypothesis? To approach these questions it was imperative to have more rigorously defined probes. This was not a trivial challenge in the early 1970's, before restriction mapping and molecular cloning were available to us.

But one potent reagent was available. In 1971, Peter Vogt reported the isolation of transformation-defective, replication-competent mutants of RSV (20). The genomic RNA subunits of these "td" mutants were shown by Duesberg and his colleagues to be about 15 percent shorter than the subunits of wildtype virus (21). The provisional interpretation was that the missing sequence (initially called "x" and later "sarc") included some or all of the viral transforming gene (v-*src*) earlier defined by temperature-sensitive mutants. Like Martin's ts mutants, the deletion mutants retained the functions required for replication, despite the extensive loss of sequence, so it was tempting to presume that the deletion was coextensive, or nearly coextensive, with the transforming gene.

Mike and I were intimately acquainted with these conjectures through a collaborative consortium of Californian laboratories, directed by Vogt, Duesberg, and us, which met every six weeks or so, in Los Angeles or the Bay Area. Through these discussions, we recognized that if we could prepare radioactive DNA specific for the sequences deleted in the td-RSV mutants, we would have a reagent that would approximate a specific probe for the transforming gene of RSV.

The strategy for doing this was straightforward in principal, but difficult in practice (Fig. 1A). In essence, single-stranded, radiolabeled DNA fragments were synthesized from a template of wild-type RSV RNA, then hybridized to td-RSV RNA to remove unwanted components by hydroxylapatite chromatography, leaving the sarc-specific DNA. Ramareddy Guntaka first put this protocol into motion with some encouraging results. But it was ultimately the ministrations of Dominique Stehelin that produced a sarc probe that met rigorous standards: nearly complete annealing to RSV RNA, no significant annealing to td-RSV RNA (Fig. 1B), and representation of

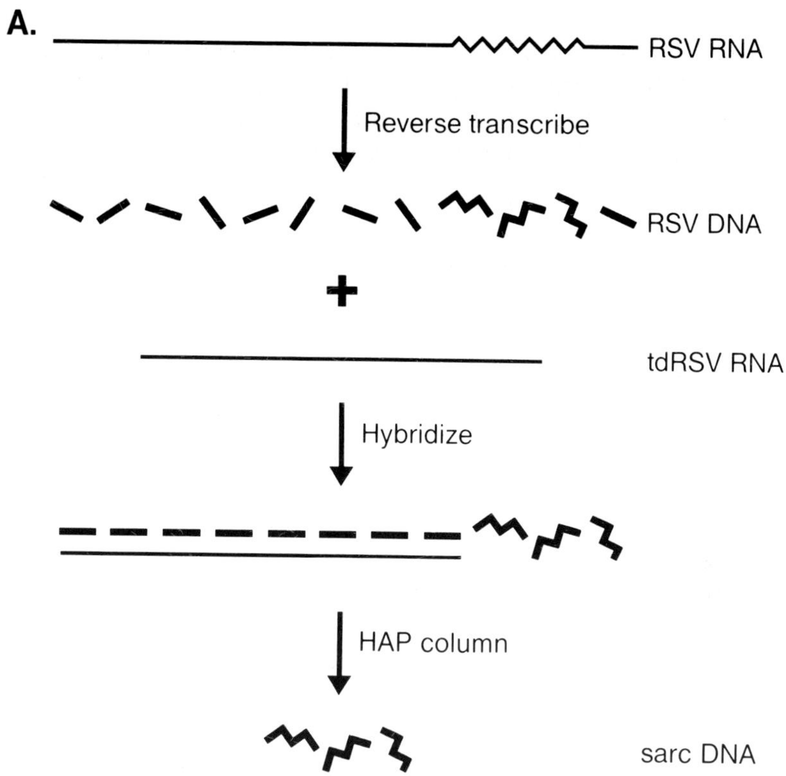

Figure 1. Schematic summary of initial experiments with sarc probe (see refs. 22 and 23 for primary data and further details). Panel A: Radiolabeled sarc-specific DNA was prepared by subtractive hybridization of the products of reverse transcription of RSV RNA to RNA from a transformation-defective deletion mutant of RSV (td RSV). The basis for the strategy is described in the text. Thin lines represent RNA, thick lines represent DNA, jagged portions represent sarc sequences (i.e. those present in RSV but not in td RSV genomes). HAP, hydroxylapatite. Panel B (next page): sarc DNA is specific for sequences that differentiate RSV and td RSV. The probe was hybridized to RSV RNA and td RSV RNA and results monitored by HAP chromatography. Panel C (next page): sarc probe (solid curves) anneals to DNA from many species of birds, whereas probe for other components of RSV genome (td RSV probe, dashed curves) anneals poorly to DNA from species other than chicken. The extent of annealing (normalized values shown here) was determined by HAP chromatography.

over 10 percent of the RSV genome, most of the sarc region, in the probe (22).

SARC PROBE DETECTS CONSERVED SEQUENCES IN AVIAN DNA

When Stehelin incubated sarc probe with normal chicken DNA, it annealed extensively (as, of course, did probe made from td-RSV RNA) (Fig. 1C). The results, unambiguously exciting, were still fully consistent with the original oncogene-virogene hypothesis. So we were yet more excited when the next results seemed to violate it: although the "virogene" probe from td-RSV annealed poorly to DNAs from several other avian species, the sarc probe annealed extensively, even to DNA from the Australian emu — a rattite (we

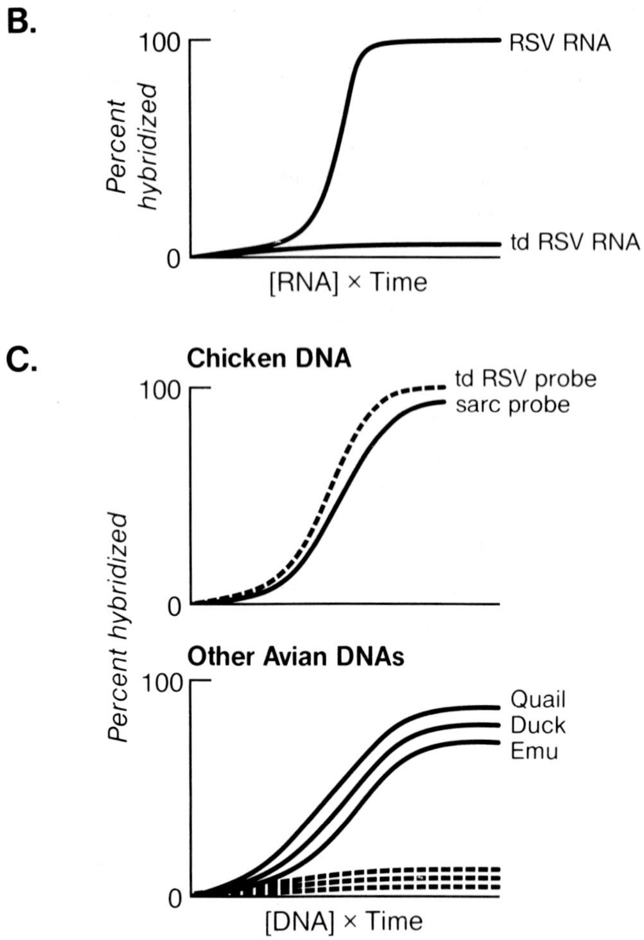

learned, from our Berkeley colleague, Allan Wilson) at a great evolutionary distance from chickens (23). The extent and fidelity of the hybrids formed with sarc probe indicated that its homologs in normal cells had diverged during avian evolution at a rate similar to that of cellular genes used in the few earlier forays into molecular evolution, suggesting that the sequences had been conserved for at least 100 million years.

From these findings, we drew conclusions that seem even bolder in retrospect, knowing they are correct, than they did at the time (23). We said that the RSV transforming gene is indeed represented in normal cellular DNA, but not in the form proposed by the virogene-oncogene hypothesis. Instead, we argued, the cellular homolog is a normal cellular gene, which was introduced into a retroviral genome in slightly altered form during the genesis of RSV. Far from being a noxious element lying in wait for a carcinogenic signal, the progenitor of the viral oncogene appeared to have a function valued by organisms, as implied by its conservation during evolution. Since the viral *src* gene allows RSV to induce tumors, we speculated that its cellular homolog normally influenced those processes gone awry in tumorigenesis, control of cell growth or development.

FIRMING UP THE RESULT: SARC REPRESENTS THE C-SRC PROTO-ONCOGENE

Despite the broad claims, the first round of experiments with sarc probe left many worrisome questions unanswered.

The most pressing question, and one foremost in the minds of our critics, seems now both essential and mundane: Was the sarc probe actually detecting a functional, protein-encoding homolog of the viral transforming gene (v-*src*)? or were the still ill-defined genetic and physical maps of the RSV genome leading us astray? Some support came from geneticists who mapped a large number of the existing transformation mutations of RSV within the region of the viral genome lost during formation of td-RSV deletion mutants (24). More exciting and stronger support came from protein biochemists: Joan Brugge and Ray Erikson discovered that the long-sought product of v-*src* was a protein of about 60,000 daltons (25), one that would require about 1600 nucleotides of coding sequence and could account for most of what was missing from td-RSV. Hermann Oppermann and others (26) then detected a protein in normal cells that seemed virtually indistinguishable from v-*src* protein, confirming the idea that sarc probe was measuring a gene (now called c-*src*) that resembled v-*src*. Ultimately, the molecular cloning and nucleotide sequencing of the RSV genome revealed how fortunate we hade been in the design of our probe (27): Most td-RSV mutants lack all of v-*src* and little else.

The second question was more subtle: Did the conservation of c-*src* during avian speciation accurately imply that it was a cellular gene? or might it still represent an inherited viral gene more conserved than other viral elements? Answers came from several quarters, all confirming the arguments based on evolution. Using chicken chromosomes fractioned according to size by Elton Stubblefield in Texas, we found that c-*src* and virogenes are unlinked; the viral genes we could detect were on one or more large chromosomes, but c-*src* was on a small chromosome (28). Steve Hughes then used restriction enzymes to gauge the diversity of sequences in and around viral genes and c-*src* in many individual chickens (29); the pattern generated with sarc probe was monotonous, as would be expected for a conserved cellular gene (and shown to be the case for genes such as globin, ovalbumin, and others). The pattern produced with a probe for viral structural genes, however, suggested variety in number and context, as though they had been introduced into the chicken genome by recent, independent germ line infections. When we examined the transcripts emanating from c-*src* and from virus-related genes, individual chicken embryos contained various amounts and types of viral RNAs but similar quantities of a single, differently-sized species of c-*src* RNA (30). The most powerful evidence for the cellular nature of c-*src* required molecular cloning. For then it was possible to show that the coding sequences of c-*src* were interrupted in many places by introns (31), in the manner recently discovered to be characteristic of cellular genes. In contrast, as described in greater detail below, endogenous virogenes have the insignia of proviruses, being composed of continuous coding domains, flanked by repeated sequences.

The third question was most informative about the mechanism by which v-*src* causes cancer: What accounted for the proposed physological differences between a beneficial proto-oncogene and a pathogenic viral oncogene derived from it? From the first measurements of *src* gene expression, it was apparent that the viral gene, controlled by a potent viral transcriptional promoter, was expressed much more vigorously than its cellular counterpart (32). But the levels of v-*src* protein required for transformation proved to be lower than non-oncogenic amounts of c-*src* protein (33), implying qualitative differences as well. Mark Collett in Erikson's laboratory (34) and Art Levinson in ours (35) had discovered that *src* proteins are protein kinases, which Tony Hunter and Bart Sefton later showed to be specific for tyrosine residues (36). Saburo Hanafusa's laboratory then defined the subtle structural and physiological differences between the viral and cellular versions of the gene (37): at least three of several aminoacid differences between $p60^{v\text{-}src}$ and $p60^{c\text{-}src}$ enhance the protein-tyrosine kinase activity of the transforming protein. Thus, quantitative and qualitative factors conspire to produce the *src* oncogene.

Finally, how well conserved is the cellular *src* gene? Early on, Deborah Spector showed that under conditions of reduced stringency most or all the sarc probe could anneal to the genomes of all vertebrates, not just birds (38). Since the implicated mammals included man, these findings helped to create a larger audience for our work, and they raised the possibility that retroviral proto-oncogenes might have a role on human cancer. New technologies ultimately extended the list of organisms that carry c-*src* to include virtually all metazoans – insects (39), worms (40) sponges (41), and hydras (42) – reminders of our evolutionary origins that are at once exhilarating and sobering.

The *src* story remains unfinished. We cannot tell you how c-*src* benefits normal organisms or cells, although recent work implicates c-*src* in both development, especially in the central nervous system (43), and in growth control during mitosis (44). We do not know the physiological targets for the *src* kinase, although numerous phosphotyrosine-containing proteins have been identified (45). And we do not know how the enzymatic activity of p60 is regulated, although phosphorylation is important (46). Nevertheless, the *src* paradigm has stimulated our field to move in several directions: to identify many new viral oncogenes and their cellular progenitors (47), to characterize a stunning variety of oncogenic proteins (48), to make unexpected connections with elements of growth regulatory networks (49), and to describe mutant proto-oncogenes in human tumors (50). These developments are recounted in the accompanying lecture by Mike Bishop. It is my mission to stay with the virus – and especially the provirus.

DECIPHERING PROVIRAL STRUCTURE

By the early 1970's the provirus was a well-accepted idea, but the organization of viral DNA and its position within chromosomes were still matters of conjecture. Several pecularities of viral RNA and the viral life cycle hinted that proviral DNA must have special attributes (19). First, the priming site for the first strand of viral DNA was near the 5' rather than at the 3' end of viral RNA (51), implying that synthesis must be a complex process and that the provirus must not be a simple copy of viral RNA. Next, a short sequence (R) was found at both ends of viral RNA, and hence present in two copies, but appeared to be copied only once during synthesis of viral DNA (52); how was the second copy of R regenerated? Finally, it was difficult to account for the efficient synthesis of viral RNA without the prospect of a strong transcriptional promoter upstream of the start site; how was that promoter provided?

These problems were solved by the unexpectedly elegant configuration of viral DNA, as worked out mainly by Peter Shank, Steve Hughes, and Hsing-Jien Kung in our group (53) and independently by John Taylor's laboratory in Philadelphia (54). Once again, RSV was the instrument of discovery, and again the results depended upon hybridization with specific probes, this time for terminal regions of the viral genome. In essence, viral genes were found to be flanked in the provirus by long terminal repeats (LTRs) derived from sequences present at both ends of viral RNA (Fig. 2). (The ends of the LTRs correspond to the priming sites for the two DNA strands and thereby helped unravel a strategy of DNA synthesis too convoluted to review here (19).) Because the R sequence is present once in each LTR, it can be reconstituted by transcribing parts of both LTRs. And viral

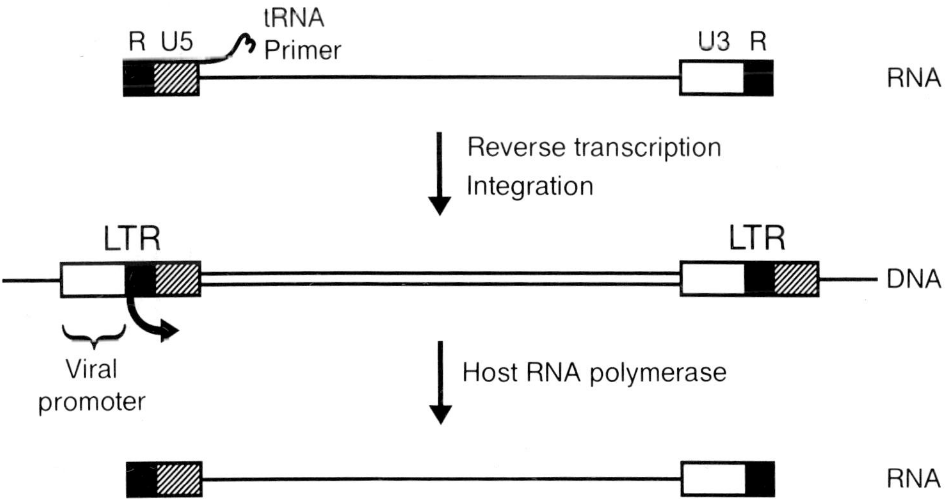

Figure 2. The organization of proviral DNA in comparison to retroviral RNA. The top line shows one subunit of a viral dimeric genome, with host tRNA positioned near the 5' end where it serves to prime synthesis of the first strand of viral DNA. R, short sequence present at both ends of viral RNA; U5 and U3 are sequences unique to the 5' and 3' regions of the RNA that are duplicated during DNA synthesis to form the long terminal repeats (LTRs). The middle line shows a provirus integrated into host cell DNA (single line). The viral coding sequences reside between the LTRs (double line). The region encompassing the viral promoter in U3 of the upstream LTRs is bracketed; the curved arrow denotes the start site and direction of transcription of the provirus by host RNA polymerase. The bottom line shows the composition of the primary viral transcript after 3' processing; the poly(A) tract at the 3' end is not illustrated.

sequences that contain strong transcriptional signals reside upstream of the RNA start site. Mapping of integration sites showed that many regions of the host genome could accommodate a provirus; thus, transcriptional self-sufficiency of the provirus allowed it to function in varied chromosomal contexts.

PROVIRUSES AS MOBILE ELEMENTS THAT CAUSE INSERTION MUTATIONS

But the structure of the provirus did more than solve some perplexities of the retrovirus life cycle. In general form and even in selected short sequences, proviruses resemble an abundant type of mobile DNA element (Fig. 3), described by now in plants, bacteria, yeast, insects, and many other organisms (55), another arresting example of conservation throughout evolution. Their connection with retroviruses has been strengthened in recent years by discoveries that several such elements are duplicated and relocated by using reverse transcriptases to make new DNA copies from RNA transcripts (56), although they never produce extracellular viruses. These properties also apply to most endogenous proviruses, cloned from the germ lines of many vertebrates (57), reemphasizing the profound differences between inherited virogenes and cellular proto-oncogenes.

One practical consequence of the startling similarities between retroviral proviruses and mobile elements was to consider the possibility that proviruses, like mobile DNA, might cause insertion mutations. In 1978, while

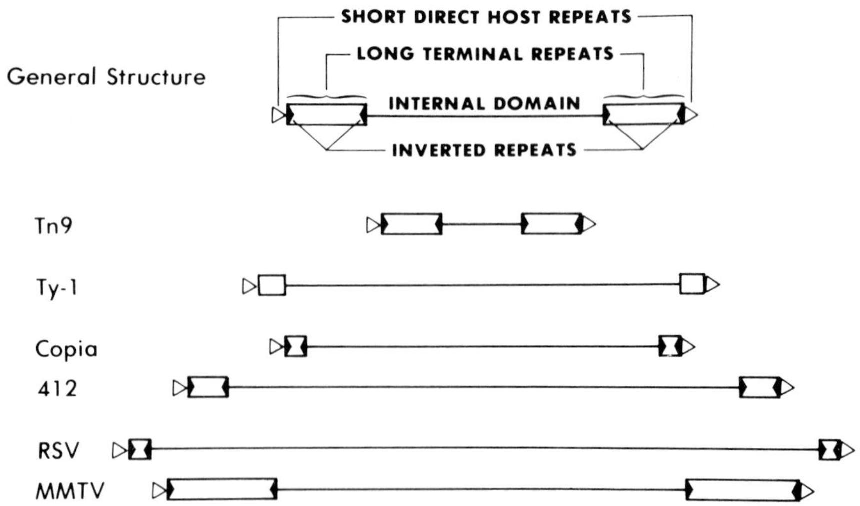

Figure 3. Many mobile DNAs are organized like proviruses. The figure demonstrates some common features of many transposable elements, including LTRs (rectangular boxes), inverted repeats within the LTRs (closed triangles), and short duplications in host DNA generated during insertion (open triangles). The illustrated mobile elements included retroviral proviruses (RSV and mouse mammary tumor virus MMTV), retrotransposons of Drosophila (copia and 412) and budding yeast (Ty1), and a conventional transposon of E.coli (Tn9). (Reprinted with permission of Academic Press; see ref. 55).

on sabbatical in Mike Fried's laboratory at the Imperial Cancer Research Fund, I designed an experiment to test this idea. John Wyke provided me with a rat cell line transformed by a single RSV provirus, which could serve as a target for insertion mutation by proviruses introduced by superinfection with mouse leukemia virus (MLV). By sifting through many clones of cells that had lost their transformed properties after infection with MLV, Suzanne Ortiz and I found two that contained an MLV provirus inserted at different sites within the pre-existing RSV provirus, interfering with the expression of v-*src* (58).

This experiment established the principle that retroviruses could serve as insertional mutagens to inactivate genes. It also had the heuristic benefit of stimulating us to think about insertion mutations that acted in a dominant fashion by activating gene expression. Greg Payne was then attempting to explain how avian leukosis virus (ALV), a virus virtually indistinguishable from td-RSV and lacking any evidence of a viral oncogene, could nevertheless induce tumors (most commonly B-cell lymphomas) within several weeks after infection of susceptible chickens (59). Might ALV proviruses occasionally integrate adjacent to a cellular proto-oncogene and augment expression through a viral LTR? Greg's evidence for this idea (60), however provocative, was nearly drowned out by the commotion caused by Hayward, Neel, and Astrin's discovery (61) that ALV DNA in B cell lymphomas was adjacent to c-*myc* — a known progenitor of a retroviral oncogene (62) — and that the viral LTR was driving c-*myc* expression.

ALV-induced tumors taught us several new principles: Retroviruses can induce neoplasia by insertionally activating proto-oncogenes (63); proviruses and their target genes can be variously arranged with similar effects on transcription (64); and proto-oncogenes do not need to be transduced to participate in oncogenesis. The last was an especially important point that presaged the later outpouring of mutant proto-oncogenes in human tumors unassociated with any virus (50).

USING PROVIRUSES AS TRANSPOSON TAGS FOR NOVEL PROTO-ONCOGENES: THE INT-1 STORY

However important, ALV has failed to introduce us to any proto-oncogenes not already known as forefathers of retroviral oncogenes. For this, we made use of another retrovirus without a viral oncogene, the mouse mammary tumor virus (MMTV). Like RSV, MMTV has a venerable history (65). Found in the milk of inbred mice with a high incidence of mammary cancer over fifty years ago in Holland (66) and at the Jackson Laboratories in Maine (67), MMTV was the first mammalian retrovirus to be discovered; it remains the only efficient viral agent of mammary carcinoma, and thus a model for one of the most common of human cancers.

MMTV-induced mammary tumors are quasi-clonal growths of virus-infected cells (68). To ask whether the tumor cells result from insertion of viral DNA near a heretofore unknown proto-oncogene, Roel Nusse examined many tumors to find one with only a single new provirus; he then cloned that provirus and its flanking cellular DNA in *E.coli*. An unfamiliar gene, which we called *int-1*, was nearby, and it was expressed in that tumor and several others with nearby insertions, but not in normal mammary glands (69).

But this was not sufficient to implicate *int-1* as a oncogene. First there was the circumstantial force of repetition: over three-quarters of mammary tumors in the C3H mouse strain harbor insertion mutations in the *int-1* locus. Then Tony Brown did what nature had not done, by placing the *int-1* gene within a retroviral genome; the resulting virus alters the growth and morphology of cultured mammary cells (70). Finally, Ann Tsukamoto followed a strategy pioneered by Ralph Brinster and Richard Palmiter and by Philip Leder (71) and introduced the *int-1* gene, linked to an MMTV LTR, into the mouse germ line (72). All the transgenic mice, male or female, experience dramatic hyperplasia of the mammary epithelium, and most of the females develop mammary carcinoma within six months. This is about as close as we can come to fulfilling Koch's postulates for a genetic disease: by placing the virally-mutated form of the gene into the germ line – ironically, much as envisioned to occur naturally in the virogene-oncogene hypothesis – we have recreated the disease.

I cannot leave our transgenic mice without making a more general point. In California and many other places, misguided efforts to abolish the use of laboratory animals seriously threaten medical science. If Peyton Rous had been denied his chickens, our field would have no past; if all of us are now denied mice and other animals, it will have little future.

A TENTATIVE SCHEME FOR TRANSDUCTION OF PROTO-ONCOGENES

int-1 is but the first entry on a now substantial list of proto-oncogenes discovered as loci repeatedly activated by proviruses in tumors (63). Thus, retroviruses usher in the genetic cast in the drama of cancer in two ways: by transduction and insertion mutation. Not surprisingly, the two phenomena appear to be mechanistically related: insertion mutation is probably the first step in the sequence of events that occasionally spawns a viral oncogene as its end product. What we can predict, but not yet fully substantiate by direct observations, is that two recombination events are required for transduction (Fig. 4; 73). The first occurs during proviral integration, placing viral DNA upstream from the activated cellular gene that will be acquired. The second occurs during virus replication in the tumor that results from the insertion mutation; the second step joins viral sequences to cellular sequences derived from the downstream region of the gene. We suppose that more or less in this fashion a close relative of ALV acquired a slightly mutated version of a chicken's *src* gene nearly a century ago, and set us on a path we are still travelling.

THE PROSPECTS FOR RETROVIROLOGY

The story thus far confirms David Baltimore's statement of thanksgiving

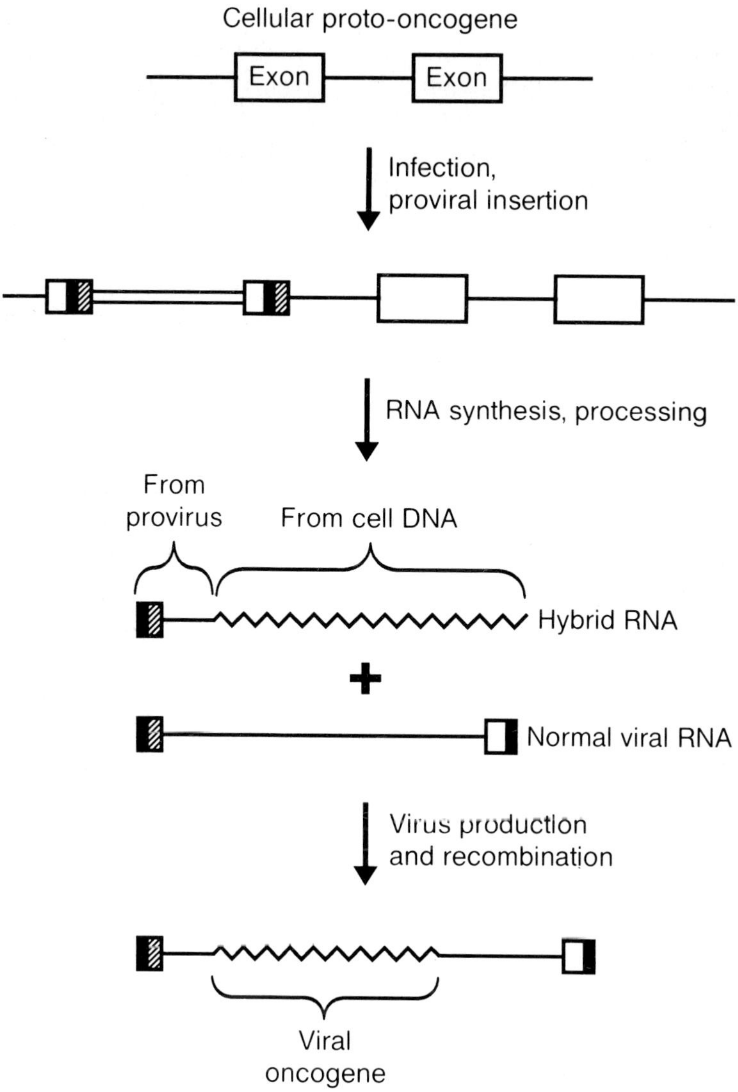

Figure 4. Model for transduction of cellular proto-oncogenes to form retroviral oncogenes. Exons of a proto-oncogene are located downstream of a retroviral provirus recently introduced by infection and denoted as in Fig. 2. Virus-host chimeric RNA, a product of the proviral insertion mutation, recombines with normal viral RNA during virus replication to join viral sequences downstream of the cellular sequences. For further details, see Ref. 73.

(74): "a virologist is among the luckiest of biologists because he can see into his chosen pet down to the details of all its molecules." Because retroviruses, our chosen pets, are such remarkable agents, it has been enough to train our sights on two brief questions – how do retroviruses grow? how do retroviruses cause cancer? – to have extended our concerns outward to the cellular host, as well as to have focused them inward upon the viruses themselves (75). As a result, we have entered into some of the liveliest arenas in modern biology: the genetic basis of cancer, the transposition of DNA through RNA intermediates, the control of gene expression in eukaryotes, and the molecular evidence for evolution.

At this point, the study of oncogenes and proto-oncogenes has attained a degree of maturity that allows it to be conducted with astonishing little virology. Yet retroviruses remain vital tools for the isolation of important new oncogenes; witness in the past few years the discoveries of the *jun* and *crk* genes (76). Likewise, since the discovery of reverse transcriptase nearly two decades ago, seemingly exhaustive attention has been given to the life cycle of retroviruses (19), yet many central features are just now coming into view (75). Cell surface receptors for viral attachment and entry have been recently identified and show a remarkable range of biochemical properties (77); the proviral integration reaction has been recapitulated in vitro with nucleoprotein complexes (78), allowing a description of integrative precursors and intermediates (79); retroviruses have been recognized as pliable genetic vectors (80) that may one day be used clinically to correct gene deficiencies, in the manner used in nature to transport host-derived oncogenes; many unexpected aspects of viral gene expression have been discovered, including translational frameshifting during the synthesis of reverse transcriptase (81) and complex viral regulatory genes that govern the behavior of two classes of human retroviruses (82); and the principles of virus assembly are emerging through physical and genetic assaults on viral structural proteins and proteases (83). These inherently fascinating problems have now taken on a special urgency, because we are all threatened by the world-wide dissemination of a lethal human retrovirus, the human immunodeficiency virus (84). Thus retroviruses continue to challenge our intellects in ways that may help us grapple with major diseases, cancer and now AIDS, while also revealing fundamental features of the lives of our cells.

ACKNOWLEDGEMENTS

My indebtedness is large: to the many coworkers and colleagues mentioned here, and to many more unnamed, for communal labor, ideas, and criticism; to several institutions (especially the National Institutes of Health, the American Cancer Society, and the University of California, San Francisco) for nurturing my career with financial support and other resources; and to my family, for affectionate tolerance. My approach to science has been most profoundly influenced by Ira Pastan, who showed me that laboratory life was preferable to the clinic; by Peter Vogt, who taught me viral genetics when I needed it; and by Mike Bishop, who has been a generous and challenging colleague for nearly two decades. My gratitude would be incomplete if I did not especially mention the friendship and assistance of Suzanne Ortiz, Nancy Quintrell, Jean Jackson, Leon Levintow, Warren Levinson, and Don Ganem.

REFERENCES

1. P. Rous, *J. Exp. Med.*, **132,** 397 (1911); *Science* **157,** 24 (1967).
2. F. Jacob, J. Monod, *J. Mol. Biol.* **3,** 318 (1961); F. Jacob, A. Ullmann, J. Monod, *ibid* **13,** 704 (1965).
3. R. L. Perlman, B. DeCrombrugghe, I. Pastan, *Nature* **223,** 810 (1969); R. Perlman, B. Chen, B. DeCrombrugghe, M. Emmer, M. Gottesman, H. Varmus, I. Pastan, in *Cold Spring Harbor Symp. Quant. Biol.* **35,** pp. 419–423 (1970).
4. G. Stent in *A Passion for Science,* L. Wolpert, A. Richards, Oxford University Press, pp. 109–119 (1988).
5. H. E. Varmus, R. L. Perlman, I. Pastan, *J. Biol. Chem.* **245,** 2259 (1970); H. E. Varmus, R. L. Perlman, I. Pastan, *J. Biol. Chem.* **245,** 6366 (1970).
6. H. M. Temin, *Nat. Cancer Inst. Monograph* **17,** 557 (1964); H. M. Temin, *Science* **192,** 1075 (1976).
7. J. P. Bader, *Virology* **26,** 253 (1965).
8. R. J. Huebner, G. J. Todaro, *Proc. Natl. Acad. Sci.* **64,** 1087 (1969).
9. H. M. Temin, S. Mizutani, *Nature* **226,** 1211, (1970).
10. D. Baltimore, *Nature* **226,** 1209 (1970).
11. G. S. Martin, *Nature* **227,** 1021 (1970).
12. L. Gelb, D. Kohne, M. Martin, *J. Mol. Biol.* **57,** 129 (1971).
13. H. E. Varmus, W. E. Levinson, J. M. Bishop, *Nat. New Biology* **223,** 19 (1971).
14. H. E. Varmus, R. A. Weiss, R. Friis, W. E. Levinson, J. M. Bishop, *Proc. Natl. Acad. Sci.* **69,** 20 (1972).
15. H. E. Varmus, S. Heasley, J. M. Bishop, *J. Virol.* **14,** 895 (1974).
16. P. E. Neiman, *Science* **178,** 750 (1972).
17. M. Hill, J. Hillova, *Nat. New Biol.* **237,** 35, (1972); M. Hill, J. Hillova, *Virology* **49,** 309 (1972).
18. H. E. Varmus, P.K. Vogt, J. M. Bishop, *J. Mol. Biol.* **74,** 613 (1973); H. E. Varmus, P. K. Vogt, J. M. Bishop, *Proc. Natl. Acad. Sci.* **70,** 3067 (1973).
19. H. Varmus, R. Swanstrom, Chapter 5 in *RNA Tumor Viruses,* R. Weiss, N. Teich, H. Varmus, J. Coffin, Eds. (Cold Spring Harbor Laboratory, Cold Spring Harbor, NY, 1982) pp. 369–512; H. Varmus, R. Swanstrom, Chapter 5S *ibid,* (1985), pp. 75–134.
20. P. K. Vogt, *Virology* **46,** 939 (1971).
21. G. S. Martin, P. H. Duesberg, *Virology* **47,** (1972); L.-H. Wang, P. Duesberg, K. Beemon, P. K. Vogt, *J. Virol.* **16,** 1051 (1975).
22. D. Stehelin, R. V. Guntaka, H. E. Varmus, J. M. Bishop, *J. Mol. Biol.* **101,** 349 (1976).
23. D. Stehelin, H. E. Varmus, J. M. Bishop, P. K. Vogt, *Nature* **260,** 170 (1976).
24. A. Bernstein, R. MacCormick, G. S. Martin, *Virology,* **70,** 206 (1976).
25. J. S. Brugge, R. L. Erikson, *Nature***269,** 346 (1977).
26. M. S. Collett, J. S. Brugge, R. L. Erikson, *Cell,* **15,** 1363, (1978); H. Oppermann, A. D. Levinson, H. E. Varmus, L. Levintow, J. M. Bishop, *Proc. Natl. Acad. Sci.* **76,** 1804 (1979); L. R. Rohrschneider, R. N. Eisenman, C. R. Leitch, *Proc. Natl. Acad. Sci.* **76,** 4479 (1979).
27. A. P Czernilofsky, A. D. Levinson, H. E. Varmus, J. M. Bishop, E. Tischer, H. M. Goodman, *Nature* **287,** 198 (1980); T. Takeya, H. Hanafusa, *J. Virol.* **44,** 12 (1982).
28. T. Padgett, E. Stublefield, H. E. Varmus, *Cell* **10,** 649 (1977); S. H. Hughes, E. Stubblefield, F. Payvar, J. D. Engel, J. B. Dodgson, D. Spector, B. Cordell, R. T. Schimke, H. E. Varmus, *Proc. Natl. Acad. Sci.* **76,** 1348 (1979).
29. S. H. Hughes, F. Payvar, D. Spector, R. T. Schimke, H. L. Robinson, J. M. Bishop, H. E. Varmus, *Cell* **18,** 347 (1979).
30. D. Spector, B. Baker, H. E. Varmus, J. M. Bishop, *Cell* **3,** 381 (1978); D.

Spector, K. Smith, T. Padgett, P. McCombe, D. Roulland- Dussoix, C. Moscovici, H. E. Varmus, J. M. Bishop, *Cell* **3,** 371 (1978).
31. R. C. Parker, H. E. Varmus, J. M. Bishop, *Proc. Natl. Acad. Sci.* **78,** 5842 (1981); T. Takeya, H. Hanafusa, R. P. Junghans, G. Ju, A. M. Skalka, *Mol. Cell. Biol.* **1,** 1024 (1981); D. Shalloway, A. D. Zelenetz, G. M. Cooper, *Cell* **24,** 531 (1981).
32. J. M. Bishop, H. Varmus, Chapter 9 in *RNA Tumor Viruses,* R. Weiss, N. Teich, H. Varmus, J. Coffin, Eds. (Cold Spring Harbor Laboratory, Cold Spring Harbor, NY, 1982) pp. 999–1108.
33. R. P. Parker, H. E. Varmus, J. M. Bishop, *Cell* **37,** 131 (1984); E. B. Jakobovits, J. E. Majors, H. E. Varmus, *Cell* **38,** 757 (1984).
34. M. S. Collett, R. L. Erikson, *Proc. Natl. Acad. Sci.* **75,** 2021 (1978).
35. A. Levinson, H. Oppermann, L. Levintow, H. E. Varmus, J. M. Bishop, *Cell* **15,** 561 (1978).
36. T. Hunter, B. M. Sefton, *Prof. Natl. Acad. Sci.* **77,** 1311 (1980).
37. T. Takeya, H. Hanafusa, *Cell* **32,** 881 (1983).
38. D. H. Spector, H. E. Varmus, J. M. Bishop, *Proc. Natl. Acad. Sci.* **75,** 4102 (1978).
39. B.-Z. Shilo, R. A. Weinberg, *Proc. Natl. Acad. Sci.* **78,** 6789 (1981); M. A. Simon, B. Drees, T. Kornberg, J. M. Bishop, *Cell* **42,** 831 (1985).
40. A. Kamb, M. Weir, B. Rudy, H. E. Varmus, C. Kenyon, *Nature* **337,** 364 (1989).
41. A. Barnekow, M. Schartl, *Mol. Cell. Biol.* **4,** 1179 (1984).
42. T. C. Bosch, T. F. Unger, D. A. Fisher, R. E. Steele, *Mol. Cell. Biol.* **9,** 4141 (1989).
43. J. S. Brugge, P. C. Cotton, A. E. Queral, J. N. Barrett, D. Nonner, R. W. Kenne, *Nature* **316,** 554 (1985); P. F. Maness, M. Aubry, C. G. Shores, L. Frame, K. H. Pfenninger, *Proc. Natl. Acad. Sci.* **85,** 5001 (1988); R. Martinez, B. Mathey-Provot, A. Bernards, D. Baltimore, *Science* **237,** 411 (1987).
44. D. O. Morgan, J. M. Kaplan, J. M. Bishop, H. E. Varmus, *Cell* **57,** 775 (1989); S. Shenoy, J.-K. Choi, S. Bagrodia, T. D. Copeland, J. L. Maller, D. Shalloway, *Cell* **57,** 763 (1989).
45. T. Hunter, J. A. Cooper, *Annu. Rev. Biochem.* **54,** 897 (1984).
46. T. Hunter, *Cell* **49,** 1 (1987).
47. J. M. Bishop, H. Varmus, Chapter 9S in *RNA Tumor Viruses,* R. Weiss, N. Teich, H. Varmus, J. Coffin, Eds. (Cold Spring Harbor Laboratory, Cold Springs Harbor, NY, 1985), pp. 249–356.
48. H. E. Varmus, J. M. Bishop, Eds., *Cancer Surveys — Proteins Encoded by Oncogenes,* Volume 5, Number 2, (Oxford University Press, England, 1986).
49. H. E. Varmus, Chapter 9 in *Molecular Basis of Blood Diseases,* G. Stamatoyannopoulos, A. W. Nienhuis, P. Leder, P. W. Majerus, Eds. (W. B. Saunders, 1987), pp. 271–346.
50. H. E. Varmus, *Ann. Rev. Genetics,* **18,** 553 (1984); J. M. Bishop, *Science* **235,** 305 (1987).
51. J. M. Taylor, R. Illmensee, *J. Virol.* **16,** 552, (1975).
52. J. M. Coffin, W. A. Haseltine, *Proc. Natl. Acadd Sci.* **74,** 1908, (1977); M. S. Collett, P. Dierks, J. F. Cahill, A. J. Faras, J. T. Parsons, *Proc. Natl. Acad. Sci.* **74,** 2389 (1977); W. A. Haseltine, J. M. Coffin, T. C. Hageman, *J. Virol.* **30,** 375 (1979); R. P. Junghans, S. Hu, C. A. Knight, N. Davidson, *Proc. Natl. Acad. Sci.* **74,** 477 (1977); E. Stoll, M. A. Billeter, A. Palmenberg, C. Weissman, *Cell* **12,** 57 (1977).
53. S. Hughes, P. K. Vogt, P. R. Shank, D. Spector, H.-J. Kung, M. L. Breitman, J. M. Bishop, H. E. Varmus, *Cell* **15,** 1397 (1978); P. R. Shank, S. Hughes, H.-J. Kung, J. Majors, N. Quintrell, R. V. Guntaka, J. M. Bishop, H. E. Varmus, *Cell* **15,** 1383 (1978); H. E. Varmus, P. R. Shank, S. Hughes, H.-J. Kung, S. Heasley, J. Majors, P. K. Vogt, J. M. Bishop, *Cold Spring Harbor Symp. Quant. Biol.* **43,** 851 (1979).

54. T. W. Hsu, J. L. Sabran, G. E. Mark, R. V. Guntaka, J. M. Taylor, *J. Virol.* **28,** 810 (1978); J. L. Sabran, T. W. Hsu, C. Yeater, A. Kaji, W. S. Mason, J. M. Taylor, *J. Virol.* **29,** 170 (1979).
55. H. E. Varmus, Chapter 10 in *Transposable Elements,* J. Shapiro, Ed., (Academic Press, NY, 1983), pp. 411–503; H. E. Varmus, P. O. Brown, in *Mobile DNA* M. Howe, D. Berg, Eds., (ASM Publications, New York, 1989) pp. 35–56.
56. D. Baltimore, *Cell* **40,** 481 (1985); H. E. Varmus, *Nature* **314,** 583 (1985).
57. J. Stoye, J. Coffin, Chapter 10 in *RNA Tumor Viruses,* R. Weiss, N. Teich, H. Varmus, J. Coffin, Eds. (Cold Spring Harbor Laboratory, Cold Spring Harbor, NY, 1985), pp. 357–404.
58. H. E. Varmus, N. Quintrell, S. Ortiz, *Cell* **25,** 23 (1981).
59. N. Teich, J. Wyke, T. Mak, A. Bernstein, W. Hardy, Chapter 8 in *RNA Tumor Viruses,* R. Weiss, N. Teich, H. Varmus, J. Coffin, Eds. (Cold Spring Harbor Laboratory, Cold Spring Harbor, NY, 1982), pp. 785–998.
60. G. S. Payne, S. A. Courtneidge, L. B. Crittenden, A. M. Fadly, J. M. Bishop, H. E. Varmus, *Cell* **23,** 31 (1981).
61. W. S. Hayward, B. G. Neel, S. M. Astrin, *Nature* **290,** 475 (1981); B. G. Neel, W. S. Hayward, H. L. Robinson, J. Fang, S. M. Astrin, *Cell* **23,** 323 (1981).
62. D. Sheiness, J. M. Bishop. *J. Virol.* **31,** 514, (1979); M. Roussel, S. Saule, C. Lagrou, C. Rommens, H. Beug, T. Graf, D. Stehelin, *Nature* **281,** 452 (1979).
63. H. E. Varmus, *Cancer Surv.* **2,** 301 (1982); R. Nusse, A. Berns, Chapter 3 in *Cellular Oncogene Activation,* G. Klein, Ed., (Marcel Dekker, Inc., New York, NY), 1988.
64. G. S. Payne, J. M. Bishop, H. E. Varmus, *Nature* **295,** 209 (1982).
65. J. Hilgers, P. Bentvelzen, *Adv. Cancer Res.* **26,** 143, (1978).
66. R. Korteweg, *Genetics* **18,** 350 (1946).
67. J. J. Bittner, *Science* **84,** 162, (1936).
68. J. C. Cohen, P. R. Shank, V. Morris, R. Cardiff, H. E. Varmus, *Cell* **16,** 333 (1979).
69. R. Nusse, H. E. Varmus, *Cell* **31,** 99 (1982); R. Nusse, A. van Ooyen, D. Cox, Y. K. T. Fung, H. Varmus, *Nature* **307,** 131 (1984).
70. A. M. C. Brown, R. S. Wildin, T. J. Prendergast, H. E. Varmus, *Cell* **46,** 1001 (1986).
71. T. A. Stewart, P. K. Pattengale, P. Leder, *Cell* **38,** 627 (1984); R. D. Palmiter, R. L. Brinster, *Annu. Rev. Genetics,* **20,** 465 (1986); S. Cory, J. M. Adams, *Annu. Rev. Immunology* **6,** 25 (1988); D. Hanahan, *Science* **246,** 1265 (1989).
72. A. S. Tsukamoto, R. Grosschedl, R. C. Guzman, T. Parslow, H. E. Varmus, *Cell* **55,** 619 (1988).
73. H. E. Varmus, *Science* **216,** 812 (1982); J. M. Bishop, *Annu. Rev. Biochem.* **52,** 301 (1983); R. Swanstrom, R. C. Parker, H. E. Varmus, J. M. Bishop, *Proc. Nat. Acad. Sci.* **80,** 2519 (1983).
74. D. Baltimore, *Science* **192,** 632 (1976).
75. H. E. Varmus, *Science* **240,** 1427 (1988).
76. Y. Maki, T. J. Bos, C. Davis, M. Starbuck, P. K. Vogt, *Proc. Natl. Acad. Sci.* **84,** 2848 (1987); P. K. Vogt, T. J. Bos, R. F. Doolittle, *Proc. Natl. Acad. Sci.* **84,** 3316 (1978); B. J. Mayer, M. Hamaguchi, H. Hanafusa, *Nature* **332,** 272 (1988).
77. P. J. Maddon, A. G. Dalgleish, J. S. McDougal, P. R. Clapham, R. A. Weiss, R. Axel, *Cell* **47,** 333 (1986); J. S. McDougal, M. S. Kennedy, J. M. Sligh, S. P. Cort, A. Mawle, J. K. A. Nicholson, *Science* **231,** 382 (1986); L. M. Albritton, L. Tseng, D. Scadden, J. M. Cunningham, *Cell* **57,** 659 (1989).
78. P. O. Brown, B. Bowerman, H. E. Varmus, J. M. Bishop, *Cell* **49,** 347 (1987).
79. T. Fujiwara, K. Mizuuchi, *Cell* **55,** 497 (1988); P. O. Brown, B. Bowerman, H. E. Varmus, J. M. Bishop, *Proc. Natl. Acad. Sci.* **86,** 2525 (1989).

80. J. Coffin, Chater 4 in *RNA Tumor Viruses,* R. Weiss, N. Teich, H. Varmus, J. Coffin, Eds. (Cold Spring Harbor Laboratory, Cold Spring Harbor, NY, 1985). pp. 17–74; E. Gilboa, *BioEssays* **5,** 252 (1987).
81. T. Jacks, H. D. Madhani, F. R. Masiarz, H. E. Varmus, *Cell* **55,** 447 (1988); T. Jacks, M.D. Power, F. R. Mariarz, P. A. Luciw, P. J. Barr, H. E. Varmus, *Nature* **331,** 280 (1988).
82. H. Varmus, *Genes & Development* **2,** 1055 (1988); B. R. Franza, B. R. Cullen, F. Wong-Staal, Eds., *The Control of Human Retrovirus Gene Expression,* (Cold Spring Harbor Laboratory, Cold Spring Harbor, NY, 1988).
83. C. Dickson, R. Eisenman, H. Fan, Chapter 6 in *RNA Tumor Viruses,* R. Weiss, N. Teich, H. Varmus, J. Coffin, Eds. (Cold Spring Harbor Laboratory, Cold Spring Harbor, NY, 1982, 1985), pp. 135–146; A. M. Skalka, *Cell* **56,** 911 (1989).
84. *The Science of AIDS: Readings from Scientific American Magazine,* W. H. Freeman & Co., New York, 1988.

J. Michael Bishop

J. MICHAEL BISHOP

"And what have kings that privates have not too,
Save ceremony, save general ceremony?"

*William Shakespeare, in Henry the Fifth, IV.*1, 243–244

My youth held little forecast of a career in biomedical research. I was born on February 22, 1936, in York, Pennsylvania, and spent my childhood in a rural area on the west bank of the Susquehanna River. Those years were pastoral in two regards: I saw little of metropolitan life until I was past the age of twenty-one; and my youth was permeated with the concerns of my father's occupation as a Lutheran minister, tending to two small parishes. My most tangible legacy from then is a passion for music, sired by the liturgy of the church, fostered by my parents through piano, organ and vocal lessons. I am deeply grateful for the legacy, albeit apostate from the church.

I obtained eight years of elementary education in a two-room school, where I encountered a stern but engaging teacher who awakened my intellect with instruction that would seem rigorous today in many colleges. History figured large in the curriculum, exciting for me what was to become an enduring interest. But I heard little of science, and what I did hear was exemplified by the collection and pressing of wild flowers. My high school was also small: eighty students graduated with me, few of whom eventually completed college. Tests conducted before I graduated predicted a future for me in journalism, forestry or the teaching of music; persons who know me well could recognize some truth in those seemingly errant prognoses.

I had taken naturally to school and was an excellent student from the beginning. But my aspirations for the future were formed outside the classroom. During the summer months of my high-school years, I befriended Dr. Robert Kough, a physician who cared for members of my family. Although he was practicing general medicine in a rural community when I met him, he was well equipped to arouse in me an interest not only in the life of a physician but in the fundaments of human biology. His influence was to have a lasting impact.

I entered Gettysburg College intent on preparing for medical school. But my ambition was far from resolute. Every new subject that I encountered in college proved a siren song. I imagined myself an historian, a philosopher, a novelist, rarely a scientist. But I stayed the course, completing my major in chemistry with diffidence but academic laurels. I met the woman who was to become my only wife. I have never been happier before or since.

I graduated from college still knowing nothing of original research in science. I knew that I would be going to medical school, but I had little interest in practicing medicine. Instead, under the influence of my college faculty, I had formed a vague hope of becoming an educator — by what means and in what subject, I knew not. Learning of this hope, an associate dean at the University of Pennsylvania recommended that I decline my admission to medical school there and, instead, accept an offer from Harvard Medical School. I followed the advice. My pastoral years were at an end.

Boston was a revelation and a revel. I could for the first time sate my burgeoning appetite for the fine arts. Harvard, on the other hand, was a revelation and a trial. I discovered that the path to an academic career in the biomedical sciences lay through research, not through teaching, and that I was probably least among my peers at Harvard in my preparation to travel that path. During my first two years of medical school, I acquired a respect for research from new-found friends among my classmates, particularly John Menninger (now at the University of Iowa) and Howard Berg (now at Harvard University). I sought summer work in a neurobiology laboratory at Harvard but was rebuffed because of my inexperience. I became ambivalent about continuing in medical school, yet at a loss for an alternative.

Two pathologists rescued me. Benjamin Castleman offered me a year of independent study in his department at the Massachusetts General Hospitality, and Edgar Taft of that department took me into his research laboratory. There was little hope that I could do any investigation of substance during that year, and I did not. But I became a practiced pathologist, which gave me an immense academic advantage in the ensuing years of medical school. I found the leisure to marry. And I was riotously free to read and think, which lead me to a new passion: molecular biology. The passion was to remain an abstraction for another four years, but my course was now set.

I was slowly becoming shrewd. I recognized that the inner sanctum of molecular biology was not accessible to me, that I would have to find an outer chamber in which to pursue my passion. I found animal virology, in the form of an elective course taken when I returned to my third year of medical school, and in the person of Elmer Pfefferkorn. From the course, I learned that the viruses of animal cells were ripe for study with the tools of molecular biology, yet still accessible to the likes of me. From Elmer, I learned the inebriation of research, the practice of rigor, and the art of disappointment.

I began my work with Elmer in odd hours snatched from the days and nights of my formal curriculum. But an enlightened dean gave me a larger opportunity when he approved my outrageous proposal to ignore the curriculum of my final year in medical school, to spend most of my time in the research laboratory. In the end, I completed only one of the courses normally required of fourth year students. Flexibility of this sort in the affaires of a medical school is rare, even now, in this allegedly more liberal age.

My work with Elmer was sheer joy, but it produced nothing of substance. I remained uncredentialed for postdoctoral work in research. So on graduation from medical school, I entered an essential interregnum of two years as a house physician at the Massachusetts General Hospital. That magnificent hospital admitted me to its prestigious training despite my woeful inexperience at the bedside, and despite my admission to the chief of service that I had no intention of ever practicing medicine. I have no evidence that they ever regretted their decision. Indeed, years later, I was privileged to receive their Warren Triennial Prize, one of my most treasured recognitions. I cherish the memories of my time there: I learned much of medicine, society and myself.

Clinical training behind me, I began research in earnest as a postdoctoral fellow in the Research Associate Training Program at the National Institute of Health in Bethesda, Maryland, a program designed to train mere physicians like myself in fundamental research. In its prime, the Program was a unique resource, providing U.S. medical schools with many of the most accomplished faculty. Without the Program, it is unlikely that I could have found my way into the community of science.

My mentor at N.I.H. was Leon Levintow, who has continued as my friend and alter ego throughout the ensuing years. My subject was the replication of poliovirus, which had a test case for the view that the study of animal viruses could tease out the secrets of the vertebrate cell. I managed my first publishable research: my feet were now thoroughly wet; I had become confident of a future in research.

Midway through my postdoctoral training, Levintow departed for the faculty at the University of California, San Francisco (known to its devotees as UCSF). In his stead came Gebhard Koch, who soon lured me to his home base in Hamburg, Germany, for a year. And again, I had an enlightened benefactor: Karl Habel, who agreed to have N.I.H. pay my salary in Germany, even though I would be in only the first year of a permanent appointment. I repaid the benefaction by never returning to Bethesda.

My year in Germany saw little success in the laboratory, but I learned the joys of Romanesque architecture and German Expressionism. As my year in Germany drew to a close, I had two offers of faculty positions in hand: one at a prestigious university on the East coast of the United States, the other from Levintow and his departmental chairman, Ernest Jawetz, at UCSF. I chose the latter, easily, because the opportunities seemed so much greater: I would have been a mere embellishment on the East Coast; I was genuinely needed in San Francisco. In February of 1968, my wife and I moved from Hamburg to San Francisco, where we remain ensconced to this day.

I continued my work on poliovirus. But new departures were also in the offing. In the laboratory adjoining mine, I found Warren Levinson, who had set up a program to study Rous Sarcoma Virus, an archetype for what we now know as retroviruses. At the time, the replication of retroviruses was one of the great puzzles of animal virology. Levinson, Levintow and I joined forces in the hope of solving that puzzle. We were hardly begun before Howard Temin and David Baltimore announced that they had solved the puzzle with the discovery of reverse transcriptase.

The discovery of reverse transcriptase was sobering for me: a momentous secret of nature, mine for the taking, had eluded me. But I was also exhilarated because reverse transcriptase offered new handles on the replication of retroviruses, handles that I seized and deployed with a vengeance. I was joined in this work by a growing force of talented postdoctoral fellows and graduate students. Among our early achievements were a description of the mechanisms by which reverse transcriptase copies RNA into DNA, the characterization of viral RNA in infected cells, and the identification and description of viral DNA in both normal and infected cells.

The work on viral DNA was particular notable because it was the handicraft of Harold Varmus, who had joined me as a postdoctoral fellow in late 1970. Harold's arrival changed my life and career. Our relationship evolved rapidly to one of coequals, and the result was surely greater than the sum of the two parts. Together we decided to extend our interests beyond the problems of retroviral replication, to address the mystery of how Rous Sarcoma Virus transforms cells to neoplastic growth.

Others had shown that transformation by Rous Sarcoma Virus could be attributed to a single gene (eventually dubbed *src*) located near the 3' end of the viral genome. Two problems engaged us: what was the origin of *src;* and what was the protein product of the gene? It was not our lot to find an answer for the second question, although we later played a part in discerning the biochemical function of the *src* protein. But with experiments performed mainly by Dominique Stehelin and Deborah Spector, we found the answer to the first question: *src* is a wayward version of a normal cellular gene (which we would now call a proto-oncogene), pirated into the retroviral genome by recombination (in a sequence of events known as transduction), and converted to a cancer gene by mutation.

In the years that followed, we consolidated our evidence for retroviral transduction, generalized the finding to retroviral oncogenes other than *src,* helped elucidate the sorts of genetic damage that convert normal cellular genes into cancer genes, explored the contributions of proto-oncogenes to the genesis of human cancer, added to the repertoire of proto-oncogenes by several experimental strategies, pursued the physiological functions of proto-oncogenes in normal organisms, and shared in the discovery of the protein kinase encoded by *src.*

I began my career at UCSF as an Assistant Professor of Microbiology and Immunology. I am now a Professor in the same department and in the Department of Biochemistry and Biophysics. I serve as Director of the G. W. Hooper Research Foundation and of the Program in Biological Sciences – the latter, an effort to unify graduate education at UCSF. I am as devoted to teaching as to research: I find the two vocations equally gratifying.

I am a member of the National Academy of Sciences, U.S.A; the American Academy of Arts and Sciences; the American Association for the Advancement of Science (elected an Honorary Fellow); the American Society for Biological Chemistry and Molecular Biology; the American Society for Microbiology; the American Society for Cell Biology; the American

Society for Virology; the Federation of American Scientists; Alpha Omega Alpha; and Phi Beta Kappa.

My honors include several awards for teaching from the students and faculty of UCSF; a Doctor of Science Honoris Causa from Gettysburg College; the American Association of Medical Colleges Award for Distinguished Research; the California Scientist of the Year; the Albert Lasker Award for Basic Medical Research; the Passano Foundation Award; the Warren Triennial Prize from the Massachusetts General Hospital; the Armand Hammer Cancer Prize; the Alfred P. Sloan, Jr. Prize from the General Motors Cancer Foundation; the Gairdner Foundation International Award; the American Cancer Society National Medal of Honor; the Lila Gruber Cancer Research Award from the American Academy of Dermatology; the Dickson Prize in Medicine from the University of Pittsburgh; the American College of Physicians Award for Basic Medical Research; and the Nobel Prize in Physiology or Medicine for 1989. Most of these have been shared with Harold Varmus.

I am married to Kathryn Ione Putman and have two sons with her, Dylan Michael Dwight and Eliot John Putman. These three have given me a gift of affection and forebearance that I cannot hope to repay. My mother and father have reached their eighth and ninth decades, respectively, and were able to join us for a joyful time at the Nobel ceremonies in Stockholm. My brother, Stephen, is a distinguished solid-state physicist and now Professor at the University of Illinois; my sister, Catharine, is arguably the finest elementary school teacher in Virginia.

If offered reincarnation, I would choose the career of a performing musician with exceptional talent, preferably, in a string quartet. One life--time as a scientist is enough — great fun, but enough. I am a self-confessed book addict, an inveterate reader of virtually anything that comes to hand (with the notable exceptions of science fiction and crime novels). I enjoy writing and abhor the dreadful prose that afflicts much of the contemporary scientific literature.

RETROVIRUSES AND ONCOGENES II

Nobel Lecture, December 8, 1989

by

J. MICHAEL BISHOP

Department of Microbiology and Immunology, Department of Biochemistry and Biophysics, and The G. W. Hooper Research Foundation, University of California at San Francisco, San Francisco, U.S.A.

I am grateful to the family in whose embrace I formed my aspirations; the institutions modest and grand that helped me nurture those aspirations; the many colleagues with whom I have shared the pursuit of nature's secrets; the Nobel Foundation and the Karolinska Institute for glorious hospitality; and Swedish Medicine, which restored my larynx to some semblance of function.

The English critic Cyril Connolly once remarked that: "The true index of a man's character is the health of his wife." My life in science has been rich and rewarding. I have sacrified very little. But my partner in that life has sacrified a great deal: Kathryn, my wife of thirty years. I take this moment to speak my gratitude for her forbearance, and to acknowledge that matters are not likely to improve.

If I were asked to choose a biographical theme for today, it would have to be procrastination. Peyton Rous withdrew from medical school for a year to spend time on a ranch in Texas, ostensibly to recover from tuberculosis (1). Once back in medicine, he found himself "unfit to be a real doctor" (in his own words) and turned to Pathology for an entry into research. Harold Varmus dallied with English Literature before entering medical school, as you have heard (2). And I came close to abandoning medicine entirely at an early age.

I entered Harvard Medical School knowing nothing of research. But during my first two years there, I was awakened to research by new-found friends among my classmates, particularly John Menninger (now at the University of Iowa) and Howard Berg (now at Harvard University). I sought summer work in a neurobiology laboratory at Harvard but was rebuffed because of my inexperience. My interest in practicing medicine declined. I became ambivalent about continuing in medical school, yet at a loss for an alternative.

Like Peyton Rous, I was rescued by pathologists. Benjamin Castleman offered me a year of independent study in his department at the Massachusetts General Hospital, and Edgar Taft of that department took me into his research laboratory. There was little hope that I could do any investigation of substance during that year, and I did not. But I was riotously free to read and think, which lead me to a new passion: molecular biology. I began my efforts to consummate that passion with two novitiates.

Novitiates

I served the first novitiate with Elmer Pfefferkorn and Sindbis Virus. I had no credentials, other than my desire. Yet Elmer took me into his laboratory, and Harvard Medical School excused me from all but one of my fourth year courses so that I could try research unencumbered. Both were enlightened acts, for which I remain grateful beyond measure.

I sought out Elmer and Sindbis Virus because I had perceived that the inner sanctum of molecular biology was closed to me, that I would have to find an outer chamber in which to pursue my passion. Through Elmer, I found animal viruses, ripe for study with the tools of molecular biology, yet still accessible to the innocent.

I was innocent, but I was brash. I resolved to test the ability of the Sindbis RNA genome to serve as mRNA *in vitro,* and to trace the fate of the genome following its entry into the host cell. These were novel ventures in their time (1961). They were also technically foolish. But they sired an abiding interest in how the genomes of RNA viruses commandeer the molecular machinery of the host cell, an interest that led me eventually to retroviruses.

My work with Elmer was sheer joy, teaching me the inebriation of research, the practice of rigor and the art of disappointment. But it produced nothing of substance. Twenty five years later, on the occasion of my fiftieth birthday, Elmer recalled my first novitiate in science with a quote from T. H. Huxley: "There is great practical benefit in making a few failures early in life."

After two years of clinical training at the Massachusetts General Hospital, I began my second novitiate by entering the Research Associates Program at the National Institutes of Health in Bethesda, Maryland, where my mentor was Leon Levintow. My subject was the replication of poliovirus, which had become a test case for the view that the study of animal viruses could tease out the secrets of the vertebrate cell. In my first publishable research, I obtained evidence that the replication of polioviral RNA engendered a multi-stranded intermediate, although my description of that intermediate proved flawed in its details (3).

Midway through my postdoctoral training, Levintow departed for the faculty at the University of California, San Francisco. In his stead came Gebhard Koch, who soon lured me to his home base in Hamburg, Germany, for a year. Together Gebhard and I explored the basis for the infectivity of multi-stranded RNAs (4). I continued this work for several years to come, eventually showing that the double stranded form of polioviral RNA is infectious because the positive strand of the duplex can be expressed in mammalian cells, as if the duplex RNA might be unraveling within the cell (5). This finding perplexed us and seemed abstruse, but it appears to have been a harbinger of unanticipated enzymatic activities whose existence and functions are coming into view only now (6).

Retroviruses

As my year in Germany drew to a close, I chose to join Levintow as a member of the faculty at the University of California, San Francisco. The decision proved providential beyond all measure. In San Francisco, I found Warren Levinson, who had set up a program to study Rous Sarcoma Virus, an archetype for what we now call retroviruses. At the time, the replication of retroviruses was one of the great puzzles of animal virology. Levinson, Levintow and I joined forces in the hope of solving that puzzle. We had hardly begun before David Baltimore and Howard Temin announced that they had solved the puzzle with the discovery of reverse transcriptase (7, 8), work that brought them Nobel Prizes a scant five years later.

The discovery of reverse transcriptase was sobering: a momentous secret of nature, mine for the taking, had eluded me (and others, of course). But I was also exilarated because the DNA synthesized by reverse transcriptase *in vitro* represented an exquisite probe for viral nucleic acids, a reagent that would give us unprecedented access to the life cycle of retroviruses. To paraphrase a memorable simile from Arthur Kornberg (9), we now had a wedge with which to pry open the infected cell, and the hammer to drive that wedge would be molecular hybridization.

I had become enamored of molecular hybridization while working on poliovirus, because of the exceptional sensitivity and specificity the technique offered in the pursuit of viral nucleic acids. It was a tool made to

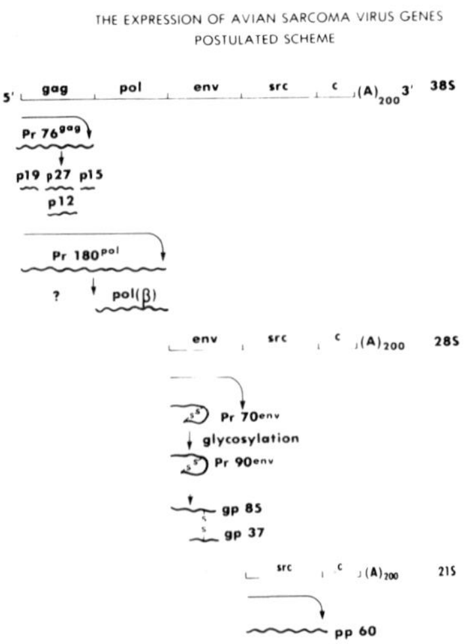

Figure 1. Expressing the genes of Rous Sarcoma Virus.

The diagram was prepared in 1980 to portray how splicing and the processing of proteins facilitate expression of the Rous Sarcoma Virus genome. In the interim, we have learned that frame shifting during translation is also required to produce the *gag-pol* polyprotein (13). Individual genes are designated by conventional nomenclature, according to the proteins they encode: *gag,* structural proteins of the capsid and nucleocapsid; *pol,* reverse transcriptase; *env,* the glycoproteins of the viral envelope; and *src,* the oncogene of the virus. Other viral functions not illustrated include the protease that mediates maturation of several viral proteins; and integrase, the enzyme that catalyzes integration of viral DNA.

order for the study of retroviral replication, which proceeds in concert with and is obscured by the normal metabolism of the host cell.

Improvising assays as we went, my colleagues and I soon had our first glimpse of viral RNA in cells infected with retroviruses (10). We were to pursue the character and genesis of those RNAs for years to come, laying out the manner of viral gene expression in considerable detail (Fig. 1). The work helped uncover an early example of RNA splicing (11, 12) and set the stage for the much later discovery of frame-shifting in the translation of retroviral mRNAs (13). But our exercises with viral RNA had a larger resonance, as well, because we had constructed the technical stage on which the discovery of retroviral transduction would eventually play out.

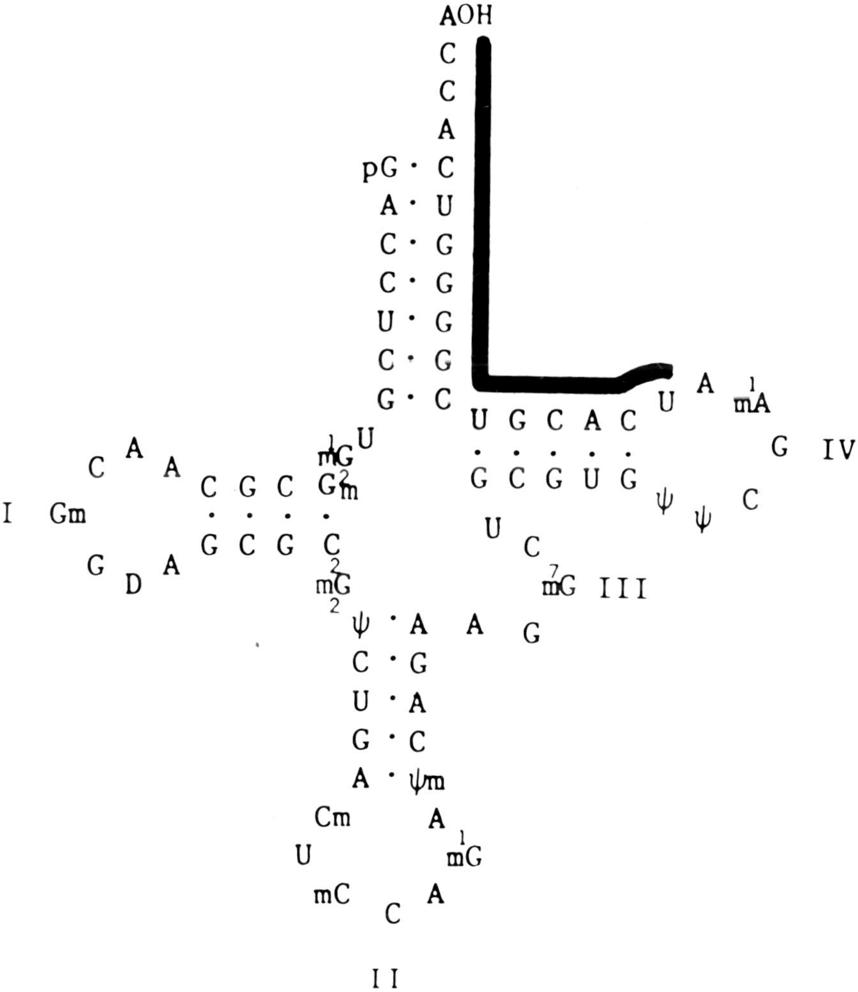

Fiure 2. The tRNA primer for reverse transcription in Rous Sarcoma Virus.

The avian tRNA for tryptophan serves as primer for reverse transcription from the genome of Rous Sarcoma Virus. The RNA is displayed here in a conventional representation of secondary structure. The first sixteen nucleotides at the 3' end of the tRNA are paired with the genome by hydrogen-bonded complementarity, as highlighted here.

I also picked up the study of reverse transcriptase itself, perhaps to exorcise my sense of failure at not discovering the enzyme in the first place. We began by working *in vitro* to explore the details of DNA synthesis by the enzyme (14). The most notable outcome was the demonstration that the reverse transcriptase of Rous Sarcoma Virus uses the cellular tRNA for tryptophan as a primer to initiate transcription from the retroviral genome (Fig. 2) (14, 15). The point was soon generalized to other retroviruses and is now a feature that serves as a signature of retrotransposons at large (16).

But what I wanted most to know was the course of events in the infected cell. Could we find the proviral DNA first imagined by Howard Temin and then foretold in substance by the discovery of reverse transcriptase? Where in the cell was this DNA synthesized following infection? What form did it take before and after integration into chromosomal DNA? Harold Varmus arrived to provide the answers (2), and to alter my life and career irrevocably. Within the year, his name became consubstantial with expertise on the synthesis and integration of retroviral DNA. In the process, I had lost a postdoctoral fellow and gained a coequal.

Oncogenes

To this point, we had thought little of cancer. But as the virus of Peyton Rous first lured me away from poliovirus, now it lured us to the study of neoplastic transformation. Peyton Rous received his Nobel Prize two years before my initial encounter with his virus. The award dramatized the great mystery of how Rous Sarcoma Virus might cause cancer. It was a mystery whose solution lay in genetics.

Soon after my arrival in San Francisco, a graduate student and I had conducted a search for temperature-sensitive conditional mutants of poliovirus and Rous Sarcoma Virus, without a particle of success. But where we had failed, others succeeded, and the study of viral tumorigenesis was transformed (17–19). The data showed with luminous clarity that a gene within Rous Sarcoma Virus is responsible for cancerous growth of infected cells, that continuous action of the gene is required to sustain cancerous growth, and that the gene probably works by directing the synthesis of a protein. The oncogene *src* had been sighted.

The genetic identification of *src* was reported in the same year as the biochemical discovery of reverse transcriptase. The two became themes that interwined and nourished one another in the daily life of our laboratory. Much as our deployment of molecular hybridization in the study of viral replication set the stage for the discovery of cellular *src*, so our ensuing success in isolating viral proteins from infected cells (20) emboldened us to seek the protein encoded by *src*.

We were in the midst of efforts to prepare antisera that would recognize the *src* protein when news of success came from Denver (21). Erikson and his colleagues had obtained persuasive evidence that the oncogene encodes a 60 kilodalton protein. Once the product of *src* was a physical reality, the puzzle of its action loomed larger than ever. How could this single protein elicit the pleiotropic change in cellular phenotype that we call neoplastic transformation?

The answer came quickly. *Src* encodes a protein kinase (22, 23), whose amino acids substrate later proved unexpectedly to be tyrosine (24). By phosphorylating numerous cellular proteins, the enzyme could rapidly change myriad aspects of cellular structure and function; by being the first exemplar of protein-tyrosine kinases, it gave notice of a previously unappreciated regulatory device that we now realize is second to none in the signalling pathways of the cell (25).

The ways in which these answers emerged are illuminating. In Denver, insight came from an inspired guess, based on the pleiotropism of *src* (22): protein phosphorylation ranks among the most versatile agents of change known to biochemists. In our laboratory, enzymological reasoning led the way (23): phosphorylation of the *src* protein in cellular extracts displayed properties suggestive of a unimolecular reaction, as if the protein were phosphorylating itself – which indeed it was (26). And at the Salk Institute, the use of an erroneous buffer led to the fortuitous separation of phosphotyrosine from phosphothreonine for the first time in recorded history, the only example of productive laziness that I have ever seen acknowledged with both candor and gratitude in the biomedical literature (25).

The sighting and subsequent characterization of *src* opened the way to a biological *cornucopia*. We now know of more than twenty retroviral oncogenes, whose diverse specificities in tumorigenesis provide experimental modes for most forms of cancer that afflict human kind (Table 1). Each of these genes encodes a protein whose biochemical action provides distinctive purchase on the mechanisms of neoplastic growth (27, 28).

Table 1. The oncogenes of retroviruses (1989)

Oncogene	Pathogenicity	Oncogene	Pathogenicity
abl	B-cell tumors and fibrosarcomas	mil/raf	sarcomas and supplemental
akt	thymomas	mos	sarcomas
cbl	B-cell and myeloid tumors	myb	myeloblastosis
		myc	carcinomas; myelocytomatosis; sarcomas
crk	sarcomas		
erb-a	supplemental		
erb-b	erythroleukemia and fibrosarcomas	ras	sarcomas; erythroleukemia
ets	supplemental	rel	b-cell tumors
fes/fps	sarcomas	ros	sarcomas
fgr	sarcomas	sea	sarcomas; leukemias
fms	sarcomas	sis	sarcomas
fos	osteosarcomas	ski	carcinomas
jun	sarcomas	src	sarcomas
kit	sarcomas	yes	sarcomas

The products of oncogenes are deployed to various reaches of the cell, including the nucleus, the cytoplasm, the plasma membrane, even the exterior beyond the cell (27, 28). And they act in different ways, which for the moment are subsumed by three genre: i) the phosphorylation of proteins, with either serine and threonine, or tyrosine as substrates – the immediate rôle of the oncogene product may be induction of the phosphorylation (as in the case of growth factors) or catalysis itself (as with the receptors for growth factors) (25); ii) the transmission of signals by GTP-binding proteins, as exemplified by the products of RAS genes – whose exact position in signalling pathways remains unresolved (29); and iii) the control of transcription from DNA (30).

Diversification of this list in the future seems likely, since the functions of many oncogenes have yet to be elucidated. But the list displays an economy of style that may survive because it reflects the need for pleiotropism: nature may have only a limited number of ways to achieve the manifold changes that create the neoplastic phenotype.

Retroviral transduction

At first it seemed that the lessons to be learned from retroviral oncogenes might apply only to the cancers induced by viruses in animals, that the oncogenes of retroviruses might be alley cats of evolution with little importance to human kind. The discovery of cellular *src* and the inference that it gave rise to the oncogene of Rous Sarcoma Virus inspired hope that this narrow view might be wrong (31). If cells contain genes capable of becoming oncogenes by transduction into retroviruses, perhaps the same genes might also become oncogenes within the cell, without ever encountering a virus. By means of accidental molecular piracy, retroviruses may have brought to view the genetic keyboard on which many different causes of cancer can play, a final common pathway to the neoplastic phenotype (Fig. 3).

The hope had its detractors (32). Even transduction itself was challenged. But for us in San Francisco, the reality of transduction seemed inescapable. The cellular homologue of *src* had been conserved through eons of evolution, whereas the other genes of Rous Sarcoma Virus could be found only in chickens and a few close kin (2). The ineluctable conclusion was that the two sorts of genes, an oncogene on the one hand and genes devoted to viral replication on the other, had separate origins.

We would eventually muster more sophisticated arguments, all of which pointed to the same conclusion: the progenitor of *src* was a conserved (and hence vital) cellular gene that found its way into Rous Sarcoma Virus by recombination (2). The images of cellular and viral *src* gained eventually from molecular cloning sustained our argument in a gratifying manner (33), but for me, they were anticlimatic.

Arguing for the cellular origins of *src* provided my first experience with the heuristic force of evolution. "Nothing in biology makes sense except in the light of evolution", to recall a famous aphorism from Dobzhansky (34).

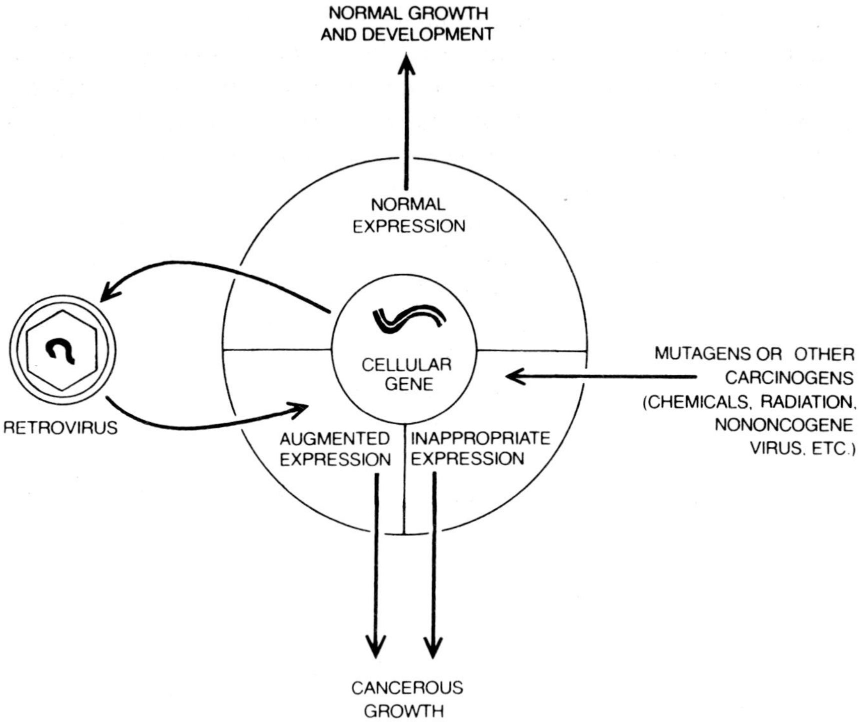

Figure 3. Transduction by retroviruses uncovers cancer genes.

The cartoon was designed in 1982 and exemplifies the now conventional view of proto-oncogenes as both important elements in the governance of normal cellular phenotype and potential substrates for various causes of cancer. The figure is reprinted here with the kind permission of *Scientific American.*

The aphorism embodies a truth that has been dishonored in the United States, where religious zealots continue their efforts to hound the teaching of evolution from public schools, and men of little learning assail the truth of evolution under the fraudulent rubric of "creation science".

The genesis of retroviral oncogenes by recombination with cellular genes had been postulated by several observers. Although it may sound self-serving, I confess my ignore of those speculations when we began our work on cellular *src.* I was motivated by a desire to test the Virogene-Oncogene Hypothesis of Huebner and Todaro (35), not by an interest in the origins of oncogenes. But in due course, Howard Temin provided a useful inspiration with his suggestion that all retroviruses arose by the cobbling together of disparate genetic elements in the cell, with intermediates that he called proto-viruses (36). The inspiration was for taxonomy, not experiment.

As transduction by retroviruses came into common discourse, the need arose for a generic term to describe the cellular progenitor of *src* and other retroviral oncogenes. The first to find general usage was "cellular oncogene". Although I was a nominal member of the responsible committee on nomenclature, I was uncomfortable with this term because of its unwarranted implication that the native cellular genes carried intrinsic tumorigenic potential, that they need not be changed to cause trouble. So in playful homage to Howard Temin, I began to use the term "proto-sarc" (37). The generic "proto-oncogene" followed in short order.

In the interim, proto-oncogene has come into general use, as the colloquial counterpoise to oncogene. It has also become an embarrassment, because the precise connotation of the word is that of prototype rather than progenitor, not far removed from the offensive connotation of cellular oncogene. But the intent of the taxonomic invention was clear: numerous investigators (ourselves included) have spent the past decade exploring the sorts of genetic damage that can convert a harmless proto-oncogene into a pathogenic oncogene (38).

The manuscript that announced our discovery of cellular *src* concluded with the speculation that the gene might be involved in "the normal regulation of cell growth and development or in the transformation of cell behavior by physical, chemical or viral agents" (39). These words were pure bravado, particularly because we then had no assurance that cellular *src* was in fact a full-fledged gene. The assurance accrued over the ensuing two years, first in the form of evidence that cellular *src* was transcribed in normal cells (40, 41), and then with the identification of the protein encoded by the gene (42, 43).

Expanding the repertoire

As our confidence in the reality of cellular *src* grew, a new challenge took shape. Could we generalize the principle of transduction? Had the oncogenes of other retroviruses also originated from cellular genes? In order to pursue the generality of transduction, we turned first to a retrovirus known as MC29 (44), which attracted our attention because it offers a model for the induction of carcinomas, the most prevalent of human cancers.

During the search for an oncogene in MC29, the impatience of molecular biologists held sway, the rigor of formal genetics was cast aside. Some investigators (ourselves included) used molecular hybridization to detect nucleotide sequences unique to the genome of MC29 (45, 46), others used chemical procedures to identify the same sequences and to map their position on the viral genome (47). Both strategies took liberties that we had not allowed ourselves with *src*. Lacking a deletion mutant that might define the oncogene, we and others made the assumption that the genomes of MC29 and its necessary helper virus were congenic except for the presence or absence of the oncogene (Fig. 4).

The same assumption had been applied previously in an effort to define the oncogene of a murine sarcoma virus and had led to a molecular quagmire (48). But now the lessons from *src* told us more precisely what we might be seeking. So we forged ahead and soon had a molecular probe that represented nucleotid sequences found in MC29 and other retroviruses with similar tumorigenicities, but not in the related helper viruses (45).

The newly found locus was taken to be the oncogene of MC29 and

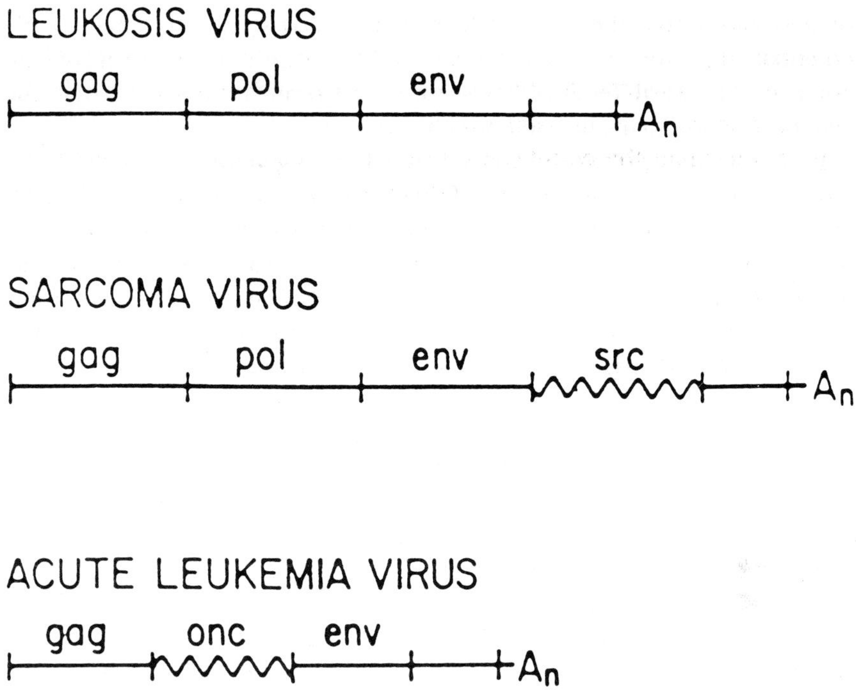

Figure 4. The genomes of Rous Sarcoma Virus and related avian retroviruses.
The genomes of Rous Sarcoma Virus, Avian Leukosis Viruses and Acute Avian Leukemia Viruses appear to have a common origin and, thus, are congenic with the exception of oncogenes, inserted into the genomes by transduction from host cells. The leukosis viruses have no oncogenes and induce malignancy by insertional mutagenesis (2, 38). The viral genes are designated as in Fig. 1, except for the use of *onc*, the generic term for retroviral oncogenes.

eventually designated *myc,* in deference to the form of leukemogenesis (myelocytomatosis) from which the virus acquired its name. Over the next several years, the authenticity of *myc* would be ascertained by molecular cloning (49), nucleotide sequencing (59), and gene-transfer (51). To this day, the gene has never been defined by the strategies of classical genetics, and there is now no need to do so. The new biology is upon us.

The lessons of *src* were powerful. We and others were able to argue that the molecular probes for the oncogene of MC29 were legitimate because they also detected nucleotide sequences in the DNA of normal vertebrates (46, 52), sequences that were transcribed into RNA in normal cells (52, 53) and that diverged among species in rough accord with phylogenetic distances (46, 52, 53). No other portion of the MC29 genome displayed these properties. Similar findings were described in parallel for the avian retroviral oncogenes *erb* (in reality, two oncogenes) and *myb* (46). The example of cellular *src* was not an exotic anomaly; it was an archetype.

In the years that followed, *myc* proved to be a great provider, a vehicle for several seminal discoveries that included the activation of cellular genes by insertional mutagenesis (54), the involvement of proto-oncogenes in chromosomal translocations (55, 56), and the amplification of proto-oncogenes in human tumors (57). These discoveries had exceptional logical force because they involved a gene whose tumorigenic potential was already known from the study of retroviruses.

No sooner had the bounty of transduction become evident than other routes to proto-oncogenes took shape, some serendipitous, others designed (58): the ability of retroviruses to mutate cellular genes, creating oncogenes at their place of residence within the cell; the dissection of chromosomal abnormalities in cancer cells, such as translocations and amplifications; the use of gene-transfer to detect mutant proto-oncogenes by means of their biological activity; and the pursuit of phylogenetic kinships. The definition of proto-oncogene had now become more expansive, subsuming any gene with the potential for conversion to an oncogene — by the hand of nature in the cell, or by the hand of the experimentalist in the test tube.

The normal function of proto-oncogenes

The tally of proto-oncogenes has now reached sixty or more. Most are genes never glimpsed before by any other means. What are these genes in their normal guise? What purposes preserved them through one billion years of evolution? Why do they harbor the potential to wreak cellular mayhem?

We formed our hypotheses from the lessons of transduction. The properties of retroviral oncogenes must echo the functions of proto-oncogenes. Three properties seemed especially telling: the stimulation of cellular proliferation; the specificity of tumorgenesis, as if each gene were designed to work in only certain cells; and the ability of many oncogenes to interrupt or sometimes reverse cellular differentiation (59). Like father, like son: it seemed possible that the actions of viral oncogenes are merely caricatures of what proto-oncogenes are normally intended to do. Retroviruses may have revealed to us not only touchstones of tumorigenesis, but clues to the nerve-center that governs the normal cell cycle and the differentiation of cellular function.

Eager to explore these thoughts in a living organism, we turned to the fruit fly, *Drosophila melanogaster*. Why the fruit fly? First, because the full collection of genes in this creature is within our scope. Second, because we have in hand a rich catalogue of normal and mutant genes from the fruit fly, the products of more than half a century of labor. Third, because the fruit fly is for the moment the only metazoan organism in which we can manipulate genes with reasonable facility, although a soil worm on the one hand and the laboratory mouse on the other are now making strong bids for our favor. Fourth, because when reduced to essentials, the fruit fly and *homo sapiens* are not very different. And fifth, because the fruit fly has a large complement of proto-oncogenes with counterparts in mammals (60).

What we have learned of proto-oncogenes from *Drosophila* indicates at least some of the bravado in our first publication on cellular *src*. Mutant alleles have been identified for the counterparts of six proto-oncogenes in the fruit fly (Table 2). In two instances, the *abl* and *myb* proto-oncogenes,

Table 2. Mutant alleles of proto-oncogenes in ***Drosophila melanogaster***

Proto-oncogene	Drosophila counterpart	Biochemical function	Mutant phenotype
abl	*abl*	tyr kinase	embryonic lethal/ defects in late development
int-1	*wingless*	? growth factor	segment polarity
raf	*pole hole*	ser/thr kinase	defective cellular proliferation
rel	*dorsal*	?	dorsal-ventral polarity
erbb-1	*flb/top*	tyr kinase	embryonic lethal/ dorsoventral patterning
myb	*myb*	transcription factor	embryonic and pupal lethality

the search for mutations was deliberate; in the reminder, mutations recognized first by their phenotypes later proved to be in *Drosophila* counterparts of mammalian proto-oncogenes. All of the mutations elicit profound disturbances of development. The work from our own hands has revealed a requirement for activity of the proto-oncogene *myb* during two distinct stages of development, embryogenesis and pupation (unpublished data of A. Katzen and J.M.B.). More precisely, a deficiency in *myb* appears to reduce the number of cell divisions that can occur in certain developmental lineages.

The actions of proto-oncogenes in development have their underpinnings in the elaborate circuitry that governs the behavior of vertebrate cells (Fig. 5). The junction boxes in this circuitry include polypeptide hormones that act on the surface of the cell, receptors for these hormones, proteins that convey signals from the receptors to the deeper recesses of the cell, and nuclear functions that orchestrate the genetic response to afferent commands (typically, by regulating transcription). Diverse lines of enquiry have brought these junction boxes to view, but the study of proto-oncogenes has been among the richest sources. Time and again, the several lines of enquiry have converged on the same junction box. We may have most of the circuitry in view. The cell is not infinitely complex; the cell can be understood.

Seeds of cancer

But what of cancer? Are proto-oncogenes the seeds of this disease in all our cells? Are they a common keyboard for many different players in tumorigenesis? As a physician, I found these questions attractive. As a scientist, I

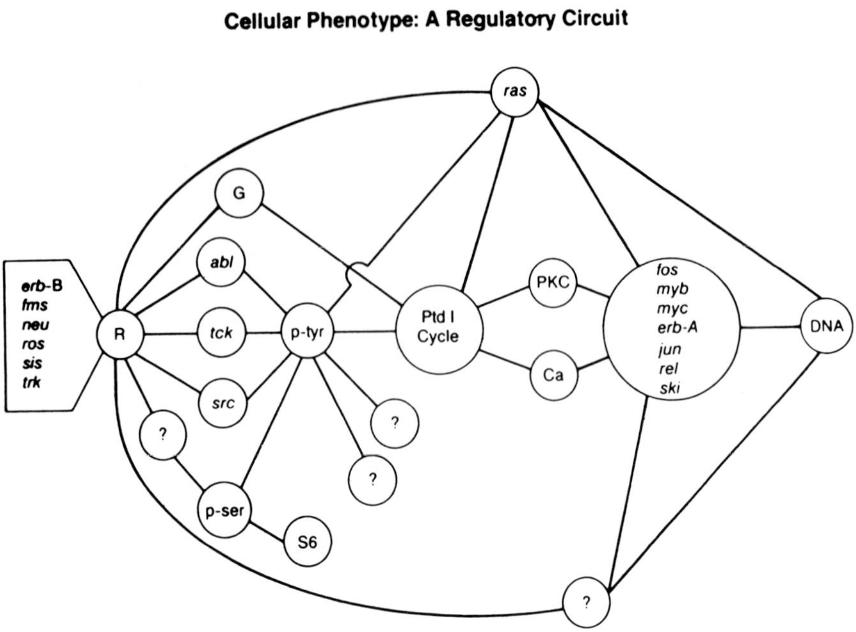

Figure 5. The biochemical circuitry that mediates cellular phenotype.

The diagram illustrates how some of the functions encoded by proto-oncogenes fit into the circuitry that regulates the phenotype of vertebrate cells. The scheme is in part hypothetical and is not intended to be comprehensive. Functions encoded by proto-oncogenes have been designated by the conventional terminology for the genes themselves. Other abbreviations include: G, GTP-binding proteins that transduce signals from cell-surface receptors; R, generic receptor; p-Ser, phosphorylation of serine in proteins; p-Tyr, phosphorylation of tyrosine in proteins; PtdI, phosphatidylinositol; S6, a ribosomal protein that is phosphorylated in response to diverse mitogenic signals; PKC, protein kinase C. Reprinted here with the kind permission of *Science*.

found them intimidating. Exploration of the cancer cell is akin to archaeology: we must infer the past from its remnants in the present, and the remnants are often cryptic.

But the first remnants to emerge from the proto-oncogenes of human cancer cells told a vivid story. Molecular dissections revealed that chromosomal translocations in human and murine tumors often affect proto-oncogenes already familiar from the study of retroviruses, with the great provider *myc* prominent among them (55, 61). Emboldened by these findings, my colleagues and I mounted a belated excavation for the genetic shards of tumorigenesis. But we chose a neglected terrain in which to dig: amplified DNA.

Focal amplification of domains within chromosomes is a scheduled and purposeful event during the life cycles of diverse organisms (62). In mammals, however, gene amplification is an unscheduled aberration that gives rise to karyotypic abnormalities known as double-minute chromosomes and homogeneously staining regions. When my colleagues and I began our excavations, gene amplification in cancer cells was known principally as a consequence of selection by chemotherapeutic agents. But the literature also contained occasional examples of double-minute chromosomes and homogeneously staining regions in untreated cancers. When a limited sampling of these were examined by ourselves and others, they proved to involve previously identified proto-oncogenes (yet again, *myc* was prevalent) (57). In due course, it became apparent that gene amplification is relatively common in untreated tumors and that it affects proto-oncogenes.

Intent on a more systematic study, we chose human neuroblastomas, in which gene amplification seemed exceptionally common. We resolved to ask whether the amplified DNA in neuroblastoma cells contained any of the proto-oncogenes then known. When we got to *myc,* we struck rich ore of an unexpected sort: a gene related to *myc,* sighted for the first time in the parallel work on neuroblastoma by ourselves and others, and eventually designated N-*myc* (63–65). In due course, it became apparent that N-*myc* is an authentic proto-oncogene, a close kin of *myc* itself that apparently evolved to serve a separate purpose in the normal organism (66–68).

As the survey of neublastomas broadened, a fertile correlation became apparent. Amplification of N-*myc* was found only in the more aggressive variants of the tumor, in perhaps one-quarter of all the specimens examined (69, 70). Moreover, within single tumors, N-*myc* was expressed abundantly only in neuroblasts, the least differentiated (and presumably most malignant) cells of the tumor (64). These correlations had two implications: first, that amplification of N-*myc* might embody a step in tumor progression, one of several events that exacerbate the malignancy of neuroblastomas; and second, that we had brought to hand a prognostic tool, a device with which to supplement the conventional staging of the disease.

The passage of time has dealt kindly with these hopes. The New England Journal of Medicine has now provided its imprimatur by arguing that "in neuroblastoma, amplification of the N-*myc* gene is of greater prognostic value than the clinical stage of the disease" (71). Thirty years after deserting the bedside, I have found clinical relevance in my research.

The excavation that unearthed N-*myc* was a pedestrian exercise in both concept and execution. Yet in one fell swoop, it served up an important proto-oncogene without the assistance of a retrovirus, hinted at a major role for gene amplification in tumorigenesis, foreshadowed the molecular dissection of tumor progression, and gave first notice that the emerging knowledge of oncogenes would eventually prove useful at the bedside.

These themes have since found wider resonance. First, we can point to a variety of human malignancies in which damage to one or another proto-oncogene has been found with some consistency (Table 3). The damage takes diverse forms, including translocations, amplifications and point mutations, all of which appear to have dominant effects on gene function. The list of malignancies containing these lesions is impressive: because of the diversity of tumors involved; because of their identities – several can be counted among the principal nemeses of humankind; and because the list has been assembled after only a few years of pursuit, with imperfect tools – there is

Tabel 3. Proto-oncogenes and human tumors: some consistent incriminations

Proto-oncogene	Neoplasm	Lesion
abl	chronic myelogenous leukemia	translocation
erbb-1	squamous cell carcinoma glioblastoma	amplification
erbb-2 *(neu)*	adenocarcinoma of breast and ovary	amplification
myc	Burkitt's lymphoma small cell carcinoma of lung carcinoma of breast carcinoma of cervix	translocation amplification
lmyc	small cell carcinoma of lung	amplification
nmyc	neuroblastoma small cell carcinoma of lung	 amplification
kras	carcinomas of colon, lung and pancreas	point mutation
nras	acute myelogenous and acute lymphoblastic leukemia; carcinoma of thyroid	point mutation
hras	carcinomas of genitourinary tract and thyroid	point mutation

doubtless more to come. Beyond this list, there lies the burgeoning repertoire of recessive lesions in human cancers, whose nature, prevalence and unquestionable importance are just now coming into focus (72).

Second, catalogues of genetic damage within individual tumors are taking shape, showing us how the malfunction of several different genes might combine to produce the malignant phenotype (58): for example, carcinomas of the colon contain no less than five different yet prevalent lesions — some genetically dominant, others recessive; carcinoma of the breast, at least five lesions; carcinoma of the lung, at least four; and neuroblastoma, at least three. Moreover, detection of the lesions is likely to provide information for prognosis, perhaps even for therapeutic management. Examples include neuroblastoma (69—71), carcinoma of the breast (73, 74) and ovary (74), and preleukemias (75—77).

For those of us who first studied cancer more than thirty years ago in medical school, then returned to the disease decades later to find it little less a mystery, the contemporary image of the cancer cell is both thrilling and an unexcelled vindication for fundamental research. The image was forged from the vantage point of molecular genetics and with the tools of that discipline. The lines of discovery trace back to minuscule columns of hydroxyapatite from which the molecular probe for *src* first flowed; to *src* itself, which even now has not been persuasively implicated in the genesis of human cancer; and to the chickens in which Peyton Rous first found his tumor. From these humble roots, a great tree of knowledge has grown. And a great truth has been reiterated: we cannot prejudge the utility of scholarship, we can only ask that it be sound.

Prospects for therapy

What of treatment? Will we acquire new antidotes for cancer from our study of oncogenes? There is little likelihood that we will be able to repair or replace damaged proto-oncogenes in the foreseeable future, particularly in the individual already burdened with countless tumor cells. There is talk of restoring functional copies of recessive oncogenes to tumors in which they are defective (78). But realization of this objective in human subjects presently seems many years distant.

If we focus on the protein handmaidens of genes, however, we can see more cause for hope. Given sufficient information about how these proteins act, the pharmaceutical chemist or the immunotherapist may be able to invent ways to interdict their action, even to exploit the specificity of genetic damage and thus to reverse the effects of oncogenes. We are not close to implementing this strategy, but it is a reasonable hope.

I am eager not to appear naive. No single therapy against an oncogene product is likely to become a panacea. We must deal with a large variety of oncogenes, whose products are deployed to the many reaches of the cell and whose actions present great chemical and enzymological diversity; and we must be prepared to cope with evolving genetic damage within cancer cells that can bring a variety of oncogenes into play sequentially. Nevertheless, the search for genetic damage in cancer cells and the explication of how that damage affects the biochemical functions of genes have become our best hope to understand and thus to thwart the ravages of cancer.

Conclusion

At the beginning of this century, the Austrian engineer and novelist Robert Musil offered a description of progress in science that foreshadowed modern views of epistemology and that now exemplifies the course of contemporary cancer research.

"... every few years, ...something that up to then was held to be error suddenly revolutionizes all views, ...an unobtrusive, despised idea becomes the ruler over a new realm ..." (79).

Harold Varmus, I and our numerous colleagues have been privileged to assist as a despised idea became the ruler over a new realm. The notion that genetic changes are important in the genesis of cancer has met strenuous resistance over the years. But now that notion has gained ascendancy. In the event, I have learned that there is no single path to creativity: we are constrained not by the necessary discipline of rigor, but by the limits to our imaginations and our intellectual courage. In the words of an American sage: "Dare to be wrong, or you may never be right" (80).

Discovery takes two forms. The first is mundane, but nevertheless legitimate: we grope our way to reality and then recognize it for what it is. The second is legitimate, but also sublime: we imagine reality as it ought to be and then find the próof for our imaginings. I have been fortunate to know the first form of discovery and am thankful for the privilege. I have miscarried opportunities to know the second and am diminished by the failure. Redemption lies in more imaginings.

"The real truths are those that can be invented" (81).

ACKNOWLEDGEMENTS

I owe a life-long debt to Harold Varmus, with whom I shared most of the events described here. The whole has been greater than the sum of the two parts, or so I believe. Numerous students and postdoctoral fellows have participated in the discovery and pursuit of proto-oncogenes with Harold and myself. Without their dedicated and talanted efforts, I would have no story to tell. Since I could not name them all in the text, I have named none. Many of their contributions are documented in the bibliography to this manuscript. I recall the memory of Richard Parker, for his special impact. I am also grateful to colleagues in the scientific community who have helped me at crucial times in my career, including David Baltimore, Howard Berg, Peter Duesberg, Howard Goodman, Robert Huebner, Malcolm Martin, John Menninger, Aaron Shatkin, Howard Temin and Peter Vogt. I acknowledge my many colleagues at the University of California, San Francisco, whose friendship and support have been vital, particularly Bruce Alberts, Mitzi Best, Herbert Boyer, Lois Fanshier, Julius Krevans, Jean Jackson, Marc Kirschner, Tom Kornberg, Suzanne Ortiz, Nancy Quintrell, Rudi Schmid, Lois Serxner, Holly Smith and Karen Smith. I thank my wife and family for a gift of forbearance that cannot be repaid. My work has been supported by both the National and California Divisions of the American Cancer Society and by the National Institutes of Health. The support of biomedical research in the United States is an act of public altruism unexcelled in history. I dedicate this manuscript to the countless individuals who sustain that support, by voluntary contributions and by the payment of state and federal taxes.

REFERENCES

1. R. Dulbecco, in *Biographical Memoirs, Vol. XLVIII,* National Academy of Sciences, Washington, D.C., pp. 275–306, 1976.
2. H. E. Varmus, in *Les Prix Nobel,* Almqvist & Wiksell Int'l., Stockholm, p. 187, 1989.
3. J. M. Bishop, D. F. Summers and L. Levintow, *Proc. Natl. Acad. Sci. USA,* **54,** 1273–1281 (1965).
4. J. M. Bishop, G. Koch, B. Evans and M. Merriman, *J. Mol. Biol.* **46,** 235–249 (1969).

5. M. Best, B. Evans and J. M. Bishop, *Virology,* **47,** 592–603 (1972).
6. B. L. Bass, H. Weintraub, R. Cattaneo and M. A. Billeter, *Cell* **56,** 331 (1989).
7. D. Baltimore, *Nature* **226,** 1209–1211 (1970).
8. H. M. Temin and S. Mizutani, *Nature* **226,** 1211–1213 (1970).
9. A. Kornberg, *Science,* **131,** 1503–1508 (1960).
10. J. Leong *et al., J. Virol.* **9,** 891–902 (1972).
11. S. R. Weiss, H. E. Varmus and J. M. Bishop, *Cell,* **12,** 983–992 (1977).
12. B. Cordell, S. R. Weiss, H. E. Varmus and J. M. Bishop, *Cell* **15,** 79–92 (1978).
13. T. Jacks and H. E. Varmus, *Science,* **230,** 1237–1246 (1985).
14. A. J. Faras, J. M. Taylor, W. Levinson, J. Goodman and J. M. Bishop, *J. Mol. Biol.* **79,** 163–183 (1973).
15. J. E. Dahlberg *et al., J. Virol.* **13,** 1126–1133 (1974).
16. H. E. Varmus, *Sci. Amer.* **257,** 56–66 (1987).
17. G. S. Martin, *Nature,* **227,** 1021–1023 (1970).
18. S. Kawai and H. Hanafusa, *Virology,* **46,** 470–479 (1971).
19. P. K. Vogt, in *Comprehensive Virology, Vol. 9* (H. Fraenkel-Conrat and R. R. Wagner, Eds.), Plenum Press, New York and London, pp. 341–455 (1977).
20. H. Oppermann, J. M. Bishop, H. E. Varmus and L. Levintow, *Cell* **12,** 993–1005 (1977).
21. J. S. Brugge and R. L. Erikson, *Nature* **269,** 346–347 (1977).
22. M. S. Collett and R. L. Erikson, *Proc. Natl. Acad. Sci. USA* **75,** 2021–2024 (1978).
23. A. Levinson, H. Oppermann, L. Levintow, H. E. Varmus and J. M. Bishop, *Cell* **15,** 561–572 (1978).
24. T. Hunter and B. M. Sefton, *Proc. Natl. Acad. Sci. USA* **77,** 1311–1315 (1980).
25. T. Hunter and J. A. Cooper, *Ann. Rev. Biochem.* **54,** 897–931 (1985).
26. A. D. Levinson, H. Oppermann, H. E. Varmus and J. M. Bishop, *J. Biol. Chem.* **255,** 11973–11980 (1980).
27. J. M. Bishop, *Cell* **42,** 23–38 (1985).
28. T. Hunter, *Sci. Amer.* **251,** 70–80 (1984).
29. M. Barbacid, *Ann. Rev. Biochem.* **56,** 779–827 (1987).
30. P. K. Vogt and R. Tjian, *Oncogene* **3,** 3–9 (1988).
31. J. M. Bishop, *Cell,* **23,** 5–6 (1981).
32. P. H. Duesberg, *Science* **228,** 669 (1985).
33. R. Jove and H. Hanafusa, *Ann. Rev. Cell Biol.* **3,** 31–57 (1987).
34. Theodosius Dobzhansky, as cited in D. J. Futuyma, *Science on Trial: The Case for Evolution,* Pantheon Books, New York, p. 114 (1983).
35. R. J. Huebner and G. J. Todaro, *Proc. Natl. Acad. Sci. USA* **64,** 1087–1094 (1969).
36. H. Temin, in *Nobel Lectures in Molecular Biology 1933–1975,* Elsevier-North Holland, Inc., New York, p. 509–529 (1977).
37. J. M. Bishop *et al.,* in *The Molecular Biology of the Mammalian Genetic Apparatus, Vol. 2* (P. O. P. Ts'o, Ed.), North-Holland Publishing Co., Amsterdam, p. 277–287 (1977).
38. H. E. Varmus, *Ann. Rev. Genet.* **18,** 553–612 (1984).
39. D. Stehelin, H. E. Varmus, J. M. Bishop and P. K. Vogt, *Nature* **260,** 170–173 (1976).
40. D. Spector *et al., Cell* **3,** 371–379 (1978).
41. D. Spector, B. Baker, H. E. Varmus and J. M. Bishop, *Cell* **3,** 381–386 (1978).
42. M. S. Collett, E. Erikson, A. F. Purchio, J. S. Brugge and R. L. Erikson, *Proc. Natl. Acad. Sci, USA* **76,** 3159–3163 (1979).
43. H. Oppermann, A. D. Levinson, H. E. Varmus, L. Levintow and J. M. Bishop, *Proc. Natl. Acad. Sci. USA* **76,** 1804–1808 (1979).
44. A. J. Langlois, S. Sankaran, P. H.-L. Hsuing and J. W. Beard, *J. Virol.* **1,** 1082–1084 (1967).

45. D. Sheiness, K. Bister, L. Fanshier, C. Moscovici and J. M. Bishop, *J. Virol.* **33,** 962–968 (1980).
46. M. Roussel *et al., Nature,* **281,** 452–455 (1979).
47. P. Mellon, A. Pawson, K. Bister, G. S. Martin and P. H. Duesberg, *Proc. Natl. Acad. Sci. USA* **75,** 5874–5878 (1978).
48. R. W. Ellis *et al., J. Virol.* **36,** 408–420 (1980).
49. B. Vennstrom, D. Sheiness, J. Zabielski and J. M. Bishop, *J. Virol.* **42,** 773–779 (1982).
50. K. Alitalo *et al., Proc. Natl. Acad. Sci. USA,* **80,** 100–104 (1983).
51. B. Vennstrom *et al., EMBO J.* **3,** 3223–3231 (1984).
52. D. Sheiness and J. M. Bishop, *J. Virol,* **3,** 514–521 (1979).
53. D. K. Sheiness, S. H. Hughes, H. E. Varmus, E. Stubblefield and J. M. Bishop, *Virology* **105,** 415–424 (1980).
54. W. S. Hayward, B. G. Neel and S. M. Astrin, *Nature* **290,** 475–480 (1981).
55. G. Klein, *Cell* **32,** 311–315 (1983).
56. M. D. Cole, *Ann. Rev. Genet.* **20,** 361–378 (1986).
57. K. Alitalo and M. Schwab *Adv. in Cancer Res.* **47,** 235–282 (1986).
58. J. M. Bishop, *Science* **235,** 305–311 (1987).
59. H. Beug, M. J. Hayman and T. Graf, in *Cancer Surveys, Vol. 1* (M. F. Greaves, Ed.,), Oxford University Press, U. K., pp. 205–230 (1982).
60. Shilo, B.-Z., *Trends in Genetics,* 69–73 (1987).
61. F. G. Haluska, Y. Tsujimoto and C. M. Croce, *Ann. Rev. Genetics,* **21**, 321–347 (1987).
62. G. R. Stark, M. Debatisse, E. Giulotto and G. M. Wahl, *Cell* **57**, 901–908 (1989).
63. M. Schwab *et al., Nature* **305,** 245–248 (1983).
64. M. Schwab *et al., Proc. Natl. Acad. Sci. USA,* **81,** 4940–4944 (1984).
65. N. E. Kohl *et al., Cell* **35,** 359–367 (1983).
66. L. W. Stanton, M. Schwab and J. M. Bishop, *Mol. Cell. Biol.* **83,** 1772–1776 (1986).
67. N. E. Kohl *et al., Nature* **319,** 73–77 (1986).
68. K. M. Downs, G. R. Martin and J. M. Bishop, *Gen. Dev.* **3,** 860–869 (1989).
69. G. M. Brodeur *et al., Science* **224,** 1121–1124 (1984).
70. R. C. Seeger *et al., New Eng. J. Med.* **313,** 1111–1116 (1985).
71. D. J. Slamon, *New Eng. J. Med.* **317,** 955–957 (1987).
72. G. Klein, *Science* **238,** 1539–1545 (1987).
73. D. J. Slamon *et al., Science* **235,** 177–182 (1987).
74. D. J. Slamon *et al., Science* **244,** 707–712 (1989).
75. E. Liu, B. Hjelle, R. Morgan, F. Hecht and J. M. Bishop, *Nature* **330,** 186–189 (1987).
76. H. Y. Hirai *et al., Nature* **327,** 430–432 (1987).
77. R. A. Padua, *Leukemia* **2,** 503–510 (1988).
78. H.-J. S. Huang *et al., Science* **242,** 1563–1566 (1988).
79. R. Musil, *The Man Without Qualities - I,* (E. Wilkins and E. Kaiser, Trans.), Pan Books Ltd., U. K., p. 41 (1979).
80. Attributed to Fats Waller, on constructing musical chords.
81. K. Kraus, in *Half-Truths & One-and-a-Half Truths* (H. Zohn, Ed. and Trans.), Carcanet Press, Ltd., Manchester, U. K., p. 61, 1986.

1990
Physiology or Medicine

JOSEPH E. MURRAY
E. DONNALL THOMAS

"for their discoveries concerning 'organ and cell transplantation in the treatment of human diseases'"

THE NOBEL PRIZE IN PHYSIOLOGY OR MEDICINE

Speech by Professor GÖSTA GAHRTON of the Karolinska Institute.
Translation from the Swedish text.

Your Majesties, Your Royal Highnesses, Ladies and Gentlemen,

During the 19th century, the association between disease symptoms and organ damage was well understood. Trouble with urine could be caused by damage of the kidney, and if the skin was yellow the cause could be in the liver. Damage to the kidney was most often incurable. Therefore, it was thought very early that perhaps a new undamaged organ from somebody else could cure the disease. Thus, at the turn of the century many heroic attempts were made to transplant kidneys from swine, sheep and goats, however without success. In 1902, attempts were made to transplant a kidney from one human being to another, again with no success. Very soon it was discovered that it was possible to transplant an organ or tissue within an individual without harm, but not between individuals. In 1912, Alexis Carrel received the Nobel Prize among other things for his discoveries concerning transplants of blood vessels and organs. However, this success was limited to transplants within an individual. Carrel concluded that there was a biological force that prevented transplantation between individuals, and he believed that it would never be possible to succeed in having an organ from one individual function in another. He received support for his belief from among others, the 1960 Nobel Prize-Winner, Peter Medawar, who discovered the role of the immune defence system in rejection of a graft and also showed that the biological force defined by Carrel was of an immunological nature.

Joseph Murray was not discouraged by this knowledge. There were reasons to believe that the immunological barrier was lacking between identical twins. Joseph Murray developed a surgical technique for kidney transplantation in dogs and showed that a kidney that was transplanted from one dog to the other could be induced to function. He used the technique in the first successful kidney transplant between identical twins in December 1954. Richard Herrick, who had incurable kidney damage was the first candidate. In order to make sure that he and his brother Ronald were identical twins, Joseph Murray asked the police in Boston to document their fingerprint patterns. During a routine review of police records, journalists found out about the investigation and its confidentiality was breached. However, Richard Herrick appeared to take this leakage to the press calmly. He became the darling of the media. The operation worked out perfectly and the kidney functioned well. Richard Herrick married his recovery-room nurse and became the father of two children. He lived happily for eight years when he died of a heart infarction. Joseph Murray later performed several other transplants between identical twins. However, most patients with incurable kidney damage had no twins, and it was therefore some time before such patients could become transplant candidates.

About two years later Donnall Thomas attempted to use bone marrow transplantation to cure terminal cancer patients, most of them with leukemias or cancers involving the bone marrow. He first treated the patients with bone-marrow-ablative total body irradiation. The goal was to cure the patients from the cancer disorder and to kill the bone marrow cells. Donnall Thomas showed that it was possible to remove about one liter of bone marrow out of the bones of a healthy individual. It was possible to give this marrow to the cancer patient by infusing it into a blood vessel. The bone marrow cells found the right spots in the new body where they could produce new normal and functioning blood cells, which soon appeared in the circulatory system. However, the healthy marrow also contained the defense cells, and these attacked their new host. The result was unfortunately a reversed and deadly rejection reaction called the graft-versus-host-reaction.

During the 1950s and 1960s some discoveries were made that were of the utmost importance for future successes in transplantion research. Jean Dausset discovered human transplantation antigens, a kind of fingerprints of the cells in the body. He was rewarded for this discovery with the Nobel Prize in 1980. About the same time George Hitchings and Gertrud Elion discovered the first cytotoxic drugs for which they were awarded the Nobel Prize in 1988. These cytotoxic drugs could also diminish the rejection reaction. Joseph Murray first used total body irradiation in attempts to prevent this reaction. Later he and others showed that azathioprin, one of the drugs discovered by Hitchings and Elion, was the most effective of these drugs in preventing rejection. This led to the first successful kidney transplantation between relatives that were not identical twins, and also to the first successful transplantation using kidneys from deceased persons. The best results were obtained when donors were selected who matched the patient's transplantation antigens. Kidney and organ transplantation was established as a treatment method. Today about 20,000 kidneys are transplanted every year, and more than 100,000 patients have gained a new and better life after transplantation.

Donnall Thomas managed to diminish the graft-versus-host reaction using the cytotoxic drug methotrexate. He showed very soon that if a donor, usually a sibling, was selected by typing for transplantation antigens, it was possible to cure leukemia, certain inherited disorders of the bone marrow, and the severe blood disorders, aplastic anemia and thalassemia. More than 10,000 patients have been cured, or have been given a normal life, with the help of bone marrow transplantation.

Dr. Joseph Murray and Dr. Donnall Thomas,

On behalf of the Nobel Assembly of the Karolinska Institute I would like to congratulate you on your outstanding accomplishments and ask you to receive the Nobel Prize in Physiology or Medicine from the hands of His Majesty the King.

Joseph E. Murray

JOSEPH E. MURRAY

I was born, as were my father and his parents, in Milford, Massachusetts, a town 30 miles southwest of Boston. My father's parents were of Southern Irish and English extraction. My mother was born in Providence, Rhode Island, soon after her parents had emigrated to the United States from Italy. Father was a lawyer and a District Court Judge, mother a school teacher. Both parents had benefited from and stressed the value of the educational opportunities this country offered. By example and precept they emphasized the need for service to others.

From earliest memory I wanted to be a surgeon, possibly influenced by the qualities of our family doctor who cared for our childhood ailments. As a second year high school chemistry student, I still have a vivid memory of my excitement when I first saw a chart of the periodic table of elements. The order in the universe seemed miraculous, and I wanted to study and learn as much as possible about the natural sciences.

I chose to attend a small liberal arts college, College of the Holy Cross, and concentrated on Latin, Greek, Philosophy and English. Assuming I'd receive ample science in medical school, I took the minimum of chemistry, physics and biology.

My four years at Harvard Medical School were all that I had dreamed they would be. The classmates and faculty were stimulating and friendly. The hospitals were filled with all varieties of patients. Although the hours of study and hospital duty were long, life was rich and full. Symphony Hall and the Gardner Museum were within walking distance, squash courts were available for daily exercise, our singing group met weekly, bicycle trips and club dances added to the variety. It was heaven.

During the final few months of medical school, while attending a Boston Symphony Orchestra concert with several classmates and their dates, I noticed a lovely young lady "far too nice" for the fellow she was with. At intermission I manipulated her towards the corridor and learned that she was Bobby Link, a music student concentrating on voice and piano. By the time the intermission had ended I realized that I had met the girl I would marry.

After intermittent dates during my internship and brief meetings during hectic wartime weekends while I was on active duty, Bobby and I were married in June 1945. We have six children, three boys and three girls. Each has contributed to society in her/his own way, in education, medicine, nursing, business and science. Bobby's music, pursued professionally for 15 years after marriage, continually adds to the richness and beauty of our family and social life.

My only medical school activity bearing any resemblance to research was a study of the then new Papanicolau smear of epithelial cells. I presented a report before the student Boyleston Society with Dr. Arthur Hertig as my faculty sponsor. Later, while a surgical intern at the Peter Bent Brigham Hospital, I introduced this technique clinically.

My interest in the biology of tissue and organ transplantation arose from my military experience at Valley Forge General Hospital in Pennsylvania. As a First Lieutenant with only a nine-month surgical internship behind me, I was randomly assigned to VFGH to await overseas duty. World War II was still raging, the Rhine River had not been crossed, the Battle of the Bulge was ahead.

VFGH was a major plastic surgical center. While there, I spent all my available spare time on the plastic surgical wards which were jammed with hundreds of battle casualties. I enjoyed talking to the patients, helping with dressings, and observing the results of the imaginative reconstructive surgical operations.

I learned only years later that Colonel James Barrett Brown, the Chief of Plastic Surgery, had noticed my day and night presence on the wards and requested that Lt. Murray be kept at VFGH and not sent overseas like the rest of the "nine-month wonders." Three years later, two years after the war ended, I finally was discharged in November 1947.

During my army service, we always had many burned patients to care for. Some were so extensively burned that donor sites for skin autografts were not available. As a life-saving measure for these patients, skin grafts were taken from other persons and used as a temporary surface cover.

The slow rejection of the foreign skin grafts fascinated me. How could the host distinguish another person's skin from his own? Colonel Brown and I often discussed this while scrubbing. In civilian life Brown had treated many severely burned patients with temporary skin allografts and observed and written about the differential dissolution of skin allografts from various donors. He tentatively postulated that the closer the genetic relationship between the skin donor and the recipient, the slower the dissolution of the graft. In 1937, he had experimentally cross skin grafted a pair of identical twins and documented permanent graft survival in both twins. This was the impetus to my study of organ transplantation, which is the subject of my Nobel Lecture.

My life as a surgeon-scientist, combining humanity and science, has been fantastically rewarding. In our daily patients we witness human nature in the raw-fear, despair, courage, understanding, hope, resignation, heroism. If alert, we can detect new problems to solve, new paths to investigate.

Our laboratory work involved close contact with many non-clinical scientists. Sir Peter Medawar, 1960 Nobel Laureate, was a frequent visitor to our lab and to the hospital. He once commented, after visiting an early renal transplant patient, that it was the first time he had been in a hospital ward. Dr. George Hitchings and Dr. Trudy Elion, 1988 Nobel Laureates, were completely at home in our lab and knew many of the dogs by name. Sir Roy

Calne, who worked in our laboratory at Harvard Medical School and Peter Bent Brigham in 1960–61 as a Surgical Research Fellow, and I frequently visited them in Tuckahoe, New York, to discuss prospective trial drugs. Billingham, Eichwald, Amos, van Rood—to mention only a few other basic investigators—also enriched the tapestry of our lives.

Medawar said it best, "This whole period was a golden age of immunology, an age abounding in synthetic discoveries all over the world, a time we all thought it was good to be alive. We, who were working on these problems, all knew each other and met as often as we could to exchange ideas and hot news from the laboratory."

For recreation, I have always been a physical enthusiast. As a family we have camped, hiked, trekked, or backpacked over portions of five continents. Competitive tennis remains fun. Our extended family, with 11 grandchildren, gets together frequently during the year, and always every summer on Martha's Vineyard Island in Massachusetts.

We have been blessed in our lives beyond my wildest dreams. My only wish would be to have ten more lives to live on this planet. If that were possible, I'd spend one lifetime each in embryology, genetics, physics, astronomy and geology. The other lifetimes would be as a pianist, backwoodsman, tennis player, or writer for the *National Geographic*. If anyone has bothered to read this far, you would note that I still have one future lifetime unaccounted for. That is because I'd like to keep open the option for another lifetime as a surgeon scientist.

THE FIRST SUCCESSFUL ORGAN TRANSPLANTS IN MAN

Nobel Lecture, December 8, 1990

by

JOSEPH E. MURRAY

Harvard Medical School, Boston, Massachusetts, U.S.A.

Preface

"If gold medals and prizes were awarded to institutions instead of individuals, the Peter Bent Brigham Hospital of 30 years ago would have qualified. The ruling board and administrative structure of that hospital did not falter in their support of the quixotic objective of treating end-stage renal disease despite a long list of tragic failures that resulted from these early efforts – leavened only by occasional encouraging notations such as those in the identical twin case. Those who were there at the time have credited Dr. George Thorn, chairman of medicine and Dr. Francis D. Moore, chairman of surgery, with the qualities of leadership, creativity, courage, and unselfishness made the Peter Bent Brigham Hospital a unique world resource for that moment of history." (1)

Introduction

Although renal transplantation had been performed sporadically during the first half of this century (2)(3), planned programs for human organ transplantation started only in the late 1940's. At that time clinicians in Paris, London, Edinburgh and Boston began renal transplantation in unmodified human recipients in spite of the warnings and pessimistic predictions of many scientists and experienced clinicians.

Many bio-scientists had difficulty understanding the determined optimism of clinicians who were willing to evaluate any type of treatment which might possibly help these terminally-ill uremic patients, most of whom were young and otherwise healthy. Tantalizing reports of functioning human renal transplants had surfaced from time to time (4)(5)(6); these hints of success were further encouragement.

In this lecture I will focus on the renal transplant program of The Peter Bent Brigham Hospital (now The Brigham & Women's Hospital) in Boston and explain how this small hospital became involved in transplantation. The medical and surgical services, along with the Department of Pathology under Dr. Gustave J. Dammin, led the way and ultimately renal transplantation involved most of the hospital in some way or other.

The full story of successful organ transplantation in man weaves together three separate pathways: the study of renal disease, skin grafting in twins, and surgical determination. A leitmotif permeates each of these pathways, i.e. a single event or report was critical for medical progress.

Renal Disease

The first two Physicians-in-Chief, the Hersey Professor at Harvard in their day, Dr. Henry Christian and Dr.Soma Weiss, had a major interest in renal disease. When Dr. Thorn succeeded Dr. Weiss in 1943 he and his associate Dr. James O'Hare continued this interest, especially the relationship of renal disease to hypertension. After World War II Dr. Thorn invited Dr. Willelm Kolff from The Netherlands to Boston to demonstrate a dialysis machine which he had developed during his forced confinement by the Germans. Dr. Carl W. Walter helped to improve the design and thus the Kolff-Brigham "artificial kidney" was devised. It was first used in patients in 1948 and set the stage for extensive new innovative approaches to both acute reversible renal disease and end-stage failure.

Because renal dialysis provided only temporary improvement for the patient, it was logical to seek a more permanent therapy.*

Skin Grafting in Twins

This thread in the story involves the biological phenomena of monozygotic and dizygotic twinning. The monozygotic (MZ), "identical", twin experience starts with the treatment of burns, the dizygotic (DZ), "non-identical", twin story begins with freemartin cattle.**

In 1932, Dr. E. Padgett of Kansas City reported the use of skin allografts from family and unrelated donors to cover severely burned patients who had insufficient unburned donor sites for the harvesting of autografts. Although none of these skin allografts survived permanently, many would remain long enough to control infection and fluid loss and thus gain time for the donor sites to re-epithelialize. It was difficult to determine accurately the duration of survival of any one allograft; some seemed to melt away slowly and be replaced by adjacent skin, others seemed to be rejected rapidly (7).

Skin grafts from family members seemed to survive longer than those from unrelated donors. But even after observing hundreds of skin allografts, one could not be certain about their survival time. One certainty was established when Dr. J.B. Brown of St. Louis in 1937 achieved permanent survival of skin grafts exchanged between MZ twins (8).

This single observation, although restricted in application, was the only ray of light in the problem of tissue and organ replacement until Gibson and Medawar demonstrated that a second allograft from the same donor was rejected more rapidly than the first (9). This clear description of the

* (Chronic dialysis was not developed until ten years later by Schribner in Seattle). In the late 1940's during a Grand Rounds at the Brigham, I was astounded to hear Dr. Thorn say, "The best way to treat hypertension is to remove both kidneys!" The entire audience gasped. The seed for the Brigham renal transplant program had been planted.

** (The spelling varies: one word, two words or hyphenated.)

"second set" phenomenon established that the rejection process was not immutable; instead it implied an allergic or immunological process which potentially might be manipulated.

The dizygotic twin story starts with John Hunter's description of freemartin cattle in 1779 (10). Freemartins are twin cattle in which the male is normal and the female sterile. Hunter cites Roman descriptions of the phenomenon and then described the physical characteristics of several pairs he had collected in England.

The trail does not appear again until 1917 when Lillie, not content with mere descriptions, dissected the placentae of several pairs of freemartin cattle and noted the placental intermingling of blood between these differently sexed twins (11). Thirty years elapsed before Owen published on the tolerogenic consequences of this placental intermingling of circulation (12). Following this Anderson in 1949 reported successful skin allografts between the freemartin and the normal male (13).

The freemartin story culminates in the report of Billingham, Brent and Medawar describing an acquired immunological tolerance produced by neonatal injection of donor cells into a future allograft recipient (14). They indicate that it was the experimental counterpart to Owen's naturally occurring model. Although not applicable to the clinical situation, their experimental breeching of the immunological barrier was another impetus for optimism in the problem so many considered hopeless.

Sir Michael Woodruff, the pioneer transplant surgeon in Edinburgh, confirmed the freemartin concept in man when he found a pair of twins, one male the other female, who shared elements of different red cell types. Postulating a shared placental circulation between the two, he cross skin grafted them successfully (15).

Surgical Determination

In 1912, Dr. A. Carrel received a Nobel Prize "in recognition of his work on vascular suture and the transplantation of blood-vessels and organs". He clearly recognized the difference in the survival times between autografts and allografts in experimental animals, but he did not conceptualize rejection as distinct from other graft-destroying processes.

Quinby in 1916 used the canine renal autograft model to study the effect of denervation on renal function (16). Mann and Williamson a decade later noted the different survival times between canine renal autografts and allografts (17)(18), but like Carrel, they did not pursue the long-term fate of the autografts. After World War II, Dempster (19) and Simonsen (20) published extensively on canine renal transplantation concentrating on the biology and biochemistry of allograft rejection. They demonstrated that skin and kidney allografts possess a common antigen which could sensitize a recipient to a subsequent allograft of either tissue from the same donor. In these reports, there was the tacit assumption that renal autograft function would deteriorate ultimately, possibly because of lack of nerve supply and/or lymphatics.

From a physiological view, if human renal transplantation were to be successful, we needed to establish that renal transplants in the absence of an immunological barrier could function permanently. In the course of many laboratory experiments on canine renal transplantation, I had developed a reproducible operation using intra-abdominal vascular anastomoses and a uretero-cystostomy for urinary drainage, placing the kidney in the lower abdomen. This has become the universal renal transplant procedure since that time. Complete functional studies of some of these autografted kidneys two years after transplantation proved them to be completely normal (21).

The Three Trails Merge

These three trails merged at the Peter Bent Brigham in the late 1940's. All the elements for a sound renal transplant program were in order: experienced knowledge in renal disease, availability of dialysis, and skilled imaginative surgeons. To minimize morbidity, the first allografts in these unmodified human recipients were added as a third kidney in the thigh under local anesthesia. Dr. David Hume was the surgeon for these patients and he anastomosed the renal vessels of the graft to the femoral vessels of the recipient. Urine was collected in a bag from a skin ureterostomy (22).

Several of these unmodified human allografts functioned better than experimental canine allografts would have predicted. Possible explanations were an immunosuppressive effect of uremia or a beneficial effect of acute tubular necrosis (ATN) which occurred regularly in these inadequately preserved donor kidneys. One thigh transplant functioned for almost six months with return of the patient's biochemical profile and blood pressure to normal, demonstrating that transplants could rectify the pathophysiologic disorder of renal insufficiency.

The very first renal transplant in 1945 at the Brigham deserves special comment. The patient was a young woman in renal failure following obstetrical complications. The purpose of the transplant was to provide temporary renal function until her own kidneys recovered from acute tubular necrosis. Dr. Thorn recalls his inability to obtain permission to have the patient transferred to a regular operating room (this was prior to Dr. Moore's tenure) so the operation was performed on the old E-Second Ward by Dr. Charles Hufnagel, then a Research Fellow working on vascular grafts, Dr. Ernest Landsteiner, then Chief Resident in Urology, and Dr. David Hume, then Assistant Resident in Surgery. The donor kidney was anastomosed in the antecubital space under local anesthesia using a cutaneous ureterostomy.

According to Dr. Robert J. Glaser, who was assistant resident on the medical service at that time, "secretion of urine was minimal, and certainly did not, 'rescue the woman from her crisis'. The kidney functioned poorly and only transiently, and the patient continued to have a stormy course, although fortunately, despite our lack of understanding at the time of how best to treat renal shutdown, she ultimately did respond and she left the hospital with normal renal function and in good health."

Dr. Glaser further reports that her happy state was short-lived because she died a few months later of fulminating hepatitis secondary to pooled plasma infusions which she had received in the course of her treatment. Interestingly, Dr. Glaser still recalls taking care of the patient whose kidney was ultimately used as the donor transplant. "The patient had disseminated lupus erythematosis and had been in the Brigham many times. Although in patients with advanced lupus the kidney is usually badly damaged, in this particular case renal manifestations were relatively limited, and when her kidney became available it was therefore used"(23).

The Identical Twin Patient

In the fall of 1954, Dr. Donald Miller of the U.S. Public Health Service telephoned Dr. Merrill in order to refer a patient with severe renal disease. Moreover, Dr. Miller suggested there might be the opportunity for transplantation of a kidney because the patient had a healthy twin brother. Needless to say, the transplant team was interested in the possibility of transplanting a genetically compatible kidney. Cross skin grafting established genetic identity, renal disease was brought under control with medications and dialysis, and we were ready to apply our laboratory-tested surgical technique to man.

The only remaining problem was the ethical decision concerning the removal of a healthy organ from a normal person for the benefit of someone else. For the first time in medical history a normal healthy person was to be subjected to a major surgical operation not for his own benefit. After many consultations with experienced physicians within and outside the Brigham and with clergy of all denominations, we felt it reasonable to offer the operations to the recipient, the donor and their family. We discussed in detail the preparations, anesthesia, operations, possible complications and anticipated result.

At the conclusion of our last pre-operative discussion, the donor asked whether the hospital would be responsible for his health care for the rest of his life if he decided to donate his kidney. Dr. Harrison, the surgeon for the donor, said, "Of course not." But he immediately followed with the question, "Ronald, do you think anyone in this room would ever refuse to take care of you if you needed help?" Ronald paused, and then understood that his future depended upon our sense of professional responsibility rather than on legal assurances.

Once the patients and the team decided to proceed with the transplant, an extra professional burden falls on the surgeon performing the donor nephrectomy because his patient is expected to survive normally. In contrast, the surgeon performing the transplant is operating on a patient otherwise doomed to die, and the nephrologist caring for these critically ill patients cannot be faulted for failure to cure.

Post-operatively the transplanted kidney functioned immediately with a dramatic improvement in the patient's renal and cardiopulmonary status. This spectacular success was a clear demonstration that organ transplantation could be life saving. In a way, it was spying into the future because we had achieved our long-term goal by bypassing, but not solving, the issue of biological incompatibility (24)(25).

Subsequent Laboratory and Clinical Study

The impact was worldwide and stimulated widespread laboratory attempts to breech the immunological barrier. Experimental protocols included total body X-ray treatment followed by marrow infusion, immunoparalysis by consecutive graftings, immunological enhancement or adaptation by prior exposure of the host or graft to antigen, matching of donor and recipient by red or white cell typing, and the use of drugs such as toluene and nitrogen mustard.

We continued with both clinical and laboratory studies. In a series of volunteer uremic patients, we noted a prolonged but not permanent survival of skin allografts, suggesting the uremic state itself was immunosuppressive (26). In several series of dogs we tried without success to establish a state of renal insuffiency by partial removal of renal mass, infusion of toxins directly into the renal artery, temporary ischemia, and/or thermal insult. Attempts to prolong graft survival by treating the host with steroids and/or anticoagulants also failed (27).

To study the "X-ray marrow" protocol, which seemed to have the best potential for human application, we used mice and rabbits. Using sublethal or lethal doses of total body X-ray, followed by marrow infusions from single or multiple donors, we were able to obtain a limited number of long surviving skin allografts (28).

Simultaneously during the 1950's we transplanted several more sets of identical twins. One twin transplanted in 1956 completed a pregnancy two years later (29). She is now a grandmother and the longest living renal transplant recipient. Her donor, also a grandmother, likewise is in perfect health. Initially regarded as a unique occurrence, the identical twin situation has continued to reappear worldwide. It is estimated that at least fifty patients have now received transplants from their identical twins.

Several patients were referred during these years suffering from accidental loss of a solitary kidney. Because we had obtained limited encouraging laboratory results, it seemed reasonable to treat some of these patients with an "X-ray-marrow" protocol, i.e. total body X-ray followed by marrow infusion and a renal allograft. In most of the patients, the transplanted kidneys functioned immediately and continued to do so for several weeks, but in only one of twelve patients did function persist beyond three months.

The one success was our third patient, a dizygotic twin who received a sublethal, non-marrow requiring, dose of total body X-ray, given by Dr. James B. Dealy, followed by a kidney graft from his twin brother in 1959. He recovered after a difficult complicated post-operative course; he subsequently led a full active normal life until he died of cardiac problems 25 years later. He was the world's first successful renal allograft and was the enticement and stimulus for us to continue this method of procedure until drugs became available (30)(31)(32). The Hamburger group in Paris subsequently had a similar success with a dizygotic twin recipient following sublethal X-ray treatment.

The First Successful Cadaveric Transplant in Man

Although we began our first experiments in rabbits using ThioTEPA as a substitute for total body X-ray treatment in 1958 (33), the real breakthrough came with the introduction of immunosuppressive drugs by Schwartz and Dameshek in 1959 (34). They prevented rabbits from producing antibody against human serum albumin by treating them for two weeks with the antimetabolite, 6-Mercaptopurine. This "drug-induced tolerance" remained after drug treatment was stopped, even though the animals could react normally against another protein antigen, bovine gamma globulin. Thus, the tolerance seemed to be specific for an antigen introduced at the time of drug administration. Roy Calne in London (35) and Charles Zukoski in Virginia (36) tested this drug in the canine renal transplant model and had encouraging results.

On the advice of Sir Peter Medawar, in 1960 Calne came to Boston to work with me in Dr. Francis D. Moore's Department at the Harvard Medical School and the Peter Bent Brigham Hospital. Calne introduced us to Dr. Hitchings and Dr. Elion of the Burroughs-Wellcome Laboratories who became enthusiastic collaborators. Following Calne's arrival and with the use of drugs supplied by Dr. Hitchings the improvement in allograft survival was rapid and dramatic. Soon we had many bilaterally nephrectomized dogs in our laboratory living on solitary renal allografts. Some survived for months, eventually for years. One produced a normal litter sired by a drug treated allografted male. One dog recovered from a severe osteomyelitis of the mandible, indicating he was not an immunological cripple, a state we feared might result from prolonged use of the drugs (37). During this time we were testing other drugs from Hitchings and Elion, who were frequent visitors and knew most of our dogs by name. The experimental drug, BW-322, the imidazole derivative of 6-MP, seemed to have the best therapeutic index. This drug is now known as azathioprine, or Imuran, and was used throughout the world to support organ transplantation for 20 years. Now newer drugs are available and under study to extend their usefulness and diminish toxicity.

Reassured by these results, we decided to use these drugs in humans for immunosuppression. The first renal transplant recipient to receive azathioprine was an adult transplanted with an unrelated kidney in March 1961. The transplant functioned well for over one month, but the patient died of drug toxicity because the dosage required in dogs was toxic for man.

Our second patient also died of drug toxicity even though we halved the dose used for our first patient. For the first time in our experience we were able to reverse the rejection process. When we discontinued the drug because of leucopenia, rejection started to occur. As his leucopenia improved, we re-started the drug which reversed the rejection process and his renal function improved. Nevertheless he did succumb to sepsis after a month (38).

Our third patient, transplanted in April 1962, was treated with azathioprine following a cadaveric renal allograft. He survived over one year and was the world's first successful unrelated renal allograft. We reported these results in the New England Journal of Medicine (39) and a case report in the Journal of the American Medical Association followed (40). Dr. Willard Goodwin, at the University of California in Los Angeles, almost immediately introduced the use of corticosteroids as a further adjunct to the treatment (41). Subsequently, several transplantation groups worldwide began their own productive transplantation programs.

By 1965, one year survival rates of allografted kidneys from living related donors were approaching 80% and from cadavers 65%. Regional and national donor procurement programs were established along with an International Renal Transplant Registry (42). Optimism and enthusiasm were high as new drugs and other methods of immune suppression were tested along with refinements in tissue typing and improved organ preservation. Anti-lymphocyte serum and globulin prepared in horse, sheep and rabbit along with thoracic duct drainage of lymphocytes were among the more promising regimens tested. Currently it is estimated that more than 200,000 human renal transplants have been performed worldwide.

Other Organs

The success with renal allografts naturally led to attempts to transplant other organs. Moore developed the surgical technique for orthotopic canine liver transplantation (43), the model procedure used by Starzl for first successful human liver allografts (44). Calne, returning to Cambridge, England, also developed an extensive human liver transplantation experience. For almost 15 years Starzl and Calne performed the vast majority of the world's human liver transplants (45). Today, the operation is second only to kidney in frequency and is performed universally .

The next organ to be transplanted was the heart. Lower and Shumway had developed the surgical technique in dogs in 1961 and were planning a careful program for cardiac transplantation in humans. After Barnard's first human cardiac transplant in 1967, many other cardiac surgeons with little or no immunological background rapidly accumulated large numbers of heart-transplanted patients, only to witness them all die of rejection within a few months. This period between 1968-1970 was undoubtedly transplantation's darkest hour. The sole redeeming feature in heart transplantation was the continuation of Shumway's program in Stanford which achieved permanent success in 1970 (46). Today, with the development of newer drugs, cardiac transplantation is a recognized and accepted form of treatment.

Single and double lung transplantation have followed, as well as combined heart-lung transplants. Transplantation of the pancreas, with or without an accompanying renal graft, is commonly done. Multiple organ transplants in combination with liver and parts of the intestinal tract have also been successful. In 1989, there were 8,890 kidney, 2,160 liver, 1,673 heart, 413 pancreas, and 67 heart-lung transplants performed in the United States alone (47).

Ironically, allografts of skin, the tissue used classically in most of the early studies of transplantation, have proven to be the most difficult to transplant. Skin is the ultimate protection of the individual against the environment and therefore over the ages has evolved into our strongest barrier against foreign proteins. The earlier conventional wisdom was that the fate of skin allografts predicted the results of other transplants. Commenting on the contrasting survival rates of skin and kidney allografts in immunosuppressed dogs, Medawar proclaimed with his customary flair that the success of organ transplantation has "overthrown the doctrinal tyranny of skin grafts" (48).

The Future

Although thousands of young lives have already been saved by the use of various immunosuppressive regimens, serious complications still occur as a result of the treatment. The ultimate aim is to achieve an immunological tolerance between donor and recipient, eliminating entirely the need for drugs. There are hints both in the laboratory (49) and in man (50) that the liver itself can produce tolerogenic factors which may reduce or eliminate the need for immunosuppressive drugs. Discovering or uncovering naturally occurring immunosuppressive substances seems likely. It surely is as probable as the prospect of obtaining successful organ transplants was 45 years ago.

REFERENCES

1. Starzl, T.E.: The Landmark Identical Twin Case. JAMA. 251:2572, 1984
2. Moore, F.D.: Transplant, The Give and Take of Tissue Transplantation. Simon and Schuster New York, 1972, p. 66.
3. Groth, C.E.: Collective Review. Landmarks in Clinical Renal Transplantation. S.G.& O. 134: 323, 1972.
4. Lawlor, R.H., West, J.W., McNulty, P.H., Clancy, E.J. and Murphy, P.P.: Homotransplantation of the Kidney in the Human: Supplemental Report of a Case. JAMA. 147:45, 1951
5. Kuss, R. Teinturier, J., and Milliez, P.: Quelques Essais de Greffe du Rein chez L'Homme. Mem. Acad. Chir. 77:755, 1951
6. Michon,L., Hamburger,J., Economos,N., Delinotte,P., Richet,G., Vaysse,J. Antoine,B.: Une Tentative de Transplantation Renale chez L'Homme: Aspects Medicolaux et Biologiques. LaPresse Medicale: 61:1419, 1953
7. Padgett, E.C.: Is Iso-skin Grafting Practicable? Southern Med. J. 25:895, 1932
8. Brown, J.B.: Homografting of Skin: With Report of Success in Identical Twins. Surgery. 1:558, 1937
9. Gibson, T. and Medawar, P.B.: Fate of Skin Homografts in Man. J. Anat. 77:299, 1942–43

10. Hunter, J.: On the Free Martin. Royal Coll. Surg. XX, Feb. 25, 1779
11. Lillie, F.R.: The Theory of the Free-Martin Science. 43:611, 1916
12. Owen, R.D.: Immunogenetic Consequences of Vascular Anastomoses Between Bovine Twins. Science. 102:400, 1945
13. Anderson, D., Billingham, R.E., Lamkin G.H., and Medawar,PB: Use of Skin Grafting to Distinguish between Monzygotic and Dizygotic Twins in Cattle: Heredity. 5:379, 1951
14. Billingham, R.E., Brent, L. and Medawar, P.B.: Actively Acquired Tolerance of Foreign Cells. Nature. 172:603, 1953
15. Woodruff, M.F.A. and Lennox, B.: Reciprocal Skin Grafts in a Pair of Twins Showing Blood Chimerism. Lancet. 2:476, 1959
16. Quinby, W.C.: The Function of the Kidney When Deprived of Its Nerves. J. Exper. Med. 23:535, 1916
17. Williamson, C.S.: Further Studies on the Transplantation of the Kidney. J. Urology. 16:231, 1926
18. Sterioff, S., Rucker-Johnson, N.: Frank C. Mann and Transplantation at the Mayo Clinic. Mayo Clinic Proceedings. 62: 1051, 1987
19. Dempster, W.J.: The Homotransplantation of Kidneys in Dogs. Brit. J. Surg. 40:477, 1953
20. Simonsen, M., Buemann, J., Gammeltoft, A., Jensen, F. and Jorgensen, K.: Biological Incompatibilty in Kidney Transplantation in Dogs. Acta. Path. Microbiol. Scand. 32:1, 1953
21. Murray, J.E., Lang, S., Miller, B.J. and Dammin, G.J.: Prolonged Functional Survival of Renal Autografts in the Dog. SGO. 103:15, 1956
22. Hume, D.M., Merrill, J.P., Miller, B.F. and Thorn, G.W.: Experiences with Renal Homotransplantation in the Human: Report of Nine Cases. J. Clin. Inv. 34:327, 1955
23. Glaser, R.J. Footnotes to Kidney Transplant History. Focus: March 31, 1988, page 8. Published by Harvard University News Office for the Medical Area, 25 Shattuck Street, Boston, MA, 02115
24. Murray, J.E., Merrill, J.P. and Harrison, J.H.: Renal Homotransplantation in Identical Twins. Surg. Forum. 6:432, 1955
25. Merrill, J.P., Murray, J.E., Harrison, J.H. and Guild, W.R.: Successful Homo transplantation of the Human Kidney Between Identical Twins. JAMA. 160:277, 1956
26. Dammin, G.J., Couch, N.P. and Murray, J.E.: Prolonged Survival of Skin Homografts in Uremic Patients. Ann. NY Acad. Sci. 64:967, 1957
27. Lang, S., Murray, J.E. and Miller, B.F.: Homotransplantation of Ischemic Kidneys into Dogs with Experimentally Produced Impairment of Renal Function. Plas. Rec. Surg. 17:211, 1956
28. Wilson, R.E., Dealy, J.B., Sadowsky, N., Corson, J.M., and Murray, J.E.: Transplantation of Homologous Bone Marrow and Skin From Common Multiple Donors Following Total Body Irradiation. Surgery. 46:261, 1959
29. Murray, J.E., Merrill, J.P., and Harrison, J.H.:Kidney Transplantation Between Seven Pairs of Identical Twins. Ann. Surg. 148:343, 1958
30. Murray, J.E., Merrill, J.P., Dammin, G.J., Dealy, J.B., Walter, C.W., Brooke, M.S. and Wilson, R.E.: Study of Transplantation Immunity After Total Body Irradiation: Clinical and Experimental Investigation. Surgery. 48:272, 1960
31. Merrill, J.P., Murray, J.E., Harrison, J.H., Friedman, E.A., Dealy, J.B. and Dammin, G.J.: Successful Homotransplantation of the Kidney Between Non-identical Twins. New Eng. J. Med. 262:1251, 1960
32. Murray, J.E., Merrill, J.P., Dammin, G.J., Dealy, J.B., Alexandre, G.P.J. and Harrison, J.H.: Kidney Transplantation in Modified Recipients. Ann. Surg. 156:337, 1962

33. Porter, K.A., and Murray, J.E.: Homologous Marrow Transplantatiom in Rabbits After Triethylenethiophosphoramide (Thio-TEPA). AMA Archives of Surgery. 76: 908, 1958
34. Schwartz, R. and Dameshek, W.: Drug-Induced Immunological Tolerance. Nature. 183: 1682, 1959
35. Calne, R.Y.: The Inhibition of Renal Homograft Rejection in Dogs by 6 Mercaptopurine. Lancet. 1: 417 1960
36. Zukoski, C., Lee, H.M., and Hume, D.M.: The Prolongation of Functional Survival of Canine Renal Homografts by 6 Mercaptopurine. Surgical Forum. 11: 470, 1960
37. Calne, R.Y., Alexandre, G.P.J., and Murray, J.E.: A Study of the Effects of Drugs in Prolonging Survival of Homologous Renal Transplants in Dogs. Ann. NY Acad. Sci. 99: 743, 1962
38. Murray, J.E., Balankura, O., Greenburg, J.B. and Dammin, G.J.: Reversibility of the Kidney Homograft Reactgion by Retransplantaion and Drug Therapy. Ann. NY Acad. Sci. 99: 768, 1962
39. Murray, J.E., Merrill, J.P., Harrison, J.H., Wilson, R.E. and Dammin, G.J.: Prolonged Survival of Human-Kidney Homografts by Immunosuppresive Drug Therapy. N. E. J. Med. 268: 1315, 1963
40. Merrill, J.P., Murray, J.E., Takacs, F., Hager, E.B., Wilson, R.E. and Dammin, G.J.: Succesful Transplantation of Kidney from a Human Cadaver. JAMA. 185: 347, 1963
41. Goodwin, W.E., Kaufman, J.J., Mims, M.M., Turner, R.D., Glassock, R., Goldman, R. and Maxwell, M.M.: Human And Renal Transplantation I. Clinical Experiences with Six cases of Renal Transplantation. J. Urol. 89: 13,1963
42. Murray, J.E., Barnes, B.A. and Atkinson, J.C.: Fifth Report of the Human Kidney Transplant Registry. Transplantation. 5: 752, 1967
43. Moore, F.D., Smith, L.L., Burnap, T.K., Dallenbeck, F.D., Dammin, G.J.,Gruber, U.F., Shoemaker, W.C., Steenburg, R.W., Ball, M.R., & Belko, J.S.: One Stage Homotransplantation of the Liver following Total Hepatectomy in Dogs. Transplantation Bulletin. 6: 103, 1959
44. Starzl,T.E., Groth,C.G., Brettschneider,L., Penn,I., Fulginiti,V.A., Moon,J.B., Blanchard,H., Martin,A.J., Porter,K.A.: Orthotopic Transplantation of the Human Liver Ann. Surgery 168: 392,1968
45. Calne, R.C. & Williams, R. Liver Transplantation in Man. 1. Observations on Technique and Organization in Five Cases. Brit. Med. J. 4: 535, 1968
46. Dong,E., Griepp,R.B., Stinson,E.B., Shumway,N.E. Clinical Transplantation of the Heart. Ann. Surg. 176: 503, 1972
47. U.S. Dept. of Health and Human Services, Division of Organ Transplantation, 5600 Fishers Lane, Rockville, MD 20857
48. Medawar, P.B.: Transplantation of Tissues & Organs: Introduction. Brit. Med. Journal. 21: 97, 1965
49. Calne, R.Y., Sells, R.A., Pena, J.R., Davis, D.R., Millard, P.R.,Herbertson, B.M., Binns,R.M., Davies,D.A.L.: Induction of Immunological Tolerance by Porcine Liver Grafts. Nature. 223: 472, 1969
50. Davies, D.R., Pollard, S.G., and Calne, R.Y.: Forum on Immune Suppression: Hellenic Transplantation Society. Athens, Greece, Nov. 6–8, 1990

ANNOTATIONS FOR REFERENCES

Ref. 2. This book provides detailed background of the earliest attempts in human renal transplantation from around the world, as well as personalized documentation along the way. The various activities within the Peter Bent Brigham Hospital are detailed in a lively informative way.

3. This excellent collective review pays particular attention to the period from 1950 to 1965, but also documents the unsuccessful attempts in the first half of this century.

4. Originally proclaimed as a "success", this report excited much publicity. However, there was no solid evidence of transplant function. When the patient died a few years later, only scar tissue was found at the site of the transplant.

5 & 6. These two talented French groups, working independently, obtained definite evidence of transient renal transplant function in several patients. We all were in frequent communication with each other and shared the optimism that the problem eventually would proved soluble.

7. The title of this report incorrectly uses the term "isografting". This indicates the confusion in terminology which existed at the time. Several titles of ours cited in this bibliography are imprecise. The concise terms "autograft", "isograft", "allograft" and "heterograft" were not established for universal usage until 1960 by Peter Gorer.

8. Brown, who had observed Padgett's patients, was never certain about the fate of any allografts used in burned patients. That is why subsequently he sought out a pair of identical twins to test with reciprocal skin grafts. These were true "isografts", i.e. grafts between genetically identical members of the same species, similar to grafts between genetically pure strains of mice.

Brown, from Washington University in St. Louis, was undoubtedly influenced by Leo Loeb of the same institution. Loeb had written a book entitled *The Basis of Individuality* in which he categorically stated that grafts between individual humans would never be possible because individual differences existed even at the cellular level.

Brown was Chief of Plastic Surgery at Valley Forge General Hospital, Pennsylvania, the hospital to which I was randomly assigned during World War II. We often discussed the biology of transplantation while scrubbing and operating.

10. Dr. Emile Holman and Dr. Brown both had noted and written that second skin grafts from the same donor 'seemed' to disappear more rapidly than the first. But they had not recognized or formulated an underlying principle on which others could build.

11. In this same issue of *Science* there is another report describing over 300 free-martin pairs without adding any new information or findings. Lilly, on the other hand, had rolled up his sleeves and dissected a few of these cattle and discovered the intermingling of placental blood vessels. Incidently, Lillie was one of the founders of the Marine Biological Laboratory at Wood's Hole, Massachusetts, and for years served at its Director.

12. Owen's work, funded by the American Guernsey Association, was part of a study of the freemartin, whose birth could be a major financial burden for an individual dairy farmer.

14. Carrel received the Nobel Prize in 1912. Medawar, together with Sir Frank MacFarlane Burnett, received the Nobel Prize in 1960 for "the discovery of acquired immunological tolerance". Others receiving the Nobel Prize in areas related to transplantation are Baruj Benacerraf, Jean Dausset, and George Snell, in 1980, "for their discoveries concerning genetically determined structures on the cell surface that regulate immunological reactions", and George Hitchings and Gertrude Elion, in 1988, "for their discoveries of 'important principles for drug treatment' ".

19. & 20. Dempster and Simonsen were imaginative and productive investigators who added much to our understanding of transplantation biology and particularly to the rejection of renal grafts. Simonsen later performed a successful renal graft between freemartin cattle.
23. & 24. Dr. Merrill and I , as respective heads of the Medical and Surgical portions of the Brigham Transplant Program, decided that, for publications generally, I would present our patients in groups as series and he as individual case reports.
31. This is a watershed report of the Brigham experience because it is our final report on the use of total body X-radiation in our twelve patients, and it also includes our first six transplant patients treated with immunosuppressive drugs. In addition, it includes experiments in dogs, selected from over 300 transplants performed within two years, which demonstrated not only prolonged survival of the allografts, but also the specificity of the drug-induced tolerance. These long-term surviving dogs were capable of rejecting skin allografts from third-party donors, as well as skin from the kidney donor if the skin graft was placed after the renal graft was well established. For a more complete "Analysis of Mechanism of Immunosuppressive Drugs in Renal Transplantation", cf. Annals of Surgery, 160: 449, 1964.
37. This report is a direct outgrowth of the earlier renal perfusion studies of Dr. Nathan Couch. After demonstrating post-perfusion viability of an extra-corporeally perfused kidney, he attempted to alter its antigenicity by perfusing it with blood components from a future recipient. After observing several unsuccessful attempts, Dr. David Hume suggested that we use the future recipient dog itself as the pump because it would eliminate the problems both of the pump and the perfusate. Following this suggestion, we found that by attaching the extra-corporeal kidney to a prospective recipient by means of canulae, we were able to initiate the rejection process within a day or two. But, if we reconnected the rejecting kidney to its original owner in time, the rejection process could be reversed.
43. This reports the first successful human organ transplant other than the kidney. In the discussion Dr. Moore says, "This is a magnificent achievement and liver surgery as of this day has an entirely new look from hither forward." Reflecting on this remarkable contribution of Starzl's, it is important to recall that his success did not come easily. It was preceded by years of disappointment and discouragement, and success resulted only from his persistent logical dedication to the project.
46. Shumway, like Starzl, achieved success only by persistence and dedication. Neither aimed for the spectacular "overnight" breakthrough; both realized that solid progress requires time, basic knowledge and unselfish teamwork.
46. Although Calne reported more than 25 years ago that the liver might possess an intrinsic immunosuppressive factor, it has been dismissed as an idiosyncrasy of the pig and not relevant for man. However, there are emerging impressions from the simultaneous transplantation of multiple organs in man that the presence of a liver allograft seems to improve the survival of the accompanying allografts. This seems to apply especially for intestinal transplants.

E. Donnall Thomas

E. DONNALL THOMAS

My father, Dr. Edward E. Thomas was born in 1870 and moved to Texas with his family in a covered wagon in 1874. He grew up in frontier Texas and, with almost no formal schooling went to the University of Louisville, Kentucky, where he received his M. D. His first wife died of tuberculosis, and I was the only child of his second wife. He was 50 years old when I was born on March 15, 1920. He was a solo general practitioner in our small Texas village. Thus, together we span the time from horse and buggy house calls to modern high-tech medicine.

My high school class consisted of about 15 people. I was not an outstanding student even in this small group. I entered the University of Texas in Austin in 1937. In my first semester I made only B grades, but as time went on and the courses became more difficult and challenging I began to enjoy the studies, mainly in chemistry and chemical engineering. I received a B. A. in 1941 and an M. A. in 1943.

During my undergraduate years at the end of the depression money was almost non-existent so I worked at a number of odd jobs. One of the jobs was waiting tables at a girls' dormitory. One January morning it snowed, a rare event in Texas. As I emerged from the girls' dormitory, an attractive young woman hit me in the face with a snow ball. I naturally had to catch her and avenge the insult to my male ego. Thus, I meet Dorothy Martin, the Dottie who has participated in all my endeavors up to the present time. We have 3 children, Don Jr. who practices internal medicine in Montana, Jeffrey who is in business in Seattle and Elaine who is a Fellow in infectious diseases at the University of Washington. We have eight grandchildren.

I entered Harvard Medical School in 1943. During medical school Dottie abandoned her journalism work to enter training as a laboratory technician while working to help support us. Her training in writing, laboratory technology and library science has been invaluable in our work. I received the M. D. in 1946.

There followed an internship, a year of hematology training under my life-long friend Dr. Clement Finch, two years in the army, a year of post-doctoral work at Massachusetts Institute of Technology, two years of medical residency, the last as the chief medical resident at the Peter Bent Brigham Hospital in Boston. During that time Dr. Joseph Murray was a surgical resident and we have been friends and colleagues over the years because of our common interests in transplantation. I was on the wards of the Brigham and helped care for his first kidney transplant patient.

During medical school I became interested in the bone marrow and in leukemia. This interest was intensified by my early association with Dr.

Sydney Farber who gave me my first laboratory in the new Jimmy Fund Building. I was fortunate to see the first child with acute lymphoblastic leukemia (ALL) whose remission was induced with an anti-folate drug. I became interested in factors that stimulate marrow function in part due to Allan Erslev's attempt to demonstrate erythropoietin. During my year at M.I.T. I worked under Dr. John Loofborrow on stimulating factors released from irradiated yeast. I hoped to apply this knowledge to marrow stimulating factors. Fortunately I left the field of stimulating factors because it is only in recent years, with recombinant technology, that great strides have been made in this area.

I had been intrigued by the studies of Dr. Leon Jacobsen et al. who demonstrated that shielding the spleen would protect mice against otherwise lethal irradiation and the subsequent demonstration by Egon Lorenz et al. that a marrow infusion was also protective. These observations were initially thought to be the result of stimulating factors. In 1955, Main and Prehn published their paper showing that a mouse protected against lethal irradiation by a marrow infusion would accept a skin graft from the marrow donor. Their study and the demonstration by Ford et al. using cytogenetic technology of donor chromosomes in such mice made it suddenly clear that the irradiation protection effect was due to the survival of living bone marrow cells.

In 1955, at the invitation of Dr. Joseph Ferrebee I went to the Mary Imogene Basset Hospital in Cooperstown, N. Y., an affiliate of Columbia University. Immediately, we began to work on marrow transplantation in human patients and in the dog, as an outbred animal suitable for clinical care comparable to human patients. Except for an occasional patient with an identical twin, we quickly learned that allogeneic marrow transplants in man were going to be very difficult. Joe Ferrebee and I and our young colleagues concentrated on working with our dogs on many aspects of marrow transplantation. The long cold winters, absence of commuting problems and opportunity for long discussions were conducive to our work. Those years had a deep and abiding influence on subsequent work since most of the basic concepts were laid out during that time.

In 1963 I moved to Seattle at the invitation of Dr. Robert Williams, a famous endocrinologist and first chairman of the Department of Medicine at the University of Washington. Professor Williams recognized that our School of Medicine was in its infancy and rather isolated in the Pacific Northwest. He envisioned the affiliation of all the relevant institutions in the area with the School of Medicine in order to create the critical mass necessary for academic excellence. Within that concept I established my program in the Seattle Public Health Hospital.

The rest of the story seems short in retrospect. The recruitment of some brilliant young co-workers who still work with me, studies of immunology and irradiation biology in the dog, borrowing knowledge of human histocompatibility from Amos, Payne and Dausset, the assembly and training of a critical care team of nurses, and, finally, the demonstration that some patients with advanced leukemia, aplastic anemia or genetic diseases could be cured by marrow transplantation.

Our team of physicians and nurses proved to be stable and dedicated. We did face problems that at times seemed almost insurmountable. In 1972, the Seattle Public Health Hospital was faced with closure by the federal government. After many conferences with the Dean of the School of Medicine, it was apparent that we could not move to the University of Washington. We found temporary space at Providence Hospital for a two year period. In 1975 our team moved into the Fred Hutchinson Cancer Research Center which provided superb facilities and the opportunity to expand the program with the cooperation of the Swedish Hospital Medical Center. While continuing laboratory and animal research, our team has now carried out more than 4,000 human marrow transplants.

It is always difficult to identify the many threads that make up the fabric of a life's work. I know that my philosophy and ideas have been heavily influenced by more than 20 years of daily interaction with a small group of colleagues, all of whom are now distinguished scientists in their own right. Bob Epstein, Rainer Storb, Dean Buckner, Reg Clift, Paul Neiman and Alex Fefer were with me at the start of the Seattle adventure, and all except Bob are still my daily companions. Ted Graham moved with me from Cooperstown and has played an essential role in our animal research. Along the way we were joined by Joel Meyers, Fred Appelbaum, John Hansen and many others who made major contributions to the achievements honored by this award.

BONE MARROW TRANSPLANTATION – PAST, PRESENT AND FUTURE

Nobel Lecture, December 8, 1990

by

E. DONNALL THOMAS

The Fred Hutchinson Cancer Research Center and the University of Washington, Seattle, Washington, U.S.A.

During my time in medical school in the 1940's it was well known that grafts of organs between individuals of different genetic origin always result in rejection. Burnett's hypothesis and Medawar's experiments began to shed some light on transplantation immunology and the possibility of tolerance (Nobel Laureates, 1960). Inbred mice and human identical twins appeared to accept grafts from each other. Dr. Murray and his colleagues were the first to carry out a kidney transplant between human identical twins, a feat of heroic proportions at the time but now almost routine. Grafting of bone marrow was an even more remote possibility since marrow cells could not be sewn into place.

In 1939, Osgood et al.(1) infused a few ml of marrow into patients with aplastic anemia without benefit. Rekers et al.(2) worked in the classified laboratories of the Atomic Energy Commission in Rochester, N. Y. attempting to reconstitute marrow function in irradiated dogs by marrow infusions. These studies were published in 1950. In retrospect, these experiments apparently failed because the irradiation exposures, although lethal, were not great enough to produce the immunosuppression necessary for allogeneic engraftment.

In 1949, Jacobson et al.(3) found that mice could be protected from otherwise lethal irradiation by shielding the spleen with lead. It was initially thought that the protective effect was due to humoral stimulants released by the protected spleen. Soon thereafter Lorenz et al.(4) showed that a similar protection could be achieved by an intravenous infusion of marrow from a mouse of the same strain. At that time I was intrigued by these experiments because of my own work with stimulating factors released by irradiated yeast (5) and by in vitro studies of marrow metabolism (6,7).

In 1955, Main and Prehn published a paper showing that mice protected from lethal irradiation by marrow infusion would accept a skin graft from the donor strain, an observation strongly suggesting that a transfer of living donor cells had occurred to account for the apparent tolerance(8). Also in 1955, Ford et al.(9) used cytogenetics to show the presence of donor cells in irradiated mice protected by a marow infusion. In the summer of 1955 I moved to the Mary Imogene Bassett Hospital in Cooperstown, N.Y. where I found that Dr. Joseph Ferrebee had been following these and related experiments very closely. We decided to begin studies of marrow grafting in outbred species, especially the canine model, and to begin cautious exploration of marrow infusion in human patients in need of a marrow graft because of disease or its treatment.

In 1957, we published our first paper on human marrow grafting which we called "intravenous infusion" because only one patient had a transient graft which hardly constituted true marrow transplantation (10). We learned two things from those studies: 1) Large quantities of human marrow could be infused intravenously without harm when properly prepared; and 2) allogeneic marrow grafting in our species would be very difficult. Because of concern about irradiation exposure our early funding came through the Atomic Energy Commission. In the 1957 paper we stated "The studies presented here show that human bone marrow can be collected and stored in significant quantities and can be administered with safety. After administration it may grow even under disadvantageous competitive circumstances in completely irradiated hosts afflicted with marrow neoplasia. In an atomic age, with reactor accidents not to mention stupidities with bombs, somebody is going to get more radiation than is good for him. If infusion of marrow can induce recovery in a mouse or monkey after lethal radiation, one had best be prepared with this form of treatment in man. The leukemic patient who needs radiation and bone marrow and the uremic patient who needs a spare kidney are people who deserve immediate consideration. From helping them one will be preparing for the atomic disaster of tomorrow and it is high time one did." The reference to a spare kidney was based on the possibility that a marrow graft might be necessary to permit permanent acceptance of a kidney from the same donor. Immunosuppressive drugs for organ grafting were unknown at that time.

Allogeneic marrow grafting continued to be unsuccessful, but we did have the opportunity to carry out marrow grafts between a few sets of identical twins. Our first two patients, reported in 1959, with refractory leukemia who had an identical twin were given supralethal irradiation and a syngeneic marrow infusion (11). Their prompt hematologic recovery and well being demonstrated that an intravenous infusion of marrow could protect against lethal irradiation. The recurrence of their leukemia in a few months prompted our speculation about how marrow grafting might cure leukemia as follows: "Evidently something more than radiation is needed to eradicate leukemia. Two possible approaches are suggested. First, one may transplant homologous marrow after lethal irradiation and depend on the homologous marrow to provide an immunologic environment unsuitable for survival of the leukemia (Barnes et al. 1956). This approach has apparently eradicated leukemia in some mice (Barnes et al., 1957, Mathé and Bernard, 1958). However, these mice subsequently have a high incidence of death from delayed foreign marrow disease due to a reaction of the graft against the host. Whether delayed foreign marrow disease will be either serious or useful in man and whether it can be controlled by clinical supportive measures available are questions currently being studied." "The second approach to the problem of eradicating leukemia lies in the observation that with chemotherapy and x-ray, the cure rate of transplantable leukemia in the mouse is an inverse function of the number of cells present – the smaller the number of leukemic cells, the greater the possibility of cure (Burchenal et al., 1951, Mathé et al., 1959). This suggests that the patient in remission, with a relatively small mass of leukemic cells, is an advantageous subject for radiation. It further suggests that chemotherapeutic agents may be more effective if administered during an immediate postradiation period when the number of leukemic cells is relatively small." Unfortunately, 15 years were to go by before we were able to carry out marrow grafts for patients with leukemia in remission.

During the late 1950's other investigators attempted allogeneic grafts in human beings. Mathé and his colleagues reported the Yugoslavian radiation accident cases in 1959, several of whom were treated with marrow infusions (12). A retrospective review of these case suggested little benefit (13). Mathé did achieve a durable allogeneic marrow graft in a patient with leukemia, only to have that patient develop chronic graft-versus-host disease and die of infectious complications (14).

Beginning in 1955, Dr. Joseph Ferrebee, Dr. Harry Lochte, Jr. and I and our colleagues carried out marrow grafting studies in the dog. The dog is a readily available outbred animal often used in transplantation research and amenable to clinical procedures comparable to those used for human patients. When Dr. John Mannick worked with us as a fellow, we found that dogs could be given three times the lethal dose of total body irradiation and recover promptly if given an infusion of their own marrow set aside before irradiation (15). Marrow grafts could be obtained with peripheral blood as well as marrow (16) and the cells responsible for recovery could be frozen and kept for long periods of time (17,18).

However, in the dog as in our human patients marrow from an allogeneic donor almost always resulted either in failure of engraftment or in successful engraftment followed by lethal graft-versus-host disease (19,20). We were encouraged by the fact that an occasional dog, usually with a littermate donor, went through the grafting procedure successfully. The persistence of donor marrow was confirmed by cytogenetic studies when donor and recipient were of opposite sex (21) and many of these dogs proved to live a normal canine life span (22). Evidently it could be done – we just had to find out how.

Delta Uphoff had reported that methotrexate ameliorated graft-versus-host disease in some strains of mice (23). We found methotrexate given post grafting to be of help in reducing the incidence and severity of the graft-versus-host reaction, and a great deal of work went into the study of various methotrexate regimens (20,24). These and other studies in the canine model produced a wealth of information, summarized in 1972 (25). Most importantly, it was clear that a successful allogeneic graft depended upon close histocompatibility matching between donor and recipient, and we developed techniques for histocompatibility typing in the dog (26). We were finally able to detect DL-A antigens and to show that marrow grafts between matched littermates were almost always successful (27). These studies pointed the way for marrow grafting in man using patients with an HL-A matched sibling donor.

The many failures of allogeneic marrow grafting in human patients caused most investigators to abandon such studies in the 1960's. However, under the impetus of kidney grafting, the knowledge of human histocompatibility antigens progressed rapidly. As we developed our knowledge of DL-A matching, we followed closely the work of Dausset (Nobel Laureate, 1980), van Rood, Payne, Bodmer and Amos in the human, the HL-A system. By 1967, we thought that the time was right to return to allogeneic marrow grafting in humans. Recognizing that the care of patients with advanced leukemia undergoing allogeneic grafts would be difficult, we began to assemble the necessary team. In 1967 we wrote a program project grant application which was funded by the National Cancer Institute in 1968. We began to assemble and train a team of nurses familiar with the care of patients without marrow function and subject to opportunistic infections. In November of 1968 Dr. Robert Good and his colleagues carried out the first marrow transplant from a matched sibling for an infant with an immunological deficiency disease (28). Our team carried out our first transplant using a matched sibling donor for a patient with advanced leukemia in March 1969.

These studies marked the beginning of the "modern" era of human allogeneic marrow grafting. A comprehensive review of the experimental background, the early clinical successes and the deliniation of problems was presented in the New England Journal of Medicine in 1975 (29). As follow-up times increased for patients transplanted for end-stage leukemia, it became apparent that a plateau was developing on a Kaplan-Meier plot of survival so that it became possible to use the term "cure" for these patients (30).

Allogeneic marrow grafts are now carried out at more than 200 centers around the world, and the number of diseases for which marrow grafting may be considered continues to increase. Currently, approximately 5,500 allogeneic and 4,000 autologous marrow transplants are performed annually (Mary Horowitz, International Bone Marrow Transplant Registry, Personal communication). The longest survivors of these otherwise lethal diseases are now 20 years post-grafting. Recent articles summarize the experience of the Seattle team with patients given allogeneic marrow grafts for acute myeloid leukemia (31,32), chronic myeloid leukemia (33,34) and aplastic anemia and thalassemia major (35). Lucarelli et al. (36) have described their extensive experience with thalassemia major.

Among the early problems in allogeneic marrow grafting, perhaps the greatest was immunologic reactivity of the host (graft rejection) and/or of the graft (graft-versus-host disease). Irradiation of the host with "supralethal" exposures was necessary for retention of the marrow graft. Obviously, irradiation could not be used after the graft. The use of methotrexate was mentioned above, but better agents were on the way. Schwartz and Dameshek (37) noted the immunosuppresive properties of 6-mercaptopurine and Hitchings and Elion (Nobel laureates, 1988) developed Immuran. Santos and Owens introduced cyclophosphamide as an immunosuppressive agent for transplantation (38). More recently, cyclosporine has proved invaluable for organ grafts. Cyclosporine was not superior to methotrexate in our randomized clinical trials of marrow grafter patients, but the regimen of a combination of short methotrexate combined with 6 months of cyclosporine proved effective and is now our standard regimen (39,40). Chronic graft-versus-host disease is severe in a small fraction of patients, but can sometimes be controlled by prolonged corticosteroid therapy (41). T-cell depletion of the marrow graft has resulted in a reduced incidence of graft-versus-host disease but at an increased risk of graft failure or recurrence of malignancy (42). Newer agents are being investigated.

Recurrence of malignant disease following an otherwise successful allogeneic graft continues to be a problem. Efforts to kill a greater fraction of the malignant cells have involved a variety of high dose chemotherapy regimens with or without total body irradiation. Efforts to increase the intensity of the pre-transplant regimen have been limited by life-threatening damage to other organs, most notably the liver and lung (43). The role of biological response modifiers such as interferon is being investigated.

Because of the graft-versus-host reaction and its treatment, patients are profoundly immuno-suppressed following an allogeneic marrow graft and therefore at great risk for all kinds of opportunistic infections (44). Bacterial and some fungal diseases can be controlled by antibiotics, sometimes with the aid of granulocyte transfusions (45). Prophylactic acyclovir has prevented Herpes simplex and zoster infections (46). Pneumonia due to cytomegalovirus (CMV) has been difficult to treat and is a major cause of death (47). For patient and donor pairs serologically negative for CMV, the use of blood products from CMV negative donors has prevented infection (48). The use of prophylactic ganciclovir seems to prevent CMV infection even when donor and/or recipient are CMV positive. Prevention of CMV infection should result in an appreciable increment in long-term survivors of allogeneic marrow grafts.

Only about one-fourth to one-third of patients will have an HL-A identical sibling. Examination of the extended family will identify a non-sibling donor in about 10 percent of patients and results with these donors are comparable to those with an identical sibling (49). With national and international cooperation large panels of volunteer donors whose tissue type is known are now being established. More than 300 transplants using volunteer donors matched to the recipient by computer search have now been carried out. The results using phenotypically matched donors or donors differing by only one HL-A haplotype seem comparable to matched sibling donors (50,51).

In the absence of a suitable matched donor the patient's own marrow may be removed, stored and given back after intensive therapy. The general principles of long-term marrow storage and autologous marrow transplantation have been known for almost 30 years (52). Recently, there has been a striking increase in the use of autologous marrow grafts as more effective measures for destruction of the tumor in the patient have been developed. Methods for destruction of tumor in the marrow graft are being developed. An autologous graft avoids the problems of graft-versus-host disease but may be associated with a greater relapse rate due to the loss of the graft-versus-leukemia effect and the possibility of tumor cells in the stored marrow.

Monoclonal antibodies (Köhler and Milstein, Nobel laureates, 1984) are being used in many ways in marrow grafting. Anti T-cell antibodies for many T-cell epitopes are being used in vitro to remove normal or malignant T-cells from marrow and in vivo to prevent or treat graft-versus-host disease. Monoclonal antibodies coupled with a toxin are being used to treat graft-versus-host disease and, coupled to radioactive isotopes, for selective irradiation exposure of marrow cavities or of tumors so that exposure of the total body to irradiation can be reduced.

Recently, hematopoietic growth factors produced by recombinant molecular biology techniques are being used in marrow grafting. Clinical trials have shown that G-CSF and GM-CSF can accelerate marrow recovery after either allogeneic or autologous marrow grafts. Other growth factors are now entering clinical trial. Biological response modifiers including IL-2, IL-6, cloned T-cells and interferon are being explored for acceleration of marrow graft recovery, for better antibacterial and antifungal effects and greater antitumor effects.

Progress has been made in the identification and purification of the hematopoietic stem cell, long a goal of experimental hematologists. Purified stem cells, free of tumor cells, may be of value in autologous marrow grafting. Retroviral vectors have increased the efficiency of gene transfer and purified stem cells are ideal targets for gene transfer therapy for many diseases. Sustained expression of genes transferred into hematopoietic stem cells has not yet been achieved and the application of gene transfer technology to diseases such as thalassemia major must await much further research.

In summary, marrow grafting has progressed from a highly experimental procedure to being accepted as the preferred form of treatment for a wide variety of diseases at many varying stages of disease. Table 1 shows the approximate 5 year disease-free survival for the most common diseases treated by marrow transplantation. Progress has been slow but steady. Important new developments are showing the way to a further improvement in results so that many more patients with otherwise incurable diseases will have a reasonable chance of long survival and cure.

Finally, it should be noted that marrow grafting could not have reached clinical application without animal research, first in inbred rodents and then in outbred species, particularly the dog. Application to human patients depended upon developments in many branches of science including understanding of the human histocompatibility system, knowledge of immunosuppressive drugs, blood transfusion technology, especially the ability to transfuse platelets, the creation of a repertoire of broad spectrum antibiotics and the development of effective anticancer chemotherapeutic agents. I echo the sentiments of many previous Nobel laureates when I say that the success we celebrate today was made possible by the work of many others in this and in related fields.

Table 1

Diseases	*Survival*
Acute leukemia in relapse	0.10–0.30
ALL,first or second remission	0.30–0.60
AML, first remission	0.45–0.70
CML, chronic phase	0.60–0.90
CML, accelerated or blastic phase	0.10–0.30
Lymphoma, Hodgkin's disease	
after failure of first line therapy	0.40–0.60
after failure of second line therapy	0.10–0.30
Immunological deficiency disease	0.50–0.90
Aplastic anemia, transfused	0.50–0.70
Aplastic anemia, untransfused	0.80–0.90
Thalassemia major	
without liver damage	0.85–0.95
with liver damage	0.60–0.85

Abbreviations:
AML, acute myeloid leukemia
ALL, acute lymphoblastic leukemia
CML, chronic myeloid leukemia

REFERENCES

1. E.E. Osgood, M.C. Riddle and T.J. Mathews, *Ann. Intern. Med.* **13**, 357–367 (1939).
2. P.E. Rekers, M.P. Coulter and S. Warren, *Arch. Surg.* **60**, 635–667 (1950).
3. L.O. Jacobson, E.K. Marks, M.J. Robson, E.O. Gaston and R.E. Zirkle, *J. Lab. Clin. Med.* **34**, 1538–1543 (1949).
4. E. Lorenz, D. Uphoff, T.R. Reid and E. Shelton, *J. Natl. Cancer Inst.* **12**, 197–201 (1951).
5. E.D. Thomas, F.B. Hershey, A.M. Abbate and J.R. Loofbourow, *J. Biol. Chem.* **196**, 575–582 (1952).
6. E.D. Thomas, *Blood* **10**, 600–611 (1955).
7. E.D. Thomas and H.L. Lochte, Jr., *Blood* **12**, 1086–1095 (1957).
8. J.M. Main and R.T. Prehn, *J. Natl. Cancer Inst.* **15**, 1023–1029 (1955).
9. C.E. Ford, J.L. Hamerton, D.W.H. Barnes and J.F. Loutit, *Nature* **177**, 452–454 (1956).
10. E.D. Thomas, H.L. Lochte, Jr., W.C. Lu and J.W. Ferrebee, *N. Engl. J. Med.* **257**, 491–496 (1957).

11. E.D. Thomas, H.L. Lochte,Jr., J.H. Cannon, O.D. Sahler and J.W. Ferrebee, *J. Clin. Invest.* **38**, 1709–1716 (1959).
12. G. Mathé, H. Jammet, B. Pendic, et al., *Rev. Franc. Etudes Clin. et Biol.* **IV**, 226–238 (1959).
13. G.A. Andrews, *Am. J. Roentgenol. Radium Ther. Nucl. Med.* **93**, 56–74 (1965).
14. G. Mathé, *Diagnostic et Traitement des Radiolésions aigues*, O.M.S., Geneva, p. 197–230 (1964).
15. J.A. Mannick, H.L. Lochte, Jr., C.A. Ashley, E.D. Thomas and J.W. Ferrebee, *Blood* **15**, 255–266 (1960).
16. J.A. Cavins, S.C. Scheer, E.D. Thomas and J.W. Ferrebee, *Blood* **23**, 38–43 (1964).
17. E.D. Thomas and J.W. Ferrebee, *Transfusion* **2**, 115–117 (1962).
18. J.A. Cavins, S. Kasakura, E.D. Thomas and J.W. Ferrebee, *Blood* **20**, 730–734 (1962).
19. E.D. Thomas, C.A. Ashley, H.L. Lochte,Jr., A. Jaretzki,III, O.D. Sahler and J.W. Ferrebee, *Blood* **14**, 720–736 (1959).
20. E.D. Thomas, J.A. Collins, E.C. Herman, Jr., and J.W. Ferrebee, *Blood* **19**, 217–228 (1962).
21. R.B. Epstein and E.D. Thomas, *Transplantation* **5**, 267–272 (1967).
22. E.D. Thomas, G.L. Plain, T.C. Graham and J.W. Ferrebee, *Blood* **23**, 488–493 (1964).
23. D.E. Uphoff, *Proc. Soc. Exp. Biol. Med.* **99**, 651–653 (1958).
24. R. Storb, R.B. Epstein, T.C. Graham and E.D. Thomas, *Transplantation* **9**, 240–246 (1970).
25. R. Storb and E.D. Thomas, in *Proceedings of the Sixth Leukocyte Culture Conference*, Ed.M.R. Schwarz). Academic Press, New York, pp. 805–840 (1972).
26. R.B. Epstein, R. Storb, H. Ragde and E.D. Thomas, *Transplantation* **6**, 45–58 (1968).
27. R. Storb, R.H. Rudolph and E.D. Thomas, *J. Clin. Invest.* **50**, 1272–1275 (1971).
28. R.A. Gatti, H.J. Meuwissen, H.D. Allen, R. Hong and R.A. Good, *Lancet* **ii**, 1366–1369 (1968).
29. E.D. Thomas, R. Storb, R.A. Clift, et al., *N. Engl. J. Med.* **292**, 832–843,-895–902 (1975).
30. E.D. Thomas, N. Flournoy, C.D. Buckner, et al., *Leuk. Res.* **1**, 67–70 (1977).
31. R.A. Clift, C.D. Buckner, E.D. Thomas, et al., *Bone Marrow Transplantation* **2**, 243–258 (1987).
32. R.A. Clift, C.D. Buckner, F.R. Appelbaum, et al., *Blood, 76:* 1867–1871 (1990).
33. E.D. Thomas, R.A. Clift, A. Fefer, et al., *Ann. Intern. Med.* **104**, 155–163 (1986).
34. E.D. Thomas and R.A. Clift, *Blood* **73**, 861–864 (1989).
35. R. Storb, C. Anasetti, F. Appelbaum, et al., *Seminars in Hematology,* (in press).
36. G. Lucarelli, M. Galimberti, P. Polchi, et al., *N. Engl. J. Med.* **322**, 417–421 (1990).
37. R. Schwartz and W. Dameshek, *Nature* **183**, 1682–1683 (1959).
38. G.W. Santos and A.H. Owens, Jr., *Transplant. Proc.* **1**, 44–46 (1969).
39. R. Storb, H.J. Deeg, L.D. Fisher, et al., *Blood* **71**, 293–298 (1988).
40. R. Storb, H.J. Deeg, M. Pepe, et al., *Br. J. Haematol.* **72**, 567–572 (1989).
41. K.M. Sullivan, R.P. Witherspoon, R. Storb, et al., *Blood* **72**, 555–561 (1988).
42. P.J. Martin, J.A. Hansen, B. Torok-Storb, et al., *Bone Marrow Transplantation* **3**, 445–456 (1988).
43. S.I. Bearman, F.R. Appelbaum, A. Back, et al., *J. Clin. Oncol.* **7**, 1288–1294 (1989).
44. R. Witherspoon, D. Noel, R. Storb, H.D. Ochs and E.D. Thomas, *Transplant. Proc.* **10**, 233–235 (1978).

45. C.D. Buckner, R.A. Clift, J.E. Sanders and E.D. Thomas, *Transplant. Proc.* **10**, 255–257 (1978).
46. J.D. Meyers, *Scand. J. Infect. Dis.* Suppl. **47**, 128–136 (1985).
47. J.D. Meyers, N. Flournoy and E.D. Thomas, *Rev. Infect. Dis.* **4**, 1119–1132 (1982).
48. R.A. Bowden, M. Sayers, N. Flournoy, et al., *N. Engl. J. Med.* **314**, 1006–1010 (1986).
49. P.G. Beatty, R.A. Clift, E.M. Mickelson, et al., *N. Engl. J. Med.* **313**, 765–771 (1985).
50. J.A. Hansen, C. Anasetti, P.G. Beatty, et al., *Bone Marrow Transplantation* **6**, 108–111 (1990).
51. P.G. Beatty, J.A. Hansen, E.D. Thomas, et al., *Transplantation*, (in press).
52. C.D. Buckner, F.R. Appelbaum and E.D. Thomas, in *Organ Preservation for Transplantation, Chapter 16*, (A.M. Karow, Jr. and D.E. Pegg, Eds.). Marcel Dekker Inc., New York, pp. 355–375 (1981).